Computational Fluid and Solid Mechanics

Available Volumes

D. Chapelle, K.J. Bathe
The Finite Element Analysis of Shells - Fundamentals
2003

D. Drikakis, W. Rider
High-Resolution Methods for Incompressible and Low-Speed Flows
2005

M. Kojic, K.J. Bathe
Inelastic Analysis of Solids and Structures
2005

E.N. Dvorkin, M.B. Goldschmit
Nonlinear Continua
2005

J. Iannelli
Characteristics Finite Element Methods in Computational Fluid Dynamics
2006

Joe Iannelli

Characteristics Finite Element Methods in Computational Fluid Dynamics

With 384 Figures

 Springer

Author:

Professor Dr. Joe Iannelli
Director, Centre for Aeronautics
School of Engineering and Mathematical Sciences
City University
10 Northampton Square
London, EC1V 0HB
United Kingdom

Library of Congress Control Number: 2006925962

ISBN-10   3-540-25181-2 Springer Berlin Heidelberg New York
ISBN-13   978-3-540-25181-1 Springer Berlin Heidelberg New York

Springer is a part of Springer Science+Business Media
springer.com

© Springer-Verlag Berlin Heidelberg 2006
Printed in Germany

Typesetting: Dataconversion by author
Final processing by PTP-Berlin Protago-TEX-Production GmbH, Germany (www.ptp-berlin.com)
Cover-Design: deblik, Berlin
Printed on acid-free paper   62/3141/Yu – 5 4 3 2 1 0

We agreed then on the good things we have in common, on the advantage of being able to test yourself, not depending on others in the test, reflecting yourself in your work, on the pleasure of seeing your creature grow, beam after beam, bolt after bolt, solid, necessary, symmetrical, suited to its purpose. When it's finished, you look at it and you think that perhaps it will live longer than you, and perhaps it will be of use to someone you don't know, who doesn't know you. Maybe, as an old man you'll be able to come back and look at it, and it will seem beautiful, and it doesn't really matter so much that it will seem beautiful only to you, and you can say to yourself: "maybe another man wouldn't have brought it off".

Primo Levi

Preface

Now, when all these studies reach the point of inter-communication and connection with one another and come to be considered in their mutual affinities, then, I think, but not until then, will the pursuit of them have a value for our objects.
Plato

Whenever there is a number there is beauty.
Proclus

In any particular theory there is only as much real Science as there is Mathematics.
I. Kant

Thus number may be said to rule the whole world of quantity.
J. C. Maxwell

One must regard nature reasonably and naturally as one would the truth, and be contented only with a representation of it, which errs to the smallest possible extent.
J. Bolyai

Mathematical theories from the happy hunting grounds of pure mathematicians are found suitable to describe the airflow produced by aircraft with such excellent accuracy that they can be applied directly to airplane design.
T. von Karman

Fluid Dynamics governs the function and design of myriad systems, from cooling towers to aircraft. Computational Fluid Dynamics allows investigating the fluid flows in these systems via computational solutions of the mathematical models of fluid dynamics. This book details a systematic, efficient, and stable characteristics-based finite element procedure computationally to investigate incompressible, free-surface and compressible flows. Inducing by design a controllable multi-dimensional upwind bias that can be locally optimized, the procedure crisply captures contact discontinuities, normal as well as oblique shocks, and generates essentially non-oscillatory solutions for incompressible, subsonic, transonic, supersonic, and hypersonic inviscid and viscous flows, adiabatic and non-adiabatic, with shaft work, heat as well as mass transfer, and chemical reactions.

This procedure has emerged from my wish to contribute a versatile unifying method that relies on the multi-dimensional theory of characteristics, employs the finite element method, and generates discrete characteristics-based formulae automatically on arbitrary grids. The theory of characteristics is beautiful, with its physical description of wave propagation within fluid flows, and the finite element method is appealing, with its systematic process of transforming a partial differential equation into a set of algebraic or ordinary differential equations. Yet the automatic generation of arbitrary-grid characteristics-based discrete formulae is far from straightforward. After almost two decades of research that began at the von Karman Institute for Fluid Dynamics and has continued at the University of Tennessee, Knoxville, NASA Lewis, (now Glenn) and the historically-first British Center for Aeronautics, at the City University of London, a solution for this challenge establishes a characteristics-based approximation of the continuum partial differential equations themselves, before the discretization in space. A conventional centered discretization of this characteristics-based approximation then automatically generates for the selected grid a corresponding characteristics-based discrete formula for the original equations.

As detailed in this book, the procedure develops authentically multi-dimensional and infinite directional upwind resolutions of the Euler and Navier-Stokes equations, with an anisotropic upwinding that correlates with the spatial distribution of the Euler characteristic speeds. Employing characteristic lines and surfaces, the procedure relies upon the mathematics and physics of multi-dimensional acoustics and convection to introduce this kind of upwinding directly at the partial differential equation level, by way of a Flux Divergence Decomposition, a multi-dimensional generalization of flux-vector splitting. Uniformly applicable to incompressible, free-surface, and compressible flows, such a process generates a characteristics-bias system that couples the Euler or Navier-Stokes equations with a regularizing hyperbolic-parabolic perturbation. This perturbation non-linearly induces a solution-dependent upwinding that provides minimal upwinding, hence diffusion, in regions of smooth flow, for accuracy, and locally increases this upwinding at solution discontinuities, for essentially non-oscillatory shock capturing of both normal and oblique shock waves. This perturbation also introduces a spatial upwinding on the time derivative of the solution and source term, a feature that results in a Galilean invariant system and leads to a crisp capturing of contact discontinuities. For the incompressible-flow Navier-Stokes equations, the procedure allows a direct solution of the coupled continuity and linear-momentum equations, with equal-order interpolation for velocity and pressure and without any time derivative of pressure in the continuity equation.

An integral formulation of the characteristics-bias system directly provides non-discrete generalized Discontinuous Galerkin (DG) and Streamline Upwind Petrov Galerkin (SUPG) statements. The most basic Galerkin finite element discretization of this integral formulation on arbitrary grids directly yields an intrinsically multi-dimensional and infinite directional upwind algorithm for computationally solving the original Euler and Navier-Stokes equations, an algorithm that does not require any further ad-hoc shock capturing terms, numerical fluxes, or extrapolations. In particular, the algorithm induces a consistent upwinding not only along the streamline direction, as in SUPG and derivatives, but also along all directions originating from each flow-field point. On occasion, this algorithm is called ACURA, an acronym that stands for Acoustics-Convection Upstream Resolution Algorithm, due to its reliance on multi-dimensional acoustics and convection.

One of the reasons for selecting the finite element method is due to its versatility in generating reliable approximations in complex flow domains, on non-uniform and unstructured grids. The method enjoys optimal approximation properties for elliptic and parabolic governing equations. It also produces unambiguous, consistent, and accurate arbitrary-grid approximations of source terms as well as the second-order partial derivatives in the heat-conduction and shear-stress expressions. As another of its unique attributes, the method directly and naturally yields several surface integrals for the straightforward and effective enforcement of physical boundary conditions such as wall tangency, surface tractions, or outlet pressure, for multi-dimensional flows and without requiring any additional localized extrapolation techniques. The methods developed in this book also show how to use the efficient linear element accurately in arbitrary geometries, without any need for Gaussian quadratures.

Recent efforts have shown increasing interest in the adoption of finite element methods in CFD, especially when the methods incorporate results from characteristics-based approximation techniques. Early finite element methods, in fact, generated approximations independently from characteristic theory and required addition of extra diffusion terms to generate stable solutions. Genuine characteristic approximations too induce diffusion, but they do so inherently in the approximation of partial derivatives along characteristic directions. These approximations with emphasis, or bias, on characteristic directions model the physical situation that the flow state at a point is affected by waves that arrive at that point from characteristic directions. As a result, characteristic-based approximations generate discrete models that can become intrinsically stable.

Invariably, computational solutions of time dependent partial differential equations involve approximations of the space and time derivatives. The approximation in space yields a system of non-linear ordinary differential equations (ODE), which may be solved using any of the well established numerical methods to solve ODE's; this specific sequence is also known as the method of lines. This book adopts implicit Runge-Kutta methods for the numerical integration of these ODE's, because such methods are proven stable for non-linear differential equations.

The book is arranged in six parts that cover the derivation of the chief fluid dynamics equations, CFD approximations, the finite element method, and computational investigations of incompressible, free-surface, and compressible flows. Part I in Chapters 1-5 derives the governing equations for unsteady multi-dimensional and quasi-one-dimensional inviscid and viscous flows. These equations are the heart of fluid dynamics. The author feels so very strongly about their importance that he has detailed how these equations directly originate from the physical principles of conservation of mass, the second law of mechanics and the first principle of thermodynamics. These first 5 chapters may also be regarded as a condensed course in continuum mechanics. Part II contains Chapter 6-8; Chapter 6 examines the spatial-discretization and time-integration processes needed in the development of CFD algorithms; Chapter 7 develops the finite element method and Chapter 8 details several implicit non-linearly stable Runge-Kutta time integrations. Part III details in Chapters 9-10 the characteristics-bias companion system for quasi-one-dimensional formulations. This system is then employed in Part IV, in Chapters 11-13, numerically to study quasi-one-dimensional incompressible, free-surface, and compressible flows. These numerical investigations cover the traditional isentropic and adiabatic flows, but also non-isentropic and non-adiabatic flows with friction, shaft work and heat as well as mass transfer. Part V in Chapter 14 details the establishment of the characteristics-bias system for two- and three-dimensional formulations. This system is further developed in Part VI, covering Chapters 15-17, computationally to investigate multi-dimensional incompressible, free-surface, and compressible flows. These chapters also detail a set of boundary conditions and practical enforcement methods.

The book assumes an expertise in vector calculus, ordinary and partial differential equations, numerical methods, and fluid mechanics, as possessed by proficient senior engineering and science undergraduate students in US universities. While preserving rigor, the presentation avoids lengthy mathematical proofs and directly focuses on chief ideas. Through the use of a few pedagogically helpful calibrated repetitions, but with each cycle incorporating additional developments, several chapters remain essentially self contained, so the reader with a specific interest does not have to follow the various chapters in sequence. For a focus on quasi-one-dimensional non-reacting incompressible and compressible flows, it suffices to consider Chapters 1-2, 4-11, 13. Those readers with an interest in quasi-one-dimensional free-surface flows need only read Chapters 5-10, 12; for two-dimensional free-surface flows, Chapters 14 and 16 are also necessary. The author recommends Chapters 1-2, 4-11, 14-15 for an emphasis on incompressible flows, and Chapters 1-10, 13, 14, 17 for compressible flows.

Over the years I have enjoyed and benefited from professional associations and discussions with numerous colleagues, whom I thank with appreciation. Their work, opinions, and encouragement have contributed to my views of CFD and stimulated my wish to present an organic, systematic and unified presentation of characteristics-bias Finite Element CFD in this book. Interested readers' remarks will reveal whether the book has achieved this goal. I will welcome all comments and reply.

I am grateful for their support to Dr. John Wendt, professor and former director of the von Karman Institute for Fluid Dynamics, Dr. Louis Povinelli, Chief Propulsion and Turbomachinery Scientist at NASA Glenn, Drs. Jerry Baker, Don Dareing, Richard Jendrucko, Jack Wasserman, and Jack Weitsman, professors in the Mechanical, Aerospace and Biomedical Engineering department at the University of Tennessee, Knoxville, and Prof. Constantine Arcoumanis, fellow of the Royal Engineering Society and dean of the School of Engineering and Mathematical Sciences at the City University of London. I am also grateful to Springer for publishing this Characteristics Finite Element book. Springer has enthusiastically welcomed this project and I could not have found a more friendly, helpful, and understanding editor than Dr. Dieter Merkle, director of Springer Engineering publishing, whose support is a pleasure to acknowledge.

Several books conclude their prefaces with a dedication like: " To my spouse for supporting me in the writing of this book". Without limits, I dedicate this book to my wife Kimberly with love.

Joe Iannelli
January 1, 2006

Contents

A Brief History of Theoretical and Computational Fluid Dynamics

Fluid Mechanics ideas had already emerged in the 3rd century BC in Archimedes' treatise: "On Floating Bodies". Yet, the rational foundations for experimental Fluid Dynamics only evolved in the 17th century, in France and England, with Europe remaining the cradle of theoretical Fluid Dynamics, in the 18th and 19th centuries. Sir Isaac Newton can be credited with the initiation, in 1687, of scientific investigation of fluid flow in his "Principia". The famous Bernoulli's theorem emerged in 1738 in Daniel Bernoulli's investigations. The continuity and linear-momentum equations were developed in 1755 by Leonard Euler for both incompressible and compressible flows, under the fundamental Euler viewpoint. Navier in 1822 and Poisson in 1829 were the first to derive the viscous-flow linear-momentum equations and the stress tensor, in terms of velocity gradients. Beginning from fundamental mechanics principles, in 1845, Stokes independently developed the same linear-momentum equations, which are now known as the Navier-Stokes equations. However, it was Saint-Venant who first produced these equations in 1843 based on a rational consideration of the internal viscous stresses. Based on the kinetic theory of gases, in 1866, Maxwell then developed expressions for the coefficients of viscosity and heat conductivity.

Although the equations governing the flow of a viscous, heat-conducting fluid were available, theoretical investigations of fluid flows continued under the simplifying assumption of inviscid non heat conducting flows. These investigations, however, significantly, contributed to the development of Gas Dynamics. In 1808, Poisson studied simple waves and in 1839 Saint-Venant and Wantzel reported their study on the outflow of gases from a highly pressurized vessel. In 1860, one of Riemann's papers detailed the propagation of finite-amplitude waves within air, while the period 1870-1881 witnessed the evolution of compression shock wave theory at the hands of Rankine and Hugoniot, with E. Mach describing the oblique shock waves generated by projectiles flying at supersonic speeds. At the dawn of the 20th century, in 1904, the role of viscosity in the calculation of drag was eventually clarified in Prandtl's seminal boundary layer theory.

The mathematical models of Fluid Mechanics are thus the equations that mathematically express the second law of Newtonian mechanics, conservation of mass and first principle of Thermodynamics. For the case of a viscous, heat-conducting fluid, these non-linear partial differential equations are frequently labeled as the Navier-Stokes system, even though the original Navier-Stokes equations only correspond to the linear momentum equations. The equations governing the flow of an inviscid non heat-conducting fluid are the energy and Euler equations, which encompass both the conservation-of-mass (continuity) and linear-momentum equations. Providing a lucid synthesis of centuries of investigations, the Navier- Stokes or Euler equations, like Maxwell's equations of Electromagnetism, represent the theoretical foundations of Fluid Dynamics.

CFD began to rise at the horizon of Fluid Dynamics in the early 20th century. In 1910 Richardson introduced his point iterative schemes for numerically solving the potential and biharmonic equations. The first reported attempt to investigate fluid flows, related to weather analysis, by a manual numerical solution of partial differential equations took place in 1917, again at the hands of Richardson. Albeit unsuccessful, his brave attempt contributed to the birth of CFD. In a 1928 celebrated paper, Courant, Friedrichs, and Lewy detailed their work on the numerical solution of partial differential equations and introduced the CFL stability requirement for the explicit computational solution of hyperbolic equations. This requirement constrains the so called Courant or CFL number. Computational solutions of viscous-flow problems began to appear in the 1930's and Southwell's 1940 relaxation scheme was extensively used to solve fluid dynamics problems.

Beginning with the wartime effort of the 1940's and the emergence of the electronic computer, CFD has been successfully used to investigate a variety of flows. In was in these years that von Neumann introduced his practical method for assessing the stability of discrete schemes for the solution of time dependent partial differential equations; only in 1950 was a detailed account of this method published by O'Brien, Hyman, and Kaplan. With reference to the calculation of shocked flows, in 1954 Lax introduced the concept of shock-capturing. This procedure consists in numerically solving the governing equations in conservation law form in order to let a shock wave appear automatically within the computational domain, without any special fitting procedure.

The increase of the number of unknowns in discrete models required faster and more efficient solvers. Peaceman and Rachford in 1955 and Douglas and Rachford in 1956 described their alternating direction implicit, or ADI, schemes, in which two- and three-dimensional problems are solved implicitly by a sequence of one-dimensional implicit solutions along grid lines. In 1960 Lax and Wendroff described a second-order accurate method for computing shocked flows and MacCormack's 1969 version of this algorithm enjoyed considerable popularity. Briley and MacDonald in 1973 and then Beam and Warming in 1976, 1978 extended the ADI method to the solution of the fluid dynamics Euler and Navier-Stokes equations. A decisive event that firmly established the usefulness of CFD for calculating flows with embedded shock waves took place in 1966 when Moretti and Abbett calculated the steady shocked flow field about a blunt body as the asymptotic steady-state solution of the time dependent Euler equations. Their solution followed years and millions of research dollars that had been invested in the USA to solve this intricate problem, hardly tractable by other means.

For the solution of the Euler equations, in 1974 and 1979, van Leer described a process to develop high-order schemes based on the original 1959 research of Godunov's, which employs the solution of a Riemann problem at grid-cell sides. The computational work required in this method led Roe, in 1980, to introduce the concept of an approximate Riemann solver linked with the flux-difference splitting scheme. Steger and Warming in 1979 and then van Leer in 1982 introduced their flux-vector splittings. All of these splitting schemes have led to efficient calculations of convection dominated flows and several production codes.

Concerning the Finite Element Method, precursor developments, in 1909-1915, include the functional minimization procedure by Ritz and the special weighted residual method by Galerkin, in which the approximation functions equal the weight functions. In its discrete approximation form, the Finite Element Method originated with Engineers in the 1950's in their analysis of aircraft structures. Turner et al. in 1956, and Argyris and Clough in the 1960s were among the first to publish papers on the application of the method, with Clough coining the phrase "Finite Element Method" to denote this systematic approximation procedure. Significant books on the theory and applications of the FEM include those of Aubin, Aziz, Babuska, Baker, Carey, Ciarlet, Fix, Hughes, Oden, Pepper, Pironneau, Raviart, Reddy, Strang, and Zienkiewicz. In particular, Bathe's wide-ranging contributions have illuminated the field of finite element analysis, as exemplified by his notable 1976, 1982 and 1996 books.

Newton asserted we can see far because we climb on giants' shoulders; by the same token, CFD scientists can see far into the remarkable possibilities of this third fluid dynamics dimension because we have all benefitted from the contributions of acknowledged computational mechanics giants. Progressing exponentially in the past 25 years, CFD has now risen to a stature equal to theoretical and experimental Fluid Dynamics, as an invaluable discipline for investigating fluid flows.

Chapter 1

Governing Equations
of Fluid Mechanics

This chapter develops the main Fluid Mechanics equations in order to describe concisely in one book the foundations of the mathematical models for Computational Fluid Dynamics. These equations originate from continuum mechanics, first thermomechanical principles, and a definition of a fluid as a substance that deforms continuously under a shear force of any magnitude.

In the mathematical description of a fluid, the continuum mechanics point of simplifies the analysis, hence a fluid will consist of a continuous distribution of matter, even though open spaces exist among the molecules of a fluid, either a liquid or a gas. On the other hand, 1 mm^3 of air, or any ideal gas, at 1 atm and 298.15 K already contains 2.46×10^{16} molecules; the same volume contains an even greater number of molecules of a liquid, which binds molecules more tightly than a gas. A sufficient number of molecules of a fluid thus exists in even smaller volumes for meaningful statistical averages for macroscopic fluid properties. As a result, the continuum specification emerges as an accurate model for a mathematically convenient investigation of Fluid Mechanics.

The fundamental Fluid Mechanics field equations emanate from the principle of conservation of mass, the second law of Newtonian mechanics, and the first law of Thermodynamics as applied to a continuous distribution of fluid. On the basis of these first principles, the equations can all be systematically developed in both integral and differential form using the transport theorem. These developments encompass the subtle difference between mechanical and thermodynamical pressure, various forms of the energy equation, and a compact form of the field equation governing entropy variations, which are shown to depend not on the expansion work of pressure, but on the heat exchanged with a fluid particle and the deformation work of the viscous stresses.

1.1 Fluid Particle

The notion of a fluid particle occupies a central position in Fluid Dynamics. A fluid particle consists of a collection of a number of molecules that is sufficiently high to apply the continuum model, while the particle size approaches the infinitesimal volume element. Following the mechanics of systems of elementary masses, we can then identify a mass center for each fluid particle. The velocity of a fluid particle therefore signifies the velocity of the mass center of that particle.

With "ℓ" denoting a generic molecule with mass δm_ℓ, the mass as well as position and velocity of the mass center "G" of a particle containing "N" molecules are

$$\delta m = \sum_{\ell=1}^{N} \delta m_\ell, \qquad \boldsymbol{x}_G = \frac{1}{\delta m} \sum_{\ell=1}^{N} \delta m_\ell \boldsymbol{x}_\ell, \qquad \boldsymbol{V}_G = \frac{1}{\delta m} \sum_{\ell=1}^{N} \delta m_\ell \boldsymbol{V}_\ell \qquad (1.1)$$

The inertial velocity of a molecule of fluid can then be expressed in terms of the velocity of the fluid particle and the velocity of the molecule relative to the particle mass center, as illustrated in Figure 1.1. The third expression in (1.1), furthermore, allows determining the relative velocity of the fluid particle mass center with respect to itself, which obviously vanishes. These results are mathematical expressed as

$$\boldsymbol{V}_\ell = \boldsymbol{V}_G + \boldsymbol{V}_{\ell/G}, \qquad \boldsymbol{0} = \boldsymbol{V}_{G/G} = \frac{1}{\delta m} \sum_{\ell=1}^{N} \delta m_\ell \boldsymbol{V}_{\ell/G} \qquad (1.2)$$

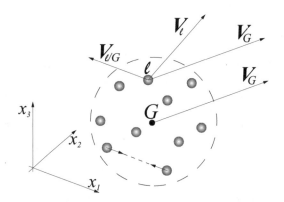

Figure 1.1: Particle Molecules and Mass Center G

From these conclusions, the kinetic energy KE of a fluid particle accrues as

$$\text{KE} = \frac{1}{2} \sum_{\ell=1}^{N} \delta m_\ell \boldsymbol{V}_\ell \cdot \boldsymbol{V}_\ell = \frac{1}{2} \delta m \boldsymbol{V}_G \cdot \boldsymbol{V}_G + \frac{1}{2} \sum_{\ell=1}^{N} \delta m_\ell \boldsymbol{V}_{\ell/G} \cdot \boldsymbol{V}_{\ell/G} \qquad (1.3)$$

which states that the kinetic energy of a fluid particle equals the kinetic energy of the particle mass center, as though all the particle mass were concentrated at the mass center, in addition to the kinetic energy of the molecular motion about the particle mass center.

The kinetic energy of a particle mass center can only be varied by the work of forces external to the particle, as the following developments show. Upon using expressions (1.1), the motion of a fluid molecule "ℓ" and a fluid particle are governed by the second law of Newtonian mechanics as

$$\delta \boldsymbol{F}_\ell = \delta \boldsymbol{F}_{E_\ell} + \delta \boldsymbol{F}_{I_\ell} = \delta m_\ell \frac{d\boldsymbol{V}_\ell}{dt} \quad \Rightarrow \quad \delta \boldsymbol{F}_E = \sum_{\ell=1}^{N} \delta \boldsymbol{F}_\ell = \delta m \frac{d\boldsymbol{V}_G}{dt} \tag{1.4}$$

where $\delta \boldsymbol{F}_E$ indicates the total external force on a particle, while $\delta \boldsymbol{F}_{E_\ell}$ and $\delta \boldsymbol{F}_{I_\ell}$ respectively denote the external and internal forces acting on molecule "ℓ". The external force is for instance due to a gravitational or electromagnetic field, whereas the internal force originates from the other molecules within the fluid particle. As shown in Figure 1.1, internal forces among any two molecules, therefore, occur in pairs, one acting on each of any two interacting molecules. Internal forces also obey the third law of mechanics, or the principle of action and reaction, according to which the forces in each pair posses equal magnitude and act in opposite senses along their common direction; as a consequence of this law, the vector sum of internal forces vanishes. The elementary work performed by all the $\delta \boldsymbol{F}_\ell$ forces as the corresponding molecules move in the interval δt is expressed as

$$\sum_{\ell=1}^{N} \delta \boldsymbol{F}_\ell \cdot \delta \boldsymbol{x}_\ell = \sum_{\ell=1}^{N} \delta \boldsymbol{F}_\ell \cdot \boldsymbol{V}_\ell \delta t = \sum_{\ell=1}^{N} \delta m_\ell \frac{d\boldsymbol{V}_\ell}{dt} \cdot \boldsymbol{V}_\ell \delta t = d \left(\sum_{\ell=1}^{N} \frac{1}{2} \delta m_\ell \boldsymbol{V}_\ell \cdot \boldsymbol{V}_\ell \right) \tag{1.5}$$

as results from using (1.4). Upon expressing the velocity of molecule "ℓ" as in (1.2), the work (1.5) becomes

$$\sum_{\ell=1}^{N} \delta \boldsymbol{F}_\ell \cdot \boldsymbol{V}_\ell \delta t = \sum_{\ell=1}^{N} \left(\delta \boldsymbol{F}_{E_\ell} + \delta \boldsymbol{F}_{I_\ell} \right) \cdot \left(\boldsymbol{V}_G + \boldsymbol{V}_{\ell/G} \right) \delta t =$$

$$\delta \boldsymbol{F}_E \cdot \boldsymbol{V}_G \delta t + \sum_{\ell=1}^{N} \delta \boldsymbol{F}_{E_\ell} \cdot \boldsymbol{V}_{\ell/G} \delta t + \underbrace{\left(\sum_{\ell=1}^{N} \delta \boldsymbol{F}_{I_\ell} \right) \cdot \boldsymbol{V}_G}_{=0} + \sum_{\ell=1}^{N} \delta \boldsymbol{F}_{I_\ell} \cdot \boldsymbol{V}_{\ell/G} \delta t =$$

$$= \delta \boldsymbol{F}_E \cdot \boldsymbol{V}_G \delta t + \sum_{\ell=1}^{N} \delta \boldsymbol{F}_\ell \cdot \boldsymbol{V}_{\ell/G} \delta t \tag{1.6}$$

where $\delta \boldsymbol{F}_E$ denotes the total external force on the particle, while the sum of internal forces vanishes because of the third law of mechanics. Upon combining (1.3), (1.5), and (1.6), the work kinetic-energy principle emerges as

$$\delta \boldsymbol{F}_E \cdot \boldsymbol{V}_G \delta t + \sum_{\ell=1}^{N} \delta \boldsymbol{F}_\ell \cdot \boldsymbol{V}_{\ell/G} \delta t = d \left(\frac{1}{2} \delta m \boldsymbol{V}_G \cdot \boldsymbol{V}_G \right) + d \left(\sum_{\ell=1}^{N} \frac{1}{2} \delta m_\ell \boldsymbol{V}_{\ell/G} \cdot \boldsymbol{V}_{\ell/G} \right) \tag{1.7}$$

On the other hand, based on the second expression in (1.4), the calculation of the work of the total external force on the particle $\delta \boldsymbol{F}_E$ through the elementary displacement $\boldsymbol{V}_G \delta t$ of the particle leads to

$$\delta \boldsymbol{F}_E \cdot \boldsymbol{V}_G \delta t = d \left(\frac{1}{2} \delta m \boldsymbol{V}_G \cdot \boldsymbol{V}_G \right) \tag{1.8}$$

which through (1.7) in turn yields

$$\sum_{\ell=1}^{N} \delta \boldsymbol{F}_\ell \cdot \boldsymbol{V}_{\ell/G} \delta t = d \left(\sum_{\ell=1}^{N} \frac{1}{2} \delta m_\ell \boldsymbol{V}_{\ell/G} \cdot \boldsymbol{V}_{\ell/G} \right) \tag{1.9}$$

These two results posses a fundamental physical significance. The first result indicates that the work of the external force by means of the elementary displacement of the fluid particle is the only agent that can vary the particle mass-center kinetic energy. The second result corresponds to the work as observed from a reference frame that moves along with the fluid particle with the particle mass-center velocity. In this reference frame, the only motion is that of the particle molecules about the particle mass center; hence, the work corresponding to the displacements about the mass center can only vary the kinetic energy of the motion about the mass center. If the particle were rigid, there would not be any motion about the mass center and the only non vanishing work would be only due to the motion of the particle mass center. Expression (1.9) thus corresponds to the work and energy of deformation.

For each fluid particle, the second law of Newtonian mechanics governs the motion of the particle mass center, as indicated in (1.4), and the first law of thermodynamics governs the evolution of the particle total energy, which includes the kinetic energy of the molecular motion about the particle mass center, as elaborated in Section 1.10.2. Since the reference point for each fluid particle is always its mass center, the subscript "G" will be implicitly understood in every expression dealing with the motion of a fluid particle.

1.2   Mathematical Description of Fluid Flow

The motion of a fluid particle can in principle be investigated by determining the time variation of the particle coordinates $x_p = x_p(t)$. This description of Fluid Dynamics corresponds to the Lagrangian viewpoint, according to which the dynamics of a mass of fluid is investigated by following the motion of all the particles. The sheer number of these particles makes this approach a formidable task. The Eulerian viewpoint allows more convenient investigations with its emphasis on velocity and acceleration at space points rather than of particles. In this viewpoint, the velocity and acceleration at a space point at any given time respectively signify the velocity and acceleration of that fluid particle whose mass center travels through the specified point at the given time. While both concerned with the motion of fluid particles, these viewpoints inherently differ, for distinct particles at different times can cross any given space point.

In the Eulerian description, a fluid flow mathematically corresponds to a time - dependent coordinate transformation, [134, 174]. At a reference initial time t_0, let x_0 denote the associated fluid particle position. As time elapses the flow field is represented as

$$x = x(x_0, t) \tag{1.10}$$

where x denote the Eulerian coordinates of points within the domain of the fluid flow investigated. For a fixed x_0, hence for a specific particle, these coordinates depend upon t because the fluid particle generally moves. At any given time level t, these coordinates depend on x_0 to identify the position at time t of that particle that was at position x_0 at time t_0.

Transformation (1.10) has to satisfy three conditions in order to describe a physically possible fluid flow. With reference to Figure 1.2, these conditions are

$$x(x_0, t_0) = x_0 \tag{1.11}$$

$$x_1 = x(x_{0_1}, t) \neq x_2 = x(x_{0_2}, t) \quad \text{whenever} \quad x_{0_1} \neq x_{0_2} \tag{1.12}$$

$$\boldsymbol{x} = \boldsymbol{x}\left(\boldsymbol{x}_0, t\right) \quad \text{has a smooth inverse} \quad \Rightarrow \boldsymbol{x}_0 = \boldsymbol{x}_0\left(\boldsymbol{x}, t\right) \tag{1.13}$$

With these conditions, transformation (1.10) maps the boundary of a fluid domain at time t_0 into the boundary the fluid domain achieves at time t.

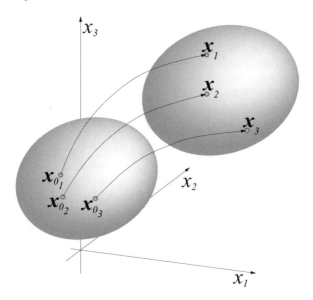

Figure 1.2: Flow of Fluid Particles

The first condition corresponds to the initial position condition. The second condition stipulates the fluid particle identity condition, hence two fluid particles $\boldsymbol{x}_{0_1}$ and $\boldsymbol{x}_{0_2}$ that are initially distinct will keep their individual masses and remain separate, without merging into each other. The third condition conforms to a traceability specification, according to which no fluid particle duplication ever takes place during the flow, so that a fluid particle at a given position at a specified time level corresponds to a single particle whose position at the initial time can always be uniquely determined.

With these specifications, the velocity field relates to transformation (1.10) as

$$\boldsymbol{V}\left(\boldsymbol{x}, t\right) \equiv \left.\frac{\partial \boldsymbol{x}}{\partial t}\right|_{\text{constant } \boldsymbol{x}_0} \tag{1.14}$$

which provides the velocity at any given time of the fluid particles within the fluid flow domain under investigation. This partial derivative obviously yields an initial function of $\boldsymbol{x}_0$ and t; by virtue of (1.13), subsequently, the eventual velocity field $\boldsymbol{V} = \boldsymbol{V}\left(\boldsymbol{x}, t\right)$ depends on $\boldsymbol{x}$.

The Eulerian viewpoint naturally leads to the idea of a "steady", that is time-independent", flow field in which $\boldsymbol{V} = \boldsymbol{V}\left(\boldsymbol{x}\right)$ only. Certainly, transformation (1.10) has to depend upon time, otherwise it would correspond to no motion of fluid particles and $\boldsymbol{V}$ from (1.14) would identically vanish. On the other hand $\boldsymbol{V}$ can remain independent of t, which signifies that fluid particles will transit by position $\boldsymbol{x}$ with the same velocity at every time level. This is precisely the meaning of a steady field. The acceleration of a steady velocity field, however, does not necessarily vanish. In fact it evolves by remembering that acceleration corresponds to the time rate of change of the velocity of a fluid particle, hence in determining acceleration from $\boldsymbol{V}$, $\boldsymbol{x}$ in $\boldsymbol{V}$ depends itself on t through (1.10). This consideration corresponds to the substantive time derivative, detailed in Section 1.8.1.

To exemplify the concepts of a coordinate transformation that represents a fluid flow and of a steady velocity field from a time-dependent transformation consider the following example of a steady two-dimensional flow field

$$\boldsymbol{V}(x, y) = y\boldsymbol{e}_1 + x\boldsymbol{e}_2 \tag{1.15}$$

where $\boldsymbol{e}_1$ and $\boldsymbol{e}_2$ respectively denote unit vectors in the x and y directions.

From the definition of velocity (1.14), the coordinate transformation (1.10) satisfies the differential system

$$\begin{cases} \dfrac{dx}{dt} = y \\[2mm] \dfrac{dy}{dt} = x \end{cases} \tag{1.16}$$

with solution

$$\begin{cases} x(x_0, y_0, t) = \dfrac{1}{2}\left[(x_0 + y_0)\exp(t) + (x_0 - y_0)\exp(-t)\right] \\[3mm] y(x_0, y_0, t) = \dfrac{1}{2}\left[(x_0 + y_0)\exp(t) - (x_0 - y_0)\exp(-t)\right] \end{cases} \tag{1.17}$$

where x_0 and y_0 denote initial-position coordinates. This solution conforms to a specific instance of transformation (1.10) that depends upon t and possesses the smooth inverse

$$\begin{cases} x_0(x, y, t) = \dfrac{1}{2}\left[(x + y)\exp(-t) + (x - y)\exp(t)\right] \\[3mm] y_0(x, y, t) = \dfrac{1}{2}\left[(x + y)\exp(-t) - (x - y)\exp(t)\right] \end{cases} \tag{1.18}$$

The partial derivatives of (1.17), in agreement with (1.14) and in conjunction with this smooth inverse then identically returns the steady velocity field (1.15).

This example can also be used to show how a fluid domain deforms with time and how a deforming time-varying fluid domain can be mapped onto a time- invariant domain. Since (1.16) corresponds to a two-dimensional velocity field, consider the flow of a fluid area Ω from an initial instant t_0 to a representative instant t, as all the fluid particles within Ω flow and consequently Ω deforms from its initial shape Ω_0. The boundary $\partial\Omega$ of Ω corresponds to a closed curve and each point on $\partial\Omega$ corresponds to one fluid particle as $\partial\Omega$ deforms, following the boundary particles that initially constituted $\partial\Omega_0$. For simplicity, take for $\partial\Omega_0$

a unit - radius circumference with center of coordinates $(-2; 2)$; the corresponding area Ω_0 is thus a circle. The parametric equation of $\partial\Omega_0$ are

$$x_0 = -2 + \cos\theta, \qquad y_0 = 2 + \sin\theta \tag{1.19}$$

where the parameter θ varies in the range $0 \le \theta \le 2\pi$. The area Ω_0 correspondingly contains all the particles for which $(x_0 + 2)^2 + (y_0 - 2)^2 < 1$. The parametric equation of $\partial\Omega_t$, as time t elapses, results from inserting (1.19) into (1.17) which yields

$$\begin{cases} x\left(\overline{x}_0, \overline{y}_0, t\right) &= \dfrac{1}{2}\left[\left(\sin\theta + \cos\theta\right)\exp(t) + \left(\cos\theta - \sin\theta - 4\right)\exp(-t)\right] \\[2mm] y\left(\overline{x}_0, \overline{y}_0, t\right) &= \dfrac{1}{2}\left[\left(\sin\theta + \cos\theta\right)\exp(t) - \left(\cos\theta - \sin\theta - 4\right)\exp(-t)\right] \end{cases} \tag{1.20}$$

where the bar over both x_0 and y_0 signifies that these variables are no longer independent of each other on $\partial\Omega_0$, but are correlated through (1.19). Figure 1.3 presents the flow of $\partial\Omega_t$ at the three instants of time $t = 0$, $t = 1/2$, and $t = 1$. The equation of each curve in this figure

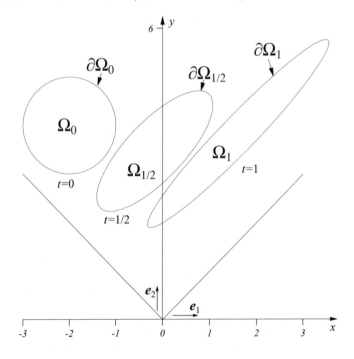

Figure 1.3: Flow of $\partial\Omega_t$

results from (1.20) by setting t equal to a time level and letting θ vary within its range; each flow particle thus corresponds to one value of θ. This figure vividly illustrates the concept of fluid motion, deformation, and fluid-flow coordinate transformation. The representative flow field (1.16) convects and deforms the initial circle on the flow plane, as each particle on the circle flows according to its own velocity, which can be obtained by differentiating (1.17) with respect to time. In particular, note how the initially closed curve is continuously mapped onto curves that remain closed but smoothly change shape. Transformation (1.18) can then be used to demonstrate how to map a time-dependent domain Ω_t onto a time-invariant domain Ω_0. It suffices to insert into (1.18) the coordinates x and y from (1.20) and verify that the resulting expressions for x_0 and y_0 no longer depend upon time, but naturally revert to (1.19). This result is obvious because (1.20) initially evolved from (1.19); however this sequence exemplifies the general process of mapping a time-varying boundary onto a time-invariant one when the process directly begins with the equations of $\partial\Omega_t$ and generates the corresponding expressions for x_0 and y_0 at the reference initial time.

1.3   Transport Theorem

Mass $\mathcal{M}$ and density ρ within a fluid volume Ω relate as

$$\mathcal{M} = \int_{\mathcal{M}} d\mathcal{M} = \int_{\Omega} \rho \, d\Omega \qquad (1.21)$$

They exemplify a pair of an extensive variable, $\mathcal{M}$ and a volume - specific intensive one, ρ, where the extensive variable depends on the spatial extension of the fluid volume, whereas the intensive variable has an "intensity" that only depends on time and coordinates of each point within Ω. Pairs of extensive - intensive quantities may be defined for scalar, vector and tensor variables. For generality's sake, they are respectively denoted as B and β.

For an arbitrary fluid volume Ω, any extensive fluid property B is thus connected to its corresponding intensive quantity $\beta\rho$ through the integral

$$B(t) = \int_{\Omega} \beta\rho \, d\Omega = \int_{\Omega(\, \boldsymbol{x}(\boldsymbol{x}_0,t)\,)} \beta\left(\boldsymbol{x},t\right)\rho\left(\boldsymbol{x},t\right) \, d\Omega = \int_{\Omega} \beta_\rho \, d\Omega \qquad (1.22)$$

where the continuous $\beta_\rho \equiv \beta\rho$ for convenience in the following proofs. The quantity β specifies a mass - specific intensive quantity and $\beta\rho$ indicates the corresponding volume - specific intensive quantity. The fundamental transport theorem, which corresponds to Leibnitz's rule for the differentiation of a multidimensional integral, provides the time rate of change of the extensive property B in terms of the time rates of change of β_ρ and Ω. By virtue of its importance, this theorem is proved in the following sections.

Figure 1.4 presents the flow of Ω from an initial instant t_0 to a representative instant t. Each point on the boundary $\partial\Omega$ of Ω corresponds to one fluid particle, as all the particles within Ω flow and Ω deforms. The flow of Ω corresponds to the motion of the fluid particles initially within Ω_0, as the boundary $\partial\Omega$ deforms following the boundary particles that were initially on $\partial\Omega_0$. The fluid domain $\Omega = \Omega(\, \boldsymbol{x}(\boldsymbol{x}_0,t)\,)$ only implicitly depends upon t, through $\boldsymbol{x}$, precisely because its motion follows the flowing particles; if Ω explicitly depended upon t, it would move independently of the fluid and fluid particles would thus cross its boundary. In this representation, however, no fluid ever crosses the boundary $\partial\Omega$.

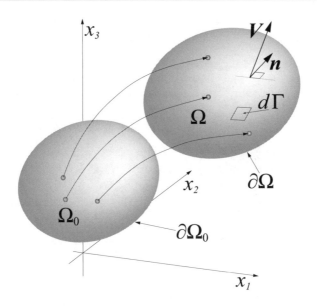

Figure 1.4: Flow of a Fluid Mass Ω

1.3.1 Surface Integral Form

The surface integral form of the transport theorem is

$$\frac{DB}{Dt} = \int_{\Omega} \frac{\partial(\beta\rho)}{\partial t}\, d\Omega + \oint_{\partial\Omega} \beta\rho \boldsymbol{V} \cdot \boldsymbol{n}\, d\Gamma = \int_{\Omega} \frac{\partial\beta_{\rho}}{\partial t}\, d\Omega + \oint_{\partial\Omega} \beta_{\rho}\boldsymbol{V} \cdot \boldsymbol{n}\, d\Gamma \qquad (1.23)$$

where with reference to Figure 1.4, $\boldsymbol{V}$, $\boldsymbol{n}$ and $d\Gamma$ respectively denote the local fluid velocity, outward pointing unit vector and surface element.

To prove (1.23), consider the definition of time rate of change

$$\frac{DB}{Dt} = \lim_{\Delta t \to 0} \frac{B(t+\Delta t) - B(t)}{\Delta t} = \lim_{\Delta t \to 0} \frac{1}{\Delta t} \left(\int_{\Omega(\boldsymbol{x}(\boldsymbol{x}_0, t+\Delta t))} \beta_{\rho}(\boldsymbol{x}, t+\Delta t)\, d\Omega - \int_{\Omega(\boldsymbol{x}(\boldsymbol{x}_0, t))} \beta_{\rho}(\boldsymbol{x}, t)\, d\Omega \right) \tag{1.24}$$

With reference to Figure 1.5, the domains of integration in (1.24) are cast as

$$\begin{aligned} \Omega(\,\boldsymbol{x}(\boldsymbol{x}_0, t+\Delta t)\,) &= \Omega_{\text{core}} + \Delta\Omega_{\text{out}} \\ \Omega(\,\boldsymbol{x}(\boldsymbol{x}_0, t)\,) &= \Omega_{\text{core}} + \Delta\Omega_{\text{in}} \end{aligned} \tag{1.25}$$

with

$$\lim_{\Delta t \to 0} \Delta\Omega_{\text{out}} = 0, \quad \lim_{\Delta t \to 0} \Omega_{\text{core}} = \Omega(\,\boldsymbol{x}(\boldsymbol{x}_0, t)\,) \quad \Rightarrow \quad \lim_{\Delta t \to 0} \Delta\Omega_{\text{in}} = 0 \tag{1.26}$$

With these specifications, (1.23) can be expressed as

$$\frac{DB}{Dt} = \lim_{\Delta t \to 0} \frac{1}{\Delta t} \left(\int_{\Omega(\,\boldsymbol{x}(\boldsymbol{x}_0, t+\Delta t)\,)} \beta_{\rho}(\boldsymbol{x}, t+\Delta t)\, d\Omega - \int_{\Omega(\,\boldsymbol{x}(\boldsymbol{x}_0, t)\,)} \beta_{\rho}(\boldsymbol{x}, t)\, d\Omega \right)$$

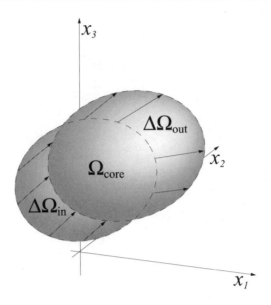

Figure 1.5: Representative Flow

$$= \lim_{\Delta t \to 0} \frac{1}{\Delta t} \left(\int_{\Omega_{core}} \beta_\rho\left(\boldsymbol{x}, t + \Delta t\right) \, d\Omega - \int_{\Omega_{core}} \beta_\rho\left(\boldsymbol{x}, t\right) \, d\Omega + \int_{\Delta\Omega_{out}} \beta_\rho\left(\boldsymbol{x}, t + \Delta t\right) \, d\Omega - \int_{\Delta\Omega_{in}} \beta_\rho\left(\boldsymbol{x}, t\right) \, d\Omega \right)$$

$$(1.27)$$

The limit of the difference of the first two integrals in (1.27) becomes

$$\lim_{\Delta t \to 0} \frac{1}{\Delta t} \left(\int_{\Omega_{core}} \beta_\rho\left(\boldsymbol{x}, t + \Delta t\right) \, d\Omega - \int_{\Omega_{core}} \beta_\rho\left(\boldsymbol{x}, t\right) \, d\Omega \right) = \int_{\Omega\left(\boldsymbol{x}(\boldsymbol{x}_0, t)\right)} \frac{\partial \beta_\rho}{\partial t} \, d\Omega \qquad (1.28)$$

which corresponds to the domain integral in (1.23). With reference to Figure 1.6, consider next the mean-value theorem and two local surface coordinate transformations, satisfying the same properties in (1.11)-(1.13); each of these local coordinate systems (s_1, s_2, s_3) is orthogonal, with the s_3 axis remaining perpendicular to the surface facet containing the (s_1, s_2) With these specifications, the remaining two integrals in (1.27) become

$$\int_{\Delta\Omega_{out}} \beta_\rho\left(\boldsymbol{x}, t + \Delta t\right) \, d\Omega = \int_{\partial\Omega_{out}} \int_0^{\Delta s_3(s_1, s_2)} \beta_\rho\left(\boldsymbol{x}(s_1, s_2, s_3), t + \Delta t\right) \, d\Gamma ds_3 =$$

$$\int_{\partial\Omega_{out}} \beta_\rho\left(\boldsymbol{x}(s_1, s_2, 0 + \alpha\Delta s_3), t + \Delta t\right) \, d\Gamma \Delta s_3 \qquad (1.29)$$

and

$$\int_{\Delta\Omega_{in}} \beta_\rho\left(\boldsymbol{x}, t\right) \, d\Omega = \int_{\partial\Omega_{in}} \int_0^{\Delta s_3(s_1, s_2)} \beta_\rho\left(\boldsymbol{x}(s_1, s_2, s_3), t\right) \, d\Gamma ds_3 =$$

$$\int_{\partial\Omega_{in}} \beta_\rho\left(\boldsymbol{x}(s_1, s_2, 0 + \alpha\Delta s_3), t\right) \, d\Gamma \Delta s_3 \qquad (1.30)$$

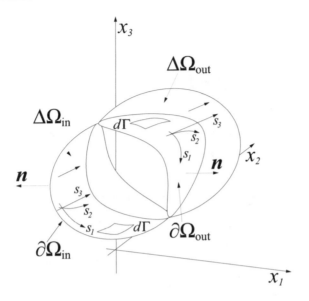

Figure 1.6: Representative Flow

where α denotes a scalar between zero and one. The limits of these two integrals become

$$\lim_{\Delta t \to 0} \int_{\partial \Omega_{\text{out}}} \beta_\rho \left(\boldsymbol{x}(s_1, s_2, 0 + \alpha \Delta s_3), t + \Delta t \right) \, d\Gamma \frac{\Delta s_3}{\Delta t} = \int_{\partial \Omega_{\text{out}}} \beta_\rho \boldsymbol{V} \cdot \boldsymbol{n} \, d\Gamma \tag{1.31}$$

and

$$\lim_{\Delta t \to 0} \int_{\partial \Omega_{\text{in}}} \beta_\rho \left(\boldsymbol{x}(s_1, s_2, 0 + \alpha \Delta s_3), t \right) \, d\Gamma \left(-\frac{\Delta s_3}{\Delta t} \right) = \int_{\partial \Omega_{\text{in}}} \beta_\rho \boldsymbol{V} \cdot \boldsymbol{n} \, d\Gamma \tag{1.32}$$

and thus the sum of these two integrals returns the surface integral in (1.23)

1.3.2   Divergence Form

When the product $\beta \rho \boldsymbol{V}$ is differentiable, then, by virtue of Gauss' theorem, (1.23) becomes

$$\frac{DB}{Dt} = \int_\Omega \left[\frac{\partial (\beta \rho)}{\partial t} + \operatorname{div} (\beta \rho \boldsymbol{V}) \right] d\Omega = \int_\Omega \left[\frac{\partial \beta_\rho}{\partial t} + \operatorname{div} (\beta_\rho \boldsymbol{V}) \right] d\Omega \tag{1.33}$$

This expression can also be proven directly. Consider the inverse coordinate transformation in (1.13) to map $\Omega(\boldsymbol{x}(\overline{\boldsymbol{x}_0}, t))$ into $\Omega_0 = \Omega(\overline{\boldsymbol{x}_0})$; the bar over $\boldsymbol{x}_0$ emphasizes that on the boundary $\partial \Omega$ of Ω the boundary coordinates are no longer independent of one another, but satisfy the equation of the boundary surface, which maps onto the time- independent coordinates $\overline{\boldsymbol{x}_0}$ of $\partial \Omega_0$. The domain $\Omega_0 = \Omega(\overline{\boldsymbol{x}_0})$, therefore, no longer depends upon t, for as soon as the coordinates $\boldsymbol{x} = \boldsymbol{x}(\overline{\boldsymbol{x}_0}, t)$ of the boundary $\partial \Omega$ of Ω are substituted into

$\boldsymbol{x}_0 = \boldsymbol{x}_0(\boldsymbol{x}, t)$ the dependence on t vanishes, because the boundary of Ω evolves from the fixed domain of Ω_0, as also exemplified by the two-dimensional flow at the end of Section 1.2. Using this inverse coordinate transformation, the domain Ω itself may be expressed as

$$\Omega = \int_{\Omega(\,\boldsymbol{x}(\overline{\boldsymbol{x}}_0, t)\,)} d\Omega = \int_{\Omega_0} \det J \, d\Omega_0 \tag{1.34}$$

where J denotes the Jacobian matrix of the coordinate transformation (1.10). Since Ω_0 does not depend upon time, the derivative of Ω with respect to time is expressed as

$$\frac{D\Omega}{Dt} = \int_{\Omega_0} \frac{\partial}{\partial t} \Big[\det J \Big]_{\text{constant } \boldsymbol{x}_0} d\Omega_0 \tag{1.35}$$

The time derivative of Ω may also be obtained directly from (1.23), which yields

$$\frac{D\Omega}{Dt} = \oint_{\partial\Omega} \boldsymbol{V} \cdot \boldsymbol{n} \, d\Gamma = \int_{\Omega} \operatorname{div}\boldsymbol{V} d\Omega = \int_{\Omega_0} \operatorname{div}\boldsymbol{V} \det J \, d\Omega_0 \tag{1.36}$$

The equality of these two integral statements for arbitrary Ω_0 yields the following expression for the derivative of $\det J$

$$\frac{\partial}{\partial t} \Big[\det J \Big]_{\text{constant } \boldsymbol{x}_0} = \operatorname{div} \boldsymbol{V} \det J \tag{1.37}$$

Consider next the integral (1.22) for the extensive property B, which may be expressed as

$$B(t) = \int_{\Omega(\,\boldsymbol{x}(\overline{\boldsymbol{x}}_0, t))} \beta_\rho(\boldsymbol{x}, t) \, d\Omega = \int_{\Omega_0} \beta_\rho(\,\boldsymbol{x}(\boldsymbol{x}_0, t), t\,) \det J \, d\Omega_0 \tag{1.38}$$

Again, since Ω_0 does not depend upon time, the derivative of B with respect to time can be brought inside the integral, which generates

$$\frac{DB}{Dt} = \int_{\Omega_0} \frac{\partial}{\partial t} \Big[\beta_\rho(\,\boldsymbol{x}(\boldsymbol{x}_0, t), t\,) \det J \Big] d\Omega_0 \tag{1.39}$$

The derivatives of β_ρ with respect to time equal

$$\frac{\partial}{\partial t} \Big[\beta_\rho(\,\boldsymbol{x}(\boldsymbol{x}_0, t), t\,) \Big]_{\text{constant } \boldsymbol{x}_0} = \frac{\partial \beta_\rho}{\partial \boldsymbol{x}} \Big|_t \frac{\partial \boldsymbol{x}}{\partial t} \Big|_{\boldsymbol{x}} + \frac{\partial \beta_\rho}{\partial t} \Big|_{\boldsymbol{x}} = \boldsymbol{V} \cdot \operatorname{grad} \beta_\rho + \frac{\partial \beta_\rho}{\partial t} \tag{1.40}$$

Using this result and (1.37), the time rate of change of B in (1.33) thus becomes

$$\frac{DB}{Dt} = \int_{\Omega_0} \Big[\frac{\partial \beta_\rho}{\partial t} + \boldsymbol{V} \cdot \operatorname{grad} \beta_\rho + \beta_\rho \operatorname{div} \boldsymbol{V} \Big] \det J \, d\Omega_0 = \int_{\Omega} \Big[\frac{\partial \beta_\rho}{\partial t} + \operatorname{div}(\beta_\rho \boldsymbol{V}) \Big] d\Omega \tag{1.41}$$

which expresses the transport theorem in divergence form.

1.4 Flowing and Stationary Fluid Volumes

Myriad Fluid Mechanics books, especially those for Engineering, rely on fixed fluid volumes, called control volumes, when they employ the transport theorem in its two forms (1.23) and (1.33). The end results are correct, yet inherently moving fluid volumes permeate the background of these applications. With reference to Figure 1.7

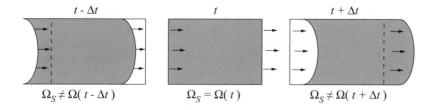

Figure 1.7: Flowing and Fixed Volumes

the fluid-mass volume $\Omega(t)$ is approaching, coinciding, and leaving the stationary control volume Ω_S. At time t, the fluid-mass volume is indistinguishable from the control volume and thus the integrals in (1.23) and (1.33) will yield the same results for Ω_S and $\Omega(t)$.

Conceiving the fluid in Ω_S as belonging to a vaster fluid mass, there will be fluid continuously flowing within Ω_S. This conception entails that the functions $\beta\rho = \beta(\boldsymbol{x},t)\rho(\boldsymbol{x},t)$ and $\boldsymbol{V} = \boldsymbol{V}(\boldsymbol{x},t)$ will remain physically meaningful when, for elapsing t, the range of $\boldsymbol{x}$ is confined to Ω_S. In this mode, the analysis no longer considers the same aggregate of fluid particles. Therefore, the analysis "boundary conditions", to be detailed in Chapters 9-17, mathematically express the effect on the fluid within Ω_S of the remaining fluid outside Ω_S.

When integrals (1.23) and (1.33) correlate with a "fixed" control volume, they thus correspond to a flowing-mass volume that wholly fills the control volume. From this perspective, the surface-integral form of the transport theorem is amenable to a different interpretation. In this case, the time rate of change of B results from two contributions: the integrated rate of change of $\beta\rho$, as if the boundaries of Ω_S were impermeable, added to the amount of $\beta\rho$ convected through the boundaries of Ω_S.

1.5 Stress Tensor

This section establishes the properties of the fluid dynamics stress tensor.

1.5.1 Geometry of a Tetrahedron

The fundamental geometric relationships among the four faces of an arbitrary tetrahedron simplify the analysis of the stress tensor on a fluid particle. These relationships are usually taken for granted, yet it is useful to see their origin. In the representative tetrahedron in Figure 1.8, three faces of this tetrahedron respectively lie on the $x_2 - x_3$, $x_3 - x_1$, and $x_1 - x_2$

planes. The fourth face corresponds to the triangle PQR with normal the unit vector $\boldsymbol{n}$, which points outside of the tetrahedron. The areas of these faces respectively are δA_1, δA_2,

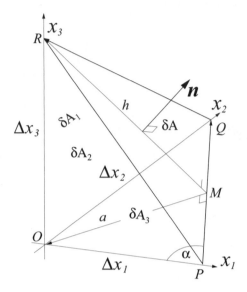

Figure 1.8: Representative Tetrahedron

δA_2, and δA. The first three faces are right triangles, and their areas are

$$\delta A_1 = \frac{1}{2}\Delta x_2 \Delta x_3, \qquad \delta A_2 = \frac{1}{2}\Delta x_3 \Delta x_1, \qquad \delta A_3 = \frac{1}{2}\Delta x_1 \Delta x_2 \tag{1.42}$$

The area δA corresponds to one half the magnitude of the cross product of the vectors $\overline{PQ}$ and $\overline{QR}$; this area may be calculated differently. From the figure, this area is $\delta A = bh/2$, where b and h respectively denote the length of the segments PQ and MR. The length of the segment PQ is $b = \sqrt{\Delta x_1^2 + \Delta x_2^2}$ and the sine of the angle α thus becomes $\sin\alpha = \Delta x_2/\sqrt{\Delta x_1^2 + \Delta x_2^2}$. Since the triangle PMO is right at M, the length a of the segment OM results as $a = \Delta x_1 \sin\alpha = \Delta x_1 \Delta x_2/\sqrt{\Delta x_1^2 + \Delta x_2^2}$. Since the vector $\overline{MO}$ is orthogonal to $\overline{PQ}$ their vector "dot" product vanishes, hence $\overline{PQ} \cdot \overline{MO} = 0$; it follows that the vector $\overline{MR}$ is also orthogonal to $\overline{PQ}$ because $\overline{PQ} \cdot \overline{MR} = \overline{PQ} \cdot \overline{MO} + \overline{PQ} \cdot \overline{OR} = 0$. The segment MR, therefore, is the height of the triangle PQR. Since the triangle ROM is right at O, the length h equals $h = \sqrt{\Delta x_3^2 + a^2}$. The area of the triangle PQR thus follows as the elegant formula

$$\delta A = \frac{1}{2}bh = \frac{1}{2}\sqrt{\Delta x_1^2 \Delta x_3^2 + \Delta x_2^2 \Delta x_1^2 + \Delta x_3^2 \Delta x_2^2} \tag{1.43}$$

The outward-pointing unit vector $\boldsymbol{n}$, orthogonal to triangle PQR, can be calculated as the cross product $\boldsymbol{n} = \overline{PQ} \times \overline{QR}/\|\overline{PQ} \times \overline{QR}\|$. From figure 1.8, the vectors $\overline{PQ}$ and $\overline{QR}$ are expressed as $\overline{PQ} = -\Delta x_1 e_1 + \Delta x_2 e_2 + 0e_3$ and $\overline{QR} = 0e_1 - \Delta x_2 e_2 + \Delta x_3 e_3$, where e_1, e_2, and e_3 denote unit vectors respectively parallel to the x_1, x_2 and x_3 axes. The unit vector n thus results as

$$n = \frac{\Delta x_2 \Delta x_3 e_1 + \Delta x_3 \Delta x_1 e_2 + \Delta x_1 \Delta x_2 e_3}{\sqrt{\Delta x_1^2 \Delta x_3^2 + \Delta x_2^2 \Delta x_1^2 + \Delta x_3^2 \Delta x_2^2}} \tag{1.44}$$

Apart from showing that $\delta A = \|\overline{PQ} \times \overline{QR}\|/2$, this result directly connects the components of n with the four areas of the tetrahedron. Comparing (1.42), (1.43), and (1.44) leads to the equalities

$$n_1 = \frac{\delta A_1}{\delta A}, \quad n_2 = \frac{\delta A_2}{\delta A}, \quad n_3 = \frac{\delta A_3}{\delta A} \tag{1.45}$$

which find important use in the analysis of the stress tensor on a fluid particle.

1.5.2 Surface Traction

With reference to Figure 1.9, the second law of mechanics along the x_1 direction for the flow of the tetrahedron yields

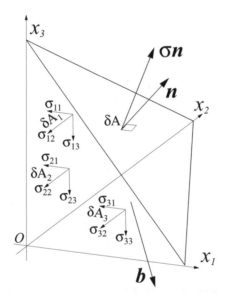

Figure 1.9: Tetrahedron Stresses

$$-\sigma_{11}\delta A_1 - \sigma_{21}\delta A_2 - \sigma_{31}\delta A_3 + \sigma_{n_1}\delta A + b_1\rho\Delta\Omega = \rho\Delta\Omega\frac{du_1}{dt} \tag{1.46}$$

where b_1, u_1, and σ_{n_1} respectively denote the fluid particle velocity and body force components in the x_1 direction and x_1 component of the surface traction on the surface with normal vector $\boldsymbol{n}$. Upon dividing through by δA, observing that $\Delta\Omega/\delta A$ vanishes as the tetrahedron shrinks to a point, and invoking results (1.45), expression (1.46) yields

$$\sigma_{n_1} = \sigma_{11}n_1 + \sigma_{21}n_2 + \sigma_{31}n_3 \tag{1.47}$$

Since similar results hold for the x_2 and x_3 directions, these results provide the Cartesian components of the traction $\boldsymbol{\sigma n}$.

1.5.3   Stress Tensor Symmetry

A certain symmetry exists among the components of $\boldsymbol{\sigma} \equiv \{\sigma_{ij}\}$, which emanates from the axial angular momentum equation. The axes considered have their origin at the mass center of the particle so that the axial moment of the body force identically vanishes. With reference to Figure 1.10 the axial moment about the x_1 axis of the surface traction equals the time

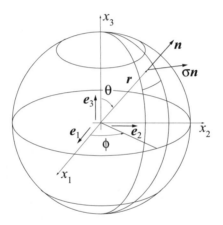

Figure 1.10: Traction $\boldsymbol{\sigma n}$ on Surface Facet

rate of change of the particle angular momentum about this axis

$$\oint_{\partial\Delta\Omega} \boldsymbol{e}_1 \cdot (\boldsymbol{r} \times (\boldsymbol{\sigma n}))\, d\Gamma = \int_{\Delta\Omega} \boldsymbol{e}_1 \cdot \left(\boldsymbol{r} \times \frac{d\boldsymbol{V}}{dt} \right) \rho\, d\Omega \tag{1.48}$$

where $\boldsymbol{r} \equiv x_1\boldsymbol{e}_1 + x_2\boldsymbol{e}_2 + x_3\boldsymbol{e}_3$ and $r_{x_1}^2 = x_2^2 + x_3^2$ respectively denote the position vector of the elementary surface element $d\Gamma$ and the square of the distance between the x_1 axis and the element of mass within $\Delta\Omega$. Since the Cartesian components of $\boldsymbol{\sigma n}$ are provided by (1.47), the surface integral in (1.48) may be expressed as

$$\oint_{\partial\Delta\Omega} \boldsymbol{e}_1 \cdot (\boldsymbol{r} \times (\boldsymbol{\sigma n}))\, d\Gamma = \oint_{\partial\Delta\Omega} (x_2\sigma_{j3}n_j - x_3\sigma_{j2}n_j)\, d\Gamma = \int_{\Delta\Omega} \left(\frac{\partial(x_2\sigma_{j3})}{\partial x_j} - \frac{\partial(x_3\sigma_{j2})}{\partial x_j} \right) d\Omega$$

$$= \int_{\Delta\Omega} (\sigma_{23} - \sigma_{32})\, d\Omega + \int_{\Delta\Omega} \left(x_2\frac{\partial\sigma_{j3}}{\partial x_j} - x_3\frac{\partial\sigma_{j2}}{\partial x_j} \right) d\Omega \tag{1.49}$$

where the domain integrals emanate from a transformation of the surface integral through Gauss' theorem. The mean value theorem allows writing (1.48) in the form

$$\overline{(\sigma_{23} - \sigma_{32})} + \overline{\left(x_2 \frac{\partial \sigma_{j3}}{\partial x_j} - x_3 \frac{\partial \sigma_{j2}}{\partial x_j}\right)} = \overline{\boldsymbol{e}_1 \cdot \left(\boldsymbol{r} \times \frac{d\boldsymbol{V}}{dt}\rho\right)} \qquad (1.50)$$

As the elementary volume $\Delta\Omega$ shrinks to a point, $\boldsymbol{r}$ and both x_2 and x_3 vanish and thus (1.50) yields $\sigma_{23} = \sigma_{32}$; another two angular momentum equations analogous to (1.48), expressing the angular momentum about the x_2 and x_3 axes, lead to similar results on the remaining shear stress components. These components, therefore satisfy the three symmetry relations

$$\sigma_{23} = \sigma_{32}, \qquad \sigma_{31} = \sigma_{13}, \qquad \sigma_{12} = \sigma_{21} \qquad (1.51)$$

with these relations, (1.47) becomes

$$\sigma_{n_i} = \sigma_{i1}n_1 + \sigma_{i2}n_2 + \sigma_{i3}n_3 = \sigma_{ij}n_j \qquad (1.52)$$

where an implied summation is understood over repeated subscript indices.

1.5.4   Normal Stresses

The normal stress in the $\boldsymbol{n}$ direction, which corresponds to the component of the surface traction $\boldsymbol{\sigma n}$ in the direction of $\boldsymbol{n}$ itself, results from the vector "dot" product of $\boldsymbol{\sigma n}$ and $\boldsymbol{n}$ as

$$\begin{aligned}
\boldsymbol{\sigma n} \cdot \boldsymbol{n} = \sigma_{nn} &= \sigma_{n_1}n_1 + \sigma_{n_2}n_2 + \sigma_{n_3}n_3 \\
&= \sigma_{11}n_1 n_1 + \sigma_{12}n_2 n_1 + \sigma_{13}n_3 n_1 \\
&+ \sigma_{12}n_1 n_2 + \sigma_{22}n_2 n_2 + \sigma_{23}n_3 n_2 \\
&+ \sigma_{13}n_1 n_3 + \sigma_{23}n_2 n_3 + \sigma_{33}n_3 n_3 = \sigma_{ij}n_i n_j
\end{aligned} \qquad (1.53)$$

Form this result, therefore, follows the normal stress along any given normal unit vector $\boldsymbol{n}$. For any three mutually orthogonal normal unit vectors $\boldsymbol{n}^1$, $\boldsymbol{n}^2$, $\boldsymbol{n}^3$ (1.53) leads to three mutually perpendicular normal stresses. With reference to Figure 1.11, consider the following convenient forms of three mutually orthogonal unit vectors

$$\begin{aligned}
\boldsymbol{n}^1 &= \sin\theta\cos\phi\boldsymbol{e}_1 + \sin\theta\sin\phi\boldsymbol{e}_2 + \cos\theta\boldsymbol{e}_3 \\
\boldsymbol{n}^2 &= \cos\theta\cos\phi\boldsymbol{e}_1 + \cos\theta\sin\phi\boldsymbol{e}_2 - \sin\theta\boldsymbol{e}_3 \\
\boldsymbol{n}^3 &= -\sin\phi\boldsymbol{e}_1 + \cos\phi\boldsymbol{e}_2 + 0\boldsymbol{e}_3
\end{aligned} \qquad (1.54)$$

As the spherical-coordinate angles $0 \leq \theta \leq \pi$ and $0 \leq \phi \leq 2\pi$ vary in their respective ranges, the unit vector $\boldsymbol{n}^1$ will point in all possible directions, while $\boldsymbol{n}^2$ and $\boldsymbol{n}^3$ will remain orthogonal to it. Given an arbitrary direction, therefore, (1.54) will provide a set of mutually orthogonal unit vectors, with $\boldsymbol{n}^1$ along the given direction. These three vectors and (1.53), therefore, allow determining three mutually perpendicular normal stresses σ_{nn^1}, σ_{nn^2}, and σ_{nn^3} along any set of mutually perpendicular directions.

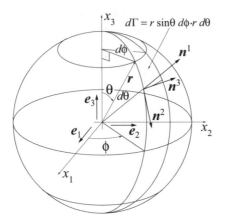

Figure 1.11: Mutually Orthogonal Unit Vectors

1.5.5   Stress Invariant, Mechanical Pressure and Deviatoric Stresses

The insertion of (1.53) each of the three unit vectors in (1.54) and then summation of the three resulting normal stresses lead to the elegant result

$$\sigma_{nn^1} + \sigma_{nn^2} + \sigma_{nn^3} = \sigma_{11} + \sigma_{22} + \sigma_{33} \tag{1.55}$$

which stipulates the fundamental invariance of the sum of three mutually orthogonal normal stresses.

This invariant possesses a profound physical significance. Consider the average normal stress $\overline{\sigma_{nn}}$ obtained by integrating (1.53) over the entire surface of a sphere with radius $r \rightarrow 0$ and centered at any given point $\boldsymbol{x}$, where σ_{ij} is evaluated; the variables in this integration, therefore, are the components of the unit vector $\boldsymbol{n} = \boldsymbol{n}^1$, from (1.54). With these specifications, $\overline{\sigma_{nn}}$ becomes

$$\overline{\sigma_{nn}} \equiv \frac{1}{\partial \Delta \Omega} \oint_{\partial \Delta \Omega} \sigma_{nn} d\Gamma = \frac{1}{4\pi r^2} \int_{\phi=0}^{\phi=2\pi} \int_{\theta=0}^{\theta=\pi} \sigma_{ij} n_i^1 n_j^1 r^2 \sin\theta \, d\theta \, d\phi = \frac{1}{3} \left(\sigma_{11} + \sigma_{22} + \sigma_{33} \right) \tag{1.56}$$

The invariant (1.55), therefore, corresponds to three times the average normal stress at any flow field point. Since this average normal stress does not depend on any particular direction, it defines the mechanical-pressure isotropic scalar field as

$$p_{\text{mech}} \equiv -\overline{\sigma_{nn}} = \frac{1}{3} \left(\sigma_{11} + \sigma_{22} + \sigma_{33} \right) \tag{1.57}$$

where the minus signifies that p_{mech} corresponds to compression while a positive $\overline{\sigma_{nn}}$ denotes a traction. A fluid motion may induce departures from local thermodynamical equilibrium. The mechanical pressure defined by (1.57) may not necessarily coincide with the thermodynamical pressure "p". This thermodynamical pressure certainly contributes to the normal stresses. The stress components σ_{ij} are thus expressed as

$$\sigma_{ij} = -p\delta_j^i + \sigma_{ij}^{\mu} \tag{1.58}$$

where σ_{ij}^{μ} denotes stresses other than pressure, known as deviatoric stresses. From this expression, the connection between mechanical and thermodynamical pressure becomes

$$p_{\text{mech}} = p - \sigma_{ii}^{\mu} \tag{1.59}$$

where σ_{ii}^{μ} may or may not vanish, depending on the fluid and the flow. This notion is amplified in Section 2.1, which presents the Newtonian constitutive relations for the stress components.

1.6   Flow State Variables

The dynamical state of a flow field is completely described by the density ρ, mass-specific linear momentum $\boldsymbol{V}$, and total energy e fields, which thus become flow state variables. In analogy to the definition of the Eulerian velocity, which equals the mass-specific linear momentum, the density and mass-specific total energy, at a location and time, signify the density and total energy of that particle that transits by the specified location.

These state variables correspond to intensive quantities. With $d\mathcal{M} = \rho d\Omega$ denoting the elementary particle mass, the corresponding extensive quantities of mass $\mathcal{M}$, linear momentum $\boldsymbol{L}$ and total energy $\mathcal{E}$ for a fluid volume Ω of any size are additively calculated as

$$\mathcal{M} = \int_{\mathcal{M}} d\mathcal{M} = \int_{\Omega} \rho d\Omega, \quad \boldsymbol{L} = \int_{\mathcal{M}} \boldsymbol{V} d\mathcal{M} = \int_{\Omega} \rho \boldsymbol{V} d\Omega, \quad \mathcal{E} = \int_{\mathcal{M}} e\, d\mathcal{M} = \int_{\Omega} \rho e\, d\Omega \tag{1.60}$$

Using the mean-value theorem, these integrals lead to the result

$$\Delta\mathcal{M} = \int_{\Delta\Omega} \rho d\Omega = \overline{\rho}\Delta\Omega, \quad \Delta\boldsymbol{L} = \int_{\Delta\Omega} \rho \boldsymbol{V} d\Omega = \overline{\rho\boldsymbol{V}}\Delta\Omega, \Delta\mathcal{E} = \int_{\Omega} \rho e\, d\Omega = \overline{\rho e}\,\Delta\Omega \tag{1.61}$$

which imply the results

$$\frac{\partial\mathcal{M}}{\partial\Omega} = \rho, \quad \frac{\partial\boldsymbol{L}}{\partial\Omega} = \rho\boldsymbol{V}, \quad \frac{\partial\mathcal{E}}{\partial\Omega} = \rho e \tag{1.62}$$

There expressions respectively indicate that ρ, $\rho\boldsymbol{V}$, and ρe correspond to the volume-specific mass, linear momentum and total energy. Since the volume-specific linear momentum and total energy recur in all the equations and analyses in the following chapters, these variables receive the following dedicated symbols

$$\boldsymbol{m} \equiv \rho\boldsymbol{V}, \quad E \equiv \rho e \tag{1.63}$$

With respect to $\boldsymbol{m}$, m_i will indicate the ith Cartesian component of linear momentum.

With the given definition of fluid particle, these state variables may not be uniquely defined at each location. Naturally, if the fluid particle size continuously shrinks to a mathematical zero, it eventually will contain no molecules and thus the state variables can no longer approach unique limits. This issue is resolved via a basic specification. As the fluid particle shrinks, it will achieve a certain threshold size at which the state variables in (1.62) approach a definite limit. From this size down, the analysis will rely on reference total particle mass, linear momentum and energy that will decrease linearly with the particle volume, hence from this size down the volume-specific variables in (1.62) will remain constant and equal to the definite limit achieved at the threshold size. In this manner, even when the particle volume reaches a mathematical zero, the magnitudes of density and volume specific linear momentum and total energy will remain uniquely defined.

1.7 Fundamental Equations

The fundamental field equations of Fluid Dynamics emanate from the principles of conservation of mass, second law of Newtonian mechanics, and first principle of thermodynamics. On the basis of these first principles, these equations can all be formulated in both integral and differential form using the transport theorem. A systematic method exists to use the transport theorem to generate field equations. This method starts with a physical principle that governs the time rate of change of an extensive quantity. The physical principle provides this rate of change and the transport theorem expresses this rate of change in terms of intensive field variables, which will directly lead to the corresponding equations.

1.7.1 Conservation of Mass: The Continuity Equation

The principle of conservation of mass simply states:

- Mass is conserved

Therefore, the mass encompassed by a fluid volume Ω with boundary $\partial\Omega$ each point of which flows with a fluid particle, does not change with time. For any two instants of time t_1 and t_2, this principle leads to

$$\mathcal{M} = \int_{\Omega(t_1)} \rho(\boldsymbol{x}, t_1) d\Omega = \int_{\Omega(t_2)} \rho(\boldsymbol{x}, t_2) d\Omega \tag{1.64}$$

stipulating the invariance of this mass with time. The time rate of change of this mass consequently vanishes and with $B \equiv \mathcal{M}$ and $\beta = 1$ the transport theorem yields

$$0 = \frac{D\mathcal{M}}{Dt} = \int_{\Omega} \frac{\partial \rho}{\partial t} d\Omega + \oint_{\partial\Omega} \rho \boldsymbol{V} \cdot \boldsymbol{n} d\Gamma \tag{1.65}$$

hence

$$\int_{\Omega} \frac{\partial \rho}{\partial t} d\Omega + \oint_{\partial\Omega} \rho \boldsymbol{V} \cdot \boldsymbol{n} d\Gamma = 0 \tag{1.66}$$

which is the continuity equation, so named to emphasize the continuity of flowing mass.

In the following form

$$-\int_{\Omega} \frac{\partial \rho}{\partial t} d\Omega = \oint_{\partial\Omega} \rho \boldsymbol{V} \cdot \boldsymbol{n} d\Gamma \tag{1.67}$$

the continuity equation yields a clear physical significance. For a flowing volume, this equation states that the decrease of density within Ω corresponds to a flow of the boundary. For a fixed volume, this equation states that the decrease of density within Ω corresponds to mass flowing away from Ω across its boundary.

When the product $\rho \boldsymbol{V}$ is differentiable, conservation of mass leads to

$$\int_\Omega \left[\frac{\partial \rho}{\partial t} + \operatorname{div}(\rho \boldsymbol{V}) \right] d\Omega = 0 \tag{1.68}$$

No assumption whatsoever constrains Ω, which implies that this integral must vanish for arbitrarily sized Ω. This integral can vanish for arbitrary Ω only when the expression between brackets itself vanishes, which leads to the continuity equation in differential form

$$\frac{\partial \rho}{\partial t} + \operatorname{div}(\rho \boldsymbol{V}) = 0 \tag{1.69}$$

Naturally, integrating this equation within a fixed volume and transforming the integral of $\operatorname{div}(\rho \boldsymbol{V})$ through Gauss' theorem returns (1.67) in its fixed-volume interpretation.

Expression (1.64) applies for any arbitrary amount of mass, including an infinitesimal mass $d\mathcal{M}$. For this elementary mass, the mass conservation principle becomes

$$\frac{D d\mathcal{M}}{Dt} = 0 \tag{1.70}$$

which mathematically stipulates conservation of the mass of a fluid particle.

1.7.2   Second Law of Mechanics: The Linear Momentum Equation

With respect to an inertial reference frame, the second law of Newtonian mechanics states:

- The total external force on a mass equals the time rate of change of the mass' absolute total linear momentum

The mathematical form of this principle is

$$\boldsymbol{F} = \frac{D\boldsymbol{L}}{Dt}, \quad \boldsymbol{L} = \int_\mathcal{M} \boldsymbol{V} d\mathcal{M} = \int_\Omega \rho \boldsymbol{V} d\Omega \tag{1.71}$$

where $\boldsymbol{F}$ and $\boldsymbol{L}$ respectively denote the total external force and linear momentum. For the case of one single particle of mass m, the linear momentum becomes $\boldsymbol{L} = m\boldsymbol{V}$ and (1.71) simplifies to the familiar expression $\boldsymbol{F} = m\boldsymbol{a}$ where $\boldsymbol{a}$ denotes the inertial acceleration of the particle center of mass. For a system of particles, including a continuum, result (1.71) is obtained on the basis of the time invariance of mass and the third law of mechanics, or the principle of action and reaction. According to this principle sets of internal forces are equal and opposite, hence they cancel each other out when the second law of mechanics for single particles is integrated over the entire distribution of mass to generate (1.71).

In terms of the velocity and density fields and using index notation, (1.71) becomes

$$F_i = \frac{D}{Dt} \int_\Omega \rho u_i d\Omega \tag{1.72}$$

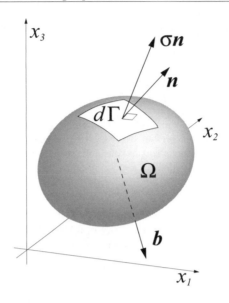

Figure 1.12: Surface and Body Forces

where u_i and F_i respectively denote the ith Cartesian component of $\mathbf{V}$ and $\mathbf{F}$. With reference to Figure 1.12 the force component F_i results from both body and surface forces in the form

$$F_i = \oint_{\partial\Omega} \sigma_{ij} n_j d\Gamma + \int_\Omega b_i \rho d\Omega \tag{1.73}$$

where $\boldsymbol{\sigma} \equiv \{\sigma_{ij}\}$, $\sigma_{ij}n_j$, and b_i respectively denote the total mechanical stress tensor, and the ith Cartesian component of surface traction and body force, with n_j indicating the jth Cartesian of a locally orthogonal outward pointing unit vector. The time rate of change of total linear momentum follows from the transport theorem as

$$\frac{D}{Dt} \int_\Omega \rho u_i d\Omega = \int_\Omega \frac{\partial \rho u_i}{\partial t} d\Omega + \oint_{\partial\Omega} \rho u_i \mathbf{V} \cdot \mathbf{n} d\Gamma \tag{1.74}$$

hence

$$\int_\Omega \frac{\partial \rho u_i}{\partial t} d\Omega + \oint_{\partial\Omega} \rho u_i u_j n_j d\Gamma = \oint_{\partial\Omega} \sigma_{ij} n_j d\Gamma + \int_\Omega b_i \rho d\Omega \tag{1.75}$$

which corresponds to the linear-momentum equation. The following form of the linear-momentum equation

$$\oint_{\partial\Omega} \sigma_{ij} n_j d\Gamma + \int_\Omega b_i \rho d\Omega - \int_\Omega \frac{\partial \rho u_i}{\partial t} d\Omega = \oint_{\partial\Omega} \rho u_i u_j n_j d\Gamma \tag{1.76}$$

leads to one physical interpretation. For a flowing volume, this equation stipulates that the effect of the external surface tractions and body force along with a decrease of internal linear momentum produce an increase of the boundary linear-momentum. For a fixed volume this equation states that the boundary surface tractions and body force, coupled with a decrease of internal linear momentum generate an outflow of linear momentum across the fixed boundary.

When the stress tensor components σ_{ij} and specific momentum flow rate components $\rho u_i u_j$ are differentiable, the second law of mechanics generates the equation

$$\int_\Omega \left[\frac{\partial \rho u_i}{\partial t} + \frac{\partial \rho u_i u_j}{\partial x_j} - \frac{\partial \sigma_{ij}}{\partial x_j} - \rho b_i \right] d\Omega = 0 \tag{1.77}$$

Once again, no assumption whatsoever constrains Ω, which implies that this equation will hold for arbitrarily sized Ω. This integral can vanish for arbitrary Ω only when the expression between brackets itself vanishes, which leads to the linear-momentum equation in differential form

$$\frac{\partial \rho u_i}{\partial t} + \frac{\partial \rho u_i u_j}{\partial x_j} - \frac{\partial \sigma_{ij}}{\partial x_j} - \rho b_i = 0 \tag{1.78}$$

Logically, integrating this equation within a fixed volume and transforming the integrals of both $\partial \rho u_i u_j / \partial x_j$ and $\partial \sigma_{ij} / \partial x_j$, through Gauss' theorem, returns (1.76) in its fixed-volume interpretation.

Expression (1.73) can also be written as

$$F_i = \int_\Omega dF_i = \oint_{\partial \Omega} \sigma_{ij} n_j d\Gamma + \int_\Omega b_i \rho d\Omega = \int_\Omega \left(\frac{\partial \sigma_{ij}}{\partial x_j} + b_i \rho \right) d\Omega \tag{1.79}$$

Consequently, the elementary force component dF_i can be expressed as

$$dF_i = \left(\frac{\partial \sigma_{ij}}{\partial x_j} + b_i \rho \right) d\Omega \tag{1.80}$$

which provides the total external force component on a fluid particle.

1.7.3 The First Law of Thermodynamics: The Total Energy Equation

The first law of Thermodynamics states:

- Total energy can only be varied through heat and work exchanges.

In Thermodynamics books, the traditional differential form of this principle for a constant-mass system reads:

$$d\mathcal{E} = \delta Q - \delta W_{\text{by}} = \delta Q + \delta W \tag{1.81}$$

where $\mathcal{E}$, Q, W_{by}, and W respectively denote total energy, heat transferred to the system, work performed "*by*" the system, and work performed "*on*" the system As usual, the differentials of energy, heat and work are indicated differently to emphasize the notion that energy $\mathcal{E}$ is a process-independent state variable with perfect differential $d\mathcal{E}$, whereas heat transferred and work performed are process dependent quantities that are not state variables, hence their elementary counterparts δQ and δW are not perfect differentials.

For the derivation of the energy equation, (1.81) is written in the form of a rate of change as

$$\frac{D\mathcal{E}}{Dt} = \frac{\delta Q}{\delta t} + \frac{\delta W}{\delta t}$$ (1.82)

where $\delta Q/\delta t$ and $\delta W/\delta t$ respectively denote the amount of heat transferred to and amount of work performed on the system per unit time. The total energy $\mathcal{E}$ within the flowing fluid mass Ω results from the integral

$$\mathcal{E} = \int_\Omega \rho e \, d\Omega$$ (1.83)

where "e" denotes the mass-specific total energy. With this expression for $\mathcal{E}$, the statement of the first principle of thermodynamics becomes

$$\frac{D}{Dt} \int_\Omega \rho e \, d\Omega = \frac{\delta Q}{\delta t} + \frac{\delta W}{\delta t}$$ (1.84)

With reference to Figure 1.13 The unit-time amount of heat is expressed as

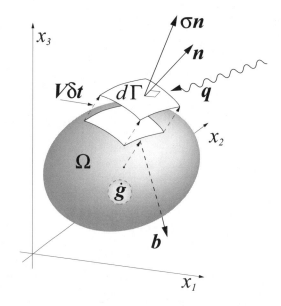

Figure 1.13: Heat Transfer and Force Displacement

$$\frac{\delta Q}{\delta t} = \oint_{\partial\Omega} \delta_\Gamma\left(\frac{\delta Q_\Gamma}{\delta t}\right) + \int_\mathcal{M} \delta_\mathcal{M}\left(\frac{\delta Q_\mathcal{M}}{\delta t}\right) = \oint_{\partial\Omega} q_i^F(-n_i)d\Gamma + \int_\Omega \dot{g}\rho d\Omega$$ (1.85)

In this expression, $\dot{g}$ denotes mass-specific unit-time body heating, as induced for instance by absorption or radiation. The surface-integral term $q_i^F(-n_i)$ denotes the unit-time and unit-area heat due to temperature gradients and transferred across the surface of Ω. The minus sign in front of n_i has the following justification. Both $\delta Q_\Gamma / \delta t$ and q_i^F correspond to heat amounts transferred *into* the system, whereas n_i denotes the component of a unit vector $\boldsymbol{n}$ pointing *away* from the system; the expression $(-n_i)$ thus provides the component of a unit vector pointing into the system; as a result, $q_i^F(-n_i)$ correctly corresponds to a heat amount transferred into the system.

The total unit-time amount of work is expressed as

$$\frac{\delta W}{\delta t} = \oint_{\partial \Omega} \delta_\Gamma \left(\frac{\delta W_\Gamma}{\delta t} \right) + \int_{\mathcal{M}} \delta_{\mathcal{M}} \left(\frac{\delta W_{\mathcal{M}}}{\delta t} \right) \tag{1.86}$$

where $\delta_\Gamma(\delta W_\Gamma / \delta t)$ and $\delta_{\mathcal{M}}(\delta W_{\mathcal{M}} / \delta t)$ denote the elementary unit-time work respectively performed by surface and body forces. These elementary unit-time work contributions emanate from basic mechanics principles as follows. The components of the elementary surface and body forces $(\delta F_i)_\Gamma$ and $(\delta F_i)_{\mathcal{M}}$ are expressed as

$$(\delta F_i)_\Gamma = \frac{\partial F_i}{\partial \Gamma} d\Gamma = \sigma_{ij} n_j d\Gamma, \quad (\delta F_i)_{\mathcal{M}} = \frac{\partial F_i}{\partial \mathcal{M}} d\mathcal{M} = b_i \rho \, d\Omega \tag{1.87}$$

With reference to Figure 1.13, the corresponding elementary work contributions, as vector dot products of forces and elementary displacements $\delta x_i = u_i \delta t$, evolve as

$$\delta_\Gamma(\delta W_\Gamma) = (\delta F_i)_\Gamma \delta x_i = \sigma_{ij} n_j d\Gamma \delta x_i, \quad \delta_{\mathcal{M}}(\delta W_{\mathcal{M}}) = (\delta F_i)_{\mathcal{M}} \delta x_i = b_i \rho d\Omega \delta x_i \tag{1.88}$$

The elementary work and displacement respectively correspond to the unit-time work and velocity component in the elementary time interval δt, hence (1.88) becomes

$$\delta_\Gamma \left(\frac{\delta W_\Gamma}{\delta t} \delta t \right) = (\delta F_i)_\Gamma \delta x_i = \sigma_{ij} n_j d\Gamma u_i \delta t, \quad \delta_{\mathcal{M}} \left(\frac{\delta W_{\mathcal{M}}}{\delta t} \delta t \right) = (\delta F_i)_{\mathcal{M}} \delta x_i = b_i \rho d\Omega u_i \delta t \tag{1.89}$$

The elementary unit-time work respectively performed by surface and body forces, therefore, result as

$$\delta_\Gamma \left(\frac{\delta W_\Gamma}{\delta t} \right) = u_i \sigma_{ij} n_j d\Gamma, \quad \delta_{\mathcal{M}} \left(\frac{\delta W_{\mathcal{M}}}{\delta t} \right) = u_i b_i \rho d\Omega \tag{1.90}$$

and the corresponding total unit-time work in (1.86) is thus expressed as

$$\frac{\delta W}{\delta t} = \oint_{\partial \Omega} u_i \sigma_{ij} n_j d\Gamma + \int_\Omega u_i b_i \rho d\Omega \tag{1.91}$$

The time rate of change of total energy follows from the transport theorem as

$$\frac{D}{Dt} \int_\Omega \rho e \, d\Omega = \int_\Omega \frac{\partial \rho e}{\partial t} d\Omega + \oint_{\partial \Omega} \rho e \boldsymbol{V} \cdot \boldsymbol{n} d\Gamma \tag{1.92}$$

hence expressions (1.84), (1.85), and (1.91) lead to

$$\int_\Omega \frac{\partial \rho e}{\partial t} d\Omega + \oint_{\partial \Omega} \rho e \boldsymbol{V} \cdot \boldsymbol{n} d\Gamma = \oint_{\partial \Omega} u_i \sigma_{ij} n_j d\Gamma + \int_\Omega u_i b_i \rho d\Omega + \oint_{\partial \Omega} q_i^F(-n_i) d\Gamma + \int_\Omega \dot{g} \rho d\Omega \tag{1.93}$$

which corresponds to the energy equation. For a flowing fluid volume, this equation stipulates that the unit-time work and heat transfer change the energy inside the volume and on its boundaries. For a fixed volume, this equation signifies that the cumulative effect of body work and heat transfer in conjunction with work and heat transfer at the boundary induce an energy increase inside the volume and an energy outflow across the fixed boundary.

When the convected energy, unit-time surface work and heat transfer expressions are differentiable, the first law of thermodynamics generates the equation

$$\int_\Omega \left[\frac{\partial \rho e}{\partial t} + \frac{\partial \rho e u_j}{\partial x_j} - \frac{\partial u_i \sigma_{ij}}{\partial x_j} + \frac{\partial q_j^F}{\partial x_j} - \rho u_i b_i - \rho \dot{g} \right] d\Omega = 0 \tag{1.94}$$

As discussed for the continuity and linear-momentum equations, no assumption whatsoever constrains Ω, which implies that this equation will hold for arbitrarily sized Ω. This integral can vanish for arbitrary Ω only when the expression between brackets itself vanishes which leads to the energy equation in differential form

$$\frac{\partial \rho e}{\partial t} + \frac{\partial \rho e u_j}{\partial x_j} - \frac{\partial u_i \sigma_{ij}}{\partial x_j} + \frac{\partial q_j^F}{\partial x_j} - \rho u_i b_i - \rho \dot{g} = 0 \tag{1.95}$$

Again, integrating this equation within a fixed volume and transforming the appropriate integrals through Gauss' theorem returns (1.93) in its fixed-volume interpretation.

Expressions (1.85) and (1.91) can be collectively expressed as

$$\frac{\delta Q}{\delta t} + \frac{\delta W}{\delta t} = \int_\Omega \delta \left(\frac{\delta Q}{\delta t} + \frac{\delta W}{\delta t} \right) = \oint_{\partial \Omega} q_i^F(-n_i) d\Gamma + \int_\Omega \dot{g} \rho d\Omega + \oint_{\partial \Omega} u_i \sigma_{ij} n_j d\Gamma + \int_\Omega u_i b_i \rho d\Omega =$$

$$= \int_\Omega \left(-\frac{\partial q_j^F}{\partial x_j} + \dot{g}\rho + \frac{\partial u_i \sigma_{ij}}{\partial x_j} + u_i b_i \rho \right) d\Omega \tag{1.96}$$

It follows that the elementary unit-time heat transfer and work can be expressed as

$$\delta \left(\frac{\delta Q}{\delta t} + \frac{\delta W}{\delta t} \right) = \left(-\frac{\partial q_j^F}{\partial x_j} + \dot{g}\rho + \frac{\partial u_i \sigma_{ij}}{\partial x_j} + u_i b_i \rho \right) d\Omega \tag{1.97}$$

which provides the total unit-time heat transfer and work on a fluid particle.

1.8   Consequences of Mass Conservation

The principle of mass conservation, hence the continuity equation, directly leads to several significant results, including the expression for the substantive derivative.

1.8.1   Substantive Derivative

The continuity equation in divergence form inherently contains the substantial derivative for all mass-specific flow properties. Consider the extensive property B as expressed in terms of the mass specific intensive property β

$$B = \int_{\mathcal{M}} \beta d\mathcal{M} = \int_\Omega \beta \rho d\Omega \tag{1.98}$$

Since mass $\mathcal{M}$ does not depend upon time, an expression for a total time derivative operator evolves from requiring that it should provide the derivative of B in the form

$$\frac{DB}{Dt} = \frac{D}{Dt} \int_{\mathcal{M}} \beta d\mathcal{M} = \int_{\mathcal{M}} \frac{D(\beta d\mathcal{M})}{Dt} = \int_{\mathcal{M}} \frac{D\beta}{Dt} d\mathcal{M} = \int_{\Omega} \frac{D\beta}{Dt} \rho d\Omega \qquad (1.99)$$

where Ω denotes the flowing fluid volume that continuously contains the same time-invariant amount of mass $\mathcal{M}$.

On the other hand, the divergence form (1.33) of the transport theorem yields

$$\frac{DB}{Dt} = \int_{\Omega} \left[\left(\frac{\partial \beta}{\partial t} + \boldsymbol{V} \cdot \operatorname{grad} \beta \right) \rho + \beta \left(\frac{\partial \rho}{\partial t} + \operatorname{div}(\rho \boldsymbol{V}) \right) \right] d\Omega = \int_{\Omega} \left(\frac{\partial \beta}{\partial t} + \boldsymbol{V} \cdot \operatorname{grad} \beta \right) \rho d\Omega$$
$$(1.100)$$

where the last equality follows from the continuity equation (1.69). Expressions (1.99) and (1.100) provide the same total rate of change of B. Subtracting (1.100) from (1.99), therefore, yields

$$\int_{\Omega} \left[\rho \frac{D\beta}{Dt} - \rho \left(\frac{\partial \beta}{\partial t} + \boldsymbol{V} \cdot \operatorname{grad} \beta \right) \right] d\Omega = 0 \qquad (1.101)$$

This equation must hold for arbitrary $\mathcal{M}$ hence Ω. As a consequence, this equation can only vanish for arbitrary Ω when the expression between brackets vanishes, which yields

$$\frac{D\beta}{Dt} = \frac{\partial \beta}{\partial t} + \boldsymbol{V} \cdot \operatorname{grad} \beta \qquad (1.102)$$

This is the well know the substantive-derivative expression, which, therefore, results as a natural consequence of the principle of mass conservation.

1.8.2 Substantive-Derivative Form of The Transport Theorem

With a clearly defined expression for the substantive derivative, (1.99) shows that the transport theorem can be equivalently cast as

$$\frac{DB}{Dt} = \frac{D}{Dt} \int_{\Omega} \beta \rho d\Omega = \int_{\Omega} \frac{D\beta}{Dt} \rho d\Omega \qquad (1.103)$$

1.8.3 Interchange of Differentiation Order for Mass and Time

From the connection between extensive and intensive quantities

$$B = \int_{\mathcal{M}} \beta d\mathcal{M} \quad \Rightarrow \quad \frac{\partial B}{\partial \mathcal{M}} = \beta \qquad (1.104)$$

the time rate of change of the mass derivative of B equals

$$\frac{D}{Dt} \left(\frac{\partial B}{\partial \mathcal{M}} \right) = \frac{D\beta}{Dt} \qquad (1.105)$$

On the other hand, (1.99) shows that the derivative of B can be expressed as

$$\frac{DB}{Dt} = \int_{\Omega} \frac{D\beta}{Dt} \rho d\Omega = \int_{\mathcal{M}} \frac{D\beta}{Dt} d\mathcal{M} \quad \Rightarrow \quad \frac{\partial}{\partial \mathcal{M}} \left(\frac{DB}{Dt} \right) = \frac{D\beta}{Dt} \qquad (1.106)$$

Expressions (1.105) and (1.106) thus sanction the interchange in the order of differentiation with respect to mass and time.

1.9   Velocity Divergence

The divergence of velocity V supplies the specific time rate of change of Ω. In conjunction with the fundamental differential equations (1.69), (1.78), and (1.95), it also leads to the equations governing the motion of an elementary fluid particle.

1.9.1   Physical Significance

With B set equal to Ω, the corresponding intensive property β equal $1/\rho$ and (1.22) and (1.33) become

$$\Omega = \int_\Omega d\Omega, \quad \frac{D\Omega}{Dt} = \int_\Omega \operatorname{div} V \, d\Omega \quad \Rightarrow \quad \frac{D\Delta\Omega}{Dt} = \int_{\Delta\Omega} \operatorname{div} V \, d\Omega \tag{1.107}$$

The mean value theorem provides for the existence of a point $\overline{x}$ within $\Delta\Omega$ for which the integral of $\operatorname{div} V$ over $\Delta\Omega$ in (1.107) becomes

$$\frac{D\Delta\Omega}{Dt} = \Delta\Omega \operatorname{div} V_{\overline{x}}, \quad \overline{x} \in \Delta\Omega \tag{1.108}$$

Conversely, and significantly, given a point x, this theorem stipulates the existence of a $\Delta\Omega$, encircling x, for which (1.108) holds true. No restriction constrains the size of $\Delta\Omega$. Hence (1.108) also applies for an infinitesimal volume $d\Omega$, which leads to the differential expression

$$\frac{Dd\Omega}{Dt} = \operatorname{div} V \, d\Omega \tag{1.109}$$

The limit of (1.108) yields

$$\lim_{\Delta\Omega \to 0} \frac{D\Delta\Omega/Dt}{\Delta\Omega} = \operatorname{div} V \tag{1.110}$$

The divergence of velocity at a point, therefore, yields the rate of change of a fluid volume encircling that point, as this volume indefinitely shrinks. Physically, then, $\operatorname{div} V$ provides the time rate of change of the volume of a flowing fluid particle per unit volume.

1.9.2   Equations for a Fluid Particle

In conjunction with (1.69), (1.78), and (1.95), the result

$$\frac{1}{\Delta\Omega} \frac{D\Delta\Omega}{Dt} = \operatorname{div} V \tag{1.111}$$

leads to the equations governing the motion of a fluid particle with volume $\Delta\Omega$, depicted in Figure 1.14

The continuity equation (1.69), linear momentum equation (1.78), and energy equation (1.95), can be expressed as

$$\frac{\partial \rho}{\partial t} + u_j \frac{\partial \rho}{\partial x_j} + \rho \operatorname{div} V = 0 \tag{1.112}$$

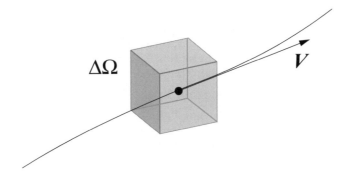

Figure 1.14: Fluid Particle $\Delta\Omega$

$$\frac{\partial \rho u_i}{\partial t} + u_j \frac{\partial \rho u_i}{\partial x_j} + \rho u_i \operatorname{div} \boldsymbol{V} = \frac{\partial \sigma_{ij}}{\partial x_j} + \rho b_i \tag{1.113}$$

$$\frac{\partial \rho e}{\partial t} + u_j \frac{\partial \rho e}{\partial x_j} + \rho e \operatorname{div} \boldsymbol{V} = \frac{\partial u_i \sigma_{ij}}{\partial x_j} - \frac{\partial q_j^F}{\partial x_j} + \rho u_i b_i + \rho \dot{g} \tag{1.114}$$

Substituting the substantive derivative and (1.111) into each of these equations yields

$$\Delta\Omega \frac{D\rho}{Dt} + \rho \frac{D\Delta\Omega}{Dt} = 0 \quad \Rightarrow \quad \frac{D(\rho\Delta\Omega)}{Dt} = \frac{D(\Delta\mathcal{M})}{Dt} = 0 \tag{1.115}$$

$$\Delta\Omega \frac{D\rho u_i}{Dt} + \rho u_i \frac{D\Delta\Omega}{Dt} = \left(\frac{\partial \sigma_{ij}}{\partial x_j} + \rho b_i \right) \Delta\Omega \tag{1.116}$$

and

$$\frac{D(u_i \Delta\mathcal{M})}{Dt} = \left(\frac{\partial \sigma_{ij}}{\partial x_j} + \rho b_i \right) \Delta\Omega \tag{1.117}$$

as well as

$$\frac{D(e\,\Delta\mathcal{M})}{Dt} = \left(\rho \dot{g} - \frac{\partial q_j^F}{\partial x_j} \right) \Delta\Omega + \left(\rho u_i b_i + \frac{\partial u_i \sigma_{ij}}{\partial x_j} \right) \Delta\Omega \tag{1.118}$$

These equations respectively express the conservation of mass, second law of mechanics and first law of thermodynamics for the elementary flowing fluid particle $\Delta\mathcal{M}$.

1.10 Energy Equations

The differential equation

$$\frac{\partial \rho e}{\partial t} + \frac{\partial \rho e u_j}{\partial x_j} - \frac{\partial u_i \sigma_{ij}}{\partial x_j} + \frac{\partial q_j^F}{\partial x_j} - \rho u_i b_i - \rho \dot{g} = 0 \tag{1.119}$$

governs the evolution of a fluid particle total energy. This section presents the equations governing the evolution of total enthalpy, kinetic as well as internal energy, internal enthalpy, temperature and entropy. These equations emanate from the governing equations already established and from the thermodynamical definitions of entropy and enthalpy. These equations lucidly show how these kinds of energy interact with one another as fluid particles move and deform.

1.10.1   Kinetic Energy Equation

From (1.8) the time rate of change of kinetic energy for a continuous distribution of mass can be cast as

$$\frac{D}{Dt} \int_\Omega \frac{1}{2} \rho u_i u_i \, d\Omega = \int_\Omega u_i \delta F_i \tag{1.120}$$

In this expression, δF_i and u_i now respectively denote the "i^{th}" Cartesian component of the total force acting on the representative fluid particle within Ω and particle velocity, which corresponds to the velocity of the particle mass center. Upon restating this expression through the transport theorem and using (1.80) for δF_i lead to the kinetic energy equation

$$\int_\Omega \left[\frac{\partial}{\partial t} \left(\frac{1}{2} \rho u_i u_i \right) + \frac{\partial}{\partial x_j} \left(\frac{1}{2} \rho u_i u_i u_j \right) \right] d\Omega = \int_\Omega \left(u_i \frac{\partial \sigma_{ij}}{\partial x_j} + \rho b_i u_i \right) d\Omega \tag{1.121}$$

which yields the following equation in differential form

$$\frac{\partial}{\partial t} \left(\frac{1}{2} \rho u_i u_i \right) + \frac{\partial}{\partial x_j} \left(\frac{1}{2} \rho u_i u_i u_j \right) - u_i \frac{\partial \sigma_{ij}}{\partial x_j} - \rho b_i u_i = 0 \tag{1.122}$$

Through (1.146), the kinetic-energy equation in terms of pressure becomes

$$\frac{\partial}{\partial t} \left(\frac{1}{2} \rho u_i u_i \right) + \frac{\partial}{\partial x_j} \left(\frac{1}{2} \rho u_i u_i u_j \right) + u_i \frac{\partial p}{\partial x_i} - u_i \frac{\partial \sigma_{ij}^\mu}{\partial x_j} - \rho b_i u_i = 0 \tag{1.123}$$

By virtue of the continuity equation, this result is also cast as

$$\frac{\partial}{\partial t} \left(\frac{1}{2} u_i u_i \right) + u_j \frac{\partial}{\partial x_j} \left(\frac{1}{2} u_i u_i \right) + \frac{u_i}{\rho} \frac{\partial p}{\partial x_i} - \frac{u_i}{\rho} \frac{\partial \sigma_{ij}^\mu}{\partial x_j} - b_i u_i = 0 \tag{1.124}$$

Naturally, this equation, or (1.123), is not independent of the governing equations (1.69) and (1.78) but can be directly obtained from them. To this end, the following identities simplify the derivation of the kinetic-energy equation from (1.69)

$$\frac{\partial}{\partial t} \left(\frac{1}{2} \rho u_i u_i \right) = u_i \frac{\partial \rho u_i}{\partial t} - \frac{1}{2} u_i u_i \frac{\partial \rho}{\partial t}, \quad \frac{\partial}{\partial x_j} \left(\frac{1}{2} \rho u_i u_i u_j \right) = u_i \frac{\partial \rho u_i u_j}{\partial x_j} - \frac{1}{2} u_i u_i \frac{\partial \rho u_j}{\partial t} \tag{1.125}$$

Upon summing up these two identities and inserting into the result the continuity equation, the following convenient identity emerges

$$u_i \frac{\partial \rho u_i}{\partial t} + u_i \frac{\partial \rho u_i u_j}{\partial x_j} = \frac{\partial}{\partial t} \left(\frac{1}{2} \rho u_i u_i \right) + \frac{\partial}{\partial x_j} \left(\frac{1}{2} \rho u_i u_i u_j \right) \tag{1.126}$$

On the basis of this identity, the product and contraction of the momentum equation components (1.78) with u_i identically returns (1.122), which shows that the kinetic-energy directly originates from the momentum equation components.

Upon inserting into (1.121) the stress tensor decomposition (1.58), and transforming the resulting rhs integrals by parts yields the time rate of change of kinetic energy in the form

$$\frac{D}{Dt} \int_\Omega \frac{1}{2} \rho u_i u_i \, d\Omega = \int_\Omega \left(p \frac{\partial u_j}{\partial x_j} + \rho b_i u_i \right) d\Omega + \oint_{\partial\Omega} u_i \sigma_{ij} n_j d\Gamma - \int_\Omega \sigma_{ij}^\mu \frac{\partial u_i}{\partial x_j} \, d\Omega \qquad (1.127)$$

As shown in Section 2.4, the deviatoric stress term $\sigma_{ij}^\mu \frac{\partial u_i}{\partial x_j}$ invariably dissipates kinetic energy.

1.10.2 Rigid-Motion and Deformation Work

The total and kinetic energy equations respectively contain the expressions for the volume-specific work per unit time of the stress tensor as

$$\frac{\delta}{\delta\Omega}\left(\frac{\delta W}{\delta t}\right) = \frac{\partial u_i \sigma_{ij}}{\partial x_j} + \rho b_i u_i, \quad \frac{\delta}{\delta\Omega}\left(\frac{\delta W_{RM}}{\delta t}\right) = u_i \frac{\partial \sigma_{ij}}{\partial x_j} + \rho b_i u_i \qquad (1.128)$$

where subscript RM indicates rigid motion. The work in the total energy equation is expanded as

$$\frac{\delta}{\delta\Omega}\left(\frac{\delta W}{\delta t}\right) = \sigma_{ij} \frac{\partial u_i}{\partial x_j} + u_i \frac{\partial \sigma_{ij}}{\partial x_j} + \rho b_i u_i \qquad (1.129)$$

which obviously incorporates the rigid-motion contribution. The first term in this expansion correlates with the work (1.9) of the external force delivered through the molecular motion about a fluid particle. The following derivation illustrates this interpretation and clarifies the rigid-body origin of the second expression in (1.128). With reference to Figure 1.15, consider the elementary work of the representative normal stresses along the x_2 axis; the work of all the shear stresses and remaining normal stresses can be calculated similarly. In this model, σ_{22} and u_2 respectively denote the normal stress and velocity component along the x_2 axis at the particle mass center G. As indicated in the figure, the normal stress and displacement to the left of G respectively are $(\sigma_{22} - \partial\sigma_{22}/\partial x_2 \Delta x_2/2)\Delta x_1 \Delta x_3$ and $(u_2 - \partial u_2/\partial x_2 \Delta x_2/2)\delta t$; the corresponding expressions for the cube face to the right of G become $(\sigma_{22} + \partial\sigma_{22}/\partial x_2 \Delta x_2/2)\Delta x_1 \Delta x_3$ and $(u_2 + \partial u_2/\partial x_2 \Delta x_2/2)\delta t$. The elementary work of these stresses thus becomes

$$\delta W = -\left(\sigma_{22} - \frac{\partial\sigma_{22}}{\partial x_2}\frac{\Delta x_2}{2}\right)\Delta x_1 \Delta x_3 \left(u_2 - \frac{\partial u_2}{\partial x_2}\frac{\Delta x_2}{2}\right)\delta t +$$

$$+ \left(\sigma_{22} + \frac{\partial\sigma_{22}}{\partial x_2}\frac{\Delta x_2}{2}\right)\Delta x_1 \Delta x_3 \left(u_2 + \frac{\partial u_2}{\partial x_2}\frac{\Delta x_2}{2}\right)\delta t = \delta W_{RM} + \delta W_D \qquad (1.130)$$

where δW_{RM} and δW_D respectively indicate rigid-motion and deformation work.

The elementary work δW_{RM} is expressed as

$$\delta W_{RM} = -\left(\sigma_{22} - \frac{\partial\sigma_{22}}{\partial x_2}\frac{\Delta x_2}{2}\right)\Delta x_1 \Delta x_3 \left(u_2\right)\delta t + \left(\sigma_{22} + \frac{\partial\sigma_{22}}{\partial x_2}\frac{\Delta x_2}{2}\right)\Delta x_1 \Delta x_3 \left(u_2\right)\delta t =$$

$$u_2 \frac{\partial\sigma_{22}}{\partial x_2} \Delta x_1 \Delta x_2 \Delta x_3 \delta t = u_2 \frac{\partial\sigma_{22}}{\partial x_2} \Delta\Omega \delta t \qquad (1.131)$$

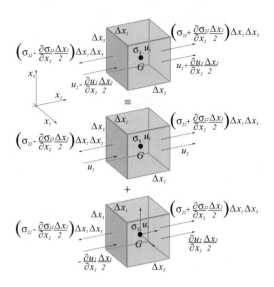

Figure 1.15: Stresses on a Fluid Particle

and it thus corresponds to the work of the external force $\partial\sigma_{22}/\partial x_2 \Delta\Omega$ as the fluid particle uniformly displaces by the single displacement $u_2\delta t$ of the particle mass center, which explains why this work corresponds to a rigid motion. This work corresponds to the motion of the particle mass center. It is this work that thus features in the kinetic-energy equation, for this equation governs the kinetic energy of the mass center of each fluid particle.

The elementary work δW_D is expressed as

$$\delta W_D = -\left(\sigma_{22} - \frac{\partial\sigma_{22}}{\partial x_2}\frac{\Delta x_2}{2}\right)\Delta x_1 \Delta x_3 \left(-\frac{\partial u_2}{\partial x_2}\frac{\Delta x_2}{2}\right)\delta t +$$

$$+\left(\sigma_{22} + \frac{\partial\sigma_{22}}{\partial x_2}\frac{\Delta x_2}{2}\right)\Delta x_1 \Delta x_3 \left(+\frac{\partial u_2}{\partial x_2}\frac{\Delta x_2}{2}\right)\delta t = \sigma_{22}\Delta x_1 \Delta x_3 \frac{\partial u_2}{\partial x_2}\Delta x_2\delta t = \sigma_{22}\frac{\partial u_2}{\partial x_2}\Delta\Omega\delta t$$

$$(1.132)$$

This result corresponds to the first term in expansion (1.129) and it corresponds to the work of the normal stresses to the left and right of G in the motion about the particle center of mass, which explains why this work corresponds to the deformation of the fluid particle. As anticipated towards the end of Section 1.1, this deformation work only affects the kinetic energy of the molecular motion about the particle mass center. It is the only work, therefore, that can feature in the internal-energy equation.

1.10.3 Internal Energy Equation

The volume specific internal energy $\rho\epsilon$ results from the total energy ρe as

$$\rho\epsilon = \rho e - \frac{1}{2}\rho u_i u_i \quad \Rightarrow \quad de = d\epsilon + d\left(\frac{1}{2}\rho u_i u_i\right) \tag{1.133}$$

The corresponding internal energy equation readily follows by subtracting the kinetic-energy equation (1.122) from the total energy equation (1.119), which yields

$$\frac{\partial\rho\epsilon}{\partial t} + \frac{\partial u_j \rho\epsilon}{\partial x_j} - \sigma_{ij}\frac{\partial u_j}{\partial x_j} + \frac{\partial q_j^F}{\partial x_j} - \rho\dot{g} = 0 \tag{1.134}$$

The effect of pressure can be separated from the other stresses by expressing σ_{ij} as in (1.146), which yields

$$\frac{\partial\rho\epsilon}{\partial t} + \frac{\partial u_j \rho\epsilon}{\partial x_j} + p\frac{\partial u_j}{\partial x_j} - \sigma_{ij}^\mu\frac{\partial u_j}{\partial x_j} + \frac{\partial q_j^F}{\partial x_j} - \rho\dot{g} = 0 \tag{1.135}$$

This equation no longer contains the work of the body force, but only the deformation work, which, therefore, directly affects the particle internal energy. Upon inserting the continuity equation into the internal energy equation, the resulting equation is

$$\rho\left(\frac{\partial\epsilon}{\partial t} + u_j\frac{\partial\epsilon}{\partial x_j}\right) + p\frac{\partial u_j}{\partial x_j} - \sigma_{ij}^\mu\frac{\partial u_i}{\partial x_j} + \frac{\partial q_j^F}{\partial x_j} - \rho\dot{g} = 0 \tag{1.136}$$

The substantive rate of change of the mass specific internal energy ϵ, therefore, results from the isotropic expansion work of pressure, the deformation work of the stresses σ_{ij}^μ, the conduction of heat $\partial q_j^F/\partial x_j$ and the heat source $\rho\dot{g}$. From the continuity equation, $\partial u_j/\partial x_j = -D\rho/\rho Dt$; upon inserting this result into the internal energy equation, the new form is

$$\rho\frac{D\epsilon}{Dt} - \frac{p}{\rho}\frac{D\rho}{Dt} = \sigma_{ij}^\mu\frac{\partial u_i}{\partial x_j} - \frac{\partial q_j^F}{\partial x_j} + \rho\dot{g} \tag{1.137}$$

which simplifies the derivation of the internal enthalpy, temperature and entropy equations.

1.10.4 Entropy Equation

The right-hand side of (1.137) corresponds to dissipative effects per unit time, which can thus be related to the unit-time variation of the mass-specific entropy s. With a specific volume $v = 1/\rho$, the classical thermodynamics differential of internal energy ϵ in terms of entropy s is expressed as

$$d\epsilon = Tds - pdv = Tds + \frac{p}{\rho^2}d\rho \tag{1.138}$$

which may constitute a definition for ds. Significantly, this expression only involves the reversible work of pressure p, which implies that the effect of the deformation work of the deviatoric stress is taken into account through entropy. The usefulness of an entropy variable s fundamentally results from the second law of thermodynamics, which constrains the variations of s, as amplified in Section 2.4. In substantive-derivative form, expression (1.138) becomes

$$\rho\frac{D\epsilon}{Dt} - \frac{p}{\rho}\frac{D\rho}{Dt} = \rho T\frac{Ds}{Dt} \tag{1.139}$$

and a comparison of this result with (1.137) leads to the following dimensionally-consistent entropy equation

$$\rho T\frac{Ds}{Dt} = \rho T\left(\frac{\partial s}{\partial t} + u_j\frac{\partial s}{\partial x_j}\right) = \sigma_{ij}^\mu\frac{\partial u_i}{\partial x_j} - \frac{\partial q_j^F}{\partial x_j} + \rho\dot{g} \tag{1.140}$$

This equation indicates that changes in entropy are induced by deformation work, heat transfer and heat generation. On the other hand, entropy remains unaffected by the expansion work by pressure, which is therefore regarded as reversible work. The entropy of fluid particles, therefore, does not change in the absence of deviatoric-stress deformation work, heat conduction, and heat generation. Conversely, substantial increases of entropy emerge when any or all of these three contributions are of considerable magnitude, as takes place, for instance, across a shock wave.

The entropy equation can be written differently by adding to it a "zero" in the form of the continuity equation multiplied by Ts. The resulting form becomes

$$\frac{\partial\rho s}{\partial t} + \frac{\partial u_j\rho s}{\partial x_j} = \frac{1}{T}\left(\sigma_{ij}^\mu\frac{\partial u_i}{\partial x_j} - \frac{\partial q_j^F}{\partial x_j} + \rho\dot{g}\right) \tag{1.141}$$

Next, integrate this equation over a fluid volume Ω and transform the integral of the entropy expression through the transport theorem. This sequence yields

$$\frac{D}{Dt}\int_\Omega \rho s\, d\Omega = \int_\Omega \frac{1}{T}\left(\sigma_{ij}^\mu\frac{\partial u_i}{\partial x_j} - \frac{\partial q_j^F}{\partial x_j} + \rho\dot{g}\right) d\Omega \tag{1.142}$$

where the lhs integral denotes the total entropy of the fluid within Ω. This equation can be recast by transforming the heat conduction term through Gauss' theorem. This transformation leads to

$$\frac{D}{Dt}\int_\Omega \rho s\, d\Omega = \int_\Omega \frac{1}{T}\left(\sigma_{ij}^\mu\frac{\partial u_i}{\partial x_j} + \rho\dot{g}\right) d\Omega - \int_\Omega \frac{1}{T^2}\frac{\partial T}{\partial x_j}q_j^F\, d\Omega - \oint_{\partial\Omega} \frac{1}{T}\boldsymbol{q}^F\cdot\boldsymbol{n}\, d\Gamma \tag{1.143}$$

which allows determining some basic constraints on the transport properties, based on the second law of thermodynamics, as shown in Section 2.4.

From the very definition of entropy variation, the lhs of (1.142) may also be expressed as

$$\frac{D}{Dt}\int_\Omega \rho s\, d\Omega = \int_\Omega \frac{\delta Q_{\text{tot}}}{T} \tag{1.144}$$

where δQ_{tot} denotes all the heat mechanisms that can change entropy. Upon comparing this equation with (1.142) shows that the total heat for a fluid particle is expressed as

$$\delta Q_{\text{tot}} = \left(\sigma_{ij}^\mu\frac{\partial u_i}{\partial x_j} - \frac{\partial q_j^F}{\partial x_j} + \rho\dot{g}\right) d\Omega \tag{1.145}$$

which shows that the deviatoric-stress deformation work contributes to the total heat that varies entropy, although, as shown in the derivation of the total energy equation in Section 1.7.3, this work physically originates separately from any external heating.

1.10.5 Total Enthalpy Equation

The effect of pressure is separated from the other stresses through decomposition (1.58) because the work of pressure is reversible. Repeated here for convenience, this decomposition is

$$\sigma_{ij} = -p\delta_j^i + \sigma_{ij}^\mu \tag{1.146}$$

Once this decomposition is inserted into (1.119), the resulting equation becomes

$$\frac{\partial \rho e}{\partial t} + \frac{\partial \rho e u_j}{\partial x_j} + \frac{\partial u_j p}{\partial x_j} - \frac{\partial u_i \sigma_{ij}^\mu}{\partial x_j} + \frac{\partial q_j^F}{\partial x_j} - \rho u_i b_i - \rho \dot{g} = 0 \tag{1.147}$$

The mass specific total enthalpy is defined as

$$H = e + \frac{p}{\rho} \quad \Rightarrow \quad \rho dH = \rho de + dp - \frac{p}{\rho} d\rho, \quad \rho \frac{DH}{Dt} = \rho \frac{De}{Dt} + \frac{Dp}{Dt} - \frac{p}{\rho} \frac{D\rho}{Dt} \tag{1.148}$$

With this definition, (1.147) becomes

$$\frac{\partial \rho e}{\partial t} + \frac{\partial \rho u_j H}{\partial x_j} - \frac{\partial u_i \sigma_{ij}^\mu}{\partial x_j} + \frac{\partial q_j^F}{\partial x_j} - \rho u_i b_i - \rho \dot{g} = 0 \tag{1.149}$$

The definition of enthalpy also leads to the equality

$$\frac{\partial \rho e}{\partial t} = \rho \frac{\partial H}{\partial t} + H \frac{\partial \rho}{\partial t} - \frac{\partial p}{\partial t} = \rho \frac{\partial H}{\partial t} - H \frac{\partial \rho u_j}{\partial x_j} - \frac{\partial p}{\partial t} \tag{1.150}$$

where the second form follows from the continuity equation. The elimination of the time partial derivative of ρe between (1.147) and (1.150) leads to the total enthalpy equation in the form

$$\rho \frac{DH}{Dt} = \frac{\partial p}{\partial t} + \frac{\partial u_i \sigma_{ij}^\mu}{\partial x_j} - \frac{\partial q_j^F}{\partial x_j} + \rho u_i b_i + \rho \dot{g} \tag{1.151}$$

A solution of this equation is presented in Section 1.11

1.10.6 Internal Enthalpy and Temperature Equations

The internal enthalpy, or just enthalpy, "h" is defined as

$$h = \epsilon + \frac{p}{\rho} \quad \Rightarrow \quad dh = d\epsilon + \frac{dp}{\rho} - \frac{p}{\rho^2} d\rho \tag{1.152}$$

which leads to the substantive-derivative expression

$$\rho \frac{Dh}{Dt} - \frac{Dp}{Dt} = \rho \frac{D\epsilon}{Dt} - \frac{p}{\rho} \frac{D\rho}{Dt} \tag{1.153}$$

The enthalpy equation, therefore, follows from (1.137) and (1.153) as

$$\rho \frac{Dh}{Dt} - \frac{Dp}{Dt} = \sigma_{ij}^\mu \frac{\partial u_i}{\partial x_j} - \frac{\partial q_j^F}{\partial x_j} + \rho \dot{g} \tag{1.154}$$

For a homogeneous fluid, a thermodynamic state variable depends on only two other state variables. Enthalpy, can thus be conceived as depending on temperature T and pressure p. The substantive derivative of enthalpy can thus be expressed as

$$\frac{Dh}{Dt} = \left(\frac{\partial h}{\partial T}\right)_p \frac{DT}{Dt} + \left(\frac{\partial h}{\partial p}\right)_T \frac{Dp}{Dt} = c_p \frac{DT}{Dt} + \left(\frac{\partial h}{\partial p}\right)_T \frac{Dp}{Dt} \qquad (1.155)$$

where c_p denotes the constant pressure specific heat, which by definition equals the partial derivative of enthalpy with respect to temperature, holding pressure constant. On the basis of this expression for the substantive derivative of enthalpy, the temperature equation emerges from the enthalpy equation as

$$\rho c_p \frac{DT}{Dt} + \left(\rho \left(\frac{\partial h}{\partial p}\right)_T - 1\right) \frac{Dp}{Dt} = \sigma_{ij}^\mu \frac{\partial u_i}{\partial x_j} - \frac{\partial q_j^F}{\partial x_j} + \rho \dot{g} \qquad (1.156)$$

Section 4.2 shows the conditions under which the expression containing the substantive derivative of pressure becomes negligible.

1.11   Special Solutions of the Internal-Energy, Enthalpy, and Momentum Equations

This section presents special exact solutions of the multi-dimensional internal-energy, total-enthalpy and momentum equations. These solutions arise from either isentropic- or adiabatic-flow conditions, or both, and remain valid for both general equilibrium gases and, in particular, perfect gases.

1.11.1   Isentropic Solution of the Internal Energy Equation

The internal-energy equation can also accrue by inserting the total energy equation into the kinetic-energy equation. The resulting partial- and substantive-derivative forms of the resulting internal-energy equation become

$$\frac{\partial \rho \epsilon}{\partial t} + \frac{\partial \rho \epsilon u_j}{\partial x_j} + p \frac{\partial u_j}{\partial x_j} - \sigma_{ij}^\mu \frac{\partial u_i}{\partial x_j} + \frac{\partial q_j^F}{\partial x_j} - \rho \dot{g} = 0 \qquad (1.157)$$

$$\rho \frac{D\epsilon}{Dt} - \frac{p}{\rho} \frac{D\rho}{Dt} = \sigma_{ij}^\mu \frac{\partial u_i}{\partial x_j} - \frac{\partial q_j^F}{\partial x_j} + \rho \dot{g} \qquad (1.158)$$

which imply that internal energy is not conserved across a normal shock.

Under conditions of negligible deformation work, heat conduction, and heat generation, this equation simplifies to

$$\rho \frac{D\epsilon}{Dt} - \frac{p}{\rho} \frac{D\rho}{Dt} = 0 \qquad (1.159)$$

and from the entropy equation (1.140), these conditions yield $Ds/Dt = 0$. This situation, therefore, corresponds to an isentropic flow, which thus results from both an adiabatic flow, $\left(\frac{\partial q_j^F}{\partial x_j} - \rho \dot{g}\right) = 0$ and an inviscid flow $\left(\sigma_{ij}^\mu \frac{\partial u_i}{\partial x_j}\right) = 0$.

For an equilibrium gas, the internal energy ϵ only depends upon two other thermodynamic variables, like pressure p and density ρ, hence $\epsilon = \epsilon(p, \rho)$. The general solution of the internal-energy equation thus emerges as $p = p(\rho)$. This result accrues by expressing the substantive derivative of ϵ as

$$\frac{D\epsilon}{Dt} = \frac{\partial \epsilon}{\partial p}\frac{Dp}{Dt} + \frac{\partial \epsilon}{\partial \rho}\frac{D\rho}{Dt} \tag{1.160}$$

Moreover, the substantive derivative of pressure can be expressed as

$$\rho \frac{Dp}{Dt} = \rho^2 \frac{Dp/\rho}{Dt} + p \frac{D\rho}{Dt} \tag{1.161}$$

With these two results, the terminal form of the internal-energy equation becomes

$$\frac{1}{p/\rho}\frac{Dp/\rho}{Dt} = \frac{1 - \left(p\dfrac{\partial \epsilon}{\partial p} + \rho\dfrac{\partial \epsilon}{\partial \rho}\right)}{\rho^2 \dfrac{\partial \epsilon}{\partial p}}\frac{D\rho}{Dt} \tag{1.162}$$

where every term only depends upon p and ρ. The solution of this equation is thus

$$p = p(\rho) \tag{1.163}$$

For a perfect gas, an explicit form for this general solution emerges from specific equations of state. These equations are expressed as

$$p = \rho R T, \quad \epsilon = c_v T, \quad \frac{c_p}{R} = \frac{\gamma}{\gamma - 1}, \quad \frac{c_v}{R} = \frac{1}{\gamma - 1}, \quad \Rightarrow \quad \frac{p}{\rho} = \frac{\gamma - 1}{\gamma}c_p T, \quad \epsilon = \frac{p}{(\gamma - 1)\rho} \tag{1.164}$$

where R, c_v, c_p, and $\gamma \equiv c_v/c_p$ respectively denote the gas constant, constant-volume as well as constant-pressure specific heats, and specific-heat ratio.

With these expressions, equation (1.162) becomes

$$\frac{1}{p/\rho}\frac{Dp/\rho}{Dt} = \frac{\gamma - 1}{\rho}\frac{D\rho}{Dt} \tag{1.165}$$

with solution

$$\ln\left(\frac{p}{\rho}\right) = (\gamma - 1)\ln\rho + \text{constant} \quad \Rightarrow \quad \frac{p}{\rho^\gamma} = \text{constant} \tag{1.166}$$

This well-known expression, therefore, naturally arises as an exact solution of the unsteady multi-dimensional internal-energy equation under isentropic-flow conditions.

1.11.2 Adiabatic- and Inviscid-Flow Solution of the Total-Enthalpy Equation

The total enthalpy $H = (\rho e + p)/\rho$ satisfies a governing equation that is conveniently repeated here in partial- and substantive-derivative forms

$$\frac{\partial \rho e}{\partial t} + \frac{\partial \rho H u_j}{\partial x_j} - \frac{\partial u_i \sigma_{ij}^\mu}{\partial x_j} + \frac{\partial q_j^F}{\partial x_j} - \rho u_i b_i - \rho \dot{g} = 0 \tag{1.167}$$

$$\rho\frac{DH}{Dt} = \frac{\partial p}{\partial t} + \frac{\partial u_i \sigma_{ij}^\mu}{\partial x_j} - \frac{\partial q_j^F}{\partial x_j} + \rho u_i b_i + \rho\dot{g} \tag{1.168}$$

In the absence of body-force and deviatoric-stress work, as well as heat conduction and generation, the partial-derivative form of this equation remains in divergence form. Total enthalpy, therefore, remains conserved across a shock wave. When the flow is also steady, hence $\partial p/\partial t = 0$, the total-enthalpy equation becomes

$$\rho\frac{DH}{Dt} = 0 \tag{1.169}$$

which shows that the total enthalpy remains constant for every fluid particle and thus throughout the flow field. With H_0 denoting the stagnation total enthalpy, the solution of (1.169) becomes

$$H_0 = H \quad\Rightarrow\quad \frac{\rho_0 e_0 + p_0}{\rho_0} = \frac{\rho e + p}{\rho} \tag{1.170}$$

which applies for both general equilibrium as well as particular perfect gases.

For a perfect gas, from (1.133), (1.164), the total energy ρe directly relates to pressure as $\rho e = p/(\gamma - 1) + \rho u_i u_i/2$. With this expression, solution (1.170) specializes as

$$\frac{\gamma}{\gamma - 1}\frac{p_0}{\rho_0} = \frac{\gamma}{\gamma - 1}\frac{p}{\rho} + \frac{1}{2}u_i u_i \tag{1.171}$$

With (1.164), this expression becomes

$$c_p T_0 = c_p T + \frac{1}{2}u_i u_i \tag{1.172}$$

This result, which is known as a solution of the 1-D or quasi-1D Euler equations, thus solves the multi-dimensional steady total-enthalpy equation.

1.11.3   Isentropic Solution of the Momentum Equation

The steady non-conservation form of the momentum equations connects the flow rotation with the flow enthalpy and entropy. The momentum equation in this form shows that its isentropic solution is an irrotational flow.

The rotation within a flow corresponds to the "curl" of the velocity vector $\boldsymbol{V}$. Vector analysis connects this rotation with the gradient of the kinetic energy as

$$\boldsymbol{V} \times \text{curl}\boldsymbol{V} = \text{grad}\left(\frac{1}{2}\boldsymbol{V} \cdot \boldsymbol{V}\right) - (\boldsymbol{V} \cdot \text{grad})\,\boldsymbol{V} \tag{1.173}$$

With this equivalence, curl$\boldsymbol{V}$ features in the momentum equation (1.78). In vector and pressure-gradient form, this equation becomes

$$\frac{\partial \boldsymbol{V}}{\partial t} + (\boldsymbol{V} \cdot \text{grad})\,\boldsymbol{V} + \frac{1}{\rho}\text{grad}p - \frac{1}{\rho}\text{div}\boldsymbol{\sigma}^\mu - \boldsymbol{b} = 0 \tag{1.174}$$

Result (1.173) transforms (1.174) into

$$\frac{\partial \boldsymbol{V}}{\partial t} - \boldsymbol{V} \times \text{curl}\boldsymbol{V} + \text{grad}\left(\frac{1}{2}\boldsymbol{V} \cdot \boldsymbol{V}\right) + \frac{1}{\rho}\text{grad}p - \frac{1}{\rho}\text{div}\boldsymbol{\sigma}^\mu - \boldsymbol{b} = 0 \tag{1.175}$$

The inner, ("dot"), product of this equation with the arbitrary displacement vector $d\boldsymbol{x}$ yields

$$\frac{\partial \boldsymbol{V}}{\partial t} \cdot d\boldsymbol{x} - (\boldsymbol{V} \times \mathrm{curl}\boldsymbol{V}) \cdot d\boldsymbol{x} + d_x \left(\frac{1}{2}\boldsymbol{V} \cdot \boldsymbol{V}\right) + \frac{d_x p}{\rho} - \left(\frac{1}{\rho}\mathrm{div}\boldsymbol{\sigma}^\mu + \boldsymbol{b}\right) \cdot d\boldsymbol{x} = 0 \qquad (1.176)$$

where d_x indicates a variation in the arbitrary direction of $d\boldsymbol{x}$ within the flow field. When $d\boldsymbol{x}$ denotes a streamline displacement, $d\boldsymbol{x} \equiv \boldsymbol{V}dt$, this equation coincides with the kinetic-energy equation (1.124), because $(\boldsymbol{V} \times \mathrm{curl}\boldsymbol{V}) \cdot \boldsymbol{V} \equiv 0$, since $(\boldsymbol{V} \times \mathrm{curl}\boldsymbol{V})$ remains perpendicular to both $\boldsymbol{V}$ and $\mathrm{curl}\boldsymbol{V}$.

For an arbitrary displacement $d\boldsymbol{x}$, the variations of pressure and kinetic energy within (1.176) depend on the variations of total enthalpy and entropy, as obtained from the differentials of total enthalpy as well as total and internal energy

$$\begin{cases} dH &= de + \dfrac{dp}{\rho} - \dfrac{p}{\rho^2}d\rho \\[2mm] de &= d\epsilon + d\left(\dfrac{1}{2}\boldsymbol{V} \cdot \boldsymbol{V}\right) \\[2mm] d\epsilon &= Tds + \dfrac{p}{\rho^2}d\rho \end{cases} \quad \Rightarrow \quad d_x H - T d_x s = d_x\left(\frac{1}{2}\boldsymbol{V} \cdot \boldsymbol{V}\right) + \frac{d_x p}{\rho} \qquad (1.177)$$

Equation (1.176) thus becomes

$$\frac{\partial \boldsymbol{V}}{\partial t} \cdot d\boldsymbol{x} - (\boldsymbol{V} \times \mathrm{curl}\boldsymbol{V}) \cdot d\boldsymbol{x} + d_x H - T d_x s - \left(\frac{1}{\rho}\mathrm{div}\boldsymbol{\sigma} + \boldsymbol{b}\right) \cdot d\boldsymbol{x} \qquad (1.178)$$

For a flow that is steady, iso-enthalpic, isentropic, and devoid of deviatoric stresses and body forces, therefore, this equation yields $(\boldsymbol{V} \times \mathrm{curl}\boldsymbol{V}) \cdot d\boldsymbol{x} = 0$; since $d\boldsymbol{x}$ is arbitrary and $\boldsymbol{V}$ is not the zero vector, it follows that

$$\mathrm{curl}\boldsymbol{V} = \boldsymbol{0} \qquad (1.179)$$

This theorem of Crocco's thus reveals that when the flow is steady, devoid of body forces, inviscid, adiabatic, and without shocks, it is irrotational. Since $\mathrm{curl}(\mathrm{grad}\phi) \equiv 0$, in an irrotational flow the velocity vector can thus be expressed as the gradient of a single scalar function ϕ

$$\boldsymbol{V} = \mathrm{grad}\phi \qquad (1.180)$$

an expression that leads to the development of potential flow.

Consider next the line integration of (1.176) between any two flow-field points $\boldsymbol{x}_1$ and $\boldsymbol{x}_2$, which yields

$$\int_{x_1}^{x_2} \frac{d_x p}{\rho} + \left(\frac{1}{2}\boldsymbol{V} \cdot \boldsymbol{V}\right)_2 - \left(\frac{1}{2}\boldsymbol{V} \cdot \boldsymbol{V}\right)_1$$

$$= \int_{x_1}^{x_2} \left((\boldsymbol{V} \times \mathrm{curl}\boldsymbol{V}) \cdot d\boldsymbol{x} - \frac{\partial \boldsymbol{V}}{\partial t} \cdot d\boldsymbol{x} + \left(\frac{1}{\rho}\mathrm{div}\boldsymbol{\sigma} + \boldsymbol{b}\right) \cdot d\boldsymbol{x}\right) \qquad (1.181)$$

For a flow that is steady and inviscid, with a body force that is only due to gravity, this result becomes

$$\int_{x_1}^{x_2} \frac{d_x p}{\rho} + \left(\frac{1}{2}\boldsymbol{V} \cdot \boldsymbol{V}\right)_2 - \left(\frac{1}{2}\boldsymbol{V} \cdot \boldsymbol{V}\right)_1 + gz_2 - gz_1 = \int_{x_1}^{x_2} (\boldsymbol{V} \times \mathrm{curl}\boldsymbol{V}) \cdot d\boldsymbol{x} \qquad (1.182)$$

When $d\boldsymbol{x}$ denotes a displacement along a streamline, $d\boldsymbol{x} \equiv \boldsymbol{V} dt$, this equation provides Bernoulli's theorem

$$\left(\frac{1}{2}\boldsymbol{V} \cdot \boldsymbol{V}\right)_1 + gz_1 = \int_{x_1}^{x_2} \frac{d_x p}{\rho} + \left(\frac{1}{2}\boldsymbol{V} \cdot \boldsymbol{V}\right)_2 + gz_2 \qquad (1.183)$$

where $\boldsymbol{x}_1$ and $\boldsymbol{x}_2$ denote any two points along this streamline. This result, therefore, corresponds to an exact solution of the kinetic-energy equation.

With reference to (1.182), when the flow is irrotational, this theorem applies for any two arbitrary points $\boldsymbol{x}_1$ and $\boldsymbol{x}_2$ within the flow field. For an adiabatic flow of a perfect gas, hence $p_1/\rho_1^\gamma = p_2/\rho_2^\gamma$, this equation can be exactly integrated, which yields

$$\frac{\gamma}{\gamma-1}\frac{p_1}{\rho_1} + \left(\frac{1}{2}\boldsymbol{V} \cdot \boldsymbol{V}\right)_1 + gz_1 = \frac{\gamma}{\gamma-1}\frac{p_2}{\rho_2} + \left(\frac{1}{2}\boldsymbol{V} \cdot \boldsymbol{V}\right)_2 + gz_2 \qquad (1.184)$$

Under the stated conditions, therefore, this result, which is a known solution of the 1-D or quasi-1-D Euler equations, thus solves the contraction of the steady multi-dimensional Euler equations. In the absence of body forces, in particular, this results coincides with the constant-enthalpy solution developed in the previous section.

Chapter 2

Constitutive and State Equations

This chapter presents the constitutive and state equations for a perfect gas and a liquid. Included are also temperature-dependent transport coefficients as well as their limits, which originate from the second law of thermodynamics.

2.1 Navier-Stokes Stress Tensor

This section presents the relations between stresses and derivatives of velocity components for the so-called Newtonian fluids. Stresses arise when a state of relative motion exists among layers of fluid. With reference to the simple one-dimensional velocity distribution in Figure 2.1, Newton thus postulated a linear relation between the shear stress σ_{12} and the velocity gradient $\partial u_1/\partial x_2$ in the form

$$\sigma_{12} = \mu \frac{\partial u_1}{\partial x_2} \tag{2.1}$$

where the coefficient of dynamic viscosity μ depends on the specific fluid. This relation remains consistent with a vanishing shear stress when the fluid is in equilibrium or uniform motion, because in this case $\partial u_1/\partial x_2 = 0$. From (2.1), the expression $\sigma_{21} = \mu \partial u_2/\partial x_1$ would result; the symmetry condition $\sigma_{21} = \sigma_{12}$, however, must also hold. Additionally, the stresses must vanish when the fluid undergoes uniform rotation with constant angular velocity $\boldsymbol{\omega}$, because no relative motion among layers of fluid will exist in this type of motion. In this case, the fluid velocity is $\boldsymbol{V} = \boldsymbol{\omega} \times \boldsymbol{r} = \boldsymbol{\omega} \times (x_1\boldsymbol{e}_1 + x_2\boldsymbol{e}_2 + x_3\boldsymbol{e}_3)$, where $\boldsymbol{e}_j$, $1 \leq j \leq 3$, denotes the unit vector in the positive direction of the x_j axis. For this velocity, the vanishing combinations of velocity component derivatives are $(\partial u_i/\partial x_j + \partial u_j/\partial x_i)$. For an isotropic fluid, μ does not depend on any direction and thus one generalization of (2.1) emerges as

$$\sigma_{ij} = \mu \left(\frac{\partial u_i}{\partial x_j} + \frac{\partial u_j}{\partial x_i} \right) \tag{2.2}$$

which yields $\sigma_{ij} = \sigma_{ji}$, returns (2.1) for $u_2 = 0$, and vanishes under both rotation and uniform translation. For the normal stresses, (2.2) also predicts a reasonable dependence on the rate of elongation $\partial u_i/\partial x_i$ along the x_i axis, but does not yet include any dependence on local pressure p, which affects normal stresses.

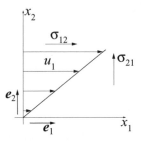

Figure 2.1: One-Dimensional Velocity Profile

Because of the fluid flow itself, the local mechanical pressure p_{mech} may not necessarily coincide with the equilibrium thermodynamic pressure p, but may be lower in proportion to the local fluid particle expansion $\lim_{\Delta\Omega\to 0}(D(\Delta\Omega)/Dt)/\Delta\Omega$, hence

$$p_{\mathrm{mech}} = p - \eta_B \lim_{\Delta\Omega\to 0} \frac{D\Delta\Omega}{\Delta\Omega Dt} = p - \eta_B \mathrm{div}\mathbf{V} = p - \eta_B \frac{\partial u_\ell}{\partial x_\ell} \tag{2.3}$$

where η_B denotes a coefficient of bulk viscosity. On the other hand, the mechanical pressure is related to the normal stresses through (1.57), hence

$$\sigma_{ii} = -3p_{\mathrm{mech}} = -3p + 3\eta_B \frac{\partial u_\ell}{\partial x_\ell} \tag{2.4}$$

a scalar expression. A further generalization of (2.2) thus follows as

$$\sigma_{ij} = \mu\left(\frac{\partial u_i}{\partial x_j} + \frac{\partial u_j}{\partial x_i}\right) + c\delta_j^i \tag{2.5}$$

with "c" a scalar function to be determined such that σ_{ii} from this expression returns (2.4). Under this condition, (2.5) becomes the Navier-Stokes expression

$$\sigma_{ij} = -p\delta_j^i + \mu\left(\frac{\partial u_i}{\partial x_j} + \frac{\partial u_j}{\partial x_i}\right) + \lambda\frac{\partial u_\ell}{\partial x_\ell}\delta_j^i, \quad \lambda = -\frac{2}{3}\mu + \eta_B \tag{2.6}$$

where λ denotes the second coefficient of viscosity. The bulk viscosity η_B plays a role in explaining some observed attenuation of acoustic waves, which cannot be attributed to thermal and shear- stress effects alone. For monatomic gases η_B vanishes, while for other fluids it remains somewhat irrelevant in ordinary fluid mechanics. In these cases, Stokes' hypothesis applies, for which $\lambda = -\frac{2}{3}\mu$. This hypothesis corresponds to a relaxation time that exceeds that of a typical fluid mechanics time, so that any local departures from equilibrium pressure virtually instantaneously lead to another thermodynamic equilibrium state, so that the mechanical pressure is essentially equal to the local thermodynamic pressure and from (2.3) $p_{\mathrm{mech}} = p$. Based on a phenomenological approach, the constitutive relations (2.6) predict results that agree with experimental results for fluids with relatively simple molecular structures, like that of air and water for instance; this agreement provides the physical justification for (2.6). When (2.6) accurately predicts the viscous stresses within a fluid, such a fluid is called a Newtonian fluid.

2.2   Heat Conduction Law

The governing equations (1.69), (1.78), and (1.95) involve density ρ, velocity components u_i, $1 \leq i \leq 3$, and total volume specific energy e, which can collectively correspond to five state variables. Additionally, the governing equations involve pressure p, through (2.6), and the heat flux vector components q_i^F, $1 \leq i \leq 3$. These components are cast in terms of static temperature through Fourier's heat conduction law

$$q_i^F = -k\frac{\partial T}{\partial x_i} \tag{2.7}$$

where the negative sign indicates that heat flows from regions of high temperature into regions of lower temperature. In this expression, "k" denotes the coefficient of thermal conductivity.

2.3   Transport and Thermo-Mechanical Properties

Figure 2.2 presents the variation of both μ/μ_0 and k/k_0 for water.

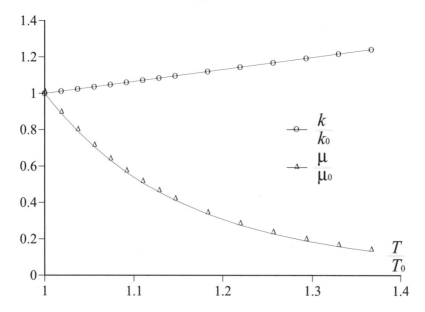

Figure 2.2: Variations of Water Viscosity and Thermal Conductivity

For increasing temperature, the coefficient of dynamic viscosity μ moderately decreases in liquids, but increases in gases. This coefficient decreases in liquids because, for increasing temperature, cohesive forces weaken, hence the liquid more readily flows; in gases, this coefficient increases, because in this case the frequency and intensity of molecular collisions increase with increasing temperature, which increases shear stress and heat conduction.

For water μ and k may be expressed through the relations

$$\frac{\mu}{\mu_0} = \frac{1 + \frac{T}{T_0}}{2\left(\frac{T}{T_0}\right)^7}, \quad \frac{k}{k_0} = 1 + 0.6584\left(\frac{T}{T_0} - 1\right)$$

$$T_0 = 273.15\text{K}, \quad \mu_0 = 1.787 \times 10^{-3}\frac{\text{kg}}{\text{m}\,\text{s}}, \quad k_0 = 5.6 \times 10^{-1}\frac{\text{J}}{\text{m}\,\text{s}\,\text{K}} \tag{2.8}$$

for $273.15\text{K} \leq T \leq 373.15\text{K}$.

For air, μ and k can be expressed through Sutherland's relations as

$$\frac{\mu}{\mu_0} = \left(\frac{T}{T_0}\right)^{\frac{3}{2}}\frac{1.4041}{0.4041 + \frac{T}{T_0}}, \quad \frac{k}{k_0} = \left(\frac{T}{T_0}\right)^{\frac{3}{2}}\frac{1.41}{0.41 + \frac{T}{T_0}}$$

$$T_0 = 273.15\text{K}, \quad \mu_0 = 1.7161 \times 10^{-5}\frac{\text{kg}}{\text{m}\,\text{s}}, \quad k_0 = 2.3360 \times 10^{-2}\frac{\text{J}}{\text{m}\,\text{s}\,\text{K}} \tag{2.9}$$

The relation for μ/μ_0 remains independent of pressure for $T < 3,000\text{K}$ and the relation for k/k_0 remains valid for $T < 2,000\text{K}$, approximately corresponding to the temperature range at which molecular oxygen does not dissociate. Figure 2.3 presents the variation of both

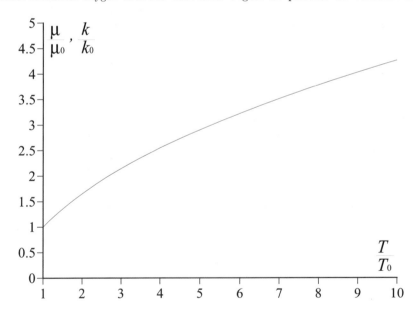

Figure 2.3: Variations of Air Viscosity and Thermal Conductivity

μ/μ_0 and k/k_0 for air, over a temperature range corresponding to $273.15\text{K} \leq T \leq 3,000\text{K}$. The variations for these non-dimensional coefficients are virtually indistinguishable within the scale of the graph.

Select thermo-mechanical properties for air and water feature in the following table. These properties correspond to a reference pressure of 1 atmosphere and a reference temperature of 20° Celsius.

Table 2.1: Thermo-Mechanical Properties of Air and Water

	ρ (Kg / m^3)	μ (N s / m^2)	k (W / m K)	c_p (J K / Kg)	β (1 / K)
Air	1.2	18.000×10^{-6}	26.3×10^{-3}	$1,007$	$1/T$
Water	998	1.003×10^{-3}	613×10^{-3}	$4,179$	276.1×10^{-6}

2.4   Second-Law Constraints on Transport Coefficients

The second law of thermodynamics imposes simple constraints on bulk viscosity η_B, viscosity μ and thermal conductivity k. The entropy of the entire universe or, more modestly, of an adiabatic system can never decrease. The time rate of change of the entropy of an adiabatic fluid mass, therefore, can only be non-negative. From (1.143), the rate of entropy change is expressed as

$$\frac{D}{Dt}\int_\Omega \rho s\, d\Omega = \int_\Omega \frac{1}{T}\left(\sigma_{ij}^\mu \frac{\partial u_i}{\partial x_j} + \rho \dot{g}\right) d\Omega - \int_\Omega \frac{1}{T^2}\frac{\partial T}{\partial x_j}q_j^F\, d\Omega - \oint_{\partial\Omega} \frac{1}{T}\boldsymbol{q}^F \cdot \boldsymbol{n}\, d\Gamma \qquad (2.10)$$

For an adiabatic system, the integrals of heat generation $\rho \dot{g}$ and heat flux $\boldsymbol{q}^F \cdot \boldsymbol{n}$ will vanish; consequently, only the deformation-work and heat-conduction integrals remain in (2.10). From (2.6) and (2.7), the deformation work and heat conduction respectively become

$$\sigma_{ij}^\mu \frac{\partial u_i}{\partial x_j} = \mu \frac{\partial u_i}{\partial x_j}\left(\frac{\partial u_i}{\partial x_j} + \frac{\partial u_j}{\partial x_i} - \frac{2}{3}\frac{\partial u_\ell}{\partial x_\ell}\delta_j^i\right) + \eta_B \frac{\partial u_i}{\partial x_j}\frac{\partial u_\ell}{\partial x_\ell}\delta_j^i =$$

$$= \frac{\mu}{2}\left(\frac{\partial u_i}{\partial x_j} + \frac{\partial u_j}{\partial x_i} - \frac{2}{3}\frac{\partial u_\ell}{\partial x_\ell}\delta_j^i\right)\left(\frac{\partial u_i}{\partial x_j} + \frac{\partial u_j}{\partial x_i} - \frac{2}{3}\frac{\partial u_\ell}{\partial x_\ell}\delta_j^i\right) + \eta_B\left(\frac{\partial u_\ell}{\partial x_\ell}\right)^2 \qquad (2.11)$$

and

$$\frac{\partial T}{\partial x_j}q_j^F = -k\frac{\partial T}{\partial x_j}\frac{\partial T}{\partial x_j} \qquad (2.12)$$

The expression multiplying $\mu/2$, in (2.11), and the rhs of (2.12) correspond to sums of squares. Inserting both (2.11) and (2.12) into the form of (2.10) for an adiabatic system yields

$$\frac{D}{Dt}\int_\Omega \rho s\, d\Omega = \int_\Omega \frac{\mu}{2T}\left(\frac{\partial u_i}{\partial x_j} + \frac{\partial u_j}{\partial x_i} - \frac{2}{3}\frac{\partial u_\ell}{\partial x_\ell}\delta_j^i\right)\left(\frac{\partial u_i}{\partial x_j} + \frac{\partial u_j}{\partial x_i} - \frac{2}{3}\frac{\partial u_\ell}{\partial x_\ell}\delta_j^i\right) d\Omega+$$

$$+ \int_\Omega \frac{\eta_B}{T} \left(\frac{\partial u_\ell}{\partial x_\ell} \right)^2 d\Omega + \int_\Omega \frac{k}{T^2} \frac{\partial T}{\partial x_j} \frac{\partial T}{\partial x_j} d\Omega \tag{2.13}$$

Hence, the total entropy of the fluid within Ω will not decrease when μ, k, and η_B are all non-negative. With these constraints on the coefficients of viscosity, the deformation work in the time rate of change of kinetic energy (1.127)

$$\frac{D}{Dt} \int_\Omega \frac{1}{2} \rho u_i u_i \, d\Omega = \int_\Omega \left(p \frac{\partial u_j}{\partial x_j} + \rho b_i u_i \right) d\Omega + \oint_{\partial\Omega} u_i \sigma_{ij} n_j d\Gamma - \int_\Omega \sigma_{ij}^\mu \frac{\partial u_i}{\partial x_j} \, d\Omega \tag{2.14}$$

always provides a negative contribution, which corresponds to a dissipative effect on kinetic energy.

2.5   Equations of State

From a mathematical standpoint, (2.7) expresses three variables through one, temperature. In order to have the same number of variables as governing equations, both T and p have to be expressed in terms of density ρ, linear-momentum $\boldsymbol{m}$, and total energy E, the state variables in the governing equations.

For a homogeneous fluid, any state variable can be expressed in terms of other two independent thermodynamic variables. In fluid mechanics, two convenient independent thermodynamic variables are density ρ and mass-specific internal energy ϵ, because ρ results from the continuity equation and ϵ follows from e as

$$\epsilon = e - \frac{1}{2} u_i u_i = \frac{E}{\rho} - \frac{1}{2\rho^2} \left(m_1^2 + m_2^2 + m_3^2 \right) \tag{2.15}$$

which indicates that the internal energy ϵ equals the total energy decreased by the mass-specific particle kinetic energy.

The expressions for p and T thus follow as

$$p = p(\rho, \epsilon), \quad T = T(\rho, \epsilon) \tag{2.16}$$

hence

$$p = p(\rho, \epsilon) = p \left(\rho, \frac{E}{\rho} - \frac{m_i m_i}{2\rho^2} \right), \quad T = T(\rho, \epsilon) = T \left(\rho, \frac{E}{\rho} - \frac{m_i m_i}{2\rho^2} \right) \tag{2.17}$$

the first expression is known as the equation of state, whereas the second equation corresponds to an internal-energy temperature equation.

Applying the chain rule to the first expression in (2.17) leads to the following expressions for the Jacobian partial derivatives of pressure

$$\left(\frac{\partial p}{\partial \rho} \right)_{\boldsymbol{m},E} = \left(\frac{\partial p}{\partial \rho} \right)_\epsilon + \left(\frac{\partial p}{\partial \epsilon} \right)_\rho \left(\frac{\partial \epsilon}{\partial \rho} \right)_{\boldsymbol{m},E} = \left(\frac{\partial p}{\partial \rho} \right)_\epsilon + \left(\frac{\partial p}{\partial \epsilon} \right)_\rho \cdot \frac{1}{\rho^2} \left(\frac{1}{\rho} \sum_{i=1}^n m_i m_i - E \right) \tag{2.18}$$

$$\left(\frac{\partial p}{\partial m_i} \right)_{\rho, m_j, E, i \neq j} = \left(\frac{\partial p}{\partial \epsilon} \right)_\rho \left(\frac{\partial \epsilon}{\partial m_i} \right)_{\rho, m_j, E, i \neq j} = \left(\frac{\partial p}{\partial \epsilon} \right)_\rho \cdot \left(-\frac{m_i}{\rho^2} \right) \tag{2.19}$$

$$\left(\frac{\partial p}{\partial E}\right)_{\rho,\boldsymbol{m}} = \left(\frac{\partial p}{\partial \epsilon}\right)_{\rho}\left(\frac{\partial \epsilon}{\partial E}\right)_{\rho,\boldsymbol{m}} = \left(\frac{\partial p}{\partial \epsilon}\right)_{\rho}\cdot\left(\frac{1}{\rho}\right) \tag{2.20}$$

The partial derivatives of T follow from these results by replacing p with T. These results also show that the derivatives of pressure with respect to m_i, $1 \le i \le 3$, and E linearly depend upon one another as

$$\left(\frac{\partial p}{\partial m_i}\right)_{\rho,m_j,E,i\ne j} + \frac{m_i}{\rho}\left(\frac{\partial p}{\partial E}\right)_{\rho,\boldsymbol{m}} = 0 \tag{2.21}$$

hence

$$\frac{\partial p}{\partial m_1} = -\frac{m_1}{\rho}\frac{\partial p}{\partial E}, \quad \frac{\partial p}{\partial m_2} = -\frac{m_2}{\rho}\frac{\partial p}{\partial E}, \quad \frac{\partial p}{\partial m_3} = -\frac{m_3}{\rho}\frac{\partial p}{\partial E} \tag{2.22}$$

as obtained by expressing the derivatives of p with respect to m_1, m_2, m_3 and E in terms of the thermodynamic derivative of p with respect to ϵ, from the first expression in (2.17). In the following sections, for simplicity, the abridged notation

$$p_\rho \equiv \frac{\partial p}{\partial \rho}, \quad p_{m_1} \equiv \frac{\partial p}{\partial m_1}, \quad p_{m_2} \equiv \frac{\partial p}{\partial m_2}, \quad p_{m_3} \equiv \frac{\partial p}{\partial m_3}, \quad p_E \equiv \frac{\partial p}{\partial E} \tag{2.23}$$

will denote the Jacobian derivatives of pressure.

The expressions for a thermally and calorically perfect gas for (2.15) and (2.16) follow from the equation of state and internal energy equation as

$$p = \rho RT, \quad \epsilon = c_v T, \quad \frac{c_p}{R} = \frac{\gamma}{\gamma-1}, \quad \frac{c_v}{R} = \frac{1}{\gamma-1}, \quad \Rightarrow \quad \frac{p}{\rho} = \frac{\gamma-1}{\gamma}c_p T, \quad \epsilon = \frac{p}{(\gamma-1)\rho} \tag{2.24}$$

where c_p, c_v, T, R and γ, as stated in the previous chapter, respectively denote the constant-pressure specific heat, the constant-volume specific heat, static temperature, gas constant and specific-heat ratio, with $\gamma = 1.4$ for perfect air. The elimination of T from these two expressions, along with (2.15), leads to the following familiar expressions for the equation of state for p.

$$p = (\gamma-1)\rho\epsilon = (\gamma-1)\left(E - \frac{1}{2\rho}\left(m_1^2 + m_2^2 + m_3^2\right)\right), \quad T = \frac{p}{\rho R} \tag{2.25}$$

With this equation of state, the corresponding Jacobian derivatives of pressure become

$$p_\rho = (\gamma-1)\frac{m_1^2 + m_2^2 + m_3^2}{2\rho^2}, \quad p_{m_i} = -(\gamma-1)\frac{m_i}{\rho}, \quad p_E = (\gamma-1) \tag{2.26}$$

which satisfy (2.22).

2.6 Speed of Sound

The speed of sound results from an isentropic variation of pressure with respect to density. On the basis of $p = p(\rho, \epsilon(\rho, s))$, the expression for the square of the speed of sound is

$$c^2 = \left(\frac{\partial p}{\partial \rho}\right)_s = \left(\frac{\partial p}{\partial \rho}\right)_\epsilon + \left(\frac{\partial p}{\partial \epsilon}\right)_\rho\left(\frac{\partial \epsilon}{\partial \rho}\right)_s = \left(\frac{\partial p}{\partial \rho}\right)_\epsilon + \left(\frac{\partial p}{\partial \epsilon}\right)_\rho\frac{p}{\rho^2} \tag{2.27}$$

Expressions (1.139) and (1.138) imply the partial derivatives of ϵ

$$\left(\frac{\partial \epsilon}{\partial s}\right)_\rho = T, \quad \left(\frac{\partial \epsilon}{\partial \rho}\right)_s = \frac{p}{\rho^2} \tag{2.28}$$

The square of the speed of sound thus becomes

$$c^2 = \left(\frac{\partial p}{\partial \rho}\right)_\epsilon + \frac{p}{\rho^2}\left(\frac{\partial p}{\partial \epsilon}\right)_\rho \tag{2.29}$$

By virtue of the Jacobian derivatives of pressure (2.18)-(2.20), this expression can be recast as

$$
\begin{aligned}
c^2 &= \left(\frac{\partial p}{\partial \rho}\right)_{\boldsymbol{m},E} + \left(\frac{\partial p}{\partial E}\right)_{\rho,\boldsymbol{m}}\left(\frac{E+p}{\rho} - \frac{1}{\rho^2}\sum_{i=1}^{n} m_i m_i\right) \\
&= p_\rho + p_E \left(\frac{E+p}{\rho} - \frac{1}{\rho^2}\left(m_1^2 + m_2^2 + m_3^2\right)\right)
\end{aligned} \tag{2.30}
$$

which corresponds to the square of the isentropic component of all the eigenvalues of the Euler inviscid flux vector. This result, in particular, allows expressing the mass-specific total enthalpy H as

$$H = \frac{E+p}{\rho} = \frac{c^2\left(1 + p_E M^2\right) - p_\rho}{p_E} = \frac{c^2 - p_\rho}{p_E} + u_i u_i \tag{2.31}$$

where $M \equiv \|\boldsymbol{u}\|/c$ denotes the Mach number. From (2.30), the perfect-gas speed of sound can be expressed as

$$c^2 = (\gamma - 1)\left[\frac{E+p}{\rho} - \frac{m_1^2 + m_2^2 + m_3^2}{2\rho^2}\right] \tag{2.32}$$

directly in terms of the dependent variables.

2.7 Isentropic Constraint on Derivatives of Density and Energy

The previous chapter showed that for an isentropic flow, the general solution of the time-dependent multi-dimensional internal-energy equation expresses pressure only as a function of density, hence $p = p(\rho)$. This solution leads to a useful constraint among the derivatives of density ρ and total energy E. This constraint emerges from the following components of the pressure gradient. In terms of the derivatives of ρ, m_j, E, the ith component of the gradient of pressure becomes

$$
\begin{aligned}
\frac{\partial p}{\partial x_i} &= \frac{\partial p}{\partial \rho}\frac{\partial \rho}{\partial x_i} + \frac{\partial p}{\partial m_j}\frac{\partial m_j}{\partial x_i} + \frac{\partial p}{\partial E}\frac{\partial E}{\partial x_i} \\
&= \frac{\partial p}{\partial \rho}\frac{\partial \rho}{\partial x_i} - \frac{\partial p}{\partial E}\frac{m_j}{\rho}\frac{\partial m_j}{\partial x_i} + \frac{\partial p}{\partial E}\frac{\partial E}{\partial x_i} \\
&= \frac{\partial p}{\partial \rho}\frac{\partial \rho}{\partial x_i} + \frac{\partial p}{\partial E}\left(\frac{\partial E}{\partial x_i} - \frac{m_j}{\rho}\frac{\partial m_j}{\partial x_i}\right)
\end{aligned} \tag{2.33}
$$

Following (2.31), the Jacobian derivative of pressure with respect to density can be expressed as

$$\frac{\partial p}{\partial \rho} = \left.\frac{\partial p}{\partial \rho}\right|_S - \frac{\partial p}{\partial E}\left(\frac{E+p}{\rho} - u_i u_i\right), \quad \left.\frac{\partial p}{\partial \rho}\right|_S = c^2 \tag{2.34}$$

With this result, (2.33) becomes

$$\frac{\partial p}{\partial x_i} = \left.\frac{\partial p}{\partial \rho}\right|_S \frac{\partial \rho}{\partial x_i} - \frac{\partial p}{\partial E}\left(\frac{E+p}{\rho} - u_i u_i\right)\frac{\partial \rho}{\partial x_i} + \frac{\partial p}{\partial E}\left(\frac{\partial E}{\partial x_i} - \frac{m_j}{\rho}\frac{\partial m_j}{\partial x_i}\right) \tag{2.35}$$

Since $p = p(\rho)$, for an isentropic flow, the first term at the right-hand side of this result coincides with the left-hand side. This recognition, leads to the constraint

$$-\left(\frac{E+p}{\rho} - u_i u_i\right)\frac{\partial \rho}{\partial x_i} + \frac{\partial E}{\partial x_i} - \frac{m_j}{\rho}\frac{\partial m_j}{\partial x_i} = 0 \tag{2.36}$$

which by virtue of the second expression in (2.31), becomes

$$\left(\frac{c^2 - p_\rho}{p_E}\right)\frac{\partial \rho}{\partial x_i} = \frac{\partial E}{\partial x_i} - \frac{m_j}{\rho}\frac{\partial m_j}{\partial x_i} \tag{2.37}$$

For a perfect gas, both sides of (2.37) equal $\dfrac{\gamma}{\gamma-1}\dfrac{p}{\rho}\dfrac{\partial \rho}{\partial x_i}$. For an acoustic field, the components m_j of linear momentum become negligible and (2.37) specializes to

$$\left(\frac{c^2 - p_\rho}{p_E}\right)\frac{\partial \rho}{\partial x_i} = \frac{\partial E}{\partial x_i} \tag{2.38}$$

Chapter 3

State Equations for Reacting Air

The chemical dissociations within reacting air absorb energy and thus lead to lower static temperatures, higher static densities, and different shock-wave positions in comparison to perfect-air predictions. These effects must therefore be modeled accurately for reliable Euler and Navier-Stokes CFD investigation of aerothermodynamic and high-temperature flows.

This chapter presents a computationally efficient solution procedure of an exact thermodynamic model of a five-species, electrically-neutral and chemically reacting air. The model involves an explicit pressure equation of state coupled to a non-linear system of six chemical-equilibrium thermodynamic equations for five mass fractions and temperature. The resulting equations then revert to the familiar perfect-gas expressions in the appropriate temperature and pressure ranges. The developed solution procedure remains valid for any exponential form of the equilibrium constants and, without introducing spurious solutions, succeeds in algebraically reducing this six-equation system to a two-equation system for nitric-oxide mass fraction and temperature, which are then numerically determined through a rapidly converging Newton's method solution. The remaining mass fractions and then pressure are explicitly calculated using this solution and the thermodynamic and Jacobian derivatives of pressure and temperature are exactly determined through an analytical differentiation of the model chemical-equilibrium equations. This procedure is employed in Chapter 17 to calculate a chemically-reacting hypersonic flow.

Classical equilibrium thermodynamics for homogeneous fluids shows that any one thermodynamic variable depends on only two other thermodynamic variables. For CFD applications, the selected two thermodynamic variables are density and mass specific internal energy, for these variables are directly available from the continuity and volume-specific total energy equations in the Euler and Navier-Stokes systems. In the equilibrium thermodynamic system, therefore, density and mass-specific internal energy become assigned parameters at each grid point. The procedure can also be used with any other two independent thermodynamic variables and this chapter presents a method and results that correspond to calculations with pressure and temperature as independent variables.

The neutral air in the procedure encompasses perfect air and consists of a mixture of five non-ionized species: nitric oxide NO and molecular as well as atomic oxygen O_2 and O and nitrogen N_2 and N. Air-equivalent molecular masses are also developed to account for the effect of carbon dioxide, argon, and the other inert noble gases within atmospheric air, without having to involve variable mass fractions of these species within the calculations.

The choice of neutral air is justified by the fact that only negligible traces of electrons and hence ionic species exist within equilibrium air for temperatures up to about 9,000 K. Each species independently behaves as a perfect-gas for which the familiar perfect-gas law applies. The mixture pressure equation of state is then obtained through Dalton's law as a sum of species partial pressures. The mixture mass-specific internal energy results from the sum of translational and rotational kinetic energies, potential vibrational energy, and formation energy, all at the single equilibrium static temperature. Two additional equations correspond to the conservation of species mass and the mole ratio of oxygen and nitrogen nuclei. The law of mass action provides the completing three equations, which express the equilibrium of any three linearly independent chemical reactions for the five species in equilibrium air. The equilibrium thermodynamic equations are then made non-dimensional by way of a versatile single reference state that makes the procedure uniformly applicable to flows ranging from shock-tube flows with zero initial velocity, to aerothermodynamic flows with supersonic / hypersonic free stream Mach numbers.

Over a temperature range of more than 10,000 K and pressure and density ranges corresponding to an increase in altitude of 30,000 m, slightly greater than 98,000 ft, above sea level, the procedure converges in two or three iterations. Agreeing with independently published results, [7, 163], the predicted distributions of pressure and temperature, as well as their partial derivatives, mass fractions, specific heats, and speed of sound remain continuous, smooth and physically meaningful, and this chapter details all the thermodynamic and chemical parameters and equations and their partial derivatives for immediate reproduction of the procedure and results.

3.1   Thermo-Chemical Equations

The equations in this section will represent equilibrium, electrically-neutral and chemically-reacting dry air. This type of air encompasses perfect air and consists of a mixture of five non-ionized species: nitric oxide NO and molecular as well as atomic oxygen O_2 and O and nitrogen N_2 and N. In the following derivations, subscript i, $1 \leq i \leq 5$, indicates the five ordered species O, N, NO, O_2, and N_2. The molecular mass of species i is indicated with M_i. Owing to the species that compose this dry-air model, the species molecular masses satisfy the constraints $M_3 = M_1 + M_2$, $M_4 = 2M_1$, and $M_5 = 2M_2$. Numerical values for these masses are presented in Section 3.1.6

3.1.1   Species Masses and Moles

The mass $\delta \mathcal{M}$ and number of moles $\mathcal{N}$ of mixture species within a fluid particle satisfy the fundamental relations

$$\delta \mathcal{M} = \sum_{i=1}^{5} \delta \mathcal{M}_i, \quad \mathcal{N} = \sum_{i=1}^{5} \mathcal{N}_i \tag{3.1}$$

where $\delta \mathcal{M}_i$ and $\mathcal{N}_i$ denote the mass and number of moles of species i. The first statement in (3.1) connotes conservation of mass, whereas the second emerges from the atomic, hence discrete, nature of species. The two statements in (3.1) establish constraints between the total mixture density ρ and species densities ρ_i, $1 \leq i \leq 5$. These densities are defined as

$$\rho = \frac{\delta \mathcal{M}}{\delta \mathcal{V}}, \quad \rho_i = \frac{\delta \mathcal{M}_i}{\delta \mathcal{V}} \tag{3.2}$$

where $\delta \mathcal{V}$ denotes the volume occupied by the mixture. The first statement in (3.1) thus yields

$$\rho = \sum_{i=1}^{5} \rho_i \tag{3.3}$$

which corresponds to conservation of mass in terms of densities.

The masses $\delta \mathcal{M}$ and $\delta \mathcal{M}_i$ can also be expressed in terms of number of moles and molecular masses as

$$\delta \mathcal{M} = \mathcal{N} M, \quad \delta \mathcal{M}_i = \mathcal{N}_i M_{\underline{i}} \tag{3.4}$$

The second expression in (3.1) and (3.2) thus lead to

$$\frac{\rho}{M} = \sum_{i=1}^{5} \frac{\rho_i}{M_i} \tag{3.5}$$

3.1.2 Mass and Mole Fractions

The mass and mole fractions of species i are defined as

$$Y_i \equiv \frac{\rho_i}{\rho}, \quad X_i \equiv \frac{\mathcal{N}_i}{\mathcal{N}} \tag{3.6}$$

From (3.3) and the second expression in (3.1), therefore, the mass and mole fractions satisfy the equations

$$\sum_{i=1}^{5} Y_i = 1, \quad \sum_{i=1}^{5} X_i = 1 \tag{3.7}$$

From the definition of mass fraction, result (3.5) becomes

$$\frac{1}{M} = \sum_{i=1}^{5} \frac{Y_i}{M_i} \tag{3.8}$$

which provides the mixture molecular mass in terms of species molecular masses and mass fractions.

The mass and mole fraction of species i can be expressed in terms of each other. Without any implied summation on an underlined repeated subscript index, the definition of mole fraction and (3.2) yield

$$X_i = \frac{\mathcal{N}_i}{\mathcal{N}} = \frac{\rho_i \delta \mathcal{V}/M_{\underline{i}}}{\rho \delta \mathcal{V}/M} = \frac{Y_i/M_{\underline{i}}}{1/M} = \frac{Y_i/M_{\underline{i}}}{\sum_{j=1}^{5} Y_j/M_j} \tag{3.9}$$

which provides each mole fraction in terms of the mass fractions. Upon multiplying this expression by M_i and summing over the model five species, the following result emerges

$$\sum_{i=1}^{5} X_i M_i = \frac{\sum_{i=1}^{5} Y_i}{\sum_{j=1}^{5} Y_j/M_j} = \frac{1}{1/M} = M \quad \Rightarrow \quad \sum_{i=1}^{5} X_i M_i = \frac{1}{\sum_{j=1}^{5} Y_j/M_j} \tag{3.10}$$

With this result, (3.9) leads to

$$Y_i = \frac{X_i M_i}{\sum_{j=1}^{5} X_j M_j} \tag{3.11}$$

which expresses each mass fraction in terms of the mole fractions.

3.1.3   Pressure, Internal-Energy, Mass-Fraction Equations

Each species i independently behaves as a perfect gas with individual EOS

$$p_i = \mathcal{R}T\frac{\rho_i}{M_i} = \mathcal{R}\rho T\frac{\rho_i/\rho}{M_i} = \mathcal{R}\rho T\frac{Y_i}{M_i} \tag{3.12}$$

where $\mathcal{R}$ and p_i respectively denote the universal gas constant and partial pressure of species i. For a mixture of perfect species, Dalton's pressure mixing law leads to the mixture pressure equation of state as

$$p = \sum_{i=1}^{5} p_i = \mathcal{R}\rho T \sum_{i=1}^{5} \frac{Y_i}{M_i} = \mathcal{R}\frac{\rho}{M}T \tag{3.13}$$

The acceptance of Dalton's law thus implies that the mixture EOS has the right-most form indicated in (3.13). On the other hand, if this form of EOS is posited, then Dalton's law deductively follows from (3.5).

The mixture mass-specific internal energy ϵ results from the sum of translational and rotational kinetic energies, potential vibrational energy, and formation energy, all at the equilibrium static temperature T,

$$\epsilon = T\sum_{i=1}^{5} c_{v_i}Y_i + \sum_{i=3}^{5} \frac{Y_i \mathcal{R}\theta_i^v/M_i}{\exp\left(\theta_i^v/T\right) - 1} + \sum_{i=1}^{3} Y_i h_i^0 \tag{3.14}$$

The second term at the right hand side of (3.14) relies upon the rigid-rotor harmonic oscillator model, which implies the perfect gas equation of state for each species, which in turn is consistent with Dalton's pressure mixing rule. Therefore, the contributions from the ground electronic state are properly modeled for NO, O_2, and N_2. In (3.14), c_{v_i} denotes the translational/rotational-mode contributions to the ith species constant-volume specific heat, while θ_i^v and h_i^0 are the vibrational temperature and formation enthalpy at 0 K; specific numerical data for these quantities appear in Section 3.1.6. The constant-volume specific heat c_v then follows from (3.14) through the derivative $c_v = (\partial\epsilon/\partial T)_\rho$, as amplified in Section 3.9. A gas is thermally perfect when M in (3.13) remains constant, hence in the absence of chemical reactions. A gas is calorically perfect when c_v remains constant, hence in the absence of both chemical reactions and vibrational effects. It is possible for a gas, therefore, to be thermally perfect and calorically not perfect, which happens, for instance, in the absence of chemical reactions when vibrational modes become significant.

Unlike a formulation in terms of mole fractions, a formulation of equations (3.13)-(3.14) in terms of mass fractions simplifies analysis because the expressions for pressure and internal energy are linear with respect to the mass fractions. Considering that both ρ and ϵ are available at each grid point from the governing equations (1.69), (1.78), (1.95) and the internal-energy expression (2.15), equations (3.13)-(3.14) will directly allow the determination of static temperature T and pressure p for the given thermodynamic state (ρ, ϵ).

Fundamental to this determination are the five variable mass fractions Y_i, $1 \leq i \leq 5$, for which five additional equations are needed.

One equation corresponds to conservation of species mass (3.3), which results in the mass-fraction conservation equation in (3.7), repeated here for convenience

$$\sum_{i=1}^{5} Y_i = 1 \tag{3.15}$$

Another equation corresponds to the conservation of the mole proportion 21/79 between oxygen and nitrogen nuclei. In terms of mole fractions, the conservation of this proportion is expressed as $79(X_1 + X_3 + 2X_4) = 21(X_2 + X_3 + 2X_5)$, which, by virtue of (3.9), corresponds to the following equation in terms of mass fractions

$$\frac{1}{21}\left(\frac{Y_1}{M_1} + \frac{Y_3}{M_3} + 2\frac{Y_4}{M_4}\right) = \frac{1}{79}\left(\frac{Y_2}{M_2} + \frac{Y_3}{M_3} + 2\frac{Y_5}{M_5}\right) \tag{3.16}$$

The law of mass action provides 3 further equations, which express the equilibrium of any 3 linearly independent chemical reactions for the five species in equilibrium air. The following chemical reactions, two dissociations and one recombination, lead to computationally convenient mass-action equations

$$O_2 \rightleftarrows 2O, \quad N_2 \rightleftarrows 2N, \quad O_2 + N_2 \rightleftarrows 2NO \tag{3.17}$$

In terms of partial pressures, the mass-action equations for reactions (3.17) are

$$\frac{p_1^2}{p_4} = k_1(T), \quad \frac{p_2^2}{p_5} = k_2(T), \quad \frac{p_3^2}{p_4 p_5} = k_3(T) \tag{3.18}$$

and according to statistical thermodynamics, the partial-pressure equilibrium functions $k_i(T)$ only depend on the static temperature T; the stoichiometric coefficients in (3.17) then become the exponents in (3.18). Replacing the partial pressure p_i using (3.12), yields the mass-fraction law of mass action

$$\frac{Y_1^2}{Y_4} = \frac{M_1^2}{M_4}\frac{k_1(T)}{\mathcal{R}T\rho} = \frac{M_1}{2}\frac{K_1(T)}{\rho} \tag{3.19}$$

$$\frac{Y_2^2}{Y_5} = \frac{M_2^2}{M_5}\frac{k_2(T)}{\mathcal{R}T\rho} = \frac{M_2}{2}\frac{K_2(T)}{\rho} \tag{3.20}$$

$$\frac{Y_3^2}{Y_4 Y_5} = \frac{M_3^2}{M_4 M_5}K_3(T) \tag{3.21}$$

where $K_i(T)$, $1 \leq i \leq 3$, represent the mass-fraction equilibrium functions, which are traditionally cast as exponential relations, as exemplified in Section 3.1.7. According to the dimensions of terms in (3.19)- (3.21), the SI units of $K_1(T)$ and $K_2(T)$ are kg-mol m^{-3}, whereas $K_3(T)$ remains dimensionless. The following analytical developments will remain valid for any exponential form of these equilibrium functions.

3.1.4   Perfect-Air Equations

Atmospheric dry air consists of a mixture of oxygen, nitrogen, carbon dioxide, and argon, in addition to traces of other noble gases. The proportions of the chief air constituents are summarized in Table I

Table I. Composition of Atmospheric Dry Air

species		molecular mass	mole fraction	mass fraction
oxygen	O_2	31.99880 kg kg-mol^{-1}	0.20950	0.231447
nitrogen	N_2	28.01348 kg kg-mol^{-1}	0.78080	0.755163
argon	Ar	39.94800 kg kg-mol^{-1}	0.00930	0.012827
carbon dioxide	CO_2	44.00980 kg kg-mol^{-1}	0.00030	0.000456

Argon and carbon dioxide feature within atmospheric air with negligible mole fractions, yet their molecular masses are greater than those of oxygen and nitrogen to the extent that they contribute about 0.4% of the total molecular mass of non-reacting atmospheric air M_{air}. This molecular mass is calculated as

$$M_{\mathrm{air}} = \sum_{i=1}^{n_{\mathrm{air}}} X_i M_i = \frac{\mathcal{R}\rho_{\mathrm{sc}} T_{\mathrm{sc}}}{p_{\mathrm{sc}}} = 28.9644827716 \text{ kg kg-mol}^{-1}$$

$$\rho_{\mathrm{sc}} = 1.225 \text{ kg m}^{-3}, \quad T_{\mathrm{sc}} = 288.15 \text{ K}, \quad p_{\mathrm{sc}} = 101{,}325.024 \text{ Pa} \qquad (3.22)$$

where n_{air} denotes the total number of atmospheric air species, and ρ_{sc}, p_{sc} and T_{sc} denote standard-condition density, temperature, and pressure.

    Reported calculations of high-temperature air flows model air as a mixture of oxygen, nitrogen, and nitric oxide, in their various forms. In non-reacting conditions, the mole fractions of oxygen and nitrogen in this mixture respectively become 0.21 and 0.79; this model, therefore, accepts the presence of 1.18% more nitrogen moles than there are within atmospheric air. The determination of pressure, temperature or density from the perfect gas law (3.13) with these mole fractions approximate the corresponding data of standard atmosphere with an error of about 0.4%.

    In atmospheric air, carbon dioxide is present in minute amounts in comparison to oxygen and nitrogen, while argon and the other noble gases in air remain inert towards all other elements. The essential effect of carbon dioxide, argon and the other noble gases in air, accordingly, is to catalyze the chemical reactions of oxygen and nitrogen through collisions and make air heavier than a mixture of only oxygen and nitrogen, reacting or not. This situation is mathematically reflected in the following pressure and internal energy equations, expressed similarly to (3.13) and (3.14), but in terms of mole fractions from (3.11) and accounting for all inert air species

$$p = \frac{\mathcal{R}\rho T}{\sum_{j=1}^{5} X_j M_j + M_{\mathrm{oth}}}, \quad \epsilon = T\left(\frac{\sum_{i=1}^{5} \bar{c}_i}{\sum_{j=1}^{5} X_j M_j + M_{\mathrm{oth}}} + c_{v_{\mathrm{oth}}} \right) +$$

$$+ \sum_{i=3}^{5} \frac{\mathcal{R}\theta_i^v X_i}{(\exp\left(\theta_i^v/T\right) - 1)(\sum_{j=1}^{5} X_j M_j + M_{\mathrm{oth}})} + \sum_{i=1}^{3} \frac{\mathcal{R}\bar{h}_i X_i}{\sum_{j=1}^{5} X_j M_j + M_{\mathrm{oth}}} \qquad (3.23)$$

where $\bar{c}_i$, $\bar{h}_i$, $c_{v_{\text{oth}}}$, and M_{oth} respectively denote the constant coefficients within c_{v_i} and h_i^0, detailed in Section 3.1.6, and the contribution to the translation / rotational specific heat and molecular mass of the other species besides nitric oxide and atomic and molecular oxygen and nitrogen. The effect of the presence of other chemicals within air, besides oxygen and nitrogen, can thus be modeled by establishing air-equivalent molecular masses for oxygen and nitrogen so that the molecular mass of a non-reacting mixture with mole ratio 21/79 of these two species equals the molecular mass of non-reacting air, hence

$$0.21 M_4 + 0.79 M_5 = M_{\text{air}} \tag{3.24}$$

where M_4 and M_5 respectively denote the air-equivalent molecular masses of oxygen and nitrogen. The second constraint to determine both M_4 and M_5 is chosen to be preservation of the mass-fraction ratio

$$\frac{Y_5^{\text{air}}}{Y_4^{\text{air}}} = \frac{Y_5}{Y_4} = \frac{X_5 M_5}{X_4 M_4} = \frac{0.79 M_5}{0.21 M_4} = 3.26279010234 \tag{3.25}$$

where Y_4^{air} and Y_5^{air} are listed in Table I. The errors induced by the 21/79 mole ratio are summarized in Table II for both non-equivalent and equivalent masses

Table II. Errors Induced by the 21/79 Mole Ratio

variable	non air-equivalent	air-equivalent
$X_{4,1}$	0.24%	0.24%
$X_{5,2}$	1.18%	1.18%
$Y_{4,1}$	0.64%	1.36%
$Y_{5,2}$	1.58%	1.36%
M_{air}	-0.39%	0.00%
R	0.39%	0.00%
$c_{v_{\text{air}}}$	0.39%	0.00%
$c_{p_{\text{air}}}$	0.39%	0.00%
γ	0.00%	0.00%

where R, $c_{v_{\text{air}}}$, $c_{p_{\text{air}}}$ and γ respectively denote the specific, not universal, gas constant, constant-volume and constant-pressure specific heats of non-reacting air, and specific heat ratio. With the air-equivalent masses M_4 and M_5, equation of state (3.13) for non-reacting conditions exactly yields the pressure and temperature of non-reacting atmospheric air.

Consider, accordingly, perfect air as a mixture of thermally and calorically perfect oxygen and nitrogen with constant chemical composition, hence constant mass fractions. From the works of Boyle, Mariotte, Charles, Gay-Lussac, and Joule, the EOS and TE for a perfect gas are

$$p = \frac{\mathcal{R}}{M_{\text{air}}} \rho T = R \rho T, \quad \epsilon = c_v T + \epsilon^0 \tag{3.26}$$

The term ϵ^0 vanishes for standard-condition perfect air; for high-temperature perfect air, it corresponds to the energy from the formation enthalpies of atomic oxygen and nitrogen, the high-temperature conditions where molecular oxygen and nitrogen completely dissociate. These equations result from (3.13)-(3.14) for a constant chemical composition and in the absence of vibrational effects. The parameters R and c_v are constant and related by Mayer's equality $c_p - c_v = R$, where c_p indicates the constant-pressure specific heat. Through the ratio $\gamma = c_p/c_v$, Mayer's equality then becomes $\gamma - 1 = R/c_v$. From this equality and (2.15), the elimination of T from (3.26) leads to the following forms of standard-condition perfect-gas EOS and TE

$$p = (\gamma - 1)\rho\epsilon = (\gamma - 1)\left(E - \frac{1}{2\rho}\sum_{i=1}^{n} m_i m_i\right), \quad T = \frac{p}{R\rho} \tag{3.27}$$

two well known expressions in CFD.

From (3.14), the temperature equation for standard-condition and high-temperature perfect air becomes

$$\epsilon = T\left(c_{v_4}Y_4 + c_{v_5}Y_5\right), \quad \epsilon = T\left(c_{v_1}Y_1 + c_{v_2}Y_2\right) + \epsilon^0 \tag{3.28}$$

Since the RHS of the TE in (3.26) and (3.28) equal one another, the expressions for the constant-volume specific heat for standard-condition and high-temperature perfect air become

$$c_{v_{\mathrm{air}}} = c_{v_4}Y_4 + c_{v_5}Y_5, \quad c_{v_{\mathrm{air}}} = c_{v_1}Y_1 + c_{v_2}Y_2 \tag{3.29}$$

For the reference perfect-air mole fractions $X_1 = X_2 = X_3 = 0$, $X_4 = 0.21$, $X_5 = 0.79$, for standard-condition molecular air, and $X_3 = X_4 = X_5 = 0$, $X_1 = 0.21$, $X_2 = 0.79$, for high- temperature atomic air, along with air-equivalent masses and the chemical and thermodynamic constants in Section 3.1.6, the corresponding numerical values of mass fractions, molecular or atomic mass M_{air}, gas constant R, constant-volume and constant-pressure specific heats $c_{v_{\mathrm{air}}}$ and $c_{p_{\mathrm{air}}}$, and specific heat ratio are

Table III. Perfect-Air Parameters

	molecular		atomic	
$Y_{4,1}$	0.23459		0.23459	
$Y_{5,2}$	0.76541		0.76541	
M_{air}	28.96448	kg kg-mol^{-1}	14.48224	kg kg-mol^{-1}
R	287.05294	J kg^{-1} K^{-1}	574.10588	J kg^{-1} K^{-1}
$c_{v_{\mathrm{air}}}$	717.63236	J kg^{-1} K^{-1}	861.15883	J kg^{-1} K^{-1}
$c_{p_{\mathrm{air}}}$	1004.68530	J kg^{-1} K^{-1}	1435.26471	J kg^{-1} K^{-1}
γ	$\frac{7}{5}$		$\frac{5}{3}$	

The numerical value of the molecular-air constant R equals the one that is used in classical gas dynamics and the values of molecular $c_{v_{\mathrm{air}}}$ and $c_{p_{\mathrm{air}}}$ correspond to those of air at about 300 K. Not surprisingly, the numerical values for R, $c_{v_{\mathrm{air}}}$, $c_{p_{\mathrm{air}}}$ for atomic air are greater than the corresponding values for molecular air. Atomic oxygen and nitrogen are less massive than their molecular counterparts, hence their kinetic energy more readily increases when temperature increases.

3.1.5   Non-dimensional Reacting-Air Equations of State

The system of equations (3.13)-(3.14) and (3.19)-(3.21) are made non-dimensional by way of a versatile single reference state that makes this system uniformly applicable to conditions ranging from flows with a specified fixed initial state with zero velocity, typical of shock-tube flows, to flows with an identifiable free stream state, typical of supersonic/hypersonic aerothermodynamic flows.

For an available free stream state with representative constant density ρ_∞, pressure p_∞, temperature T_∞, and Mach number M_∞, the reference molecular mass, density, mass-specific energy (speed squared), pressure and temperature are expressed as

$$
\begin{aligned}
M_r &= \frac{\mathcal{R}\rho_\infty T_\infty}{p_\infty} \\
\rho_r &= \rho_\infty \\
U_r^2 &= U_\infty^2 = \gamma_\infty \mathrm{M}_\infty^2 (\mathcal{R}/M_r)T_\infty, \quad \gamma_\infty \equiv \frac{c_\infty^2}{p_\infty/\rho_\infty}
\end{aligned}
\tag{3.30}
$$

$$
\begin{aligned}
p_r &= \rho_r U_r^2 = \gamma_\infty \mathrm{M}_\infty^2 p_\infty \\
T_r &= \frac{p_r}{\rho_r(\mathcal{R}/M_r)} = \frac{U_r^2}{(\mathcal{R}/M_r)} = \gamma_\infty \mathrm{M}_\infty^2 T_\infty
\end{aligned}
\tag{3.31}
$$

where c_∞^2 denotes the square of the free-stream speed of sound. When the free-stream air behaves as a standard-condition perfect air, then $\gamma_\infty = 7/5$ and when ρ_∞, p_∞ and T_∞ already satisfy the perfect-gas law, then the reference molecular mass M_r corresponds to the air molecular mass. Otherwise it simply represents a scaling factor by which to divide the species molecular masses.

For a typical shock-tube initial state with representative constant pressure p_∞, temperature T_∞, and density ρ_∞ the reference variables are

$$
\begin{aligned}
M_r &= \frac{\mathcal{R}\rho_\infty T_\infty}{p_\infty} \\
\rho_r &= \rho_\infty \\
U_r^2 &= (\mathcal{R}/M_r)T_\infty = \gamma_\infty \mathrm{M}_\infty^2 (\mathcal{R}/M_r)T_\infty, \quad \mathrm{M}_\infty^2 \equiv \frac{1}{\gamma_\infty}
\end{aligned}
\tag{3.32}
$$

$$
\begin{aligned}
p_r &= p_\infty = \gamma_\infty \mathrm{M}_\infty^2 p_\infty \\
T_r &= T_\infty = \gamma_\infty \mathrm{M}_\infty^2 T_\infty
\end{aligned}
\tag{3.33}
$$

where γ_∞ is calculated as in (3.30). Heed that for $\mathrm{M}_\infty^2 = 1/\gamma_\infty$, this reference state formally coincides with (3.30)-(3.31). Therefore, by setting M_∞ equal to either the free-stream Mach number or $1/\gamma_\infty$, depending on the flow class, a unique set of non-dimensional equations of state emerges for both reference sets (3.30)-(3.31) and (3.32)-(3.33)

With definitions (3.30)-(3.31) and (3.32)-(3.33), the reference pressure, density, and temperature then satisfy the perfect gas law. For high Mach numbers they constitute sizable constants that conveniently scale down the large pressure, density and temperature across a normal shock, like the stagnation-streamline normal section of supersonic and hypersonic aerodynamic-flow bow shocks.

Using either the reference states (3.30)-(3.31) or (3.32)-(3.33), the non-dimensional density, mass-specific internal energy, pressure and temperature are then expressed as

$$\tilde{\rho} = \frac{\rho}{\rho_r}, \quad \tilde{\epsilon} = \frac{\epsilon}{U_r^2} = \frac{\epsilon}{\gamma_\infty M_\infty^2 (\mathcal{R}/M_r) T_\infty}, \quad \tilde{p} = \frac{p}{p_r} = \frac{p}{\gamma_\infty M_\infty^2 p_\infty}, \quad \tilde{T} = \frac{T}{T_r} = \frac{T}{\gamma_\infty M_\infty^2 T_\infty} \tag{3.34}$$

The corresponding non-dimensional pressure equation is

$$\tilde{p} = \frac{\mathcal{R}\tilde{\rho}\rho_r \tilde{T} T_r}{p_r} \sum_{i=1}^{5} \frac{Y_i}{M_i} = \frac{\mathcal{R}\tilde{\rho}\tilde{T}}{(\mathcal{R}/M_r)} \sum_{i=1}^{5} \frac{Y_i}{M_i} = \tilde{\rho}\tilde{T} \sum_{i=1}^{5} \frac{Y_i}{(M_i/M_r)} = \tilde{\rho}\tilde{T} \sum_{i=1}^{5} \frac{Y_i}{\widetilde{M_i}} \tag{3.35}$$

where $\widetilde{M_i} = M_i/M_r$. It follows that with this non-dimensionalization, the universal gas constant $\mathcal{R}$ no longer appears in this non-dimensional equation of state. The non-dimensional energy equation is

$$\tilde{\epsilon} = \frac{T}{\gamma_\infty M_\infty^2 (\mathcal{R}/M_r) T_\infty} \sum_{i=1}^{5} c_{v_i} Y_i +$$

$$+ \sum_{i=3}^{5} \frac{Y_i \mathcal{R}\theta_i^v / M_i}{(\gamma_\infty M_\infty^2 (\mathcal{R}/M_r) T_\infty) \cdot (\exp(\theta_i^v/T) - 1)} + \sum_{i=1}^{3} Y_i \frac{h_i^0}{\gamma_\infty M_\infty^2 (\mathcal{R}/M_r) T_\infty} \tag{3.36}$$

which becomes

$$\tilde{\epsilon} = \frac{T}{\gamma_\infty M_\infty^2 T_\infty} \sum_{i=1}^{5} \left(c_{v_i} \frac{M_r}{\mathcal{R}} \right) Y_i + \sum_{i=3}^{5} \frac{Y_i \left(\dfrac{\theta_i^v}{\gamma_\infty M_\infty^2 T_\infty} \right) \cdot \dfrac{1}{M_i/M_r}}{\exp\left(\left(\dfrac{\theta_i^v}{\gamma_\infty M_\infty^2 T_\infty} \right) \dfrac{\gamma_\infty M_\infty^2 T_\infty}{T} \right) - 1} + \sum_{i=1}^{3} Y_i \tilde{h}_i^0 \tag{3.37}$$

and is simplified as

$$\tilde{\epsilon} = \tilde{T} \sum_{i=1}^{5} \tilde{c}_{v_i} Y_i + \sum_{i=3}^{5} \frac{Y_i \tilde{\theta}_i^v / \widetilde{M_i}}{\exp(\tilde{\theta}_i^v/\tilde{T}) - 1} + \sum_{i=1}^{3} Y_i \tilde{h}_i^0 \tag{3.38}$$

and the universal gas constant $\mathcal{R}$ no longer multiplies any term in this equation. Furthermore, as indicated in Section 3.1.6, the magnitude of the non-dimensional specific heat $\tilde{c}_{v_i}$, characteristic vibrational temperature $\tilde{\theta}_i^v$, and formation enthalpy $\tilde{h}_i^0$ decrease with respect to their dimensional values.

Since the mass fractions Y_i are already dimensionless variables, the non-dimensional species conservation equations are

$$\sum_{i=1}^{5} Y_i = 1, \quad \frac{1}{21} \left(\frac{Y_1}{\widetilde{M_1}} + \frac{Y_3}{\widetilde{M_3}} + 2\frac{Y_4}{\widetilde{M_4}} \right) = \frac{1}{79} \left(\frac{Y_2}{\widetilde{M_2}} + \frac{Y_3}{\widetilde{M_3}} + 2\frac{Y_5}{\widetilde{M_5}} \right) \tag{3.39}$$

The non-dimensional mass-action equations are then

$$\frac{Y_1^2}{Y_4} = \frac{\widetilde{M_1}}{2} \frac{K_1(\tilde{T})}{\tilde{\rho}(\rho_r/M_r)} = \frac{\widetilde{M_1}}{2} \frac{\widetilde{K_1}(\tilde{T})}{\tilde{\rho}}, \quad \widetilde{K_1}(\tilde{T}) \equiv \frac{K_1(\tilde{T})}{\rho_r/M_r} \tag{3.40}$$

$$\frac{Y_2^2}{Y_5} = \frac{\widetilde{M_2}}{2} \frac{K_2(\tilde{T})}{\tilde{\rho}(\rho_r/M_r)} = \frac{\widetilde{M_2}}{2} \frac{\widetilde{K_2}(\tilde{T})}{\tilde{\rho}}, \quad \widetilde{K_2}(\tilde{T}) \equiv \frac{K_2(\tilde{T})}{\rho_r/M_r} \tag{3.41}$$

$$\frac{Y_3^2}{Y_4 Y_5} = \frac{\widetilde{M_3}^2}{\widetilde{M_4}\widetilde{M_5}} \widetilde{K_3}(\tilde{T}) \tag{3.42}$$

where $\widetilde{K}_i$, $1 \leq i \leq 3$, denote the non-dimensional equilibrium functions, as exemplified in Section 3.1.7.

3.1.6 Chemical and Thermodynamics Parameters

The molecular masses for the chosen five species are

$$M_1^{na} = 15.99940 \text{ kg kg-mol}^{-1} , \quad M_2^{na} = 14.00674 \text{ kg kg-mol}^{-1}$$

$$M_3^{na} = 30.00614 \text{ kg kg-mol}^{-1} , \quad M_4^{na} = 31.99880 \text{ kg kg-mol}^{-1} , \quad M_5^{na} = 28.01348 \text{ kg kg-mol}^{-1}$$
$$(3.43)$$

where superscript "na" signifies non air-equivalent masses. The corresponding air-equivalent masses are

$$M_1 = 16.17791 \text{ kg kg-mol}^{-1} , \quad M_2 = 14.03149 \text{ kg kg-mol}^{-1}$$

$$M_3 = 30.20940 \text{ kg kg-mol}^{-1} , \quad M_4 = 32.35582 \text{ kg kg-mol}^{-1} , \quad M_5 = 28.06298 \text{ kg kg-mol}^{-1}$$
$$(3.44)$$

Both non equivalent and equivalent masses satisfy the constraints $M_3 = M_1 + M_2$, $M_4 = 2M_1$, and $M_5 = 2M_2$.

In SI units, the universal gas constant is

$$\mathcal{R} = 8,314.34 \text{ J kg-mol}^{-1} \text{ K}^{-1} \tag{3.45}$$

For the selected five species, the vibrational temperatures in (3.14) are expressed as

$$\theta_3^v = 2,740 \text{ K}, \quad \theta_4^v = 2,270 \text{ K}, \quad \theta_5^v = 3,390 \text{ K} \tag{3.46}$$

and the formation enthalpies are

$$h_1^0 = 29,682.446 \frac{\mathcal{R}}{M_1} \text{ J kg}^{-1}, \quad h_2^0 = 56,627.830 \frac{\mathcal{R}}{M_2} \text{ J kg}^{-1}, \quad h_3^0 = 10,797.780 \frac{\mathcal{R}}{M_3} \text{ J kg}^{-1}$$
$$(3.47)$$

Further, each c_{v_i} in (3.14) is determined as

$$c_{v_1} = \frac{3}{2} \frac{\mathcal{R}}{M_1} \text{ J kg}^{-1} \text{ K}^{-1} , \quad c_{v_2} = \frac{3}{2} \frac{\mathcal{R}}{M_2} \text{ J kg}^{-1} \text{ K}^{-1}$$

$$c_{v_3} = \frac{5}{2} \frac{\mathcal{R}}{M_3} \text{ J kg}^{-1} \text{ K}^{-1} , \quad c_{v_4} = \frac{5}{2} \frac{\mathcal{R}}{M_4} \text{ J kg}^{-1} \text{ K}^{-1} , \quad c_{v_5} = \frac{5}{2} \frac{\mathcal{R}}{M_5} \text{ J kg}^{-1} \text{ K}^{-1} \tag{3.48}$$

These expressions result from the kinetic theory of gases dictating that a mono-atomic gas, like O, possesses three independent energy absorption mechanisms (degrees of freedom), which correspond to the translation kinetic energy of the motion along three mutually perpendicular directions; a diatomic gas, like N_2, has five such degrees, three of translation and two of rotation about two mutually perpendicular axes that are orthogonal to the axis through the molecule nuclei.

The non-dimensional form of the internal energy equations (3.38) thus originates from the following expressions

$$\tilde{c}_{v_1} = \frac{3}{2} \frac{1}{\widetilde{M_1}}, \quad \tilde{c}_{v_2} = \frac{3}{2} \frac{1}{\widetilde{M_2}}, \quad \tilde{c}_{v_3} = \frac{5}{2} \frac{1}{\widetilde{M_3}}, \quad \tilde{c}_{v_4} = \frac{5}{2} \frac{1}{\widetilde{M_4}}, \quad \tilde{c}_{v_5} = \frac{5}{2} \frac{1}{\widetilde{M_5}} \tag{3.49}$$

for each specific heat,

$$\tilde{\theta}_3^v = \frac{2,740}{\gamma_\infty M_\infty^2 T_\infty}, \quad \tilde{\theta}_4^v = \frac{2,270}{\gamma_\infty M_\infty^2 T_\infty}, \quad \tilde{\theta}_5^v = \frac{3,390}{\gamma_\infty M_\infty^2 T_\infty} \tag{3.50}$$

for each vibrational temperature, and

$$\tilde{h}_1^0 = \frac{29,682.446}{\gamma_\infty M_\infty^2 T_\infty \widetilde{M}_1}, \quad \tilde{h}_2^0 = \frac{56,627.830}{\gamma_\infty M_\infty^2 T_\infty \widetilde{M}_2}, \quad \tilde{h}_3^0 = \frac{10,797.780}{\gamma_\infty M_\infty^2 T_\infty \widetilde{M}_3} \tag{3.51}$$

for each formation enthalpy, which collectively indicate that either (3.30)-(3.31) or (3.32)-(3.33) can significantly scale down these parameters.

3.1.7 Chemical Equilibrium Functions

While the procedure developed to solve system (3.35), (3.38)-(3.42) is valid for any exponential form of equilibrium functions $K_i(T)$, $i = 1, 2, 3$, the numerical results presented in Section 3.11 were generated using the following specific expressions for $K_i(T)$ from [155, 156]

$$K_i(T) = \exp\left(A_{i4}Z^4 + A_{i3}Z^3 + A_{i2}Z^2 + A_{i1}Z + A_{i0}\right), \quad Z \equiv \log\left(\frac{10,000}{T}\right) \tag{3.52}$$

where A_{ij} denote constant coefficients that also ensure a smooth asymptotic convergence of the equilibrium-air system (3.35), (3.38)-(3.42) to either a standard-condition perfect gas equation of state, at lower temperatures, or a high-temperature equation of state, at higher temperatures.

According to (3.19)-(3.21), this asymptotic convergence is achieved when $K_i(T)$, $i = 1, 2, 3$ approach zero at lower temperatures, since Y_1, Y_2, and Y_3 vanish in this temperature range. Furthermore, $K_1(T)$ and $K_2(T)$ have to increase monotonically at higher temperatures, since Y_4 and Y_5 vanish at these temperatures. Conversely, $K_3(T)$ has to remain bounded at higher temperatures as a sufficient condition for attaining from (3.21) a vanishing Y_3. The corresponding coefficients A_{ij} for these equilibrium functions are

Table IV. Coefficients for Equilibrium Functions (49)

i	A_{i4}	A_{i3}	A_{i2}	A_{i1}	A_{i0}
1	- 0.466031	- 1.78672	- 1.24877	- 5.15926	+ 2.97975
2	- 1.007340	- 2.74128	- 2.93912	- 11.0496	- 2.06856
3	- 0.196317	- 0.42724	- 0.86667	- 1.65356	+ 0.61473

Heed that the coefficients A_{i0}, $i = 1, 2$, in this table result by adding $\log(1,000)$ to those in [155, 156], which corresponds to a change of units from mol cm^{-3} to kg-mol m^{-3}. The equilibrium functions in (3.52) are compactly expressed as

$$K_i(T) = \exp\left(G_i(T)\right) \tag{3.53}$$

Consequently, the derivative of $K_i(T)$ with respect to T is cast as

$$\frac{dK_i(T)}{dT} = \exp\left(G_i(T)\right)\frac{dG_i(T)}{dT} = K_i\frac{dG_i}{dT} \tag{3.54}$$

Expressions (3.40)-(3.41) feature the non-dimensional equilibrium functions $\widetilde{K}_i(\widetilde{T})$ corresponding to (3.52). These non-dimensional equilibrium functions depend upon $\widetilde{T}$ by expressing the variable Z in (3.52) as

$$\exp(Z) = \frac{10,000}{T} = \left(\frac{10,000}{\gamma_\infty \, M_\infty^2 T_\infty}\right)\left(\frac{\gamma_\infty \, M_\infty^2 T_\infty}{T}\right) = \frac{T}{\widetilde{T}}, \quad \overline{T} = \frac{10,000}{\gamma_\infty \, M_\infty^2 T_\infty} \tag{3.55}$$

where $\overline{T}$ can be made substantially smaller than $10,000$.

3.2 Algebraic Solution

Dropping the tildes for simplicity, yet remembering that all variables in the following expressions are non-dimensional ones, the non-dimensional equations of state are

$$p = \rho T \sum_{i=1}^{5} \frac{Y_i}{M_i} \tag{3.56}$$

$$\epsilon = T \sum_{i=1}^{5} c_{v_i} Y_i + \sum_{i=3}^{5} \frac{Y_i \theta_i^v / M_i}{\exp\left(\theta_i^v / T\right) - 1} + \sum_{i=1}^{3} Y_i h_i^0 \tag{3.57}$$

$$\sum_{i=1}^{5} Y_i = 1 \tag{3.58}$$

$$\frac{1}{21}\left(\frac{Y_1}{M_1} + \frac{Y_3}{M_3} + 2\frac{Y_4}{M_4}\right) = \frac{1}{79}\left(\frac{Y_2}{M_2} + \frac{Y_3}{M_3} + 2\frac{Y_5}{M_5}\right) \tag{3.59}$$

$$\frac{Y_1^2}{Y_4} = \frac{M_1}{2}\frac{K_1(T)}{\rho} \tag{3.60}$$

$$\frac{Y_2^2}{Y_5} = \frac{M_2}{2}\frac{K_2(T)}{\rho} \tag{3.61}$$

$$\frac{Y_3^2}{Y_4 Y_5} = \frac{M_3^2}{M_4 M_5}K_3(T) \tag{3.62}$$

The five mass fractions Y_i, $1 \le i \le 5$ and static temperature T can then be theoretically obtained by solving the non-linear system of the last six equations given above and practically determined through a numerical iterative procedure, which for CFD applications can become a daunting proposition if such a 6×6 system required even a few iterations at each of hundreds of thousands of grid points. In the following developments, however, the four variable mass fractions Y_i, $i = 1, 2, 4, 5$, of the six variables in this 6×6 system are explicitly expressed algebraically in terms of only two variables: nitric-oxide mass fraction Y_3 and temperature T. The complete solution of system (3.56)-(3.62) is then obtained from the solution for these two variables of a far less daunting 2×2 system. The selection of the specific relations leading to this system was predicated on the elimination of the non-physical spurious solutions that can solve a non-linear algebraic system of equations. This criterion ensured that the resulting equations only possess a physically meaningful solution.

The two mass fractions Y_1 and Y_2 in the two linear equations (3.58)-(3.59) can be explicitly solved in terms of Y_3, Y_4, and Y_5. The expression for Y_1 results by summing (3.59) to the product of (3.58) and $1/(79M_2)$; Y_2 then follows from (3.58). This sequence of operations yields the expressions

$$Y_1 = \frac{0.21M_1}{0.21M_1 + 0.79M_2} - \frac{M_1}{M_3}Y_3 - Y_4 \tag{3.63}$$

$$Y_2 = \frac{0.79M_2}{0.21M_1 + 0.79M_2} - \frac{M_2}{M_3}Y_3 - Y_5 \tag{3.64}$$

For conciseness, these two expressions are then cast as

$$Y_1 = \alpha_{10} - \alpha_{13}Y_3 - Y_4 \tag{3.65}$$

$$Y_2 = \alpha_{20} - \alpha_{23}Y_3 - Y_5 \tag{3.66}$$

where the constants α_{10}, α_{13}, α_{20}, and α_{23} follow from inspection of (3.63)-(3.64).

The explicit relations for Y_4 and Y_5 result from inserting (3.65) and (3.66) into (3.60) and (3.61) respectively. This operation yields the quadratic equations

$$Y_4^2 - 2\left(\alpha_{10} - \alpha_{13}Y_3 + \frac{M_1}{2}\frac{K_1(T)}{2\rho}\right)Y_4 + (\alpha_{10} - \alpha_{13}Y_3)^2 = 0 \tag{3.67}$$

$$Y_5^2 - 2\left(\alpha_{20} - \alpha_{23}Y_3 + \frac{M_2}{2}\frac{K_2(T)}{2\rho}\right)Y_5 + (\alpha_{20} - \alpha_{23}Y_3)^2 = 0 \tag{3.68}$$

which remain valid for any form of the equilibrium function $K_i(T)$, $i = 1, 2$, and intrinsically depend upon Y_3. Of the two mathematical solutions for each of (3.67) and (3.68), the solution devoid of physical significance is discarded and the physically meaningful solutions are then established as

$$Y_4 = \alpha_{10} - \alpha_{13}Y_3 + \frac{M_1}{2}\frac{K_1(T)}{2\rho} - \left(\frac{M_1}{2}\frac{K_1(T)}{\rho}\left(\alpha_{10} - \alpha_{13}Y_3 + \frac{M_1}{2}\frac{K_1(T)}{4\rho}\right)\right)^{1/2} \tag{3.69}$$

$$Y_5 = \alpha_{20} - \alpha_{23}Y_3 + \frac{M_2}{2}\frac{K_2(T)}{2\rho} - \left(\frac{M_2}{2}\frac{K_2(T)}{\rho}\left(\alpha_{20} - \alpha_{23}Y_3 + \frac{M_2}{2}\frac{K_2(T)}{4\rho}\right)\right)^{1/2} \tag{3.70}$$

According to the algebraic sign of the coefficients in (3.69)-(3.70), Y_4 and Y_5 are positive if so are the expressions $\alpha_{i0} - \alpha_{i3}Y_{13}$, $i = 1, 2$, since K_1 and K_2 are intrinsically positive. Given the definition of the coefficients α_{10}, α_{13}, α_{20}, α_{23} in (3.65)-(3.66), this condition is always met since Y_3 remains below one. For increasing temperature, furthermore, both Y_4 and Y_5 from (3.69)-(3.70) consistently approach zero, as physically correct, owing to the monotonic increase of K_1 and K_2. Finally, at lower temperatures, K_1, K_2 and Y_3 approach zero. Hence, Y_4 and Y_5 from (3.69)-(3.70) converge to their respective perfect-gas numerical values in Table III, while from (3.63)-(3.64) both Y_1 and Y_2 vanish.

By virtue of (3.69)-(3.70), Y_4 and Y_5 are functionally expressed as

$$Y_4 = Y_4\left(Y_3(\rho, \epsilon), T(\rho, \epsilon), \rho\right), \quad Y_5 = Y_5\left(Y_3(\rho, \epsilon), T(\rho, \epsilon), \rho\right) \tag{3.71}$$

and the functional relations for Y_1 and Y_2 from (3.63)-(3.64) are then

$$Y_1 = Y_1 \left(Y_3(\rho, \epsilon), Y_4 \left(Y_3(\rho, \epsilon), T(\rho, \epsilon), \rho \right) \right), \quad Y_2 = Y_2 \left(Y_3(\rho, \epsilon), Y_5 \left(Y_3(\rho, \epsilon), T(\rho, \epsilon), \rho \right) \right) \quad (3.72)$$

which show that these four mass fractions explicitly depend upon ρ and ϵ as well as Y_3 and T. For the thermodynamic state (ρ, ϵ), existing at each grid node from (1.69), (1.78), (1.95), and (1.95), along with the associated Y_3 and T, therefore, both Y_4 and Y_5 are directly obtained from (3.69)-(3.70), which then allows determining both Y_1 and Y_2 from (3.63)-(3.64). The corresponding pressure is then determined from (3.56). A complete solution for the non-linear 6- equation system (3.57)-(3.62), is thus obtained when both Y_3 and T are determined.

3.3 Calculation of Y_3 and T

The two equations that remain to be solved for system (3.57)-(3.62) are the nitric-oxide mass-action equation (3.62) and the mass-specific internal energy equation (3.57). The solution for T and Y_3 is thus determined by solving the two-equation system

$$f_1(\rho, Y_3, T) \equiv Y_3 - M_3 \left(\frac{Y_4}{M_4} \frac{Y_5}{M_5} K_3(T) \right)^{1/2} = 0 \quad (3.73)$$

$$f_2(\rho, \epsilon, Y_3, T) \equiv \epsilon - T \sum_{i=1}^{5} c_{v_i} Y_i - \sum_{i=3}^{5} \frac{Y_i \theta_i^v / M_i}{\exp\left(\theta_i^v / T \right) - 1} - \sum_{i=1}^{3} Y_i h_i^0 = 0 \quad (3.74)$$

with Y_i, $i = 1, 2, 4, 5$ expressed via (3.65)-(3.66) and (3.69)-(3.70). For a positive thermodynamic state (ρ, ϵ), each term in (3.73)-(3.74), is a non-positive monotone function of temperature T and nitric oxide mass fraction Y_3. Furthermore, the square root expression in (3.73) is also a non-negative monotone function of T. Therefore, a solution of system (3.73)-(3.74) with positive Y_3 and T exists and is unique. This solution is numerically determined by solving this system through Newton's method

3.3.1 Numerical Solution

In system (3.73)-(3.74), the thermodynamic state (ρ, ϵ) is known and fixed at each grid point. For the auxiliary variable $Q \equiv \{Y_3, T\}^T$, therefore, this system is cast as

$$F(\rho, \epsilon, Q) \equiv F(\rho, \epsilon, Y_3, T) \equiv \left\{ \begin{array}{c} f_1(\rho, Y_3, T) \\ f_2(\rho, \epsilon, Y_3, T) \end{array} \right\} = 0 \quad (3.75)$$

The Newton's algorithm to solve this system is

$$Q^{i+1} = Q^i - \left[\left(\frac{\partial F}{\partial Q} \right)^i_{\rho, \epsilon} \right]^{-1} \{F^s\} \quad (3.76)$$

where superscript i denotes the iteration index. The initial estimate Q^0 at a grid node can coincide with the value of Q at an adjacent grid point where this system has already been solved. If no solution for Q is available at an adjacent node, as is typical when (3.73)-(3.74) is solved at the first grid node, then an initial estimate for Q^0 can correspond to a perfect-gas low T and consequently $Y_3 \equiv 0$. These two selections are consistent with each other because K_3, Y_1 and Y_2 approach zero at lower temperatures, which leads to a vanishing Y_3, as a solution of (3.73). Under the same low-temperature conditions, equation (3.74) then asymptotically converges to the corresponding perfect gas expressions. With symmetrized f_1 and f_2 with respect to the f axis, that is $f_1(-Y_3, -T) = f_1(Y_3, T)$ and $f_2(-Y_3, -T) = f_2(Y_3, T)$, the absolute value of both Y_3^i and T^i, at the end of each iteration, will equally lead to the solution of (3.73)-(3.74). Iteration (3.76) is cast in closed form, since the 2×2 Jacobian is analytically determined in the following section. Hence, evaluating (3.76) is relatively inexpensive, leads to a quadratically convergent process, and directly yields Y_3 and T.

3.3.2   Iteration Partial Derivatives

For a given flow state $q \equiv \{\rho, \boldsymbol{m}, E\}^T$, and hence for a fixed (ρ, ϵ), equations (3.73)-(3.74) are functionally cast as

$$f_1\left(\rho, Y_3, Y_4\left(Y_3, T, \rho\right), Y_5\left(Y_3, T, \rho\right), T\right) = 0 \tag{3.77}$$

$$f_2\left(\epsilon, T, Y_1\left(Y_3, Y_4\left(Y_3, T, \rho\right)\right), Y_2\left(Y_3, Y_5\left(Y_3, T, \rho\right)\right), Y_3, Y_4\left(Y_3, T, \rho\right), Y_5\left(Y_3, T, \rho\right)\right) = 0 \tag{3.78}$$

Therefore, the partial derivatives in the Jacobian matrix in (3.76) are expressed as

$$\left[\left(\frac{\partial F}{\partial Q}\right)_q\right] \equiv \left[\begin{array}{cc} \left(\dfrac{\partial f_1}{\partial Y_3}\right)_{q,T} & \left(\dfrac{\partial f_1}{\partial T}\right)_{q,Y_3} \\[3mm] \left(\dfrac{\partial f_2}{\partial Y_3}\right)_{q,T} & \left(\dfrac{\partial f_2}{\partial T}\right)_{q,Y_3} \end{array} \right] \tag{3.79}$$

where subscripts denote the variables held constant in the partial differentiation. The partial derivatives in (3.79) are thus determined by application of the chain rule to (3.73)-(3.74), which yields the unabridged forms

$$\left(\frac{\partial f_1}{\partial Y_3}\right)_{q,T} = \left(\frac{\partial f_1}{\partial Y_3}\right)_{q,Y_4,Y_5,T} + \left(\frac{\partial f_1}{\partial Y_4}\right)_{q,Y_3,Y_5,T} \cdot \left(\frac{\partial Y_4}{\partial Y_3}\right)_{q,T}$$
$$+ \left(\frac{\partial f_1}{\partial Y_5}\right)_{q,Y_3,Y_4,T} \cdot \left(\frac{\partial Y_5}{\partial Y_3}\right)_{q,T} \tag{3.80}$$

$$\left(\frac{\partial f_1}{\partial T}\right)_{q,Y_3} = \left(\frac{\partial f_1}{\partial T}\right)_{q,Y_4,Y_5,Y_3} + \left(\frac{\partial f_1}{\partial Y_4}\right)_{q,Y_3,Y_5,T} \cdot \left(\frac{\partial Y_4}{\partial T}\right)_{q,Y_3}$$
$$+ \left(\frac{\partial f_1}{\partial Y_5}\right)_{q,Y_3,Y_4,T} \cdot \left(\frac{\partial Y_5}{\partial T}\right)_{q,Y_3} \tag{3.81}$$

$$\left(\frac{\partial f_2}{\partial Y_3}\right)_{q,T} = \left(\frac{\partial f_2}{\partial Y_3}\right)_{q,Y_1,Y_2,Y_4,Y_5,T} + \left(\frac{\partial f_2}{\partial Y_1}\right)_{q,Y_2,Y_3,Y_4,Y_5,T} \cdot \left(\frac{\partial Y_1}{\partial Y_3}\right)_{q,T}$$
$$+ \left(\frac{\partial f_2}{\partial Y_2}\right)_{q,Y_1,Y_3,Y_4,Y_5,T} \cdot \left(\frac{\partial Y_2}{\partial Y_3}\right)_{q,T}$$

$$\left(\frac{\partial f_2}{\partial T}\right)_{q,Y_3} = \left(\frac{\partial f_2}{\partial T}\right)_{q,Y_1,Y_2,Y_3,Y_4,Y_5} \begin{aligned} &+ \left(\frac{\partial f_2}{\partial Y_4}\right)_{q,Y_1,Y_2,Y_3,Y_5,T} \cdot \left(\frac{\partial Y_4}{\partial Y_3}\right)_{q,T} \\ &+ \left(\frac{\partial f_2}{\partial Y_5}\right)_{q,Y_1,Y_2,Y_3,Y_4,T} \cdot \left(\frac{\partial Y_5}{\partial Y_3}\right)_{q,T} \qquad (3.82) \\ &+ \left(\frac{\partial f_2}{\partial Y_1}\right)_{q,Y_2,Y_3,Y_4,Y_5,T} \cdot \left(\frac{\partial Y_1}{\partial T}\right)_{q,Y_3} \\ &+ \left(\frac{\partial f_2}{\partial Y_2}\right)_{q,Y_1,Y_3,Y_4,Y_5,T} \cdot \left(\frac{\partial Y_2}{\partial T}\right)_{q,Y_3} \\ &+ \left(\frac{\partial f_2}{\partial Y_4}\right)_{q,Y_1,Y_2,Y_3,Y_5,T} \cdot \left(\frac{\partial Y_4}{\partial T}\right)_{q,Y_3} \\ &+ \left(\frac{\partial f_2}{\partial Y_5}\right)_{q,Y_1,Y_2,Y_3,Y_4,T} \cdot \left(\frac{\partial Y_5}{\partial T}\right)_{q,Y_3} \qquad (3.83) \end{aligned}$$

Despite their deceptive complexity, these expressions become peculiarly simple, as detailed next.

The Jacobian partial derivatives (3.80)-(3.83) depend upon the four derivatives

$$\left(\frac{\partial Y_4}{\partial Y_3}\right)_{q,T}, \quad \left(\frac{\partial Y_4}{\partial T}\right)_{q,Y_3}, \quad \left(\frac{\partial Y_5}{\partial Y_3}\right)_{q,T}, \quad \left(\frac{\partial Y_5}{\partial T}\right)_{q,Y_3} \qquad (3.84)$$

which are directly computed by differentiating the mass-action equations (3.60)-(3.61) and mass- fraction equations (3.65)-(3.66) and then solving for the required derivatives. With such a procedure, these derivatives become

$$\left(\frac{\partial Y_4}{\partial Y_3}\right)_{q,T} = -\frac{2\alpha_{13}Y_4}{Y_1 + 2Y_4}, \quad \left(\frac{\partial Y_4}{\partial T}\right)_{q,Y_3} = -\frac{Y_1 Y_4}{Y_1 + 2Y_4}\frac{dG_1}{dT} \qquad (3.85)$$

and

$$\left(\frac{\partial Y_5}{\partial Y_3}\right)_{q,T} = -\frac{2\alpha_{23}Y_5}{Y_2 + 2Y_5}, \quad \left(\frac{\partial Y_5}{\partial T}\right)_{q,Y_3} = -\frac{Y_2 Y_5}{Y_2 + 2Y_5}\frac{dG_2}{dT} \qquad (3.86)$$

These simple expressions depend upon the mass fractions Y_1, Y_4, Y_2 and Y_5 and never become indeterminate. This is because the denominators constantly remain positive since Y_1 and Y_4 as well as Y_2 and Y_5 never vanish simultaneously.

With these results, the four partial derivatives

$$\left(\frac{\partial Y_1}{\partial Y_3}\right)_{q,T}, \quad \left(\frac{\partial Y_1}{\partial T}\right)_{q,Y_3}, \quad \left(\frac{\partial Y_2}{\partial Y_3}\right)_{q,T}, \quad \left(\frac{\partial Y_2}{\partial T}\right)_{q,Y_3} \qquad (3.87)$$

originate from differentiating (3.65)-(3.66) as

$$\left(\frac{\partial Y_1}{\partial Y_3}\right)_{q,T} = \left(\frac{\partial Y_1}{\partial Y_3}\right)_{Y_4} + \left(\frac{\partial Y_1}{\partial Y_4}\right)_{Y_3} \cdot \left(\frac{\partial Y_4}{\partial Y_3}\right)_{q,T} = -\frac{\alpha_{13}Y_1}{Y_1 + 2Y_4} \qquad (3.88)$$

$$\left(\frac{\partial Y_1}{\partial T}\right)_{q,Y_3} = \left(\frac{\partial Y_1}{\partial Y_4}\right)_{Y_3} \cdot \left(\frac{\partial Y_4}{\partial T}\right)_{q,Y_3} = \frac{Y_1 Y_4}{Y_1 + 2Y_4} \cdot \frac{dG_1}{dT} \qquad (3.89)$$

$$\left(\frac{\partial Y_2}{\partial Y_3}\right)_{q,T} = \left(\frac{\partial Y_2}{\partial Y_3}\right)_{Y_5} + \left(\frac{\partial Y_2}{\partial Y_5}\right)_{Y_3} \cdot \left(\frac{\partial Y_5}{\partial Y_3}\right)_{q,T} = -\frac{\alpha_{23}Y_2}{Y_2 + 2Y_5} \tag{3.90}$$

$$\left(\frac{\partial Y_2}{\partial T}\right)_{q,Y_3} = \left(\frac{\partial Y_2}{\partial Y_5}\right)_{Y_3} \cdot \left(\frac{\partial Y_5}{\partial T}\right)_{q,Y_3} = \frac{Y_2 Y_5}{Y_2 + 2Y_5} \cdot \frac{dG_2}{dT} \tag{3.91}$$

Consequently, the expressions for the partial derivatives (3.80)-(3.83) of f_1 and f_2 with respect to Y_3 and T become

$$\left(\frac{\partial f_1}{\partial Y_3}\right)_{q,T} = 1 \;-\; \frac{Y_3 - f_1}{2Y_4}\left(-\frac{2\alpha_{13}Y_4}{Y_1 + 2Y_4}\right) - \frac{Y_3 - f_1}{2Y_5}\left(-\frac{2\alpha_{23}Y_5}{Y_2 + 2Y_5}\right)$$

$$= 1 \;+\; (Y_3 - f_1) \cdot \left(\frac{\alpha_{13}}{Y_1 + 2Y_4} + \frac{\alpha_{23}}{Y_2 + 2Y_5}\right) \tag{3.92}$$

$$\left(\frac{\partial f_1}{\partial T}\right)_{q,Y_3} = -\frac{Y_3 - f_1}{2}\frac{dG_3}{dT} - \frac{Y_3 - f_1}{2Y_4}\left(-\frac{Y_1 Y_4}{Y_1 + 2Y_4}\frac{dG_1}{dT}\right) - \frac{Y_3 - f_1}{2Y_5}\left(-\frac{Y_2 Y_5}{Y_2 + 2Y_5}\frac{dG_2}{dT}\right)$$

$$= \frac{(Y_3 - f_1)}{2} \cdot \left(\frac{Y_1}{Y_1 + 2Y_4}\frac{dG_1}{dT} + \frac{Y_2}{Y_2 + 2Y_5}\frac{dG_2}{dT} - \frac{dG_3}{dT}\right) \tag{3.93}$$

$$\left(\frac{\partial f_2}{\partial Y_3}\right)_{q,T} = -\; Tc_{v_3} - \frac{\theta_3^v/M_3}{\exp\left(\theta_3^v/T\right) - 1} - h_3^0$$

$$+ \left(Tc_{v_1} + h_1^0\right)\frac{\alpha_{13}Y_1}{Y_1 + 2Y_4} + \left(Tc_{v_2} + h_2^0\right)\frac{\alpha_{23}Y_2}{Y_2 + 2Y_5}$$

$$+ \left(Tc_{v_4} + \frac{\theta_4^v/M_4}{\exp\left(\theta_4^v/T\right) - 1}\right)\frac{2\alpha_{13}Y_4}{Y_1 + 2Y_4}$$

$$+ \left(Tc_{v_5} + \frac{\theta_5^v/M_5}{\exp\left(\theta_5^v/T\right) - 1}\right)\frac{2\alpha_{23}Y_5}{Y_2 + 2Y_5} \tag{3.94}$$

$$\left(\frac{\partial f_2}{\partial T}\right)_{q,Y_3} = -\; \sum_{i=1}^{5} c_{v_i}Y_i - \sum_{i=3}^{5}\frac{Y_i(\theta_i^v/M_i)(\theta_i^v/T^2)\exp\left(\theta_i^v/T\right)}{\left(\exp\left(\theta_i^v/T\right) - 1\right)^2}$$

$$-\; \left(Tc_{v_1} + h_1^0\right)\frac{Y_1 Y_4}{Y_1 + 2Y_4}\frac{dG_1}{dT} - \left(Tc_{v_2} + h_2^0\right)\frac{Y_2 Y_5}{Y_2 + 2Y_5}\frac{dG_2}{dT}$$

$$+\; \left(Tc_{v_4} + \frac{\theta_4^v/M_4}{\exp\left(\theta_4^v/T\right) - 1}\right)\frac{Y_1 Y_4}{Y_1 + 2Y_4}\frac{dG_1}{dT}$$

$$+\; \left(Tc_{v_5} + \frac{\theta_5^v/M_5}{\exp\left(\theta_5^v/T\right) - 1}\right)\frac{Y_2 Y_5}{Y_2 + 2Y_5}\frac{dG_2}{dT} \tag{3.95}$$

With these analytical partial derivatives, the procedure for determining temperature, and the five mass fractions is theoretically complete. Therefore, a practical implementation utilizes expressions (3.69)-(3.70) and (3.65)-(3.66) to compute Y_i, $i \neq 3$, for a given state (ρ, ϵ) at each grid point and associated (Y_3, T). All of these variables are then employed to evaluate functions (3.73)-(3.74) and all the corresponding partial derivatives for the Newton's-algorithm determination of Y_3 and T. The computational results discussed in Section 3.11 indicate that this procedure quadratically converges, in two or three iterations, and directly yields temperature and the five mass fractions. Pressure is then explicitly computed using (3.56).

3.4   Partial Derivatives of T, Y_1, Y_2, Y_3, Y_4, Y_5 with Respect to ρ and ϵ

From the EOS (3.56), the thermodynamic and Jacobian partial derivatives of pressure depend on the partial derivatives of temperature T and mass fractions Y_i, $1 \leq i \leq 5$ with respect to ρ and ϵ. This section presents these derivatives. The derivatives of Y_3 and T are exactly determined from differentiating system (3.75). The derivatives of Y_1, Y_2, Y_4, Y_5 then follow from differentiating their defining expressions.

3.4.1   Derivatives of T and Y_3

Considering that $Q = Q(\rho, \epsilon)$, the differential of both sides of expression (3.75) yields

$$\left(\left[\left(\frac{\partial F}{\partial Q} \right)_{\rho,\epsilon} \right] \left(\frac{\partial Q}{\partial \rho} \right)_\epsilon + \left(\frac{\partial F}{\partial \rho} \right)_{Q,\epsilon} \right) d\rho + \left(\left[\left(\frac{\partial F}{\partial Q} \right)_{\rho,\epsilon} \right] \left(\frac{\partial Q}{\partial \epsilon} \right)_\rho + \left(\frac{\partial F}{\partial \epsilon} \right)_{Q,\rho} \right) d\epsilon = 0 \quad (3.96)$$

This constitutes a linear combination of the linearly independent differentials $d\rho$ and $d\epsilon$, which holds true if and only if their coefficients independently vanish, which defines the two linear systems

$$\left(\frac{\partial Q}{\partial \rho} \right)_\epsilon = - \left[\left(\frac{\partial F}{\partial Q} \right)_{\rho,\epsilon} \right]^{-1} \left(\frac{\partial F}{\partial \rho} \right)_{Q,\epsilon}, \quad \left(\frac{\partial Q}{\partial \epsilon} \right)_\rho = - \left[\left(\frac{\partial F}{\partial Q} \right)_{\rho,\epsilon} \right]^{-1} \left(\frac{\partial F}{\partial \epsilon} \right)_{Q,\rho} \quad (3.97)$$

The Jacobian in both of these expressions is invariant and coincides with that in (3.76) at convergence. Systems (3.97) thus directly supply the partial derivatives

$$\left(\frac{\partial Y_3}{\partial \rho} \right)_\epsilon, \quad \left(\frac{\partial Y_3}{\partial \epsilon} \right)_\rho, \quad \left(\frac{\partial T}{\partial \rho} \right)_\epsilon, \quad \left(\frac{\partial T}{\partial \epsilon} \right)_\rho \quad (3.98)$$

Systems (3.97) depend on the partial derivatives

$$\left(\frac{\partial F}{\partial \rho} \right)_{Q,\epsilon}, \quad \left(\frac{\partial F}{\partial \epsilon} \right)_{Q,\rho} \quad (3.99)$$

which are expressed as

$$\left(\frac{\partial f_1}{\partial \rho} \right)_{Y_3,T,\epsilon} = \left(\frac{\partial f_1}{\partial Y_4} \right)_{q,Y_3,Y_5,T} \cdot \left(\frac{\partial Y_4}{\partial \rho} \right)_{Y_3,T} + \left(\frac{\partial f_1}{\partial Y_5} \right)_{q,Y_3,Y_4,T} \cdot \left(\frac{\partial Y_5}{\partial \rho} \right)_{Y_3,T} \quad (3.100)$$

$$\left(\frac{\partial f_1}{\partial \epsilon} \right)_{Y_3,T,\rho} = 0.0 \quad (3.101)$$

$$\left(\frac{\partial f_2}{\partial \rho} \right)_{Y_3,T,\epsilon} = \left(\frac{\partial f_2}{\partial Y_1} \right)_{q,Y_2,Y_3,Y_4,Y_5,T} \cdot \left(\frac{\partial Y_1}{\partial \rho} \right)_{Y_3,T} + \left(\frac{\partial f_2}{\partial Y_2} \right)_{q,Y_1,Y_3,Y_4,Y_5,T} \cdot \left(\frac{\partial Y_2}{\partial \rho} \right)_{Y_3,T}$$

$$+ \left(\frac{\partial f_2}{\partial Y_4} \right)_{q,Y_1,Y_2,Y_3,Y_5,T} \cdot \left(\frac{\partial Y_4}{\partial \rho} \right)_{Y_3,T} + \left(\frac{\partial f_2}{\partial Y_5} \right)_{q,Y_1,Y_2,Y_3,Y_4,T} \cdot \left(\frac{\partial Y_5}{\partial \rho} \right)_{Y_3,T} \quad (3.102)$$

$$\left(\frac{\partial f_2}{\partial \epsilon} \right)_{Y_3,T,\rho} = \left(\frac{\partial f_2}{\partial \epsilon} \right)_{Y_1,Y_2,Y_3,Y_4,Y_5,T,\rho} = 1.0 \quad (3.103)$$

These relationships depend on the partial derivatives

$$\left(\frac{\partial f_1}{\partial Y_i}\right)_{q,Y_j,T}, \quad \left(\frac{\partial f_2}{\partial Y_i}\right)_{q,Y_j,T}, \quad i \neq j \tag{3.104}$$

and

$$\left(\frac{\partial Y_4}{\partial \rho}\right)_{Y_3,T}, \quad \left(\frac{\partial Y_5}{\partial \rho}\right)_{Y_3,T}, \quad \left(\frac{\partial Y_1}{\partial \rho}\right)_{Y_3,T}, \quad \left(\frac{\partial Y_2}{\partial \rho}\right)_{Y_3,T} \tag{3.105}$$

At convergence of (3.76), the derivatives of f_1 and f_2 in (3.104) directly follow from the corresponding ones within (3.92)-(3.93), after setting $f_1 = 0$ and $f_2 = 0$. The derivatives of Y_4 and Y_5 in (3.105) are obtained from directly differentiating the mass-action equations (3.60)-(3.61). With this procedure, the mass-fraction derivatives (3.105) become

$$\left(\frac{\partial Y_4}{\partial \rho}\right)_{Y_3,T} = \frac{1}{\rho}\frac{Y_1 Y_4}{Y_1 + 2Y_4}, \quad \left(\frac{\partial Y_5}{\partial \rho}\right)_{Y_3,T} = \frac{1}{\rho}\frac{Y_2 Y_5}{Y_2 + 2Y_5} \tag{3.106}$$

The partial derivatives of Y_1 and Y_2 then directly follow from differentiating the mass-fraction equations (3.65)- (3.66) as

$$\left(\frac{\partial Y_1}{\partial \rho}\right)_{Y_3,T} = -\frac{1}{\rho}\frac{Y_1 Y_4}{Y_1 + 2Y_4}, \quad \left(\frac{\partial Y_2}{\partial \rho}\right)_{Y_3,T} = -\frac{1}{\rho}\frac{Y_2 Y_5}{Y_2 + 2Y_5} \tag{3.107}$$

With expressions (3.106)-(3.107), the partial derivatives of f_1 and f_2 become

$$\left(\frac{\partial f_1}{\partial \rho}\right)_{Y_3,T,\epsilon} = -\frac{1}{2\rho}\frac{Y_1 Y_3}{Y_1 + 2Y_4} - \frac{1}{2\rho}\frac{Y_2 Y_3}{Y_2 + 2Y_5} \tag{3.108}$$

$$\left(\frac{\partial f_1}{\partial \epsilon}\right)_{Y_3,T,\rho} = 0.0 \tag{3.109}$$

$$\left(\frac{\partial f_2}{\partial \rho}\right)_{Y_3,T,\epsilon} = \left(Tc_{v_1} + h_1^0\right)\frac{1}{\rho}\frac{Y_1 Y_4}{Y_1 + 2Y_4} + \left(Tc_{v_2} + h_2^0\right)\frac{1}{\rho}\frac{Y_2 Y_5}{Y_2 + 2Y_5}$$
$$- \left(Tc_{v_4} + \frac{\theta_4^v/M_4}{\exp\left(\theta_4^v/T\right) - 1}\right)\frac{1}{\rho}\frac{Y_1 Y_4}{Y_1 + 2Y_4}$$
$$- \left(Tc_{v_5} + \frac{\theta_5^v/M_5}{\exp\left(\theta_5^v/T\right) - 1}\right)\frac{1}{\rho}\frac{Y_2 Y_5}{Y_2 + 2Y_5} \tag{3.110}$$

$$\left(\frac{\partial f_2}{\partial \epsilon}\right)_{Y_3,T,\rho} = 1.0 \tag{3.111}$$

3.4.2 Derivatives of Y_1, Y_2, Y_4, Y_5

Once the derivatives of T and Y_3 with respect to ρ and ϵ have been calculated, the partial derivatives

$$\left(\frac{\partial Y_4}{\partial \rho}\right)_\epsilon, \quad \left(\frac{\partial Y_4}{\partial \epsilon}\right)_\rho, \quad \left(\frac{\partial Y_5}{\partial \rho}\right)_\epsilon, \quad \left(\frac{\partial Y_5}{\partial \epsilon}\right)_\rho \tag{3.112}$$

can then be exactly determined. Given the functional relations (3.71)-(3.72) for Y_i, $i \neq 3$, these derivatives are expressed as

$$\left(\frac{\partial Y_4}{\partial \rho}\right)_\epsilon = \left(\frac{\partial Y_4}{\partial Y_3}\right)_{\rho,T} \cdot \left(\frac{\partial Y_3}{\partial \rho}\right)_\epsilon + \left(\frac{\partial Y_4}{\partial T}\right)_{\rho,Y_3} \cdot \left(\frac{\partial T}{\partial \rho}\right)_\epsilon + \left(\frac{\partial Y_4}{\partial \rho}\right)_{Y_3,T} \tag{3.113}$$

$$\left(\frac{\partial Y_4}{\partial \epsilon}\right)_\rho = \left(\frac{\partial Y_4}{\partial Y_3}\right)_{\rho,T} \cdot \left(\frac{\partial Y_3}{\partial \epsilon}\right)_\rho + \left(\frac{\partial Y_4}{\partial T}\right)_{\rho,Y_3} \cdot \left(\frac{\partial T}{\partial \epsilon}\right)_\rho \tag{3.114}$$

$$\left(\frac{\partial Y_5}{\partial \rho}\right)_\epsilon = \left(\frac{\partial Y_5}{\partial Y_3}\right)_{\rho,T} \cdot \left(\frac{\partial Y_3}{\partial \rho}\right)_\epsilon + \left(\frac{\partial Y_5}{\partial T}\right)_{\rho,Y_3} \cdot \left(\frac{\partial T}{\partial \rho}\right)_\epsilon + \left(\frac{\partial Y_5}{\partial \rho}\right)_{Y_3,T} \tag{3.115}$$

$$\left(\frac{\partial Y_5}{\partial \epsilon}\right)_\rho = \left(\frac{\partial Y_5}{\partial Y_3}\right)_{\rho,T} \cdot \left(\frac{\partial Y_3}{\partial \epsilon}\right)_\rho + \left(\frac{\partial Y_5}{\partial T}\right)_{\rho,Y_3} \cdot \left(\frac{\partial T}{\partial \epsilon}\right)_\rho \tag{3.116}$$

where the partial derivatives of Y_4 and Y_5 with respect to Y_3 and T are detailed in Section 3.4.1. Hence, expressions (3.112) become

$$\left(\frac{\partial Y_4}{\partial \rho}\right)_\epsilon = -\frac{2\alpha_{13}Y_4}{Y_1 + 2Y_4}\left(\frac{\partial Y_3}{\partial \rho}\right)_\epsilon - \frac{Y_1 Y_4}{Y_1 + 2Y_4}\frac{dG_1}{dT}\left(\frac{\partial T}{\partial \rho}\right)_\epsilon + \frac{1}{\rho}\frac{Y_1 Y_4}{Y_1 + 2Y_4} \tag{3.117}$$

$$\left(\frac{\partial Y_4}{\partial \epsilon}\right)_\rho = -\frac{2\alpha_{13}Y_4}{Y_1 + 2Y_4}\left(\frac{\partial Y_3}{\partial \epsilon}\right)_\rho - \frac{Y_1 Y_4}{Y_1 + 2Y_4}\frac{dG_1}{dT}\left(\frac{\partial T}{\partial \epsilon}\right)_\rho \tag{3.118}$$

$$\left(\frac{\partial Y_5}{\partial \rho}\right)_\epsilon = -\frac{2\alpha_{23}Y_5}{Y_2 + 2Y_5}\left(\frac{\partial Y_3}{\partial \rho}\right)_\epsilon - \frac{Y_2 Y_5}{Y_2 + 2Y_5}\frac{dG_2}{dT}\left(\frac{\partial T}{\partial \rho}\right)_\epsilon + \frac{1}{\rho}\frac{Y_2 Y_5}{Y_2 + 2Y_5} \tag{3.119}$$

$$\left(\frac{\partial Y_5}{\partial \epsilon}\right)_\rho = -\frac{2\alpha_{23}Y_5}{Y_2 + 2Y_5}\left(\frac{\partial Y_3}{\partial \epsilon}\right)_\rho - \frac{Y_2 Y_5}{Y_2 + 2Y_5}\frac{dG_2}{dT}\left(\frac{\partial T}{\partial \epsilon}\right)_\rho \tag{3.120}$$

With these expressions, along with (3.65)-(3.66) the partial derivatives

$$\left(\frac{\partial Y_1}{\partial \rho}\right)_\epsilon, \quad \left(\frac{\partial Y_1}{\partial \epsilon}\right)_\rho, \quad \left(\frac{\partial Y_2}{\partial \rho}\right)_\epsilon, \quad \left(\frac{\partial Y_2}{\partial \epsilon}\right)_\rho \tag{3.121}$$

are developed as

$$\left(\frac{\partial Y_1}{\partial \rho}\right)_\epsilon = -\left(\frac{\partial Y_4}{\partial \rho}\right)_\epsilon - \alpha_{13}\left(\frac{\partial Y_3}{\partial \rho}\right)_\epsilon$$

$$= \frac{-\alpha_{13}Y_1}{Y_1 + 2Y_4}\left(\frac{\partial Y_3}{\partial \rho}\right)_\epsilon + \frac{Y_1 Y_4}{Y_1 + 2Y_4}\frac{dG_1}{dT}\left(\frac{\partial T}{\partial \rho}\right)_\epsilon - \frac{1}{\rho}\frac{Y_1 Y_4}{Y_1 + 2Y_4} \tag{3.122}$$

$$\left(\frac{\partial Y_1}{\partial \epsilon}\right)_\rho = -\left(\frac{\partial Y_4}{\partial \epsilon}\right)_\rho - \alpha_{13}\left(\frac{\partial Y_3}{\partial \epsilon}\right)_\rho$$

$$= \frac{-\alpha_{13}Y_1}{Y_1 + 2Y_4}\left(\frac{\partial Y_3}{\partial \epsilon}\right)_\rho + \frac{Y_1 Y_4}{Y_1 + 2Y_4}\frac{dG_1}{dT}\left(\frac{\partial T}{\partial \epsilon}\right)_\rho \tag{3.123}$$

$$\left(\frac{\partial Y_2}{\partial \rho}\right)_\epsilon = -\left(\frac{\partial Y_5}{\partial \rho}\right)_\epsilon - \alpha_{23}\left(\frac{\partial Y_3}{\partial \rho}\right)_\epsilon$$

$$= \frac{-\alpha_{23} Y_2}{Y_2 + 2Y_5} \left(\frac{\partial Y_3}{\partial \rho} \right)_\epsilon + \frac{Y_2 Y_5}{Y_2 + 2Y_5} \frac{dG_2}{dT} \left(\frac{\partial T}{\partial \rho} \right)_\epsilon - \frac{1}{\rho} \frac{Y_2 Y_5}{Y_2 + 2Y_5} \quad (3.124)$$

$$\left(\frac{\partial Y_2}{\partial \epsilon} \right)_\rho = - \left(\frac{\partial Y_5}{\partial \epsilon} \right)_\rho - \alpha_{23} \left(\frac{\partial Y_3}{\partial \epsilon} \right)_\rho$$

$$= \frac{-\alpha_{23} Y_2}{Y_2 + 2Y_5} \left(\frac{\partial Y_3}{\partial \epsilon} \right)_\rho + \frac{Y_2 Y_5}{Y_2 + 2Y_5} \frac{dG_2}{dT} \left(\frac{\partial T}{\partial \epsilon} \right)_\rho \quad (3.125)$$

3.5   Pressure and its Thermodynamic Derivatives

With the calculated temperature and mass fractions, pressure is readily calculated via the EOS

$$p = \rho T \sum_{i=1}^{5} \frac{Y_i}{M_i} \quad (3.126)$$

The thermodynamic derivatives of pressure are then exactly determined by differentiating this EOS in the form

$$\left(\frac{\partial p}{\partial \rho} \right)_\epsilon = T \sum_{i=1}^{5} \frac{Y_i}{M_i} + \rho \left(\frac{\partial T}{\partial \rho} \right)_\epsilon \sum_{i=1}^{5} \frac{Y_i}{M_i} + \rho T \sum_{i=1}^{5} \frac{1}{M_i} \left(\frac{\partial Y_i}{\partial \rho} \right)_\epsilon \quad (3.127)$$

$$\left(\frac{\partial p}{\partial \epsilon} \right)_\rho = \rho \left(\frac{\partial T}{\partial \epsilon} \right)_\rho \sum_{i=1}^{5} \frac{Y_i}{M_i} + \rho T \sum_{i=1}^{5} \frac{1}{M_i} \left(\frac{\partial Y_i}{\partial \epsilon} \right)_\rho \quad (3.128)$$

which shows dependence on the thermodynamic derivatives of both T and mass fractions Y_i, $1 \leq i \leq 5$, with respect to ρ and ϵ.

3.6   Jacobian Partial Derivatives
of Pressure and Temperature

The convergence of implicit Euler and Navier-Stokes CFD algorithms also depends upon accurate and continuous Jacobians of pressure with respect to the state variable q. Implicit Navier-Stokes CFD algorithms also require accurate and continuous Jacobians of temperature with respect to q. This section exactly determines these derivatives.

Using (2.15), an application of the chain rule to the EOS $p = p(\rho, \epsilon(q))$, leads to the partial derivatives of pressure with respect to the flow variable $q \equiv \{\rho, \boldsymbol{m}, E\}^T$ as

$$\left(\frac{\partial p}{\partial \rho} \right)_{\boldsymbol{m},E} = \left(\frac{\partial p}{\partial \rho} \right)_\epsilon + \left(\frac{\partial p}{\partial \epsilon} \right)_\rho \left(\frac{\partial \epsilon}{\partial \rho} \right)_{\boldsymbol{m},E} = \left(\frac{\partial p}{\partial \rho} \right)_\epsilon + \left(\frac{\partial p}{\partial \epsilon} \right)_\rho \cdot \frac{1}{\rho^2} \left(\frac{1}{\rho} \sum_{i=1}^{n} m_i m_i - E \right) \quad (3.129)$$

$$\left(\frac{\partial p}{\partial m_i} \right)_{\rho, m_j, E, i \neq j} = \left(\frac{\partial p}{\partial \epsilon} \right)_\rho \left(\frac{\partial \epsilon}{\partial m_i} \right)_{\rho, m_j, E, i \neq j} = \left(\frac{\partial p}{\partial \epsilon} \right)_\rho \cdot \left(-\frac{m_i}{\rho^2} \right) \quad (3.130)$$

$$\left(\frac{\partial p}{\partial E} \right)_{\rho, \boldsymbol{m}} = \left(\frac{\partial p}{\partial \epsilon} \right)_\rho \left(\frac{\partial \epsilon}{\partial E} \right)_{\rho, \boldsymbol{m}} = \left(\frac{\partial p}{\partial \epsilon} \right)_\rho \cdot \left(\frac{1}{\rho} \right) \quad (3.131)$$

The corresponding Jacobian partial derivatives of T are directly obtained by replacing p with T in (3.129)-(3.131). As noted in Section 3.2, equations (3.65)-(3.66), (3.69)-(3.70), and (3.73)-(3.74) asymptotically approach perfect-gas expressions. Therefore (3.129)-(3.131) will converge for low temperatures to perfect-gas partial derivatives. In any case, these Jacobian derivatives depend on the thermodynamic derivatives $\left(\frac{\partial p}{\partial \rho}\right)_\epsilon$ and $\left(\frac{\partial p}{\partial \epsilon}\right)_\rho$. Similarly, the Jacobian partial derivatives of T depend on the thermodynamic derivatives $\left(\frac{\partial T}{\partial \rho}\right)_\epsilon$ and $\left(\frac{\partial T}{\partial \epsilon}\right)_\rho$.

3.7 Derivatives of p and Y_i with Respect to ρ, T, Derivatives of ρ with Respect to p and T, and Derivatives of T with Respect to ρ and p

These derivatives are established in terms of the thermodynamic derivatives of pressure and temperature. The expressions for the differentials of $p = p(\rho, \epsilon)$ and $T = T(\rho, \epsilon)$ are

$$dp = \left(\frac{\partial p}{\partial \rho}\right)_\epsilon d\rho + \left(\frac{\partial p}{\partial \epsilon}\right)_\rho d\epsilon \tag{3.132}$$

$$dT = \left(\frac{\partial T}{\partial \rho}\right)_\epsilon d\rho + \left(\frac{\partial T}{\partial \epsilon}\right)_\rho d\epsilon \tag{3.133}$$

Similarly, the expression for the differential of $p = p(\rho, T)$ is

$$dp = \left(\frac{\partial p}{\partial \rho}\right)_T d\rho + \left(\frac{\partial p}{\partial T}\right)_\rho dT \tag{3.134}$$

Replacing in this expression the differential of T via (3.133) yields

$$dp = \left(\left(\frac{\partial p}{\partial \rho}\right)_T + \left(\frac{\partial p}{\partial T}\right)_\rho \cdot \left(\frac{\partial T}{\partial \rho}\right)_\epsilon\right) d\rho + \left(\frac{\partial p}{\partial T}\right)_\rho \cdot \left(\frac{\partial T}{\partial \epsilon}\right)_\rho d\epsilon \tag{3.135}$$

This equation has to coincide with (3.132), hence the required derivatives of pressure become

$$\left(\frac{\partial p}{\partial T}\right)_\rho = \left(\frac{\partial p}{\partial \epsilon}\right)_\rho / \left(\frac{\partial T}{\partial \epsilon}\right)_\rho, \quad \left(\frac{\partial p}{\partial \rho}\right)_T = \left(\frac{\partial p}{\partial \rho}\right)_\epsilon - \left(\frac{\partial p}{\partial \epsilon}\right)_\rho \left(\frac{\partial T}{\partial \rho}\right)_\epsilon / \left(\frac{\partial T}{\partial \epsilon}\right)_\rho \tag{3.136}$$

The partial derivatives of the mass fractions Y_i, $1 \leq i \leq 5$, as well as any other thermodynamic variable, with respect to T and ρ can be obtained similarly to the derivative of pressure with respect to T and ρ. Following the procedure that generated these pressure derivatives, therefore, the derivatives of all Y_i's, $1 \leq i \leq 5$, with respect to T and ρ become

$$\left(\frac{\partial Y_i}{\partial T}\right)_\rho = \left(\frac{\partial Y_i}{\partial \epsilon}\right)_\rho / \left(\frac{\partial T}{\partial \epsilon}\right)_\rho, \quad \left(\frac{\partial Y_i}{\partial \rho}\right)_T = \left(\frac{\partial Y_i}{\partial \rho}\right)_\epsilon - \left(\frac{\partial Y_i}{\partial \epsilon}\right)_\rho \left(\frac{\partial T}{\partial \rho}\right)_\epsilon / \left(\frac{\partial T}{\partial \epsilon}\right)_\rho \tag{3.137}$$

where the partial derivatives $\left(\frac{\partial Y_i}{\partial \epsilon}\right)_\rho$ and $\left(\frac{\partial Y_i}{\partial \rho}\right)_\epsilon$ have already been determined in Section 3.4.

The differential of pressure (3.132) also allows determining relationships between the derivatives of pressure and those of density. The expression for the differential of $\rho = \rho(p, T)$ is expressed as

$$d\rho = \left(\frac{\partial \rho}{\partial T}\right)_p dT + \left(\frac{\partial \rho}{\partial p}\right)_T dp \qquad (3.138)$$

Upon substituting for $d\rho$ in (3.134) with this differential, the following linear combination emerges

$$\left(\left(\frac{\partial p}{\partial \rho}\right)_T \left(\frac{\partial \rho}{\partial p}\right)_T - 1\right) dp + \left(\left(\frac{\partial p}{\partial \rho}\right)_T \left(\frac{\partial \rho}{\partial T}\right)_p + \left(\frac{\partial p}{\partial T}\right)_T\right) dT = 0 \qquad (3.139)$$

Considering that the differentials dp and dT are linearly independent in such a combination, this combination can vanish if and only if its coefficients vanish, which yields

$$\left(\frac{\partial \rho}{\partial p}\right)_T = \left(\frac{\partial p}{\partial \rho}\right)_T^{-1}, \quad \left(\frac{\partial \rho}{\partial T}\right)_p = -\left(\frac{\partial p}{\partial T}\right)_T / \left(\frac{\partial p}{\partial \rho}\right)_T \qquad (3.140)$$

The expression for the differential of $T = T(\rho, p)$ is expressed as

$$dT = \left(\frac{\partial T}{\partial \rho}\right)_p d\rho + \left(\frac{\partial T}{\partial p}\right)_\rho dp \qquad (3.141)$$

Replacing in this expression the differential of p via (3.132) yields

$$dT = \left(\left(\frac{\partial T}{\partial \rho}\right)_p + \left(\frac{\partial T}{\partial p}\right)_\rho \cdot \left(\frac{\partial p}{\partial \rho}\right)_\epsilon\right) d\rho + \left(\frac{\partial T}{\partial p}\right)_\rho \cdot \left(\frac{\partial p}{\partial \epsilon}\right)_\rho d\epsilon \qquad (3.142)$$

This equation has to coincide with (3.133), hence the required derivatives of temperature become

$$\left(\frac{\partial T}{\partial p}\right)_\rho = \left(\frac{\partial T}{\partial \epsilon}\right)_\rho / \left(\frac{\partial p}{\partial \epsilon}\right)_\rho, \quad \left(\frac{\partial T}{\partial \rho}\right)_p = \left(\frac{\partial T}{\partial \rho}\right)_\epsilon - \left(\frac{\partial T}{\partial \epsilon}\right)_\rho \left(\frac{\partial p}{\partial \rho}\right)_\epsilon / \left(\frac{\partial p}{\partial \epsilon}\right)_\rho \qquad (3.143)$$

A comparison of these findings with (3.136), in particular, reiterates the general result $\left(\frac{\partial T}{\partial p}\right)_\rho = \left(\frac{\partial p}{\partial T}\right)_\rho^{-1}$

3.8 Thermodynamic Properties with p and T as Independent Variables

The procedure described in Sections 3.2-3.4 generates the thermodynamic properties of air with density and internal energy as independent variables. This set of independent variables is eminently suited for CFD applications, for, as indicated in Section 3.1.3, these variables are readily available from the dependent state variable q in the Euler and Navier-Stokes conservation law systems. The same procedure, however, can also deliver thermodynamic properties for any other set of independent variables because the procedure also delivers partial derivatives. This section presents a rapidly converging Newton's method to use the procedure in Sections 3.2-3.4 to generate thermodynamic properties with p and T as independent variables. The method can also be swiftly modified to generate thermodynamic properties for any other set of independent thermodynamic variables.

The procedure in Sections 3.2-3.4 delivers the functions $p = p(\rho, \epsilon)$, $T = T(\rho, \epsilon)$ and their partial derivatives with respect to ρ and ϵ. Repeated here for convenience, the expressions for the differentials of $p = p(\rho, \epsilon)$ and $T = T(\rho, \epsilon)$ are

$$dp = \left(\frac{\partial p}{\partial \rho}\right)_\epsilon d\rho + \left(\frac{\partial p}{\partial \epsilon}\right)_\rho d\epsilon, \quad dT = \left(\frac{\partial T}{\partial \rho}\right)_\epsilon d\rho + \left(\frac{\partial T}{\partial \epsilon}\right)_\rho d\epsilon \qquad (3.144)$$

Upon setting the differentials in these expressions equal to first-order variations and solving for the variations $\Delta \rho$ and $\Delta \epsilon$, the following expressions emerge

$$\Delta \rho = \frac{\left(\frac{\partial T}{\partial \epsilon}\right)_\rho \Delta p - \left(\frac{\partial p}{\partial \epsilon}\right)_\rho \Delta T}{\left(\frac{\partial p}{\partial \rho}\right)_\epsilon \left(\frac{\partial T}{\partial \epsilon}\right)_\rho - \left(\frac{\partial T}{\partial \rho}\right)_\epsilon \left(\frac{\partial p}{\partial \epsilon}\right)_\rho}, \quad \Delta \epsilon = \frac{-\left(\frac{\partial T}{\partial \rho}\right)_\epsilon \Delta p + \left(\frac{\partial p}{\partial \rho}\right)_\epsilon \Delta T}{\left(\frac{\partial p}{\partial \rho}\right)_\epsilon \left(\frac{\partial T}{\partial \epsilon}\right)_\rho - \left(\frac{\partial T}{\partial \rho}\right)_\epsilon \left(\frac{\partial p}{\partial \epsilon}\right)_\rho} \qquad (3.145)$$

The variations $\Delta \rho$ and $\Delta \epsilon$ remain well defined since the denominator in these results never vanishes. Set all the variations in these results equal to the expressions

$$\Delta p = p_0 - p^i, \quad \Delta T = T_0 - T^i, \quad \Delta \rho = \rho^{i+1} - \rho^i, \quad \Delta \epsilon = \epsilon^{i+1} - \epsilon^i \qquad (3.146)$$

In these expressions, p_0 and T_0 denote prescribed pressure and temperature; p^i and T^i correspond to the pressure and temperature the procedure delivers in relation to ρ^i and ϵ^i. Hence ρ^{i+1} and ϵ^{i+1} calculated through (3.145) represent the refinements on ρ and ϵ as p^i and T^i respectively converge toward p_0 and T_0. Since the expressions in (3.145) correspond to Newton's method on the system of functions $p(\rho, \epsilon) - p_0 = 0$ and $T(\rho, \epsilon) - T_0 = 0$, the method has been found to converge quadratically in two or three iterations, provided that the initial estimate for ρ and ϵ lie within the range of convergence and the absolute value of ρ^{i+1} and ϵ^{i+1} are used in each cycle, following the indications in Section 3.3. With the procedure described in Sections 3.2-3.4 encapsulated in a subroutine, the iteration corresponding to equations (3.145) finds its realization with two or three calls to such a subroutine. At convergence, $\Delta \rho \rightarrow 0$, $\Delta \epsilon \rightarrow 0$, and $p^i \rightarrow p_0$, $T^i \rightarrow T_0$, hence the procedure effectively provides thermodynamic properties for prescribed p and T.

3.9 Specific Heats

The mixture specific heat at constant volume c_v evolves from the derivative of the internal energy ϵ with respect to temperature. The specific heat at constant pressure c_p follows in terms of c_v.

3.9.1   Specific Heat c_v

The differential of $\epsilon = \epsilon(\rho, T)$ is expressed as

$$d\epsilon = \left(\frac{\partial \epsilon}{\partial \rho}\right)_T d\rho + \left(\frac{\partial \epsilon}{\partial T}\right)_\rho dT \tag{3.147}$$

Upon inserting this result into (3.133), the following linear combination emerges

$$\left(\left(\frac{\partial T}{\partial \rho}\right)_\epsilon + \left(\frac{\partial T}{\partial \epsilon}\right)_\rho \left(\frac{\partial \epsilon}{\partial \rho}\right)_T\right) d\rho + \left(\left(\frac{\partial T}{\partial \epsilon}\right)_\rho \left(\frac{\partial \epsilon}{\partial T}\right)_\rho - 1\right) dT = 0 \tag{3.148}$$

Considering that the differentials $d\rho$ and dT are linearly independent in such a combination, this combination can vanish if and only if its coefficients vanish, which yields

$$\left(\frac{\partial \epsilon}{\partial \rho}\right)_T = -\left(\frac{\partial T}{\partial \rho}\right)_\epsilon \bigg/ \left(\frac{\partial T}{\partial \epsilon}\right)_\rho, \quad \left(\frac{\partial \epsilon}{\partial T}\right)_\rho = \left(\frac{\partial T}{\partial \epsilon}\right)_\rho^{-1} \tag{3.149}$$

By definition, c_v is expressed as

$$c_v = \left(\frac{\partial \epsilon}{\partial T}\right)_\rho \tag{3.150}$$

This specific heat is thus obtained in terms of the derivative of temperature as

$$c_v = \left(\frac{\partial T}{\partial \epsilon}\right)_\rho^{-1} \tag{3.151}$$

which derivative is calculated as detailed in Section 3.4.

3.9.2   Specific Heat c_p

The determination of this specific heat in terms of c_v requires the partial derivatives of internal energy ϵ and entropy s, both with respect to ρ at constant temperature. From (1.138) the differential of ϵ is expressed as

$$d\epsilon = Tds - pdv = Tds + \frac{p}{\rho^2}d\rho \tag{3.152}$$

where s and v respectively denote entropy and specific volume. This expression thus implies the partial derivatives of ϵ

$$\left(\frac{\partial \epsilon}{\partial s}\right)_\rho = T, \quad \left(\frac{\partial \epsilon}{\partial \rho}\right)_s = \frac{p}{\rho^2} \tag{3.153}$$

The differential of $s = s(\rho, T)$ equals

$$ds = \left(\frac{\partial s}{\partial \rho}\right)_T d\rho + \left(\frac{\partial s}{\partial T}\right)_\rho dT \tag{3.154}$$

Upon substitution of this ds into (3.152), the following differential of ϵ is obtained

$$d\epsilon = \left(T\left(\frac{\partial s}{\partial \rho}\right)_T + \frac{p}{\rho^2}\right)d\rho + T\left(\frac{\partial s}{\partial T}\right)_\rho dT \qquad (3.155)$$

This differential has to coincide with (3.147) which yields

$$\left(\frac{\partial \epsilon}{\partial T}\right)_\rho = T\left(\frac{\partial s}{\partial T}\right)_\rho, \quad \left(\frac{\partial \epsilon}{\partial \rho}\right)_T = T\left(\frac{\partial s}{\partial \rho}\right)_T + \frac{p}{\rho^2} \qquad (3.156)$$

Furthermore, substituting for $d\rho$ in (3.147) with differential (3.138) yields

$$d\epsilon = \left(\left(\frac{\partial \epsilon}{\partial T}\right)_\rho + \left(\frac{\partial \epsilon}{\partial \rho}\right)_T \left(\frac{\partial \rho}{\partial T}\right)_p\right)dT + \left(\frac{\partial \epsilon}{\partial \rho}\right)_T \left(\frac{\partial \rho}{\partial p}\right)_T dp \qquad (3.157)$$

which for $\epsilon = \epsilon(T, p)$ leads to the following derivative of ϵ

$$\left(\frac{\partial \epsilon}{\partial T}\right)_p = \left(\frac{\partial \epsilon}{\partial T}\right)_\rho + \left(\frac{\partial \epsilon}{\partial \rho}\right)_T \left(\frac{\partial \rho}{\partial T}\right)_p = c_v + T\left(\frac{\partial s}{\partial \rho}\right)_T \left(\frac{\partial \rho}{\partial T}\right)_p + \frac{p}{\rho^2}\left(\frac{\partial \rho}{\partial T}\right)_p \qquad (3.158)$$

To determine the derivative of entropy with respect to density consider Helmholtz function H_e and its differential

$$H_e = \epsilon - Ts \quad \Rightarrow \quad dH_e = d\epsilon - d(Ts) = \frac{p}{\rho^2}d\rho - sdT \qquad (3.159)$$

Since $H_e = H_e(\rho, T)$ is a function of state, its mixed second-order derivatives must equal each other whenever H_e is differentiable, hence

$$\frac{\partial^2 H_e}{\partial T \partial \rho} = \frac{\partial^2 H_e}{\partial \rho \partial T} \quad \Rightarrow \quad \left(\frac{\partial s}{\partial \rho}\right)_T = -\left(\frac{\partial}{\partial T}\left(\frac{p}{\rho^2}\right)\right)_\rho = -\frac{1}{\rho^2}\left(\frac{\partial p}{\partial T}\right)_\rho \qquad (3.160)$$

which provides the derivative of s in terms of the derivative of p, determined in Section 3.7; incidentally, this result is one of Maxwell's thermodynamic equations, but cast in terms of density.

With derivatives (3.158) and (3.160), the expression for c_p follows from its definition as the derivative of enthalpy h as

$$
\begin{aligned}
c_p &= \left(\frac{\partial h}{\partial T}\right)_p = \left(\frac{\partial}{\partial T}\left(\epsilon + \frac{p}{\rho}\right)\right)_p = \left(\frac{\partial \epsilon}{\partial T}\right)_p - \frac{p}{\rho^2}\left(\frac{\partial \rho}{\partial T}\right)_p \\
&= c_v + T\left(\frac{\partial s}{\partial \rho}\right)_T \left(\frac{\partial \rho}{\partial T}\right)_p + \frac{p}{\rho^2}\left(\frac{\partial \rho}{\partial T}\right)_p - \frac{p}{\rho^2}\left(\frac{\partial \rho}{\partial T}\right)_p \\
&= c_v - \frac{T}{\rho^2}\left(\frac{\partial p}{\partial T}\right)_\rho \left(\frac{\partial \rho}{\partial T}\right)_p
\end{aligned}
\qquad (3.161)
$$

This result can then be expressed in terms of the derivatives of only one thermodynamic variable as

$$c_p = c_v - \frac{T}{\rho^2 \left(\frac{\partial T}{\partial p}\right)_\rho \left(\frac{\partial T}{\partial \rho}\right)_p}, \quad c_p = c_v + \frac{T\left(\frac{\partial p}{\partial T}\right)_\rho^2}{\rho^2 \left(\frac{\partial p}{\partial \rho}\right)_T} \qquad (3.162)$$

where the second result follows from the second expression in (3.140). The ratio of specific heats then follows as

$$\gamma = \frac{c_p}{c_v} = \left(\frac{\partial h}{\partial T}\right)_p \left(\frac{\partial T}{\partial \epsilon}\right)_\rho = 1 - \frac{T\left(\frac{\partial T}{\partial \epsilon}\right)_\rho}{\rho^2 \left(\frac{\partial T}{\partial p}\right)_\rho \left(\frac{\partial T}{\partial \rho}\right)_p} \tag{3.163}$$

In the absence of chemical reactions, $M = $ constant and the derivatives of pressure from (3.13) reduce (3.162) to the well known result

$$c_p = c_v + \frac{\mathcal{R}}{M} \tag{3.164}$$

3.10   Speed of Sound

The square of the speed of sound calculated from $p = p(\rho, \epsilon(\rho, s))$ is expressed as

$$c^2 = \left(\frac{\partial p}{\partial \rho}\right)_s = \left(\frac{\partial p}{\partial \rho}\right)_\epsilon + \left(\frac{\partial p}{\partial \epsilon}\right)_\rho \left(\frac{\partial \epsilon}{\partial \rho}\right)_s = \left(\frac{\partial p}{\partial \rho}\right)_\epsilon + \frac{p}{\rho^2}\left(\frac{\partial p}{\partial \epsilon}\right)_\rho \tag{3.165}$$

where the final step derives from (3.153). By virtue of the Jacobian derivatives of pressure (3.129)-(3.131), this expression becomes

$$c^2 = \left(\frac{\partial p}{\partial \rho}\right)_{\boldsymbol{m},E} + \left(\frac{\partial p}{\partial E}\right)_{\rho,\boldsymbol{m}} \left(\frac{E+p}{\rho} - \frac{1}{\rho^2}\sum_{i=1}^n m_i m_i\right) \tag{3.166}$$

which coincides with the square of the isotropic component of all the eigenvalues of the inviscid flux vector, as determined in Chapters 13, 17.

The speed of sound, is calculated via

$$c^2 = \left(\frac{\partial p}{\partial \rho}\right)_\epsilon + \frac{p}{\rho^2}\left(\frac{\partial p}{\partial \epsilon}\right)_\rho \tag{3.167}$$

for the thermodynamic derivatives of pressure are determined as shown in Section 3.5.

This expression for the square of speed of sound can also be expressed in terms of the specific heat ratio γ using the pressure derivatives (3.136) and the following equality from the second expression in (3.162)

$$\left(\frac{\partial p}{\partial \rho}\right)_T = \gamma \left(\frac{\partial p}{\partial \rho}\right)_T - \frac{T}{c_v \rho^2}\left(\frac{\partial p}{\partial T}\right)_\rho^2 \tag{3.168}$$

The expression for the square of the speed of sound thus becomes

$$
\begin{aligned}
c^2 &= \left(\frac{\partial p}{\partial \rho}\right)_T + \left(\frac{\partial p}{\partial T}\right)_\rho \left(\frac{\partial T}{\partial \rho}\right)_\epsilon + \frac{p}{\rho^2}\left(\frac{\partial p}{\partial T}\right)_\rho \left(\frac{\partial T}{\partial \epsilon}\right)_\rho \\
&= \gamma\left(\frac{\partial p}{\partial \rho}\right)_T + \left(\frac{\partial p}{\partial T}\right)_\rho \left(\left(\frac{\partial T}{\partial \rho}\right)_\epsilon + \frac{p}{\rho^2}\left(\frac{\partial T}{\partial \epsilon}\right)_\rho - \frac{T}{c_v\rho^2}\left(\frac{\partial p}{\partial T}\right)_\rho\right) \\
&= \gamma RT + \gamma\rho T\left(\frac{\partial R}{\partial \rho}\right)_T + \left(\frac{\partial p}{\partial T}\right)_\rho \left(\left(\frac{\partial T}{\partial \rho}\right)_\epsilon + \frac{p}{c_v\rho^2} - \frac{T}{c_v\rho^2}\left(\frac{\partial p}{\partial T}\right)_\rho\right) \tag{3.169}
\end{aligned}
$$

where the EOS (3.13), with $R \equiv \mathcal{R}/M$, contributed the expression for the derivative of pressure with respect to ρ. In the presence of chemical reactions, therefore, the speed of sound differs from the "frozen" speed of sound $\sqrt{\gamma R T}$. In the absence of chemical reactions $R = $ constant and with $T = T(\epsilon)$ and (3.13) for EOS, the additional terms beyond $\gamma R T$ in (3.169) vanish and the speed of sound then coincides with $\sqrt{\gamma R T}$.

3.11   Computational Results

The procedure detailed in this chapter has generated the thermodynamic properties of equilibrium air using as independent thermodynamic variables both density and internal energy, suitable for CFD applications, and pressure and temperature, following the developments in Section 3.8. The results are generated for the pressure and density ranges that correspond to an increase of altitude in standard atmosphere of 30,000 m, slightly over 98,000 ft, above sea level. Table V summarizes the reference pressure and density

Table V. Reference Density and Pressure

altitude (km)	density (kg m^{-3})	pressure (Pa)
30	0.0184	1196.968507
24	0.0469	2971.756296
18	0.1216	7565.246266
12	0.3119	19399.475779
6	0.6601	47217.594506
0	1.2250	101325.024000

The symbols ρ_∞ and p_∞ in the following discussion denote sea-level density and pressure.

An electrically neutral five-species model for air remains accurate for temperatures up to about 9,000 K. The temperature range used for the results in this section, however, reaches over 10,000 K, in order to verify that at high temperatures the procedure yields results that correspond to a perfect gas of fully dissociated, hence atomic, species.

All variables in this section remain dimensional and are arranged in non-dimensional groupings. The results are presented in sets of isochors for pressure and sets of isobars for mass and mole fractions, mixture molecular mass, temperature, thermodynamic derivatives of pressure and temperature, constant-volume and constant-pressure specific heats, specific heat ratio, and sound speed squared.

Figures 3.1-3.2 present the distributions versus temperature of species mass and mole fractions. At $p = 1$ atm, the dissociation of molecular oxygen begins above 2,000 K and is virtually complete above 4,000 K; the dissociation of molecular nitrogen begins above 4,000 K, when that of oxygen completes, and is virtually complete over 9,000 K; nitric oxide begins to form at about 2,000 K; its mole fraction increases, reaches a peak at about 3,500 K and then decreases. These features and all the $p = 1$ atm curves virtually coincide with the results reported in [7]. This reference, also confirms the correctness of the shift of the mole fraction curves observed in the figure. A decrease in pressure favors dissociations; they can thus initiate at comparatively lower temperatures, which explains the shift to the left of the mass and mole fraction curves.

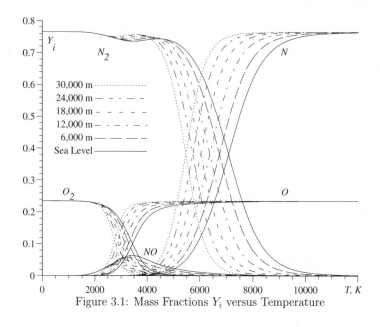

Figure 3.1: Mass Fractions Y_i versus Temperature

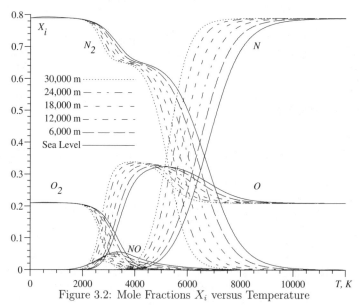

Figure 3.2: Mole Fractions X_i versus Temperature

These curves show a rapid decrease of N_2. This effect is due to the formation of NO, which requires nitrogen atoms. As dictated by conservation of mass, a relative minimum in N_2 mass fraction, for $T < 4,000$ K, corresponds to a maximum in NO mass fraction. The mass fraction of O increases monotonically, whereas the mole fraction decreases from a relative maximum for $T < 5,000$ K. Since a mole fraction is the ratio of species moles and mixture moles, this decrease of O mole fraction is not so much due to a decrease in the number of atoms and molecules of oxygen, but rather to a drastic increase of the total number of mixture moles due to the rapid dissociation of molecular nitrogen. The results reported in [163] then indicate that electrons and ionic species are virtually absent, for their mole fractions are less than 0.008 for $T < 9,000K$, and $p = 0.01$ atm; even smaller mole fractions of these charged species are present at higher pressure. This observation justifies the selection of an electrically neutral reacting-air model for this temperature range.

Figure 3.3 presents the variation of mixture molecular mass, which directly correlates with the variations of mass and mole fractions

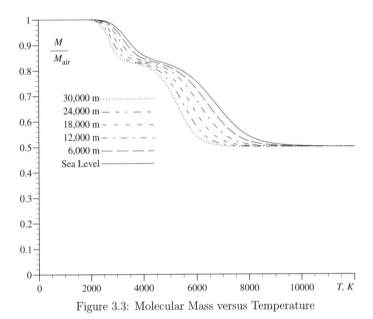

Figure 3.3: Molecular Mass versus Temperature

The decrease in this mass for rising temperature is expected because as temperature increases several atomic species join the existing molecular species, the number of moles, accordingly, increases, and hence less massive particles are present within each mixture mole. The variation of N_2 mole fraction reflects the variation of molecular mass for $T < 4,000$ K, for N_2 is the dominant species in this temperature range. As temperature increases, the chemical dissociations are complete and the ratio $M/M_{\rm air}$ approaches $1/2$, because air becomes a mixture of atomic oxygen and nitrogen, the molecular mass of which is half that of their molecular counterparts.

For the various density levels in Table V, Figures 3.4-3.5 show the variation of pressure versus internal energy and temperature.

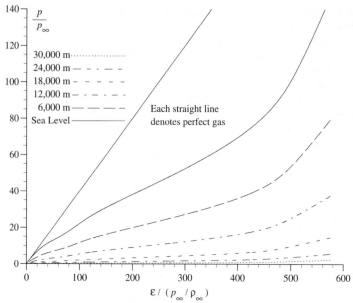

Figure 3.4: Pressure versus Internal Energy

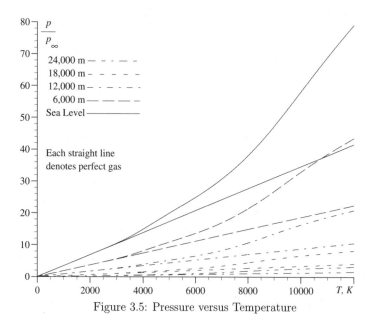

Figure 3.5: Pressure versus Temperature

Pressure is charted versus ϵ because from the first EOS in (3.27), pressure linearly increases with ϵ. This representation clearly shows the difference between perfect- and reacting-air predictions. For a given density, the chart illustrates the incorrect high pressure predictions of the perfect-gas model, which become wholly unrealistic as ϵ increases. This noticeable discrepancy between perfect- and reacting-air predictions originates in the significance of internal energy. In a perfect gas, the internal energy is due to rotation and translation kinetic energy, which determines species collision intensity, hence pressure. Within a reacting gas, on the other hand, internal energy also encompasses vibrational and formation energy, hence only a portion of ϵ contributes to the pressure-causing molecular kinetic energy, which corresponds to a reduction in both pressure and pressure increase in comparison to the perfect-gas case. As ϵ increases further, the curves tend to approach straight lines, because at completion of the chemical reactions, the equilibrium air begins to behave for a certain range of temperature as a perfect gas of atomic oxygen and nitrogen, corresponding to the first EOS in (3.27) with $\gamma = 5/3$.

At low temperature, when the mixture molecular mass decreases only modestly, see Figure 3.3, pressure is observed to increase linearly with temperature. This correct trend follows the EOS (3.13), which, with approximately constant M, leads to pressure remaining directly proportional to temperature. As temperature increases, the molecular mass decreases, as illustrated in Figure 3.3. As a consequence, pressure increases more rapidly versus temperature in the reacting than in the non-reacting case. Mechanically, more species, hence particles, that are less massive and faster contribute to particle collisions, which increases pressure in comparison to the non-reacting case. smooth.

Figures 3.6-3.7 present the variations of thermodynamic partial derivatives of pressure $\left(\frac{\partial p}{\partial \rho}\right)_\epsilon$ and $\left(\frac{\partial p}{\partial \epsilon}\right)_\rho$ versus temperature. The kinetic-theory interpretation of pressure as collisional variation of linear momentum explains these variations. Concerning the variation of the constant-density derivative $\left(\frac{\partial p}{\partial \epsilon}\right)_\rho$, as temperature increases the mixture species possess greater kinetic energy, hence this derivative is positive. An increase in temperature with constant pressure, however, leads to a decrease in density, hence fewer particles collide, pressure increases less rapidly and $\left(\frac{\partial p}{\partial \epsilon}\right)_\rho$ decreases as temperature increases. In the presence of chemical reactions, on the other hand, more particles collide hence this derivative is greater than in the perfect-gas case.

For a fixed ϵ, as temperature increases, an increase in mixture density corresponds to more species involved in collisions, which explains the increase in $\left(\frac{\partial p}{\partial \rho}\right)_\epsilon$ for both the non-reacting and reacting cases. This thermodynamic derivative varies linearly for low temperature, because for a perfect gas, hence from (3.27), this derivative equals ϵ, which from (3.26) increases linearly with temperature. As temperature increases, the reacting-air values of this derivative are greater than those for a perfect-air, because the chemical reactions lead to more colliding species than in the perfect-air case. For both derivatives, a dissociation-favoring decrease in pressure then further reduces the increase in molecular kinetic energy and increases the number of colliding species, which explains the observed respective variations of these derivatives with respect to pressure. As temperature increases further, the chemical dissociations are complete, the mixture becomes a perfect gas of atomic species and the isochors converge to a single straight line. This line corresponds to the derivative of the first expression in (3.27), but with $\gamma = 5/3$, rather than the smaller $\gamma = 7/5$ which explains the

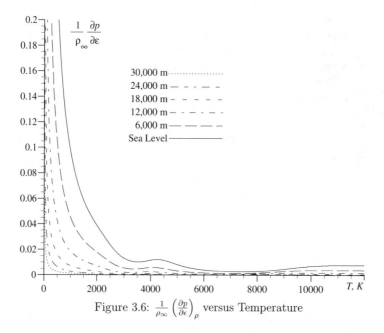

Figure 3.6: $\frac{1}{\rho_\infty}\left(\frac{\partial p}{\partial \epsilon}\right)_\rho$ versus Temperature

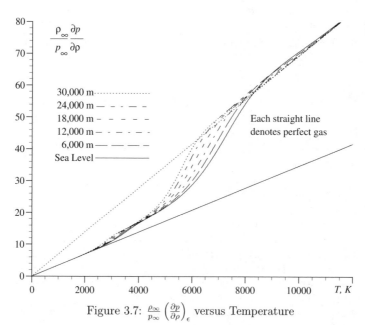

Figure 3.7: $\frac{\rho_\infty}{p_\infty}\left(\frac{\partial p}{\partial \rho}\right)_\epsilon$ versus Temperature

steeper slope of the atomic-air line versus that for molecular air. Observe that for both these thermodynamic derivatives, the procedure has generated continuous and smooth distributions.

Figure 3.8 shows the variation of temperature versus non-dimensional specific volume $1/(\rho/\rho_\infty)$.

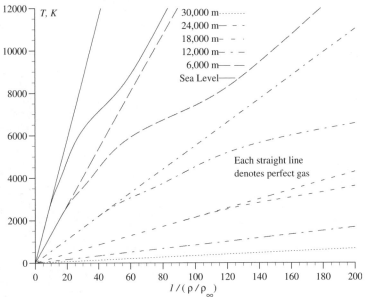

Figure 3.8: Temperature versus Specific Volume

Temperature is plotted versus specific volume $1/(\rho/\rho_\infty)$ because from the second EOS in (3.27), temperature increases linearly with $1/(\rho/\rho_\infty)$ for a perfect gas. This representation clearly shows the difference between perfect- and reacting-air temperature predictions. For a given density, hence specific volume, the chart illustrates the well known incorrect high temperature predictions of the perfect-gas model, which become totally unrealistic as $1/(\rho/\rho_\infty)$ increases. This discrepancy results from the mechanism of specific volume increase. In the non- reacting case, the specific volume increases in direct proportion to a rise in molecular kinetic energy, hence a rise in temperature. In the reacting case, for a given temperature, specific volume rapidly increases because the atomic species that emerge from the chemical dissociations occupy larger spaces when they translate freely than when they remain bound within molecules. As temperature increases further, the curves tend to approach straight lines, because at completion of the chemical reactions involved, the equilibrium air begins to behave for a certain range of temperature as perfect gas of atomic oxygen and nitrogen. Overall, the temperature distributions in the figure remain continuous and smooth.

Another clear indication of the significant differences between perfect- and reacting-air behavior is provided by Figure 3.9, which correlates with Figure 3.8 and presents the variation of temperature versus internal energy ϵ. For low energies, no chemical reactions occur nor are the vibrational modes fully excited. Consequently, temperature remains independent of pressure and increases linearly with internal energy in this range. This correct trend follows the perfect-gas TE in (3.26), which shows both independence from pressure and a constant specific heat for the constant slope in the curve. Temperature is a measure of the molecular kinetic-energy mode. Hence, further increases in energy, accompanied by chemical reactions, reduce the rise in temperature, in comparison to the perfect-air case, because only part of the internal energy raises molecular kinetic energy, while the rest triggers chemical reactions, and the larger ϵ the greater the discrepancy between perfect-air and reacting-air predictions. As the graphs show, the rise in temperature is further retarded by a decrease in pressure, which increases the rate of dissociations and thus further reduces the increase in molecular kinetic energy. For continuing increase of internal-energy, the curves then indicate a diminishing dependence on pressure and concurrent convergence towards a single straight line. This convergence results from completion of the chemical reactions, hence the equilibrium-air behaves for a certain range of energies as a perfect-gas of atomic oxygen and nitrogen with pressure-independent constant specific heat, hence the constant slope in the corresponding curve.

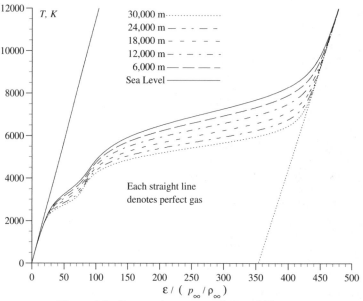

Figure 3.9: Temperature versus Internal Energy

Figures 3.10-3.11 present the thermodynamic partial derivatives of temperature $\left(\frac{\partial T}{\partial \rho}\right)_{\epsilon}$ and $\left(\frac{\partial T}{\partial \epsilon}\right)_{\rho}$ versus temperature. For low temperatures, $\left(\frac{\partial T}{\partial \rho}\right)_{\epsilon}$ has to vanish, for temperature remains constant for fixed ϵ, according to the perfect-air TE in (3.26).

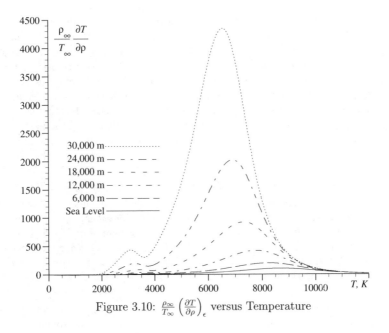

Figure 3.10: $\frac{\rho_\infty}{T_\infty}\left(\frac{\partial T}{\partial \rho}\right)_\epsilon$ versus Temperature

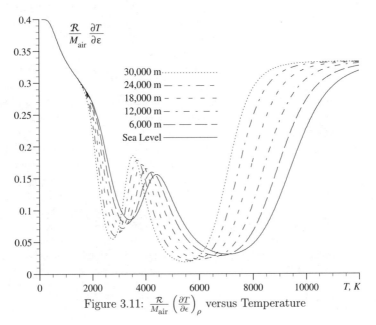

Figure 3.11: $\frac{\mathcal{R}}{M_{\mathrm{air}}}\left(\frac{\partial T}{\partial \epsilon}\right)_\rho$ versus Temperature

As temperature increases, the presence of chemical dissociations induce a strong dependence of T and $\left(\frac{\partial T}{\partial \rho}\right)_\epsilon$ on ρ because an increase in mixture density provides for more atoms in the mixture that become free from molecular bonds and travel faster than molecules because they are less massive. The number and intensity of particle collisions, therefore, increases, impact linear momentum is exchanged more rapidly, and kinetic energy and temperature increase more swiftly than in the non-reacting case. This effect is accentuated by a dissociation-favoring pressure decrease. As temperature increases further, however, the dependence of $\left(\frac{\partial T}{\partial \rho}\right)_\epsilon$ on T rapidly decreases because at completion of the chemical reactions, no additional atomic particles emerge within the mixture and internal energy no longer depends on density, but only on temperature. For constant ϵ, it thus follows that $\left(\frac{\partial T}{\partial \rho}\right)_\epsilon \simeq 0$, in these conditions.

The variation of $\left(\frac{\partial T}{\partial \epsilon}\right)_\rho$ versus T directly correlates with the variation of T versus ϵ and mole fractions versus T. Even before chemical dissociations begin, this derivative is seen to decrease, as determined by the non-linear increase of the vibrational energy, which equals kinetic energy at equilibrium. Therefore, temperature, a measure of one mode of molecular energy among the translational, rotational and vibrational modes, increases less rapidly than ϵ, the sum of all the molecular energy modes. As the chemical reactions progress, they require increasing amounts of energy that will not be present as kinetic energy, hence $\left(\frac{\partial T}{\partial \epsilon}\right)_\rho$ decreases further.

When the peak in nitric oxide mass fraction is reached, an increase in internal energy contributes to a proportional increase in molecular kinetic energy, which explains the increase in $\left(\frac{\partial T}{\partial \epsilon}\right)_\rho$ until a state where a further increase in temperature initiates the dissociation of molecular nitrogen. This dissociation will then require increasing amounts of energy that will not be present as kinetic energy, hence $\left(\frac{\partial T}{\partial \epsilon}\right)_\rho$ begins to decrease again. As the chemical dissociations near completion, an increase in ϵ induces a proportional increase in molecular kinetic energy, which explains the renewed increase in $\left(\frac{\partial T}{\partial \epsilon}\right)_\rho$. A thermodynamic state is then reached where the reactions are complete, the mixture begins to behave as a perfect air, within an appropriate range of T, and thus $\left(\frac{\partial T}{\partial \epsilon}\right)_\rho$ becomes equal to a constant, corresponding to the inverse of a constant specific-heat. It is important to observe that for both derivatives, the procedure generated smooth results.

Figures 3.12-3.13 present the variations of the specific heats c_v and c_p versus temperature. These variations directly correlate with the variation of $\left(\frac{\partial T}{\partial \epsilon}\right)_\rho$ and with each other, following (3.162). At low temperature, the mixture behaves as a perfect gas of molecular oxygen and nitrogen. From (3.48)-(3.49) and (3.164), therefore, $c_v/(\mathcal{R}/M_{\text{air}})$ and $c_p/(\mathcal{R}/M_{\text{air}})$ approach their respective perfect-gas magnitudes of $5/2$ and $7/2$. Before any chemical reactions, pressure does not influence c_v and c_p, but as temperature increases, c_v and c_p rise above corresponding perfect-gas magnitudes because the internal energy experiences an increase in both kinetic energy, contributing perfect-gas magnitudes for these specific heats, and vibrational energy. With the onset of chemical reactions, the rate of change of Y_3 initially increases with temperature, while internal energy and enthalpy rapidly increase through NO formation enthalpy, which leads to the observed rapid rise in c_v and c_p.

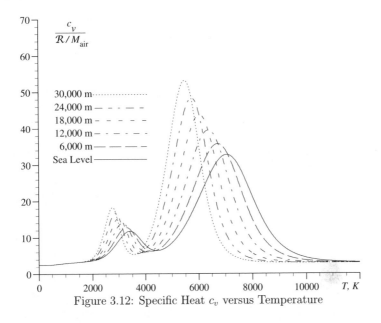

Figure 3.12: Specific Heat c_v versus Temperature

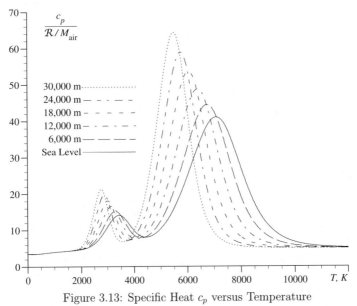

Figure 3.13: Specific Heat c_p versus Temperature

When the NO mass fraction decreases with temperature, so does the corresponding formation-enthalpy contribution to internal energy and enthalpy. Consequently both c_v and

c_p decrease, but remain greater than perfect gas levels owing to vibrational energy modes and the increasing contribution of atomic-oxygen formation enthalpy. As temperature increases further, the mass and mole fraction of N rapidly rise, hence the internal energy significantly increases due to the increasing contribution from the formation enthalpy of N. A decrease in pressure favors the mixture chemical reactions and it thereby accentuates the effects of NO and N formation-enthalpy contributions to internal energy, which intensifies the rises in c_v and c_p. For $T \leq 9,000$ K, hence in the virtual absence of electrons and ionic species, the distribution of $c_v/(\mathcal{R}/M_{\mathrm{air}})$ agrees with the one reported in [7]. For higher temperatures, the mass fraction of N increases less rapidly until it reaches a constant plateau. Accordingly, c_v and c_p begin to decrease and a further temperature rise merely increases atomic kinetic energy. The mixture thus behaves as a non-reacting gas that consists of atomic oxygen and nitrogen. The atomic mass of this mixture equals half of M_{air}, and from (3.48)-(3.49) and (3.164), $c_v/(\mathcal{R}/M_{\mathrm{air}})$ and $c_p/(\mathcal{R}/M_{\mathrm{air}})$ approach their respective perfect-gas magnitudes of 3 and 5.

Figures 3.14-3.15 show the variation of specific heat ratio γ and non-dimensional sound speed squared $c^2/(p/\rho)$.

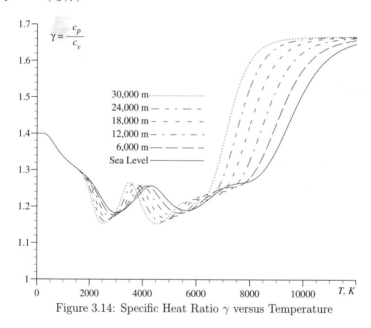

Figure 3.14: Specific Heat Ratio γ versus Temperature

Without any chemical reactions or vibrational-energy effects, both γ and $c^2/(p/\rho)$ equal the molecular perfect-gas magnitude of $7/5$. From the first expression in (3.163), γ equals the product of the derivative of enthalpy with respect to temperature and the derivative of temperature with respect to internal energy. The variation of γ, therefore, directly correlates with the variation of $\left(\frac{\partial T}{\partial \epsilon}\right)_\rho$, portrayed in Figure 3.11.

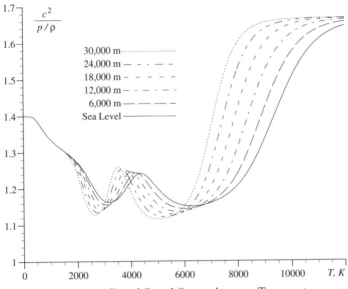

Figure 3.15: Sound Speed Squared versus Temperature

In the absence of chemical reactions, all terms beyond $\gamma R T$ vanish in expression (3.169) for the square of the speed of sound. In these conditions, therefore, γ and $c^2/(p/\rho) = c^2/(RT)$ equal each other, as Figures 3.14-3.15 indicate. When the mixture chemical reactions take place, expression (3.169) shows that $c^2/(p/\rho)$ no longer coincides with γ, as the charts indicate. For $T \leq 9{,}000$ K, hence in the virtual absence of electrons and ionic species, the distribution of $c^2/(p/\rho)$ agrees with the one reported in [7]. When the reactions are completed, γ and $c^2/(p/\rho)$ coincide with each other again and approach the perfect-gas magnitude of $5/3$, following expressions from (3.48)-(3.49) and (3.164).

3.12   Performance Summary

The procedure detailed in this chapter directly generates pressure, and temperature, as well as their thermodynamic and Jacobian partial derivatives, mass and mole fractions, molecular mass, specific heats and speed of sound for a five-species neutral equilibrium air. This procedure algebraically reduces the six-equation chemical- equilibrium thermodynamic system and explicitly expresses four variables in terms of nitric-oxide mass fraction and temperature. These two variables are then numerically determined by solving the internal-energy and nitric-oxide mass-action equations through a Newton's method solution, which is observed to converge rapidly in two to three iterations. The procedure then exactly determines the partial derivatives of pressure, temperature and mass fractions analytically.

All of the computational results over a temperature range of more than 10,000 K and pressure range corresponding to an increase in altitude of 30,000 m, about 98,000 ft, above sea level, remain physically meaningful. Independently published results are available for mole fractions, constant-volume specific-heat and speed of sound and the distributions of these variables predicted by the procedure agree with these independent results. The thermodynamic properties so generated, including the critical thermodynamic partial derivatives of pressure and temperature are then observed to remain continuous and smooth. These results, accordingly, support the procedure as an attractive alternative both to generate the thermodynamic properties of electrically neutral chemically reacting equilibrium air and to incorporate these properties directly and efficiently within Euler and Navier-Stokes CFD algorithms.

Chapter 4

Euler and Navier Stokes Systems

This chapter specializes the fundamental governing equations for the case of compressible and incompressible flows of Newtonian fluids and then establishes corresponding non-dimensional forms. This chapter also clarifies the subtle yet important difference between the incompressible- and compressible-flow temperature equations.

The governing equations in differential form

$$
\begin{cases}
\dfrac{\partial \rho}{\partial t} + \dfrac{\partial \rho u_j}{\partial x_j} = 0 \\[2mm]
\dfrac{\partial \rho u_i}{\partial t} + \dfrac{\partial (\rho u_i u_j + p\delta_j^i)}{\partial x_j} - \dfrac{\partial \sigma_{ij}^\mu}{\partial x_j} - \rho b_i = 0 \\[2mm]
\dfrac{\partial \rho e}{\partial t} + \dfrac{\partial (u_j(\rho e + p))}{\partial x_j} - \dfrac{\partial u_i \sigma_{ij}^\mu}{\partial x_j} + \dfrac{\partial q_j^F}{\partial x_j} - \rho u_i b_i - \rho \dot{g} = 0
\end{cases}
\tag{4.1}
$$

show the separate effects of pressure and deviatoric stress. This non-linear time dependent partial differential system in the independent variables (t, x_1, x_2, x_3) and dependent variables $(\rho, \rho u_1, \rho u_2, \rho u_3, \rho e)$ provides 5 equations for three-dimensional flows and 4 equations for two-dimensional flows.

The Euler equations derive from this system when the contributions from deviatoric stress σ_{ij}^μ and heat conduction q_i^F components can be neglected. The corresponding flow is an "inviscid" flow. In this case, the only additional equation to close the system is an equation of state that relates pressure p to density ρ and energy e.

The Navier Stokes equations, instead, emerge from this system upon specification of constitutive relations for σ_{ij}^μ and q_i^F in addition to an equation of state for pressure, as all presented in Chapters 2-3.

System (4.1) presents the governing equations in the so-called conservation form. In the absence of the source term for body forces and heat generation, this system is abridged as

$$
\frac{\partial q}{\partial t} + \frac{\partial F_j}{\partial x_j} = 0, \quad 1 \le j \le 3
\tag{4.2}
$$

In integral form, system (4.2) is expressed as

$$
\int_\Omega \left(\frac{\partial q}{\partial t} + \frac{\partial F_j}{\partial x_j} \right) d\Omega = 0
\tag{4.3}
$$

This equation can be transformed through Gauss' theorem, which for a steady flow leads to

$$\oint_{\partial\Omega} F_j n_j \, d\Gamma = 0 \qquad (4.4)$$

This result stipulates the absence of any net "flow" of the quantity $\boldsymbol{F}$ across the boundary $\partial\Omega$ of Ω, which corresponds to conservation of this quantity. In CFD, the conservation form of the equations leads to accurate calculations of shock-wave position and strength. The arrays q and F_j, $1 \leq j \leq 3$, are defined as

$$q \equiv \left\{ \begin{array}{c} \rho \\ \rho u_1 \\ \rho u_2 \\ \rho u_3 \\ \rho e \end{array} \right\}, \quad F_1 \equiv \left\{ \begin{array}{c} \rho u_1 \\ \rho u_1 u_1 + p - \sigma_{11}^{\mu} \\ \rho u_2 u_1 - \sigma_{21}^{\mu} \\ \rho u_3 u_1 - \sigma_{31}^{\mu} \\ u_1 (\rho e + p) - u_i \sigma_{i1}^{\mu} + q_1^F \end{array} \right\}$$

$$F_2 \equiv \left\{ \begin{array}{c} \rho u_2 \\ \rho u_1 u_2 - \sigma_{12}^{\mu} \\ \rho u_2 u_2 + p - \sigma_{22}^{\mu} \\ \rho u_3 u_2 - \sigma_{32}^{\mu} \\ u_2 (\rho e + p) - u_i \sigma_{i2}^{\mu} + q_2^F \end{array} \right\}, \quad F_3 \equiv \left\{ \begin{array}{c} \rho u_3 \\ \rho u_1 u_3 - \sigma_{13}^{\mu} \\ \rho u_2 u_3 - \sigma_{23}^{\mu} \\ \rho u_3 u_3 + p - \sigma_{33}^{\mu} \\ u_3 (\rho e + p) - u_i \sigma_{i3}^{\mu} + q_3^F \end{array} \right\} \qquad (4.5)$$

The following chapters amplify on this conservation concept for specific flows.

The so-called "non-conservation form" of the governing equations follow from subtracting the continuity equation from the momentum and total energy equations. With this transformation, system (4.1) becomes

$$\left\{ \begin{array}{l} \dfrac{\partial \rho}{\partial t} + u_j \dfrac{\partial \rho}{\partial x_j} + \rho \dfrac{\partial u_j}{\partial x_j} = 0 \\[2ex] \rho \dfrac{\partial u_i}{\partial t} + \rho u_j \dfrac{\partial u_i}{\partial x_j} + \dfrac{\partial p}{\partial x_i} - \dfrac{\partial \sigma_{ij}^{\mu}}{\partial x_j} - \rho b_i = 0 \\[2ex] \rho \dfrac{\partial e}{\partial t} + \rho u_j \dfrac{\partial e}{\partial x_j} + \dfrac{\partial u_j p}{\partial x_j} - \dfrac{\partial u_i \sigma_{ij}^{\mu}}{\partial x_j} + \dfrac{\partial q_j^F}{\partial x_j} - \rho u_i b_i - \rho \dot{g} = 0 \end{array} \right. \qquad (4.6)$$

These equations are said to be in non-conservation form only and simply because they do not directly lead to (4.4). It goes without saying, they still express conservation of mass and energy and the second law of mechanics.

4.1 Compressible Flows

Expansion, compression, and shock waves can all emerge within compressible flows. The conservation form of the equations will lead to accurate calculations of these flows, which is the form of the equations used for a CFD investigation of compressible flows.

4.1.1 Conservation Form

The time derivative terms in the momentum and energy equations in (4.1) contain products of variables; these variables are replaced by the volume-specific linear momentum and total energy in order to generate a single variable within each of these partial derivatives, which in turn will permit more convenient calculations. The volume-specific linear momentum components and total energy are defined as

$$m_i \equiv \rho u_i, \quad E \equiv \rho e \tag{4.7}$$

In particular, the velocity components become derived variables as linear momentum components per unit density. A certain balance emerges in the governing equations with the introduction of these variables because density itself is a volume specific quantity, mass in this case.

In terms of volume specific mass, linear momentum components and total energy, the Navier-Stokes system thus becomes

$$
\begin{cases}
\dfrac{\partial \rho}{\partial t} + \dfrac{\partial m_j}{\partial x_j} = 0 \\[2mm]
\dfrac{\partial m_i}{\partial t} + \dfrac{\partial}{\partial x_j}\left(\dfrac{m_j}{\rho}m_i + p\delta_j^i - \sigma_{ij}^\mu\right) - \rho b_i = 0 \\[2mm]
\dfrac{\partial E}{\partial t} + \dfrac{\partial}{\partial x_j}\left(\dfrac{m_j}{\rho}(E+p) - \dfrac{m_i}{\rho}\sigma_{ij}^\mu + q_j^F\right) - m_i b_i - \rho \dot{g} = 0
\end{cases}
\tag{4.8}
$$

and the Euler system obtains by neglecting the deviatoric stress components σ_{ij}^μ and heat conduction term q_j^F.

From (2.6) and (2.7) the constitutive equations for stress tensor and heat conduction terms are expressed as

$$\sigma_{ij}^\mu = \mu\left(\frac{\partial m_i/\rho}{\partial x_j} + \frac{\partial m_j/\rho}{\partial x_i} - \frac{2}{3}\frac{\partial m_\ell/\rho}{\partial x_\ell}\delta_j^i\right) + \eta_B\frac{\partial m_\ell/\rho}{\partial x_\ell}, \quad q_j^F = -k\frac{\partial T}{\partial x_j} \tag{4.9}$$

Stokes' hypothesis is practically always used and consequently the bulk viscosity η_B is set to zero. The pressure and temperature state equations in terms of volume - specific variables are expressed as

$$p = p(\rho,\epsilon) = p\left(\rho, \frac{1}{\rho}\left(E - \frac{1}{2\rho}m_i m_i\right)\right), \quad T = T(\rho,\epsilon) = T\left(\rho, \frac{1}{\rho}\left(E - \frac{1}{2\rho}m_i m_i\right)\right) \tag{4.10}$$

as shown in Chapters 2-3.

In abridged notation, the Navier-Stokes system becomes

$$\frac{\partial q}{\partial t} + \frac{\partial f_j}{\partial x_j} - \frac{\partial f_j^v}{\partial x_j} - \phi = 0 \tag{4.11}$$

where the arrays q, f_j, f_j^v, $1 \leq j \leq 3$, and ϕ are defined as

$$q \equiv \left\{ \begin{array}{c} \rho \\ m_1 \\ m_2 \\ m_3 \\ E \end{array} \right\}, \quad f_j \equiv \left\{ \begin{array}{c} m_j \\ \dfrac{m_j}{\rho} m_1 + p\delta_j^1 \\ \dfrac{m_j}{\rho} m_2 + p\delta_j^2 \\ \dfrac{m_j}{\rho} m_3 + p\delta_j^3 \\ \dfrac{m_j}{\rho}(E + p) \end{array} \right\}, \quad f_j^v \equiv \left\{ \begin{array}{c} 0 \\ \sigma_{1j}^\mu \\ \sigma_{2j}^\mu \\ \sigma_{3j}^\mu \\ \dfrac{m_i}{\rho}\sigma_{ij}^\mu + q_j^F \end{array} \right\}, \quad \phi \equiv \left\{ \begin{array}{c} 0 \\ \rho f_1 \\ \rho f_2 \\ \rho f_3 \\ \dfrac{m_i}{\rho} m_i b_i + \rho \dot{g} \end{array} \right\}$$

$$\tag{4.12}$$

The so-called inviscid flux array with components f_j depends on the state array q only, hence $f_j = f_j(q)$. In many gas dynamics applications, the magnitude of the source term ϕ remains negligible in comparison to the other terms in the governing equations and is thus stricken out of the equations.

4.1.2 Non-Dimensional System

Computationally convenient non-dimensional forms of the governing equations emerge from a set of reference variables for the Cartesian coordinates x_j, $1 \leq j \leq n$, with n the number of flow dimensions, and the dependent variables of density ρ and volume-specific linear momentum components m_j, $1 \leq j \leq n$, and total energy E. The reference time rests on a reference length and a reference velocity, which in turn depends on the reference pressure and density. With these specifications, the non-dimensional Euler equations, retain the same form as their dimensional counterparts, whereas the non-dimensional Navier-Stokes equations display the Reynolds, Prandtl, and Eckert numbers.

Consider the following reference length, density, pressure, and temperature

$$L, \quad \rho_{\mathrm{re}}, \quad p_{\mathrm{re}}, \quad T_{\mathrm{re}} \tag{4.13}$$

Depending on the flow investigated, these parameters can correspond to a dominant geometric dimension as well as free-stream or stagnation conditions. On the basis of their respective dimensions, the corresponding reference parameters for speed, time, and volume-specific total energy obtain as

$$V_{\mathrm{re}}^2 = \frac{p_{\mathrm{re}}}{\rho_{\mathrm{re}}}, \quad t_{\mathrm{re}} = \frac{L}{V_{\mathrm{re}}}, \quad E_{\mathrm{re}} = p_{\mathrm{re}}, \tag{4.14}$$

With a tilde "~" denoting a non-dimensional variable, corresponding dimensional and dimensionless variables depend upon one another through (4.13)-(4.14) as follows

$$t = t_{\mathrm{re}}\tilde{t}, \quad x_j = L\tilde{x}_j$$

$$\rho = \rho_{\mathrm{re}}\widetilde{\rho} \quad , \quad m_j = \rho_{\mathrm{re}}V_{\mathrm{re}}\widetilde{m}_j, \qquad E = E_{\mathrm{re}}\widetilde{E} \tag{4.15}$$

These specifications lead to the following form of the compressible Navier-Stokes equations

$$\begin{cases} \dfrac{\rho_{\mathrm{re}}V_{\mathrm{re}}}{L}\dfrac{\partial \widetilde{\rho}}{\partial \widetilde{t}} + \dfrac{\rho_{\mathrm{re}}V_{\mathrm{re}}}{L}\dfrac{\partial \widetilde{m}_j}{\partial \widetilde{x}_j} = 0 \\[2mm] \dfrac{\rho_{\mathrm{re}}V_{\mathrm{re}}^2}{L}\dfrac{\partial \widetilde{m}_i}{\partial \widetilde{t}} + \dfrac{\rho_{\mathrm{re}}V_{\mathrm{re}}^2}{L}\dfrac{\partial}{\partial \widetilde{x}_j}\left(\dfrac{\widetilde{m}_j}{\widetilde{\rho}}\widetilde{m}_i\right) + \dfrac{p_{\mathrm{re}}}{L}\dfrac{\partial \widetilde{p}\delta_j^i}{\partial \widetilde{x}_j} - \dfrac{\mu V_{\mathrm{re}}}{L^2}\dfrac{\partial \widetilde{\sigma}_{ij}^{\mu}}{\partial \widetilde{x}_j} = 0 \\[2mm] \dfrac{\rho_{\mathrm{re}}V_{\mathrm{re}}}{L}\dfrac{\partial \widetilde{E}}{\partial \widetilde{t}} + \dfrac{\rho_{\mathrm{re}}V_{\mathrm{re}}}{L}\dfrac{\partial}{\partial \widetilde{x}_j}\left(\dfrac{\widetilde{m}_j}{\widetilde{\rho}}(\widetilde{E}+\widetilde{p})\right) - \dfrac{\mu V_{\mathrm{re}}^2}{L^2}\dfrac{\partial}{\partial \widetilde{x}_j}\left(\dfrac{\widetilde{m}_i}{\widetilde{\rho}}\widetilde{\sigma}_{ij}^{\mu}\right) - \dfrac{k_0 T_{\mathrm{re}}}{L^2}\dfrac{\partial}{\partial \widetilde{x}_j}\left(\dfrac{k}{k_0}\dfrac{\partial \widetilde{T}}{\partial \widetilde{x}_j}\right) = 0 \end{cases} \tag{4.16}$$

The following non-dimensional groups of reference parameters are rearranged as

$$\frac{L}{\rho_{\mathrm{re}}V_{\mathrm{re}}^2}\frac{\mu V_{\mathrm{re}}}{L^2} = \frac{\mu}{\rho_{\mathrm{re}}V_{\mathrm{re}}L} = \frac{1}{Re}, \quad \frac{L}{p_{\mathrm{re}}V_{\mathrm{re}}}\frac{\mu V_{\mathrm{re}}^2}{L^2} = \frac{\mu}{\rho_{\mathrm{re}}V_{\mathrm{re}}L} = \frac{1}{Re}$$

The reference parameters multiplying the second-order derivatives of temperature in (4.16) become

$$\frac{L}{p_{\mathrm{re}}V_{\mathrm{re}}}\frac{k_0 T_{\mathrm{re}}}{L^2} = \frac{k_0 T_{\mathrm{re}}}{L}\frac{1}{V_{\mathrm{re}}^2\rho_{\mathrm{re}}V_{\mathrm{re}}} = \frac{1}{\rho_{\mathrm{re}}V_{\mathrm{re}}L}\frac{k_0 T_{\mathrm{re}}}{V_{\mathrm{re}}^2} = \frac{\mu}{\rho_{\mathrm{re}}V_{\mathrm{re}}L}\frac{k_0}{\mu c_p}\frac{c_p T_{\mathrm{re}}}{V_{\mathrm{re}}^2} = \frac{1}{Ec\,Pr\,Re}$$

where "Ec", "Pr", and "Re" respectively denote the Eckert, Prandtl, and Reynolds numbers. With these numbers, the non-dimensional compressible Navier-Stokes equations become

$$\begin{cases} \dfrac{\partial \widetilde{\rho}}{\partial \widetilde{t}} + \dfrac{\partial \widetilde{m}_j}{\partial \widetilde{x}_j} = 0 \\[2mm] \dfrac{\partial \widetilde{m}_i}{\partial \widetilde{t}} + \dfrac{\partial}{\partial \widetilde{x}_j}\left(\dfrac{\widetilde{m}_j}{\widetilde{\rho}}\widetilde{m}_i\right) + \dfrac{\partial \widetilde{p}\delta_j^i}{\partial \widetilde{x}_j} - \dfrac{1}{Re}\dfrac{\partial \widetilde{\sigma}_{ij}^{\mu}}{\partial \widetilde{x}_j} = 0 \\[2mm] \dfrac{\partial \widetilde{E}}{\partial \widetilde{t}} + \dfrac{\partial}{\partial \widetilde{x}_j}\left(\dfrac{\widetilde{m}_j}{\widetilde{\rho}}(\widetilde{E}+\widetilde{p})\right) - \dfrac{1}{Re}\dfrac{\partial}{\partial \widetilde{x}_j}\left(\dfrac{\widetilde{m}_i}{\widetilde{\rho}}\widetilde{\sigma}_{ij}^{\mu}\right) - \dfrac{1}{Ec\,Pr\,Re}\dfrac{\partial}{\partial \widetilde{x}_j}\left(\dfrac{k}{k_0}\dfrac{\partial \widetilde{T}}{\partial \widetilde{x}_j}\right) = 0 \end{cases} \tag{4.17}$$

With a reference speed expressed in terms reference of pressure and density, the local Mach number "M" for a perfect gas becomes

$$M \equiv \frac{V}{\sqrt{\gamma p/\rho}} = \frac{V/V_{\mathrm{re}}}{(\sqrt{\gamma p/\rho})/V_{\mathrm{re}}} = \frac{V/V_{\mathrm{re}}}{(\sqrt{\gamma p/\rho})/\sqrt{p_{\mathrm{re}}/\rho_{\mathrm{re}}}} = \frac{\widetilde{V}}{\sqrt{\gamma \widetilde{p}/\widetilde{\rho}}} \tag{4.18}$$

which directly depends on non-dimensional variables. Since the reference volume-specific total energy coincides with the reference pressure, the non-dimensional perfect-gas equation of state becomes

$$\frac{p}{p_{\mathrm{re}}} = (\gamma - 1)\left(\frac{E}{p_{\mathrm{re}}} - \frac{\rho V^2}{2p_{\mathrm{re}}}\right) = (\gamma - 1)\left(\frac{E}{p_{\mathrm{re}}} - \frac{\gamma}{2}\frac{p}{p_{\mathrm{re}}}\frac{V^2}{\gamma p/\rho}\right) =$$

$$= (\gamma - 1)\left(\frac{E}{p_{\mathrm{re}}} - \frac{\gamma}{2}\frac{p}{p_{\mathrm{re}}}M^2\right) \quad \Rightarrow \quad \frac{E}{p_{\mathrm{re}}} = \frac{p}{p_{\mathrm{re}}}\left(\frac{1}{\gamma - 1} + \frac{\gamma}{2}M^2\right) \tag{4.19}$$

where "$\gamma = c_p/c_v$" denotes the specific-heat ratio. Since the specific heats and gas constant R are related by Mayer's equality $c_p - c_v = R$, the non-dimensional ratio c_p/R depends on γ only as $c_p/R = \gamma/(\gamma - 1)$. Like the reference speed V_{re}, the reference temperature T_{re} too can depend upon the reference density and pressure. In this case:

$$T_{\mathrm{re}} = \frac{p_{\mathrm{re}}}{\rho_{\mathrm{re}} R} \quad \Rightarrow \quad Ec = \frac{V_{\mathrm{re}}^2}{c_p T_{\mathrm{re}}} = \frac{V_{\mathrm{re}}^2 R}{c_p p_{\mathrm{re}}/\rho_{\mathrm{re}}} = \frac{R}{c_p} = \frac{\gamma - 1}{\gamma} \tag{4.20}$$

4.1.3   Variational Formulation

In the variational weighted formulation introduced in Chapter 7, the Navier-Stokes equations are conveniently expressed as

$$\int_\Omega w \frac{\partial \rho}{\partial t} d\Omega - \int_\Omega \frac{\partial w}{\partial x_j} m_j d\Omega + \oint_{\partial \Omega} w m_j n_j d\Gamma = 0 \tag{4.21}$$

$$\int_\Omega w \frac{\partial m_i}{\partial t} d\Omega - \int_\Omega \frac{\partial w}{\partial x_j} \left(\frac{m_j}{\rho} m_i + p\delta_j^i - \frac{\sigma_{ij}^\mu}{Re} \right) d\Omega +$$

$$+ \oint_{\partial \Omega} w \frac{m_i}{\rho} m_j n_j d\Gamma + \oint_{\partial \Omega} w \left(p\delta_j^i - \frac{\sigma_{ij}^\mu}{Re} \right) n_j d\Gamma = 0 \tag{4.22}$$

$$\int_\Omega w \frac{\partial E}{\partial t} d\Omega - \int_\Omega \frac{\partial w}{\partial x_j} \left(\frac{m_j}{\rho} (E + p) - \frac{m_i \sigma_{ij}^\mu}{\rho \, Re} - \frac{k/k_o}{EcPrRe} \frac{\partial T}{\partial x_j} \right) d\Omega +$$

$$+ \oint_{\partial \Omega} w E m_j n_j d\Gamma + \oint_{\partial \Omega} w \frac{m_i}{\rho} \left(p\delta_j^i - \frac{\sigma_{ij}^\mu}{Re} \right) n_j d\Gamma - \oint_{\partial \Omega} w \frac{k/k_o}{EcPrRe} \frac{\partial T}{\partial x_j} n_j d\Gamma = 0 \tag{4.23}$$

where the tilde has been dropped for simplicity, even though all variables in these statements remain non dimensional. The surface integrals in these results arise from a transformation of the corresponding domain integrals through Gauss' theorem. These surface integrals provide quite a convenient setting to enforce physically meaningful boundary conditions on surface linear momentum flux $m_j n_j$, traction components $(p\delta_j^i - \sigma_{ij}^\mu/Re)n_j$ and heat flux $\partial T/\partial n$. The boundary conditions that must be enforced on these quantities can thus be directly inserted into the corresponding surface integrals in these variational equations. Additional boundary conditions, of course, constrain q on appropriate facets of the total boundary, as detailed in Chapters 13, 17.

4.2   Low Speed and Incompressible Flows

The flow of gases too can be characterized as incompressible when the flow speed does not exceed about 30% of the speed of sound. The difference between an incompressible fluid and an incompressible flow is the following. In an incompressible fluid, like a liquid, e.g. water, the density practically remains constant irrespective of the state of motion. An incompressible flow involves the motion of either a liquid or a gas within which the variations of density are negligible in comparison to gradients of velocity in the continuity and momentum equations. In this sense it is possible to speak of incompressible flows of highly compressible fluids like gases.

4.2.1 Continuity Equation

The continuity equation in system (4.1) can be expressed as

$$\frac{1}{\rho}\frac{D\rho}{Dt} + \frac{\partial u_j}{\partial x_j} = 0 \tag{4.24}$$

The case of each fluid particle maintaining practically invariant density implies

$$\frac{D\rho}{Dt} \simeq 0 \quad \Rightarrow \quad \frac{\partial u_j}{\partial x_j} = 0 \tag{4.25}$$

As a result, if the fluid particle density does not change, the divergence of velocity vanishes, which, with reference to Section 1.9, signifies that the volume of the particle will not change, as expected. The vanishing of the substantive derivative of density does not directly imply, however, that the density field is a uniform constant. In fact, this field can vary with both time and space while possessing a vanishing substantive derivative.

4.2.2 Momentum Equation

The non-conservation momentum equation in system (4.6)

$$\rho\frac{\partial u_i}{\partial t} + \rho u_j\frac{\partial u_i}{\partial x_j} + \frac{\partial p}{\partial x_i} - \frac{\partial \sigma_{ij}^\mu}{\partial x_j} - \rho b_i = 0 \tag{4.26}$$

applies for any flow, compressible or incompressible. The form for incompressible flows results from the following specifications.

With a vanishing velocity divergence $\partial u_j/\partial x_j$, the constitutive expressions for the deviatoric stress components from (2.6) become

$$\sigma_{ij}^\mu = \mu\left(\frac{\partial u_i}{\partial x_j} + \frac{\partial u_j}{\partial x_i}\right) \tag{4.27}$$

Since the velocity divergence vanishes, the specific numerical value of the bulk viscosity η_B no longer matters in an incompressible flow, for it multiplies the velocity divergence, which vanishes and thus no longer features in the deviatoric stress. Within an incompressible flow and with reference to Chapter 2, as a consequence of a vanishing velocity divergence, mechanical and thermodynamical pressure always equal each other.

If variations of density are induced by temperature gradients, the ρ multiplying the body force is left to vary, whereas the ρ multiplying the substantial derivative of velocity is set to a constant, for its slight variations would not induce substantial changes in the velocity gradients. This specification corresponds to Bousinnesq's approximation, in which the density multiplying the body force is expressed as

$$\rho \simeq \rho_o + \Delta\rho = \rho_o + \left(\frac{\partial \rho}{\partial T}\right)_p (T - T_o) = \rho_o\left(1 - \beta_{\mathrm{te}}\left(T - T_o\right)\right), \qquad \beta_{\mathrm{te}} \equiv -\frac{1}{\rho_o}\left(\frac{\partial \rho}{\partial T}\right)_p \tag{4.28}$$

where β_{te} denotes the coefficient of thermal expansion. The momentum equation for an incompressible flow in this case becomes

$$\frac{\partial u_i}{\partial t} + u_j \frac{\partial u_i}{\partial x_j} + \frac{1}{\rho_o}\frac{\partial p}{\partial x_i} - \frac{1}{\rho_o}\frac{\partial}{\partial x_j}\left(\mu\left(\frac{\partial u_i}{\partial x_j} + \frac{\partial u_j}{\partial x_i}\right)\right) - \frac{\rho}{\rho_o}b_i = 0 \qquad (4.29)$$

where ρ_o denotes a reference constant density. The insertion of (4.28) into the moment equation yields

$$\frac{\partial u_i}{\partial t} + u_j \frac{\partial u_i}{\partial x_j} + \frac{1}{\rho_o}\frac{\partial p}{\partial x_i} - \frac{1}{\rho_o}\frac{\partial}{\partial x_j}\left(\mu\left(\frac{\partial u_i}{\partial x_j} + \frac{\partial u_j}{\partial x_i}\right)\right) - b_i + \beta_{te}\left(T - T_o\right)b_i = 0 \qquad (4.30)$$

Let b_i denote the ith component of a constant body-force vector. In this case, a "potential" pressure "P" exists, which satisfies the expressions

$$\frac{\partial P}{\partial x_i} = \frac{1}{\rho_o}\frac{\partial p}{\partial x_i} - b_i \quad \Rightarrow \quad P = \frac{p}{\rho_o} - b_i x_i = \frac{p}{\rho_o} + gx_3 \qquad (4.31)$$

the latter expression applying when the body force is the weight, with "g" denoting the acceleration of gravity parallel to the x_3 axis. With this potential pressure, the momentum equation becomes

$$\frac{\partial u_i}{\partial t} + u_j \frac{\partial u_i}{\partial x_j} + \frac{\partial P}{\partial x_i} - \frac{1}{\rho_o}\frac{\partial}{\partial x_j}\left(\mu\left(\frac{\partial u_i}{\partial x_j} + \frac{\partial u_j}{\partial x_i}\right)\right) + \beta_{te}\left(T - T_o\right)b_i = 0 \qquad (4.32)$$

The conservation form of the momentum equation for incompressible flow arises by adding to (4.29) $u_i \partial u_j/\partial x_j = 0$, which yields

$$\frac{\partial u_i}{\partial t} + \frac{\partial u_j u_i}{\partial x_j} + \frac{\partial P}{\partial x_i} - \frac{1}{\rho_o}\frac{\partial}{\partial x_j}\left(\mu\left(\frac{\partial u_i}{\partial x_j} + \frac{\partial u_j}{\partial x_i}\right)\right) + \beta_{te}\left(T - T_o\right)b_i = 0 \qquad (4.33)$$

4.2.3 Temperature Equation

The derivation of the temperature equation for incompressible flows presents an interesting riddle. The straightforward insertion of the incompressibility condition into the internal-energy equation

$$\rho\frac{D\epsilon}{Dt} - \frac{p}{\rho}\frac{D\rho}{Dt} = \sigma_{ij}^{\mu}\frac{\partial u_i}{\partial x_j} - \frac{\partial q_j^F}{\partial x_j} + \rho\dot{g} \qquad (4.34)$$

leads to a temperature equation that is correct for liquids, but quite incorrect for gases, because such a direct substitution would neglect the important temperature variations that result from a density change, however minute.

While in a liquid, density remains practically constant under temperature variations, in a gas, density readily varies under temperature changes. In the incompressible flow of a gas, slight density variations minimally affect the linear momentum of a fluid particle, but significantly affect its temperature. A temperature equation for deriving the corresponding correct equation for the incompressible flow of both liquids and gases is the following equation (1.156) that accounts for the variation of temperature due to density changes

$$\rho c_p \frac{DT}{Dt} + \left(\rho \left(\frac{\partial h}{\partial p}\right)_T - 1\right) \frac{Dp}{Dt} = \sigma_{ij}^\mu \frac{\partial u_i}{\partial x_j} - \frac{\partial q_j^F}{\partial x_j} + \rho \dot{g} \qquad (4.35)$$

This equation applies for both compressible and incompressible flows and does not explicitly contain any density gradients; the velocity divergence only implicitly features within σ_{ij}^μ. The determination of an incompressible-flow temperature equation for liquids and gases from this equation thus amounts to a characterization of these fluids through the substantive derivative of pressure and the thermodynamic derivatives of enthalpy h.

For a homogeneous fluid, a thermodynamic state variable depends on only two other independent state variables. Internal energy and enthalpy can thus be expressed in terms of pressure and temperature as follows

$$\epsilon = \epsilon(T, \rho(p, T)), \quad h = h(T, \rho(p, T)) = \epsilon(T, \rho(p, T)) + \frac{p}{\rho(p, T)} \qquad (4.36)$$

By definition, the specific heats at constant specific-volume, hence constant density, and constant pressure are

$$c_v \equiv \left(\frac{\partial \epsilon}{\partial T}\right)_\rho, \quad c_p \equiv \left(\frac{\partial h}{\partial T}\right)_p \qquad (4.37)$$

where the subscripts ρ and p signify derivatives calculated respectively at constant density and constant pressure. The substantive derivatives of ϵ and h can thus be expressed as

$$\frac{D\epsilon}{Dt} = \left(\frac{\partial \epsilon}{\partial T}\right)_\rho \frac{DT}{Dt} + \left(\frac{\partial \epsilon}{\partial \rho}\right)_T \frac{D\rho}{Dt} = c_v \frac{DT}{Dt} + \left(\frac{\partial \epsilon}{\partial \rho}\right)_T \frac{D\rho}{Dt}$$

$$\frac{Dh}{Dt} = \left(\frac{\partial h}{\partial T}\right)_p \frac{DT}{Dt} + \left(\frac{\partial h}{\partial p}\right)_T \frac{Dp}{Dt} = c_p \frac{DT}{Dt} + \left(\frac{\partial h}{\partial p}\right)_T \frac{Dp}{Dt} \qquad (4.38)$$

The connection between c_p and c_v and the partial derivative of enthalpy with respect to pressure originate from (4.36) and (4.37) as

$$c_p = c_v + \left(\frac{\partial \epsilon}{\partial \rho}\right)_T \left(\frac{\partial \rho}{\partial T}\right)_p - \frac{p}{\rho^2}\left(\frac{\partial \rho}{\partial T}\right)_p, \quad \left(\frac{\partial h}{\partial p}\right)_T = \left(\frac{\partial \epsilon}{\partial \rho}\right)_T \left(\frac{\partial \rho}{\partial p}\right)_T + \frac{1}{\rho} - \frac{p}{\rho^2}\left(\frac{\partial \rho}{\partial p}\right)_T \qquad (4.39)$$

Liquids

For a liquid, $\rho \simeq$ constant, hence (4.38) and (4.39) become

$$\frac{D\epsilon}{Dt} = c_v \frac{DT}{Dt}, \quad \frac{Dh}{Dt} = c_p \frac{DT}{Dt} + \frac{1}{\rho}\frac{Dp}{Dt} \qquad (4.40)$$

$$c_p = c_v, \quad \left(\frac{\partial h}{\partial p}\right)_T = \frac{1}{\rho} \qquad (4.41)$$

which imply

$$\left(\rho \left(\frac{\partial h}{\partial p}\right)_T - 1\right) \frac{Dp}{Dt} = 0 \qquad (4.42)$$

Gases

For a sufficiently dilute gas, the internal energy ϵ practically remains independent of density hence (4.38) and (4.39) become

$$\frac{D\epsilon}{Dt} = c_v \frac{DT}{Dt}, \quad \frac{Dh}{Dt} = c_p \frac{DT}{Dt} + \left(\frac{1}{\rho} - \frac{p}{\rho^2}\left(\frac{\partial\rho}{\partial p}\right)_T\right)\frac{Dp}{Dt} \tag{4.43}$$

$$c_p = c_v - \frac{p}{\rho^2}\left(\frac{\partial\rho}{\partial T}\right)_p, \quad \left(\frac{\partial h}{\partial p}\right)_T = \frac{1}{\rho} - \frac{p}{\rho^2}\left(\frac{\partial\rho}{\partial p}\right)_T \tag{4.44}$$

It follows, that both liquids and gases posses the same substantive derivative of internal energy ϵ. For either a dilute or perfect gas, when the local flow speed is lower than the speed of sound, pressure variations on a fluid particle are negligible in determining its temperature in comparison to density variations; in this case $Dp/Dt \simeq 0$, which implies

$$\left(\rho\left(\frac{\partial h}{\partial p}\right)_T - 1\right)\frac{Dp}{Dt} \simeq 0 \tag{4.45}$$

Certainly, this model does not imply that pressure would remain a constant throughout the flow field. The perfect gas model with $p = \rho RT$ then simplifies these expressions to

$$\frac{Dh}{Dt} = c_p \frac{DT}{Dt}, \quad c_p = c_v + R, \quad \left(\frac{\partial h}{\partial p}\right)_T = 0 \tag{4.46}$$

Equation for Liquids and Gases

With results (4.42), (4.45), (2.7) and (4.27) the temperature equation (4.35) for the incompressible flow of both liquids and gases becomes

$$\rho c_p\left(\frac{\partial T}{\partial t} + u_j \frac{\partial T}{\partial x_j}\right) = \frac{\partial}{\partial x_j}\left(k\frac{\partial T}{\partial x_j}\right) + \rho\dot{g} + \mu\frac{\partial u_i}{\partial x_j}\left(\frac{\partial u_i}{\partial x_j} + \frac{\partial u_j}{\partial x_i}\right) \tag{4.47}$$

The specific heat c_v instead of c_p would appear in a temperature equation which originates from inserting both the incompressibility condition and the first of (4.40) or (4.43) into the internal-energy equation (4.34). While this is correct for a liquid, for $c_p = c_v$, it is not for a gas, as indicated by (4.41) and (4.44).

The conservation form of the temperature equation for incompressible flows arises by adding to (4.47) $\rho c_p T \partial u_j/\partial x_j = 0$, which yields

$$\frac{\partial T}{\partial t} + \frac{\partial u_j T}{\partial x_j} = \frac{k_o}{\rho c_p}\frac{\partial}{\partial x_j}\left(\frac{k}{k_o}\frac{\partial T}{\partial x_j}\right) + \frac{\dot{g}}{c_p} + \frac{\mu}{\rho c_p}\frac{\partial u_i}{\partial x_j}\left(\frac{\partial u_i}{\partial x_j} + \frac{\partial u_j}{\partial x_i}\right) \tag{4.48}$$

With a constant reference temperature T_o and for an incompressible flow, the following expressions all vanish

$$\frac{\partial T_o}{\partial t} = 0, \quad T_o\frac{\partial u_j}{\partial x_j} = 0, \quad \frac{\partial T_o}{\partial x_j} = 0 \tag{4.49}$$

The subtraction of these expressions from the temperature equation leads to the following energy equation, which depends on the temperature variation $\Delta T \equiv T - T_o$

$$\frac{\partial \Delta T}{\partial t} + \frac{\partial u_j \Delta T}{\partial x_j} - \frac{k_o}{\rho c_p}\frac{\partial}{\partial x_j}\left(\frac{k}{k_o}\frac{\partial \Delta T}{\partial x_j}\right) - \frac{\dot{g}}{c_p} - \frac{\mu}{\rho c_p}\frac{\partial u_i}{\partial x_j}\left(\frac{\partial u_i}{\partial x_j} + \frac{\partial u_j}{\partial x_i}\right) = 0 \tag{4.50}$$

4.2.4   Governing System

The Navier-Stokes system for incompressible flows can thus be expressed as

$$
\left\{
\begin{array}{l}
\dfrac{\partial u_j}{\partial x_j} = 0 \\[2mm]
\dfrac{\partial u_i}{\partial t} + \dfrac{\partial u_j u_i}{\partial x_j} + \dfrac{\partial P}{\partial x_i} - \dfrac{\mu_o}{\rho_o}\dfrac{\partial}{\partial x_j}\left(\dfrac{\mu}{\mu_o}\left(\dfrac{\partial u_i}{\partial x_j}+\dfrac{\partial u_j}{\partial x_i}\right)\right) + \beta_{\text{te}}\left(\Delta T\right)b_i = 0 \\[2mm]
\dfrac{\partial \Delta T}{\partial t} + \dfrac{\partial u_j \Delta T}{\partial x_j} - \dfrac{k_o}{\rho_o c_p}\dfrac{\partial}{\partial x_j}\left(\dfrac{k}{k_o}\dfrac{\partial \Delta T}{\partial x_j}\right) - \dfrac{\dot{g}}{c_p} - \dfrac{\mu_o}{\rho_o c_p}\dfrac{\mu}{\mu_o}\dfrac{\partial u_i}{\partial x_j}\left(\dfrac{\partial u_i}{\partial x_j}+\dfrac{\partial u_j}{\partial x_i}\right) = 0
\end{array}
\right.
\tag{4.51}
$$

In contrast with its compressible-flow counterpart, the continuity equation in this system does not contain any partial derivative with respect to time. This particular feature calls for special numerical methods for solving this system computationally.

4.2.5   Non-Dimensional System

Computationally convenient non-dimensional forms of the incompressible-flow Navier-Stokes equations emerge from a set of reference variables for the Cartesian coordinates x_j, $1 \leq j \leq n$, and the dependent variables of pressure p, velocity components u_j, $1 \leq j \leq n$, and temperature variation ΔT. The reference time rests on a reference length and a reference velocity, which in turn depends on the reference pressure and density. With these specifications, the non-dimensional Euler equations, retain the same form as their dimensional counterparts, whereas the non-dimensional Navier-Stokes equations display the Reynolds, Prandtl, and Eckert numbers, as well as the Grashof or Rayleigh numbers for thermally induced flows.

Consider the following reference length, density, pressure, and temperature variation

$$
L, \quad \rho_{\text{re}}, \quad p_{\text{re}}, \quad \Delta T_{\text{re}}
\tag{4.52}
$$

Depending on the flow investigated, these parameters can correspond to a dominant geometric dimension as well as free-stream or stagnation conditions. On the basis of their respective dimensions, the corresponding reference parameters for speed, time, and volume-specific total energy become

$$
V_{\text{re}}^2 = \frac{p_{\text{re}}}{\rho_{\text{re}}}, \quad t_{\text{re}} = \frac{L}{V_{\text{re}}}
\tag{4.53}
$$

With a tilde "~" denoting a non-dimensional variable, corresponding dimensional and dimensionless variables depend upon one another through (4.52)-(4.53) as follows

$$
t = t_{\text{re}}\tilde{t}, \quad x_j = L\tilde{x}_j
$$

$$
p = p_{\text{re}}\tilde{p}, \quad u_j = V_{\text{re}}\tilde{u}_j, \quad \Delta T = \Delta T_{\text{re}}\widetilde{\Delta T}
\tag{4.54}
$$

On the basis of these expressions, the incompressible Navier-Stokes equations become

$$
\begin{cases}
\dfrac{V_{\mathrm{re}}}{L}\dfrac{\partial \tilde{u}_j}{\partial \tilde{x}_j} = 0 \\[2ex]
\dfrac{V_{\mathrm{re}}^2}{L}\dfrac{\partial \tilde{u}_i}{\partial \tilde{t}} + \dfrac{V_{\mathrm{re}}^2}{L}\dfrac{\partial \tilde{u}_j \tilde{u}_i}{\partial \tilde{x}_j} + \dfrac{p_{\mathrm{re}}}{\rho_{\mathrm{re}} L}\dfrac{\partial \tilde{P}}{\partial \tilde{x}_i} - \dfrac{V_{\mathrm{re}}\,\mu_o}{L^2\,\rho_o}\dfrac{\partial}{\partial \tilde{x}_j}\left(\dfrac{\mu}{\mu_o}\left(\dfrac{\partial \tilde{u}_i}{\partial \tilde{x}_j} + \dfrac{\partial \tilde{u}_j}{\partial \tilde{x}_i}\right)\right) + \beta_{\mathrm{te}}(\Delta T)_{\mathrm{re}} g \Delta \tilde{T}\dfrac{g_i}{g} = 0 \\[2ex]
\dfrac{V_{\mathrm{re}}(\Delta T)_{\mathrm{re}}}{L}\dfrac{\partial \Delta \tilde{T}}{\partial \tilde{t}} + \dfrac{V_{\mathrm{re}}(\Delta T)_{\mathrm{re}}}{L}\dfrac{\partial \tilde{u}_j \Delta \tilde{T}}{\partial \tilde{x}_j} \\[2ex]
\quad - \dfrac{(\Delta T)_{\mathrm{re}}}{L^2}\dfrac{k_o}{\rho c_p}\dfrac{\partial}{\partial \tilde{x}_j}\left(\dfrac{k}{k_o}\dfrac{\partial \Delta \tilde{T}}{\partial \tilde{x}_j}\right) - \dfrac{V_{\mathrm{re}}^2}{L^2}\dfrac{\mu}{\rho c_p}\dfrac{\partial \tilde{u}_i}{\partial \tilde{x}_j}\left(\dfrac{\partial \tilde{u}_i}{\partial \tilde{x}_j} + \dfrac{\partial \tilde{u}_j}{\partial \tilde{x}_i}\right) = 0
\end{cases}
$$

$$(4.55)$$

The following non-dimensional groups of reference parameters are rearranged as

$$
\frac{L}{V_{\mathrm{re}}^2}\frac{\mu_o V_{\mathrm{re}}}{\rho_o L^2} = \frac{\mu_o}{\rho_o V_{\mathrm{re}} L} = \frac{1}{Re}
$$

$$
\frac{L}{V_{\mathrm{re}}^2}\beta_{\mathrm{te}}(\Delta T)_{\mathrm{re}}g = \frac{L\beta_{\mathrm{te}}(\Delta T)_{\mathrm{re}}g}{\dfrac{\rho_o^2 V_{\mathrm{re}}^2 L^2}{\mu_o^2}\dfrac{\mu_o^2}{\rho_o^2 L^2}} = \frac{1}{Re^2}\frac{L^3\rho_o^2\beta_{\mathrm{te}}(\Delta T)_{\mathrm{re}}g}{\mu_o^2} = \frac{Gr}{Re^2}
$$

$$
\frac{L}{V_{\mathrm{re}}(\Delta T)_{\mathrm{re}}}\frac{(\Delta T)_{\mathrm{re}}}{L^2}\frac{k_o}{\rho_o c_p} = \frac{k_o}{\mu_o c_p}\frac{\mu_o}{\rho_o V_{\mathrm{re}} L} = \frac{1}{Pr Re}
$$

$$
\frac{L}{V_{\mathrm{re}}(\Delta T)_{\mathrm{re}}}\frac{V_{\mathrm{re}}^2}{L^2}\frac{\mu}{\rho c_p} = \frac{V_{\mathrm{re}}^2}{(\Delta T)_{\mathrm{re}}c_p}\frac{\mu}{\rho_o V_{\mathrm{re}} L} = \frac{Ec}{Re}
$$

where "Re", "Gr", "Pr", and "Ec" respectively denote the Reynolds, Grashof, Prandtl, and Eckert numbers. With these numbers, the incompressible Navier-Stokes system becomes

$$
\begin{cases}
\dfrac{\partial \tilde{u}_j}{\partial \tilde{x}_j} = 0 \\[2ex]
\dfrac{\partial \tilde{u}_i}{\partial \tilde{t}} + \dfrac{\partial \tilde{u}_j \tilde{u}_i}{\partial \tilde{x}_j} + \dfrac{\partial \tilde{P}}{\partial \tilde{x}_i} - \dfrac{1}{Re}\dfrac{\partial}{\partial \tilde{x}_j}\left(\dfrac{\mu}{\mu_o}\left(\dfrac{\partial \tilde{u}_i}{\partial \tilde{x}_j} + \dfrac{\partial \tilde{u}_j}{\partial \tilde{x}_i}\right)\right) + \dfrac{Gr}{Re^2}\Delta \tilde{T}\dfrac{g_i}{g} = 0 \\[2ex]
\dfrac{\partial \Delta \tilde{T}}{\partial \tilde{t}} + \dfrac{\partial \tilde{u}_j \Delta \tilde{T}}{\partial \tilde{x}_j} - \dfrac{1}{Pr Re}\dfrac{\partial}{\partial \tilde{x}_j}\left(\dfrac{k}{k_o}\dfrac{\partial \Delta \tilde{T}}{\partial \tilde{x}_j}\right) - \dfrac{Ec}{Re}\dfrac{\partial \tilde{u}_i}{\partial \tilde{x}_j}\left(\dfrac{\partial \tilde{u}_i}{\partial \tilde{x}_j} + \dfrac{\partial \tilde{u}_j}{\partial \tilde{x}_i}\right) = 0
\end{cases}
$$

$$(4.56)$$

The Rayleigh number "Ra" can replace the Grashof number in the momentum equations as

$$Ra = Gr Pr \quad \Rightarrow \quad \frac{Gr}{Re^2} = \frac{Ra}{Pr Re^2} \tag{4.57}$$

By the same token, the Eckert number in the energy equation can be expressed in terms of either the Grashof or Rayleigh number, through the reference temperature variation, as follows

$$Gr = \frac{L^3\rho_o^2\beta_{\mathrm{te}}(\Delta T)_{\mathrm{re}}g}{\mu_o^2} \quad \Rightarrow \quad (\Delta T)_{\mathrm{re}} = \frac{\mu_o^2}{L^3\rho_o^2\beta_{\mathrm{te}}g}$$

$$\frac{Ec}{Re} = \frac{1}{Re} \frac{V_{\text{re}}^2}{(\Delta T)_{\text{re}} c_p} = \frac{1}{Re} \frac{V_{\text{re}}^2 L^3 \rho_o^2 \beta_{\text{te}} g}{c_p Gr \mu_o^2}$$

$$= \frac{1}{Re} \frac{V_{\text{re}}^2 L^2 \rho_o^2}{Gr \mu_o^2} \frac{L \beta_{\text{te}} g}{c_p} = \frac{L \beta_{\text{te}} g / c_p}{Re} \frac{Re^2}{Gr} = \frac{L \beta_{\text{te}} g / c_p}{Re} \frac{Pr Re^2}{Ra} \tag{4.58}$$

For temperature-driven flows, where no established inlet or outlet pressure and velocity exist, dimensional analysis reveals that

$$\frac{p_{\max}}{\rho g L} = f\left(\frac{\rho L \sqrt{gL}}{\mu}, \beta_{\text{te}}(\Delta T)_{\text{re}}\right) \tag{4.59}$$

which leads to the following reference pressure and speed

$$p_{\text{re}} = \rho g L \quad \Rightarrow \quad V_{\text{re}} = \sqrt{\frac{p_{\text{re}}}{\rho}} = \sqrt{gL} \tag{4.60}$$

The Reynolds number, in this case, becomes

$$Re = \frac{\rho V_{\text{re}} L}{\mu} = \frac{\rho L \sqrt{gL}}{\mu} \tag{4.61}$$

and the ratio Gr/Re^2 thus simplifies as

$$\frac{Gr}{Re^2} = \frac{1}{Re^2} \frac{L^3 \rho_o^2 \beta_{\text{te}}(\Delta T)_{\text{re}} g}{\mu_o^2} = \frac{\beta_{\text{te}}(\Delta T)_{\text{re}}}{Re^2} \left(\frac{\rho L \sqrt{gL}}{\mu}\right)^2 = \beta_{\text{te}}(\Delta T)_{\text{re}} \tag{4.62}$$

which provides the order of magnitude for such a ratio.

4.2.6 Variational Formulation

In the variational weighted formulation, the incompressible-flow Navier-Stokes equations are conveniently expressed as

$$\int_\Omega \frac{\partial w}{\partial x_j} u_j d\Omega - \oint_{\partial \Omega} w u_j n_j d\Gamma = 0 \tag{4.63}$$

$$\int_\Omega w \left(\frac{\partial u_i}{\partial t} - \frac{Gr}{Re^2} \Delta T \frac{g_i}{g}\right) d\Omega - \int_\Omega \frac{\partial w}{\partial x_j} \left(u_i u_j + \left(P\delta_j^i - \frac{\sigma_{ij}^\mu}{Re}\right)\right) d\Omega +$$

$$+ \oint_{\partial \Omega} w u_i u_j n_j d\Gamma + \oint_{\partial \Omega} w \left(P\delta_j^i - \frac{\sigma_{ij}^\mu}{Re}\right) n_j d\Gamma = 0 \tag{4.64}$$

$$\int_\Omega w \left(\frac{\partial \Delta T}{\partial t} + \frac{Ec}{Re} \frac{\partial u_i}{\partial x_j} \left(\frac{\partial u_i}{\partial x_j} + \frac{\partial u_j}{\partial x_i}\right)\right) d\Omega$$

$$- \int_\Omega \frac{\partial w}{\partial x_j} u_j \Delta T d\Omega + \int_\Omega \frac{\partial w}{\partial x_j} \frac{k/k_o}{Pr Re} \frac{\partial \Delta T}{\partial x_j} d\Omega +$$

$$+ \oint_{\partial \Omega} w \Delta T u_j n_j d\Gamma - \oint_{\partial \Omega} w \frac{k/k_o}{Pr Re} \frac{\partial \Delta T}{\partial x_j} n_j d\Gamma = 0 \tag{4.65}$$

where the tilde has been dropped for simplicity, even though all variables in these statements remain non dimensional. The surface integrals in these results arise from a transformation of the corresponding domain integrals through Gauss' theorem. For $w = 1$, the continuity-equation integral statement becomes

$$\oint_{\partial\Omega} \boldsymbol{V} \cdot \boldsymbol{n} \, d\Gamma = 0 \qquad (4.66)$$

which corresponds to a global mass conservation statement. As in the compressible-flow counterparts, these surface integrals provide quite a convenient setting to enforce physically meaningful boundary conditions on surface velocity flux $u_j n_j$, traction components $(p\delta_j^i - \sigma_{ij}^\mu/Re)n_j$ and heat flux $\partial \Delta T/\partial n$. Boundary conditions that must be enforced on these quantities can thus directly inserted into the corresponding surface integrals in these variational equations. Additional boundary conditions, of course, constrain u_i and ΔT on appropriate facets of the total boundary, as detailed in Chapters 11, 15.

Chapter 5

Quasi One-Dimensional and Free-Surface Equations

The quasi-one-dimensional and free-surface sets of equations correspond to simplified forms of the Euler and incompressible Navier-Stokes equations. Both of these sets emerge from the application of Reynolds' transport theorem to integral definitions of mass, linear momentum as well as total energy, and subsequent introduction of averages for density, linear momentum, and total energy. This chapter presents these averages and then derives the corresponding governing equations for the two cases of quasi-one-dimensional gas dynamics and free-surface open-channel flows. Progressing beyond the ideal isentropic specifications, these equations also model the presence of wall friction, heat as well as mass transfer, and mechanical shaft work.

5.1 Area Averages

The quantity of mass of an amount of fluid within a volume $\Omega \in \mathcal{R}^3$ results from the integral

$$\mathcal{M} = \int_\Omega \rho d\Omega \tag{5.1}$$

With reference to Figure 5.1, this integral becomes

$$\int_\Omega \rho \, d\Omega = \int_{x_{\text{in}}}^{x_{\text{out}}} dx \int_A \rho \, dA = \int_{x_{\text{in}}}^{x_{\text{out}}} \overline{\rho} A \, dx \tag{5.2}$$

where $\overline{\rho}$ denotes the area averaged density. Similarly, "$\overline{m}$" and "$\overline{E}$" denote the area averaged volume-specific linear momentum and total energy. These three averages are defined as

$$\overline{\rho} \equiv \frac{1}{A} \int_A \rho \, dA, \quad \overline{m} \equiv \frac{1}{A} \int_A m \, dA, \quad \overline{E} \equiv \frac{1}{A} \int_A E \, dA \tag{5.3}$$

The average speed is then based on the average linear momentum and density as

$$\overline{u} \equiv \frac{\overline{m}}{\overline{\rho}} = \frac{\int_A u\rho \, dA}{\int_A \rho \, dA} \tag{5.4}$$

which corresponds to a "mass averaged" speed. With a suitable variation of "$A = A(x)$", the area- and mass-averaged speeds remain accurate representations of each other.

5.2   Quasi One-Dimensional Euler Equations

The quasi-one-dimensional Euler equations evolve from the principles of mass conservation, 2nd law of mechanics, and first principle of thermodynamics as Reynolds' transport theorem is applied to integral expressions for mass, linear momentum, and total energy. Referring to Figure 5.1, the following derivations dispense with the overbars of averages, for simplicity, even though all dependent variables represent averaged quantities.

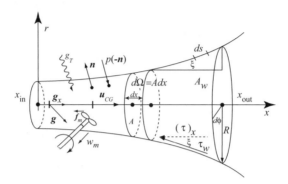

Figure 5.1: Quasi-One-Dimensional Duct

5.2.1   Continuity Equation

In a nozzle flow, the presence of both mass transfer through a side opening and longitudinal variation of duct cross-sectional area will induce two additional source terms in the one-dimensional continuity equation. Figure 5.2 illustrates the transfer of mass to a nozzle flow through a side opening of width b. This mass is transferred with density ρ_b and a velocity $\mathbf{u}_b$, at an angle θ with respect to the direction perpendicular to the nozzle x-axis, a velocity on a plane that contains the x-axis; the corresponding mass flow rate per unit length $\dot{m}_b$ will be positive when mass is added to the nozzle flow.

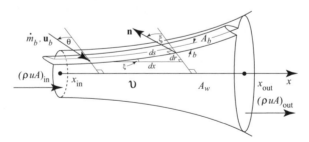

Figure 5.2: Fluid in Nozzle and Side Opening

The mass $\mathcal{M}$ of an amount of fluid that flows in the nozzle and through the side opening will not change as a result of the flow, which implies

$$\frac{D\mathcal{M}}{Dt} = \frac{D}{Dt} \int_{\mathcal{V}(t)} \rho \, d\mathcal{V} = 0 \tag{5.5}$$

where $\mathcal{V}$ denotes the instantaneous volume occupied by the fluid. With respect to a reference frame attached to the nozzle, Reynolds' transport theorem transforms this statement into

$$\int_{\mathcal{V}} \frac{\partial \rho}{\partial t} d\mathcal{V} + \oint_{\partial \mathcal{V}} \rho \mathbf{u} \cdot \mathbf{n} \, d\Gamma = 0 \tag{5.6}$$

where $\mathbf{n}$ denotes the outward pointing unit vector, perpendicular to every fluid-surface facet, and $\partial \mathcal{V} \equiv A_{\text{in}} \bigcup A_{\text{out}} \bigcup A_w \bigcup A_b$, with A_w and A_b respectively denoting the area of the lateral solid surface of the duct and the area of the lateral surface of the amount of fluid flowing through the side opening. Although for clarity Figure 5.2 shows a finite volume of fluid through the opening, for the purpose of deriving the equations for the nozzle flow it suffices to consider an infinitesimal volume of transferred mass because all the corresponding equations only depend on rates of mass transfer. For a moderate curvature of the duct and using cross-sectional area averaged variables, the area average of result (5.6) is expressed as

$$\int_{x_{\text{in}}}^{x_{\text{out}}} \frac{\partial \rho}{\partial t} A \, dx - \rho u A\big|_{\text{in}} + \rho u A\big|_{\text{out}} + \int_{A_b} \rho_b \mathbf{u}_b \cdot \mathbf{n} \, d\Gamma = 0 \tag{5.7}$$

where $A = A(x)$ denotes the time-independent cross sectional area of the flow converging-diverging duct. With reference to Figure 5.2, the mass-transfer integral is transformed as

$$\int_{A_b} \rho_b \mathbf{u}_b \cdot \mathbf{n} \, d\Gamma = \int_{A_b} -\rho_b u_b \cos(\xi - \theta) b \, ds = \int_{x_{\text{in}}}^{x_{\text{out}}} -\rho_b u_b \cos(\xi - \theta) b \frac{dx}{\cos \xi}$$

$$= \int_{x_{\text{in}}}^{x_{\text{out}}} -(\rho_b u_b b)(\cos \theta + \tan \xi \sin \theta) \, dx = -\int_{x_{\text{in}}}^{x_{\text{out}}} \dot{m}_b \left(\cos \theta + \frac{\partial A}{\partial x} \frac{\sin \theta}{2\sqrt{\pi A}} \right) dx \tag{5.8}$$

where the integrand expression corresponds to the rate of mass per unit length transferred to the nozzle flow, $\dot{m}_b = \rho_b u_b b$ denotes the mass flow rate per unit length in the direction of $\mathbf{u}_b$, and $\tan \xi = \frac{\partial r}{\partial x}$ is expressed in terms of the gradient of the cross-sectional area A owing to the expression $\frac{\partial r}{\partial x} = \frac{1}{2\pi r} \frac{\partial(\pi r^2)}{\partial x} = \frac{1}{2\sqrt{\pi A}} \frac{\partial A}{\partial x}$. Accordingly, the expression between parenthesis multiplying $\dot{m}_b$ in this integral identically reduces to unity for $\mathbf{u}_b$ perpendicular to the x-axis. With this result, statement (5.7) is further developed as

$$\int_{x_{\text{in}}}^{x_{\text{out}}} \frac{\partial \rho A}{\partial t} dx + \int_{x_{\text{in}}}^{x_{\text{out}}} \frac{\partial \rho u A}{\partial x} dx - \int_{x_{\text{in}}}^{x_{\text{out}}} \dot{m}_b \left(\cos \theta + \frac{\partial A}{\partial x} \frac{\sin \theta}{2\sqrt{\pi A}} \right) dx = 0 \tag{5.9}$$

This expression remains unaltered for arbitrary intervals $(x_{\text{in}}, x_{\text{out}})$. This recognition leads to the equation

$$\frac{\partial \rho}{\partial t} + \frac{\partial \rho u}{\partial x} + \frac{\rho u}{A} \frac{\partial A}{\partial x} = \frac{\dot{m}_b}{A} \left(\cos \theta + \frac{\partial A}{\partial x} \frac{\sin \theta}{2\sqrt{\pi A}} \right) \tag{5.10}$$

which corresponds to the continuity equation for flows in a variable-cross section duct with mass transfer.

5.2.2   Generalized Linear-Momentum Equation

With reference to Figure 5.1, the generalized quasi-one-dimensional linear-momentum equation arises from the second law of Newtonian mechanics as

$$\frac{d(\mathcal{M}u_{CG})}{dt} = F_x \Rightarrow \frac{D}{Dt} \int_{\mathcal{V}(t)} \rho u \, d\mathcal{V} = F_x \tag{5.11}$$

In this expression, F_x denotes the total external force on the fluid within $\mathcal{V}$, while $\mathcal{M}$, u_{CG}, and $\partial\mathcal{V}$ respectively denote the mass of the flowing amount of fluid, the x-axis component of the velocity of its center of gravity, and the surface that entirely envelops a nozzle-shaped mass of fluid flowing in the x-direction.

Using Reynolds' transport theorem, the substantive derivative of the integral of ρu in (5.11) becomes

$$\frac{D}{Dt} \int_{\mathcal{V}} \rho u \, d\mathcal{V} = \int_{\mathcal{V}} \frac{\partial \rho u}{\partial t} d\mathcal{V} + \oint_{\partial\mathcal{V}} \rho u \mathbf{u} \cdot \mathbf{n} \, d\Gamma \tag{5.12}$$

With reference to Figures 5.2-5.1, reflecting developments (5.6)-(5.9) and employing cross-sectional area averaged variables, the area average of this integral statement is expressed as

$$\int_{x_{in}}^{x_{out}} \frac{\partial \rho u}{\partial t} A \, dx - \rho u^2 A \Big|_{in} + \rho u^2 A \Big|_{out} + \int_{A_b} \rho_b (u_b)_x \mathbf{u}_b \cdot \mathbf{n} \, d\Gamma =$$

$$\int_{x_{in}}^{x_{out}} \frac{\partial \rho u A}{\partial t} dx + \int_{x_{in}}^{x_{out}} \frac{\partial \rho u^2 A}{\partial x} dx - \int_{x_{in}}^{x_{out}} \frac{\dot{m}_b^2}{b \rho_b} \sin\theta \left(\cos\theta + \frac{\partial A}{\partial x} \frac{\sin\theta}{2\sqrt{\pi A}} \right) dx \tag{5.13}$$

The total external force F_x in the x-direction results from surface stress, gravity, and the action of any mechanical shaft interacting with the flow. This total force is thus expressed as

$$F_x = \oint_{\partial\mathcal{V}} \sigma_{1j} n_j d\Gamma + \int_{\Omega} \rho g_x \, d\Omega + \int_{x_{in}}^{x_{out}} f_{mx} \, dx \tag{5.14}$$

where $\sigma_{1j} n_j$, g_x, and f_{mx} respectively denote the multi-dimensional traction in the $x_1 \equiv x$ direction, the x-component of the acceleration of gravity, and the shaft force per unit length. The stress tensor component $\sigma_{1j} n_j$, $1 \le j \le 3$, is decomposed as $\sigma_{1j} = -(p)_w \delta_j^1 + \tau_{1j}$, with δ_j^1 and τ_{1j} respectively denoting the Kronecker delta and shear-stress tensor components. With reference to $(p)_w$, this term indicates the nozzle-wall pressure, which equals the mass-transfer pressure p_b, at the side opening, and a circumferentially continuous duct-flow pressure P, at all other wall locations; this pressure is indicated with a capital letter to distinguish it from the area averaged pressure p. The surface-traction force thus becomes

$$\oint_{\partial\mathcal{V}} \sigma_{1j} n_j d\Gamma = \oint_{\partial\mathcal{V}} (p)_w \delta_j^1 (-n_j) d\Gamma + \oint_{\partial\mathcal{V}} \tau_{1j} n_j d\Gamma \tag{5.15}$$

Pressure-Gradient Force

The contribution to F_x from pressure is expressed as

$$\oint_{\partial\mathcal{V}} (p)_w \delta_j^1 (-n_j) d\Gamma = \int_{\partial\mathcal{V} \backslash A_b} P \delta_j^1 (-n_j) d\Gamma + \int_{A_b} p_b \delta_j^1 (-n_j) d\Gamma$$

$$= \oint_{\partial\mathcal{V}} P \delta_j^1 (-n_j) d\Gamma + \int_{A_b} (p_b - P) \, \delta_j^1 (-n_j) d\Gamma \tag{5.16}$$

With reference to Figures 5.2-5.1 and results (5.6)-(5.9), the surface integral of $(p_b - P)$ is expressed as

$$\int_{A_b} (p_b - P)\,\delta_j^1(-n_j)d\Gamma = \int_{x_{\text{in}}}^{x_{\text{out}}} b(\Delta p)_{b_x}\,dx \qquad (5.17)$$

with

$$(\Delta p)_{b_x} \equiv \frac{(p_b - P)}{2\sqrt{\pi A}}\frac{\partial A}{\partial x},\quad P|_{A_b} \simeq p \qquad (5.18)$$

where p denotes the area average of P. The effect of boundary-layer growth may be modelled by adding to $b(\Delta p)_{b_x}$ a term of the form $-f(x)(1-\eta)pA/L$, where $f(x)$ depends on duct inlet conditions and shape, and L and η respectively denote duct length and efficiency, with $0.8 < \eta < 1.0$, [79]. In respect of the integral of P, this is expressed as

$$\oint_{\partial v} P\delta_j^1(-n_j)d\Gamma = -\int_\Omega \frac{\partial(P\delta_j^1)}{\partial x_j}d\Omega = -\int_\Omega \frac{\partial P}{\partial x_1}d\Omega = -\int_{x_{\text{in}}}^{x_{\text{out}}} \overline{\frac{\partial P}{\partial x}}A\,dx,\quad x \equiv x_1 \qquad (5.19)$$

which depends on the average pressure gradient

$$\overline{\frac{\partial P}{\partial x}} \equiv \frac{1}{A}\int_A \frac{\partial P}{\partial x}dA = \frac{1}{A}\int_A \frac{\partial P}{\partial x}r\,d\phi\,dr = \frac{2}{R^2}\int_0^R \frac{\partial \tilde{p}}{\partial x}r\,dr \qquad (5.20)$$

With reference to Figure 5.3, $\tilde{p}$ in this result indicates a circumferentially averaged pressure.

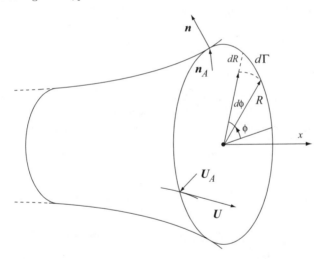

Figure 5.3: Velocity and Normal Vectors on Cross-Sectional Plane

By virtue of Leibnitz's theorem for the differentiation of an integral, the derivative of the area average $\overline{P}$ of pressure becomes

$$\frac{\partial(\overline{P})}{\partial x} \equiv \frac{\partial}{\partial x}\left[\frac{2}{R^2}\int_0^R \tilde{p}r\,dr\right] = \frac{\partial}{\partial x}\left(\frac{2}{R^2}\right)\left[\int_0^R \tilde{p}r\,dr\right] +$$

$$\left(\frac{2}{R^2}\right)\left[\int_0^R \frac{\partial \tilde{p}}{\partial x} r\,dr\right] + \left(\frac{2}{R^2}\right)\frac{\partial R}{\partial x}\tilde{p}(x,R)R \tag{5.21}$$

From this result, the correlation between $\overline{\partial P/\partial x}$ and $\partial(\overline{P})/\partial x$ becomes

$$\overline{\frac{\partial P}{\partial x}} = \frac{\partial(\overline{P})}{\partial x} + \frac{1}{A}\frac{\partial A}{\partial x}\left[\frac{2}{R^2}\int_0^R \tilde{p}(x,r)r\,dr - \tilde{p}(x,R)\right] \tag{5.22}$$

For a small $\partial A/\partial x$, or a small $\partial \tilde{p}/\partial r$, or both, the expression between brackets becomes negligible and this average pressure gradient can be accurately represented by the gradient of the average pressure as

$$\overline{\frac{\partial P}{\partial x}} \simeq \frac{\partial(\overline{P})}{\partial x}, \quad \overline{P} = p = \frac{1}{A}\int_A P\,dA \tag{5.23}$$

Denoting with $\gamma \equiv c_p/c_v$ the ratio of specific heats, the static pressure p in the linear-momentum equation is expressed by way of the equation of state

$$p = (\gamma - 1)\left(E - \frac{m^2}{2\rho}\right) \tag{5.24}$$

Shear-Stress Force

With implied summation on repeated subscript indices, the expression $\tau_{1j}n_j$ denotes the x-axis component of the shear surface traction on a surface facet with outward pointing unit vector $\boldsymbol{n}$ with components n_j, $1 \leq j \leq 3$. The contribution to F_x from this traction becomes

$$\oint_{\partial\Omega} \tau_{1j}n_j\,d\Gamma = \oint_{A_w} -(\tau)_x\,dA_w \tag{5.25}$$

with $(\tau)_x$ indicating the x-axis component of the shear surface traction on the lateral surface, with area A_w, of the fluid in the nozzle, as shown in Figure 5.1. With τ_w denoting a circumferentially averaged wall shear stress, the shear-stress force integral becomes

$$\oint_{A_w}(\tau)_x\,dA_w = \int_{x_{in}}^{x_{out}}(\tau_w\cos\xi)\,2\pi r\,ds = \int_{x_{in}}^{x_{out}}(\tau_w\cos\xi)\,2\pi r\sqrt{dx^2 + dr^2} =$$

$$\int_{x_{in}}^{x_{out}}(\tau_w\cos\xi)\,2\pi r\sqrt{1 + \left(\frac{dr}{dx}\right)^2}\,dx = \int_{x_{in}}^{x_{out}}(\tau_w\cos\xi)\,2\pi r\sqrt{1 + \tan^2\xi}\,dx =$$

$$\int_{x_{in}}^{x_{out}}(\tau_w\cos\xi)\,2\pi r\frac{dx}{\cos\xi} = \int_{x_{in}}^{x_{out}}2\pi r\tau_w\,dx \tag{5.26}$$

On the basis of Darcy's friction factor "f_D" and the corresponding correlation between the wall shear stress and the flow specific kinetic energy, the expression $2\pi r\tau_w$ becomes

$$2\pi r\tau_w = \frac{2\pi r^2}{r}\tau_w = \frac{4A}{D}\tau_w, \quad \tau_w = \frac{f_D}{4}\frac{\rho u|u|}{2} \Rightarrow \frac{4A}{D}\tau_w = f_D\frac{A}{D}\frac{\rho u|u|}{2} \tag{5.27}$$

where u in this expression signifies the average flow velocity component relative to the duct walls and D denotes the hydraulic radius of the duct. This result thus models the wall shear-stress force on the flow.

Force of Gravity

The gravitational contribution to F_x results from the component of the gravitational force along the x-axis as

$$\int_\Omega \rho g_x \, d\Omega = \int_{x_{\text{in}}}^{x_{\text{out}}} \rho g_x A \, dx \tag{5.28}$$

Mechanical-Work Force

Whenever an exchange of work takes place between a flow and a mechanical system, a corresponding force, hence a force per unit length f_{mx} develops that also depends on the system stage efficiency. Because of this correlation, it is convenient to express this force in terms of the originating work w_m per unit mass of flowing fluid per unit length of duct so that the shaft-work contributions to both the linear-momentum and energy equations depend on the single originating work w_m. Following thermodynamics usage, w_m will be positive or negative depending on whether the work is performed by or on the fluid system, for any direction of the flow velocity. In terms of w_m, the work exchanged with the flow per unit time per unit length of duct may be expressed as

$$\dot{w}_m = \rho|u|Aw_m \tag{5.29}$$

which depends on the flow speed $|u|$ so that it is the algebraic sign of w_m that determines whether work is performed by or on the system. These work specifications correspond to a force that is respectively opposite or concurrent with the flow direction. Based on these considerations, the algebraic sign of f_{mx} will have to depend on the sign of both w_m and u. The work $\dot{w}_m$ may then be expressed in terms of the force component f_{mx} as

$$\dot{w}_m = -\chi f_{mx} u \tag{5.30}$$

where χ correlates with the system stage efficiency, with $0 < \chi = \eta_{\text{c}} \leq 1$ or $\chi = 1/\eta_{\text{T}} \geq 1$ respectively for a compressor or a turbine, with $\eta_{\text{C}}, \eta_{\text{T}}$ denoting the corresponding efficiencies. The minus sign in (5.30) is included so that a negative component f_{mx} both corresponds to a positive $\dot{w}_m$ and generates a force that opposes a flow in the positive x-direction. For instance, when the flow progresses in this direction, hence $u > 0$, and performs work on a mechanical shaft, hence $w_m > 0$, f_{mx} must then be negative so that the corresponding work from (5.30) remains positive as in (5.29). By comparing (5.29) and (5.30), the force f_{mx} is expressed in terms of w_m as

$$f_{mx} = -\chi \rho \operatorname{sgn}(u) A w_m \tag{5.31}$$

a force that opposes or favors the flow for all combinations of flow directions and work.

Governing Equation

Since (5.11) applies for arbitrary intervals $(x_{\text{in}}, x_{\text{out}})$ and in view of (5.12), (5.17), (5.19), (5.23), (5.27), (5.28), as well as (5.31), the corresponding differential equation is expressed as

$$\frac{\partial \rho u}{\partial t} + \frac{\partial}{\partial x}\left(\rho u^2 + p\right) + \frac{\rho u^2}{A}\frac{\partial A}{\partial x} =$$

$$-f_D \frac{\rho u|u|}{2D} + \rho g_x - \chi \rho \, \mathrm{sgn}(u) w_m + \frac{b(\Delta p)_{bx}}{A} + \frac{\dot{m}_b^2}{b\rho_{b_*} A} \sin\theta \left(\cos\theta + \frac{\partial A}{\partial x} \frac{\sin\theta}{2\sqrt{\pi A}} \right) \tag{5.32}$$

which corresponds to the generalized linear-momentum equation.

5.2.3 Energy Equation

This section determines the energy equation with specific source terms that correspond to shaft work and mass as well as energy transferred to the flow. The general energy equation arises from the first principle of thermodynamics as

$$\frac{d\mathcal{E}}{dt} = \frac{D}{Dt} \int_{\mathcal{V}(t)} E \, d\mathcal{V} = \frac{\delta \mathcal{Q}}{\delta t} - \frac{\delta \mathcal{W}}{\delta t} \tag{5.33}$$

where $\mathcal{E}$, $\mathcal{Q}$, and $\mathcal{W}$ respectively denote the fluid-system total energy as well as heat transferred to the system and work performed by the system, both positive following thermodynamics specifications.

Through Reynolds' transport theorem and reflecting developments (5.12)-(5.13), the substantive derivative of the integral of E in (5.33) is transformed into

$$\frac{D}{Dt} \int_{\mathcal{V}} E \, d\mathcal{V} = \int_{\mathcal{V}} \frac{\partial E}{\partial t} d\mathcal{V} + \oint_{\partial \mathcal{V}} E \mathbf{u} \cdot \mathbf{n} \, d\Gamma \tag{5.34}$$

Using cross-sectional area averaged variables, the area average of this expression becomes

$$\int_{x_{\mathrm{in}}}^{x_{\mathrm{out}}} \frac{\partial E}{\partial t} A \, dx - EuA\big|_{\mathrm{in}} + EuA\big|_{\mathrm{out}} + \int_{A_b} E_b \mathbf{u}_b \cdot \mathbf{n} \, d\Gamma$$

$$= \int_{x_{\mathrm{in}}}^{x_{\mathrm{out}}} \frac{\partial EA}{\partial t} \, dx + \int_{x_{\mathrm{in}}}^{x_{\mathrm{out}}} \frac{\partial EuA}{\partial x} \, dx - \int_{x_{\mathrm{in}}}^{x_{\mathrm{out}}} \frac{E}{\rho}\bigg|_b \dot{m}_b \left(\cos\theta + \frac{\partial A}{\partial x} \frac{\sin\theta}{2\sqrt{\pi A}} \right) dx \tag{5.35}$$

The total work is expressed as

$$\frac{\delta \mathcal{W}}{\delta t} = \frac{\delta \mathcal{W}_\sigma}{\delta t} + \frac{\delta \mathcal{W}_m}{\delta t} + \frac{\delta \mathcal{W}_g}{\delta t} \tag{5.36}$$

where $\delta \mathcal{W}_\sigma / \delta t$, $\delta \mathcal{W}_m / \delta t$ and $\delta \mathcal{W}_g / \delta t$ indicate the work per unit time respectively due to the effect of the pressure gradient as well as shear stress, mechanical shaft, and gravitational force.

Pressure and Shear-Stress Work

With U_i, $1 \leq i \leq 3$, denoting the components of a circumferentially continuous flow velocity with $U_i \equiv u_i = 0$, on $\partial \mathcal{V} \setminus A_b$, and $U_i n_i = 0$, on A_b, the work per unit time $\delta \mathcal{W}_\sigma / \delta t$ is expressed as

$$-\frac{\delta \mathcal{W}_\sigma}{\delta t} = \oint_{\partial \mathcal{V}} (p)_w \delta_j^i u_i (-n_j) \, d\Gamma + \oint_{\partial \mathcal{V}} \tau_{ij} u_i n_j \, d\Gamma$$

$$= \int_{\partial \mathcal{V} \setminus A_b} P \delta_j^i U_i (-n_j) \, d\Gamma + \int_{A_b} \left(-p_b \delta_j^i + \tau_{ij} \right) u_i n_j \, d\Gamma$$

$$= \oint_{\partial V} P\delta_j^i U_i(-n_j)\, d\Gamma + \int_{A_b} \left(-p_b\delta_j^i + \tau_{ij}\right) u_i n_j\, d\Gamma \tag{5.37}$$

With reference to Figures 5.2- 5.1 and results (5.6)-(5.9), the work per unit time along A_b is expressed as

$$\int_{A_b} \left(p_b\delta_j^i - \tau_{ij}\right)(u_i)_b\, n_j\, d\Gamma = \int_{x_{in}}^{x_{out}} \left(\frac{p}{\rho}\Big|_b\, \dot{m}_b\left(\cos\theta + \frac{\partial A}{\partial x}\frac{\sin\theta}{2\sqrt{\pi A}}\right) - \frac{|\dot{m}_b|}{b}Aw_\tau\right) dx \tag{5.38}$$

this work will vanish in the absence of any mass transfer. In this expression, τ_{w_b} and w_τ respectively indicate the shear stress at the mass-transfer side opening and the corresponding work, per unit mass per unit duct length, expressed as

$$\tau_{w_b} = \frac{f_{D_s}\,\rho_b}{8}(u - u_b\sin\theta)|u - u_b\sin\theta|, \quad w_\tau \equiv \frac{b\,\tau_{w_b}}{A\rho_b}\left(\sin\theta - \frac{\partial A}{\partial x}\frac{\cos\theta}{2\sqrt{\pi A}}\right) \tag{5.39}$$

As was stipulated for pressure in the previous section, the capital letter U denotes in this section the multi-dimensional velocity component, whereas the unsubscripted lower case u indicates an average velocity component. With these specifications the work per unit time of pressure P becomes

$$\oint_{\partial V} Pn_j U_j d\Gamma = \int_\Omega \frac{\partial(PU_j)}{\partial x_j} d\Omega = \int_{x_{in}}^{x_{out}} \overline{\frac{\partial(PU_j)}{\partial x_j}} A\, dx = \int_{x_{in}}^{x_{out}} \frac{\partial(puA)}{\partial x_1} dx \tag{5.40}$$

The last equality is proven beginning with the expression for the average divergence

$$\overline{\frac{\partial(PU_j)}{\partial x_j}} \equiv \frac{1}{A}\int_A \frac{\partial(PU_j)}{\partial x_j} dA = \frac{1}{A}\int_A \frac{\partial(PU_1)}{\partial x_1} dA + \sum_{j=2}^{3}\frac{1}{A}\int_A \frac{\partial(PU_j)}{\partial x_j} dA \tag{5.41}$$

By virtue of Leibnitz's theorem for the differentiation of an integral and with reference to Figure 5.3, the divergence of the area average $\overline{\{PU_j\}}$ is cast as

$$\frac{\partial(\overline{PU_j})}{\partial x_j} \equiv \frac{\partial}{\partial x_j}\underbrace{\left[\frac{1}{A(x_1)}\int_0^{2\pi}\int_0^{R(\phi,x_1)} PU_j\, r\, d\phi dr\right]}_{\text{function of } x_1 \text{ only}} = \frac{\partial(\overline{PU_1})}{\partial x_1} =$$

$$\frac{\partial}{\partial x_1}\left(\frac{1}{A}\right)\left[\int_A PU_1 dA\right] + \frac{1}{A}\int_0^{2\pi}\left[\int_0^{R(\phi,x_1)}\frac{\partial(PU_1)}{\partial x_1}r\, dr + \frac{\partial R}{\partial x_1}RPU_1\right]d\phi =$$

$$-\frac{1}{A}\frac{\partial A}{\partial x_1}\left[\frac{1}{A}\int_A PU_1 dA\right] + \frac{1}{A}\int_A \frac{\partial(PU_1)}{\partial x_1}dA + \int_0^{2\pi}\frac{\partial R}{\partial x_1}RPU_1 d\phi =$$

$$-\frac{1}{A}\frac{\partial A}{\partial x_1}\overline{PU_1} + \overline{\frac{\partial(PU_1)}{\partial x_1}} + \frac{1}{A}\int_0^{2\pi}\frac{\partial R}{\partial x_1}RPU_1 d\phi \tag{5.42}$$

From this result, the correlation between $A(\overline{\partial PU_j/\partial x_j})$ and $A\partial(\overline{PU_j})/\partial x_j$ becomes

$$A\left(\overline{\frac{\partial PU_j}{\partial x_j}}\right) = A\left(\overline{\frac{\partial PU_1}{\partial x_1}}\right) + \sum_{j=2}^{3}\int_A \frac{\partial PU_j}{\partial x_j}dA =$$

$$\frac{\partial A}{\partial x_1}\overline{PU_1} + A\frac{\partial(\overline{PU_1})}{\partial x_1} - \int_0^{2\pi}\frac{\partial R}{\partial x_1}RPU_1 d\phi + \oint_{\Gamma_A} PU_A \cdot n_A d\Gamma =$$

$$\frac{\partial(\overline{PU_1}A)}{\partial x_1} + \int_0^{2\pi}\left[PU_1\left(-\frac{\partial R}{\partial x_1}R\right) + PU_A \cdot n_A\sqrt{\left(\frac{\partial R}{\partial\phi}\right)^2 + R^2}\right]d\phi =$$

$$\frac{\partial(\overline{PU_1}A)}{\partial x_1} + \int_0^{2\pi} PU \cdot n d\phi \tag{5.43}$$

In this expression, with reference to Figure 5.3, U_A and the unit vector n_A respectively denote the vector components, on a plane A, of velocity U and outward-pointing normal vector n, with plane and spatial vectors correlated as

$$U = U_1 i + U_A$$

$$n = \left(-\frac{\partial R}{\partial x_1}R\right)i + n_A\sqrt{\left(\frac{\partial R}{\partial\phi}\right)^2 + R^2}, \quad \Gamma_A \equiv x_2^2 + x_3^2 - R^2(\phi(x_2, x_3), x_1) = 0 \tag{5.44}$$

where i denotes the unit vector in the positive direction of the x-axis. The component of velocity $U \cdot n$ in the direction of the outward-pointing normal vector n vanishes on the lateral surfaces of the duct in Figure 5.1, because of the no-penetration boundary condition for U on the surface of the duct. Because of this condition, the terminal correlation between $A(\overline{\partial PU_j/\partial x_j})$ and $A\partial(\overline{PU_j})/\partial x_j$ exactly becomes

$$A\left(\overline{\frac{\partial PU_j}{\partial x_j}}\right) = \frac{\partial(\overline{PU_1}A)}{\partial x_1} \tag{5.45}$$

This result is further simplified using the area-averaged pressure discussed in the previous section, which implies a pressure-averaged speed u as follows

$$\overline{PU}A = \int_A PU dA = puA \quad \Rightarrow \quad \left(\int_A P dA\right)u = \int_A PU dA \quad \Rightarrow \quad u \equiv \frac{\int_A PU dA}{\int_A P dA} \tag{5.46}$$

This pressure-averaged speed generally differs from the mass-averaged speed in Section 5.1. The variation of $A = A(x)$ must, therefore, allow these two average velocities to remain accurate representations of each other. Under the same conditions, the average of the speed squared and the square of the average speed also accurately represent each other. For a perfect gas: $p = \rho RT$, where T indicates a mass-averaged temperature; hence, the pressure- and mass-averaged speeds can remain accurate representations of each other when T inappreciably varies over A.

Work of Gravity

With reference to the work due to the force of gravity, this is expressed as

$$-\frac{\delta W_g}{\delta t} = \int_\Omega g_x u\rho \, d\Omega = \int_{x_{\text{in}}}^{x_{\text{out}}} g_x mA \, dx \tag{5.47}$$

where the minus sign indicates work performed on the flow.

Mechanical Shaft Work

The shaft work is directly expressed as

$$\frac{\delta \mathcal{W}_m}{\delta t} = \int_{x_{\text{in}}}^{x_{\text{out}}} \ddot{w}_m \, dx = \int_{x_{\text{in}}}^{x_{\text{out}}} \rho |u| A w_m \, dx \tag{5.48}$$

where $\ddot{w}_m$ and w_m respectively indicate the work per unit time per unit duct length and the work per unit mass of flow per unit duct length. In this expression, $\ddot{w}_m$ depends not on u but on the absolute value $|u|$ of u so that independently of the flow direction, hence sign of u, this work will correctly contribute to a decrease or increase of total energy $\mathcal{E}$ in (5.33) depending on whether the work is performed by or on the flow.

Heat Transfer

The total exchanged heat is cast as

$$\frac{\delta \mathcal{Q}}{\delta t} = \oint_A \left. \frac{\delta Q}{\delta t} \right|_S dA + \int_V \left. \frac{\delta Q}{\delta t} \right|_V d\mathcal{V} = \int_{x_{\text{in}}}^{x_{\text{out}}} s_g \, dx \tag{5.49}$$

where the heats transferred to the system $\left. \frac{\delta Q}{\delta t} \right|_S$, $\left. \frac{\delta Q}{\delta t} \right|_V$, and s_g respectively denote heat per unit time and unit surface, heat per unit time and unit volume, and corresponding heating per unit time and unit length of duct.

The connection between s_g and Rayleigh heating is established by developing the generalized energy equation from (5.33) for a steady flow and comparing the result with the energy equation of the classical Rayleigh flow, a non-adiabatic flow with vanishing mass transfer as well as shaft work. In this classical kind of flow, the steady-state energy equation may be expressed as

$$\text{sgn}(u)d\left(c_p T_0\right) = \text{sgn}(u)d\left(c_p T + \frac{u^2}{2}\right) = d\tilde{g} = \frac{d\tilde{g}}{dx}dx = g_\text{T}dx \tag{5.50}$$

where c_p, T_0, T, $\tilde{g}$ and g_T respectively denote the constant-pressure specific heat, stagnation temperature, static temperature, the energy transferred per unit mass and the energy transferred per unit mass per unit length of duct; the sign of u enters this expression so that for a positive g, i.e. heating, the stagnation enthalpy $c_p T_0$ increases or decreases along the positive x-axis depending on whether the flow progresses in the positive or negative x-direction.

In the absence of mass transfer and shaft as well as gravity work, the energy equation from (5.33), in view of (5.35) and (5.40), is expressed as

$$\int_{x_{\text{in}}}^{x_{\text{out}}} \left(\frac{\partial (EA)}{\partial t} + \frac{\partial u A (E+p)}{\partial x} \right) dx = \int_{x_{\text{in}}}^{x_{\text{out}}} s_g dx \tag{5.51}$$

The specialization of this equation for Rayleigh flow requires equation (5.10), which in steady differential form becomes $d(\rho u A) = 0$, and the following expressions: $E \equiv \rho \epsilon + \rho u^2/2$, $\epsilon = c_v T$, $p = \rho R T$, where ϵ, c_v, and $R = c_p - c_v$ denote internal energy, constant-volume specific heat, and gas constant. With these expressions, the differential steady-flow form of equation (5.51) becomes

$$d\left(\rho u A \frac{E+p}{\rho}\right) = s_g \, dx \;\Rightarrow\; d\left(\rho u A\right) \frac{E+p}{\rho} + (\rho u A)\, d\left(\epsilon + \frac{u^2}{2} + \frac{p}{\rho}\right) =$$

$$(\rho u A)\, d\left(c_v T + RT + \frac{u^2}{2}\right) = (\rho u A)\, d\left(c_p T + \frac{u^2}{2}\right) = s_g \, dx \qquad (5.52)$$

Comparing this result with (5.50) thus shows that g_T and s_g are connected as

$$s_g = (\rho |u| A) g_T \qquad (5.53)$$

In this expression, s_g depends not on u but on the absolute value $|u|$ of u so that independently of the flow direction, hence sign of u, this heat-transfer term will correctly contribute to an increase or decrease of total energy $\mathcal{E}$ in (5.33) depending on whether the flow is heated or cooled.

Governing Equation

Since (5.33) applies for arbitrary intervals $(x_{\mathrm{in}}, x_{\mathrm{out}})$ and in view of (5.35), (5.38), (5.40), (5.47), (5.48), (5.49), as well as (5.53), the associated differential equation is expressed

$$\frac{\partial E}{\partial t} + \frac{\partial u(E+p)}{\partial x} + \frac{u(E+p)}{A} \frac{\partial A}{\partial x} =$$

$$+ \rho |u| \left(g_T - w_m\right) + \rho u g_x - \frac{|\dot{m}_b|}{b} w_\tau + \left.\frac{E+p}{\rho}\right|_b \frac{\dot{m}_b}{A}\left(\cos\theta + \frac{\partial A}{\partial x}\frac{\sin\theta}{2\sqrt{\pi A}}\right) \qquad (5.54)$$

which corresponds to the generalized energy equation.

5.2.4   Compressible-Flow System

The governing equations are made non-dimensional via the reference pressure p_{ref}, reference density ρ_{ref}, reference speed $u_{\mathrm{ref}} = \sqrt{p_{\mathrm{ref}}/\rho_{\mathrm{ref}}}$, duct length L, reference time $t_{\mathrm{ref}} = L/u_{\mathrm{ref}}$, and reference cross-sectional throat area A_*, typically the straight-duct area or, for a nozzle, the upstream sonic-flow area.

With subscript "dim" denoting a dimensional variable, the non-dimensional mass-transfer rate, duct diameter, width of mass-transfer port, total energy, shaft as well as shear-stress work, and heat transfer terms in these equations correlate with dimensional counterparts as

$$\dot{m}_b = \frac{L(\dot{m}_b)_{\mathrm{dim}}}{\rho_{\mathrm{ref}} u_{\mathrm{ref}} A_*}, \quad D = \frac{(D)_{\mathrm{dim}}}{\sqrt{A_*}} = \frac{2}{\sqrt{\pi}}\sqrt{\frac{(A)_{\mathrm{dim}}}{A_*}}$$

$$b = \frac{L(b)_{\mathrm{dim}}}{A_*}, \quad E = \frac{(E)_{\mathrm{dim}}}{p_{\mathrm{ref}}}, \quad w_{m,\tau} = \frac{L(w_{m,\tau})_{\mathrm{dim}}}{p_{\mathrm{ref}}/\rho_{\mathrm{ref}}}, \quad g_{x,\mathrm{T}} = \frac{L(g_{x,\mathrm{T}})_{\mathrm{dim}}}{p_{\mathrm{ref}}/\rho_{\mathrm{ref}}} \qquad (5.55)$$

where $w_{m,\tau}$ synthetically stands for w_m and w_τ. The total mass transfer $\widetilde{m}_b$ is then defined as

$$\widetilde{m}_b \equiv \dot{m}_b \left(\cos\theta + \frac{\partial A}{\partial x} \frac{(\sqrt{A_*}/L)\sin\theta}{2\sqrt{\pi A}} \right) \tag{5.56}$$

With these specifications and $s \equiv \mathrm{sgn}(m)$, the non-dimensional generalized Euler system becomes

$$\begin{cases} \dfrac{\partial\rho}{\partial t} + \dfrac{\partial m}{\partial x} + \dfrac{m}{A}\dfrac{\partial A}{\partial x} = \dfrac{\widetilde{m}_b}{A} \\[2ex] \dfrac{\partial m}{\partial t} + \dfrac{\partial}{\partial x}\left(\dfrac{m^2}{\rho}+p\right) + \dfrac{m^2}{\rho A}\dfrac{\partial A}{\partial x} = \\[2ex] \quad -f_D \dfrac{m|m|}{2\rho D\sqrt{A_*}/L} + \rho g_x - s\chi\,\rho w_m + \dfrac{b(\Delta p)_{b_x}}{A} + \dfrac{\dot{m}_b\widetilde{m}_b}{b\rho_b A}\sin\theta \\[2ex] \dfrac{\partial E}{\partial t} + \dfrac{\partial}{\partial x}\left(m\dfrac{(E+p)}{\rho}\right) + \dfrac{m(E+p)}{\rho A}\dfrac{\partial A}{\partial x} = \\[2ex] \quad + |m|\,(g_{\mathrm{T}} - w_m) + mg_x - \dfrac{|\dot{m}_b|}{b}w_\tau + \left.\dfrac{E+p}{\rho}\right|_b \dfrac{\widetilde{m}_b}{A} \end{cases} \tag{5.57}$$

With respect to an inertial reference frame, the generalized quasi-1D Euler conservation law system becomes

$$\frac{\partial q}{\partial t} + \frac{\partial f(q)}{\partial x} = \phi \tag{5.58}$$

where the independent variable (x,t) varies in the domain $D \equiv \Omega \times [t_o, T]$, $\Omega \equiv [x_{\mathrm{in}}, x_{\mathrm{out}}]$. This system consists of the continuity, linear-momentum and total-energy equations, with arrays of dependent variables $q = q(x,t)$, $f = f(q)$ and $\phi = \phi(x,q)$ defined as

$$q \equiv \left\{\begin{array}{c} \rho \\ m \\ E \end{array}\right\}, \quad f(q) \equiv \left\{\begin{array}{c} m \\[1ex] \dfrac{m^2}{\rho}+p \\[1ex] \dfrac{m}{\rho}(E+p) \end{array}\right\}$$

$$\phi \equiv -\frac{m}{\rho A}\frac{\partial A}{\partial x}\left\{\begin{array}{c} \rho \\ m \\ E+p \end{array}\right\} + \frac{\widetilde{m}_b}{A}\left\{\begin{array}{c} 1 \\[1ex] \dfrac{\dot{m}_b}{b\rho_b}\sin\theta \\[1ex] \left.\dfrac{E+p}{\rho}\right|_b \end{array}\right\}$$

$$-\left\{\begin{array}{c} 0 \\[1ex] f_D\dfrac{m|m|}{2\rho D\sqrt{A_*}/L} - \rho g_x + s\,\rho w_m - \dfrac{b(\Delta p)_{b_x}}{A} \\[1ex] |m|(w_m - g_{\mathrm{T}}) - mg_x + |\dot{m}_b|w_\tau/b \end{array}\right\} \tag{5.59}$$

The fluid and its thermodynamic state transferred to the nozzle flow may generally differ from the nozzle fluid and local thermodynamic conditions, so that the transferred fluid pressure p_b, density ρ_b, and enthalpy $((E + p)/\rho)_b$ may not coincide with the local state. The results discussed in this book, however, correspond to equality of fluids and thermodynamic conditions with negligible shear-stress work. The resulting generalized system (5.57) is computationally solved using a Galilean invariant formulation for conservation - law systems with a source term, with computational results that reflect the available independently derived analytical solutions.

5.2.5   Incompressible-Flow System

Using the previous reference variables for incompressible flows, the governing non-dimensional system becomes

$$M\frac{\partial q}{\partial t} + \frac{\partial f(q)}{\partial x} = \phi, \quad M = \begin{pmatrix} 0 & 0 & 0 \\ 0 & 1 & 0 \\ 0 & 0 & 1 \end{pmatrix} \tag{5.60}$$

and system (5.60) consists of the continuity, linear-momentum, and energy equations. For a constant density ρ, the arrays $q = q(x, t)$, $f = f(q)$ and $\phi = \phi(x, q)$ are then defined as

$$q \equiv \left\{ \begin{array}{c} p \\ u \\ E/\rho \end{array} \right\}, \quad f(q) \equiv \left\{ \begin{array}{c} u \\ u^2 + p/\rho \\ u\dfrac{E+p}{\rho} \end{array} \right\}, \quad \phi \equiv -\frac{u}{A}\frac{\partial A}{\partial x}\left\{ \begin{array}{c} 1 \\ u \\ \dfrac{E+p}{\rho} \end{array} \right\}$$

$$+ \frac{\widetilde{\dot{m}_b}}{\rho A} \left\{ \begin{array}{c} 1 \\ \dfrac{\dot{m}_b}{b\rho_b}\sin\theta \\ \dfrac{E+p}{\rho}\Big|_b \end{array} \right\} - \left\{ \begin{array}{c} 0 \\ f_D\dfrac{u|u|}{2D\sqrt{A_*}/L} - g_x + s\chi\,w_m - \dfrac{b(\Delta p)_{b_x}}{\rho A} \\ |u|(w_m - g_{\text{T}}) - ug_x + |\dot{m}_b|w_{\text{T}}/(\rho b) \end{array} \right\} \tag{5.61}$$

where ρ now stands for a constant reference density, whereas u, p and E respectively denote the area averaged speed, static pressure and volume-specific total energy. No completing equation of state for pressure is needed, for the first two governing equations yield both p and u. With these two dependent variables, the third equation provides the distribution of energy. Thermodynamically expressed as

$$E = \rho c T + \frac{\rho u^2}{2}, \quad T = \frac{E - \rho u^2/2}{\rho c}, \quad E > \frac{\rho u^2}{2} \tag{5.62}$$

with "c" the specific heat, the total energy E in this expression leads to the temperature within the flow. The constraint on the kinetic energy limits the maximum amount of the combination of work and heat that can be extracted from the system.

5.3 Open-Channel Configuration

The following developments relate to the open-channel flows that occur over a variously shaped channel bed, as illustrated in Figure 5.4. This figure shows the vertically measured free-surface height $h = h(x, y, t)$ and contoured-channel bathymetry $b(x, y, t)$; h also depends upon time because of the flow, whereas b may depend upon time due to seismic events. The figure also shows the outward pointing unit vectors n_s and n_b, respectively normal to the free surface and channel bed, vectors that are needed in subsequent derivations. These unit vectors may be expressed in terms of the constant surface functions F_s and F_b, respectively describing the free surface and channel bed, as

$$n_s = \frac{-\dfrac{\partial(b+h)}{\partial x_1}i - \dfrac{\partial(b+h)}{\partial x_2}j + k}{\sqrt{\left(\dfrac{\partial(b+h)}{\partial x_1}\right)^2 + \left(\dfrac{\partial(b+h)}{\partial x_2}\right)^2 + 1}}, \quad n_b = \frac{\dfrac{\partial b}{\partial x_1}i + \dfrac{\partial b}{\partial x_2}j - k}{\sqrt{\left(\dfrac{\partial b}{\partial x_1}\right)^2 + \left(\dfrac{\partial b}{\partial x_2}\right)^2 + 1}} \tag{5.63}$$

where i, j, k respectively denote the unit vectors in the positive x, y, z directions. The following derivations will also use the convenient equivalences: $x \equiv x_1$, $y \equiv x_2$, $z \equiv x_3$.

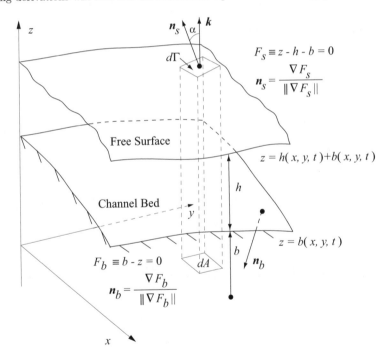

Figure 5.4: Channel Bed and Free Surface

With reference to this figure, the connection between the element of surface area $d\Gamma$ and its projection dA onto the xy plane is expressed as

$$dA = d\Gamma \cos\alpha = d\Gamma\, \boldsymbol{n}_s \cdot \boldsymbol{k} = \frac{d\Gamma}{\sqrt{\left(\frac{\partial(b+h)}{\partial x_1}\right)^2 + \left(\frac{\partial(b+h)}{\partial x_2}\right)^2 + 1}} \tag{5.64}$$

The z component of velocity of a free-surface fluid particle is then expressed as

$$w\big|_{\text{FS}} \equiv \frac{Dz_{\text{FS}}}{Dt} = \frac{D(b+h)}{Dt} = \frac{\partial(b+h)}{\partial x_1}\, u\big|_{\text{FS}} + \frac{\partial(b+h)}{\partial x_2}\, v\big|_{\text{FS}} + \underbrace{\frac{\partial b}{\partial t}}_{=\,0} + \frac{\partial h}{\partial t} \tag{5.65}$$

These two results are employed in the derivation of the quasi-one-dimensional open-channel flow equations in Section 5.4.

5.3.1   Depth Averages

Open-channel free-surface flows are practically investigated by way of averaged dependent variables in the generalized continuity, linear-momentum, energy and species governing equations. Figure 5.5 depicts a fluid region of depth "h". With reference to this figure, the amount of mass within the region may be expressed as

$$\mathcal{M} = \int_\Omega \rho\, d\Omega = \int_A dA \int_b^{b+h} \rho\, dz = \int_A \bar{\rho}h\, dA \tag{5.66}$$

where $\bar{\rho}$ denotes the depth-averaged density, which remains constant for incompressible flows. Similarly, $\overline{\rho\boldsymbol{u}}$ and $\widehat{\overline{E}}$ denote the depth-averaged volume-specific linear momentum vector and total energy. These averages are defined as

$$\bar{\rho} \equiv \frac{1}{h}\int_b^{b+h} \rho\, dz, \quad \overline{\rho\boldsymbol{u}} \equiv \frac{1}{h}\int_b^{b+h} \rho\boldsymbol{u}\, dz, \quad \widehat{\overline{E}} \equiv \frac{1}{h}\int_b^{b+h} \widehat{E}\, dz \tag{5.67}$$

The average velocity is then based on the average linear momentum vector and density as

$$\bar{\boldsymbol{u}} \equiv \frac{\overline{\rho\boldsymbol{u}}}{\bar{\rho}} = \frac{\displaystyle\int_b^{b+h} \boldsymbol{u}\rho\, dz}{\displaystyle\int_b^{b+h} \rho\, dz} \tag{5.68}$$

which corresponds to a "mass averaged" velocity. Considering that ρ remains constant, for any variation of "$h = h(x, y, t)$", the depth and mass averaged velocities remain identical.

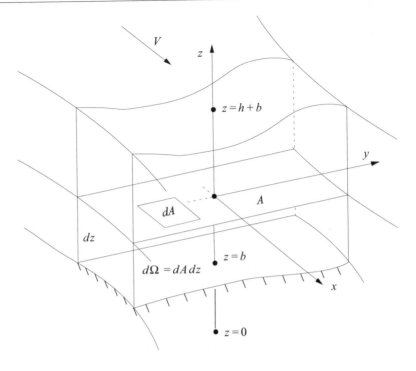

Figure 5.5: Free-Surface Flow Region

5.3.2 Area Averages

For quasi-one-dimensional open-channel flows, with reference to Figure 5.6, the cross-sectional area $S = S(x,t)$, channel-width averaged height $\overline{h}$, and reference channel width W are expressed as

$$S(x,t) = \int_{-W_-}^{W_+} h(x,y,t)\,dy$$

$$\overline{h} = \frac{1}{W_+ + W_-} \int_{-W_-}^{W_+} h(x,y,t)\,dy, \quad W \equiv \frac{S(x,t)}{\overline{h}} = W_+ + W_- \tag{5.69}$$

where W_- and W_+ not only correspond to the distances from respectively the left and right channel banks to the origin of the y axis, but also lead to a variable channel width. In a flow that involves measurable changes in free surface height, this variable width can remain independent of time, $W = W(x)$, when the vertical banks of the flow open channel are sufficiently tall, as represented in Figure 5.9.

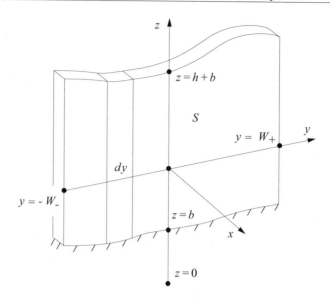

Figure 5.6: Flow Cross Area

Figure 5.7 depicts a fluid layer within a channel flow with channel width W. With reference to this figure, the amount of mass (5.66) within the region may be further expressed as

$$M = \int_A \bar{\rho} h \, dA = \int_{x_{\text{in}}}^{x_{\text{out}}} dx \int_{-W_-}^{W_+} \bar{\rho} h \, dy = \int_{x_{\text{in}}}^{x_{\text{out}}} \bar{\bar{\rho}} dx \int_{-W_-}^{W_+} h \, dy = \int_{x_{\text{in}}}^{x_{\text{out}}} \bar{\bar{\rho}} h W \, dx \qquad (5.70)$$

where $\bar{\bar{\rho}}$ denotes an average density. Likewise, $\widehat{\bar{E}}$ represents an average volume-specific total energy. With reference to (5.70), these averages are expressed as

$$\bar{\bar{\rho}} \equiv \frac{\displaystyle\int_{-W_-}^{W_+} \bar{\rho} h \, dy}{\displaystyle\int_{-W_-}^{W_+} h \, dy} = \frac{1}{S} \int_{-W_-}^{W_+} \int_{b}^{b+h} \rho \, dy dz, \qquad \widehat{\bar{E}} \equiv \frac{\displaystyle\int_{-W_-}^{W_+} \widehat{E} h \, dy}{\displaystyle\int_{-W_-}^{W_+} h \, dy} = \frac{1}{S} \int_{-W_-}^{W_+} \int_{b}^{b+h} \widehat{E} \, dy dz$$

$$(5.71)$$

which show that $\bar{\bar{\rho}}$ and $\widehat{\bar{E}}$ correspond to the area-averaged density and total energy over the cross-sectional area S.

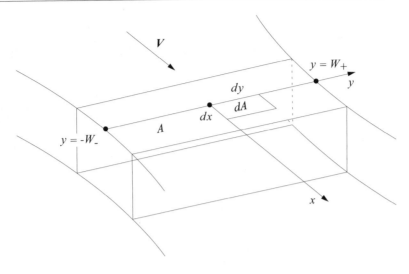

Figure 5.7: Flow Layer

Similarly, the total linear momentum within the region becomes

$$\mathcal{M}\boldsymbol{V}_G = \int_A \overline{\rho\boldsymbol{u}}h\,dA = \int_{x_{in}}^{x_{out}} dx \int_{-W_-}^{W_+} \overline{\rho\boldsymbol{u}}h\,dy = \int_{x_{in}}^{x_{out}} \overline{\overline{\rho\boldsymbol{u}}}dx \int_{-W_-}^{W_+} h\,dy = \int_{x_{in}}^{x_{out}} \overline{\overline{\rho\boldsymbol{u}}}hW\,dx$$
(5.72)

where $\overline{\overline{\rho\boldsymbol{u}}}$ denotes the area-averaged linear momentum. This average and the corresponding mass-averaged velocity $\overline{\overline{\boldsymbol{u}}}$ are expressed as

$$\overline{\overline{\rho\boldsymbol{u}}} \equiv \frac{\int_{-W_-}^{W_+} \overline{\rho\boldsymbol{u}}h\,dy}{\int_{-W_-}^{W_+} h\,dy} = \frac{1}{S}\int_{-W_-}^{W_+}\int_b^{b+h} \rho\boldsymbol{u}\,dydz, \qquad \overline{\overline{\boldsymbol{u}}} = \frac{\overline{\overline{\rho\boldsymbol{u}}}}{\overline{\overline{\rho}}} = \frac{\int_S \rho\boldsymbol{u}\,dydz}{\int_S \rho\,dydz} \qquad (5.73)$$

5.4 Quasi One-Dimensional Free-Surface Open-Channel Flow Equations

The quasi one-dimensional free-surface open-channel flow equations evolve from the principles of mass conservation, 2nd law of mechanics, and first principle of thermodynamics, as Reynolds' transport theorem is applied to the integral expressions for mass, linear momentum, energy, and species partial density. Flows with a free surface involve a liquid, hence the density in these flows remains constant. The equations in this section model not only the traditional effects of wall friction and variable bathymetry, but also the presence of heat transfer, shaft work and mass transfer due to a tributary or distributary flow. The following derivations dispense with the overbars of averages, for simplicity, even though all dependent variables represent averaged quantities.

Figure 5.8 displays the transfer of mass to a channel flow through a side branch with cross-sectional area A_b. This perspective figure simply illustrates the flow and its geometry, although no transversal variations of depth-averaged height h and other variables take place in a one-dimensional flow.

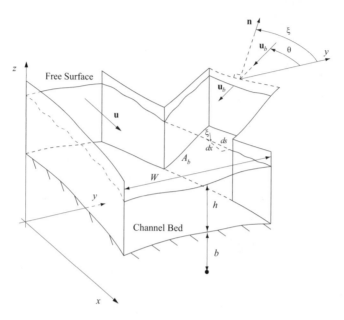

Figure 5.8: Free-Surface Open-Channel and Tributary Flow

This mass is transferred with free-surface height h_b and a velocity $\mathbf{u}_b$, at an angle θ with respect to the direction perpendicular to the channel x-axis; the corresponding mass flow rate per unit length $\dot{m}_b$ will be positive when mass is added to the channel flow.

5.4.1   Continuity Equation

The mass $\mathcal{M}$ of an amount of fluid that flows in a channel and through a lateral tributary or distributary will not change as a result of the flow, which implies

$$\frac{D\mathcal{M}}{Dt} = \frac{D}{Dt}\int_{\mathcal{V}(t)} \rho \, d\mathcal{V} = 0 \tag{5.74}$$

where $\mathcal{V}$ denotes the instantaneous volume occupied by the fluid. With respect to a reference frame fixed to the channel bed, Reynolds' transport theorem transforms this statement into

$$\int_{\mathcal{V}} \frac{\partial \rho}{\partial t} d\mathcal{V} + \oint_{\partial \mathcal{V}} \rho \mathbf{u} \cdot \mathbf{n} \, d\Gamma = 0 \tag{5.75}$$

where $\mathbf{n}$ denotes the outward pointing unit vector, perpendicular to every fluid-surface facet, and $\partial \mathcal{V} \equiv A_{\mathrm{in}} \cup A_{\mathrm{out}} \cup A_w \cup A_{\mathrm{FS}} \cup A_b$, with A_w, A_{FS}, and A_b respectively denoting the area of the lateral solid surface of the channel, the area of the free surface, and the area of the lateral surface of the amount of fluid flowing through the branch channel. For a moderate curvature of the channel and using depth- as well as width-averaged variables, the average-variable counterpart of result (5.75) is expressed as

$$\int_{x_{\mathrm{in}}}^{x_{\mathrm{out}}} \frac{\partial \rho}{\partial t} h W \, dx - \rho u h W \big|_{\mathrm{in}} + \rho u h W \big|_{\mathrm{out}} + \int_{A_{\mathrm{FS}}} \rho \mathbf{u} \cdot \mathbf{n}_s \, d\Gamma + \int_{A_b} \rho_b \mathbf{u}_b \cdot \mathbf{n} \, d\Gamma = 0 \qquad (5.76)$$

where $W = W(x)$ denotes the time-independent channel width, while the integral over A_w vanishes because of the no-penetration boundary condition $\mathbf{u} \cdot \mathbf{n} = 0$ on this area.

On the basis of (5.63), (5.64), (5.65) and using averaged variables, the integral over the free-surface area A_{FS} is expressed as

$$\int_{A_{\mathrm{FS}}} \rho \mathbf{u} \cdot \mathbf{n}_s \, d\Gamma = \int_{A_{\mathrm{FS}}} \rho \frac{\left(-\frac{\partial (b+h)}{\partial x_1} u \big|_{\mathrm{FS}} - \frac{\partial (b+h)}{\partial x_2} v \big|_{\mathrm{FS}} + w \big|_{\mathrm{FS}} \right) d\Gamma}{\sqrt{\left(\frac{\partial (b+h)}{\partial x_1} \right)^2 + \left(\frac{\partial (b+h)}{\partial x_2} \right)^2 + 1}}$$

$$= \int_A \rho \frac{\partial h}{\partial t} dA \Rightarrow \int_{x_{\mathrm{in}}}^{x_{\mathrm{out}}} \rho \frac{\partial h}{\partial t} W \, dx \qquad (5.77)$$

With reference to Figure 5.8 and using depth-averaged variables, the mass-transfer integral is transformed as

$$\int_{A_b} \rho_b \mathbf{u}_b \cdot \mathbf{n} \, d\Gamma \Rightarrow \int_{A_b} -\rho_b u_b \cos(\xi - \theta) h_b \, ds = \int_{x_{\mathrm{in}}}^{x_{\mathrm{out}}} -\rho_b u_b h_b \cos(\xi - \theta) \frac{dx}{\cos \xi}$$

$$= \int_{x_{\mathrm{in}}}^{x_{\mathrm{out}}} -(\rho_b u_b h_b)(\cos \theta + \tan \xi \sin \theta) \, dx = -\int_{x_{\mathrm{in}}}^{x_{\mathrm{out}}} (\dot{m}_b) \left(\cos \theta + \frac{\partial W}{\partial x} \frac{\sin \theta}{2} \right) dx \qquad (5.78)$$

where the integrand expression corresponds to the rate of mass per unit length transferred to the channel flow, $(\dot{m}_b) \equiv \rho_b u_b h_b$, $\rho_b \simeq \rho$, denotes the mass flow rate per unit length in the direction of $\mathbf{u}_b$, and $\tan \xi = \frac{\partial y}{\partial x}$ is expressed in terms of the gradient of the cross-sectional channel width W as $\frac{\partial y}{\partial x} = \frac{\partial W}{\partial x}/2$. The expression between parenthesis multiplying $\dot{m}_b$ in this integral identically reduces to unity for $\mathbf{u}_b$ perpendicular to the x-axis. With these results for the mass-flux integrals, statement (5.76) is further developed as

$$\int_{x_{\mathrm{in}}}^{x_{\mathrm{out}}} \frac{\partial \rho h W}{\partial t} \, dx + \int_{x_{\mathrm{in}}}^{x_{\mathrm{out}}} \frac{\partial \rho u h W}{\partial x} \, dx - \int_{x_{\mathrm{in}}}^{x_{\mathrm{out}}} \dot{m}_b \left(\cos \theta + \frac{\partial W}{\partial x} \frac{\sin \theta}{2} \right) dx = 0 \qquad (5.79)$$

Applying for arbitrary intervals $(x_{\mathrm{in}}, x_{\mathrm{out}})$, this expression implies the equation

$$\frac{\partial h}{\partial t} + \frac{\partial h u}{\partial x} + \frac{h u}{W} \frac{\partial W}{\partial x} = \frac{\dot{m}_b}{\rho W} \left(\cos \theta + \frac{\partial W}{\partial x} \frac{\sin \theta}{2} \right) \qquad (5.80)$$

which corresponds to the continuity equation for free-surface incompressible flows in a variable-width channel with mass transfer.

5.4.2   Momentum Equation

The depth-averaged quasi-one-dimensional linear-momentum equation arises from the second law of Newtonian mechanics as

$$\frac{d(\mathcal{M}u_{CG})}{dt} = F_x \Rightarrow \frac{D}{Dt} \int_{\mathcal{V}(t)} \rho u \, d\mathcal{V} = F_x \tag{5.81}$$

In this expression, F_x denotes the total external force on the fluid within $\mathcal{V}$, while $\mathcal{M}$, u_{CG}, and $\partial\mathcal{V}$ respectively denote the mass of the flowing amount of fluid, the x-axis component of the velocity of its center of gravity, and the surface that entirely envelops a nozzle-shaped mass of fluid flowing in the x-direction.

Using Reynolds' transport theorem, the substantive derivative of the integral of ρu in (5.81) becomes

$$\frac{D}{Dt} \int_{\mathcal{V}} \rho u \, d\mathcal{V} = \int_{\mathcal{V}} \frac{\partial \rho u}{\partial t} d\mathcal{V} + \oint_{\partial\mathcal{V}} \rho u \mathbf{u} \cdot \mathbf{n} \, d\Gamma \tag{5.82}$$

Reflecting developments (5.75)-(5.79) and employing depth- and width-averaged variables, this integral statement in terms of average variables is expressed as

$$\int_{x_{in}}^{x_{out}} \frac{\partial \rho u}{\partial t} hW \, dx - \rho u^2 hW \Big|_{in} + \rho u^2 hW \Big|_{out} + \int_{A_{FS}} \rho(u)_x \mathbf{u} \cdot \mathbf{n}_s \, d\Gamma + \int_{A_b} \rho(u_b)_x \mathbf{u}_b \cdot \mathbf{n} \, d\Gamma \Rightarrow$$

$$\int_{x_{in}}^{x_{out}} \frac{\partial \rho u hW}{\partial t} dx + \int_{x_{in}}^{x_{out}} \frac{\partial \rho u^2 hW}{\partial x} dx - \int_{x_{in}}^{x_{out}} \frac{\dot{m}_b^2}{\rho h_b} \sin\theta \left(\cos\theta + \frac{\partial W}{\partial x} \frac{\sin\theta}{2} \right) dx \tag{5.83}$$

The total external force F_x in the x-direction results from the pressure gradient, viscous shear stress, and the action of any mechanical shaft interacting with the flow. With reference to the effect of gravity, this is incorporated within pressure as shown in this section. The force F_x is thus expressed as

$$F_x = \oint_{\partial\Omega} P\delta_j^1(-n_j) d\Gamma + \oint_{\partial\Omega} \tau_{1j} n_j d\Gamma + \int_{x_{in}}^{x_{out}} f_{mx} \, dx \tag{5.84}$$

where $\partial\Omega$, P, τ_{1j}, and f_{mx} respectively denote the surface that entirely envelops an amount of mass of fluid flowing in the x-direction, the multi-dimensional pressure, indicated in this section with a capital letter to distinguish it from the average pressure p, the components of the viscous shear stress tensor in the x-direction, and the shaft force per unit length.

Shaft-Work Force

As noted in Section 5.2.2, whenever an exchange of work takes place between a flow and a mechanical system, a corresponding force, hence a force per unit length f_{mx} develops. Because of this correlation, it is convenient to express this force in terms the originating work w_m per unit mass of flowing fluid per unit length of channel so that the shaft-work contributions to both the linear-momentum and energy equations will depend on the single originating work w_m. Following the developments in Section 5.2.2, the expressions for the work exchanged with the flow per unit time per unit length of duct and force component f_{mx} are cast as

$$\dot{w}_m = \rho|u|hW w_m, \quad \dot{w}_m = -\chi f_{mx} u \Rightarrow f_{mx} = -\chi\rho \, \mathrm{sgn}(u)hW w_m \tag{5.85}$$

Pressure Force

In respect of the force due to pressure, this force is expressed as

$$F_x = \oint_{\partial\Omega} P\delta_j^1(-n_j)d\Gamma = -\int_\Omega \frac{\partial P}{\partial x_1}d\Omega = -\int_\Omega \frac{\partial P}{\partial x}dAdz = -\int_A \overline{\frac{\partial P}{\partial x}}hdA \tag{5.86}$$

where

$$\overline{\frac{\partial P}{\partial x}}h = \int_b^{b+h} \frac{\partial P}{\partial x}dz \tag{5.87}$$

In open-channel flows, the components of acceleration and viscous force in the z direction remain negligible. The third component of the incompressible-flow linear-momentum Navier-Stokes equations thus yields the hydrostatics expression

$$\frac{\partial P}{\partial z} = -\rho g \tag{5.88}$$

where g denotes the acceleration of gravity. The distribution of pressure arises from this equation as

$$P(x, y, z, t) = P_{\text{atm}} - \rho g(z - h(x, y, t) - b(x, y, t)) \tag{5.89}$$

which leads to

$$\int_b^{b+h} \frac{\partial P}{\partial x}dz = \int_b^{b+h} \rho g\left(\frac{\partial h}{\partial x} + \frac{\partial b}{\partial x}\right)dz = \rho gh\left(\frac{\partial h}{\partial x} + \frac{\partial b}{\partial x}\right) = \frac{\partial}{\partial x}\left(\rho g\frac{h^2}{2}\right) + \rho gh\frac{\partial b}{\partial x} \tag{5.90}$$

The expression $\rho gh^2/2$ is related to an average gauge pressure as follows

$$h\overline{P} = \int_b^{b+h} Pdz = hP_{\text{atm}} + \frac{\rho gh^2}{2} \quad \Rightarrow \quad h\left(\overline{P} - P_{\text{atm}}\right) = \frac{\rho gh^2}{2} \tag{5.91}$$

Integral (5.86) becomes

$$F_x = -\int_A \left[\frac{\partial}{\partial x}\left(\rho g\frac{h^2}{2}\right) + \rho gh\frac{\partial b}{\partial x}\right]dA \tag{5.92}$$

For a channel-width averaged formulation, this integral is further expressed as

$$F_x = -\int_A \frac{\partial}{\partial x}\left[\left(\rho g\frac{h^2}{2}\right) + \rho gh\frac{\partial b}{\partial x}\right]dA = -\int_{x_{in}}^{x_{out}} \overline{\left[\left(\rho g\frac{h^2}{2}\right) + \rho gh\frac{\partial b}{\partial x}\right]}Wdx$$

$$\overline{\left[\left(\rho g\frac{h^2}{2}\right) + \rho gh\frac{\partial b}{\partial x}\right]} \equiv \frac{1}{W}\int_{-W_-}^{W_+} \left[\left(\rho g\frac{h^2}{2}\right) + \rho gh\frac{\partial b}{\partial x}\right]dy \tag{5.93}$$

and for a relatively small $\partial W/\partial x$, or for a small $\partial h/\partial y$, or both, this result simplifies as

$$\overline{\left[\left(\rho g\frac{h^2}{2}\right) + \rho gh\frac{\partial b}{\partial x}\right]} \simeq \frac{\partial}{\partial x}\left(\rho g\frac{\overline{h}^2}{2}\right) + \rho g\overline{h}\frac{\partial \overline{b}}{\partial x} \tag{5.94}$$

Shear-Stress Force

With implied summation on repeated subscript indices, the expression $\tau_{1j}n_j$ denotes the x-axis component of the shear surface traction on a surface facet with outward pointing unit vector $\boldsymbol{n}$ with components n_j, $1 \leq j \leq 3$. The contribution to F_x from this traction becomes

$$\oint_{\partial\Omega} \tau_{1j}n_j d\Gamma = \oint_{A_w} -(\tau)_x \, dA = \int_{A_{\text{bed}}} -(\tau)_x \, dA + \int_{A_{\text{sides}}} -(\tau)_x \, dA + \int_{A_{\text{free surf.}}} \tau_x \, dA \qquad (5.95)$$

with $(\tau)_x$ indicating the x-axis component of the shear surface traction on the lateral surface, with area $A_w = A_{\text{bed}} + A_{\text{sides}} + A_{\text{free surf.}}$, of the fluid in the open channel, as represented in Figure 5.9.

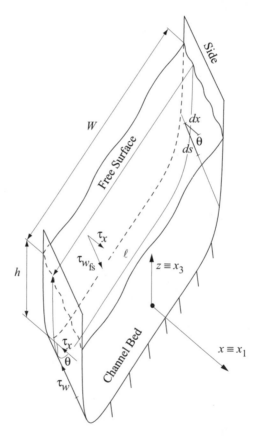

Figure 5.9: Shear Stress on Channel Walls

With τ_w denoting a circumferentially averaged wall shear stress, the shear-stress force integral on the channel bed and free surface become

$$\int_{A_{\text{bed}}} (\tau)_x \, dA = \int_{x_{\text{in}}}^{x_{\text{out}}} \int_{\ell} (\tau_w \cos \theta) \, d\ell \, ds =$$

$$\int_{x_{\text{in}}}^{x_{\text{out}}} \int_{\ell} (\tau_w \cos \theta) \, d\ell \, \frac{dx}{\cos \theta} = \int_{x_{\text{in}}}^{x_{\text{out}}} \int_{\ell} \tau_w \, d\ell \, dx = \int_{x_{\text{in}}}^{x_{\text{out}}} \tau_w \, W \, dx \qquad (5.96)$$

and

$$\int_{A_{\text{free surf.}}} \tau_x \, dA = \int_{x_{\text{in}}}^{x_{\text{out}}} \int_{\ell_{\text{free surf.}}} (\tau_w \cos \theta) \, d\ell \, ds =$$

$$\int_{x_{\text{in}}}^{x_{\text{out}}} \int_{\ell_{\text{free surf.}}} (\tau_w \cos \theta) \, d\ell \, \frac{dx}{\cos \theta} = \int_{x_{\text{in}}}^{x_{\text{out}}} \int_{\ell_{\text{free surf.}}} \tau_w \, d\ell \, dx = \int_{x_{\text{in}}}^{x_{\text{out}}} \tau_{w_{\text{free surf.}}} \, W \, dx \qquad (5.97)$$

The corresponding result for the channel sides becomes

$$\int_{A_{\text{sides}}} (\tau)_x \, dA = \int_{x_{\text{in}}}^{x_{\text{out}}} \int_{\ell_{\text{sides}}} \tau_w \, d\ell \, dx = 2 \int_{x_{\text{in}}}^{x_{\text{out}}} \tau_w \, h \, dx \qquad (5.98)$$

On the basis of Darcy's factor f and the corresponding correlation between the wall shear stress and the flow specific kinetic energy, the expression $(W + 2h)\tau_w$ becomes

$$\tau_w = \frac{f_D}{4} \frac{\rho u |u|}{2} \quad \Rightarrow \quad (W + 2h)\tau_w = (W + 2h)\frac{f_D}{4} \frac{\rho h^2 u |u|}{2h^2} \qquad (5.99)$$

The shear force on the free surface results from wind action. Accordingly, this force may be expressed as

$$\tau_{w_{\text{free surf.}}} = \frac{f_{\text{air}}}{4} \frac{\rho_{\text{air}} (u_{\text{wind}} - u)|u_{\text{wind}} - u|}{2} \qquad (5.100)$$

which depends on the relative velocity $(u_{\text{wind}} - u)$ of the wind with respect to the free surface, where u_{wind} denotes the wind velocity component. These results model the effect of shear-stress forces on an open-channel flow.

Governing Equation

For h denoting a width-averaged free-surface height, integrals (5.83) and (5.84) together with results (5.85), (5.94), (5.99), and (5.100) lead to

$$\frac{\partial m}{\partial t} + \frac{\partial}{\partial x}\left(\frac{m^2}{h} + g\frac{h^2}{2}\right) + \frac{m^2}{hW}\frac{\partial W}{\partial x} = -gh\frac{\partial b}{\partial x} - \text{sgn}(m)hw_m$$

$$-\left(1 + 2\frac{h}{W}\right)\tilde{f}_D \frac{m|m|}{8h^2} + \frac{f_{\text{air}}}{8}\frac{\rho_{\text{air}}}{\rho}(u_{\text{wind}} - u)|u_{\text{wind}} - u| + \frac{\dot{m}_b^2}{\rho^2 h_b W}\sin\theta\left(\cos\theta + \frac{\partial W}{\partial x}\frac{\sin\theta}{2}\right)$$
$$(5.101)$$

which corresponds to the generalized differential linear momentum equation.

5.4.3   Energy Equation

The energy equation for depth-averages flows from the first principle of thermodynamics as

$$\frac{d\mathcal{E}}{dt} = \frac{D}{Dt}\int_{\mathcal{V}(t)} \tilde{E}\, d\mathcal{V} = \frac{\delta\mathcal{Q}}{\delta t} - \frac{\delta\mathcal{W}}{\delta t} \tag{5.102}$$

where $\tilde{E}$, $\mathcal{Q}$, and $\mathcal{W}$ respectively denote the total flow energy per unit volume, heat transferred to the flow system and work performed by the system, both positive following thermodynamics specifications. The total work is expressed as

$$\frac{\delta\mathcal{W}}{\delta t} = \frac{\delta\mathcal{W}_p}{\delta t} + \frac{\delta\mathcal{W}_\tau}{\delta t} + \frac{\delta\mathcal{W}_m}{\delta t} \tag{5.103}$$

with $\delta\mathcal{W}_p/\delta t$, $\delta\mathcal{W}_m/\delta t$ and $\delta\mathcal{W}_m/\delta t$ respectively indicating the pressure-gradient, shear-stress, and mechanical-shaft work per unit time.

Through Reynolds' transport theorem and reflecting developments (5.82)-(5.83), the substantive derivative of the integral of E in (5.102) is transformed into

$$\frac{D}{Dt}\int_{\mathcal{V}} \tilde{E}\, d\mathcal{V} = \int_{\mathcal{V}} \frac{\partial\tilde{E}}{\partial t}\, d\mathcal{V} + \oint_{\partial\mathcal{V}} \tilde{E}\mathbf{u}\cdot\mathbf{n}\, d\Gamma \tag{5.104}$$

Using depth- and width-averaged variables, this expression becomes

$$\int_{x_{\text{in}}}^{x_{\text{out}}} \frac{\partial\tilde{E}}{\partial t} hW\, dx - \tilde{E}uhW\Big|_{\text{in}} + \tilde{E}uhW\Big|_{\text{out}} + \int_{A_{\text{FS}}} \tilde{E}\mathbf{u}\cdot\mathbf{n}_s\, d\Gamma + \int_{A_b} \tilde{E}_{_b}\mathbf{u}_{_b}\cdot\mathbf{n}\, d\Gamma$$

$$= \int_{x_{\text{in}}}^{x_{\text{out}}} \frac{\partial EW}{\partial t}\, dx + \int_{x_{\text{in}}}^{x_{\text{out}}} \frac{\partial EuW}{\partial x}\, dx - \int_{x_{\text{in}}}^{x_{\text{out}}} \frac{E}{\rho h}\Big|_b \dot{m}_b \left(\cos\theta + \frac{\partial W}{\partial x}\frac{\sin\theta}{2}\right) dx \tag{5.105}$$

where $E \equiv \overline{\tilde{E}\tilde{h}}$ denotes the total energy per unit horizontal area.

Mechanical Shaft Work

The shaft work is directly expressed as

$$\frac{\delta\mathcal{W}_m}{\delta t} = \int_{x_{\text{in}}}^{x_{\text{out}}} \dot{w}_m\, dx = \int_{x_{\text{in}}}^{x_{\text{out}}} \rho|u| hW w_m\, dx \tag{5.106}$$

where $\dot{w}_m$ and w_m respectively indicate the work per unit time per unit length of duct and the work per unit mass of flow per unit channel length. In the expression, $\dot{w}_m$ depends not on u but on the absolute value $|u|$ of u so that independently of the flow direction, hence sign of u, this work will correctly contribute to a decrease or increase of total energy $\mathcal{E}$ in (5.102) depending on whether the work is performed by or on the flow.

Pressure Work

The work per unit time due to pressure is calculated by benefitting from the 3-D continuity equation and the distribution of gauge pressure. As it was stipulated for pressure in the previous section, the capital letter U_i, $1 \leq i \leq 3$, denotes in this section the multi-dimensional velocity components, whereas the lower case u_i indicates an average velocity component. Since free-surface flows are incompressible, the corresponding 3-D continuity equation is

$$\frac{\partial U_i}{\partial x_i} = 0 \tag{5.107}$$

for a constant atmospheric pressure P_{atm}, this equation leads to the integral

$$-\int_{\Omega} P_{\text{atm}} \frac{\partial U_i}{\partial x_i} \, d\Omega = \oint_{\partial \mathcal{V}} P_{\text{atm}} U_i(-n_i) \, d\Gamma = 0 \tag{5.108}$$

which is used in the expression for the pressure-gradient work. This type of work is thus cast as

$$\frac{\delta W_p}{\delta t} = \oint_{\partial \mathcal{V}} U_i P \delta_j^i(-n_j) d\Gamma = \oint_{\partial \mathcal{V}} (P - P_{\text{atm}}) \, U_i(-n_i) d\Gamma$$

$$= \int_{\partial \mathcal{V} \backslash A_b} (P - P_{\text{atm}}) \, U_i(-n_i) d\Gamma + \int_{A_b} (P - P_{\text{atm}})_b \, \mathbf{u}_b \cdot (-\mathbf{n}) \, d\Gamma \tag{5.109}$$

as results from using (5.108). With reference to the integral over A_{FS}, this contribution vanishes because $(P - P_{\text{atm}})_{\text{FS}} = (P_{\text{atm}} - P_{\text{atm}})_{\text{FS}} = 0$. The component of velocity $U_i n_i$ in the direction of the outward pointing unit vector also vanishes on the lateral surfaces of the channel configuration in Figure 5.5, whereas $U_i n_i = -U_1$, on S_{in}, and $U_i n_i = U_1$, on S_{out}. With this specification, the pressure work per unit time over $\partial \mathcal{V} \backslash A_b$ becomes

$$\int_{\partial \mathcal{V} \backslash A_b} (P - P_{\text{atm}}) \, U_i(-n_i) d\Gamma = -\left[-\int_{S_{\text{in}}} (P - P_{\text{atm}}) \, U_1 dS + \int_{S_{\text{out}}} (P - P_{\text{atm}}) \, U_1 dS \right] =$$

$$-\left[-\int_{-W_-}^{W_+} \int_b^{b+h} (P - P_{\text{atm}}) \, U_1 dz dy \bigg|_{\text{in}} + \int_{-W_-}^{W_+} \int_b^{b+h} (P - P_{\text{atm}}) \, U_1 dz dy \bigg|_{\text{out}} \right] \tag{5.110}$$

Since $(P - P_{\text{atm}}) > 0$, each of these integrals may be further transformed via an average speed u as follows

$$\int_b^{b+h} (P - P_{\text{atm}}) \, U_1 dz = u \int_b^{b+h} (P - P_{\text{atm}}) \, dz \quad \Rightarrow \quad u \equiv \frac{\int_b^{b+h} (P - P_{\text{atm}}) \, U_1 dz}{\int_b^{b+h} (P - P_{\text{atm}}) \, dz} \tag{5.111}$$

This pressure-averaged speed generally differs from the depth-averaged speed in Section 5.3. With suitable variations of $S = S(x, t)$, these two average velocities remain accurate representations of each other. By virtue of (5.89), the remaining pressure integral in (5.110) becomes

$$\int_b^{b+h} (P - P_{\text{atm}}) \, dz = \rho g \frac{h^2}{2} \tag{5.112}$$

which leads to

$$\int_{-W_-}^{W_+} \int_b^{b+h} (P - P_{\text{atm}}) U dz dy = \int_{-W_-}^{W_+} u \left(\rho g \frac{h^2}{2} \right) dy = \overline{u \rho g \frac{h^2}{2} W} \simeq u \rho g \frac{h^2}{2} W \qquad (5.113)$$

The pressure-work (5.110) thus becomes

$$-\left[-\int_{-W_-}^{W_+} \int_b^{b+h} (P - P_{\text{atm}}) U dz dy \Big|_{\text{in}} + \int_{-W_-}^{W_+} \int_b^{b+h} (P - P_{\text{atm}}) U dz dy \Big|_{\text{out}} \right] =$$

$$- \int_{x_{\text{in}}}^{x_{\text{out}}} \frac{\partial}{\partial x} \left(u \rho g \frac{h^2}{2} W \right) dx \qquad (5.114)$$

With reference to the pressure integral over A_b in (5.109), by virtue of (5.112) and reflecting developments (5.82)-(5.83), this integral is expressed as

$$\int_{A_b} (P - P_{\text{atm}})_b \, \mathbf{u}_b \cdot (-\mathbf{n}) \, d\Gamma = \int_{x_{\text{in}}}^{x_{\text{out}}} \left(\rho g \frac{h^2}{2} \right)_b \frac{1}{\rho h} \Big|_b \dot{m}_b \left(\cos\theta + \frac{\partial W}{\partial x} \frac{\sin\theta}{2} \right) dx \qquad (5.115)$$

Shear-Stress Work

With reference to Figure 5.9 as well as the developments leading to (5.97) and for a negligible shear work at a tributary or distributary confluence, the shear work is due to wind shear as

$$-\frac{\delta W_\tau}{\delta t} = \oint_{\partial V} u_i \tau_{ij} n_j d\Gamma \simeq \int_{x_{\text{in}}}^{x_{\text{out}}} \tau_{w\text{free surf.}} u \, W \, dx = \int_{x_{\text{in}}}^{x_{\text{out}}} \frac{f_{\text{air}}}{4} \frac{\rho_{\text{air}} u (u_{\text{wind}} - u)|u_{\text{wind}} - u|}{2} W \, dx$$

$$(5.116)$$

Heat Transfer

In respect of the total heat $\delta Q/\delta t$, this energy contribution is expressed on the analogy of the developments in Section 5.2.3 as

$$\frac{\delta Q}{\delta t} = \int_\Omega \rho |u| g_\text{T} \, d\Omega = \int_{x_{\text{in}}}^{x_{\text{out}}} \rho |u| h W g_\text{T} \, dx \qquad (5.117)$$

where g_T denotes the energy transferred per unit mass of flow per unit length of duct.

Governing Equation

Integrals (5.105), (5.117), and (5.114) together lead to

$$\frac{\partial E}{\partial t} + \frac{\partial}{\partial x} \left(\frac{m}{h} \left(E + \rho g \frac{h^2}{2} \right) \right) + \frac{m}{hW} \left(E + \rho g \frac{h^2}{2} \right) \frac{\partial W}{\partial x} - \rho |m| (g_\text{T} - w_m)$$

$$= \frac{f_{\text{air}} \rho_{\text{air}}}{8} u (u_{\text{wind}} - u)|u_{\text{wind}} - u| + \left(E + \rho g \frac{h^2}{2} \right)_b \frac{1}{\rho h} \Big|_b \frac{\dot{m}_b}{W} \left(\cos\theta + \frac{\partial W}{\partial x} \frac{\sin\theta}{2} \right) \qquad (5.118)$$

which corresponds to the generalized energy equation. This equation identifies the "enthalpy" H of free-surface flows as

$$H \equiv \frac{1}{h}\left(E + \rho g \frac{h^2}{2} \right) \qquad (5.119)$$

Once the area-specific energy E is computed, the free-surface flow temperature results from

$$\frac{E}{h} = \rho \epsilon + \frac{1}{2}\rho u^2 = \rho c T + \frac{1}{2}\rho u^2 \;\Rightarrow\; T = \frac{E/h - \rho u^2/2}{\rho c}, \quad \frac{E}{h} > \frac{1}{2}\rho u^2 \qquad (5.120)$$

where "c" denotes the specific heat of the fluid in free-surface flow. Since the free-surface flow continuity and linear-momentum equations may be solved independently of the energy equation, the chief use of this equation consists in determining the temperature of the flow. The constraint on the kinetic energy limits the maximum amount of the combination of work and heat that can be extracted from an open-channel flow.

5.4.4 Quasi One-Dimensional Free-Surface System

The governing equations are expressed in non-dimensional form. For the non-dimensional variables, the reference magnitudes are the free-surface height h_{ref}, speed $u_{\mathrm{ref}} = \sqrt{gh_{\mathrm{ref}}}$, with g indicating the acceleration of gravity, duct length L, time $t_{\mathrm{ref}} = L/u_{\mathrm{ref}}$, and channel width W_{ref}, typically the width of a straight channel, or the critical-flow width of a variable-width channel, that is the width at the first upstream location where $u = \sqrt{gh}$. With subscript "dim" denoting a dimensional variable, the non-dimensional channel-bed elevation, friction factor, reference mass transfer rate, total energy and heat transfer as well as shaft work relate to the reference variables as

$$b = \frac{b_{\mathrm{dim}}}{h_{\mathrm{ref}}}, \quad f_D = \frac{\widetilde{f}_D L}{8h_{\mathrm{ref}}}, \quad \dot{m}_b = \frac{L(\dot{m}_b)_{\mathrm{dim}}}{\rho u_{\mathrm{ref}} h_{\mathrm{ref}} W_{\mathrm{ref}}}$$

$$E = \frac{E_{\mathrm{dim}}}{\rho g h_{\mathrm{ref}}^2}, \quad g_{\mathrm{T}} = \frac{L(g_{\mathrm{T}})_{\mathrm{dim}}}{gh_{\mathrm{ref}}}, \quad w_m = \frac{L(w_m)_{\mathrm{dim}}}{gh_{\mathrm{ref}}} \qquad (5.121)$$

The Froude number is then expressed as

$$Fr \equiv \frac{|u_{\mathrm{dim}}|}{\sqrt{gh_{\mathrm{dim}}}} = \frac{|u|u_{\mathrm{ref}}}{\sqrt{gh}\sqrt{h_{\mathrm{ref}}}} = \frac{|u|}{\sqrt{h}} \qquad (5.122)$$

which can thus be directly calculated in terms of non-dimensional variables. On the basis of these specifications, the non-dimensional total mass transfer rate $\widetilde{m}_b$ is defined as

$$\widetilde{m}_b \equiv \dot{m}_b \left(\cos\theta + \sin\theta \frac{\partial W}{\partial x} \frac{W_{\mathrm{ref}}/L}{2} \right) \qquad (5.123)$$

the generalized non-dimensional depth-averaged equations are expressed as

$$
\begin{cases}
\dfrac{\partial h}{\partial t} + \dfrac{\partial m}{\partial x} + \dfrac{m}{W}\dfrac{\partial W}{\partial x} = \dfrac{\widetilde{m}_b}{W} \\[2ex]
\dfrac{\partial m}{\partial t} + \dfrac{\partial}{\partial x}\left(\dfrac{m^2}{h} + \dfrac{h^2}{2}\right) + \dfrac{m^2}{hW}\dfrac{\partial W}{\partial x} = -\left(1 + \dfrac{2h}{W(W_{\mathrm{ref}}/h_{\mathrm{ref}})}\right)f_D\dfrac{m|m|}{h^2} \\[2ex]
\quad - h\dfrac{\partial b}{\partial x} - s\chi\, h w_m + \dfrac{\dot{m}_b \widetilde{m}_b (W_{\mathrm{ref}}/L)\sin\theta}{h_b W} \\[2ex]
\dfrac{\partial E}{\partial t} + \dfrac{\partial}{\partial x}\left(\dfrac{m}{h}\left(E + \dfrac{h^2}{2}\right)\right) + \dfrac{m(E + h^2/2)}{hW}\dfrac{\partial W}{\partial x} = \\[2ex]
\quad |m|\,(g_{\mathrm{T}} - w_m) + \dfrac{f_{\mathrm{air}}}{8}\dfrac{\rho_{\mathrm{air}}}{\rho}u(u_{\mathrm{wind}} - u)|u_{\mathrm{wind}} - u| + \dfrac{(E + h^2/2)_b\,\widetilde{m}_b}{h_b W}
\end{cases}
\tag{5.124}
$$

where $s \equiv \mathrm{sgn}(m)$. With respect to an inertial reference frame, the quasi-1D free-surface conservation law system thus becomes

$$
\frac{\partial q}{\partial t} + \frac{\partial f(q)}{\partial x} = \phi
\tag{5.125}
$$

where the independent variable (x,t) varies in the domain $D \equiv \Omega_1 \times [t_o, T]$, $\Omega_1 \equiv [x_{\mathrm{in}}, x_{\mathrm{out}}]$ and the system consists of the continuity, linear-momentum, and energy equations. The arrays $q = q(x,t)$, $f = f(q)$ and $\phi = \phi(x,q)$ are defined as

$$
q \equiv \left\{\begin{array}{c} h \\ m \\ E \end{array}\right\}, \quad
f(q) \equiv \left\{\begin{array}{c} m \\ \dfrac{m^2}{h} + \dfrac{h^2}{2} \\ \dfrac{m}{h}\left(E + \dfrac{h^2}{2}\right) \end{array}\right\},
$$

$$
\phi \equiv -\left\{\begin{array}{c} 0 \\ \left(1 + \dfrac{2h}{W(W_{\mathrm{ref}}/h_{\mathrm{ref}})}\right)f_D\dfrac{m|m|}{h^2} + h\dfrac{\partial b}{\partial x} + s\chi\, h w_m \\ |m|\,(w_m - g_{\mathrm{T}}) - \dfrac{f_{\mathrm{air}}}{8}\dfrac{\rho_{\mathrm{air}}}{\rho}u(u_{\mathrm{wind}} - u)|u_{\mathrm{wind}} - u| \end{array}\right\}
$$

$$
- \frac{m}{hW}\frac{\partial W}{\partial x}\left\{\begin{array}{c} h \\ m \\ E + \dfrac{h^2}{2} \end{array}\right\} + \frac{\widetilde{m}_b}{W}\left\{\begin{array}{c} 1 \\ \dfrac{\dot{m}_b}{h_b}(W_{\mathrm{ref}}/L)\sin\theta \\ \left.\dfrac{E + h^2/2}{h}\right|_b \end{array}\right\}
\tag{5.126}
$$

which provide the free-surface open-channel flow equations. In the inviscid flux $f(q)$, the term $h^2/2$ corresponds to a gauge pressure force per unit channel width, concisely designated as a "pressure". This interpretation leads to a convenient outlet pressure boundary condition, as discussed in Chapters 12, 16. The fluid and its thermodynamic state transferred to the channel flow may generally differ from the channel fluid and local thermodynamic conditions, so that the transferred fluid pressure p_b, free-surface height h_b, and enthalpy $((E + h^2/2)/h)_b$ may not coincide with the local state. The results discussed in Chapter 12, however, correspond to equality of fluids and thermodynamic conditions with negligible shear-stress work. The resulting generalized system (5.124) is computationally solved using a Galilean invariant formulation for conservation - law systems with a source term, with computational results that reflect the available independently derived analytical solutions. For the initial boundary-value problem, this system is subject to suitable initial and boundary conditions, as amplified in Chapters 12, 16.

5.5 Two-Dimensional Free-Surface Equations

On the analogy of the previous derivations, also the two-dimensional free-surface equations evolve from the principles of mass conservation, 2nd law of mechanics, and 1st principle of thermodynamics, as Reynolds' transport theorem is applied to the depth-averaged expressions for mass, linear momentum, energy and species partial density. Flows with a free surface involve a liquid, hence the density in these flows remains constant. The following derivations dispense with the overbars of averages, for simplicity, even though all dependent variables represent averaged quantities.

5.5.1 Continuity Equation

The expressions for mass and conservation of mass are cast as

$$\mathcal{M} = \int_A \rho h \, dA, \quad \frac{D\mathcal{M}}{Dt} = 0 \tag{5.127}$$

Reynolds' transport theorem applied to $\mathcal{M}$ generates

$$\frac{D\mathcal{M}}{Dt} = \int_A \left(\frac{\partial (\rho h)}{\partial t} + \frac{\partial (\rho h u_j)}{\partial x_j} \right) dA = 0 \tag{5.128}$$

For a linear momentum component $m_j = h u_j$, $1 \leq j \leq 2$, this integral equation leads to

$$\frac{\partial h}{\partial t} + \frac{\partial m_j}{\partial x_j} = 0 \tag{5.129}$$

which corresponds to the 2-D free-surface continuity equation.

5.5.2 Momentum Equation

The expressions for linear momentum and 2nd law of mechanics are

$$L_{x_i} = \int_A \rho u_i h \, dA, \quad \frac{DL_{x_i}}{Dt} = F_{x_i} \tag{5.130}$$

Reynolds' transport theorem applied to L_{x_i} yields

$$\frac{DL_{x_i}}{Dt} = \int_A \left(\frac{\partial (\rho u_i h)}{\partial t} + \frac{\partial (\rho u_i u_j h)}{\partial x_j} \right) dA = F_{x_i} \tag{5.131}$$

The total external force component F_{x_i}, $1 \leq i \leq 2$, results from the three-dimensional distribution of stresses and shaft-work contributions as

$$F_{x_i} = \oint_{\partial \Omega} \sigma_{ij} n_j d\Gamma + \int_\Omega f_{m_i} d\Omega \tag{5.132}$$

where σ_{ij}, $1 \leq i, j \leq 3$, denote the components of the stress tensor and f_{m_i} stands for the components of the mechanical shaft-work force.

Mechanical-Shaft Force

As noted previously, whenever an exchange of work takes place between a two-dimensional flow and a mechanical system, a corresponding shaft force, hence a shaft force per unit area $\boldsymbol{f}_m \equiv \{f_{m_i}\}$ develops that also depends on the system stage efficiency. This force is conveniently correlated with the originating shaft work w_m per unit mass of flowing fluid per unit area. Through this correlation, the shaft-work contributions to both the linear-momentum and energy equations will depend on the single work w_m. This work will be positive or negative depending on whether the work is performed by or on the fluid system, for any direction of the flow velocity. In terms of w_m, the work exchanged with the flow per unit time per unit area may be expressed as

$$\dot{w}_m = \rho h w_m |\boldsymbol{r} \cdot \boldsymbol{u}| \tag{5.133}$$

where the unit vector $\boldsymbol{r}$ lies on the action line of the specific shaft force, in such a way that $w_m \boldsymbol{u} \cdot \boldsymbol{r} \leq 0$, as established next. The expression for $\dot{w}_m$ depends on the magnitude of the dot product $\boldsymbol{u} \cdot \boldsymbol{r}$ so that it is the algebraic sign of w_m that determines whether work is performed by or on the system. These work specifications correspond to a force that respectively favors or opposes the flow. The work $\dot{w}_m$ may then be expressed in terms of the shaft force $\boldsymbol{f}_m$ as

$$\dot{w}_m = -\chi \boldsymbol{f}_m \cdot \boldsymbol{u} = -\chi \left(\|\boldsymbol{f}_m\| \boldsymbol{r} \right) \cdot \boldsymbol{u} \tag{5.134}$$

so that a negative work corresponds to a shaft force that favors the flow. For instance, when the flow progresses in the x_1-direction, hence $u_1 > 0$, and receives work from a mechanical shaft, hence $w_m < 0$, the force component f_{m_1} must then be positive so that the corresponding work from (5.134) remains negative as in (5.133). By comparing (5.133) and (5.134), the magnitude of the shaft force $\boldsymbol{f}_m$ is expressed in terms of w_m as

$$-\chi \|\boldsymbol{f}_m\| \boldsymbol{r} \cdot \boldsymbol{u} = \rho h w_m |\boldsymbol{r} \cdot \boldsymbol{u}| \tag{5.135}$$

Since this magnitude must remain non-negative for any "h" and "w_m", the expression $w_m \boldsymbol{r} \cdot \boldsymbol{u}$ must remain non-positive, as anticipated. The expression for the shaft force $\boldsymbol{f}_m$ thus follows from (5.134)-(5.135) as

$$\boldsymbol{f}_m = \chi \rho h w_m \boldsymbol{r} \tag{5.136}$$

a force that opposes or favors the flow for all combinations of flow directions and work.

Stress Force

The three-dimensional distribution of stresses contain the multi-dimensional pressure P, indicated again with a capital letter to distinguish it from an average pressure p. These stresses lead to the following force

$$F_{x_i} = \oint_{\partial\Omega} \sigma_{ij} n_j \, d\Gamma = \int_\Omega \frac{\partial \sigma_{ij}}{\partial x_j} \, d\Omega = \int_\Omega \frac{\partial}{\partial x_j} \left(-P\delta_j^i + \tau_{ij} \right) d\Omega =$$

$$= -\int_\Omega \frac{\partial P}{\partial x_i} \, dA \, dz + \int_\Omega \frac{\partial \tau_{ij}}{\partial x_j} \, dA \, dz = -\int_A \overline{\frac{\partial P}{\partial x_i}} h \, dA + \int_A \frac{\partial \tau_{ij}}{\partial x_j} h \, dA \qquad (5.137)$$

where τ_{ij} denote the components of the deviatoric stress tensor. Concerning the average pressure gradient, Section 5.4.2 has shown that this gradient is related to the gradients of both the free-surface height h and channel-bed height b as

$$\overline{\frac{\partial P}{\partial x_i}} h = \frac{\partial}{\partial x_i} \left(\rho g \frac{h^2}{2} \right) + \rho g h \frac{\partial b}{\partial x_i} \qquad (5.138)$$

As an advancement over other formulations, the average divergence of the deviatoric stress tensor

$$\overline{\sum_{j=1}^3 \frac{\partial \tau_{ij}}{\partial x_j}} \equiv \frac{1}{h} \int_b^{b+h} \sum_{j=1}^2 \frac{\partial \tau_{ij}}{\partial x_j} \, dz + \frac{1}{h} \int_b^{b+h} \frac{\partial \tau_{i3}}{\partial x_3} \, dz \qquad (5.139)$$

is correlated to both the divergence of the average tensor and the channel-bed and free-surface shear stress as follows. According to Leibnitz's differentiation rule, which coincides with the one-dimensional form of Reynolds' transport theorem, the divergence of the average deviatoric stress tensor, for $1 \le j \le 2$, is expressed as

$$h\frac{\partial (\overline{\tau_{ij}})}{\partial x_j} \equiv h\frac{\partial}{\partial x_j} \left(\frac{1}{h} \int_b^{b+h} \tau_{ij} \, dz \right) = h\frac{\partial}{\partial x_j} \left(\frac{1}{h} \right) \int_b^{b+h} \tau_{ij} \, dz +$$

$$h \left(\frac{1}{h} \int_b^{b+h} \frac{\partial \tau_{ij}}{\partial x_j} \, dz \right) + \frac{\partial (b+h)}{\partial x_j} \tau_{ij}(x,y,b+h) - \frac{\partial b}{\partial x_j} \tau_{ij}(x,y,b) =$$

$$-\frac{\partial h}{\partial x_j} \overline{\tau_{ij}} + h \left(\overline{\frac{\partial \tau_{ij}}{\partial x_j}} \right) + \frac{\partial (b+h)}{\partial x_j} \tau_{ij}(x,y,b+h) - \frac{\partial b}{\partial x_j} \tau_{ij}(x,y,b) \qquad (5.140)$$

These developments show that for $1 \le j \le 2$ the average divergence multiplied by h equals

$$h \left(\overline{\frac{\partial \tau_{ij}}{\partial x_j}} \right) = h\frac{\partial (\overline{\tau_{ij}})}{\partial x_j} + \frac{\partial h}{\partial x_j} \overline{\tau_{ij}} - \frac{\partial (b+h)}{\partial x_j} \tau_{ij}(x,y,b+h) + \frac{\partial b}{\partial x_j} \tau_{ij}(x,y,b) =$$

$$\frac{\partial (h\overline{\tau_{ij}})}{\partial x_j} - \frac{\partial (b+h)}{\partial x_j} \tau_{ij}(x,y,b+h) + \frac{\partial b}{\partial x_j} \tau_{ij}(x,y,b) \qquad (5.141)$$

which significantly shows that the free-surface height h in the end occurs not outside, but inside the divergence. With reference to $\partial \tau_{i3}/\partial x_3$, the average of this derivative becomes

$$h \left(\frac{1}{h} \int_b^{b+h} \frac{\partial \tau_{i3}}{\partial x_3} \, dz \right) = \tau_{i3}(x,y,b+h) - \tau_{i3}(x,y,b) \qquad (5.142)$$

Based on results (5.141) and (5.142), the average divergence of τ_{ij} multiplied by h in (5.137), for $1 \leq j \leq 3$ is cast as

$$\int_A \left(\sum_{j=1}^{3} \frac{\partial \tau_{ij}}{\partial x_j} \right) h \, dA = \int_A \left(-\sum_{j=1}^{2} \frac{\partial (b+h)}{\partial x_j} \tau_{ij}(x,y,b+h) + \tau_{i3}(x,y,b+h) \right) dA +$$

$$\int_A \sum_{j=1}^{2} \frac{\partial \left(h \overline{\tau}_{ij} \right)}{\partial x_j} \, dA + \int_A \left(\sum_{j=1}^{2} \frac{\partial b}{\partial x_j} \tau_{ij}(x,y,b) - \tau_{i3}(x,y,b) \right) dA \qquad (5.143)$$

The average shear-stress components in this integral are cast as

$$\overline{\tau}_{ij} \equiv \overline{\mu \left(\frac{\partial u_i}{\partial x_j} + \frac{\partial u_j}{\partial x_i} \right)} = \frac{1}{h} \int_b^{b+h} \left[\mu \left(\frac{\partial u_i}{\partial x_j} + \frac{\partial u_j}{\partial x_i} \right) \right] dz \qquad (5.144)$$

where μ represents a constant coefficient of dynamic viscosity. The following developments highlight physically meaningful conditions for which the derivative of an average velocity component provides an accurate representation for the average of the derivative of the same velocity component. Through Leibnitz's theorem, the partial derivative of an average velocity component is expressed as

$$\frac{\partial \left(\overline{u_i} \right)}{\partial x_j} = \frac{\partial}{\partial x_j} \left(\frac{1}{h} \int_b^{b+h} u_i \, dz \right) = \frac{\partial}{\partial x_j} \left(\frac{1}{h} \right) \int_b^{b+h} u_i \, dz + \frac{1}{h} \left[\int_b^{b+h} \frac{\partial u_i}{\partial x_j} dz + \frac{\partial h}{\partial x_j} u_i(x,y,h) \right] =$$

$$\frac{\overline{\partial u_i}}{\partial x_j} + \frac{1}{h} \frac{\partial h}{\partial x_j} \left[-\frac{1}{h} \int_b^{b+h} u_i(x,y,z) \, dz + u_i(x,y,h) \right] \qquad (5.145)$$

This result shows that when the derivatives of h are negligible, or the derivative of u_i with respect to z is negligible, or both, then the last term in this expression becomes negligible and the derivative of an average velocity component accurately represents the average of a derivative of the same velocity component.

Based on results (5.63) and with reference to Figure 5.10, the integrals in (5.143) of the inner products of the gradients of "b" as well as "h" and "τ_{ij}" provide the shear-traction contribution to the total force F_{x_i} in (5.132) as follows

$$\int_A \left(-\sum_{j=1}^{2} \frac{\partial (b+h)}{\partial x_j} \tau_{ij}(x,y,b+h) + \tau_{i3}(x,y,b+h) \right) dA =$$

$$\int_A \left(\tau_{ij} n_j \right) \left(\sqrt{\left(\frac{\partial (b+h)}{\partial x_1} \right)^2 + \left(\frac{\partial (b+h)}{\partial x_2} \right)^2 + 1} \right) dA \Bigg|_{\text{free surface}} \qquad (5.146)$$

and

$$\int_A \left(\sum_{j=1}^{2} \frac{\partial b}{\partial x_j} \tau_{ij}(x,y,b) - \tau_{i3}(x,y,b) \right) dA =$$

$$\int_A \left(\tau_{ij} n_j \right) \left(\sqrt{\left(\frac{\partial b}{\partial x_1} \right)^2 + \left(\frac{\partial b}{\partial x_2} \right)^2 + 1} \right) dA \Bigg|_{\text{channel bed}} \qquad (5.147)$$

which respectively correspond to the effects of free-surface and channel-bed shear stress.

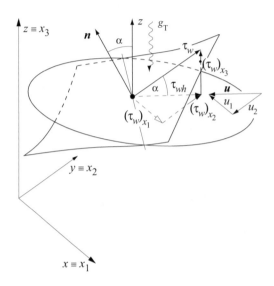

Figure 5.10: Shear Stress on Channel Bed Facet

The correlations between wall as well as wind shear stress τ_w and τ_{ws} and the respective flow specific kinetic energy are expressed as

$$\tau_w = \frac{f_w}{4} \frac{\rho \|\boldsymbol{u}\|^2}{2}, \quad \tau_{ws} = \frac{f_{ws}}{4} \frac{\rho_{ws} \|\boldsymbol{u}_w - \boldsymbol{u}\|^2}{2} \tag{5.148}$$

where $\boldsymbol{u}_w$ denotes the wind velocity, with a corresponding shear force that depends on the relative velocity $\boldsymbol{u}_w - \boldsymbol{u}$ of the wind with respect to the free surface. Next, consider the condition that in the vicinity of the channel bed the flow takes place in the direction that approaches the steepest descent direction, locally on a bed-surface facet. This descent direction then becomes the direction of the local wall shear stress τ_w, as shown in Figure 5.10. Accordingly, the projection τ_{wh} of the channel-bed stress τ_w onto a horizontal x_1-x_2 plane becomes $\tau_{wh} = \tau_w \cos(\alpha) = \tau_w / \sqrt{\left(\frac{\partial b}{\partial x_1}\right)^2 + \left(\frac{\partial b}{\partial x_2}\right)^2 + 1}$, with x_1-x_2-plane components $(\tau_w)_{x_i} \equiv \tau_{ij} n_j = -\tau_{wh} u_i / \|\boldsymbol{u}\|$. With these results, the channel-bed integral (5.147) becomes

$$\int_A (\tau_{ij} n_j) \left(\sqrt{\left(\frac{\partial b}{\partial x_1}\right)^2 + \left(\frac{\partial b}{\partial x_2}\right)^2 + 1} \right) dA = -\int_A \tau_w \frac{u_i}{\|\boldsymbol{u}\|} dA = -\int_A \rho \frac{f_w}{8} u_i \|\boldsymbol{u}\| dA \tag{5.149}$$

Similarly, consider the condition in the vicinity of the free surface that the steepest descent direction on a free-surface facet approaches the direction of the relative wind velocity $\boldsymbol{u}_w - \boldsymbol{u}$. This descent direction then becomes the direction of the local wind shear stress τ_{ws}. Accordingly, the projection $(\tau_{ws})_{z=0}$ of τ_{ws} onto a horizontal plane becomes $(\tau_{ws})_{z=0} = \tau_{ws}\cos(\alpha_{ws}) = \tau_{ws}/\sqrt{\left(\frac{\partial(b+h)}{\partial x_1}\right)^2 + \left(\frac{\partial(b+h)}{\partial x_2}\right)^2 + 1}$, with components that are expressed as $\tau_{ij}n_j = (\tau_{ws_{z=0}})\,(u_{w_i} - u_i)/\|\boldsymbol{u}_w\|$. With these results, the free-surface integral (5.146) becomes

$$\int_A (\tau_{ij}n_j)\left(\sqrt{\left(\frac{\partial(b+h)}{\partial x_1}\right)^2 + \left(\frac{\partial(b+h)}{\partial x_2}\right)^2 + 1}\right)\,dA =$$

$$\int_A \tau_{ws}\frac{u_{w_i}-u_i}{\|\boldsymbol{u}_w-\boldsymbol{u}\|}\,dA = \int_A \rho_{ws}\frac{f_{ws}(u_{w_i}-u_i)}{8}\|\boldsymbol{u}_w-\boldsymbol{u}\|\,dA \qquad (5.150)$$

Governing Equation

With all of these results, in terms of averaged variables, integral (5.137) becomes

$$F_{x_i} = \int_A \left[-\frac{\partial}{\partial x_i}\left(\rho g\frac{h^2}{2}\right) - \rho g h\frac{\partial b}{\partial x_i} + \frac{\partial}{\partial x_j}\left(\mu h\left(\frac{\partial u_i}{\partial x_j} + \frac{\partial u_j}{\partial x_i}\right)\right)\right.$$

$$\left. -\rho\frac{f_w}{8}u_i\|\boldsymbol{u}\| + \rho_{ws}\frac{f_{ws}(u_{w_i}-u_i)}{8}\|\boldsymbol{u}_{ws}-\boldsymbol{u}\| + \chi\rho h w_m r_i\right]\,dA \qquad (5.151)$$

For $m_i \equiv hu_i$ denoting a depth-averaged linear-momentum component, integrals (5.131) and (5.151) together lead to the result

$$\frac{\partial m_i}{\partial t} + \frac{\partial}{\partial x_j}\left(\frac{m_j m_i}{h} + g\frac{h^2}{2}\delta_j^i\right) + gh\frac{\partial b}{\partial x_i} - \chi h w_m r_i$$

$$-\frac{\partial}{\partial x_j}\left(h\frac{\mu}{\rho}\left(\frac{\partial(m_i/h)}{\partial x_j} + \frac{\partial(m_j/h)}{\partial x_i}\right)\right) + \frac{f_w}{8}u_i\|\boldsymbol{u}\| - \frac{\rho_{ws}}{\rho}\frac{f_{ws}(u_{w_i}-u_i)}{8}\|\boldsymbol{u}_w-\boldsymbol{u}\| = 0 \quad (5.152)$$

which corresponds to the generalized differential linear-momentum equation.

5.5.3 Energy Equation

The expressions for total energy and first principle of thermodynamics become

$$\mathcal{E} = \int_A E\,dA, \qquad \frac{D\mathcal{E}}{Dt} = \frac{\delta Q}{\delta t} - \frac{\delta W}{\delta t} \qquad (5.153)$$

Reynolds' transport theorem to evaluate the derivative of $\mathcal{E}$ yields

$$\frac{D\mathcal{E}}{Dt} = \int_A \left(\frac{\partial(E)}{\partial t} + \frac{\partial(Eu_j)}{\partial x_j}\right)\,dx = \frac{\delta Q}{\delta t} - \frac{\delta W}{\delta t} \qquad (5.154)$$

where $\delta Q/\delta t$ and $\delta W/\delta t$ respectively denote the heat and work exchanged per unit time with the flow, with a positive $\delta Q/\delta t$ indicating heat transferred to the flow per unit time and a positive $\delta W/\delta t$ indicating work performed by the flow. The total work is expressed as

$$\frac{\delta W}{\delta t} = \frac{\delta W_p}{\delta t} + \frac{\delta W_\tau}{\delta t} + \frac{\delta W_m}{\delta t} \qquad (5.155)$$

with $\delta W_\tau/\delta t$, $\delta W_p/\delta t$ and $\delta W_m/\delta t$ respectively indicating the shear-stress, pressure-gradient and mechanical-shaft, work per unit time.

Mechanical-Shaft Work

Following the developments in Section 5.5.2, the shaft work is directly expressed as

$$\frac{\delta \mathcal{W}_m}{\delta t} = \int_\Omega \dot{w}_m \, d\Omega = \int_A \rho |\boldsymbol{u} \cdot \boldsymbol{r}| h w_m \, dA \qquad (5.156)$$

where $\dot{w}_m$ and w_m respectively indicate the work per unit time per unit volume and the work per unit mass of flow per unit area. In the expression, $\dot{w}_m$ depends on $|\boldsymbol{u} \cdot \boldsymbol{r}|$ so that independently of the flow direction this work will correctly contribute to a decrease or increase of total energy $\mathcal{E}$ in (5.154) depending on whether the work is performed by or on the flow.

Pressure Work

The work per unit time due to pressure is calculated by benefitting from the 3-D continuity equation and the distribution of gauge pressure. As previously stipulated, the capital letter U_i denotes in this section the multi-dimensional velocity component, whereas the lower case u_i indicates an average velocity component. Since free-surface flows are incompressible, the corresponding 3-D continuity equation is

$$\frac{\partial U_i}{\partial x_i} = 0 \qquad (5.157)$$

for a constant atmospheric pressure P_{atm}, this equation leads to the integral

$$-\int_\Omega P_{\text{atm}} \frac{\partial U_i}{\partial x_i} \, d\Omega = \oint_{\partial \Omega} P_{\text{atm}} U_i (-n_i) \, d\Gamma = 0 \qquad (5.158)$$

which is used in the expression for the pressure-gradient work. This type of work is thus cast as

$$\frac{\delta \mathcal{W}_p}{\delta t} = \oint_{\partial \Omega} U_i P \delta_j^i (-n_j) d\Gamma = \oint_{\partial \Omega} (P - P_{\text{atm}}) U_i (-n_i) d\Gamma = \oint_{\partial \Omega} \Delta P U_i (-n_i) d\Gamma \qquad (5.159)$$

as results from using (5.158). With these specifications, the pressure work per unit time becomes

$$\oint_{\partial \Omega} \Delta P \, U_i n_i d\Gamma = \sum_{i=1}^3 \int_\Omega \frac{\partial (\Delta P U_i)}{\partial x_i} \, d\Omega =$$

$$\sum_{i=1}^2 \int_A \frac{\overline{\partial (\Delta P U_i)}}{\partial x_i} h \, dA + \int_A \int_b^{b+h} \frac{\partial (\Delta P U_3)}{\partial x_3} dz \, dA =$$

$$\sum_{i=1}^2 \int_A \frac{\overline{\partial (\Delta P U_i)}}{\partial x_i} h \, dA + \int_A \left(\Delta P U_3 |_{b+h} - \Delta P U_3 |_b \right) dA \qquad (5.160)$$

The average of the divergence of $\{(\Delta P U_i)\}$ in this result is correlated with the divergence of the average of $\{(\Delta P U_i)\}$ as follows. According to Leibnitz's differentiation rule, the divergence of the average of $\{(\Delta P U_i)\}$, for $1 \leq i \leq 2$, is expressed as

$$\frac{\partial \left(\overline{\Delta P U_i} \right)}{\partial x_i} \equiv \frac{\partial}{\partial x_i} \left(\frac{1}{h} \int_b^{b+h} \Delta P U_i \, dz \right) = \frac{\partial}{\partial x_i} \left(\frac{1}{h} \right) \int_b^{b+h} \Delta P U_i \, dz +$$

$$
\frac{1}{h}\int_b^{b+h} \frac{\partial \Delta P\,U_i}{\partial x_i}\,dz + \frac{1}{h}\left(\frac{\partial(b+h)}{\partial x_i}\,\Delta P\,U_i|_{b+h} - \frac{\partial b}{\partial x_i}\,\Delta P\,U_i|_b\right) =
$$

$$
-\frac{1}{h}\frac{\partial h}{\partial x_i}\overline{\Delta P\,U_i} + \overline{\frac{\partial(\Delta P\,U_i)}{\partial x_i}} + \frac{1}{h}\left(\frac{\partial(b+h)}{\partial x_i}\,\Delta P\,U_i|_{b+h} - \frac{\partial b}{\partial x_i}\,\Delta P\,U_i|_b\right) \tag{5.161}
$$

which leads to

$$
\sum_{i=1}^{2}\int_A \left(\overline{\frac{\partial(\Delta P\,U_i)}{\partial x_i}}\right) h\,dA =
$$

$$
\int_A \frac{\partial}{\partial x_i}\left[h\,\overline{(\Delta P\,U_i)}\right]dA - \int_A \left(\frac{\partial(b+h)}{\partial x_i}\,\Delta P\,U_i|_{b+h} - \frac{\partial b}{\partial x_i}\,\Delta P\,U_i|_b\right)dA \tag{5.162}
$$

With these results, the pressure work integral (5.160) becomes

$$
\sum_{i=1}^{3}\oint_{\partial\Omega}\Delta P\,U_i n_i d\Gamma = \sum_{i=1}^{2}\int_A \frac{\partial}{\partial x_i}\left[h\,\overline{(\Delta P\,U_i)}\right]dA +
$$

$$
\int_A \Delta P|_{b+h}\left(-\frac{\partial(b+h)}{\partial x_i}\,U_i|_{b+h} + U_3|_{b+h}\right)dA + \int_A \Delta P|_b\left(\frac{\partial b}{\partial x_i}\,U_i|_b - U_3|_b\right)dA \tag{5.163}
$$

With reference to results (5.63), the vector with components $(\partial b/\partial x_1, \partial b/\partial x_2, -1)$, normal to a channel-bed facet, remains perpendicular to the channel-bed velocity, independently of the no-slip boundary condition. Analogously, under the condition that the free surface approaches a stream surface, the vector with components $(-\partial(b+h)/\partial x_1, -\partial(b+h)/\partial x_2, 1)$, normal to a free-surface facet, remains perpendicular to the free-surface velocity. Accordingly, the inner products of each of these normal vectors and the respective velocity in (5.163) will vanish, which leads to the pressure-work result

$$
\sum_{i=1}^{3}\oint_{\partial\Omega}\Delta P\,U_i n_i d\Gamma = \sum_{i=1}^{2}\int_A \frac{\partial}{\partial x_i}\left[h\,\overline{(\Delta P\,U_i)}\right]dA \tag{5.164}
$$

Since $\Delta P \equiv (P - P_{\text{atm}}) > 0$, the area integral in this result may be further transformed via an average velocity component u_i as follows

$$
\int_b^{b+h}(P - P_{\text{atm}})\,U_i dz = u_i \int_b^{b+h}(P - P_{\text{atm}})\,dz \quad \Rightarrow \quad u_i \equiv \frac{\int_b^{b+h}(P - P_{\text{atm}})\,U_i dz}{\int_b^{b+h}(P - P_{\text{atm}})\,dz} \tag{5.165}
$$

This pressure-averaged velocity component generally differs from the mass-averaged velocity component in Section 5.1. With a suitable variation of $h = h(x, y, t)$ these two average velocities will remain accurate representations of each other. By virtue of (5.89) the pressure integral in (5.165) becomes

$$
\int_b^{b+h}(P - P_{\text{atm}})\,U_i dz = u_i \int_b^{b+h}(P - P_{\text{atm}})\,dz = u_i \rho g \frac{h^2}{2} \tag{5.166}
$$

which leads to the pressure-work expression

$$
\sum_{i=1}^{3}\oint_{\partial\Omega}\Delta P\,U_i n_i d\Gamma = \sum_{i=1}^{2}\int_A \frac{\partial}{\partial x_i}\left[h\,\overline{(\Delta P\,U_i)}\right]dA = \sum_{i=1}^{2}\int_A \frac{\partial}{\partial x_i}\left(u_i \rho g \frac{h^2}{2}\right)dA \tag{5.167}
$$

Shear-Stress Work

The work per unit time due to the shear-stress components $\{\tau_{ij}\}$ is then expressed as

$$\frac{\delta W_\tau}{\delta t} = \sum_{i,j=1}^{3} \oint_{\partial\Omega} U_i(-\tau_{ij})n_j \, d\Gamma = -\sum_{i,j=1}^{3} \int_{\Omega} \frac{\partial(U_i\tau_{ij})}{\partial x_j} \, d\Omega =$$
$$-\sum_{i=1}^{3}\sum_{j=1}^{2} \int_A \overline{\frac{\partial(U_i\tau_{ij})}{\partial x_j}} h \, dA - \sum_{i=1}^{3} \int_A \int_b^{b+h} \frac{\partial(U_i\tau_{i3})}{\partial x_3} \, dz \, dA \qquad (5.168)$$

Following developments (5.145), (5.160)-(5.163), the final expression for $1 \le i, j \le 2$ for the shear-work exactly becomes

$$\frac{\delta W_\tau}{\delta t} = -\int_A \frac{\partial}{\partial x_j} \left(u_i \mu h \left(\frac{\partial u_i}{\partial x_j} + \frac{\partial u_j}{\partial x_i} \right) \right) dA - \int_A \dot{W}_{\text{wind}} \, dA \qquad (5.169)$$

where $\dot{W}_{\text{wind}}$ denotes the work per unit time of the wind shear stress. When the product of the vertical components of velocity and the corresponding wind shear traction is negligible, the wind-shear stress work becomes

$$\dot{W}_{\text{wind}} = \int_A \rho_{ws} \frac{f_{ws}}{8} \boldsymbol{u} \cdot (\boldsymbol{u}_w - \boldsymbol{u}) \|\boldsymbol{u}_w - \boldsymbol{u}\| \, dA \qquad (5.170)$$

Heat Transfer

In respect of the total heat per unit time $\delta Q/\delta t$, this energy contribution is expressed as

$$\frac{\delta Q}{\delta t} = \oint_{\partial\Omega} \left(-\boldsymbol{q}^F \cdot \boldsymbol{n} \right) d\Gamma \qquad (5.171)$$

where $\boldsymbol{q}^F$ denotes the Fourier heat-flux vector. The minus sign in this expression is justified by the thermodynamics convention that an amount of heat transferred to the flow, for which $\boldsymbol{q}^F \cdot \boldsymbol{n} \le 0$, corresponds to a positive $\delta Q/\delta t$, for the correctness of (5.153). The expression for $\delta Q/\delta t$ is further transformed as

$$\frac{\delta Q}{\delta t} = -\sum_{j=1}^{3} \oint_{\partial\Omega} q_j^F n_j \, d\Gamma = -\sum_{j=1}^{3} \int_{\Omega} \frac{\partial q_j^F}{\partial x_j} \, d\Omega = -\int_A \sum_{j=1}^{3} \frac{\partial q_j^F}{\partial x_j} h \, dA \qquad (5.172)$$

Following developments (5.139)-(5.150), this result becomes

$$-\int_A \overline{\sum_{j=1}^{3} \frac{\partial q_j^F}{\partial x_j}} h \, dA = -\int_A \sum_{j=1}^{2} \frac{\partial(h\bar{q}_j^F)}{\partial x_j} \, dA + \int_A \rho\|\boldsymbol{u}\| h g_\text{T} \, dA \qquad (5.173)$$

where g_T denotes the energy transferred from both the free surface and channel bed, per unit mass of flow per unit area, orthogonally to the area "A", with "A" the area indicated in Figure 5.5. On the basis of Fourier's heat-transfer law

$$\boldsymbol{q}^F = -k\nabla T \qquad (5.174)$$

with k a constant coefficient of heat conductivity, together with derivations (5.143)-(5.145), the final expression for $\delta Q/\delta t$ in terms of temperature "T" and surface heating "g_T" becomes

$$\frac{\delta Q}{\delta t} \simeq \int_A \sum_{j=1}^{2} \frac{\partial}{\partial x_j} \left(kh \frac{\partial T}{\partial x_j} \right) dA + \int_A \rho\|\boldsymbol{u}\| h g_\text{T} \, dA \qquad (5.175)$$

Governing Equation

The generalized energy equation thus becomes

$$\frac{\partial E}{\partial t} + \frac{\partial}{\partial x_i}\left(\frac{m_i}{h}\left(E + \rho g\frac{h^2}{2}\right)\right) = \frac{\partial}{\partial x_j}\left(u_i\mu h\left(\frac{\partial u_i}{\partial x_j} + \frac{\partial u_j}{\partial x_i}\right)\right) +$$

$$\frac{\partial}{\partial x_j}\left(kh\frac{\partial T}{\partial x_j}\right) + \rho\|\boldsymbol{m}\|\,g_{\mathrm{T}} - \rho|\boldsymbol{m}\cdot\boldsymbol{r}|w_m + \rho_{ws}\frac{f_{ws}}{8}\boldsymbol{u}\cdot(\boldsymbol{u}_w - \boldsymbol{u})\|\boldsymbol{u}_w - \boldsymbol{u}\| \qquad (5.176)$$

This expression identifies the "enthalpy" H of free-surface flows as

$$H \equiv \frac{1}{h}\left(E + \rho g\frac{h^2}{2}\right) \qquad (5.177)$$

Once the area-specific energy E is computed, the free-surface flow temperature results as

$$\frac{E}{h} = \rho\epsilon + \frac{1}{2}\rho\|\boldsymbol{m}\|^2 = \rho cT + \frac{1}{2}\rho\|\boldsymbol{m}\|^2 \;\Rightarrow\; T = \frac{E/h - \rho\|\boldsymbol{m}\|^2/2}{\rho c}, \quad \frac{E}{h} > \frac{1}{2}\rho\|\boldsymbol{m}\|^2 \quad (5.178)$$

where "c" denotes the specific heat of the fluid in free-surface flow. As in the quasi-one dimensional case, since the free-surface flow continuity and linear-momentum equations may be solved independently of the energy equation, the chief use of this equation consists in determining the temperature of the flow. With T_{ref} a constant reference temperature, the total energy may also be expressed in terms of the temperature variation ΔT, by writing $T = T_{\mathrm{ref}} + \Delta T$; owing to the continuity equation, the corresponding energy equation simply replaces T with ΔT, with E only depending on ΔT. The constraint on the kinetic energy limits the maximum amount of the combination of work and heat that can be extracted from an open-channel flow.

5.5.4   Non-Dimensional 2-D Free-Surface Equations

The system of 2-D free-surface open-channel flow equations is thus expressed as

$$\begin{cases} \dfrac{\partial h}{\partial t} + \dfrac{\partial m_j}{\partial x_j} = 0 \\[2mm] \dfrac{\partial m_i}{\partial t} + \dfrac{\partial}{\partial x_j}\left(\dfrac{m_j m_i}{h} + g\dfrac{h^2}{2}\delta_j^i\right) + gh\dfrac{\partial b}{\partial x_i} + \dfrac{f_w}{8}u_i\|\boldsymbol{u}\| - \chi h w_m r_i - \\[2mm] \dfrac{\partial}{\partial x_j}\left(h\dfrac{\mu}{\rho}\left(\dfrac{\partial(m_i/h)}{\partial x_j} + \dfrac{\partial(m_j/h)}{\partial x_i}\right)\right) - \dfrac{f_{ws}}{8}\dfrac{\rho_{ws}}{\rho}(u_{w_i} - u_i)\|\boldsymbol{u}_w - \boldsymbol{u}\| = 0 \\[2mm] \dfrac{\partial E}{\partial t} + \dfrac{\partial}{\partial x_i}\left(\dfrac{m_i}{h}\left(E + \rho g\dfrac{h^2}{2}\right)\right) - \dfrac{\partial}{\partial x_j}\left(\mu m_i\left(\dfrac{\partial(m_i/h)}{\partial x_j} + \dfrac{\partial(m_j/h)}{\partial x_i}\right)\right) - \rho\|\boldsymbol{m}\|\,g_{\mathrm{T}} - \\[2mm] \dfrac{\partial}{\partial x_j}\left(kh\dfrac{\partial T}{\partial x_j}\right) + \rho|\boldsymbol{m}\cdot\boldsymbol{r}|w_m - \dfrac{f_{ws}}{8}\rho_{ws}\boldsymbol{u}\cdot(\boldsymbol{u}_w - \boldsymbol{u})\|\boldsymbol{u}_w - \boldsymbol{u}\| = 0 \end{cases}$$

$$(5.179)$$

which consists of the four equations of continuity, linear-momentum components, and energy, one species for the dependent variable $q \equiv (h, m_1, m_2, E)$.

Both the dependent and independent variables in this system are expressed in terms of non dimensional variables as

$$h = h_0 h_{\mathrm{nd}}, \quad E = \rho g h_0^2 E_{\mathrm{nd}}, \quad V^2 \equiv g h_0, \quad u_j = V u_{j_{\mathrm{nd}}}$$

$$t = \frac{L}{V} t_{\mathrm{nd}}, \quad x_j = L x_{j_{\mathrm{nd}}}, \quad w_m = \frac{V^2}{L} w_{m_{\mathrm{nd}}}, \quad g_{\mathrm{T}} = \frac{V^2}{L} g_{\mathrm{T \, nd}}, \quad T = \frac{V^2}{c} T_{\mathrm{nd}} \tag{5.180}$$

where subscript "nd" indicates a non-dimensional variable and L, V, h_0, and c respectively indicate reference length, speed, free-surface height and specific heat. From (5.91), the non-dimensional form of the average gauge pressure becomes

$$h(\overline{P} - P_{\mathrm{atm}}) = \frac{\rho g h^2}{2} \;\Rightarrow\; h_0 \rho V^2 h_{\mathrm{nd}} (\overline{P} - P_{\mathrm{atm}})_{\mathrm{nd}} = \rho g h_0^2 \frac{h_{\mathrm{nd}}^2}{2} \;\Rightarrow\; h_{\mathrm{nd}} (\overline{P} - P_{\mathrm{atm}})_{\mathrm{nd}} = \frac{h_{\mathrm{nd}}^2}{2} \tag{5.181}$$

which corresponds to a non-dimensional gauge pressure force per unit channel width "p_g" with "equation of state"

$$p_g = \frac{h_{\mathrm{nd}}^2}{2} \tag{5.182}$$

With a reference speed expressed in terms of reference free-surface height h_0, the local Froude number "Fr" is expressed as

$$Fr \equiv \frac{\|\boldsymbol{u}\|}{\sqrt{gh}} = \frac{\|\boldsymbol{u}\|/V}{(\sqrt{gh})/V} = \frac{\|\boldsymbol{u}\|/V}{(\sqrt{gh})/\sqrt{gh_0}} = \frac{\|\boldsymbol{u}\|_{\mathrm{nd}}}{\sqrt{h_{\mathrm{nd}}}} \tag{5.183}$$

which, again, directly depends on non-dimensional variables. The transport coefficients "μ" and "k" in (5.179), together with (5.180) in turn lead to the Prandtl and Reynolds numbers

$$Pr \equiv \frac{\mu c}{k}, \quad Re \equiv \frac{\rho V L}{\mu} \tag{5.184}$$

Dispensing with the subscript "nd", for simplicity, but recognizing each variable is non dimensional, these numbers and (5.180) lead to

$$\begin{cases}
\dfrac{\partial h}{\partial t} + \dfrac{\partial m_j}{\partial x_j} = 0 \\[2mm]
\dfrac{\partial m_i}{\partial t} + \dfrac{\partial}{\partial x_j}\left(\dfrac{m_j m_i}{h} + \dfrac{h^2}{2}\delta_j^i\right) + h\dfrac{\partial b}{\partial x_i} + \dfrac{f_w}{8}\dfrac{u_i \|\boldsymbol{u}\|}{h_0/L} - \chi h w_m r_i - \\[2mm]
\quad \dfrac{1}{Re}\dfrac{\partial}{\partial x_j}\left(h\left(\dfrac{\partial(m_i/h)}{\partial x_j} + \dfrac{\partial(m_j/h)}{\partial x_i}\right)\right) - \dfrac{f_{ws}}{8}\dfrac{\rho_{ws}}{\rho}\dfrac{(u_{w_i} - u_i)\|\boldsymbol{u}_w - \boldsymbol{u}\|}{h_0/L} = 0 \\[2mm]
\dfrac{\partial E}{\partial t} + \dfrac{\partial}{\partial x_i}\left(\dfrac{m_i}{h}\left(E + \dfrac{h^2}{2}\right)\right) - \dfrac{1}{Re}\dfrac{\partial}{\partial x_i}\left(m_i\left(\dfrac{\partial(m_i/h)}{\partial x_j} + \dfrac{\partial(m_j/h)}{\partial x_i}\right)\right) - \|\boldsymbol{m}\|\,g_{\mathrm{T}} - \\[2mm]
\quad \dfrac{1}{Pr\,Re}\dfrac{\partial}{\partial x_j}\left(h\dfrac{\partial T}{\partial x_j}\right) + |\boldsymbol{m}\cdot\boldsymbol{r}|w_m - \dfrac{f_{ws}}{8}\dfrac{\rho_{ws}}{\rho}\dfrac{\boldsymbol{u}\cdot(\boldsymbol{u}_w - \boldsymbol{u})\|\boldsymbol{u}_w - \boldsymbol{u}\|}{h_0/L} = 0
\end{cases} \tag{5.185}$$

which corresponds to the non-dimensional 2-D free-surface open-channel flow equations.

Chapter 6

Overview of CFD Algorithm Development

This overview delineates the chief phases in the development of CFD algorithms. Computational Fluid Dynamics is the art and science of investigating the flows of fluids by means of computer solutions of the mathematical models of fluid dynamics. CFD, accordingly, begins with one such mathematical model, which corresponds to a set of time-dependent partial differential equations.

These equations require auxiliary data in the form of suitable boundary and initial conditions. Independently of any approximation, an inappropriate selection of these conditions may lead to a mathematical solution that may not correspond to the physical system of interest.

With a chosen model that may lead to a physically meaningful solution, the development of a CFD algorithm calls for the approximation of the partial derivatives with respect to the space variables. This process generates a system of non-linear ordinary differential equations with respect to the time variable, a process known as the method of lines. This approximation critically impacts the stability of the eventual CFD solution in the sense that an incorrect approximation may lead to a system of ordinary differential equations with solutions that grow exponentially, even when the original partial differential equations with bounded initial and boundary conditions admit a physically meaningful solution. This instability may ensue independently of the approximation of the derivatives with respect to time.

If it were possible exactly to solve arbitrary systems of non-linear ordinary differential equations (ODE's), in closed form, the solution of these systems would correspond to the eventual CFD solution. CFD hardly affords this luxury! This system must then be solved numerically and myriad numerical methods for solving ODE's can be chosen. An inappropriate method may yield a numerical solution that grows unboundedly, even when the solution of the ordinary differential equations does not. A simple example clarifies this phenomenon.

Both explicit and implicit methods are available to solve these ordinary differential equations numerically. Explicit methods yield the solution without requiring the solution of systems of coupled algebraic equations; on the other hand, solution stability only exists under frequently severe restrictions on the integration time step. Implicit methods do not impose such stability restrictions, but generate systems of coupled algebraic equations for computing the eventual solution. General non-linear systems are invariably solved via Newton's method, in full or linearized form.

6.1 Mathematical Model

The mathematical model of a CFD investigation involves a set of partial differential equations, which can depend on both time and space variables. These partial differential equations usually admit myriad solutions. Among them, the solution that describes the physical system under investigation corresponds to additional information, which completes the description of the system. This additional information, or auxiliary data, will consist of initial and boundary conditions. Initial conditions prescribe the system state at a reference initial time, and boundary conditions specify the state of the system at all times on surface regions of the system boundary. A mathematical model of a physical system thus consists of both a partial differential equation set and auxiliary data.

In Computational Fluid Dynamics, as well as mathematical physics, the mathematical models that lead to physically meaningful solutions are termed well posed problems. A well posed problem possesses three properties:

1. A solution exists

2. The solution is unique

3. The solution continuously depends on the auxiliary data

The third property signifies that slight changes in the auxiliary data can only induce correspondingly slight changes in the solution. A sign of ill-posedness would, for instance, be a solution for a temperature distribution that changes by 1,000K inside a domain when a 0.01K degree temperature change occurs on the domain boundary. This continuous dependence on the auxiliary data, which the investigator controls and specifies, corresponds to the idea of stability of a solution with respect to the auxiliary data.

To exemplify the importance of these three properties, consider the parabolic equation

$$\frac{\partial q}{\partial t} + a\frac{\partial q}{\partial x} - \frac{1}{\text{Pe}}\frac{\partial^2 q}{\partial x^2} = 0 \tag{6.1}$$

where the constant "a" and positive $1/\text{Pe}$ respectively denote a convection velocity component and diffusion coefficient. This equation can model the distribution of temperature q within a fluid that flows with constant speed a within a segment of a straight tube. In this case, t denotes time, whereas x represents the spatial position along the tube axis. From a mathematical standpoint, the solution of (6.1) is sought in the form $q = q(x,t)$, for $x \in [0,1]$ and $t \in [0, t_\infty)$, with a possibly unbounded t_∞. As an initial condition, the state of temperature $q^0(x) = q(x,0)$ at $t = 0$ must be prescribed in order to determine this temperature at subsequent instants of time.

Without imposing any boundary conditions, consider the problem of finding solutions of (6.1) corresponding to the following initial conditions

$$q^0(x) = 10\exp(-\ell x), \quad q^0(x) = 10 + 10\sin(\ell\pi x) \tag{6.2}$$

where "ℓ" denotes a positive integer. The associated solutions are either counterintuitive or physically meaningless. For instance, the solution corresponding to the first initial condition, which decreases with respect to x, is:

$$q(x,t) = 10 \exp\left(\frac{\ell^2}{Pe}t\right) \exp(\ell(x - at)) \tag{6.3}$$

and this solution, as time elapses is anything but decaying; in fact it grows unboundedly with time, hence is devoid of any physical significance. Instead, the mathematical solution corresponding to the second undecaying initial condition is

$$q(x,t) = 10 + 10 \exp\left(-\frac{(\ell\pi)^2}{Pe}t\right) \sin\left(\ell\pi(x - at)\right) \tag{6.4}$$

which actually decays as time elapses! These two examples stress the notion that while mathematical solutions of physics partial differential equations can be found, they do not necessarily correspond to physically meaningful solutions. For a solution that corresponds to a physical situation, equation (6.1) must be complemented by all auxiliary data.

A "well-posed" problem associated with equation (6.1), therefore, consists of both the equation and auxiliary data in the form of initial and boundary conditions. Consider, for instance, the problem of determining the steady-state temperature distribution corresponding to the following initial and boundary conditions

$$q(0,t) = 0 \quad , \quad q(1,t) = 1$$
$$q(x,0) \;\; = \;\; x \tag{6.5}$$

In this case the solution is

$$q(x,t_\infty) = \frac{\exp(aPe\,x) - 1}{\exp(aPe) - 1} \tag{6.6}$$

which remains bounded. Figure 6.1 shows the graph of this solution for several magnitudes of Pe. A "boundary layer" develops with this solution in the neighborhood of $x = 1$ as Pe grows unboundedly.

The time-dependent temperature distribution that asymptotically reverts to (6.6), for increasing t, evolves as a solution of (6.1) along with the following initial and boundary conditions

$$q(0,t) = 0, \quad q(1,t) = 1$$
$$q(x,0) = f(x), \quad f(0) = 0, \quad f(1) = 1 \tag{6.7}$$

where $f(x)$ expresses the initial distribution of temperature. With these auxiliary data, the solution of (6.1) is found through separation of variables by expressing q as $q(x,t) = \phi(x)\psi(t) + \gamma(x)$, where $\gamma(x)$ satisfies the steady form of (6.1) with boundary conditions $\gamma(0) = 0$ and $\gamma(1) = 1$. The solution thus becomes

$$q(x,t) = \exp\left(\frac{aPe}{2}\left(x - \frac{a}{2}t\right)\right) \sum_{\ell=1}^{\infty} C_\ell \exp\left(-\frac{(\ell\pi)^2}{Pe}t\right) \sin\left(\ell\pi x\right) + \frac{\exp(aPe\,x) - 1}{\exp(aPe) - 1} \tag{6.8}$$

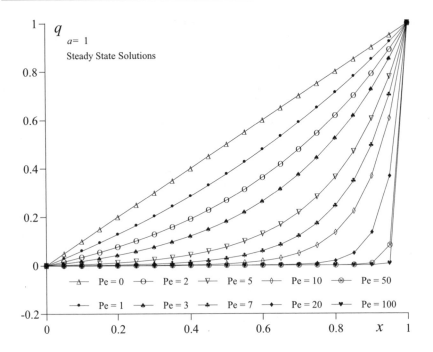

Figure 6.1: "Boundary Layer" Solution for Growing Pe

where C_ℓ denotes the ℓth Fourier's series coefficient, which is found by setting (6.8) at $t = 0$ equal to the initial condition $f(x)$. With this specification, the expression for C_ℓ becomes

$$C_\ell = 2 \int_0^1 f(x) \exp\left(-\frac{a\text{Pe}}{2} x\right) \sin(\ell\pi x) dx - 2 \int_0^1 \frac{\exp\left(\frac{a\text{Pe}\, x}{2}\right) - \exp\left(\frac{-a\text{Pe}\, x}{2}\right)}{\exp(a\,\text{Pe}) - 1} \sin(\ell\pi x) dx$$

$$= 2 \int_0^1 f(x) \exp\left(-\frac{a\,\text{Pe}}{2} x\right) \sin(\ell\pi x) dx + \frac{8\ell\pi \cos(\ell\pi) \exp\left(-\frac{a\text{Pe}}{2}\right)}{a^2\,\text{Pe}^2 + 4\ell^2\pi^2} \qquad (6.9)$$

For the case $f(x) = x$, the integral $\mathcal{I}_f$ of $f(x)$ in this expression equals

$$\mathcal{I}_f = \frac{8}{\left(a^2\,\text{Pe}^2 + 4\ell^2\pi^2\right)^2} \left[4a\text{Pe}\,\ell\pi - \exp\left(-\frac{a\text{Pe}}{2}\right) \cos(\ell\pi) \left(4a\text{Pe}\,\ell\pi + \ell\pi\left(a^2\text{Pe}^2 + 4\ell^2\pi^2\right)\right)\right]$$
$$(6.10)$$

With these expressions for C_ℓ, the exact solution (6.8) is complete and can thus be used to assess correctness of a software program that numerically solves (6.1) under the same initial and boundary conditions.

Each Fourier series component in (6.8) is itself a solution of (6.1) and the rapidity of series convergence depends on $a\text{Pe}$. For $a = 1$ and $\text{Pe} = 10$, a 15-term expansion yields for (6.8) at $t = 0$ an expression that coincides with $f(x) = x$ within four significant digits.

Figure 6.2 graphs (6.8) for various time levels and shows that the solution smoothly varies with time, with terminal steady state that arises for $t \geq 2.5$. The time-dependent exponen-

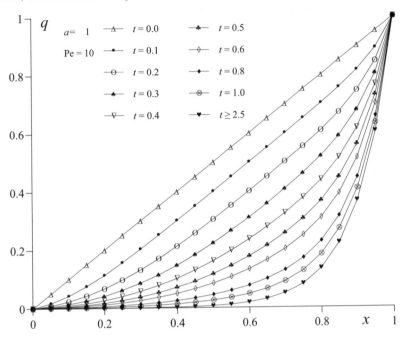

Figure 6.2: Solutions at Several Time Levels

tial components of this solution rapidly decay as time t elapses, which corresponds to the smoothing action of the second order derivative in (6.1).

Consider, a solution of (6.1) in the form of a general M-term Fourier series with time-dependent coefficients and boundary conditions

$$q(x,t) = \sum_{\ell=1}^{M} v_\ell(t) \exp(j\omega_\ell x) \tag{6.11}$$

where $j \equiv \sqrt{-1}$, $\exp(j\omega_\ell x) \equiv \cos(\omega_\ell x) + j\sin(\omega_\ell x)$, and ω_ℓ denotes a component spatial frequency.

Insertion of this series into (6.1) yields the following ordinary differential equation for the ℓth time-dependent coefficient in (6.11)

$$\frac{dv_\ell}{dt} = -\left(j\omega_\ell a + \frac{\omega_\ell^2}{\text{Pe}} \right) v_\ell \tag{6.12}$$

hence the Fourier-series solution of (6.1) becomes

$$q(x,t) = \sum_{\ell=1}^{N} C_\ell \exp\left(-\frac{\omega_\ell^2}{\text{Pe}} t \right) \exp(j\omega_\ell(x - at)) \tag{6.13}$$

which exhibits an analogous exponential smoothing as time elapses. An approximate solution of (6.1), therefore, is acceptable when it too displays a similar exponential damping under correspondingly bounded initial and boundary conditions.

6.2   Spatial Approximation: Discretization

The process of approximating partial derivatives with respect to the space variables is called discretization because it replaces a continuum partial derivative with an algebraic expression that depends on solution values at a discrete set of points. This discretization significantly impacts CFD stability.

Even when the continuum problem is well posed and the solution exhibits an exponential damping, the solution of a corresponding spatial approximation of the governing equation (6.1) may explosively diverge, regardless of the boundedness of initial and boundary conditions. This phenomenon is entirely due to the approximation of the odd-order derivatives with respect to the space variables, as the following example delineates.

For simplicity, consider a uniform discretization of the solution interval for (6.1) such that $x = i\Delta x$, with i denoting an integer number; also consider the following approximations for the first- and second-order derivatives in (6.1)

$$\frac{\partial^2 q}{\partial x^2} \simeq \frac{q_{i-1} - 2q_i + q_{i+1}}{\Delta x^2} \tag{6.14}$$

$$\frac{\partial q}{\partial x} \simeq \frac{q_i - q_{i-1}}{\Delta x}, \qquad \frac{\partial q}{\partial x} \simeq \frac{q_{i+1} - q_{i-1}}{2\Delta x}, \qquad \frac{\partial q}{\partial x} \simeq \frac{q_{i+1} - q_i}{\Delta x} \tag{6.15}$$

Each of these three discretizations for the corresponding first-order partial derivative is algebraically correct, yet depending on the sign of the convection velocity component "a", one of these choices can destabilize the approximation. For convenience, all these three discretization can be combined as

$$\frac{\partial q}{\partial x} \simeq \beta \frac{q_i - q_{i-1}}{\Delta x} + (1 - \beta)\frac{q_{i+1} - q_i}{\Delta x} \tag{6.16}$$

where β denotes a positive parameter such that $\beta = 1$, $\beta = \frac{1}{2}$, or $\beta = 0$ respectively corresponds to the first, second, or third discretization in (6.15), which are respectively known as a fully upwind, centered, and fully downwind discretization. After the spatial discretization of (6.1), the resulting expressions no longer depend on the continuum space variable x, but only on t. The partial derivative with respect to time thus becomes the following ordinary derivative

$$\left(\frac{\partial}{\partial t}\right) q(x_i, t) = \left(\frac{\partial}{\partial t}\right) q(i\Delta x, t) \simeq \left(\frac{\partial}{\partial t}\right) q_i(t) = \frac{dq_i}{dt} \tag{6.17}$$

With approximations (6.16)-(6.17) the model partial differential equation becomes

$$\frac{dq_i}{dt} = -a\beta \frac{q_i - q_{i-1}}{\Delta x} - a(1 - \beta)\frac{q_{i+1} - q_i}{\Delta x} + \frac{1}{\text{Pe}}\frac{q_{i-1} - 2q_i + q_{i+1}}{\Delta x^2} \tag{6.18}$$

for $i = 0, 1, 2, \ldots, N$. The spatial discretization thus replaces the single partial differential equation (6.1) with (6.18), which constitutes a set of $N + 1$ coupled ordinary differential equations in the continuum time variable t. Since (6.1) is linear, (6.18) is also linear.

6.2.1   Steady-Solution Monotonicity

The steady form of (6.18) becomes

$$q_{i+1}\left(\frac{1}{\mathrm{Pe}\Delta x} + \frac{a}{2}(\psi - 1)\right) - 2q_i\left(\frac{1}{\mathrm{Pe}\Delta x} + \frac{a\psi}{2}\right) + q_{i-1}\left(\frac{1}{\mathrm{Re}\Delta x} + \frac{a}{2}(\psi + 1)\right) = 0 \quad (6.19)$$

which, for a fixed $\psi \equiv 2\beta - 1$, constitutes a linear, constant-coefficient, second-order difference equation. This difference equation may be equivalently expressed as

$$\left(\frac{1}{\mathrm{Pe}\Delta x} + \frac{a\psi}{2}\right)(q_{i+1} - 2q_i + q_{i-1}) - \frac{a}{2}(q_{i+1} - q_{i-1}) = 0 \quad (6.20)$$

This result shows the presence of an artificial dissipation, depending on ψ, which leads to the effective Peclet number Pe^E

$$\frac{1}{\mathrm{Pe}^E \Delta x} \equiv \frac{1}{\mathrm{Pe}\Delta x} + \frac{a\psi}{2} \quad (6.21)$$

For various magnitudes of the coefficients a, ψ and Pe, the exact solutions of the difference equation (6.19) can be expressed in closed form, leading to the chief conclusion that a monotone as well as accurate discrete solution requires a discretization interval Δx of order $1/\mathrm{Pe}$. The general solution of this difference equation is expressed as

$$q_i = C_1\left(\lambda_1\right)^i + C_2\left(\lambda_2\right)^i \quad (6.22)$$

where superscript "i" is an exponent, the constants C_1 and C_2 depend on the boundary conditions for (6.1), hence (6.18), and λ_1 and λ_2 denote the two distinct roots of the characteristic equation

$$\lambda^2\left(\frac{1}{\mathrm{Pe}\Delta x} + \frac{a}{2}(\psi - 1)\right) - 2\lambda\left(\frac{1}{\mathrm{Pe}\Delta x} + \frac{a\psi}{2}\right) + \left(\frac{1}{\mathrm{Re}\Delta x} + \frac{a}{2}(\psi + 1)\right) = 0 \quad (6.23)$$

When this equation has two coincident roots λ_1, the solution of the corresponding difference equation becomes

$$q_i = C_1\left(\lambda_1\right)^i + iC_2\left(\lambda_1\right)^i \quad (6.24)$$

For a purely convection equation, hence for $\mathrm{Pe} \to \infty$, the fully upwind scheme for $\psi = 1$, hence $\beta = 1$ though only first-order accurate, yields the exact continuum solution

$$q_i = C_1 \quad (6.25)$$

This depends on the single constant C_1 that is determined using one boundary condition, in complete agreement with the associated first order of the resulting convection equation. In the absence of low order artificial dissipation, with $\psi = 0$, hence $\beta = \frac{1}{2}$, the solution becomes

$$q_i = C_1 + C_2(-1)^i \quad (6.26)$$

For the determination of the two constants C_1 and C_2, this solution requires two boundary conditions, one physical and one mathematical, even though the associated partial differential equation depends upon only one boundary condition. Significantly, the physical and mathematical boundary conditions must lead to a vanishing C_2, otherwise the solution will exhibit a daunting oscillatory behavior.

For a purely diffusion equation, hence $a = 0$, the characteristic equation (6.23) has one double root. Accordingly, the solution becomes

$$q_i = C_1 + iC_2 \tag{6.27}$$

which corresponds to the exact continuum solution $q(x_i) = C_1 + C_2 x_i$, since (6.18) corresponds to a second order accurate discretization.

For the general convection diffusion equation with artificial dissipation, the semi-discrete solution becomes

$$q_i = C_1 + C_2 \left(\frac{\dfrac{1}{\mathrm{Pe}\Delta x} + \dfrac{a}{2}(\psi + 1)}{\dfrac{1}{\mathrm{Pe}\Delta x} + \dfrac{a}{2}(\psi - 1)} \right)^i \tag{6.28}$$

For a positive convection speed a, solution monotonicity can be preserved by forcing the denominator to become positive

$$\frac{1}{\mathrm{Pe}\Delta x} + \frac{a}{2}(\psi - 1) > 0 \tag{6.29}$$

For a given Δx, this positiveness constraint is satisfied by any ψ, hence β, constrained by the inequality

$$\psi > 1 - \frac{2}{a\mathrm{Pe}\Delta x}, \quad \beta > 1 - \frac{1}{a\mathrm{Pe}\Delta x} \tag{6.30}$$

Thus, a formally second-order monotone linear scheme can be obtained when (6.30) yields $\psi = \mathcal{O}(\Delta x)$, $\beta = \frac{1}{2} + \mathcal{O}(\Delta x)$. However, this result leads to a regrettable reduction of the effective Peclet number, which from (6.21) is expressed as

$$\mathrm{Pe}^E = \frac{\mathrm{Pe}}{1 + a\psi\mathrm{Pe}/(2\Delta x)} < \mathrm{Pe} \tag{6.31}$$

Depending on the magnitude of Pe, the effective Pe^E may be much smaller than Pe. For example, for Pe$=1 \times 10^6$, $a = 1$, $\psi = \Delta x = 1 \times 10^{-3}$, the effective Peclet number becomes $\mathrm{Pe}^E = 666,666$, which corresponds to a 33% reduction. This predicament cannot be remedied by increasing the nominal Peclet number, because the effective Peclet number reaches the limit

$$\mathrm{Pe}^E = \frac{2}{a\psi\Delta x} \tag{6.32}$$

One avenue to counter this reduction consists in selecting a suitably small Δx such that (6.30) is satisfied for a fixed ψ. This selection leads to the expression

$$\Delta x < \frac{2}{a(1 - \psi)\mathrm{Pe}} \tag{6.33}$$

which entails a large number of mesh nodes, eventuating in a computationally expensive procedure. A viable resolution is attained when the algorithm uses an adapted mesh that clusters nodes in boundary layer regions, with Δx satisfying (6.33), and employs elsewhere within the computational domain a variable ψ that depends on semi-discrete solution smoothness, as discussed in Chapter 10.

6.2.2 Unsteady-Solution Stability

The discretization of the first-order partial derivative, hence the parameter β, also critically impacts the stability of the unsteady solutions of (6.18). This section also addresses an apparent confusion in reported research, related to the precise magnitude of β required for stability. The previous section has shown that a monotone steady solution results when

$$\beta > 1 - \frac{1}{a\mathrm{Pe}\Delta x} \tag{6.34}$$

a frequently-cited stability constraint. This section shows that solution stability requires

$$\beta > \frac{1}{2} - \frac{1}{a\mathrm{Pe}\Delta x} \tag{6.35}$$

another frequently reported stability result. The resolution of the seeming confusion revolves around the interpretation of the word "stability". A solution of (6.18) will remain stable in the sense of monotone, hence non-oscillatory, when constraint (6.34) is satisfied. Instead, a solution of (6.18) will remain stable in the sense of boundedness, but not necessarily monotone, when constraint (6.35) is satisfied, as developed in this section.

The stability, in the sense of boundedness, of the time-dependent solutions of (6.18) is conveniently investigated by means of the following general M-term Fourier series, analogous to (6.11),

$$q_i(t) = \sum_{\ell=1}^{M} v_\ell(t) \exp(\boldsymbol{j}\omega_\ell \, i\Delta x) \tag{6.36}$$

Insertion of this series into (6.18) yields the following ordinary differential equation for the time-dependent series coefficient v_ℓ

$$\frac{dv_\ell}{dt} = -\left(a\left((2\beta - 1)\omega_\ell \frac{1 - \cos(\omega_\ell \Delta x)}{\omega_\ell \Delta x} + \boldsymbol{j}\omega_\ell \frac{\sin(\omega_\ell \Delta x)}{\omega_\ell \Delta x} \right) + \frac{\omega_\ell^2}{\mathrm{Pe}} \frac{2(1 - \cos(\omega_\ell \Delta x))}{(\omega_\ell \Delta x)^2} \right) v_\ell \tag{6.37}$$

which correlates with (6.12). The solution of this equation is

$$v_\ell(t) = C_\ell \exp\left[-\left(a\left((2\beta - 1)\omega_\ell \frac{1 - \cos(\omega_\ell \Delta x)}{\omega_\ell \Delta x} + \boldsymbol{j}\omega_\ell \frac{\sin(\omega_\ell \Delta x)}{\omega_\ell \Delta x} \right) \right. \right.$$
$$\left. \left. + \frac{\omega_\ell^2}{\mathrm{Pe}} \frac{2(1 - \cos(\omega_\ell \Delta x))}{(\omega_\ell \Delta x)^2} \right) t \right] \tag{6.38}$$

which leads to the following Fourier-series solution of the ordinary differential equations (6.18)

$$q_i(t) = \sum_{\ell=1}^{M} \left\{ C_\ell \exp\left[-\left(a(2\beta - 1)\omega_\ell \frac{1 - \cos(\omega_\ell \Delta x)}{\omega_\ell \Delta x} \right) t \right] \right.$$
$$\times \exp\left[-\left(\frac{\omega_\ell^2}{\mathrm{Pe}} \frac{2(1 - \cos(\omega_\ell \Delta x))}{(\omega_\ell \Delta x)^2} \right) t \right] \exp\left[\boldsymbol{j}\omega_\ell \left(i\Delta x - a\frac{\sin(\omega_\ell \Delta x)}{\omega_\ell \Delta x} t \right) \right] \right\} \tag{6.39}$$

where the coefficients C_ℓ may be determined as in (6.9)- (6.10). This solution may also be expressed as

$$
q_i(t) = \sum_{\ell=1}^{M} C_\ell \left\{ \exp\left[-\left(\left(a(2\beta - 1)\Delta x + \frac{2}{\text{Pe}} \right) \frac{\omega_\ell^2(1 - \cos(\omega_\ell \Delta x))}{(\omega_\ell \Delta x)^2} \right) t \right] \right.
$$

$$
\left. \times \exp\left[j\omega_\ell \left(i\Delta x - a\frac{\sin(\omega_\ell \Delta x)}{\omega_\ell \Delta x} t \right) \right] \right\} \tag{6.40}
$$

which remains bounded in time when

$$
a(2\beta - 1)\Delta x + \frac{2}{\text{Pe}} \geq 0 \quad \Rightarrow \quad \beta > \frac{1}{2} - \frac{1}{a\text{Pe}\Delta x} \tag{6.41}
$$

the boundedness result previously mentioned. As Δx approaches zero, with a finite $x_i = i\Delta x$, this solution (6.39) converges to (6.13) by virtue of the limits

$$
\lim_{\Delta x \to 0} \frac{1 - \cos(\omega_\ell \Delta x)}{\omega_\ell \Delta x} = 0, \quad \lim_{\Delta x \to 0} \frac{2(1 - \cos(\omega_\ell \Delta x))}{(\omega_\ell \Delta x)^2} = 1, \quad \lim_{\Delta x \to 0} \frac{\sin(\omega_\ell \Delta x)}{\omega_\ell \Delta x} = 1 \tag{6.42}
$$

Solution (6.39) thus correlates with (6.13) and lucidly exposes the effect of the spatial discretization on solution accuracy and stability. A comparison of (6.13) with (6.39) reveals that the spatial discretization alters the convection speed and diffusion coefficient, respectively from a to $a\frac{\sin(\omega_\ell \Delta x)}{\omega_\ell \Delta x}$ and from $\frac{1}{\text{Pe}}$ to $\frac{1}{\text{Pe}}\frac{2(1-\cos(\omega_\ell \Delta x))}{(\omega_\ell \Delta x)^2}$. This approximate diffusion coefficient, in particular, always remains non-negative, which indicates that the discretization in (6.14) of the second-order partial derivative in (6.1) is intrinsically stable, for the corresponding second time-exponential in (6.39) will never grow as time elapses.

The discretization (6.16) of first-order spatial derivatives, however, is not unconditionally stable. In fact, this discretization also induces the exponential term

$$
\mathcal{D} = \exp\left[-\left(a(2\beta - 1)\omega_\ell \frac{1 - \cos(\omega_\ell \Delta x)}{\omega_\ell \Delta x} \right) t \right] \tag{6.43}
$$

This term will not grow as time elapses if $(2\beta - 1) \leq 0$, for a negative "a", and $(2\beta - 1) \geq 0$, for a positive "a". The case $(2\beta - 1) = 0$, hence $\beta = \frac{1}{2}$, corresponds to a traditional centered discretization, which generally induces no intrinsic dissipation. In practice, however, it is convenient to take $\|2\beta - 1\| > 0$ for a dissipative discretization.

While in this linear case, $\beta \geq 0.5$ leads to a bounded solution, for non-linear equations, only the constraint $\beta > 0.5$ may generally yield bounded solutions. To illustrate the type of instability that can result from a non-dissipative spatial discretization, consider the time evolution of solution (6.39) for an unsuitable numerical value of β. For instance, when $a = 1$, $\text{Pe} = 500$, and $\Delta x = 0.05$, solution (6.39) will grow unboundedly when $\beta < 0.46$. Figure 6.3 compares solution (6.39) for $\beta = 0.45$ and $\beta = 0.55$ at several time levels. When $\beta = 0.55$ the solution remains bounded at all times and follows the exact solution. Conversely, for $\beta = 0.45$, instability rapidly sets in as time elapses and obviously originates from the solution exponential components. Since the integration in time that leads to (6.39) is exact, this instability entirely results from the discretization in space. Independently of any approximation of the time derivative, the spatial discretization impacts solution stability, in this

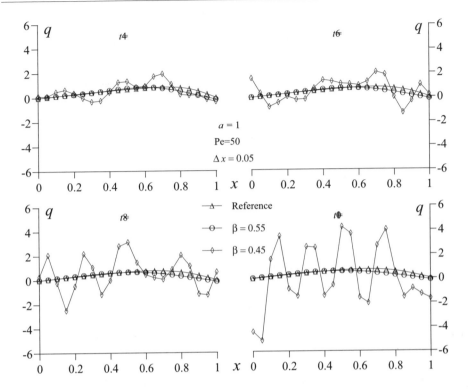

Figure 6.3: Stable and Unstable Spatially Discrete Solutions

representative case through the discretization of first-order space derivatives. An unsuitable discretization of first-order derivatives, especially in non-linear problems, will induce catastrophic exponential growth, even for well posed problems and bounded initial and boundary conditions. Independently of any time integration technique, a spatial discretization is acceptable when it leads to a system of continuum-time ordinary differential equations with a solution that evolves in time analogously to the spatially continuum solution.

Figure 6.3 indicates that in this example $\beta = 0.55$ leads to a bounded solution. The inherently stable discretization resulting from the mathematical constraint $\|2\beta - 1\| > 0$ corresponds to a fundamental physical situation. Consider, as a representative example, the case of a positive "a". This case corresponds to a flow from left to right, hence from node "$i-1$" to nodes "i" and "$i+1$". The inherent stability condition for this case is $(2\beta - 1) > 0$ hence $\beta > \frac{1}{2}$. This condition indicates that an inherently stable discretization of a first-order partial derivative results when the weight, or bias, is greater on the approximation that involves nodes "$i-1$" and "i". These are the nodes on the side of "i" from where the flow arrives and for this reason a discretization with such a bias is called an upstream-bias discretization. Upstream bias discretizations, accordingly, can be intrinsically stable, because they reflect the physical situation that the flow state at a point depends on flow conditions convected to that point along a specific direction. Form this perspective, upstream-bias discretizations are acceptable, in the sense defined previously.

Mathematically, upstream discretizations are inherently stable because they intrinsically induce an amount of dissipation. In the case of (6.16), the induced dissipation can be exposed by expressing (6.16) in the following form

$$\beta \frac{q_i - q_{i-1}}{\Delta x} + (1 - \beta) \frac{q_{i+1} - q_i}{\Delta x} = \frac{q_{i+1} - q_{i-1}}{2\Delta x} - (2\beta - 1) \frac{\Delta x}{2} \frac{q_{i-1} - 2q_i + q_{i+1}}{(\Delta x)^2} \tag{6.44}$$

This result consists of a non-dissipative centered discretization added, most importantly, to the discretization of a second-order partial derivative, which is intrinsically diffusive when its coefficient $(2\beta - 1)\frac{\Delta x}{2} \equiv \psi \frac{\Delta x}{2}$ remains positive. Expression (6.44) also indicates that an upstream discretization can equivalently result from a traditional centered discretization of a companion, or upstream bias, equation associated with (6.1). On the basis on (6.44), an upstream bias equation for (6.1) is

$$\frac{\partial q}{\partial t} + a \left(\frac{\partial q}{\partial x} - \varepsilon \psi \frac{\partial^2 q}{\partial x^2} \right) - \frac{1}{\text{Pe}} \frac{\partial^2 q}{\partial x^2} = 0 \tag{6.45}$$

for a three-node centered discretization of this equation along with $\varepsilon = \frac{\Delta x}{2}$ identically returns (6.44). The following chapters show that centered discretizations of characteristics-bias continuum equations directly and conveniently yield consistent upstream discretizations for the original governing equations.

6.3 Time Integration

The spatial-discretization system of ordinary differential equations, like (6.18), is then numerically integrated in time, which introduces additional accuracy and stability issues. Consider the following general form of an ordinary differential equation system

$$\frac{dq}{dt} = f(q(t), t) \tag{6.46}$$

Both explicit and implicit methods are available to integrate this system numerically. A representative explicit method is

$$\frac{q^{n+1} - q^n}{\Delta t} = f(q^n, t_n) \tag{6.47}$$

whereas an implicit method can be cast as

$$\frac{q^{n+1} - q^n}{\Delta t} = f(q^{n+1}, t_{n+1}) \tag{6.48}$$

where

$$t_n = n\Delta t, \qquad q^n \simeq q(t_n) \tag{6.49}$$

As (6.47) indicates, an explicit method allows the direct determination of the solution array q^{n+1} at time level t_{n+1} without solving any equations, but through a direct calculation of the right-hand-side at each node x_i and time level t_n. Conversely, an implicit method requires the solution of a system of generally non-linear coupled equations for computing the solution array q^{n+1} simultaneously at all discretization nodes at time level t_{n+1}.

Even for intrinsically stable time-continuum ordinary differential equations, an explicit method generates stable solutions only if the time step Δt does not exceed a certain stability limit that usually remains proportional to the discretization interval Δx; hence the smaller Δx the more minute Δt becomes. Implicit methods, on the other hand, generate stable solutions for much larger time steps Δt.

To illustrate these fundamental stability concepts, consider the integration of the model system (6.18) using methods (6.47) and (6.49), which respectively yield the two systems

$$\frac{q_i^{n+1} - q_i^n}{\Delta t} = -a\beta \frac{q_i^n - q_{i-1}^n}{\Delta x} - a(1-\beta)\frac{q_{i+1}^n - q_i^n}{\Delta x} + \frac{1}{Pe}\frac{q_{i-1}^n - 2q_i^n + q_{i+1}^n}{\Delta x^2} \tag{6.50}$$

$$\frac{q_i^{n+1} - q_i^n}{\Delta t} = -a\beta \frac{q_i^{n+1} - q_{i-1}^{n+1}}{\Delta x} - a(1-\beta)\frac{q_{i+1}^{n+1} - q_i^{n+1}}{\Delta x} + \frac{1}{Pe}\frac{q_{i-1}^{n+1} - 2q_i^{n+1} + q_{i+1}^{n+1}}{\Delta x^2} \tag{6.51}$$

The stability of the solutions of these algebraic systems with respect to the time step Δt is easily investigated by means of a general M-term Fourier-series solution, analogous to (6.11) and (6.36) in the form

$$q_i^n = \sum_{\ell=1}^M C_\ell v_\ell^n \exp(j\omega_\ell i \Delta x) \tag{6.52}$$

where in this case v_ℓ^n signifies that the "amplification factor" v_ℓ is actually raised to the power "n", unlike "q_i^n" where "n" quite simply denotes a time - station superscript. Insertion on this series into (6.50) yields the following equations for v_ℓ

$$\frac{v_\ell^{n+1} - v_\ell^n}{\Delta t} =$$

$$-\underbrace{\left(a\left((2\beta-1)\omega_\ell \frac{1 - \cos(\omega_\ell \Delta x)}{\omega_\ell \Delta x} + j\omega_\ell \frac{\sin(\omega_\ell \Delta x)}{\omega_\ell \Delta x}\right) + \frac{\omega_\ell^2}{Pe}\frac{2(1 - \cos(\omega_\ell \Delta x))}{(\omega_\ell \Delta x)^2}\right)}_{= \lambda_\ell} v_\ell^n \tag{6.53}$$

and

$$\frac{v_\ell^{n+1} - v_\ell^n}{\Delta t} =$$

$$-\underbrace{\left(a\left((2\beta-1)\omega_\ell \frac{1 - \cos(\omega_\ell \Delta x)}{\omega_\ell \Delta x} + j\omega_\ell \frac{\sin(\omega_\ell \Delta x)}{\omega_\ell \Delta x}\right) + \frac{\omega_\ell^2}{Pe}\frac{2(1 - \cos(\omega_\ell \Delta x))}{(\omega_\ell \Delta x)^2}\right)}_{= \lambda_\ell} v_\ell^{n+1} \tag{6.54}$$

with respective solutions

$$v_\ell = (1 - \lambda_\ell \Delta t), \qquad v_\ell = (1 + \lambda_\ell \Delta t)^{-1} \tag{6.55}$$

which lead to the following series for q_i^n

$$q_i^n = \sum_{\ell=1}^{M} C_\ell (1 - \lambda_\ell \Delta t)^n \exp(j\omega_\ell i \Delta x), \quad q_i^n = \sum_{\ell=1}^{M} C_\ell ((1 + \lambda_\ell \Delta t)^{-1})^n \exp(j\omega_\ell i \Delta x) \quad (6.56)$$

where the coefficients C_ℓ can be determined as in (6.9)- (6.10). For a vanishing Δt, with a finite $t_n = n\Delta t$, both of these solutions converge to the time-continuum solution (6.39). This is shown by remembering that

$$\lim_{m \to \pm\infty} \left(1 + \frac{1}{m}\right)^m = \exp(1); \quad \Delta t = \frac{t_n}{n} \quad \Rightarrow \quad \lim_{\Delta t \to 0} \Delta t = \lim_{n \to \infty} \frac{t_n}{n} \quad (6.57)$$

and consequently

$$\lim_{\Delta t \to 0} (1 - \lambda_\ell \Delta t)^n = \lim_{n \to \infty} \left(\left(1 + \frac{1}{\left(\frac{n}{-\lambda_\ell t_n}\right)}\right)^{\left(\frac{n}{-\lambda_\ell t}\right)} \right)^{-\lambda_\ell t_n} = \exp\left(-\lambda_\ell t_n\right) \quad (6.58)$$

$$\lim_{\Delta t \to 0} \left((1 + \lambda_\ell \Delta t)^{-1}\right)^n = \lim_{n \to \infty} \left(\left(1 + \frac{1}{\left(\frac{n}{\lambda_\ell t_n}\right)}\right)^{\left(\frac{n}{\lambda_\ell t}\right)} \right)^{-\lambda_\ell t_n} = \exp\left(-\lambda_\ell t_n\right) \quad (6.59)$$

Although both solutions in (6.56) converge to the single time continuum solution (6.39), these solutions feature profoundly distinct stability properties. Solutions (6.56) depend on an amplification factor that is raised to a positive integer power "n" greater than one. These solutions will thus remain stable when this factor does not exceed 1, which leads to the stability inequalities

$$\|1 - \lambda_\ell \Delta t\| \leq 1, \quad \left\|(1 + \lambda_\ell \Delta t)^{-1}\right\| \leq 1 \quad (6.60)$$

The first inequality, hence the explicit method, leads to the following realistic sufficient stability constraints on both Δt and β

$$\beta \geq 1 - \frac{1}{2a\mathrm{Pe}\Delta x}, \quad \Delta t \leq \frac{1}{\frac{a}{\Delta x}(2\beta - 1) + \frac{1}{\mathrm{Pe}\Delta x^2}} \quad (6.61)$$

These results indicate that in the absence of any physical diffusion, that is $\mathrm{Pe} \to \infty$, the limitation on β is $\beta = 1$. This limitation signifies this explicit numerical time integration method remains stable only in conjunction with a fully upstream spatial discretization, even though system (6.18) is stable for $\beta > \frac{1}{2}$. For the fully upstream spatial discretization, the stability constraint involving Δt becomes $\frac{a\Delta t}{\Delta x} \leq 1$ which corresponds to the fundamental Courant-Friedrichs-Lewy, or CFL, condition, a useful reference. When physical diffusion is present, hence Pe remains finite, constraints (6.61) impose stringent limitations on Δt. For instance, when $\mathrm{Pe}\Delta x \simeq 1$ then $\Delta t \simeq \Delta x$.

To illustrate the type of instability that can result from an explicit time integration, consider the time evolution of the explicit solution in (6.56) for a Δt that violates (6.61). For instance, when $a = 1$, $\mathrm{Pe} = 500$, and $\Delta x = 0.05$, this explicit solution will remain bounded

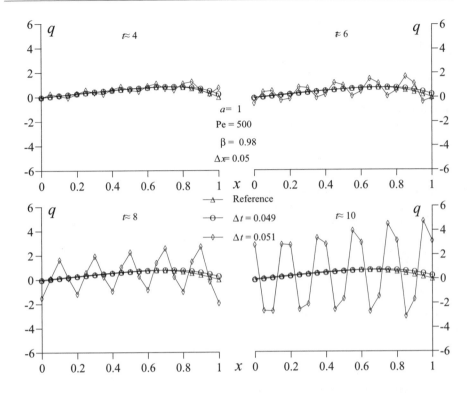

Figure 6.4: Stable and Unstable Fully Discrete Solutions

when $\beta \geq 0.98$ and $\Delta t \leq \Delta x$. Figure 6.3 compares explicit solutions for $\Delta t = 0.049$ and $\Delta t = 0.051$ at several time stations. When $\Delta t = 0.049$ the solution remains bounded at all times and follows the exact solution. Conversely, for $\Delta t = 0.051$, instability rapidly develops at subsequent time stations. Since $\beta = 0.98$ corresponds to a stable spatial discretization, as established in the previous section, this instability entirely results from the explicit time integration with a Δt that violates a stability constraint. In general, a time integration is acceptable when it leads to a system of equations with a solution that evolves at subsequent time stations analogously to the solution of the associated system of ordinary differential equations.

The second inequality in (6.60), hence the implicit method, does not impose any stability restriction on Δt. With $\Re(\lambda_\ell)$ and $\Im(\lambda_\ell)$ respectively denoting the real and imaginary parts of λ_ℓ, this inequality yields

$$\frac{1}{\left(1 + \Delta t \Re(\lambda_\ell)\right)^2 + \left(\Delta t \Im(\lambda_\ell)\right)^2} \leq 1 \tag{6.62}$$

which is unconditionally satisfied for any positive Δt.

The solution for q_i^{n+1} with implicit time-integration methods in practice originates from solving a system of equations. For the model example (6.18), this system of equations becomes

$$-\left(\beta\tfrac{a\Delta t}{\Delta x} + \tfrac{\Delta t}{\mathrm{Pe}\Delta x^2}\right) q_{i-1}^{n+1} + \left(1 + (2\beta - 1)\tfrac{a\Delta t}{\Delta x} + \tfrac{2\Delta t}{\mathrm{Pe}\Delta x^2}\right) q_i^{n+1} - \left((\beta - 1)\tfrac{a\Delta t}{\Delta x} + \tfrac{\Delta t}{\mathrm{Pe}\Delta x^2}\right) q_{i+1}^{n+1} = q_i^n$$

$$(6.63)$$

which contains $N+1$ coupled linear equations for $i = 0, 2, 3, \ldots, N-1$ and with two boundary conditions, one for $i = 0$ and the other for $i = N$.

6.4 Solution of Non-Linear Systems

An implicit method requires the solution of non-linear systems for the calculation of the solution because such a method couples all the spatial approximation nodes. Unlike explicit methods, however, implicit methods allow larger time steps in the determination of stable numerical solutions.

For the general spatial-discretization system of non-linear ordinary differential equations

$$\frac{dq}{dt} = f(q(t), t) \tag{6.64}$$

an implicit numerical time integration method like

$$\frac{q^{n+1} - q^n}{\Delta t} = f(q^{n+1}, t_{n+1}) \tag{6.65}$$

thus corresponds to a system of coupled non-linear equations. This system is invariably approximated by means of Newton's method, which in essence reduces the solution process to a sequence of solutions of linear algebraic systems. CFD procedures employ either direct or iterative methods to solve this terminal linear system of algebraic equations.

To delineate Newton's method, introduce the solution variation $\Delta q \equiv q^{n+1} - q^n$ and express (6.65) in the homogeneous form

$$F(\Delta q) \equiv \Delta q - \Delta t f(q^n + \Delta q, t_{n+1}) = 0 \tag{6.66}$$

Newton's method to solve $F(\Delta q) = 0$ for Δq is

$$0 = F(\Delta q^m) + \left.\frac{\partial F}{\partial \Delta q}\right|_{\Delta q = \Delta q^m} \left(\Delta q^{m+1} + \Delta q^m\right) \tag{6.67}$$

which for (6.66) specializes to

$$\left(I - \Delta t \left.\frac{\partial f}{\partial q}\right|_{q = q^n + \Delta q^m}\right) \left(\Delta q^{m+1} + \Delta q^m\right) = \Delta t f(q^n + \Delta q^m, t_{n+1}) - \Delta q^m \tag{6.68}$$

This linear system for determining Δq expresses an iterative process with "m" denoting not an exponent, but an iteration superscript. This iterative process can be concluded when the norm of difference between any two subsequent solutions decreases below a prescribed tolerance, like

$$\left\| \Delta q^{m+1} - \Delta q^m \right\| \leq \text{tolerance} \tag{6.69}$$

Frequently, however, the full iterative process (6.68) is replaced with a one-iteration algorithm with $\Delta q^0 = 0$, which leads to the linear system

$$\left(I - \Delta t \left. \frac{\partial f}{\partial q} \right|_{q=q^n} \right) \Delta q = \Delta t f(q^n, t_{n+1}) \tag{6.70}$$

This single system corresponds to a linearization of (6.65) in the form

$$q^{n+1} - q^n = \Delta t f(q^n, t_{n+1}) + \Delta t \left. \frac{\partial f}{\partial q} \right|_{q=q^n} \Delta q \tag{6.71}$$

which leads to the solution q^{n+1} in one step at each time level. In this case, the time step Δt can no longer be as large as desired, but must remain finite for a stable solution. The linearization stability limit on Δt, however, definitely exceeds that imposed by an explicit time- integration method.

Chapter 7

The Finite Element Method

The Finite Element method is a most systematic procedure to generate optimally accurate discrete analogues and solutions of a system of continuum partial differential equations, [17, 30, 144, 160, 214]. Minimizing arbitrariness as well as ambiguity in the synthesis of the discrete equations, the method represents each single continuum partial differential equation via a system of ordinary differential equations. Concisely presented in this chapter, the method consists of six chief sequential stages:

1. recasting of the partial differential system as an integral statement, from a generalized "least squares" minimization, and transformation of the statement into a "weak" statement;

2. restatement of each integral as a sum of integrals in subdomains;

3. execution of a local coordinate transformation within each subdomain for a practical calculation of each integral;

4. representation of each function and coordinate as a linear combination of chosen basis functions within each subdomain;

5. evaluation of each integral to generate a discrete system;

6. solution of this system to determine the computational solution of the original partial-differential-system problem.

Each integration subdomain corresponds to a finite element, which is thus a finite region where solutions and coordinates vary according to prescribed basis functions. Among several finite elements and methods, this book employs Lagrange elements and the classical Galerkin procedure, a method in which the weak-statement weight functions coincide with the solution basis functions. As well known, a Galerkin procedure generates a centered discretization, which can generate oscillatory solutions for convection dominated systems. As amplified in the following sections and chapters, the methods in this book apply the Galerkin procedure not so much directly to the Euler or Navier-Stokes system, but rather to a companion characteristics-bias system. Developed at the partial differential equation level, in the continuum and before any discretization, the characteristics-bias system augments the Euler or Navier-Stokes system with an intrinsically multidimensional upstream bias, along all the infinite wave-propagation directions radiating from each flow-field point, so that a classical Galerkin discretization of this system automatically generates a physically consistent multi-dimensional upstream discretization of the original Euler and Navier-Stokes systems. As the computational results bear out, this discretization generates monotone and essentially non-oscillatory solutions for smooth and shocked flows. Over the years, several finite element discretization procedures have been developed that both employ characteristics-theory notions and generate essentially non-oscillatory solutions for convection dominated flows. Among these procedures are the Streamline Upwind Petrov Galerkin (SUPG) method and the Discontinuous Galerkin (DG) method. This chapter shows that the integral statements from the Characteristics-Bias Galerkin (CBG) methods not only encompass but also generalize the SUPG and DG formulations.

The integral statement equivalent to a system of partial differential equations is shown also to evolve from a generalized minimization procedure. Employing the Green-Gauss divergence theorem, that is multi-dimensional integration by parts, the integral statement is then transformed into a powerful "weak" statement. This statement both provides efficient venues effectively to enforce physically meaningful boundary conditions and allows use of linear, bi-linear, and tri-linear piecewise continuous functions accurately to represent second-order stress-tensor and heat-transfer terms.

This chapter also summarizes the vector tools for formulating curvilinear-coordinate forms of the governing equations and corresponding integral statements. Basic geometric definitions generate straightforward vector expressions for finite areas and volumes, which then lead to their differential counterparts; the rules for transforming partial derivatives follow. With these significant results, the traditional derivatives and integrals with respect to the cartesian coordinates (x_1, x_2, x_3) are more conveniently and easily evaluated in terms of curvilinear coordinates (η_1, η_2, η_3).

A crucial stage in the finite element method consists in discretizing both functions and coordinate transformations in terms of prescribed functions. These functions belong to a so-called complete set of functions, which, somewhat informally, means that a sequence of these functions will converge to a member of the set. Investigated in depth in functional analysis, these fundamental approximation and convergence notions both identify this set of functions as a Hilbert space and involve such fundamental concepts as Hilbert and Sobolev spaces of functions as well as optimal approximations within inner-product spaces. This chapter highlights the main results of these investigations and then presents some of the commonly used functions. Allowing a convenient and efficient calculation of all the weak statement integrals, the completion of the discretization is the stage that leads to a system of discrete equations. The remaining sections of this chapter detail this systematic finite element discretization in arbitrarily shaped computational domains. These sections also present a set of optimal coordinate-transformation metric data that obviates the need for numerical integration and a sample "C" pseudo code that illustrates a typical sequence of operations within a finite element program for the characteristics-bias solution of the Euler equations.

7.1 Variational Weighted Integral Statements

This section develops several integral statements that are equivalent to corresponding governing partial differential equations. As shown in Sections 7.1.1-7.1.2, these statements evolve from a minimization principle that apply to fairly general equations. Section 7.1.3 presents a complementary procedure specific to the Fluid Dynamics equations.

7.1.1 Least Squares Statements

Consider the two vectors $\boldsymbol{F} = f_1\boldsymbol{i} + f_2\boldsymbol{j} + f_3\boldsymbol{k}$ and $\boldsymbol{Q} = q_1\boldsymbol{i} + q_2\boldsymbol{j} + 0\boldsymbol{k}$ with the objective of determining the best components q_1 and q_2 so that $\boldsymbol{Q}$ is closest to $\boldsymbol{F}$ in Gauss' least squares sense. This requirement leads to the minimization of the square of the error magnitude

$$2\mathcal{I} = (\boldsymbol{F} - \boldsymbol{Q}) \cdot (\boldsymbol{F} - \boldsymbol{Q}) \tag{7.1}$$

with respect to (q_1, q_2), which leads to

$$\frac{\partial \mathcal{I}}{\partial q_1} = 0, \quad \frac{\partial \mathcal{I}}{\partial q_2} = 0 \tag{7.2}$$

and consequently to

$$(\boldsymbol{F} - \boldsymbol{Q}) \cdot \boldsymbol{i} = 0, \quad (\boldsymbol{F} - \boldsymbol{Q}) \cdot \boldsymbol{j} = 0 \Rightarrow (\boldsymbol{F} - \boldsymbol{Q}) \cdot \boldsymbol{Q} = 0 \tag{7.3}$$

According to these expressions, the solution is $q_1 = f_1$ and $q_2 = f_2$. This finding means that the optimal $\boldsymbol{Q}$ corresponds to the projection of $\boldsymbol{F}$ onto the $(\boldsymbol{i}, \boldsymbol{j})$ plane and correspondingly the "error" $(\boldsymbol{F} - \boldsymbol{Q})$ is orthogonal to this optimal solution $\boldsymbol{Q}$. These basic results generalize to the determination of optimal approximating functions as follows.

Consider the smooth function $f = f(x, t)$ in terms of its Taylor's series about $x = x_0$

$$f(x, t) = f(x_0, t) + \sum_{n=1}^{\infty} \frac{d^n f}{dx}\bigg|_{x_0, t} \frac{(x - x_0)^n}{n!} \tag{7.4}$$

This exact series indicates that the expansion

$$f(x, t) = f(x_0, t) + \sum_{n=1}^{\infty} c_n(t)(x - x_0)^n \tag{7.5}$$

with each term a product of a function $c_n(t)$ of t and a function $(x - x_0)^n$ of x is neither a separability assumption nor an approximation, but an exact representation. This representation, in particular, confirms the correctness of expressing a function as a linear combination of simpler functions. Accordingly, consider the function $q = q(x, t)$ expressed as

$$q(x, t) = \sum_{n=1}^{N} a_n(t)(x - x_0)^{n-1} \tag{7.6}$$

in terms of a finite number of simpler functions, with the objective of determining the components $a_n(t)$ such that this function is closest to $f = f(x, t)$ in the least square sense.

In comparison to the previous vectors, the simple functions $(x - x_0)^i$, $i \geq 0$, thus correspond to the basis vectors $\boldsymbol{i}$, $\boldsymbol{j}$, $\boldsymbol{k}$ and are thus called basis functions. In direct analogy with the previous form of $\mathcal{I}$, the square of the error norm in this case is expressed as

$$2\mathcal{I} = \int_{\Omega} \left(f(x,t) - \sum_{n=1}^{N} a_n(t)(x - x_0)^{n-1} \right)^2 d\Omega \tag{7.7}$$

The minimization of this expression generates the system

$$\frac{\partial \mathcal{I}}{\partial a_i} = \int_{\Omega} (x - x_0)^{i-1} \left(f(x,t) - \sum_{n=1}^{N} a_n(t)(x - x_0)^{n-1} \right) d\Omega = 0, \quad 1 \leq i \leq N \tag{7.8}$$

that leads to the linear-combination components $a_n(t)$, $1 \leq n \leq N$. This integral, in particular, leads to the result

$$\sum_{i=1}^{N} a_i \frac{\partial \mathcal{I}}{\partial a_i} = \int_{\Omega} q(x,t) \left(f(x,t) - \sum_{n=1}^{N} a_n(t)(x - x_0)^{n-1} \right) d\Omega = 0 \tag{7.9}$$

With respect to the generalized "inner product" $(f, q) \equiv \int f g d\Omega$, as amplified in Section 7.5, this result indicates that the error $(f - q)$ remains "orthogonal" to the best approximation q, which, accordingly, represents the "projection" of the function f onto the "plane" defined by the basis functions $(x - x_0)^{n-1}$, $1 \leq n \leq N$. The generalization of these results to functions of several space variables $f = f(\boldsymbol{x}, t)$ and $q = q(\boldsymbol{x}, t)$ is direct, as it employs multi-dimensional integrals with an expansion $q(\boldsymbol{x}, t) = \sum_{n=1}^{N} a_n(t) W_n(\boldsymbol{x})$ in terms of prescribed functions $W_n(\boldsymbol{x})$. This minimization procedure extends to the determination of an optimally accurate discrete solution of a system of time-dependent partial differential equations.

For a function $q(\boldsymbol{x}, t) = \sum_{n=1}^{N} a_n(t) W_n(\boldsymbol{x})$ that is not an exact solution, a conservation-law system, like the Euler or Navier-Stokes system, becomes

$$\frac{\partial q}{\partial t} + \frac{\partial f_j}{\partial x_j} - \frac{\partial f_j^{\nu}}{\partial x_j} - \phi = \varepsilon_{\text{TOT}}(\boldsymbol{x}, t) = \varepsilon(\boldsymbol{x}, t) + \Delta\varepsilon(\boldsymbol{x}, t) \tag{7.10}$$

with

$$\varepsilon(\boldsymbol{x}, t) = \sum_{n=1}^{N} b_n(t) W_n(\boldsymbol{x}) \tag{7.11}$$

while $\Delta\varepsilon(\boldsymbol{x}, t)$ indicates a residual error that appears when the fluxes f_j and f_j^{ν} non-linearly depend on the basis functions, but that vanishes when these fluxes are expressed as a linear combination of basis functions, i.e. a "group" representation. The minimization procedure in this case aims to determine an optimal function q that minimizes the error ε_{TOT}. The square of the error magnitude becomes

$$2\mathcal{I} = \int_{\Omega} (\varepsilon(\boldsymbol{x}, t) + \Delta\varepsilon(\boldsymbol{x}, t))^2 d\Omega \tag{7.12}$$

and the minimization of this expression with respect to $b_i(t)$, $1 \leq i \leq N$, yields

$$\frac{\partial \mathcal{I}}{\partial b_i} = \int_{\Omega} (\varepsilon(\boldsymbol{x}, t) + \Delta\varepsilon(\boldsymbol{x}, t)) W_i(\boldsymbol{x}) d\Omega = \int_{\Omega} W_i \left(\frac{\partial q}{\partial t} + \frac{\partial f_j}{\partial x_j} - \frac{\partial f_j^{\nu}}{\partial x_j} - s \right) d\Omega = 0 \tag{7.13}$$

and

$$\sum_i^N c_i(t)\frac{\partial \mathcal{I}}{\partial b_i} = \int_\Omega W\left(\frac{\partial q}{\partial t} + \frac{\partial f_j}{\partial x_j} - \frac{\partial f_j^\nu}{\partial x_j} - s\right)d\Omega = 0 \tag{7.14}$$

These results show that an arbitrary function W, expressed via the prescribed functions W_i, $1 \le i \le N$, hence also q itself, is orthogonal to the error ε_{TOT}. Most importantly, the weighted integral statement

$$\int_\Omega W\left(\frac{\partial q}{\partial t} + \frac{\partial f_j}{\partial x_j} - \frac{\partial f_j^\nu}{\partial x_j} - s\right)d\Omega = 0 \tag{7.15}$$

defines an expansion q that minimizes the error ε_{TOT}. In particular, this integral statement generates a system of ordinary differential equations in time for each time dependent partial differential equation, as exemplified in the following section. For an infinite dimensional exact solution q, that is when $N \to \infty$, this weighted integral statement is equivalent to the original equations. Remaining valid for any function W as well as domain Ω, in this case, the statement implies the equation system. For a finite N, the expansion q becomes a finite dimensional optimal representation of the solution on Ω. Accordingly, this statement is employed to generate computational solutions.

7.1.2   Representative Optimal Solutions

This section demonstrates the determination of an optimal solution via a finite dimensional expansion and corresponding weighted integral statement for representative one- and two-dimensional convection diffusion equations. For the one dimensional convection diffusion equation

$$\frac{\partial q}{\partial t} + c\frac{\partial q}{\partial x} - \frac{1}{\text{Re}}\frac{\partial^2 q}{\partial x^2} = 0 \tag{7.16}$$

consider determining a finite dimensional solution in the form

$$q(x,t) = a_1(t) + a_2(t)x + a_3 x^2 \tag{7.17}$$

Substituting this expansion in the equation generates the "error"

$$\varepsilon_{\text{TOT}} = \left(\frac{da_1}{dt} + ca_2 - \frac{2}{\text{Re}}a_3\right) + \left(\frac{da_2}{dt} + 2ca_3\right)x + \frac{da_3}{dt}x^2 \tag{7.18}$$

With this error, for $1 \le i \le 3$, the weighted integral system (7.15) generates the ordinary differential equation system

$$\frac{da_1}{dt} + ca_2 - \frac{2}{\text{Re}}a_3 = 0, \quad \frac{da_2}{dt} + 2ca_3 = 0, \quad \frac{da_3}{dt} = 0 \tag{7.19}$$

For A_1, A_2, A_3 denoting arbitrary constants, the solution of this system is directly found as

$$a_3 = A_3, \quad a_2 = -2cA_3 t + A_2, \quad a_1 = \frac{2}{\text{Re}}A_3 t + c^2 A_3 t^2 - cA_2 t + A_1 \tag{7.20}$$

which leads to the following expression for q

$$q(x,t) = A_1 + A_2(x - ct) + A_3\left(x^2 - 2ctx + c^2t^2 + \frac{2}{Re}t\right) \tag{7.21}$$

Significantly, this is an exact solution of the original partial differential equation. The constants may then be determined by minimizing the difference between this solution and the initial and two boundary conditions. In this respect, this q is a finite-dimensional representation of the solution of a complete initial-boundary value problem; the greater N the more accurate this solution. In particular, this representative example shows that the optimization weighted integral statement (7.15) for a single time-dependent partial differential equation, generates a system of ordinary differential equations.

Consider next the determination of an optimal finite-dimensional solution for the two-dimensional convection-diffusion equation

$$\frac{\partial q}{\partial t} + u\frac{\partial q}{\partial x} + v\frac{\partial q}{\partial y} - \frac{1}{Re}\left(\frac{\partial^2 q}{\partial x^2} + \frac{\partial^2 q}{\partial y^2}\right) = 0 \tag{7.22}$$

A finite dimensional solution may be expressed in the form

$$q(x,y,t) = a_1(t) + a_2(t)x + a_3(t)y + a_4(t)xy + a_5(t)x^2 + a_5(t)y^2 \tag{7.23}$$

Substituting this expansion in the equation generates the "error"

$$\varepsilon_{TOT} = \left(\frac{da_1}{dt} + ua_2 + va_3 - \frac{2}{Re}(a_5 + a_6)\right) + \left(\frac{da_2}{dt} + 2ua_5 + va_4\right)x +$$

$$\left(\frac{da_3}{dt} + ua_4 + 2va_4\right)y + \frac{da_4}{dt}xy + \frac{da_5}{dt}x^2 + \frac{da_6}{dt}y^2 \tag{7.24}$$

With this error, for $1 \leq i \leq 6$, the weighted integral system (7.15) generates the ordinary differential equation system

$$\frac{da_1}{dt} + ua_2 + va_3 - \frac{2}{Re}(a_5 + a_6) = 0, \quad \frac{da_2}{dt} + 2ua_5 + va_4 = 0$$

$$\frac{da_3}{dt} + ua_4 + 2va_4 = 0, \quad \frac{da_4}{dt} = 0, \quad \frac{da_5}{dt} = 0, \quad \frac{da_6}{dt} = 0 \tag{7.25}$$

For A_1, A_2, A_3, A_4, A_5, A_6 denoting arbitrary constants, the solution of this system is directly found as

$$a_6 = A_6, \quad a_5 = A_5, \quad a_4 = A_4, \quad a_3 = -uA_4t - 2vA_6t + A_3, \quad a_2 = -vA_4t - 2uA_5t + A_2$$

$$a_1 = (u^2A_5 + uvA_4 + v^2A_6)t^2 + \left(\frac{2}{Re}(A_5 + A_6) - uA_2 - vA_3\right)t + A_1 \tag{7.26}$$

which leads to the following expansion for q

$$q(x,y,t) = A_1 + A_2(x - ut) + A_3(y - vt) + A_4\left(uvt^2 - vx - uty + xy\right) +$$

$$A_5 \left(x^2 - 2utx + u^2t^2 + \frac{2}{\text{Re}}t \right) + A_6 \left(y^2 - 2vty + v^2t^2 + \frac{2}{\text{Re}}t \right) \tag{7.27}$$

Similarly to the one-dimensional case, this is an exact solution of the original partial differential equation. The constants may then be determined by minimizing the difference between this solution and the initial and boundary conditions. In this respect, this q is a finite-dimensional representation of the solution of a complete initial-boundary value problem; the greater N the more accurate this solution. Again, this representative example shows that the optimization weighted integral statement (7.15) for a single time-dependent partial differential equation, generates a system or ordinary differential equations.

7.1.3 Fluid Dynamics Equations from a Variational Principle

This section shows how the Fluid Dynamics governing equations also emerge from a variational principle that yields three weighted integral statements. Consider expressing the intensive variable $\beta\rho$ of Chapter 1 as a linear combination of "N" time dependent parameters and suitable basis functions in the form

$$\beta(\boldsymbol{x},t)\rho(\boldsymbol{x},t) \simeq \sum_{j=1}^{N} \beta_j(t)\rho_j(t)w_j(\boldsymbol{x}), \quad w_j(\boldsymbol{x}_i) = \delta_j^i \tag{7.28}$$

so that

$$\beta(\boldsymbol{x}_i,t)\rho(\boldsymbol{x}_i,t) \simeq \beta_i(t)\rho_i(t) \tag{7.29}$$

The parameters $\beta_i(t)$ and $\rho_i(t)$, however, may or may not represent the values attained by $\beta(\boldsymbol{x},t)$ and $\rho(\boldsymbol{x},t)$ at $\boldsymbol{x} = \boldsymbol{x}_i$. For given trial functions $w_j(\boldsymbol{x})$, a fundamental question related to (7.28) concerns the choice of the parameters $\beta_i(t)$ and $\rho_i(t)$, so that (7.28) becomes the most accurate representation for $\beta(\boldsymbol{x},t)\rho(\boldsymbol{x},t)$. As detailed in the previous sections, Gauss' least squares method indicates that a most accurate representation emerges when these parameters are linearly related to "weighted" state variables as shown next.

The local deviation between $\beta(\boldsymbol{x},t)\rho(\boldsymbol{x},t)$ and (7.28) can be simply characterized as the difference $\varepsilon_{TOT} \equiv \beta(\boldsymbol{x},t)\rho(\boldsymbol{x},t) - \sum_{j=1}^{N} \beta_j(t)\rho_j(t)w_j(\boldsymbol{x})$. At each instant of time, the total deviation can then be expressed as the integral of the square of ε_{TOT} over the fluid domain Ω of interest as

$$2\mathcal{I} = \int_{\Omega} \varepsilon_{TOT}^2 \, d\Omega = \int_{\Omega} \left(\beta(\boldsymbol{x},t)\rho(\boldsymbol{x},t) - \sum_{j=1}^{N} \beta_j(t)\rho_j(t)w_j(\boldsymbol{x}) \right)^2 d\Omega \tag{7.30}$$

From this expression, the total deviation is always positive and depends on the parameters $\beta_i(t)$ and $\rho_i(t)$. Representation (7.28) becomes "optimal" in the least squares sense when ε_{TOT} is a minimum with respect to $\beta_i(t)\rho_i(t)$. This requirement yields the conditions

$$d\mathcal{I} = \sum_{i=1}^{N} \frac{\partial \mathcal{I}}{\partial(\beta_i(t)\rho_i(t))} d(\beta_i(t)\rho_i(t)) = 0 \quad \Rightarrow \quad \frac{\partial \mathcal{I}}{\partial(\beta_i(t)\rho_i(t))} = 0, \quad 1 \leq i \leq N \tag{7.31}$$

where the vanishing of N partial derivatives follow from the linear independence of the parameter products $(\beta_i(t)\rho_i(t))$, for $1 \leq i \leq N$. These partial derivatives yield the system

$$\int_{\Omega} w_i(\boldsymbol{x}) \sum_{j=1}^{N} w_j(\boldsymbol{x})\beta_j(t)\rho_j(t)d\Omega = \int_{\Omega} w_i(\boldsymbol{x})\beta(\boldsymbol{x},t)\rho(\boldsymbol{x},t)d\Omega, \quad 1 \leq i \leq N \tag{7.32}$$

which determines the parameter products $(\beta_j(t)\rho_j(t))$, for $1 \le j \le N$. These optimal parameters, therefore, depend on the "weighted" extensive quantity at the rhs of this linear system. For simplicity, the subscript "i" is eliminated from the basis function w for now, because the following considerations apply for any suitable basis function. For $\beta = 1$, $\beta = V$, and $\beta = e$, this weighted quantity respectively corresponds to the weighted mass, linear momentum and energy

$$\mathcal{M}_w = \int_\Omega w\rho\, d\Omega, \quad \boldsymbol{L}_w = \int_\Omega w\rho\boldsymbol{V}\, d\Omega, \quad \mathcal{E}_w = \int_\Omega w\rho e\, d\Omega \tag{7.33}$$

The prominent appearance of these weighted quantities in the variational problem of obtaining an optimal representation (7.28) justifies inquiring into the total rate of change of these quantities.

From the transport theorem, the time rate of change of a weighted quantity is

$$\frac{DB_w}{Dt} = \frac{D}{Dt}\int_\Omega w\beta\rho\, d\Omega = \int_\Omega \left[\frac{\partial(w\beta\rho)}{\partial t} + \mathrm{div}\,(w\beta\rho\boldsymbol{V})\right] d\Omega =$$

$$= \int_\Omega w\left[\frac{\partial\beta\rho}{\partial t} + \frac{\partial(\beta\rho u_j)}{\partial x_j}\right] d\Omega + \int_\Omega \beta\rho u_j \frac{\partial w}{\partial x_j}\, d\Omega \tag{7.34}$$

The presence of an integral involving $u_j\partial w/\partial x_j$, leads to a convenient interpretation of the weight function w. Retracing the steps that prove the transport theorem in divergence form, shows that $u_j\partial w/\partial x_j$ corresponds to the derivative of w with respect to time, holding the fluid particle coordinates $\boldsymbol{x}_0$ constant. This observation shows that the weight function w can be expressed as $w(\boldsymbol{x}) = w(\boldsymbol{x}(\boldsymbol{x}_0, t))$, which corresponds to a function that "follows" fluid particles as Ω deforms, in the sense that one specific value of w within the deforming Ω always corresponds to one fluid particle.

Expression (7.34) yields a different form of the rate of change of B_w upon regrouping terms as follows

$$\frac{DB_w}{Dt} = \int_\Omega w\left[\frac{\partial\beta\rho}{\partial t} + \frac{\partial(\beta\rho u_j)}{\partial x_j}\right] d\Omega + \int_\Omega \beta\rho u_j \frac{\partial w}{\partial x_j}\, d\Omega =$$

$$= \int_\Omega w\left[\left(\frac{\partial\beta\rho}{\partial t} + u_j\frac{\partial(\beta\rho)}{\partial x_j}\right) d\Omega + \beta\rho\,\mathrm{div}\boldsymbol{V}\, d\Omega\right] + \int_\Omega \beta\rho u_j \frac{\partial w}{\partial x_j}\, d\Omega =$$

$$= \int_\Omega w\left[\frac{D\beta\rho}{Dt} d\Omega + \beta\rho\frac{Dd\Omega}{Dt}\right] + \int_\Omega \beta\rho u_j \frac{\partial w}{\partial x_j}\, d\Omega \tag{7.35}$$

The time rate of change of a weighted quantity is thus expressed as

$$\frac{D}{Dt}\int_\Omega w\beta\rho\, d\Omega = \int_\Omega w\frac{D(\beta\rho d\Omega)}{Dt} + \int_\Omega \beta\rho u_j \frac{\partial w}{\partial x_j}\, d\Omega = \int_\Omega w\frac{D(\beta d\mathcal{M})}{Dt} + \int_\Omega \beta\rho u_j \frac{\partial w}{\partial x_j}\, d\Omega \tag{7.36}$$

and equals the weighted integral of the substantive derivative of a fluid particle intensive quantity augmented by the integral of $\beta\rho u_j\partial w/\partial x_j$.

Expressions (7.34) and (7.36) provide the same rate of change and upon equating them leads to the equation governing the integral of a weighted substantive rate of change as

$$\int_\Omega w \left[\frac{\partial \beta \rho}{\partial t} + \frac{\partial(\beta \rho u_j)}{\partial x_j} \right] d\Omega = \int_\Omega w \frac{D(\beta d\mathcal{M})}{Dt} \tag{7.37}$$

For $\beta = 1$, $\beta = u_i$, and $\beta = e$ the rhs of this equation contains the weighted integral of the substantive derivative of a fluid particle mass, i^{th} component of linear momentum and total energy. For the linear momentum and total energy cases these integrals can thus be recast as

$$\int_\Omega w \left[\frac{\partial u_i \rho}{\partial t} + \frac{\partial(u_i \rho u_j)}{\partial x_j} \right] d\Omega = \int_\Omega w \, dF_i \tag{7.38}$$

$$\int_\Omega w \left[\frac{\partial \rho e}{\partial t} + \frac{\partial(\rho e u_j)}{\partial x_j} \right] d\Omega = \int_\Omega w \delta \left(\frac{\delta Q}{\delta t} + \frac{\delta W}{\delta t} \right) \tag{7.39}$$

Since $D(d\mathcal{M})/Dt = 0$ and the elementary force component and unit-time heat transfer and work are expressed as shown in Chapter 1, equations (7.37)-(7.39) lead to the weighted integral equations

$$\int_\Omega w \left[\frac{\partial \rho}{\partial t} + \frac{\partial \rho u_j}{\partial x_j} \right] d\Omega = 0 \tag{7.40}$$

$$\int_\Omega w \left[\frac{\partial \rho u_i}{\partial t} + \frac{\partial \rho u_i u_j}{\partial x_j} - \frac{\partial \sigma_{ij}}{\partial x_j} - \rho b_i \right] d\Omega = 0 \tag{7.41}$$

$$\int_\Omega w \left[\frac{\partial \rho e}{\partial t} + \frac{\partial \rho e u_j}{\partial x_j} - \frac{\partial u_i \sigma_{ij}}{\partial x_j} + \frac{\partial q_j}{\partial x_j} - \rho u_i b_i - \rho \dot{g} \right] d\Omega = 0 \tag{7.42}$$

These equations can also be expressed for a fixed volume Ω. Since these equations must then be valid for arbitrary weight functions w, it then follows they can only hold true when the expressions between brackets themselves vanish, which returns the governing equations in differential form. This observation shows that these weighted integral equations are as fundamental as the differential equations themselves. On the other hand, as shown in the previous sections, the weighted integral form of these equations provides a minimization foundation for the development of a finite element computational solution.

7.2   Integrals from Gauss' Divergence Theorem

Gauss' divergence theorem is expressed as

$$\int_\Omega \nabla \cdot \boldsymbol{f} \, d\Omega = \oint_{\partial \Omega} \boldsymbol{f} \cdot \boldsymbol{n} \, d\partial\Omega \tag{7.43}$$

for a smooth function $\boldsymbol{f}$. With implied summation on repeated subscript indices, the Cartesian-coordinate form of this theorem is

$$\int_\Omega \frac{\partial f_j}{\partial x_j} \, d\Omega = \oint_{\partial \Omega} f_j n_j \, d\partial\Omega \tag{7.44}$$

This central theorem leads to the integral theorems that underpin the integral and weak-statement formulations of the governing equations.

The multidimensional rule for integration by parts emerges from Gauss's theorem, by expressing the vector $\boldsymbol{f}$ as the product of scalar and vector functions as

$$\boldsymbol{f} = w\boldsymbol{g} \tag{7.45}$$

The insertion of this form into (7.43) yields

$$\int_\Omega \nabla\cdot(w\boldsymbol{g})\,d\Omega = \oint_{\partial\Omega} w\boldsymbol{g}\cdot\boldsymbol{n}\,d\partial\Omega$$

$$\int_\Omega (w\nabla\cdot\boldsymbol{g} + \boldsymbol{g}\cdot\nabla w)\,d\Omega = \oint_{\partial\Omega} w\boldsymbol{g}\cdot\boldsymbol{n}\,d\partial\Omega$$

$$\int_\Omega w\nabla\cdot\boldsymbol{g}\,d\Omega = -\int_\Omega \boldsymbol{g}\cdot\nabla w\,d\Omega + \oint_{\partial\Omega} w\boldsymbol{g}\cdot\boldsymbol{n}\,d\partial\Omega \tag{7.46}$$

which states the multidimensional rule for integration by parts. In Cartesian coordinates, this rule becomes

$$\int_\Omega w\frac{\partial g_j}{\partial x_j}\,d\Omega = -\int_\Omega \frac{\partial w}{\partial x_j}g_j\,d\Omega + \oint_{\partial\Omega} wg_jn_j\,d\partial\Omega \tag{7.47}$$

Integration by parts also allows transforming integrals that contain second-order derivatives of a scalar function ϕ. This particular transformation emerges from the rule for integration by parts (7.46) by expressing the vector $\boldsymbol{g}$ as the gradient of a scalar function ϕ as

$$\boldsymbol{g} = \nabla\phi \;\Rightarrow\; g_j = \frac{\partial\phi}{\partial x_j} \tag{7.48}$$

The insertion of this form of $\boldsymbol{g}$ transforms (7.46) into

$$\int_\Omega w\nabla^2\phi\,d\Omega = -\int_\Omega \nabla w\cdot\nabla\phi\,d\Omega + \oint_{\partial\Omega} w\nabla\phi\cdot\boldsymbol{n}\,d\partial\Omega$$

$$\int_\Omega w\nabla^2\phi\,d\Omega = -\int_\Omega \nabla w\cdot\nabla\phi\,d\Omega + \oint_{\partial\Omega} w\frac{\partial\phi}{\partial n}\,d\partial\Omega \tag{7.49}$$

The corresponding Cartesian-coordinate form becomes

$$\int_\Omega w\frac{\partial^2\phi}{\partial x_j\partial x_j}\,d\Omega = -\int_\Omega \frac{\partial w}{\partial x_j}\frac{\partial\phi}{\partial x_j}\,d\Omega + \oint_{\partial\Omega} w\frac{\partial\phi}{\partial n}\,d\partial\Omega \tag{7.50}$$

which shows that the integral at the right of the equal sign only involve first-order derivatives. Green's symmetrical integral theorem emerges from the rule for integration by parts of second order derivatives (7.49). Considering that both w and ϕ are scalar functions, an analogous expression arises by interchanging w with ϕ, which leads to

$$\int_\Omega \phi\nabla^2 w\,d\Omega = -\int_\Omega \nabla\phi\cdot\nabla w\,d\Omega + \oint_{\partial\Omega} \phi\frac{\partial w}{\partial n}\,d\partial\Omega \tag{7.51}$$

Upon subtracting (7.51) from (7.49), Green's symmetrical theorem emerges as

$$\int_\Omega \left(w\nabla^2\phi - \phi\nabla^2 w\right)d\Omega = \oint_{\partial\Omega} \left(w\frac{\partial\phi}{\partial n} - \phi\frac{\partial w}{\partial n}\right)d\partial\Omega \tag{7.52}$$

In Cartesian coordinates, this theorem is expressed as

$$\int_\Omega \left(w\frac{\partial^2\phi}{\partial x_j\partial x_j} - \phi\frac{\partial^2 w}{\partial x_j\partial x_j}\right)d\Omega = \oint_{\partial\Omega} \left(w\frac{\partial\phi}{\partial n} - \phi\frac{\partial w}{\partial n}\right)d\partial\Omega \tag{7.53}$$

7.3 Weak Derivatives and Integral Statement

Within a domain Ω, consider a function ϕ such that itself and its derivatives of all orders are smooth in Ω and vanish on all the boundary of Ω. Such a function is said to have compact support in Ω and its properties are symbolically indicated by the expression $\phi \in C_0^\infty(\Omega)$; instead, a function "$f$" that is simply square integrable in Ω is indicated as $f \in \mathcal{H}^0(\Omega)$.

For any $\phi \in C_0^\infty(\Omega)$, consider the classical integration-by-parts statement for $f \in C^\infty(\Omega)$

$$\int_\Omega \frac{\partial f}{\partial x} \phi d\Omega = -\int_\Omega f \frac{\partial \phi}{\partial x} d\Omega \tag{7.54}$$

In respect of the right-hand side, this term exists even for non smooth functions $f \in \mathcal{H}^0(\Omega)$. Consider then a function $f_{w_1} \in \mathcal{H}^0(\Omega)$ defined as

$$\int_\Omega f_{w_1} \phi d\Omega \equiv -\int_\Omega f \frac{\partial \phi}{\partial x} d\Omega \tag{7.55}$$

On the analogy of (7.54), this function f_{w_1}, if it exists, is the "weak" first derivative of f. Generalizing this result, the expression

$$\int_\Omega f_{w_n} \phi d\Omega = (-1)^n \int_\Omega f \frac{\partial^n \phi}{\partial x^n} d\Omega \tag{7.56}$$

defines a function $f_{w_n} \in \mathcal{H}^0(\Omega)$, if it exists, that corresponds to the n-th "weak" derivative of f. When $f \in C^n(\Omega)$, then its weak derivative coincides with the classical derivative. Expression (7.56) may also exist for general $f \in \mathcal{H}^0(\Omega)$; it follows that a weak derivative corresponds to a generalized derivative for generalized functions. To exemplify the notion of weak derivative, consider determining the first weak derivative for two functions, one with continuous, the other with discontinuous slopes. These two sample functions are

$$f_1(x) = -1 + x, \quad f_2(x) = \begin{cases} -1+x & x < 0 \\ 1 & x = 0 \\ 1-x & x > 0 \end{cases}, \quad x \in \Omega \equiv [-1,1] \tag{7.57}$$

The right-hand side of (7.54) for the first function becomes

$$-\int_{-1}^1 (-1+x) \frac{\partial \phi}{\partial x} dx = \int_{-1}^1 \frac{\partial(-1+x)}{\partial x} \phi dx = \int_{-1}^1 \phi dx \tag{7.58}$$

and comparison of this result with the left-hand side of (7.54) shows that the weak derivative f_{w_1} of f_1 is $f_{w_1} = 1$, which corresponds to the classical derivative of f_1.

With reference to f_2, this function does not possess a classical first derivative at $x = 0$; nevertheless it does have a first weak derivative with multiple values at $x = 0$. The right-hand side of (7.54) for this function becomes

$$-\int_{-1}^1 f_2 \frac{\partial \phi}{\partial x} dx = \int_{-1}^0 \frac{\partial(-1+x)}{\partial x} \phi dx + \int_0^1 \frac{\partial(1-x)}{\partial x} \phi dx$$

$$= \int_{-1}^0 \phi dx - \int_0^1 \phi dx \tag{7.59}$$

and comparison of this result with the left-hand side of (7.54) shows that actually several first weak derivatives f_{w_1} of f_2 exist, such as

$$f_{w_1a}(x) = \begin{cases} 1 & x < 0 \\ -1 & x \geq 0 \end{cases}, \quad f_{w_1b}(x) = \begin{cases} 1 & x \leq 0 \\ -1 & x > 0 \end{cases} \tag{7.60}$$

Although the difference between these two weak derivatives may seem a mere technicality, the important concept they reflect is that, unlike the classical derivative, these weak derivatives exist over the entire Ω, coincide with one another almost everywhere within Ω, and may differ on sets of point of vanishing volume, area, or length measure, in this particular case the point $x = 0$. Most importantly, although generally discontinuous as in this example, weak derivatives may still be integrated without any preoccupation of potentially infinite derivatives at the discontinuity point. This observation, in particular, may be used as the definition of a generalized integral, e.g. the Lebesgue integral, that is an integral that exists for functions that may be discontinuous, but in regions of zero measure. All the integrals in this book are to be implicitly understood as Lebesgue integrals and the classical notation $f_{w_n} = \partial^n f / \partial x^n$ will always indicate a weak derivative.

7.4 Characteristics-Bias Formulation

The computational solutions of the Euler and Navier-Stokes equations are generated from a Galerkin discretization of an associated characteristics-bias system. Developed in the following chapters, this system augments the Euler and Navier-Stokes system with a multi-dimensional upstream bias term and is cast as

$$\frac{\partial q}{\partial t} - \frac{\partial}{\partial x_\ell} \left(\varepsilon \psi a_\ell \frac{\partial q}{\partial t} \right) + \frac{\partial \left(f_\ell - f_\ell^\nu \right)}{\partial x_\ell} - \frac{\partial}{\partial x_\ell} \left(\varepsilon \psi a_\ell \left(\frac{\partial (f_m^q - f_m^\nu)}{\partial x_m} + \delta \frac{\partial f_m^p}{\partial x_m} \right) \right)$$

$$- \frac{\partial}{\partial x_\ell} \left(\varepsilon \psi \left(c \left(\alpha a_\ell a_m + \alpha^N \left(a_\ell^{N_1} a_m^{N_1} + a_\ell^{N_2} a_m^{N_2} \right) \right) \frac{\partial q}{\partial x_m} \right) \right) = 0 \tag{7.61}$$

As amplified in the following chapters, the spatial derivative of $\frac{\partial q}{\partial t}$ and the second-order terms in this system result from a continuum, as opposed to discrete, multi-dimensional consistent upwind representation of the inviscid-flux divergence $\frac{\partial (f_\ell - f_\ell^\nu)}{\partial q}$ along all directions of wave propagation. In these terms, the fluxes f_ℓ^q and f_ℓ^p satisfy the equality $f_\ell = f_\ell^q + f_\ell^p$, the vector components a_ℓ, $a_\ell^{N_1}$, $a_\ell^{N_2}$, $1 \leq \ell \leq 3$, correspond to the direction cosines of three unit vectors, one parallel to the local velocity vector and the other two orthogonal to each other on a plane perpendicular to velocity, the parameters ε and ψ respectively correspond to a reference length and solution smoothness controller, the variable "c" denotes the speed of sound, and the functions δ, α, α^N respectively balance the contributions to the multi-dimensional upwind of the pressure gradient and the acoustic components in the streamline and crossflow directions.

This system is also equivalent to the integral statement

$$\int_\Omega w \left(\frac{\partial q}{\partial t} - \frac{\partial}{\partial x_\ell} \left(\varepsilon \psi a_\ell \frac{\partial q}{\partial t} \right) + \frac{\partial \left(f_\ell - f_\ell^\nu \right)}{\partial x_\ell} - \frac{\partial}{\partial x_\ell} \left(\varepsilon \psi a_\ell \left(\frac{\partial (f_m^q - f_m^\nu)}{\partial x_m} + \delta \frac{\partial f_m^p}{\partial x_m} \right) \right) \right) d\Omega$$

$$-\int_\Omega w \frac{\partial}{\partial x_\ell}\left(\varepsilon\psi c\left(\alpha a_\ell a_m + \alpha^N\left(a_\ell^{N_1} a_m^{N_1} + a_\ell^{N_2} a_m^{N_2}\right)\right)\frac{\partial q}{\partial x_m}\right) d\Omega = 0 \qquad (7.62)$$

Integrating this statement by parts yields

$$\int_\Omega w \frac{\partial q}{\partial t} d\Omega - \int_\Omega \frac{\partial w}{\partial x_\ell}(f_\ell - f_\ell^\nu)\, d\Omega + \int_\Omega \frac{\partial w}{\partial x_\ell} a_\ell \varepsilon\psi \left(\frac{\partial q}{\partial t} + \frac{\partial(f_m^q - f_m^\nu)}{\partial x_m} + \delta\frac{\partial f_m^p}{\partial x_m}\right) d\Omega$$

$$+ \int_\Omega \frac{\partial w}{\partial x_\ell}\varepsilon\psi c\left(\alpha a_\ell a_m + \alpha^N\left(a_\ell^{N_1} a_m^{N_1} + a_\ell^{N_2} a_m^{N_2}\right)\right)\frac{\partial q}{\partial x_m} d\Omega$$

$$- \oint_{\partial\Omega} w\varepsilon\psi a_\ell \left(\frac{\partial q}{\partial t} + \frac{\partial(f_m^q - f_m^\nu)}{\partial x_m} + \delta\frac{\partial f_m^p}{\partial x_m}\right) n_\ell\, d\partial\Omega$$

$$- \oint_{\partial\Omega} w\varepsilon\psi \left(c\left(\alpha a_\ell a_m + \alpha^N\left(a_\ell^{N_1} a_m^{N_1} + a_\ell^{N_2} a_m^{N_2}\right)\right)\frac{\partial q}{\partial x_m}\right) n_\ell\, d\partial\Omega$$

$$+ \oint_{\partial\Omega} w f_j n_j\, d\partial\Omega - \oint_{\partial\Omega} w f_j^\nu n_j\, d\partial\Omega = 0 \qquad (7.63)$$

which corresponds to the characteristics-bias Euler and Navier-Stokes integral weak statement. The word "weak" in this phrase signifies not so much that the statement is somehow "feeble", but rather that the differentiability requirements on q have been weakened, which allows this integral statement to admit solutions with shock. This statement is elaborated in Section 7.7.

7.4.1 Non-Discrete Discontinuous Galerkin (DG) Form

For both a weight function w with non-vanishing trace and a solution q that is discontinuous on the boundary $\partial\widehat{\Omega}$ of a subdomain $\widehat{\Omega} \subset \Omega$, an integration by parts of both the flux divergence and the characteristics-bias expression in (7.62) generates the weak statement

$$\int_{\widehat{\Omega}} w \frac{\partial q}{\partial t} d\Omega + \int_{\widehat{\Omega}} \frac{\partial w}{\partial x_\ell}\varepsilon\psi\left[\left(\frac{\partial q}{\partial t} + \frac{\partial(f_m - f_m^\nu)}{\partial x_m} + (\delta - 1)\frac{\partial f_m^p}{\partial x_m}\right)a_\ell\right.$$

$$\left. + c\left(\alpha a_\ell a_m + \alpha^N\left(a_\ell^{N_1} a_m^{N_1} + a_\ell^{N_2} a_m^{N_2}\right)\right)\frac{\partial q}{\partial x_m}\right] d\Omega$$

$$- \int_{\widehat{\Omega}} \frac{\partial w}{\partial x_\ell}(f_\ell - f_\ell^\nu)\, d\Omega + \oint_{\partial\widehat{\Omega}} w\left\{(f_\ell - f_\ell^\nu) - \varepsilon\psi\left[\left(\frac{\partial q}{\partial t} + \frac{\partial(f_m - f_m^\nu)}{\partial x_m} + (\delta - 1)\frac{\partial f_m^p}{\partial x_m}\right)a_\ell\right.\right.$$

$$\left.\left. + c\left(\alpha a_\ell a_m + \alpha^N\left(a_\ell^{N_1} a_m^{N_1} + a_\ell^{N_2} a_m^{N_2}\right)\right)\frac{\partial q}{\partial x_m}\right]\right\} n_\ell\, d\partial\Omega = 0 \qquad (7.64)$$

This statement may be viewed as a non-discrete Galilean invariant DG formulation. In respect of the second domain integral, this expression presents a direct counterpart of the "shock-capturing" term employed in the DG algorithms; rather than corresponding to a Laplacian dissipation added to the formulation, this expression naturally emerges from the characteristics-bias procedure and system. With reference to the boundary integral, this expression features not only the flux vector f_ℓ, but also an expression that corresponds to an intrinsically multi-dimensional characteristics bias resolution of f_ℓ; as a result, the entire boundary expression may be viewed as non-discrete "numerical flux" for f_ℓ. Rather than executing an *ad hoc* substitution of f_ℓ with a chosen numerical flux, as is the case in reported DG developments, the characteristics-bias system procedure directly leads to a surface integral with a specific non-discrete multi-dimensional numerical flux, in this case a characteristics-bias flux. A discrete DG algorithm may then result from a Galerkin finite element discretization of (7.64), which reiterates the idea that a classical Galerkin method, as applied to an augmented system, can generate a set of upwind-stable equations.

7.4.2 Non-Discrete Streamline Upwind Petrov-Galerkin (SUPG) Form

For a weight function w with compact support in $\widehat{\Omega} \subset \Omega$ and a continuous solution q, an integration by parts of the characteristics-bias expression in (7.62) generates the weak statement

$$\int_{\widehat{\Omega}} \left(w + \varepsilon \psi a_\ell \frac{\partial w}{\partial x_\ell} \right) \left(\frac{\partial q}{\partial t} + \frac{\partial (f_m - f_m^\nu)}{\partial x_m} \right) d\Omega$$

$$+ \int_{\widehat{\Omega}} \varepsilon \psi \frac{\partial w}{\partial x_\ell} \left(c \left(\alpha a_\ell a_m + \alpha^N \left(a_\ell^{N_1} a_m^{N_1} + a_\ell^{N_2} a_m^{N_2} \right) \right) \frac{\partial q}{\partial x_m} + a_\ell (\delta - 1) \frac{\partial f_m^p}{\partial x_m} \right) d\Omega = 0 \quad (7.65)$$

Since w enjoy compact support in $\widehat{\Omega}$, the boundary integrals from the integration by parts do not feature in this formulation; when w does not have compact support, these surface integrals still vanish because of suitable boundary conditions. This statement is recognized as a non-discrete SUPG integral statement. As an alternative to the reported SUPG formulations, the upstream-bias term in the first domain integral in (7.65) does not require any premultiplication by the transpose of the Euler flux Jacobian matrix $\frac{\partial f_\ell}{\partial q}$; as the counterparts of the τ_{SUPG}, and δ_{SUPG} SUPG stability parameters, moreover, the characteristics-bias parameters $\varepsilon \psi$, $\varepsilon \psi \alpha$, $\varepsilon \psi \alpha^N$, and $\varepsilon \psi (\delta - 1)$ naturally emerge from the characteristics-bias procedure and explicitly depend on the square or cube of a local mesh spacing, respectively for shocked and smooth flows, which dependence has contributed to the definite asymptotic convergence rates detailed in the following chapters. The second integral in this statement corresponds to the stability and shock-capturing terms of reported SUPG formulations. Unlike the common *ad-hoc* terms, the stability and shock-capturing expressions in this formulation naturally emerge from characteristic wave propagation, for they are directly generated by the characteristics-bias decomposition of the Euler flux divergence $\partial f_\ell / \partial q$. The first integral corresponds to a weighted integral of the original system by way of a weighting function that is biased in the streamline direction, for the unit vector with components a_ℓ, $1 \le \ell \le 3$, points in the velocity direction and thus the expression $a_\ell \frac{\partial w}{\partial x_\ell}$ provides the rate of change of w in the streamline direction. The spatially discrete equations in this book derive from a Galerkin finite element discretization of this integral statement, which again emphasizes the technology of applying the versatile and optimal Galerkin method to a companion system to generate both an upwind stable system and corresponding essentially non-oscillatory solutions.

7.5   Hilbert Spaces and Approximation Results

For a finite N, as introduced in Section 7.1.1, the expansion

$$q_N(\boldsymbol{x}, t) = \sum_{j=1}^{N} q_j(t) W_j(\boldsymbol{x}) \tag{7.66}$$

corresponds to a finite-dimensional approximation. As N increases, $q_N(\boldsymbol{x}, t)$ becomes a sequence of functions, which prompts the question of convergence according to a selected norm in a specified space of functions. To address these fundamental questions, especially for piece-wise continuous functions, this section summarizes some basic concepts and results.

7.5.1   Hilbert Spaces

As well known, the inner product and squared norm of two vectors $\boldsymbol{F} = f_1\boldsymbol{i} + f_2\boldsymbol{j} + f_3\boldsymbol{k}$ and $\boldsymbol{Q} = q_1\boldsymbol{i} + q_2\boldsymbol{j} + q_3\boldsymbol{k}$ are expressed as

$$\boldsymbol{F} \cdot \boldsymbol{G} = \sum_{i=1}^{3} f_i g_i, \quad \|\boldsymbol{F}\|^2 = \sum_{i=1}^{3} f_i f_i \tag{7.67}$$

For any two functions f and g belonging to a vector space of functions $\mathcal{H}$, the generalization of these finite-dimensional inner product and norm evolves as

$$(f, g)_{\mathcal{H}^0(\Omega)} = \int_{\Omega} fg\, d\Omega, \quad \|f\|^2_{\mathcal{H}^0(\Omega)} = \int_{\Omega} f^2 d\Omega \tag{7.68}$$

As a further generalization of these results, consider the following expression

$$(f, g) = \int_{\Omega} \left(fg + f'g' \right) d\Omega \tag{7.69}$$

where the integral is a Lebesgue integral and the derivatives are weak derivatives, as discussed in Section 7.3. This expression is known as a Sobolev inner product with corresponding norm

$$\|f\|^2_{\mathcal{H}^1(\Omega)} \equiv (f, f) = \int_{\Omega} \left(f^2 + (f')^2 \right) d\Omega \tag{7.70}$$

A sequence of functions $\{f_n\}$ is termed a Cauchy sequence when it satisfies the limit

$$\lim_{m, n \to \infty} \|f_n - f_m\| = 0 \tag{7.71}$$

The limit of a Cauchy sequence is then expressed as

$$\lim_{n \to \infty} \{f_n\} = f \tag{7.72}$$

A vector space of functions $\mathcal{H}$ equipped with an inner product and induced norm, like, for instance, (7.69) or (7.68), is known as a Hilbert space, when any Cauchy sequence $\{f_n\}$ of elements of $\mathcal{H}$ converges to a limit f that also belongs to $\mathcal{H}$. In this sense, a Hilbert space is a complete space.

Owing to their inner product and induced norm, Hilbert spaces provide a convenient setting for investigating questions of optimal approximations, that is approximations with minimal error. As the next section highlights, within Hilbert space an optimal approximation exists and is unique.

7.5.2   Approximation Results

Consider two function Hilbert spaces $\mathcal{H}^1$ $\mathcal{H}^1_S$, with $\mathcal{H}^1_S \subseteq \mathcal{H}^1$. The space $\mathcal{H}^1_S$ is thus a subspace of $\mathcal{H}^1$, and the following sections in this chapter will consider finite-dimensional subspaces. Since $\mathcal{H}^1_S$ is a Hilbert space, it is complete, in the sense of the previous section. The optimal approximation problem may then be stated as follows: given a function $f \in \mathcal{H}^1$ find a function $g \in \mathcal{H}^1_S$ such that $\|f - g\|_{\mathcal{H}^1}$ is minimized. Functional analysis shows that g exists, is unique, and remains "orthogonal" to the error $(f - g)$.

For any subspace $\mathcal{H}^1_S \subseteq \mathcal{H}^1$, the space $\mathcal{H}^1$ may be exactly decomposed as

$$\mathcal{H}^1 = \mathcal{H}^1_S + \mathcal{H}^{1\perp}_S \tag{7.73}$$

where $\mathcal{H}^{1\perp}_S$ is the orthogonal complement of $\mathcal{H}^1_S$ in the sense that for any $g \in \mathcal{H}^1_S$ and any $h \in \mathcal{H}^{1\perp}_S$ then $(g, h) = 0$. The best approximation g can then be expressed as $f = g + h$, where $h = f - g$ denotes the error. Accordingly, the optimal approximation remains orthogonal to the error.

Although the optimization process that employs expansion (7.66) and yields (7.15) may not directly enforce a minimization of the norm $\|f - g\|_{\mathcal{H}^0}$, it does nevertheless minimize the error induced by q in the differential equation system. In this context, the best-approximation theorem remains important because it justifies selecting the finite dimensional expansion q within a Hilbert subspace. The completeness of the subspace remains a fundamental property because as the number of dimensions of the subspace increases, an optimal solution may converge to an element of the set and eventually to the exact solution when this belongs to the subspace.

7.6   Curvilinear Coordinate Transformations

This section presents the chief tools and results for curvilinear-coordinate transformations of partial derivatives with respect to Cartesian coordinates, equation systems and weak integral statements.

7.6.1   Areas and Volumes

In Figure 7.1, the unit vector n remains perpendicular to the plane of A and B, while "h" denotes the height of the plane parallelogram. With reference to this figure,

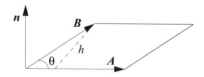

Figure 7.1: Area of Parallelogram

the area of a plane parallelogram is calculated as

$$\text{Area} = |A|h = |A|\,(|B|\sin\theta) = |A \times B| = (A \times B) \cdot n \tag{7.74}$$

which thus depends upon the three vectors $\boldsymbol{n}$, $\boldsymbol{A}$, $\boldsymbol{B}$. If the three vectors do not form a right handed frame, then

$$\text{Area} = |\boldsymbol{n} \cdot \boldsymbol{A} \times \boldsymbol{B}| \tag{7.75}$$

Similarly, with reference to Figure 7.2, the volume of a parallelepiped emerges as

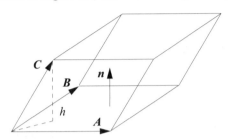

Figure 7.2: Volume of Parallelepiped

$$\text{Volume} = h(\text{Area}) = (\boldsymbol{C} \cdot \boldsymbol{n}) |\boldsymbol{A} \times \boldsymbol{B}| = (|\boldsymbol{A} \times \boldsymbol{B}| \, \boldsymbol{n}) \cdot \boldsymbol{C} = (\boldsymbol{A} \times \boldsymbol{B}) \cdot \boldsymbol{C} \tag{7.76}$$

which depends on the three vectors $\boldsymbol{A}$, $\boldsymbol{B}$, $\boldsymbol{C}$. If the three vectors do not form a right-handed frame, then

$$\text{Volume} = |(\boldsymbol{A} \times \boldsymbol{B}) \cdot \boldsymbol{C}| \tag{7.77}$$

With reference to Figure 7.3

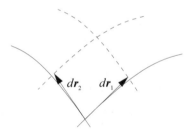

Figure 7.3: Differential Area

and following (7.74), the element of area then becomes

$$dA = \boldsymbol{n} \cdot (d\boldsymbol{r}_1 \times d\boldsymbol{r}_2) \tag{7.78}$$

Likewise, with reference to Figure 7.4, the corresponding element of volume is

$$dV = d\boldsymbol{r}_1 \cdot (d\boldsymbol{r}_2 \times d\boldsymbol{r}_3) \tag{7.79}$$

as expected.

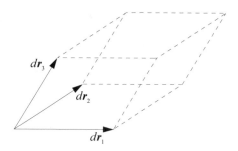

Figure 7.4: Differential Volume

7.6.2   Two-Dimensional Coordinate Transformation

Let (η_1, η_2) denote a set of 2-D curvilinear coordinates and let (x_1, x_2) depend upon (η_1, η_2) through invertible as well as differentiable functions.

$$\begin{cases} x_1 = x_1\,(\eta_1, \eta_2) \\ x_2 = x_2\,(\eta_1, \eta_2) \end{cases}, \quad \begin{cases} \eta_1 = \eta_1\,(x_1, x_2) \\ \eta_2 = \eta_2\,(x_1, x_2) \end{cases} \tag{7.80}$$

These requirements ensure the existence of all partial derivatives of (x_1, x_2) with respect to (η_1, η_2) as well as all partial derivatives of (η_1, η_2) with respect to (x_1, x_2). These derivatives support the transformation of the Cartesian-coordinate governing equations.

Partial Derivative Transformation

For a differentiable function $F = F(\eta_1, \eta_2)$, consider the issue of establishing the derivatives of F with respect to (x_1, x_2). By virtue of (7.80) and the chain rule, the derivatives of F with respect to (η_1, η_2) become

$$\begin{aligned} \frac{\partial F}{\partial \eta_1} &= \frac{\partial F}{\partial x_1}\frac{\partial x_1}{\partial \eta_1} + \frac{\partial F}{\partial x_2}\frac{\partial x_2}{\partial \eta_1} \\ \frac{\partial F}{\partial \eta_2} &= \frac{\partial F}{\partial x_1}\frac{\partial x_1}{\partial \eta_2} + \frac{\partial F}{\partial x_2}\frac{\partial x_2}{\partial \eta_2} \end{aligned} \tag{7.81}$$

which stand for the linear system for $(\partial F/\partial x_1, \partial F/\partial x_2)$

$$\begin{pmatrix} \frac{\partial x_1}{\partial \eta_1} & \frac{\partial x_2}{\partial \eta_1} \\ \frac{\partial x_1}{\partial \eta_2} & \frac{\partial x_2}{\partial \eta_2} \end{pmatrix} \begin{pmatrix} \frac{\partial F}{\partial x_1} \\ \frac{\partial F}{\partial x_2} \end{pmatrix} = \begin{pmatrix} \frac{\partial F}{\partial \eta_1} \\ \frac{\partial F}{\partial \eta_2} \end{pmatrix} \tag{7.82}$$

with solution

$$\begin{aligned} \begin{pmatrix} \frac{\partial F}{\partial x_1} \\ \frac{\partial F}{\partial x_2} \end{pmatrix} &= \begin{pmatrix} \frac{\partial x_1}{\partial \eta_1} & \frac{\partial x_2}{\partial \eta_1} \\ \frac{\partial x_1}{\partial \eta_2} & \frac{\partial x_2}{\partial \eta_2} \end{pmatrix}^{-1} \begin{pmatrix} \frac{\partial F}{\partial \eta_1} \\ \frac{\partial F}{\partial \eta_2} \end{pmatrix} \\ &= \frac{1}{\det J} \begin{pmatrix} \frac{\partial x_2}{\partial \eta_2} & -\frac{\partial x_2}{\partial \eta_1} \\ -\frac{\partial x_1}{\partial \eta_2} & \frac{\partial x_1}{\partial \eta_1} \end{pmatrix} \begin{pmatrix} \frac{\partial F}{\partial \eta_1} \\ \frac{\partial F}{\partial \eta_2} \end{pmatrix} \end{aligned} \tag{7.83}$$

The derivatives of F with respect to (x_1, x_2), accordingly, rely on those with respect to (η_1, η_2) through the derivatives of (x_1, x_2) with respect to (η_1, η_2).

Results (7.83) depend on the metric data

$$
\begin{aligned}
e_{11} &= \frac{\partial x_2}{\partial \eta_2}, \quad e_{12} = -\frac{\partial x_2}{\partial \eta_1} \\
e_{21} &= -\frac{\partial x_1}{\partial \eta_2}, \quad e_{22} = \frac{\partial x_1}{\partial \eta_1} \\
\det J &= \frac{\partial x_2}{\partial \eta_2}\frac{\partial x_1}{\partial \eta_1} - \frac{\partial x_2}{\partial \eta_1}\frac{\partial x_1}{\partial \eta_2} = e_{11}e_{22} - e_{21}e_{12}
\end{aligned}
\tag{7.84}
$$

In terms of these metric data, the derivatives $(\partial F/\partial x_1, \partial F/\partial x_2)$ become

$$
\frac{\partial F}{\partial x_i} = \frac{\partial F}{\partial \eta_j}\frac{\partial \eta_j}{\partial x_i} = \frac{e_{ij}}{\det J}\frac{\partial F}{\partial \eta_j} \Rightarrow \frac{\partial \eta_j}{\partial x_i} = \frac{e_{ij}}{\det J}
\tag{7.85}
$$

With $F \equiv x_\ell$, these expressions result in the equality

$$
\det J \delta_i^\ell = e_{ij}\frac{\partial x_\ell}{\partial \eta_j} \Rightarrow \det J = \frac{e_{ij}}{2}\frac{\partial x_i}{\partial \eta_j}
\tag{7.86}
$$

The metric data (e_{ij}) satisfy the fundamental relationship

$$
\frac{\partial e_{ij}}{\partial \eta_j} = 0
\tag{7.87}
$$

as proven by

$$
\begin{aligned}
\frac{\partial e_{11}}{\partial \eta_1} + \frac{\partial e_{12}}{\partial \eta_2} &= \frac{\partial^2 x_2}{\partial \eta_1 \partial \eta_2} - \frac{\partial^2 x_2}{\partial \eta_2 \partial \eta_1} = 0 \\
\frac{\partial e_{21}}{\partial \eta_1} + \frac{\partial e_{22}}{\partial \eta_2} &= -\frac{\partial^2 x_1}{\partial \eta_1 \partial \eta_2} + \frac{\partial^2 x_1}{\partial \eta_2 \partial \eta_1} = 0
\end{aligned}
\tag{7.88}
$$

by virtue of the equality of mixed derivatives.

Normal Vectors and Area

A set of vectors perpendicular to the coordinate lines $\eta_1 = $ constant and $\eta_2 = $ constant arise from the derivatives of (x_1, x_2) with respect to (η_1, η_2). With respect to Figure 7.5, The position vector $\boldsymbol{r}$ of a point on the (x_1, x_2) plane depends on the unit vectors $\boldsymbol{e}_1, \boldsymbol{e}_2$ as well as (x_1, x_2) as

$$
\boldsymbol{r} = x_1 \boldsymbol{e}_1 + x_2 \boldsymbol{e}_2
\tag{7.89}
$$

A set of vectors tangent to the $(\eta_1 = $ constant$)$ and $(\eta_2 = $ constant$)$ lines emerges as

$$
\begin{aligned}
\frac{\partial \boldsymbol{r}}{\partial \eta_1} &= \frac{\partial x_1}{\partial \eta_1}\boldsymbol{e}_1 + \frac{\partial x_2}{\partial \eta_1}\boldsymbol{e}_2 \\
\frac{\partial \boldsymbol{r}}{\partial \eta_2} &= \frac{\partial x_1}{\partial \eta_2}\boldsymbol{e}_1 + \frac{\partial x_2}{\partial \eta_2}\boldsymbol{e}_2
\end{aligned}
\tag{7.90}
$$

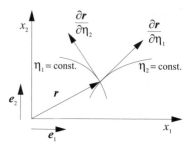

Figure 7.5: Two-Dimensional Coordinates

The vector $\boldsymbol{p}_1 = \frac{\partial \boldsymbol{r}}{\partial \eta_2} \times \boldsymbol{n}$ is then perpendicular to the line $\eta_1 = $ constant and its components are (e_{11}, e_{21}), as follows from the cross product

$$\frac{\partial \boldsymbol{r}}{\partial \eta_2} \times \boldsymbol{n} = \begin{vmatrix} \boldsymbol{e}_1 & \boldsymbol{e}_2 & \boldsymbol{e}_3 \\ \frac{\partial x_1}{\partial \eta_2} & \frac{\partial x_2}{\partial \eta_2} & 0 \\ 0 & 0 & 1 \end{vmatrix} = \frac{\partial x_2}{\partial \eta_2} \boldsymbol{e}_1 - \frac{\partial x_1}{\partial \eta_2} \boldsymbol{e}_2 + 0 \boldsymbol{e}_3 = e_{11} \boldsymbol{e}_1 + e_{21} \boldsymbol{e}_2 + 0 \boldsymbol{e}_3 \qquad (7.91)$$

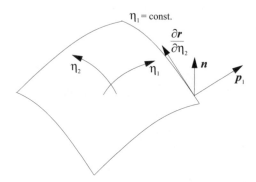

Figure 7.6: Vector Perpendicular to η_1=constant

Similarly, the vector $\boldsymbol{p}_2 = \boldsymbol{n} \times \frac{\partial \boldsymbol{r}}{\partial \eta_1}$ is perpendicular to the line $\eta_2 = $ constant and its components are (e_{12}, e_{22}), as follows from the cross product

$$\boldsymbol{n} \times \frac{\partial \boldsymbol{r}}{\partial \eta_1} = \begin{vmatrix} \boldsymbol{e}_1 & \boldsymbol{e}_2 & \boldsymbol{e}_3 \\ 0 & 0 & 1 \\ \frac{\partial x_1}{\partial \eta_1} & \frac{\partial x_2}{\partial \eta_1} & 0 \end{vmatrix} = -\frac{\partial x_2}{\partial \eta_1} \boldsymbol{e}_1 + \frac{\partial x_1}{\partial \eta_1} \boldsymbol{e}_2 + 0 \boldsymbol{e}_3 = e_{12} \boldsymbol{e}_1 + e_{22} \boldsymbol{e}_2 + 0 \boldsymbol{e}_3 \qquad (7.92)$$

The determinant $\det J$ also corresponds to the triple product

$$\det J = \boldsymbol{n} \cdot \left(\frac{\partial \boldsymbol{r}}{\partial \eta_1} \times \frac{\partial \boldsymbol{r}}{\partial \eta_2} \right) = \begin{vmatrix} 0 & 0 & 1 \\ \frac{\partial x_1}{\partial \eta_1} & \frac{\partial x_2}{\partial \eta_1} & 0 \\ \frac{\partial x_1}{\partial \eta_2} & \frac{\partial x_2}{\partial \eta_2} & 0 \end{vmatrix} \tag{7.93}$$

Following (7.74), therefore, $\det J$ at any point $P \equiv (x_1, x_2)$ equals the area of the parallelogram with sides $\frac{\partial \boldsymbol{r}}{\partial \eta_1}$ and $\frac{\partial \boldsymbol{r}}{\partial \eta_2}$ evaluated at P.

7.6.3 Three-Dimensional Coordinate Transformation

Similarly to the 2-D coordinate transformation, let (η_1, η_2, η_3) denote a set of 3-D curvilinear coordinates and let (x_1, x_2, x_3) depend upon (η_1, η_2, η_3) through invertible as well as differentiable functions.

$$\begin{cases} x_1 = x_1 (\eta_1, \eta_2, \eta_3) \\ x_2 = x_2 (\eta_1, \eta_2, \eta_3) \\ x_3 = x_3 (\eta_1, \eta_2, \eta_3) \end{cases}, \qquad \begin{cases} \eta_1 = \eta_1 (x_1, x_2, x_3) \\ \eta_2 = \eta_2 (x_1, x_2, x_3) \\ \eta_3 = \eta_3 (x_1, x_2, x_3) \end{cases} \tag{7.94}$$

These requirements ensure the existence of all partial derivatives of (x_1, x_2, x_3) with respect to (η_1, η_2, η_3) as well as all partial derivatives of (η_1, η_2, η_4) with respect to (x_1, x_2, x_3). These derivatives support the transformation of the 3-D Cartesian-coordinate governing equations.

Partial Derivative Transformation

For a differentiable function $F = F(\eta_1, \eta_2, \eta_3)$, consider the issue of establishing the derivatives of F with respect to (x_1, x_2, x_3). By virtue of (7.94) and the chain rule, the derivatives of F with respect to (η_1, η_2, η_3) become

$$\begin{aligned} \frac{\partial F}{\partial \eta_1} &= \frac{\partial F}{\partial x_1} \frac{\partial x_1}{\partial \eta_1} + \frac{\partial F}{\partial x_2} \frac{\partial x_2}{\partial \eta_1} + \frac{\partial F}{\partial x_3} \frac{\partial x_3}{\partial \eta_1} \\ \frac{\partial F}{\partial \eta_2} &= \frac{\partial F}{\partial x_1} \frac{\partial x_1}{\partial \eta_2} + \frac{\partial F}{\partial x_2} \frac{\partial x_2}{\partial \eta_2} + \frac{\partial F}{\partial x_3} \frac{\partial x_3}{\partial \eta_2} \\ \frac{\partial F}{\partial \eta_3} &= \frac{\partial F}{\partial x_1} \frac{\partial x_1}{\partial \eta_3} + \frac{\partial F}{\partial x_2} \frac{\partial x_2}{\partial \eta_3} + \frac{\partial F}{\partial x_3} \frac{\partial x_3}{\partial \eta_3} \end{aligned} \tag{7.95}$$

which stand for the linear system for $(\partial F / \partial x_1, \partial F / \partial x_2, \partial F / \partial x_3)$

$$\begin{pmatrix} \frac{\partial x_1}{\partial \eta_1} & \frac{\partial x_2}{\partial \eta_1} & \frac{\partial x_3}{\partial \eta_1} \\ \frac{\partial x_1}{\partial \eta_2} & \frac{\partial x_2}{\partial \eta_2} & \frac{\partial x_3}{\partial \eta_2} \\ \frac{\partial x_1}{\partial \eta_3} & \frac{\partial x_2}{\partial \eta_3} & \frac{\partial x_3}{\partial \eta_3} \end{pmatrix} \begin{pmatrix} \frac{\partial F}{\partial x_1} \\ \frac{\partial F}{\partial x_2} \\ \frac{\partial F}{\partial x_3} \end{pmatrix} = \begin{pmatrix} \frac{\partial F}{\partial \eta_1} \\ \frac{\partial F}{\partial \eta_2} \\ \frac{\partial F}{\partial \eta_3} \end{pmatrix} \tag{7.96}$$

with solution

$$\begin{pmatrix} \frac{\partial F}{\partial x_1} \\ \frac{\partial F}{\partial x_2} \\ \frac{\partial F}{\partial x_3} \end{pmatrix} = \begin{pmatrix} \frac{\partial x_1}{\partial \eta_1} & \frac{\partial x_2}{\partial \eta_1} & \frac{\partial x_3}{\partial \eta_1} \\ \frac{\partial x_1}{\partial \eta_2} & \frac{\partial x_2}{\partial \eta_2} & \frac{\partial x_3}{\partial \eta_2} \\ \frac{\partial x_1}{\partial \eta_3} & \frac{\partial x_2}{\partial \eta_3} & \frac{\partial x_3}{\partial \eta_3} \end{pmatrix}^{-1} \begin{pmatrix} \frac{\partial F}{\partial \eta_1} \\ \frac{\partial F}{\partial \eta_2} \\ \frac{\partial F}{\partial \eta_3} \end{pmatrix}$$

$$= \frac{1}{\det J} \begin{pmatrix} e_{11} & e_{12} & e_{13} \\ e_{21} & e_{22} & e_{23} \\ e_{31} & e_{32} & e_{33} \end{pmatrix} \begin{pmatrix} \frac{\partial F}{\partial \eta_1} \\ \frac{\partial F}{\partial \eta_2} \\ \frac{\partial F}{\partial \eta_3} \end{pmatrix} \tag{7.97}$$

$$\det J = \frac{\partial x_1}{\partial \eta_1} \left(\frac{\partial x_2}{\partial \eta_2} \frac{\partial x_3}{\partial \eta_3} - \frac{\partial x_2}{\partial \eta_3} \frac{\partial x_3}{\partial \eta_2} \right)$$

$$+ \frac{\partial x_1}{\partial \eta_2} \left(- \frac{\partial x_2}{\partial \eta_1} \frac{\partial x_3}{\partial \eta_3} + \frac{\partial x_3}{\partial \eta_1} \frac{\partial x_2}{\partial \eta_3} \right) + \frac{\partial x_1}{\partial \eta_3} \left(\frac{\partial x_2}{\partial \eta_1} \frac{\partial x_3}{\partial \eta_2} - \frac{\partial x_3}{\partial \eta_1} \frac{\partial x_2}{\partial \eta_2} \right) \tag{7.98}$$

The derivatives of F with respect to (x_1, x_2, x_3), therefore, rely on those with respect to (η_1, η_2, η_3) through the derivatives of (x_1, x_2, x_3) with respect to (η_1, η_2, η_3).

Results (7.97) depend on the metric data

$$e_{11} = \frac{\partial x_2}{\partial \eta_2} \frac{\partial x_3}{\partial \eta_3} - \frac{\partial x_2}{\partial \eta_3} \frac{\partial x_3}{\partial \eta_2}$$

$$e_{12} = -\frac{\partial x_2}{\partial \eta_1} \frac{\partial x_3}{\partial \eta_3} + \frac{\partial x_3}{\partial \eta_1} \frac{\partial x_2}{\partial \eta_3}$$

$$e_{13} = \frac{\partial x_2}{\partial \eta_1} \frac{\partial x_3}{\partial \eta_2} - \frac{\partial x_3}{\partial \eta_1} \frac{\partial x_2}{\partial \eta_2} \tag{7.99}$$

$$e_{21} = -\frac{\partial x_1}{\partial \eta_2} \frac{\partial x_3}{\partial \eta_3} + \frac{\partial x_3}{\partial \eta_2} \frac{\partial x_1}{\partial \eta_3}$$

$$e_{22} = \frac{\partial x_1}{\partial \eta_1} \frac{\partial x_3}{\partial \eta_3} - \frac{\partial x_3}{\partial \eta_1} \frac{\partial x_1}{\partial \eta_3}$$

$$e_{23} = -\frac{\partial x_1}{\partial \eta_1} \frac{\partial x_3}{\partial \eta_2} + \frac{\partial x_3}{\partial \eta_1} \frac{\partial x_1}{\partial \eta_2} \tag{7.100}$$

$$e_{31} = \frac{\partial x_1}{\partial \eta_2} \frac{\partial x_2}{\partial \eta_3} - \frac{\partial x_2}{\partial \eta_2} \frac{\partial x_1}{\partial \eta_3}$$

$$e_{32} = -\frac{\partial x_1}{\partial \eta_1} \frac{\partial x_2}{\partial \eta_3} + \frac{\partial x_2}{\partial \eta_1} \frac{\partial x_1}{\partial \eta_3}$$

$$e_{33} = \frac{\partial x_1}{\partial \eta_1} \frac{\partial x_2}{\partial \eta_2} - \frac{\partial x_2}{\partial \eta_1} \frac{\partial x_1}{\partial \eta_2} \tag{7.101}$$

In terms of these metric data, the derivatives $(\partial F/\partial x_1, \partial F/\partial x_2, \partial F/\partial x_3)$ become

$$\frac{\partial F}{\partial x_i} = \frac{\partial F}{\partial \eta_j} \frac{\partial \eta_j}{\partial x_i} = \frac{e_{ij}}{\det J} \frac{\partial F}{\partial \eta_j} \quad \Rightarrow \quad \frac{\partial \eta_j}{\partial x_i} = \frac{e_{ij}}{\det J} \tag{7.102}$$

With $F \equiv x_\ell$, these expressions result in the equality

$$\det J \delta_i^\ell = e_{ij} \frac{\partial x_\ell}{\partial \eta_j} \quad \Rightarrow \quad \det J = \frac{e_{ij}}{3} \frac{\partial x_i}{\partial \eta_j} \tag{7.103}$$

The metric data (e_{ij}) also satisfy the fundamental relationship

$$\frac{\partial e_{ij}}{\partial \eta_j} = 0 \tag{7.104}$$

as can be shown through the equality of mixed partial derivatives.

Normal Vectors and Volume

A set of vectors perpendicular to the coordinate surfaces ($\eta_1 = $ constant), ($\eta_2 = $ constant), and ($\eta_3 = $ constant) originate from the derivatives of (x_1, x_2, x_3) with respect to (η_1, η_2, η_3). With respect to Figure 7.7,

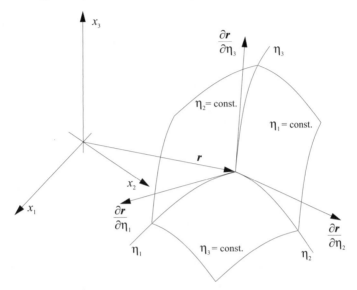

Figure 7.7: Three-Dimensional Coordinates

the position vector $\boldsymbol{r}$ of a point in the (x_1, x_2, x_3) space depends on (x_1, x_2, x_3) as

$$\boldsymbol{r} = x_1\boldsymbol{e}_1 + x_2\boldsymbol{e}_2 + x_3\boldsymbol{e}_3 \tag{7.105}$$

A set of vectors tangent to the ($\eta_2 = $ constant, $\eta_3 = $ constant), ($\eta_3 = $ constant, $\eta_1 = $ constant), and ($\eta_1 = $ constant, $\eta_2 = $ constant) lines emerge as

$$
\begin{aligned}
\frac{\partial \boldsymbol{r}}{\partial \eta_1} &= \frac{\partial x_1}{\partial \eta_1}\boldsymbol{e}_1 + \frac{\partial x_2}{\partial \eta_1}\boldsymbol{e}_2 + \frac{\partial x_3}{\partial \eta_1}\boldsymbol{e}_3 \\
\frac{\partial \boldsymbol{r}}{\partial \eta_2} &= \frac{\partial x_1}{\partial \eta_2}\boldsymbol{e}_1 + \frac{\partial x_2}{\partial \eta_2}\boldsymbol{e}_2 + \frac{\partial x_3}{\partial \eta_2}\boldsymbol{e}_3 \\
\frac{\partial \boldsymbol{r}}{\partial \eta_3} &= \frac{\partial x_1}{\partial \eta_3}\boldsymbol{e}_1 + \frac{\partial x_2}{\partial \eta_3}\boldsymbol{e}_2 + \frac{\partial x_3}{\partial \eta_3}\boldsymbol{e}_3
\end{aligned}
\tag{7.106}
$$

The vector $\boldsymbol{p}_1 = \frac{\partial \boldsymbol{r}}{\partial \eta_2} \times \frac{\partial \boldsymbol{r}}{\partial \eta_3}$ is perpendicular to the surface $\eta_1 = $ constant and its components are (e_{11}, e_{21}, e_{31}), as follows from the cross product

$$
\frac{\partial \boldsymbol{r}}{\partial \eta_2} \times \frac{\partial \boldsymbol{r}}{\partial \eta_3} =
\begin{vmatrix}
\boldsymbol{e}_1 & \boldsymbol{e}_2 & \boldsymbol{e}_3 \\
\frac{\partial x_1}{\partial \eta_2} & \frac{\partial x_2}{\partial \eta_2} & \frac{\partial x_2}{\partial \eta_2} \\
\frac{\partial x_1}{\partial \eta_3} & \frac{\partial x_2}{\partial \eta_3} & \frac{\partial x_2}{\partial \eta_3}
\end{vmatrix}
$$

$$= \left(\frac{\partial x_2}{\partial \eta_2}\frac{\partial x_3}{\partial \eta_3} - \frac{\partial x_2}{\partial \eta_3}\frac{\partial x_3}{\partial \eta_2}\right)\boldsymbol{e}_1 + \left(-\frac{\partial x_1}{\partial \eta_2}\frac{\partial x_3}{\partial \eta_3} + \frac{\partial x_3}{\partial \eta_2}\frac{\partial x_1}{\partial \eta_3}\right)\boldsymbol{e}_2 + \left(\frac{\partial x_1}{\partial \eta_2}\frac{\partial x_2}{\partial \eta_3} - \frac{\partial x_2}{\partial \eta_2}\frac{\partial x_1}{\partial \eta_3}\right)\boldsymbol{e}_3$$

$$= e_{11}\boldsymbol{e}_1 + e_{21}\boldsymbol{e}_2 + e_{31}\boldsymbol{e}_3 \tag{7.107}$$

Similarly, the vector $\boldsymbol{p}_2 = \frac{\partial \boldsymbol{r}}{\partial \eta_3} \times \frac{\partial \boldsymbol{r}}{\partial \eta_1}$, perpendicular to the surface ($\eta_2 = $ constant), has components (e_{12}, e_{22}, e_{32}), and the vector $\boldsymbol{p}_3 = \frac{\partial \boldsymbol{r}}{\partial \eta_1} \times \frac{\partial \boldsymbol{r}}{\partial \eta_2}$, perpendicular to the surface ($\eta_3 = $ constant), has components (e_{13}, e_{23}, e_{33}).

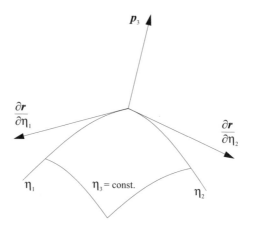

Figure 7.8: Vector Perpendicular to ($\eta_3 = $ constant) Surface

With reference to (7.96), the determinant $\det J$ also corresponds to any one of the three triple products

$$\det J = \frac{\partial \boldsymbol{r}}{\partial \eta_1} \cdot \left(\frac{\partial \boldsymbol{r}}{\partial \eta_2} \times \frac{\partial \boldsymbol{r}}{\partial \eta_3}\right) = \frac{\partial \boldsymbol{r}}{\partial \eta_2} \cdot \left(\frac{\partial \boldsymbol{r}}{\partial \eta_3} \times \frac{\partial \boldsymbol{r}}{\partial \eta_1}\right) = \frac{\partial \boldsymbol{r}}{\partial \eta_3} \cdot \left(\frac{\partial \boldsymbol{r}}{\partial \eta_1} \times \frac{\partial \boldsymbol{r}}{\partial \eta_2}\right)$$

$$= \begin{vmatrix} \frac{\partial x_1}{\partial \eta_1} & \frac{\partial x_2}{\partial \eta_1} & \frac{\partial x_3}{\partial \eta_1} \\ \frac{\partial x_1}{\partial \eta_2} & \frac{\partial x_2}{\partial \eta_2} & \frac{\partial x_3}{\partial \eta_2} \\ \frac{\partial x_1}{\partial \eta_3} & \frac{\partial x_2}{\partial \eta_3} & \frac{\partial x_3}{\partial \eta_3} \end{vmatrix} \tag{7.108}$$

Since any one of the three cross products in this result has the metric data e_{ij} as its components, the expression for $\det J$ can also be cast as

$$\det J \delta_i^\ell = e_{j\ell}\frac{\partial x_j}{\partial \eta_i} \tag{7.109}$$

Following (7.76), furthermore, $\det J$ at any point $P \equiv (x_1, x_2, x_3)$ equals the volume of the parallelepiped with sides $\frac{\partial \boldsymbol{r}}{\partial \eta_1}$, $\frac{\partial \boldsymbol{r}}{\partial \eta_2}$, and $\frac{\partial \boldsymbol{r}}{\partial \eta_3}$ evaluated at P.

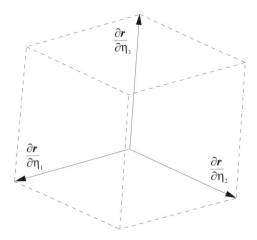

Figure 7.9: Volume from $\frac{\partial \boldsymbol{r}}{\partial \eta_1}, \frac{\partial \boldsymbol{r}}{\partial \eta_2}, \frac{\partial \boldsymbol{r}}{\partial \eta_3}$

7.6.4 Differential Arc, Area, Volume

The expressions for differential arcs, areas, and volumes directly support the transformation of line, surface, and volume integrals between any two coordinate sets.

Arc and Area in Two Dimensions

The differential displacement vector tangent to the ($\eta_2 =$ constant) and ($\eta_1 =$ constant) lines can be expressed as

$$
d\boldsymbol{r}_1 \equiv \frac{\partial \boldsymbol{r}}{\partial \eta_1} d\eta_1 = \left(\frac{\partial x_1}{\partial \eta_1} \boldsymbol{e}_1 + \frac{\partial x_2}{\partial \eta_1} \boldsymbol{e}_2 \right) d\eta_1
$$

$$
d\boldsymbol{r}_2 \equiv \frac{\partial \boldsymbol{r}}{\partial \eta_2} d\eta_2 = \left(\frac{\partial x_1}{\partial \eta_2} \boldsymbol{e}_1 + \frac{\partial x_2}{\partial \eta_2} \boldsymbol{e}_2 \right) d\eta_2 \tag{7.110}
$$

These differential expressions directly depend on the derivatives of (x_1, x_2) with respect to (η_1, η_2) and lead to the differential arc lengths

$$
ds_2 = |d\boldsymbol{r}_2 \times \boldsymbol{n}| = \left| \frac{\partial \boldsymbol{r}}{\partial \eta_2} \times \boldsymbol{n} \right| d\eta_2 = |e_{11}\boldsymbol{e}_1 + e_{21}\boldsymbol{e}_2| \, d\eta_2 = \sqrt{e_{11}^2 + e_{21}^2} d\eta_2
$$

$$
ds_1 = |\boldsymbol{n} \times d\boldsymbol{r}_1| = \left| \boldsymbol{n} \times \frac{\partial \boldsymbol{r}}{\partial \eta_1} \right| d\eta_1 = |e_{12}\boldsymbol{e}_1 + e_{22}\boldsymbol{e}_2| \, d\eta_1 = \sqrt{e_{12}^2 + e_{22}^2} d\eta_1 \tag{7.111}
$$

With reference to (7.91) and (7.92), these results correspond to the magnitudes of $\boldsymbol{p}_1 d\eta_1$ and $\boldsymbol{p}_2 d\eta_2$

With results (7.111)

$$
ds_1 = \sqrt{e_{12}^2 + e_{22}^2} d\eta_1, \quad ds_2 = \sqrt{e_{11}^2 + e_{21}^2} d\eta_2 \tag{7.112}
$$

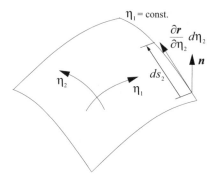

Figure 7.10: Differential Arc

the line integrals along the (η_2 = constant) and (η_1 = constant) lines transform as

$$\int_\Gamma w ds_1 = \int_{\eta_1=a}^{\eta_1=b} w\sqrt{e_{12}^2 + e_{22}^2}\, d\eta_1, \qquad \int_\Gamma w ds_2 = \int_{\eta_2=a}^{\eta_2=b} w\sqrt{e_{11}^2 + e_{21}^2}\, d\eta_2 \qquad (7.113)$$

Following (7.78) and Figure 7.11

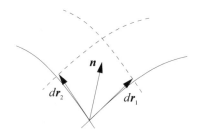

Figure 7.11: Two-Dimensional Differential Area

the elementary area dA depends on $d\boldsymbol{r}_1$ and $d\boldsymbol{r}_2$ as

$$dA = \boldsymbol{n} \cdot (d\boldsymbol{r}_1 \times d\boldsymbol{r}_2) = \boldsymbol{n} \cdot \left(\frac{\partial \boldsymbol{r}}{\partial \eta_1} \times \frac{\partial \boldsymbol{r}}{\partial \eta_2} \right) d\eta_1 d\eta_2 = \det J d\eta_1 d\eta_2 \qquad (7.114)$$

$$\frac{dA}{d\eta_1 d\eta_2} = \det J \qquad (7.115)$$

The corresponding transformation of an area integral becomes

$$\int_\Omega w d\Omega = \int_{\Omega_\eta} w \det J d\eta_1 d\eta_2 \qquad (7.116)$$

where Ω_η emerges from Ω through the coordinate transformation (7.80).

Area and Volume in Three Dimensions

The differential areas on the ($\eta_1 = $ constant), ($\eta_2 = $ constant), ($\eta_3 = $ constant) surfaces are expressed as

$$
dA_1 = |d\boldsymbol{r}_2 \times d\boldsymbol{r}_3| = \left| \frac{\partial \boldsymbol{r}}{\partial \eta_2} \times \frac{\partial \boldsymbol{r}}{\partial \eta_3} \right| d\eta_2 d\eta_3 = |e_{11}\boldsymbol{e}_1 + e_{21}\boldsymbol{e}_2 + e_{31}\boldsymbol{e}_3| \, d\eta_2 d\eta_3
$$

$$
dA_2 = |d\boldsymbol{r}_3 \times d\boldsymbol{r}_1| = \left| \frac{\partial \boldsymbol{r}}{\partial \eta_3} \times \frac{\partial \boldsymbol{r}}{\partial \eta_1} \right| d\eta_3 d\eta_1 = |e_{12}\boldsymbol{e}_1 + e_{22}\boldsymbol{e}_2 + e_{32}\boldsymbol{e}_3| \, d\eta_3 d\eta_1
$$

$$
dA_3 = |d\boldsymbol{r}_1 \times d\boldsymbol{r}_2| = \left| \frac{\partial \boldsymbol{r}}{\partial \eta_1} \times \frac{\partial \boldsymbol{r}}{\partial \eta_2} \right| d\eta_1 d\eta_2 = |e_{13}\boldsymbol{e}_1 + e_{23}\boldsymbol{e}_2 + e_{33}\boldsymbol{e}_3| \, d\eta_1 d\eta_2 \quad (7.117)
$$

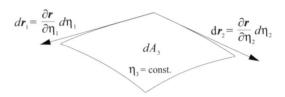

Figure 7.12: Three-Dimensional Surface Area

These differential expressions directly depend on the derivatives of (x_1, x_2, x_3) with respect to (η_1, η_2, η_3) and simplify as

$$
\frac{dA_1}{d\eta_2 d\eta_3} = \sqrt{e_{11}^2 + e_{21}^2 + e_{31}^2}, \quad \frac{dA_2}{d\eta_3 d\eta_1} = \sqrt{e_{12}^2 + e_{22}^2 + e_{32}^2}, \quad \frac{dA_3}{d\eta_1 d\eta_2} = \sqrt{e_{13}^2 + e_{23}^2 + e_{33}^2}
$$

$$(7.118)$$

With these results, the surface integrals over ($\eta_1 = $ constant), ($\eta_2 = $ constant), and ($\eta_3 = $ constant) surfaces transform as

$$
\int_{A_1} w dA_1 = \int_{A_{1\eta}} w \sqrt{e_{11}^2 + e_{21}^2 + e_{31}^2} d\eta_2 d\eta_3 \quad (7.119)
$$

$$
\int_{A_2} w dA_2 = \int_{A_{2\eta}} w \sqrt{e_{12}^2 + e_{22}^2 + e_{32}^2} d\eta_3 d\eta_1 \quad (7.120)
$$

$$
\int_{A_3} w dA_3 = \int_{A_{3\eta}} w \sqrt{e_{13}^2 + e_{23}^2 + e_{33}^2} d\eta_1 d\eta_2 \quad (7.121)
$$

Following (7.79), the elementary volume dV depends on $d\boldsymbol{r}_1$, $d\boldsymbol{r}_2$, $d\boldsymbol{r}_3$ as

$$
dV = d\boldsymbol{r}_1 \cdot (d\boldsymbol{r}_2 \times d\boldsymbol{r}_3) = \frac{\partial \boldsymbol{r}}{\partial \eta_1} \left(\frac{\partial \boldsymbol{r}}{\partial \eta_2} \times \frac{\partial \boldsymbol{r}}{\partial \eta_3} \right) d\eta_1 d\eta_2 d\eta_3 = \det J d\eta_1 d\eta_2 d\eta_3 \quad (7.122)
$$

$$
\frac{dV}{d\eta_1 d\eta_2 d\eta_3} = \det J \quad (7.123)
$$

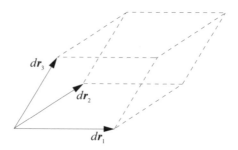

Figure 7.13: Differential Volume

The corresponding transformation of a volume integral becomes

$$\int_\Omega w d\Omega = \int_\Omega w dV = \int_{\Omega_\eta} w \det J d\eta_1 d\eta_2 d\eta_3 \tag{7.124}$$

where Ω_η emerges from Ω through the coordinate transformation (7.94).

7.7 Discretized Statement

7.7.1 Differential and Integral Curvilinear-Coordinate Statements

Using either (7.85), in 2-D, or (7.102), in 3-D, the characteristics-bias Euler and Navier-Stokes system

$$\frac{\partial q}{\partial t} - \frac{\partial}{\partial x_\ell}\left(\varepsilon\psi a_\ell \frac{\partial q}{\partial t}\right) + \frac{\partial\left(f_\ell - f_\ell^\nu\right)}{\partial x_\ell} - \frac{\partial}{\partial x_\ell}\left(\varepsilon\psi a_\ell \left(\frac{\partial\left(f_m^q - f_m^\nu\right)}{\partial x_m} + \delta\frac{\partial f_m^p}{\partial x_m}\right)\right)$$

$$-\frac{\partial}{\partial x_\ell}\left(\varepsilon\psi\left(c\left(\alpha a_\ell a_m + \alpha^N \left(a_\ell^{N_1} a_m^{N_1} + a_\ell^{N_2} a_m^{N_2}\right)\right)\frac{\partial q}{\partial x_m}\right)\right) = 0 \tag{7.125}$$

is expressed in curvilinear-coordinate form as

$$\frac{\partial q}{\partial t} - \frac{e_{\ell k}}{\det J}\frac{\partial}{\partial \eta_k}\left(\varepsilon\psi a_\ell \frac{\partial q}{\partial t}\right) + \frac{e_{\ell k}}{\det J}\frac{\partial\left(f_\ell - f_\ell^\nu\right)}{\partial \eta_k} - \frac{e_{\ell k}}{\det J}\frac{\partial}{\partial \eta_k}\left(\varepsilon\psi a_\ell \frac{e_{mn}}{\det J}\left(\frac{\partial\left(f_m^q - f_m^\nu\right)}{\partial \eta_n} + \delta\frac{\partial f_m^p}{\partial \eta_n}\right)\right)$$

$$-\frac{e_{\ell k}}{\det J}\frac{\partial}{\partial \eta_k}\left(\varepsilon\psi\left(c\left(\alpha a_\ell a_m + \alpha^N \left(a_\ell^{N_1} a_m^{N_1} + a_\ell^{N_2} a_m^{N_2}\right)\right)\frac{e_{mn}}{\det J}\frac{\partial q}{\partial \eta_n}\right)\right) = 0 \tag{7.126}$$

Moreover, $\det J$ remains independent of time t, while the metric data e_{ij} satisfy the invariance relationships (7.87), (7.104). System (7.125) can then also be expressed as

$$\frac{\partial\left(q \det J\right)}{\partial t} + \frac{\partial\left(e_{\ell k}\left(f_\ell - f_\ell^\nu\right)\right)}{\partial \eta_k}$$

$$-\frac{\partial}{\partial\eta_k}\left[\varepsilon\psi e_{\ell k}a_\ell\left(\frac{\partial q}{\partial t}+\frac{e_{mn}}{\det J}\left(\frac{\partial(f_m^q-f_m^\nu)}{\partial\eta_n}+\delta\frac{\partial f_m^p}{\partial\eta_n}\right)\right)\right]$$

$$-\frac{\partial}{\partial\eta_k}\left[\varepsilon\psi e_{\ell k}\left(c\left(\alpha a_\ell a_m+\alpha^N\left(a_\ell^{N_1}a_m^{N_1}+a_\ell^{N_2}a_m^{N_2}\right)\right)\frac{e_{mn}}{\det J}\frac{\partial q}{\partial\eta_n}\right)\right]=0 \qquad (7.127)$$

Similarly, the weak statement

$$\int_\Omega w\frac{\partial q}{\partial t}d\Omega-\int_\Omega\frac{\partial w}{\partial x_\ell}\left(f_\ell-f_\ell^\nu\right)d\Omega+\int_\Omega\frac{\partial w}{\partial x_\ell}a_\ell\varepsilon\psi\left(\frac{\partial q}{\partial t}+\frac{\partial\left(f_m^q-f_m^\nu\right)}{\partial x_m}+\delta\frac{\partial f_m^p}{\partial x_m}\right)d\Omega$$

$$+\int_\Omega\frac{\partial w}{\partial x_\ell}\varepsilon\psi c\left(\alpha a_\ell a_m+\alpha^N\left(a_\ell^{N_1}a_m^{N_1}+a_\ell^{N_2}a_m^{N_2}\right)\right)\frac{\partial q}{\partial x_m}d\Omega$$

$$-\oint_{\partial\Omega}w\varepsilon\psi a_\ell\left(\frac{\partial q}{\partial t}+\frac{\partial\left(f_m^q-f_m^\nu\right)}{\partial x_m}+\delta\frac{\partial f_m^p}{\partial x_m}\right)n_\ell\,d\partial\Omega$$

$$-\oint_{\partial\Omega}w\varepsilon\psi\left(c\left(\alpha a_\ell a_m+\alpha^N\left(a_\ell^{N_1}a_m^{N_1}+a_\ell^{N_2}a_m^{N_2}\right)\right)\frac{\partial q}{\partial x_m}\right)n_\ell\,d\partial\Omega$$

$$+\oint_{\partial\Omega}wf_jn_j\,d\partial\Omega-\oint_{\partial\Omega}wf_j^\nu n_j\,d\partial\Omega=0 \qquad (7.128)$$

is further cast in terms of the curvilinear coordinates $\boldsymbol{\eta}$ following transformations (7.85), (7.102)

$$\int_\Omega w\frac{\partial q}{\partial t}\det J\,d\Omega-\int_\Omega e_{\ell k}\frac{\partial w}{\partial\eta_k}\left(f_\ell-f_\ell^\nu\right)d\Omega$$

$$+\int_\Omega e_{\ell k}\frac{\partial w}{\partial\eta_k}a_\ell\varepsilon\psi\left(\frac{\partial q}{\partial t}+\left(\frac{\partial\left(f_m-f_m^\nu\right)}{\partial\eta_n}+(\delta-1)\frac{\partial f_m^p}{\partial\eta_n}\right)\frac{e_{mn}}{\det J}\right)d\Omega$$

$$+\int_\Omega e_{\ell k}\frac{\partial w}{\partial\eta_k}\varepsilon\psi c\left(\alpha a_\ell a_m+\alpha^N\left(a_\ell^{N_1}a_m^{N_1}+a_\ell^{N_2}a_m^{N_2}\right)\right)\frac{\partial q}{\partial\eta_n}\frac{e_{mn}}{\det J}d\Omega$$

$$-\oint_{\partial\Omega}w\varepsilon\psi a_\ell\left(\frac{\partial q}{\partial t}+\frac{\partial\left(f_m^q-f_m^\nu\right)}{\partial x_m}+\delta\frac{\partial f_m^p}{\partial x_m}\right)n_\ell\,d\partial\Omega$$

$$-\oint_{\partial\Omega}w\varepsilon\psi\left(c\left(\alpha a_\ell a_m+\alpha^N\left(a_\ell^{N_1}a_m^{N_1}+a_\ell^{N_2}a_m^{N_2}\right)\right)\frac{\partial q}{\partial x_m}\right)n_\ell\,d\partial\Omega$$

$$+\oint_{\partial\Omega}wf_jn_j\,d\partial\Omega-\oint_{\partial\Omega}wf_j^\nu n_j\,d\partial\Omega=0 \qquad (7.129)$$

which states the curvilinear-coordinate form of an integral statement for the Euler and Navier-Stokes equations. This weak statement is subject to prescribed initial conditions $q(\boldsymbol{x},0)=q_0(\boldsymbol{x})$ and boundary conditions on $\partial\Omega\equiv\overline{\Omega}\backslash\Omega$. Synthetically, these boundary conditions are expressed as

$$B(\boldsymbol{x}_{\partial\Omega})q(\boldsymbol{x}_{\partial\Omega},t)=G_v(\boldsymbol{x}_{\partial\Omega},t) \qquad (7.130)$$

where $G_v(\boldsymbol{x}_{\partial\Omega},t)$ corresponds to the array of prescribed Dirichlet boundary conditions, with a zero entry for each corresponding unconstrained component of q, and $B(\boldsymbol{x}_{\partial\Omega})$ denotes a square diagonal matrix, with a 1 for each diagonal entry, but replaced by zero for each corresponding unconstrained component of q. For a hyperbolic system, that is for $f_j^\nu \equiv 0$, the number of boundary conditions at inlet and outlet depend on the signs of the eigenvalues of the boundary Jacobian matrix $\partial f/\partial q$.

A weak statement formulation affords several advantages. This statement provides dedicated surface integrals that allow direct, effective, and efficient enforcement of prescribed boundary conditions on the surface fluxes $f_j n_j$ and $f_j^\nu n_j$. The weak statement also allows an accurate representation of the viscous flux f_ℓ^ν, which depends on the derivatives of velocity, because the statement does not require the derivative of this flux, hence it does not require that the velocity field should be twice differentiable, but only once. This significant feature allows use of efficient piece-wise continuous multi-linear basis functions w in the domain integral that involves the gradient of w and this flux, the integral that provides the important contributions from the viscous flux to the formulation. In the case of any alternative formulation, e.g. finite volume method, where the weighting function w remains constant, the viscous flux must also be approximated on the surface of every internal subdomain for the evaluation of the corresponding surface integrals. Conversely, the weak statement does not require this additional surface approximation because the surface integral identically vanishes on the surface of every internal subdomain, where w vanishes, whereas the integrals on the boundary surfaces, as noted, are employed for enforcing prescribed boundary conditions. According to the order of the derivatives in the weak statement, the weighting functions belong to an $\mathcal{H}^1$ space. The solution belongs to the same space, for a viscous flow, but need only belong to an $\mathcal{H}^0$ space for inviscid shocked or shockless flows.

7.7.2   Finite Element Galerkin Formulation

The continuum weak statement

$$
\int_\Omega w \frac{\partial q}{\partial t} \det J \, d\Omega - \int_\Omega e_{\ell k} \frac{\partial w}{\partial \eta_k} \left(f_\ell - f_\ell^\nu \right) d\Omega
$$

$$
+ \int_\Omega e_{\ell k} \frac{\partial w}{\partial \eta_k} a_\ell \varepsilon \psi \left(\frac{\partial q}{\partial t} + \left(\frac{\partial \left(f_m - f_m^\nu \right)}{\partial \eta_n} + (\delta - 1) \frac{\partial f_m^p}{\partial \eta_n} \right) \frac{e_{mn}}{\det J} \right) d\Omega
$$

$$
+ \int_\Omega e_{\ell k} \frac{\partial w}{\partial \eta_k} \varepsilon \psi c \left(\alpha a_\ell a_m + \alpha^N \left(a_\ell^{N_1} a_m^{N_1} + a_\ell^{N_2} a_m^{N_2} \right) \right) \frac{\partial q}{\partial \eta_n} \frac{e_{mn}}{\det J} d\Omega
$$

$$
- \oint_{\partial \Omega} w \varepsilon \psi a_\ell \left(\frac{\partial q}{\partial t} + \frac{\partial \left(f_m^q - f_m^\nu \right)}{\partial x_m} + \delta \frac{\partial f_m^p}{\partial x_m} \right) n_\ell \, d\partial \Omega
$$

$$
- \oint_{\partial \Omega} w \varepsilon \psi \left(c \left(\alpha a_\ell a_m + \alpha^N \left(a_\ell^{N_1} a_m^{N_1} + a_\ell^{N_2} a_m^{N_2} \right) \right) \frac{\partial q}{\partial x_m} \right) n_\ell \, d\partial \Omega
$$

$$
+ \oint_{\partial \Omega} w f_j n_j \, d\partial \Omega - \oint_{\partial \Omega} w f_j^\nu n_j \, d\partial \Omega = 0 \tag{7.131}
$$

provides the foundation for developing finite-dimensional discrete solutions. Following the error-minimization and best-approximation results cited in Section 7.1.1, the finite dimensional solution q^h is sought within a finite dimensional Hilbert subspace $\mathcal{H}^h$, where superscript h denotes a finite - dimensional discretization. Each basis function w^h will also belong to a finite-dimensional subspace; when the space of the weighting functions coincides with the space of the discrete solution, the formulation is named Galerkin formulation, which is the formulation employed in the following sections. With $\mathcal{P}_n(\Omega_e)$ and $\mathcal{P}_{nv}(\Omega_e)$ the spaces of respectively diagonal square matrix-valued and vector-valued n-th-degree polynomials within each Ω_e, for each "t", the corresponding diagonal square matrix-valued and vector-valued N-dimensional finite element discretization spaces employed in this presentation are defined as

$$\mathcal{S}^{1h}(\Omega^h) \equiv \left\{ w^h \in \mathcal{H}^1(\Omega^h) : w^h\big|_{\Omega_e} \in \mathcal{P}_n(\Omega_e), \forall \Omega_e \in \Omega^h, B(\boldsymbol{x}_{\partial\Omega^h}) w^h(\boldsymbol{x}_{\partial\Omega^h}) = 0 \right\}$$

$$\mathcal{S}_v^{0h}(\Omega^h) \equiv \left\{ w_v^h \in \mathcal{H}_v^0(\Omega^h) : w_v^h\big|_{\Omega_e} \in \mathcal{P}_{nv}(\Omega_e), \forall \Omega_e \in \Omega^h, B(\boldsymbol{x}_{\partial\Omega^h}) w_v^h(\boldsymbol{x}_{\partial\Omega^h}, t) = G_v(\boldsymbol{x}_{\partial\Omega^h}, t) \right\}$$

$$(7.132)$$

Based on these spaces, the finite element approximation $q^h \in \mathcal{S}_v^{0h}(\Omega^h)$, is determined for each "t" as the solution of the finite element weak statement

$$\int_\Omega w^h \frac{\partial q^h}{\partial t} \det J \, d\Omega - \int_\Omega e_{\ell k} \frac{\partial w^h}{\partial \eta_k} \left(f_\ell^h - f_\ell^{\nu^h} \right) d\Omega$$

$$+ \int_\Omega e_{\ell k} \frac{\partial w^h}{\partial \eta_k} a_\ell^h \varepsilon^h \psi^h \left(\frac{\partial q^h}{\partial t} + \left(\frac{\partial \left(f_m^h - f_m^{\nu^h} \right)}{\partial \eta_n} + (\delta^h - 1) \frac{\partial f_m^{p^h}}{\partial \eta_n} \right) \frac{e_{mn}}{\det J} \right) d\Omega$$

$$+ \int_\Omega e_{\ell k} \frac{\partial w^h}{\partial \eta_k} \varepsilon^h \psi^h c^h \left(\alpha^h a_\ell^h a_m^h + \alpha^{N^h} \left(a_\ell^{N_1^h} a_m^{N_1^h} + a_\ell^{N_2^h} a_m^{N_2^h} \right) \right) \frac{\partial q^h}{\partial \eta_n} \frac{e_{mn}}{\det J} d\Omega$$

$$- \oint_{\partial\Omega} w^h \varepsilon^h \psi^h a_\ell^h \left(\frac{\partial q^h}{\partial t} + \frac{\partial \left(f_m^h - f_m^{\nu^h} \right)}{\partial x_m} + (\delta^h - 1) \frac{\partial f_m^{p^h}}{\partial x_m} \right) n_\ell \, d\partial\Omega$$

$$- \oint_{\partial\Omega} w^h \varepsilon^h \psi^h \left(c^h \left(\alpha^h a_\ell^h a_m^h + \alpha^{N^h} \left(a_\ell^{N_1^h} a_m^{N_1^h} + a_\ell^{N_2^h} a_m^{N_2^h} \right) \right) \frac{\partial q^h}{\partial x_m} \right) n_\ell \, d\partial\Omega$$

$$+ \oint_{\partial\Omega} w^h f_j^h n_j \, d\partial\Omega - \oint_{\partial\Omega} w^h f_j^{\nu^h} n_j \, d\partial\Omega = 0 \qquad (7.133)$$

for every basis function $w^h \in \mathcal{S}^{1h}(\Omega^h)$. Since each of these functions results from a linear combination of the N linearly independent basis functions $w_k(\boldsymbol{x})$ of $\mathcal{S}^{1h}(\Omega^h)$, this statement is equivalent to the N equations, $1 \leq k \leq N$,

$$\int_\Omega w_k \frac{\partial q^h}{\partial t} \det J \, d\Omega - \int_\Omega e_{\ell s} \frac{\partial w_k}{\partial \eta_s} \left(f_\ell^h - f_\ell^{\nu^h} \right) d\Omega$$

$$+ \int_\Omega e_{\ell s} \frac{\partial w_k}{\partial \eta_s} a_\ell^h \varepsilon^h \psi^h \left(\frac{\partial q^h}{\partial t} + \left(\frac{\partial \left(f_m^h - f_m^{\nu^h} \right)}{\partial \eta_n} + (\delta^h - 1) \frac{\partial f_m^{p^h}}{\partial \eta_n} \right) \frac{e_{mn}}{\det J} \right) d\Omega$$

$$+ \int_\Omega e_{\ell s} \frac{\partial w_k}{\partial \eta_s} \varepsilon^h \psi^h c^h \left(\alpha^h a_\ell^h a_m^h + \alpha^{N^h} \left(a_\ell^{N_1^h} a_m^{N_1^h} + a_\ell^{N_2^h} a_m^{N_2^h} \right) \right) \frac{\partial q^h}{\partial \eta_n} \frac{e_{mn}}{\det J} d\Omega$$

$$- \oint_{\partial\Omega} w_k \varepsilon^h \psi^h a_\ell^h \left(\frac{\partial q^h}{\partial t} + \frac{\partial \left(f_m^h - f_m^{\nu^h} \right)}{\partial x_m} + (\delta^h - 1) \frac{\partial f_m^{p^h}}{\partial x_m} \right) n_\ell \, d\partial\Omega$$

$$-\oint_{\partial\Omega} w_k \varepsilon^h \psi^h \left(c^h \left(\alpha^h a_\ell^h a_m^h + \alpha^{N^h} \left(a_\ell^{N_1^h} a_m^{N_1^h} + a_\ell^{N_2^h} a_m^{N_2^h} \right) \right) \right) \frac{\partial q^h}{\partial x_m} n_\ell \, d\partial\Omega$$

$$+ \oint_{\partial\Omega} w_k f_j^h n_j \, d\partial\Omega - \oint_{\partial\Omega} w_k f_j^{\nu^h} n_j \, d\partial\Omega = 0 \tag{7.134}$$

This statement is further expressed as

$$\sum_{e=1}^{N_e} \left\{ \int_{\Omega_e} w_k \frac{\partial q^h}{\partial t} \det J \, d\Omega - \int_{\Omega_e} e_{\ell s} \frac{\partial w_k}{\partial \eta_s} \left(f_\ell^h - f_\ell^{\nu^h} \right) d\Omega \right.$$

$$+ \int_{\Omega_e} e_{\ell s} \frac{\partial w_k}{\partial \eta_s} a_\ell^h \varepsilon^h \psi^h \left(\frac{\partial q^h}{\partial t} + \left(\frac{\partial \left(f_m^h - f_m^{\nu^h} \right)}{\partial \eta_n} + (\delta^h - 1) \frac{\partial f_m^{p^h}}{\partial \eta_n} \right) \frac{e_{mn}}{\det J} \right) d\Omega$$

$$+ \int_{\Omega_e} e_{\ell s} \frac{\partial w_k}{\partial \eta_s} \varepsilon^h \psi^h c^h \left(\alpha^h a_\ell^h a_m^h + \alpha^{N^h} \left(a_\ell^{N_1^h} a_m^{N_1^h} + a_\ell^{N_2^h} a_m^{N_2^h} \right) \right) \frac{\partial q^h}{\partial \eta_n} \frac{e_{mn}}{\det J} d\Omega$$

$$- \oint_{\partial\Omega_e} w_k \varepsilon^h \psi^h a_\ell^h \left(\frac{\partial q^h}{\partial t} + \frac{\partial \left(f_m^h - f_m^{\nu^h} \right)}{\partial x_m} + (\delta^h - 1) \frac{\partial f_m^{p^h}}{\partial x_m} \right) n_\ell \, d\partial\Omega$$

$$- \oint_{\partial\Omega_e} w_k \varepsilon^h \psi^h \left(c^h \left(\alpha^h a_\ell^h a_m^h + \alpha^{N^h} \left(a_\ell^{N_1^h} a_m^{N_1^h} + a_\ell^{N_2^h} a_m^{N_2^h} \right) \right) \right) \frac{\partial q^h}{\partial x_m} n_\ell \, d\partial\Omega$$

$$\left. + \oint_{\partial\Omega_e} w_k f_j^h n_j \, d\partial\Omega - \oint_{\partial\Omega_e} w_k f_j^{\nu^h} n_j \, d\partial\Omega \right\} = 0 \tag{7.135}$$

that is the sum of contributions from each subregion Ω_e. This subregion is that which is known as a finite element. A finite element, accordingly, is a finite region of space where functions and coordinate transformations are cast in terms of locally prescribed basis functions $w_k(\boldsymbol{x})$. As shown in Section 7.1.2, this finite element formulation provides the significant advantage of a local element-by-element coordinate transformation, which is far more convenient and practical to establish and employ as opposed to a global domain-wide transformation.

For a finite dimensional q^h, this weak statement requires the evaluation of several integrals, such as

$$\int_{\Omega_e} w_k \frac{\partial q^h}{\partial t} \det J \, d\Omega, \quad \int_{\Omega_e} \frac{\partial w_k}{\partial \eta_\ell} e_{j\ell} f_j^h \, d\Omega \tag{7.136}$$

If the basis functions remain similar to the functions in the representative examples in Section 7.1.1, functions that do not identically vanish over subregions of the computational domain Ω, these integrals will couple all of the unknown coefficients to be determined in q^h, which will lead to a computationally demanding procedure with a fully-populated matrix. The formulation in the following sections will therefore introduce and employ locally defined basis functions that identically vanish over most subregions of the computational domain Ω, so that these integrals will lead to banded-matrix systems that are more efficiently solved. As the number of elements increases, the magnitude of each integral in the weak statement (7.135) correspondingly decreases, which may lead to an ill-conditioned system. This situation is prevented by dividing each nodal equation "k" by a nodal "$\det J_k$", obtained as the average of the det J's from the elements sharing the k-th node. After all the integrations with respect to the space variable $\boldsymbol{x}$ are completed, the weak statement generates the system

$$\mathcal{M}\frac{dQ}{dt} + F(Q,t) = 0 \qquad (7.137)$$

which is a system of ordinary differential equations in continuum time. In this system, $\mathcal{M}$ denotes the matrix, known as the mass matrix, which couples the time derivatives of the time dependent coefficients Q in q^h. The array F contains all other terms from the weak statement as well as boundary conditions. Since the minimization procedure described in Section 7.1.1 minimizes the error ε_{TOT} it is possible for q^h to become oscillatory, since the derivatives of such a function can still lead to a vanishing weighted-residual integral statement. The following chapters show that a characteristics-bias companion Euler or Navier-Stokes system leads to essentially non-oscillatory solutions.

7.7.3 Discretization of Space

The continuum domain Ω is represented by a finite-dimensional partition Ω^h, with either $\Omega^h \subseteq \Omega$ or $\Omega^h \supseteq \Omega$, where superscript "$h$" denotes discretization. This partition Ω^h has its boundary nodes on the boundary $\partial\Omega$ of Ω and results from the union of N_e non-overlapping elements Ω_e, $\Omega^h = \bigcup_{e=1}^{N_e} \Omega_e$, as illustrated in Figure 7.14.

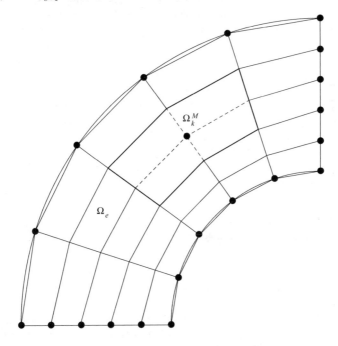

Figure 7.14: Continuum and Discrete Domains

For N mesh nodes within Ω^h, there exist clusters of "master" elements Ω_k^M, each comprising only those adjacent elements that share a mesh node $\boldsymbol{x}_k$, with $1 \leq k \leq N$; accordingly, within Ω^h there exist exactly N master elements. Note that Ω_k^M represents a "finite volume" as used in finite volume schemes.

7.7.4 Discretization of Functions

The discrete solution q^h at each time t assumes the form of the following linear combination

$$q^h (\boldsymbol{x}, t) \equiv \sum_{k=1}^{N} w_k (\boldsymbol{x}) \cdot q^h (\boldsymbol{x}_k, t) \tag{7.138}$$

of time-dependent nodal solution values $q^h (\boldsymbol{x}_k, t)$, to be determined, and basis functions. Similarly, all fluxes like $f(q(\boldsymbol{x}, t))$ are discretized through the group expression

$$f_j^h (\boldsymbol{x}, t) \equiv \sum_{k=1}^{N} w_k (\boldsymbol{x}) \cdot f_j \left(q^h (\boldsymbol{x}_k, t) \right) \tag{7.139}$$

Owing to this expression, the residual error $\Delta\varepsilon$ of Section 7.1.1 will vanish.

For a representative one-dimensional case, Figure 7.15 displays a continuum function q and a discrete function q^h corresponding to (7.138). The main objective of this section is to demonstrate for a simple Cartesian master element how a nodal basis function may be expressed in terms of local element functions. For arbitrarily shaped master elements, the basis functions are more efficiently formed using local coordinate transformations, as shown in the next section.

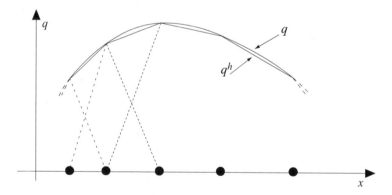

Figure 7.15: Continuum and Discrete Functions

With reference to Figures 7.16, 7.18, the discrete test function w^h within each master element Ω_k^M coincides with the "pyramid" basis function $w_k = w_k (\boldsymbol{x})$, $1 \leq k \leq N$, with compact support on Ω_k^M. Such a function equals one at node $\boldsymbol{x}_k$, zero at all other mesh nodes and also identically vanishes both on the boundary segments of Ω_k^M not containing $\boldsymbol{x}_k$ and on the computational domain outside Ω_k^M.

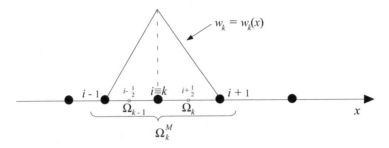

Figure 7.16: One-Dimensional Master Element Ω_k^M and Test Function $w_k = w_k(x)$

For instance, a one-dimensional piece-wise continuous linear pyramid test function w_k, can be expressed as

$$\Delta x_{k-\frac{1}{2}} \equiv x_k - x_{k-1}, \quad w_k(x) \equiv \begin{cases} \dfrac{x - x_{k-1}}{\Delta x_{k-\frac{1}{2}}} & , \quad x_{k-1} \leq x \leq x_k \\[2mm] \dfrac{x_{k+1} - x}{\Delta x_{k+\frac{1}{2}}} & , \quad x_k \leq x \leq x_{k+1} \end{cases} \tag{7.140}$$

Considering the one-dimensional pyramid functions that form w_k allows shifting the focus from node-based to element-based functions. The pyramid test functions w_k may also be expressed as

$$w_k(x) \equiv \begin{cases} w_{k_e-1,2} & , \quad x_{k-1} \leq x \leq x_k \\ w_{k_e,1} & , \quad x_k \leq x \leq x_{k+1} \end{cases} \tag{7.141}$$

where $w_{k_e-1,2}$ and $w_{k_e,1}$ denote the basis functions respectively for node 2 of element $k_e - 1$ and for node 1 of element k_e. With reference to Figure 7.17, and dispensing with subscript k_e, the linear basis functions for every element Ω_k are cast as

$$w_1(x) = \frac{x_{k+1} - x}{x_{k+1} - x_k}, \quad w_2(x) = \frac{x - x_k}{x_{k+1} - x_k} \tag{7.142}$$

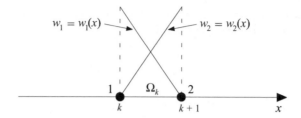

Figure 7.17: One-Dimensional Linear Basis Functions

Figure 7.18 shows a representative pyramid basis functions for a two-dimensional formulation

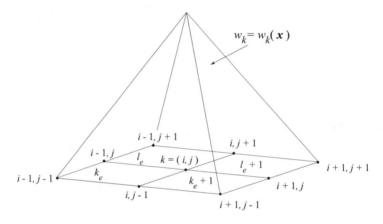

Figure 7.18: Pyramid Test Function for Ω_k^M

For uniform rectangular elements, the expression for this pyramid basis function is

$$
w_k(\boldsymbol{x}) \equiv \begin{cases} \left(\dfrac{x - x_{i-1,j}}{\Delta x_{i-\frac{1}{2}}}\right)\left(\dfrac{y - y_{i,j-1}}{\Delta y_{j-\frac{1}{2}}}\right) & , \quad \begin{array}{ccc} x_{i-1,j} & \leq x \leq & x_{i,j} \\ y_{i,j-1} & \leq y \leq & y_{i,j} \end{array} \\[3ex] \left(\dfrac{x_{i+1,j} - x}{\Delta x_{i+\frac{1}{2}}}\right)\left(\dfrac{y - y_{i,j-1}}{\Delta y_{j-\frac{1}{2}}}\right) & , \quad \begin{array}{ccc} x_{i,j} & \leq x \leq & x_{i+1,j} \\ y_{i,j-1} & \leq y \leq & y_{i,j} \end{array} \\[3ex] \left(\dfrac{x_{i+1,j} - x}{\Delta x_{i+\frac{1}{2}}}\right)\left(\dfrac{y_{i,j+1} - y}{\Delta y_{j+\frac{1}{2}}}\right) & , \quad \begin{array}{ccc} x_{i,j} & \leq x \leq & x_{i+1,j} \\ y_{i,j} & \leq y \leq & y_{i,j+1} \end{array} \\[3ex] \left(\dfrac{x - x_{i-1,j}}{\Delta x_{i-\frac{1}{2}}}\right)\left(\dfrac{y_{i,j+1} - y}{\Delta y_{j+\frac{1}{2}}}\right) & , \quad \begin{array}{ccc} x_{i-1,j} & \leq x \leq & x_{i,j} \\ y_{i,j} & \leq y \leq & y_{i,j+1} \end{array} \end{cases}
\tag{7.143}
$$

As in the one-dimensional case, considering the two-dimensional pyramid functions that form this $w_k(\boldsymbol{x})$ allows shifting the focus from node-based to element-based functions. Accordingly, this pyramid test function $w_k(\boldsymbol{x})$ may also be expressed as

$$
w_k(\boldsymbol{x}) \equiv \begin{cases} w_{k_e,3}(\boldsymbol{x}) & , & \begin{aligned} x_{i-1,j} &\leq x \leq x_{i,j} \\ y_{i,j-1} &\leq y \leq y_{i,j} \end{aligned} \\[2mm] w_{k_e+1,4}(\boldsymbol{x}) & , & \begin{aligned} x_{i,j} &\leq x \leq x_{i+1,j} \\ y_{i,j-1} &\leq y \leq y_{i,j} \end{aligned} \\[2mm] w_{l_e+1,1}(\boldsymbol{x}) & , & \begin{aligned} x_{i,j} &\leq x \leq x_{i+1,j} \\ y_{i,j} &\leq y \leq y_{i,j+1} \end{aligned} \\[2mm] w_{l_e,2}(\boldsymbol{x}) & , & \begin{aligned} x_{i-1,j} &\leq x \leq x_{i,j} \\ y_{i,j} &\leq y \leq y_{i,j+1} \end{aligned} \end{cases} \tag{7.144}
$$

where $w_{k_e,3}$, $w_{k_e+1,4}$, $w_{l_e+1,1}$, $w_{l_e,2}$ denote the basis functions respectively for node 3 of element k_e, node 4 of element k_e+1, node 1 of element l_e+1, and node 2 of element l_e. With reference to Figure 7.18 and dispensing with the element subscripts, the four bi-linear basis functions for a rectangular element Ω_e are cast as

$$
w_1(\boldsymbol{x}) = \left(\frac{x_{i+1,j}-x}{x_{i+1,j}-x_{i,j}}\right)\left(\frac{y_{i,j+1}-y}{y_{i,j+1}-y_{i,j}}\right), \quad w_2(\boldsymbol{x}) = \left(\frac{x-x_{i,j}}{x_{i+1,j}-x_{i,j}}\right)\left(\frac{y_{i,j+1}-y}{y_{i,j+1}-y_{i,j}}\right)
$$

$$
w_3(\boldsymbol{x}) = \left(\frac{x-x_{i,j}}{x_{i+1,j}-x_{i,j}}\right)\left(\frac{y-y_{i,j}}{y_{i,j+1}-y_{i,j}}\right), \quad w_4(\boldsymbol{x}) = \left(\frac{x_{i+1,j}-x}{x_{i+1,j}-x_{i,j}}\right)\left(\frac{y-y_{i,j}}{y_{i,j+1}-y_{i,j}}\right)
$$

$$\tag{7.145}$$

The corresponding basis functions for three-dimensional Cartesian master elements emerge as a generalization of these expression, by introducing the contributions from basis functions in the z directions. As all of these developments show, even for simple rectangular elements the basis functions may become quite involved; for arbitrarily shaped elements, the complexity of the Cartesian expressions for these functions will significantly increase. It is far more convenient, efficient, and practical to express these functions in terms of local coordinates, via a local coordinate transformation, as shown in the following sections.

7.7.5 A-Posteriori Accuracy Assessment

This section presents some basic error correlations in terms of a time interval Δt and a maximum element length measure ΔL. Let q denote the solution of the characteristics-bias system

$$
\frac{\partial q}{\partial t} - \frac{\partial}{\partial x_\ell}\left(\varepsilon\psi a_\ell\frac{\partial q}{\partial t}\right) + \frac{\partial(f_\ell - f_\ell^\nu)}{\partial x_\ell} - \frac{\partial}{\partial x_\ell}\left(\varepsilon\psi a_\ell\left(\frac{\partial(f_m^q - f_m^\nu)}{\partial x_m} + \delta\frac{\partial f_m^p}{\partial x_m}\right)\right)
$$

$$
- \frac{\partial}{\partial x_\ell}\left(\varepsilon\psi\left(c\left(\alpha a_\ell a_m + \alpha^N\left(a_\ell^{N_1}a_m^{N_1} + a_\ell^{N_2}a_m^{N_2}\right)\right)\frac{\partial q}{\partial x_m}\right)\right) = 0 \tag{7.146}
$$

and $\widehat{q}$ the solution of the unperturbed system, corresponding that is to $\varepsilon = 0$. The norm of the error between the fully discrete solution Q of the characteristics-bias system and the solution $\widehat{q}$ of the unperturbed system may be expressed as

$$
\|Q - \widehat{q}\|_{\mathcal{H}^\eta(\Omega^h)} \leq \left\|Q - q^h\right\|_{\mathcal{H}^\eta(\Omega^h)} + \left\|q^h - q\right\|_{\mathcal{H}^\eta(\Omega^h)} + \|q - \widehat{q}\|_{\mathcal{H}^\eta(\Omega^h)} \tag{7.147}
$$

where η equals either 1 or 0 respectively for a shockless and shocked solution, q denotes the solution of the characteristics-bias system (7.61) and q^h and Q correspond to expansion (7.138) respectively for the case of an exact solution of the ODE system (7.137) and for a numerical solution of this system, as discussed in Chapter 8; at steady state, $Q = q^h$. Each of these three norms may be bounded as

$$\left\| Q - q^h \right\|_{\mathcal{H}^\eta(\Omega^h)} \leq C_{\Delta t} \Delta t^{m_t}, \quad \left\| q^h - q \right\|_{\mathcal{H}^\eta(\Omega^h)} \leq C_{\Delta x} \Delta x^{m_h}, \quad \left\| q - \widehat{q} \right\|_{\mathcal{H}^\eta(\Omega^h)} \leq C_{\varepsilon\psi} (\varepsilon\psi)^{m_\varepsilon}$$
(7.148)

with $C_{\Delta t}$, $C_{\Delta L}$, $C_{\varepsilon\psi}$, m_t, m_h, m_ε suitable constants, totally independent of Δt, ΔL, and $\varepsilon\psi$. If these constants where available in terms of correlations with geometric measures of Ω, magnitudes of pertinent parameters in the characteristics-bias systems, and norms of associated initial and boundary conditions, these inequalities would lead to an a-priori error estimates. For general multi-dimensional, non-linear hyperbolic and parabolic systems, such correlations are not available and these norm expressions, accordingly, serve the purpose of guiding the a-posteriori determination of asymptotic spatial convergence rates.

The time step Δt is correlated to the mesh measure via a generalized Courant number as $\Delta t = \overline{C} \Delta L$. For linear elements and for a suitable time integration algorithm that is at least second order accurate, it follows that $m_t \simeq m_h$. Norm (7.147) thus becomes

$$
\begin{aligned}
\left\| Q - \widehat{q} \right\|_{\mathcal{H}^\eta(\Omega^h)} &\leq \left\| Q - q^h \right\|_{\mathcal{H}^\eta(\Omega^h)} + \left\| q^h - q \right\|_{\mathcal{H}^\eta(\Omega^h)} + \left\| q - \widehat{q} \right\|_{\mathcal{H}^\eta(\Omega^h)} \\
&\leq \max \left(C_{\Delta t} (\overline{C}\Delta L)^{m_h}, C_{\Delta L}\Delta L^{m_h}, C_{\varepsilon\psi}(\varepsilon\psi)^{m_\varepsilon} \right)
\end{aligned}
$$
(7.149)

In the characteristics-bias system, the $\varepsilon\psi$ term dominates this error and the perturbation parameters ε and ψ are cast as

$$\varepsilon = \frac{\Delta L}{2}, \quad \psi = C_\psi \Delta L^{m_\psi}$$
(7.150)

As a result, norm (7.149) becomes

$$\left\| Q - \widehat{q} \right\|_{\mathcal{H}^\eta(\Omega^h)} \leq C \left(\Delta L^{1+m_\psi} \right)^{m_\varepsilon} = C\Delta L^{m_\varepsilon(1+m_\psi)} = C\Delta L^{m_e}$$
(7.151)

for some suitable constants C, m_ε, and m_e, which may, however, differ for the $\mathcal{H}^0$ and $\mathcal{H}^1$ norms. For one-dimensional solutions $\Delta L = 1/N_e$, and (7.151) is equivalently expressed as

$$\log \left\| Q - \widehat{q} \right\|_{\mathcal{H}^\eta(\Omega^h)} \leq \log C - m_e \log N_e$$
(7.152)

to correlate the decrease of the norm with the increase in the number N_e of elements. In any case, for one- and multi-dimensional solutions, with C and m_e those constants for which (7.152) is an equality for the two solutions Q_1 and Q_2, respectively corresponding to ΔL_1 and $\Delta L_2 = \Delta L_1/2$, the exponent m_e may be calculated as

$$m_e = \frac{1}{\log 2} \log \left(\frac{\left\| Q_1 - \widehat{q} \right\|_{\mathcal{H}^\eta(\Omega^h)}}{\left\| Q_2 - \widehat{q} \right\|_{\mathcal{H}^\eta(\Omega^h)}} \right)$$
(7.153)

for a sequence of progressively denser grids. According to the computational results obtained with multi-linear elements, as discussed in the following chapters, the computed exponent m_e exceeds 2, for smooth flows, and can even be marginally greater than 3 in the $\mathcal{H}^0$ norm. This finding indicates the characteristic-bias Galerkin weak statement with linear elements produces an algorithm with accuracy between 2nd and 3rd order.

7.8 Basis Functions, Element and Coordinate Transformations

The element wise coordinate transformation conveniently results from the same finite - dimensional expansion for q. With reference to (7.138), in terms of curvilinear coordinates $\boldsymbol{\eta}$, this expansion is

$$q^h\left(\boldsymbol{x}(\boldsymbol{\eta}), t\right) \equiv \sum_{k=1}^{N} w_k\left(\boldsymbol{\eta}\right) \cdot q^h\left(\boldsymbol{x}_k, t\right) \tag{7.154}$$

Within each element, the coordinate transformation is thus expressed as

$$\boldsymbol{x} = \sum_{k=1}^{e_n} w_k\left(\boldsymbol{\eta}\right) \boldsymbol{x}_k \tag{7.155}$$

where "e_n" denotes the number of nodes within the element. The following developments are based on the concept of the "isoparametric" element, that is an element where the number of nodes for the solution q is the same as for the coordinate transformation for $\boldsymbol{x}$; additionally, the elements employed will be of the Lagrangian type, which means that the multi-dimensional basis functions result from products of suitable one-dimensional functions.

These one-dimensional functions may be systematically developed by way of a dedicated Taylor series as follows. Express the value at η_a of the one-dimensional function $q = q(\eta)$ in terms of the values of the function and its derivatives at η as

$$q(\eta_a) = q(\eta) + \sum_{n=1}^{e_n} \frac{d^n q}{d\eta}\bigg|_{\eta} \frac{(\eta_a - \eta)}{n!} \tag{7.156}$$

Upon writing this series for a number "e_n" of locations "η_a" it is possible to obtain a linear system where the unknowns are $q(\eta)$ and its derivatives. The solution of this system for $q(\eta)$ automatically provides the corresponding basis functions as exemplified next for several cases.

7.8.1 One-Dimensional Elements and Coordinate Transformations

The Taylor-series system for linear elements is expressed as

$$\begin{cases} q(\eta) + \dfrac{dq}{d\eta}(\eta_a - \eta) = q(\eta_a) \\[2mm] q(\eta) + \dfrac{dq}{d\eta}(\eta_b - \eta) = q(\eta_b) \end{cases} \tag{7.157}$$

For $\eta_a = -1$, $\eta_b = +1$, and $-1 \le \eta \le 1$, the solution of this system for $q(\eta)$ provides

$$q(\eta) = \frac{1}{2}(1 - \eta)q(\eta_a) + \frac{1}{2}(1 + \eta)q(\eta_b) \tag{7.158}$$

and the corresponding basis functions are

$$w_1(\eta) = \frac{1}{2}(1 - \eta), \quad w_2(\eta) = \frac{1}{2}(1 + \eta) \tag{7.159}$$

as illustrated in Figure 7.19

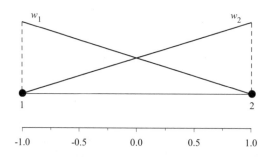

Figure 7.19: Linear Basis Functions

For quadratic elements, the Taylor-series system is expressed as

$$\begin{cases} q(\eta) + \dfrac{dq}{d\eta}(\eta_a - \eta) + \dfrac{d^2q}{d\eta^2}\dfrac{(\eta_a - \eta)^2}{2} = q(\eta_a) \\[2mm] q(\eta) + \dfrac{dq}{d\eta}(\eta_b - \eta) + \dfrac{d^2q}{d\eta^2}\dfrac{(\eta_b - \eta)^2}{2} = q(\eta_b) \\[2mm] q(\eta) + \dfrac{dq}{d\eta}(\eta_c - \eta) + \dfrac{d^2q}{d\eta^2}\dfrac{(\eta_c - \eta)^2}{2} = q(\eta_c) \end{cases} \tag{7.160}$$

For $\eta_a = -1$, $\eta_b = +1$, $\eta_c = 0$, and $-1 \le \eta \le 1$, the solution of this system for $q(\eta)$ provides

$$q(\eta) = \frac{1}{2}\eta(\eta - 1)q(\eta_a) + (1 - \eta^2)q(\eta_c) + \frac{1}{2}\eta(1 + \eta)q(\eta_b) \tag{7.161}$$

and the corresponding basis functions are

$$w_1(\eta) = \frac{1}{2}\eta(\eta - 1), \quad w_2(\eta) = (1 - \eta^2), \quad w_3(\eta) = \frac{1}{2}\eta(1 + \eta) \tag{7.162}$$

as illustrated in Figure 7.20

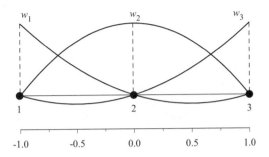

Figure 7.20: Quadratic Basis Functions

For cubic elements, the Taylor-series system is expressed as

$$
\begin{cases}
q(\eta) + \dfrac{dq}{d\eta}(\eta_a - \eta) + \dfrac{d^2q}{d\eta^2}\dfrac{(\eta_a - \eta)^2}{2} + \dfrac{d^3q}{d\eta^3}\dfrac{(\eta_a - \eta)^3}{6} = q(\eta_a) \\[2mm]
q(\eta) + \dfrac{dq}{d\eta}(\eta_b - \eta) + \dfrac{d^2q}{d\eta^2}\dfrac{(\eta_b - \eta)^2}{2} + \dfrac{d^3q}{d\eta^3}\dfrac{(\eta_b - \eta)^3}{6} = q(\eta_b) \\[2mm]
q(\eta) + \dfrac{dq}{d\eta}(\eta_c - \eta) + \dfrac{d^2q}{d\eta^2}\dfrac{(\eta_c - \eta)^2}{2} + \dfrac{d^3q}{d\eta^3}\dfrac{(\eta_c - \eta)^3}{6} = q(\eta_c) \\[2mm]
q(\eta) + \dfrac{dq}{d\eta}(\eta_c - \eta) + \dfrac{d^2q}{d\eta^2}\dfrac{(\eta_d - \eta)^2}{2} + \dfrac{d^3q}{d\eta^3}\dfrac{(\eta_d - \eta)^3}{6} = q(\eta_d)
\end{cases}
\tag{7.163}
$$

For $\eta_a = -1$, $\eta_b = +1$, $\eta_c = -\frac{1}{3}$, $\eta_d = \frac{1}{3}$, and $-1 \le \eta \le 1$, the solution of this system for $q(\eta)$ provides

$$
\begin{aligned}
q(\eta) = {}& -\frac{9}{16}(\eta^3 - \eta^2 - \frac{\eta}{9} + \frac{1}{9})q(\eta_a) + \frac{27}{16}(\eta^3 - \frac{\eta^2}{3} - \eta + \frac{1}{3})q(\eta_c) \\
& -\frac{27}{16}(\eta^3 + \frac{\eta^2}{3} - \eta - \frac{1}{3})q(\eta_d) + \frac{9}{16}(\eta^3 + \eta^2 - \frac{\eta}{9} - \frac{1}{9})q(\eta_b)
\end{aligned}
\tag{7.164}
$$

and the corresponding basis functions are

$$
w_1(\eta) = -\frac{9}{16}(\eta^3 - \eta^2 - \frac{\eta}{9} + \frac{1}{9}), \quad w_2(\eta) = \frac{27}{16}(\eta^3 - \frac{\eta^2}{3} - \eta + \frac{1}{3}),
$$

$$
w_3(\eta) = -\frac{27}{16}(\eta^3 + \frac{\eta^2}{3} - \eta - \frac{1}{3}), \quad w_4(\eta) = \frac{9}{16}(\eta^3 + \eta^2 - \frac{\eta}{9} - \frac{1}{9})
\tag{7.165}
$$

as illustrated in Figure 7.21

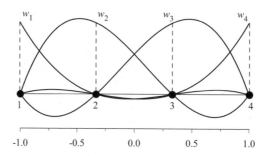

Figure 7.21: Cubic Basis Functions

The basis functions for even higher-order elements are determined following a similar procedure.

The local coordinate transformation for any of these elements is expressed as

$$x(\eta) = \sum_{k=1}^{e_n} w_k(\eta)\, x_k \qquad (7.166)$$

With this expression, an element in Cartesian space is transformed into an element in "η" space, as illustrated in Figure 7.22

Figure 7.22: Element Transformation

For these one-dimensional elements the transformation metric data $\det J = e_{22}$ becomes

$$\det J = e_{22} = \frac{\partial x}{\partial \eta} = \sum_{k=1}^{e_n} \frac{dw_k}{d\eta} x_k \qquad (7.167)$$

For the linear element, for instance, this metric data becomes

$$\det J = \frac{x_b - x_a}{2} \qquad (7.168)$$

which, in this case, remains constant.

7.8.2 Multi-Dimensional Elements

The Lagrange functions for multi-dimensional elements result from products of the one-dimensional functions presented in the previous section. The basis functions for a quadrilateral linear element are thus expressed as

$$w_1(\boldsymbol{\eta}) = \frac{1}{4}(1 - \eta_1)(1 - \eta_2), \quad w_2(\boldsymbol{\eta}) = \frac{1}{4}(1 + \eta_1)(1 - \eta_2)$$

$$w_3(\boldsymbol{\eta}) = \frac{1}{4}(1+\eta_1)(1+\eta_2), \quad w_4(\boldsymbol{\eta}) = \frac{1}{4}(1-\eta_1)(1+\eta_2) \tag{7.169}$$

as graphed in Figure 7.23

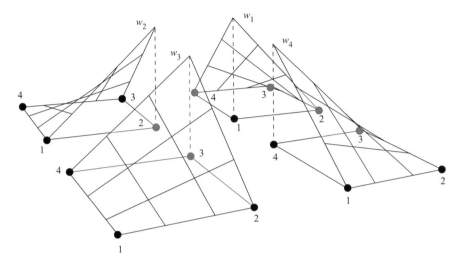

Figure 7.23: Two-Dimensional Linear Basis Functions

The basis functions for a quadrilateral quadratic element are cast as

$$w_1(\boldsymbol{\eta}) = \frac{1}{4}\eta_1\eta_2(\eta_1-1)(\eta_2-1)$$

$$w_2(\boldsymbol{\eta}) = \frac{1}{2}\eta_2\left(1-\eta_1^2\right)(\eta_2-1), \quad w_3(\boldsymbol{\eta}) = \frac{1}{4}\eta_1\eta_2(\eta_1+1)(\eta_2-1)$$

$$w_4(\boldsymbol{\eta}) = \frac{1}{2}\eta_1(\eta_1+1)\left(1-\eta_2^2\right), \quad w_5(\boldsymbol{\eta}) = \frac{1}{4}\eta_1\eta_2(\eta_1+1)(\eta_2+1)$$

$$w_6(\boldsymbol{\eta}) = \frac{1}{2}\eta_2\left(1-\eta_1^2\right)(\eta_2+1), \quad w_7(\boldsymbol{\eta}) = \frac{1}{4}\eta_1\eta_2(\eta_1-1)(\eta_2+1)$$

$$w_8(\boldsymbol{\eta}) = \frac{1}{2}\eta_1(\eta_1-1)\left(1-\eta_2^2\right), \quad w_9(\boldsymbol{\eta}) = \left(1-\eta_1^2\right)\left(1-\eta_2^2\right) \tag{7.170}$$

The corresponding basis functions for three-dimensional elements are similarly obtained by performing products of three one-dimensional functions.

For any of these elements, the local coordinate transformation is expressed as

$$\boldsymbol{x} = \sum_{k=1}^{e_n} w_k(\boldsymbol{\eta})(\boldsymbol{x}_k) \tag{7.171}$$

The metric data e_{11}, e_{12}, e_{21}, e_{22} are cast as

$$e_{11} = \frac{\partial x_2}{\partial \eta_2} = \sum_{k=1}^{e_n} \frac{\partial w_k}{\partial \eta_2}(x_2)_k, \quad e_{12} = -\frac{\partial x_2}{\partial \eta_1} = -\sum_{k=1}^{e_n} \frac{\partial w_k}{\partial \eta_1}(x_2)_k$$

$$e_{21} = -\frac{\partial x_1}{\partial \eta_2} = -\sum_{k=1}^{e_n} \frac{\partial w_k}{\partial \eta_2}(x_1)_k, \quad e_{22} = \frac{\partial x_1}{\partial \eta_1} = \sum_{k=1}^{e_n} \frac{\partial w_k}{\partial \eta_1}(x_1)_k \qquad (7.172)$$

This transformation and metric data map a rectangular region in Cartesian space into a square in the (η_1, η_2) space, as illustrated in Figure 7.24

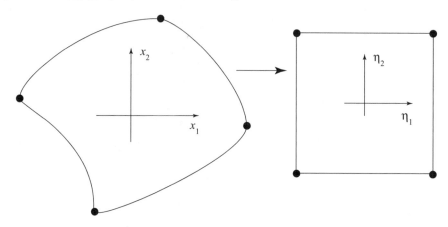

Figure 7.24: Two-Dimensional Transformation of Quadrilateral Region

For the linear element, the expressions for the metric data become

$$
\begin{aligned}
e_{11} &= \frac{1}{4}\left((x_2)_4 - (x_2)_1 + (x_2)_3 - (x_2)_2\right) + \frac{\eta_1}{4}\left((x_2)_3 - (x_2)_4 - ((x_2)_2 - (x_2)_1)\right) \\
e_{12} &= -\frac{1}{4}\left((x_2)_3 - (x_2)_4 + (x_2)_2 - (x_2)_1\right) - \frac{\eta_2}{4}\left((x_2)_3 - (x_2)_4 - ((x_2)_2 - (x_2)_1)\right) \\
e_{21} &= -\frac{1}{4}\left((x_1)_4 - (x_1)_1 + (x_1)_3 - (x_1)_2\right) - \frac{\eta_1}{4}\left((x_1)_3 - (x_1)_2 - ((x_1)_4 - (x_1)_1)\right) \\
e_{22} &= \frac{1}{4}\left((x_1)_2 - (x_1)_1 + (x_1)_3 - (x_1)_4\right) + \frac{\eta_2}{4}\left((x_1)_3 - (x_1)_2 - ((x_1)_4 - (x_1)_1)\right)
\end{aligned}
$$
$$(7.173)$$

The corresponding determinant of the transformation Jacobian is then expressed as

$$
\begin{aligned}
\det J &= e_{11}e_{22} - e_{12}e_{21} \\
&= \frac{1}{8}\left(((x_1)_3 - (x_1)_1)((x_2)_4 - (x_2)_2) + ((x_1)_2 - (x_1)_4)((x_2)_3 - (x_2)_1)\right) \\
&+ \frac{\eta_1}{8}\left(((x_1)_2 - (x_1)_1)((x_2)_3 - (x_2)_4) - ((x_1)_3 - (x_1)_4)((x_2)_2 - (x_2)_1)\right) \\
&+ \frac{\eta_2}{8}\left(((x_1)_3 - (x_1)_2)((x_2)_4 - (x_2)_1) - ((x_1)_4 - (x_1)_1)((x_2)_3 - (x_2)_2)\right) \quad (7.174)
\end{aligned}
$$

For a rectangular element, $((x_2)_4 - (x_2)_1) = ((x_2)_3 - (x_2)_2)$, $((x_1)_2 - (x_1)_1) = ((x_1)_3 - (x_1)_4)$, $((x_2)_2 = (x_2)_1)$, $(\ (x_2)_3 = (x_2)_4)$, $((x_1)_4 = (x_1)_1)$, $(\ (x_1)_3 = (x_1)_2)$ and this determinant becomes

$$\det J = \frac{1}{4}((x_1)_2 - (x_1)_1)((x_2)_3 - (x_2)_2) \tag{7.175}$$

which correctly returns the ratio of the areas of the rectangular element and of the corresponding transformed square element in the (η_1, η_2) space.

7.9 Element Integrals, Gaussian Quadratures, Optimal Metric Data

Owing to the local element-based coordinate transformation, the integrals in the transformed weak statement (7.135) are evaluated in the transformed elements, which only involve constant limits of integration. The integrals that feature in the Euler and Navier Stokes equations are now presented for one- and multi-dimensional formulations.

7.9.1 One-Dimensional Integrals

The five integrals involved in the finite element group discretization of any linear or non-linear convection diffusion equation in conservation-law form lead to the five element matrices

$$\mathcal{M} \equiv \left\{ \int_{-1}^{+1} w_i w_j \left(\frac{dx}{d\eta}\right) d\eta \right\}, \quad \mathcal{C} \equiv \left\{ \int_{-1}^{+1} w_i \frac{dw_j}{d\eta} d\eta \right\}, \quad \mathcal{C}_w \equiv \left\{ \int_{-1}^{+1} \frac{dw_i}{d\eta} w_i d\eta \right\}$$

$$\mathcal{D} \equiv \left\{ \int_{-1}^{+1} \frac{dw_i}{d\eta} \frac{dw_j}{d\eta} \frac{d\eta}{\frac{dx}{d\eta}} \right\}, \quad \mathcal{DV} \equiv \left\{ \int_{-1}^{+1} \frac{dw_i}{d\eta} w_s \frac{dw_j}{d\eta} \frac{d\eta}{\frac{dx}{d\eta}} \right\} \tag{7.176}$$

The first three integrals only involve polynomials of "η" and as such they can be integrated exactly. The fourth and fifth integral involve the ratio of polynomials, which can be exactly integrated, although for higher order elements the result becomes involved. In this case, it may be convenient to evaluate the integral using the Gaussian quadrature procedure presented in the next section.

For a linear element of length Δx_e, these integrals are exactly integrated and the results are

$$\mathcal{M} = \frac{\Delta x_e}{2} \begin{bmatrix} \frac{2}{3} & \frac{1}{3} \\ \frac{1}{3} & \frac{2}{3} \end{bmatrix}, \quad \mathcal{C} = \begin{bmatrix} -\frac{1}{2} & \frac{1}{2} \\ -\frac{1}{2} & \frac{1}{2} \end{bmatrix}$$

$$\mathcal{D} = \frac{2}{\Delta x_e} \begin{bmatrix} \frac{1}{2} & -\frac{1}{2} \\ -\frac{1}{2} & \frac{1}{2} \end{bmatrix}, \quad \mathcal{DV} = \frac{2}{\Delta x_e} \begin{bmatrix} \left[\frac{1}{4},\frac{1}{4}\right] & -\left[\frac{1}{4},\frac{1}{4}\right] \\ -\left[\frac{1}{4},\frac{1}{4}\right] & \left[\frac{1}{4},\frac{1}{4}\right] \end{bmatrix} \tag{7.177}$$

with the matrix $\mathcal{C}_w$ obtained as the transpose of $\mathcal{C}$. These matrices rapidly lead to the discrete counterpart of a prescribed time dependent partial differential equation. With reference to Figure 7.25, the discrete equation for node "i" sequentially assembles the contributions from entries $(2,1)$, $(2,2)$, $(1,1)$ and $(1,2)$ in these matrices. Following this sequence, the discretization on a non-uniform grid of the weak statement

$$\int_\Omega w \left(\frac{\partial q}{\partial t} + \frac{\partial f(q)}{\partial x}\right) d\Omega + \int_\Omega \frac{1}{\text{Re}} \frac{\partial w}{\partial x} \frac{\partial q}{\partial x} d\Omega = 0 \tag{7.178}$$

leads to the finite element statement

$$\sum_{e=1}^{N_e}\left(\int_{\Omega_e} w_i\left(w_j\frac{\partial q_j}{\partial t}+\frac{\partial w_j}{\partial x}f(q_j)\right)d\Omega+\int_{\Omega_e}\frac{1}{Re}\frac{\partial w_i}{\partial x}\frac{\partial w_j}{\partial x}q_jd\Omega\right)=0 \tag{7.179}$$

which becomes

$$\int_{\Omega_{e_{i-1}}} w_2\left(w_j\frac{\partial q_j}{\partial t}+\frac{\partial w_j}{\partial x}f(q_j)\right)d\Omega+\int_{\Omega_{e_{i-1}}}\frac{1}{Re}\frac{\partial w_2}{\partial x}\frac{\partial w_j}{\partial x}q_jd\Omega$$

$$+\int_{\Omega_{e_{i+1}}} w_1\left(w_j\frac{\partial q_j}{\partial t}+\frac{\partial w_j}{\partial x}f(q_j)\right)d\Omega+\int_{\Omega_{e_{i+1}}}\frac{1}{Re}\frac{\partial w_1}{\partial x}\frac{\partial w_j}{\partial x}q_jd\Omega=0 \tag{7.180}$$

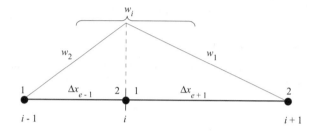

Figure 7.25: Assembly of Two Non-Uniform Elements

The corresponding discrete equation is thus expressed as

$$\frac{\Delta x_{e_{i-1}}}{2}\left(\frac{1}{3}\dot{q}_{i-1}+\frac{2}{3}\dot{q}_i\right)+\frac{\Delta x_{e_{i+1}}}{2}\left(\frac{2}{3}\dot{q}_i+\frac{1}{3}\dot{q}_{i+1}\right)$$

$$+\left(-\frac{1}{2}f\left(q_{i-1}\right)+\frac{1}{2}f\left(q_{i+1}\right)\right)+\frac{2}{Re\Delta x_{e_{i-1}}}\left(\frac{1}{2}q_{i-1}-\frac{1}{2}q_i\right)+\frac{2}{Re\Delta x_{e_{i+1}}}\left(-\frac{1}{2}q_i+\frac{1}{2}q_{i+1}\right)=0 \tag{7.181}$$

The discrete equations for other second order systems and different elements result from the same sequential assembly process.

7.9.2 Gaussian Quadratures

The one-dimensional integral of a continuous function

$$\mathcal{I}=\int_{-1}^{+1}f(\eta)d\eta \tag{7.182}$$

may be accurately evaluated using the Gaussian quadrature

$$\int_{-1}^{+1}f(\eta)d\eta\simeq\sum_{i=1}^{N_G}G_if(H_i) \tag{7.183}$$

where N_G stands for the number of Gauss abscissae, with G_i as well as H_i, $1 \leq i \leq N_G$, respectively denoting the Gauss weights and abscissae. The order of accuracy of this quadrature equals $2N_G - 1$, so that a formula with 3 abscissae exactly integrates a polynomial of 5-th degree. For such a formula, the Gauss weights and abscissae are

$$
G_1 = \frac{5}{9}, \qquad G_2 = \frac{8}{9}, \quad G_3 = \frac{5}{9}
$$
$$
H_1 = -\sqrt{\frac{3}{5}}, \quad H_2 = 0, \quad H_3 = \sqrt{\frac{3}{5}} \tag{7.184}
$$

Multi-dimensional integrals may be similarly evaluated via an iterate use of the quadrature along each coordinate. A double integral of a continuous function, for instance, is evaluated as

$$
\int_{-1}^{+1} \int_{-1}^{+1} f(\eta_1, \eta_2) \, d\eta_1 d\eta_2 \simeq \sum_{i=1}^{N_G} \sum_{j=1}^{N_G} G_i G_j f(H_i, H_j) \tag{7.185}
$$

which is the formula that may be generally employed to evaluate the two-dimensional finite element integrals for arbitrary elements.

7.9.3 Multi-Dimensional Integrals and Optimal Metric Data

With implied summation on a repeated subscript index, the integrals involved in the multi-dimensional finite element group discretization of any linear or non-linear 2nd-order convection diffusion equation in conservation-law form lead to five element matrices. For rectangular elements, the corresponding metric data are constant and these matrices are expressed as

$$
\mathcal{M} \equiv \left\{ \int_{\Omega_e} w_i w_j d\Omega \right\} \det J, \quad \mathcal{C}_\ell \equiv \left\{ \int_{\Omega_e} w_i \frac{\partial w_j}{\partial \eta_k} d\Omega \right\} e_{\ell k}, \quad \mathcal{C}_{w\ell} \equiv \left\{ \int_{\Omega_e} \frac{\partial w_i}{\partial \eta_k} w_j \, d\Omega \right\} e_{\ell k}
$$
$$
\mathcal{D}_{\ell m} \equiv \left\{ \int_{\Omega_e} \frac{\partial w_i}{\partial \eta_k} \frac{\partial w_j}{\partial \eta_n} d\Omega \right\} \frac{e_{\ell k} e_{mn}}{\det J}, \quad \mathcal{DV}_{\ell s m} \equiv \left\{ \int_{\Omega_e} \frac{\partial w_i}{\partial \eta_k} w_s \frac{\partial w_j}{\partial \eta_n} d\Omega \right\} \frac{e_{\ell k} e_{mn}}{\det J} \tag{7.186}
$$

For linear elements, the matrices of the integrals of the basis functions and their derivatives are expressed as

$$
\left\{ \int_{\Omega_e} w_i w_j d\Omega \right\} = \frac{1}{9} \begin{bmatrix} 4 & 2 & 1 & 2 \\ 2 & 4 & 2 & 1 \\ 1 & 2 & 4 & 2 \\ 2 & 1 & 2 & 4 \end{bmatrix}
$$

$$
\left\{ \int_{\Omega_e} w_i \frac{\partial w_j}{\partial \eta_1} d\Omega \right\} = \frac{1}{6} \begin{bmatrix} -2 & 2 & 1 & -1 \\ -2 & 2 & 1 & -1 \\ -1 & 1 & 2 & -2 \\ -1 & 1 & 2 & -2 \end{bmatrix}, \quad \left\{ \int_{\Omega_e} w_i \frac{\partial w_j}{\partial \eta_2} d\Omega \right\} = \frac{1}{6} \begin{bmatrix} -2 & -1 & 1 & 2 \\ -1 & -2 & 2 & 1 \\ -1 & -2 & 2 & 1 \\ -2 & -1 & 1 & 2 \end{bmatrix}
$$

$$\left\{\int_{\Omega_e} \frac{\partial w_i}{\partial \eta_1} \frac{\partial w_j}{\partial \eta_1} d\Omega\right\} = \frac{1}{6} \begin{bmatrix} 2 & -2 & -1 & 1 \\ -2 & 2 & 1 & -1 \\ -1 & 1 & 2 & -2 \\ 1 & -1 & -2 & 2 \end{bmatrix}, \quad \left\{\int_{\Omega_e} \frac{\partial w_i}{\partial \eta_1} \frac{\partial w_j}{\partial \eta_2} d\Omega\right\} = \frac{1}{4} \begin{bmatrix} 1 & 1 & -1 & -1 \\ -1 & -1 & 1 & 1 \\ -1 & -1 & 1 & 1 \\ 1 & 1 & -1 & -1 \end{bmatrix}$$

$$\left\{\int_{\Omega_e} \frac{\partial w_i}{\partial \eta_2} \frac{\partial w_j}{\partial \eta_1} d\Omega\right\} = \frac{1}{4} \begin{bmatrix} 1 & -1 & -1 & 1 \\ 1 & -1 & -1 & 1 \\ -1 & 1 & 1 & -1 \\ -1 & 1 & 1 & -1 \end{bmatrix}, \quad \left\{\int_{\Omega_e} \frac{\partial w_i}{\partial \eta_2} \frac{\partial w_j}{\partial \eta_2} d\Omega\right\} = \frac{1}{6} \begin{bmatrix} 2 & 1 & -1 & -2 \\ 1 & 2 & -2 & -1 \\ -1 & -2 & 2 & 1 \\ -2 & -1 & 1 & 2 \end{bmatrix}$$

$$\tag{7.187}$$

For elements of arbitrary non-degenerate shape in Cartesian space, the metric data are no longer constant, and the corresponding five element matrices are expressed as

$$\mathcal{M} \equiv \left\{\int_{\Omega_e} w_i w_j \det J d\Omega\right\}, \quad \mathcal{C}_\ell \equiv \left\{\int_{\Omega_e} w_i e_{\ell k} \frac{\partial w_j}{\partial \eta_k} d\Omega\right\}, \quad \mathcal{C}_{w\ell} \equiv \left\{\int_{\Omega_e} e_{\ell k} \frac{\partial w_i}{\partial \eta_k} w_j\, d\Omega\right\}$$

$$\mathcal{D}_{\ell m} \equiv \left\{\int_{\Omega_e} e_{\ell k} \frac{\partial w_i}{\partial \eta_k} e_{mn} \frac{\partial w_j}{\partial \eta_n} \frac{d\Omega}{\det J}\right\}, \quad DV_{\ell sm} \equiv \left\{\int_{\Omega_e} e_{\ell k} \frac{\partial w_i}{\partial \eta_k} w_s e_{mn} \frac{\partial w_j}{\partial \eta_n} \frac{d\Omega}{\det J}\right\} \tag{7.188}$$

These integrals may be evaluated by way of Gaussian quadratures, as described in the previous section.

For linear elements of arbitrary shape in Cartesian space, the first three matrices have been exactly calculated, and the remaining two have been accurately evaluated using metric data evaluated at optimal points in (η_1, η_2) coordinates.

With subscript point coordinates indicating evaluation at the indicated point, the exact components of the mass matrix $\mathcal{M}$ are expressed as

$$\mathcal{M}_{11} = \tfrac{4}{9} \det J\big|_{-\frac{1}{2},-\frac{1}{2}} \qquad \mathcal{M}_{21} = \tfrac{2}{9} \det J\big|_{0,-\frac{1}{2}}$$

$$\mathcal{M}_{12} = \tfrac{2}{9} \det J\big|_{0,-\frac{1}{2}} \qquad \mathcal{M}_{22} = \tfrac{4}{9} \det J\big|_{\frac{1}{2},-\frac{1}{2}}$$

$$\mathcal{M}_{13} = \tfrac{1}{9} \det J\big|_{0,0} \qquad \mathcal{M}_{23} = \tfrac{2}{9} \det J\big|_{\frac{1}{2},0}$$

$$\mathcal{M}_{14} = \tfrac{2}{9} \det J\big|_{-\frac{1}{2},0} \qquad \mathcal{M}_{24} = \tfrac{1}{9} \det J\big|_{0,0}$$

$$\mathcal{M}_{31} = \tfrac{1}{9} \det J\big|_{0,0} \qquad \mathcal{M}_{41} = \tfrac{2}{9} \det J\big|_{-\frac{1}{2},0}$$

$$\mathcal{M}_{32} = \tfrac{2}{9} \det J\big|_{\frac{1}{2},0} \qquad \mathcal{M}_{42} = \tfrac{1}{9} \det J\big|_{0,0}$$

$$\mathcal{M}_{33} = \tfrac{4}{9} \det J\big|_{\frac{1}{2},\frac{1}{2}} \qquad \mathcal{M}_{43} = \tfrac{2}{9} \det J\big|_{0,\frac{1}{2}}$$

$$\mathcal{M}_{34} = \tfrac{2}{9} \det J\big|_{0,\frac{1}{2}} \qquad \mathcal{M}_{44} = \tfrac{4}{9} \det J\big|_{-\frac{1}{2},\frac{1}{2}}$$

$$\tag{7.189}$$

Similarly, the exact components of the first-order-derivative matrix $\mathcal{C}_\ell$ are expressed as

$$\mathcal{C}_{\ell 11} = -\tfrac{1}{3}e_{\ell 1}\big|_{-\frac{1}{3},-\frac{1}{3}} - \tfrac{1}{3}e_{\ell 2}\big|_{-\frac{1}{3},-\frac{1}{3}} \qquad \mathcal{C}_{\ell 21} = -\tfrac{1}{3}e_{\ell 1}\big|_{\frac{1}{3},-\frac{1}{3}} - \tfrac{1}{6}e_{\ell 2}\big|_{\frac{1}{3},-\frac{1}{3}}$$

$$\mathcal{C}_{\ell 12} = \tfrac{1}{3}e_{\ell 1}\big|_{-\frac{1}{3},-\frac{1}{3}} - \tfrac{1}{6}e_{\ell 2}\big|_{-\frac{1}{3},-\frac{1}{3}} \qquad \mathcal{C}_{\ell 22} = \tfrac{1}{3}e_{\ell 1}\big|_{\frac{1}{3},-\frac{1}{3}} - \tfrac{1}{3}e_{\ell 2}\big|_{\frac{1}{3},-\frac{1}{3}}$$

$$\mathcal{C}_{\ell 13} = \tfrac{1}{6}e_{\ell 1}\big|_{-\frac{1}{3},-\frac{1}{3}} + \tfrac{1}{6}e_{\ell 2}\big|_{-\frac{1}{3},-\frac{1}{3}} \qquad \mathcal{C}_{\ell 23} = \tfrac{1}{6}e_{\ell 1}\big|_{\frac{1}{3},-\frac{1}{3}} + \tfrac{1}{3}e_{\ell 2}\big|_{\frac{1}{3},-\frac{1}{3}}$$

$$\mathcal{C}_{\ell 14} = -\tfrac{1}{6}e_{\ell 1}\big|_{-\frac{1}{3},-\frac{1}{3}} + \tfrac{1}{3}e_{\ell 2}\big|_{-\frac{1}{3},-\frac{1}{3}} \qquad \mathcal{C}_{\ell 24} = -\tfrac{1}{6}e_{\ell 1}\big|_{\frac{1}{3},-\frac{1}{3}} + \tfrac{1}{6}e_{\ell 2}\big|_{\frac{1}{3},-\frac{1}{3}}$$

$$\mathcal{C}_{\ell 31} = -\tfrac{1}{6}e_{\ell 1}\big|_{\frac{1}{3},\frac{1}{3}} - \tfrac{1}{6}e_{\ell 2}\big|_{\frac{1}{3},\frac{1}{3}} \qquad \mathcal{C}_{\ell 41} = -\tfrac{1}{6}e_{\ell 1}\big|_{-\frac{1}{3},\frac{1}{3}} - \tfrac{1}{3}e_{\ell 2}\big|_{-\frac{1}{3},\frac{1}{3}}$$

$$\mathcal{C}_{\ell 32} = \tfrac{1}{6}e_{\ell 1}\big|_{\frac{1}{3},\frac{1}{3}} - \tfrac{1}{3}e_{\ell 2}\big|_{\frac{1}{3},\frac{1}{3}} \qquad \mathcal{C}_{\ell 42} = \tfrac{1}{6}e_{\ell 1}\big|_{-\frac{1}{3},\frac{1}{3}} - \tfrac{1}{6}e_{\ell 2}\big|_{-\frac{1}{3},\frac{1}{3}}$$

$$\mathcal{C}_{\ell 33} = \tfrac{1}{3}e_{\ell 1}\big|_{\frac{1}{3},\frac{1}{3}} + \tfrac{1}{3}e_{\ell 2}\big|_{\frac{1}{3},\frac{1}{3}} \qquad \mathcal{C}_{\ell 43} = \tfrac{1}{3}e_{\ell 1}\big|_{-\frac{1}{3},\frac{1}{3}} + \tfrac{1}{6}e_{\ell 2}\big|_{-\frac{1}{3},\frac{1}{3}}$$

$$\mathcal{C}_{\ell 34} = -\tfrac{1}{3}e_{\ell 1}\big|_{\frac{1}{3},\frac{1}{3}} + \tfrac{1}{6}e_{\ell 2}\big|_{\frac{1}{3},\frac{1}{3}} \qquad \mathcal{C}_{\ell 44} = -\tfrac{1}{3}e_{\ell 1}\big|_{-\frac{1}{3},\frac{1}{3}} + \tfrac{1}{3}e_{\ell 2}\big|_{-\frac{1}{3},\frac{1}{3}}$$

$$(7.190)$$

The matrix $\mathcal{C}_{w\ell}$ is then obtained as the transpose of $\mathcal{C}_\ell$.

Next consider the second-order-derivative integrals

$$\int_{\Omega_e} \frac{\partial w_i}{\partial x_\ell}\frac{\partial w_j}{\partial x_m}\,d\Omega = \int_{-1}^{1}\int_{-1}^{1}\frac{\partial w_i}{\partial \eta_k}\frac{\partial w_j}{\partial \eta_n}\frac{e_{\ell k}e_{mn}}{\det J}\,d\Omega$$

$$\int_{\Omega_e} \frac{\partial w_i}{\partial x_\ell}w_s\frac{\partial w_j}{\partial x_m}\,d\Omega = \int_{-1}^{1}\int_{-1}^{1}\frac{\partial w_i}{\partial \eta_k}w_s\frac{\partial w_j}{\partial \eta_n}\frac{e_{\ell k}e_{mn}}{\det J}\,d\Omega \qquad (7.191)$$

For a rectangular element, as noted previously, the metric-data expression $e_{\ell k}e_{mn}/\det J$ reverts to a constant and these integrals become

$$\int_{-1}^{1}\int_{-1}^{1}\frac{\partial w_i}{\partial \eta_k}\frac{\partial w_j}{\partial \eta_n}\frac{e_{\ell k}e_{mn}}{\det J}\,d\Omega = \frac{e_{\ell k}e_{mn}}{\det J}\int_{-1}^{1}\int_{-1}^{1}\frac{\partial w_i}{\partial \eta_k}\frac{\partial w_j}{\partial \eta_n}\,d\Omega$$

$$\int_{-1}^{1}\int_{-1}^{1}\frac{\partial w_i}{\partial \eta_k}w_s\frac{\partial w_j}{\partial \eta_n}\frac{e_{\ell k}e_{mn}}{\det J}\,d\Omega = \frac{e_{\ell k}e_{mn}}{\det J}\int_{-1}^{1}\int_{-1}^{1}\frac{\partial w_i}{\partial \eta_k}w_s\frac{\partial w_j}{\partial \eta_n}\,d\Omega \qquad (7.192)$$

For computational efficiency, a similar expression is employed for arbitrarily shaped elements, as

$$\int_{-1}^{1}\int_{-1}^{1}\frac{\partial w_i}{\partial \eta_k}\frac{\partial w_j}{\partial \eta_n}\frac{e_{\ell k}e_{mn}}{\det J}\,d\Omega \simeq \frac{\overline{e_{\ell k}}\,\overline{e_{mn}}}{\widetilde{\det J}}\int_{-1}^{1}\int_{-1}^{1}\frac{\partial w_i}{\partial \eta_k}\frac{\partial w_j}{\partial \eta_n}\,d\Omega$$

$$\int_{-1}^{1}\int_{-1}^{1}\frac{\partial w_i}{\partial \eta_k}w_s\frac{\partial w_j}{\partial \eta_n}\frac{e_{\ell k}e_{mn}}{\det J}\,d\Omega \simeq \frac{\overline{e_{\ell k}}\,\overline{e_{mn}}}{\widetilde{\det J}}\int_{-1}^{1}\int_{-1}^{1}\frac{\partial w_i}{\partial \eta_k}w_s\frac{\partial w_j}{\partial \eta_n}\,d\Omega \qquad (7.193)$$

where $\overline{e_{\ell k}}$, $\overline{e_{mn}}$, $\widetilde{\det J}$ denote metric data computed at optimal evaluation points in (η_1,η_2) coordinates, for both sets of integrals, from the specifications

$$\int_{-1}^{1}\int_{-1}^{1}\frac{\partial w_i}{\partial \eta_k}\frac{\partial w_j}{\partial \eta_n}\frac{e_{\ell k}e_{mn}}{\det J}\,d\Omega \simeq \overline{e_{\ell k}}\int_{-1}^{1}\int_{-1}^{1}\frac{\partial w_i}{\partial \eta_k}\frac{\partial w_j}{\partial \eta_n}\frac{e_{mn}}{\det J}\,d\Omega$$

$$\simeq \frac{\overline{e_{\ell k}}}{\det \widetilde{J}} \int_{-1}^{1}\int_{-1}^{1} \frac{\partial w_i}{\partial \eta_k}\frac{\partial w_j}{\partial \eta_n} e_{mn}\, d\Omega = \frac{\overline{e_{\ell k}}\,\overline{\overline{e_{mn}}}}{\det \widetilde{J}} \int_{-1}^{1}\int_{-1}^{1} \frac{\partial w_i}{\partial \eta_k}\frac{\partial w_j}{\partial \eta_n}\, d\Omega \qquad (7.194)$$

According to these specifications, the optimal metric data $\overline{\overline{e_{mn}}}$ are determined from the equality

$$\overline{\overline{e_{mn}}} \int_{-1}^{1}\int_{-1}^{1} \frac{\partial w_i}{\partial \eta_k}\frac{\partial w_j}{\partial \eta_n}\, d\Omega = \int_{-1}^{1}\int_{-1}^{1} \frac{\partial w_i}{\partial \eta_k}\frac{\partial w_j}{\partial \eta_n} e_{mn}\, d\Omega \qquad (7.195)$$

The remaining optimal metric data $\overline{e_{\ell k}}$ and $\det \widetilde{J}$ respectively originate from contracting $(x_m)_j$, that is array of nodal values of the x_m coordinate, with the leftmost and rightmost sides of (7.194) and equating this contraction with the exact result. The contraction with the leftmost side leads to

$$(x_m)_j \int_{-1}^{1}\int_{-1}^{1} \frac{\partial w_i}{\partial \eta_k}\frac{\partial w_j}{\partial \eta_n}\frac{e_{\ell k}e_{mn}}{\det \widetilde{J}}\, d\Omega = \int_{-1}^{1}\int_{-1}^{1} \frac{\partial w_i}{\partial \eta_k}e_{\ell k}\left(\frac{\partial w_j(x_m)_j}{\partial \eta_n}\frac{e_{mn}}{\det \widetilde{J}}\right) d\Omega$$

$$= \int_{-1}^{1}\int_{-1}^{1} \frac{\partial w_i}{\partial \eta_k}e_{\ell k}\frac{\partial x_m}{\partial x_m}d\Omega = \int_{-1}^{1}\int_{-1}^{1} \frac{\partial w_i}{\partial \eta_k}e_{\ell k}\, d\Omega = \overline{e_{\ell k}}\int_{-1}^{1}\int_{-1}^{1} \frac{\partial w_i}{\partial \eta_k}\, d\Omega \qquad (7.196)$$

which determines $\overline{e_{\ell k}}$. The contraction with the rightmost side leads to

$$(x_m)_j \frac{\overline{e_{\ell k}}\,\overline{\overline{e_{mn}}}}{\det \widetilde{J}} \int_{-1}^{1}\int_{-1}^{1} \frac{\partial w_i}{\partial \eta_k}\frac{\partial w_j}{\partial \eta_n}\, d\Omega = \overline{e_{\ell k}}\int_{-1}^{1}\int_{-1}^{1} \frac{\partial w_i}{\partial \eta_k}\, d\Omega \qquad (7.197)$$

For every "i" and "k", this equality determines $\widetilde{\det J}$ as

$$\widetilde{\det J}_{i\,k}\ \overline{e_{\ell k}}\int_{-1}^{1}\int_{-1}^{1} \frac{\partial w_i}{\partial \eta_k}\, d\Omega = \overline{e_{\ell k}}\ \overline{\overline{e_{mn}}}\int_{-1}^{1}\int_{-1}^{1} \frac{\partial w_i}{\partial \eta_k}\frac{\partial w_j(x_m)_j}{\partial \eta_n}\, d\Omega$$

$$= \overline{e_{\ell k}}\ \overline{\overline{e_{mn}}}\int_{-1}^{1}\int_{-1}^{1} \frac{\partial w_i}{\partial \eta_k}(-1)^{m+n}e_{3-m,3-n}\, d\Omega = \overline{e_{\ell k}}\ (-1)^{m+n}\overline{\overline{e_{3-m,3-n}}}\int_{-1}^{1}\int_{-1}^{1} \frac{\partial w_i}{\partial \eta_k}\, d\Omega \qquad (7.198)$$

Denote the optimal evaluations points as $\boldsymbol{\eta}_0 \equiv (0,0)$, $\{\boldsymbol{\eta}_i\} \equiv \{\boldsymbol{\eta}_1,\boldsymbol{\eta}_2,\boldsymbol{\eta}_3,\boldsymbol{\eta}_4\}$, with $\boldsymbol{\eta}_1 \equiv (-\frac{1}{3},-\frac{1}{3})$, $\boldsymbol{\eta}_2 \equiv (\frac{1}{3},-\frac{1}{3})$, $\boldsymbol{\eta}_3 \equiv (\frac{1}{3},\frac{1}{3})$, $\boldsymbol{\eta}_4 \equiv (-\frac{1}{3},\frac{1}{3})$. With these points, the optimal metric data from (7.195), (7.196), (7.198) are

$$\overline{\overline{e_{mn}}}\int_{-1}^{1}\int_{-1}^{1} \frac{\partial w_i}{\partial \eta_k}\frac{\partial w_j}{\partial \eta_n}\, d\Omega = \int_{-1}^{1}\int_{-1}^{1} \frac{\partial w_i}{\partial \eta_k}\frac{\partial w_j}{\partial \eta_n} e_{mn}\, d\Omega \Rightarrow \begin{array}{l} \overline{\overline{e_{mn}}} = e_{mn}\big|_{\boldsymbol{\eta}_0}\ ,\ k = n \\[2mm] \overline{\overline{e_{mn}}} = e_{mn}\big|_{\boldsymbol{\eta}_i}\ ,\ k \neq n \end{array} \qquad (7.199)$$

$$\overline{e_{\ell s}}\int_{-1}^{1}\int_{-1}^{1} \frac{\partial w_i}{\partial \eta_k}\, d\Omega = \int_{-1}^{1}\int_{-1}^{1} \frac{\partial w_i}{\partial \eta_k}e_{\ell s}\, d\Omega \Rightarrow \begin{array}{l} \overline{e_{\ell s}} = e_{\ell s}\big|_{\boldsymbol{\eta}_0}\ ,\ s = k \\[2mm] \overline{e_{\ell s}} = e_{\ell s}\big|_{\boldsymbol{\eta}_i}\ ,\ s = 3-k \end{array} \qquad (7.200)$$

and

$$\begin{array}{rcl} \widetilde{\det J}_{i1} &=& e_{11}\big|_{\boldsymbol{\eta}_0}\, e_{22}\big|_{\boldsymbol{\eta}_i} - e_{21}\big|_{\boldsymbol{\eta}_0}\, e_{12}\big|_{\boldsymbol{\eta}_i} \\[2mm] \widetilde{\det J}_{i2} &=& e_{11}\big|_{\boldsymbol{\eta}_i}\, e_{22}\big|_{\boldsymbol{\eta}_0} - e_{21}\big|_{\boldsymbol{\eta}_i}\, e_{12}\big|_{\boldsymbol{\eta}_0} \end{array} \qquad (7.201)$$

These optimal metric data also satisfy the following constraints

$$\sum_{j=1}^{en} \int_{\Omega_e} \frac{\partial w_i}{\partial x_\ell} \frac{\partial w_j}{\partial x_m} \, d\Omega = 0, \quad \sum_{j=1}^{en} \int_{\Omega_e} \frac{\partial w_i}{\partial x_\ell} \frac{\partial w_j (x_{3-m})_j}{\partial x_m} \, d\Omega = 0 \tag{7.202}$$

that are satisfied by the exact expressions. In comparison to the iterated three-point Gaussian quadrature, these optimal matrices lead to the same order of accuracy, yet they remain at least four times more efficient computationally.

7.10   Implementation Sequence

This section discusses the salient issues in the practical implementation of the several finite element operations presented in the previous sections. The included pseudo "C" code demonstrates a representative implementation.

The finite element matrices presented in the previous sections feature not only in the Euler or Navier-Stokes equations, but also in any linear or non-linear time-dependent partial differential system that employs the group discretization within first and second order derivatives. This recognition makes such matrices akin to discrete operators and justifies forming these matrices within a dedicated subroutine, separately from the discrete equations. Employing this strategy, a structured finite element code may be rapidly adapted either to the solution of distinct equations systems or to the use of different elements, by only modifying the corresponding subroutine. To exemplify this process, consider the discretization of the weak statement for the two-dimensional characteristics-bias Euler system

$$\frac{\partial q}{\partial t} - \frac{\partial}{\partial x_\ell} \left(\varepsilon \psi a_\ell \frac{\partial q}{\partial t} \right) + \frac{\partial f_\ell}{\partial x_\ell}$$

$$-\frac{\partial}{\partial x_\ell} \left(\varepsilon \psi \left(c \left(\alpha a_\ell a_m + \alpha^N a_\ell^N a_m^N \right) \frac{\partial q}{\partial x_m} + a_\ell \left(\frac{\partial f_m^q}{\partial x_m} + \delta \frac{\partial f_m^p}{\partial x_m} \right) \right) \right) = 0 \tag{7.203}$$

The Cartesian and curvilinear form of the weak statement for this characteristic-bias system can be expressed as

$$\int_\Omega w \frac{\partial q}{\partial t} \, d\Omega + \int_\Omega \frac{\partial w}{\partial x_\ell} \varepsilon \psi a_\ell \frac{\partial q}{\partial t} \, d\Omega + \int_\Omega w \frac{\partial f_\ell}{\partial x_\ell} \, d\Omega$$

$$+ \int_\Omega \frac{\partial w}{\partial x_\ell} \left(\varepsilon \psi \left(c \left(\alpha a_\ell a_m + \alpha^N a_\ell^N a_m^N \right) \frac{\partial q}{\partial x_m} + a_\ell \left(\frac{\partial f_m^q}{\partial x_m} + \delta \frac{\partial f_m^p}{\partial x_m} \right) \right) \right) d\Omega = 0 \tag{7.204}$$

and

$$\int_\Omega w \frac{\partial q}{\partial t} \det J \, d\Omega + \int_\Omega e_{\ell k} \frac{\partial w}{\partial \eta_k} \varepsilon \psi a_\ell \frac{\partial q}{\partial t} \, d\Omega + \int_\Omega w e_{\ell k} \frac{\partial f_\ell}{\partial \eta_k} \, d\Omega$$

$$+ \int_\Omega e_{\ell k} \frac{\partial w}{\partial \eta_k} \left(\varepsilon \psi \left(c \left(\alpha a_\ell a_m + \alpha^N a_\ell^N a_m^N \right) \frac{\partial q}{\partial \eta_n} + \frac{\partial f_j^q}{\partial \eta_n} + \delta \frac{\partial f_j^p}{\partial \eta_n} \right) \right) \frac{e_{mn}}{\det J} d\Omega = 0 \tag{7.205}$$

The insertion in this statement of the nodal basis function w_i and the discrete arrays for q, f_ℓ, f_ℓ^q, f_ℓ^p leads to the finite element weak statement

$$\sum_{e=1}^{N_e} \left(\int_{\Omega_e} w_i w_j \det J \, d\Omega \frac{\partial}{\partial t} \left(q_j \right) + \int_{\Omega_e} e_{\ell k} \frac{\partial w_i}{\partial \eta_k} w_j \varepsilon \psi a_\ell \, d\Omega \frac{\partial}{\partial t} \left(q_j \right) + \int_{\Omega_e} w_i e_{\ell k} \frac{\partial w_j}{\partial \eta_k} \, d\Omega \left(f_{\ell_j} \right) \right.$$

$$+ \sum_{e=1}^{N_e} \left(\int_{\Omega_e} e_{\ell k} \frac{\partial w_i}{\partial \eta_k} \varepsilon \psi \, c \left(\alpha a_\ell \, a_m + \alpha^N a_\ell^N a_m^N \right) \frac{\partial w_j}{\partial \eta_n} \frac{e_{mn}}{\det J} \, d\Omega \, (q_j) \right)$$

$$+ \sum_{e=1}^{N_e} \left(\int_{\Omega_e} a_\ell e_{\ell k} \frac{\partial w_i}{\partial \eta_k} \varepsilon \psi \frac{\partial w_j}{\partial \eta_n} \frac{e_{mn}}{\det J} \, d\Omega \, \left(f_{m_j}^q \right) + \int_{\Omega_e} a_\ell e_{\ell k} \frac{\partial w_i}{\partial \eta_k} \varepsilon \psi \delta \frac{\partial w_j}{\partial \eta_n} \frac{e_{mn}}{\det J} d\Omega \, \left(f_{m_j}^p \right) \right) = 0$$

$$(7.206)$$

By way of the symbolic operators

$$\left. \frac{\widetilde{\partial}}{\partial t} \right|_{\Omega_{e_{ij}}} \equiv \int_{\Omega_e} w_i w_j \det J \, d\Omega \frac{\partial}{\partial t}, \quad \left. \frac{\widetilde{\partial^2}}{\partial x_\ell \partial t} \right|_{\Omega_{e_{ij}}} \equiv - \int_{\Omega_e} e_{\ell k} \frac{\partial w_i}{\partial \eta_k} w_j \, d\Omega \frac{\partial}{\partial t}$$

$$\left. \frac{\widetilde{\partial}}{\partial x_\ell} \right|_{\Omega_{e_{ij}}} \equiv \int_{\Omega_e} w_i e_{\ell k} \frac{\partial w_j}{\partial \eta_k} \, d\Omega, \quad \left. \frac{\widetilde{\partial^2}}{\partial x_\ell x_m} \right|_{\Omega_{e_{ij}}} \equiv - \int_{\Omega_e} \frac{\partial w_i}{\partial \eta_k} \frac{\partial w_j}{\partial \eta_n} \frac{e_{\ell k} e_{mn}}{\det J} \, d\Omega \qquad (7.207)$$

this statement may also be expressed in the following symbolic form

$$\sum_{e=1}^{N_e} \left(\left. \frac{\widetilde{\partial}}{\partial t} \right|_{\Omega_{e_{ij}}} (q_j) - (\varepsilon \psi a_\ell)_{\Omega_e} \left. \frac{\widetilde{\partial^2}}{\partial x_\ell \partial t} \right|_{\Omega_{e_{ij}}} (q_j) + \left. \frac{\widetilde{\partial}}{\partial x_\ell} \right|_{\Omega_{e_{ij}}} \left(f_{\ell_j} \right) \right)$$

$$- \sum_{e=1}^{N_e} \left(\left(\varepsilon \psi \, c \left(\alpha a_\ell \, a_m + \alpha^N a_\ell^N a_m^N \right) \right)_{\Omega_e} \left. \frac{\widetilde{\partial^2}}{\partial x_\ell x_m} \right|_{\Omega_{e_{ij}}} (q_j) \right)$$

$$- \sum_{e=1}^{N_e} \left((\varepsilon \psi a_\ell)_{\Omega_e} \left. \frac{\widetilde{\partial^2}}{\partial x_\ell x_m} \right|_{\Omega_{e_{ij}}} \left(f_{m_j}^q \right) + (\varepsilon \psi a_\ell \delta)_{\Omega_e} \left. \frac{\widetilde{\partial^2}}{\partial x_\ell x_m} \right|_{\Omega_{e_{ij}}} \left(f_{m_j}^p \right) \right) = 0 \qquad (7.208)$$

Remaining valid for any (q_j), (f_{ℓ_j}), $(f_{\ell_j}^q)$, $(f_{\ell_j}^p)$, this expression significantly resembles the original differential equation (7.203). Using the symbolic-operator concept, any other time dependent second-order partial differential equation is cast as a similar expression. Accordingly, a subroutine for this expression receives the symbolic operators as input and calculates this expression by way of a programming formula that mirrors the original equation, as further exemplified in the `EqJac` subroutine in the appended pseudo "C" program.

A structured finite element code requires the Cartesian-coordinate arrays `x1[]`, `x2[]`, the solution array `q[]`, and element-connectivity array `elnods[][]`. For each element `ie`, this connectivity array provides the global number `ig = elnods[ie][in]` of each node `in` within the element. These numbers are required to retrieve from the corresponding global arrays `x1[]`, `x2[]`, `q[]`, the Cartesian coordinates `xe1[]`, `xe2[]` and values `qe[]` of the solution variables at the nodes of the element.

The included pseudo "C" program has the chief purpose of showing a sample sequence of operations within a finite element program that solves time dependent partial differential equations. As this program demonstrates, the coordinates are needed to form the transformation metric data and the symbolic operators as the matrices `ma[][]`, `dx1[][]`, `dx2[][]`, as well as the "weak" statement first-order derivatives `dwx1[][]`, `dwx2[][]` and second-order derivatives `dx1x1[][]`, `dx1x2[][]`, `dx2x1[][]`, `dx2x2[][]`. Along with these matrices, the values of the nodal solution are needed to form the equation array `ff[]` and Jacobian arrays `bb[][]`, `aa[][]` of the system to be solved. These local equations and Jacobians are then stored in the global matrix band `a[][]` and rhs array `f[]` for the element-by-element assembly of each discrete nodal equation.

```
/*----------------------------------------------------------*/
/* For all the grid elements, this core subroutine forms */
/* the element matrices, discrete equations as well as   */
/* Jacobians  and global matrix band and rhs array       */
/*----------------------------------------------------------*/


/*----------------------------------------------------------*/
/*      Loop over grid elements                          */
/*      nelem = total number of elements                 */

for ( ie = 1; ie <= nelem; ie = ie + 1 ) {

    /*-------------------------------------------------------*/
    /*   Collect element nodal coordinates               */
    /*      nodel = number of nodes in element           */

    for( je = 1; je <= nodel ; je = je + 1 ) {

        ig  = elnods[ ie ][ 1 ] ) ;
        xe1[ je ] = x1[ ig ] ;
        xe2[ je ] = x2[ ig ] ;
    }
    /*-------------------------------------------------------*/


    /*-------------------------------------------------------*/
    /*      Determine element matrices and               */
    /*      centroidal parameters                        */

    finelm( xe1, xe2, ma, dx1, dx2, dwx1, dwx2, \
                dx1x1, dx1x2, dx2x1, dx2x2 ) ;

    cenpar( elnods, ie, xe1, xe2, q, par ) ;
    /*-------------------------------------------------------*/


    /*-------------------------------------------------------*/
    /*      Loop over nodal equations                    */

    for ( in = 1; in <= nodel; in = in + 1 ) {


        /*---------------------------------------------------*/
        /* For each global node "ig" form the array     */
        /* of pointers "iv[]" that contains the         */
        /* addresses of the corresponding degrees of    */
        /* freedom at the node                          */
```

```
        ig = elnods[ ie ][ in ] ;
        VarPoint( nvar, ig, iv ) ;
/*-------------------------------------------------*/


/*-------------------------------------------------*/
/*        Loop over nodal variables                 */

    for ( jn = 1; jn <= nodel; jn = jn + 1 ) {

        jg = elnods[ ie ][ jn ] ;
        VarPoint( nvar, jg, jv ) ;

        /*-----------------------------------------*/
        /* Collect nodal variables;                */
        /*  nvar = no. of variables per node       */

        for( je = 1; je <= nvar ; je = je + 1 ) {

            qe[ je ] = q[ jv[ 1 ] ] ;

        }
        /*-----------------------------------------*/

        /*-----------------------------------------*/
        /*        Form discrete equations           */

        EqJac(  ma[ in ][ jn ],    dx1[ in ][ jn ], \
              dx2[ in ][ jn ],   dwx1[ in ][ jn ], \
             dwx2[ in ][ jn ], dx1x1[ in ][ jn ], \
            dx1x2[ in ][ jn ], dx2x1[ in ][ jn ], \
            dx2x2[ in ][ jn ],                    \
              xe1, xe2, t, par, qe, ff, aa, bb ) ;
        /*-----------------------------------------*/

        /*-----------------------------------------*/
        /*  Newton's Iteration Steps.              */
        /*  In these steps, "k[]" is the global    */
        /*  array of solution variations for the   */
        /*  implicit Runge-Kutta integration       */
        /*  "aldt" is the product of "Dt" and the  */
        /*  Runge-Kutta parameter "alpha"          */

        for ( i = 1 ; i <= nvar ; i = i + 1 ){
            ff[ i ] = ff[ i ] * Dt ;

            for ( j = 1 ; j <= nvar ; j = j + 1 ){
```

```
                    ff[ i ] = ff[ i ] - \
                            bb[ i ][ j ] * k[ jv[ j ] ] ;
                    aa[ i ][ j ] = \
                    aa[ i ][ j ] * aldt + bb[ i ][ j ] ;
                }
            }
            /*----------------------------------------*/

            /*----------------------------------------*/
            /*  Insert element-based rhs              */
            /*    and Jacobian matrix in              */
            /*    global rhs and matrix band.         */
            /*    nband = semi-bandwidth              */

            for ( i = 1 ; i <= nvar ; i = i + 1 ){
                f[ iv[ i ] ] = f[ iv[ i ] ] + ff[ i ] ;
            }

            for ( i = 1 ; i <= nvar ; i = i + 1 ){
                for ( j = 1 ; j <= nvar ; j = j + 1 ){
                    ib = jv[ j ] - iv[ i ] + nband ;
                    a[ iv[i] ][ ib ] = a[ iv[i] ][ ib ] + \
                                            aa[ i ][ j ] ;
                }
            }
            /*----------------------------------------*/

        }
        /*--------------------------------------------*/

    }
    /*------------------------------------------------*/
}
return ;
/*----------------------------------------------------*/


/*--------------------------------------------------------*/
/* This subroutine forms the discrete system and          */
/* corresponding Jacobians. Notice how the symbolic        */
/* operators lead to an expression for "f[]" that          */
/* resembles the original partial differential equations */
/*--------------------------------------------------------*/

/*--------------------------------------------------------*/

void EqJac( long double ma,    long double dx1,    \
```

```
                long double dx2,    long double dwx1,  \
                long double dwx2,   long double dx1x1, \
                long double dx1x2,  long double dx2x1, \
                long double dx2x2,                     \
                long double xe1[],  long double xe2[], \
                long double t,      long double par[], \
                long double qe[],   long double f[],   \
                long double a[][ MNVAR ], long double b[][ MNVAR ] ){

                long double eps, psi, c, alf, alfN ;
                long double del, a1, a2, aN1, aN2  ;

                long double r, m1, m2, E ;
                long double p, pr, pm1, pm2, pE ;

/*-------------------------------------------------------*/
/*      Extract centroidal parameters                  */

    eps  = par[  1 ] ;
    psi  = par[  2 ] ;
      c  = par[  3 ] ;
    alf  = par[  4 ] ;
    alfN = par[  5 ] ;
    del  = par[  6 ] ;
     a1  = par[  7 ] ;
     a2  = par[  8 ] ;
    aN1  = par[  9 ] ;
    aN2  = par[ 10 ] ;
/*-------------------------------------------------------*/


/*-------------------------------------------------------*/
/*          Extract nodal variables                    */

    r  = qe[ 1 ] ;
    m1 = qe[ 2 ] ;
    m2 = qe[ 3 ] ;
    E  = qe[ 4 ] ;


/*-------------------------------------------------------*/


/*-------------------------------------------------------*/
/*   x1-Component of the linear - momentum equation   */
/*   and its Jacobians                                */



    /*-------------------------------------------------------*/
    /*    Calculate Pressure and its Jacobians            */
```

```
PresJac( r, m1, m2, E, &p, &pr, &pm1, &pm1, &pE ) ;

/*------------------------------------------------*/


b[ 2 ][ 1 ] = 0.0 ;
b[ 2 ][ 2 ] = ma - eps * psi * ( a1 * dwx1 + a2 * dwx2 )   ;
b[ 2 ][ 3 ] = 0.0 ;
b[ 2 ][ 4 ] = 0.0 ;

f[ 2 ] = -dx1 * m1 * m1 / r -dwx1 * p -dx2 * m2 * m1 / r +       \
           eps * psi * c *                                       \
         ( dx1x1 * ( alf * a1 * a1 + alfN * aN1 * aN1 )       + \
           dx1x2 * ( alf * a1 * a2 + alfN * aN1 * aN2 )       + \
           dx2x1 * ( alf * a2 * a1 + alfN * aN2 * aN1 )       + \
           dx2x2 * ( alf * a2 * a2 + alfN * aN2 * aN2 ) ) * r + \
           eps * psi *                                           \
         (      ( dx1x1 * a1 + dx2x1 * a2 ) * ( m1 * m1 / r ) + \
           del * ( dx1x1 * a1 + dx2x1 * a2 ) *   p             + \
                ( dx1x2 * a1 + dx2x2 * a2 ) * ( m2 * m1 / r ) ) ;

a[ 2 ][ 1 ] = -dx1 * ( - m1 * m1 / ( r * r ) ) - dwx1 * pr   - \
               dx2 * ( - m2 * m1 / ( r * r ) )               + \
             eps * psi * c *                                   \
           ( dx1x1 * ( alf * a1 * a1 + alfN * aN1 * aN1 )     + \
             dx1x2 * ( alf * a1 * a2 + alfN * aN1 * aN2 )     + \
             dx2x1 * ( alf * a2 * a1 + alfN * aN2 * aN1 )     + \
             dx2x2 * ( alf * a2 * a2 + alfN * aN2 * aN2 ) )   + \
             eps * psi *                                       \
           ( ( dx1x1 * a1 + dx2x1 * a2 ) * ( -m1 * m1 / ( r * r ) ) + \
             ( dx1x2 * a1 + dx2x2 * a2 ) * ( -m2 * m1 / ( r * r ) ) + \
        del * ( dx1x1 * a1 + dx2x1 * a2 ) *   pr ) ;


a[ 2 ][ 2 ] = -dx1 * ( 2.0 * m1 / r ) -dwx1 * pm1             - \
               dx2 * ( m2 / r )                                + \
             eps * psi *                                        \
           ( ( dx1x1 * a1 + dx2x1 * a2 ) * ( 2.0 * m1 / r )    + \
             ( dx1x2 * a1 + dx2x2 * a2 ) * (     m2 / r    )   + \
        del * ( dx1x1 * a1 + dx2x1 * a2 ) *   pm1 ) ;

a[ 2 ][ 3 ] =                             -dwx1 * pm2          - \
               dx2 * ( m1 / r )                                + \
             eps * psi *                                        \
           ( ( dx1x2 * a1 + dx2x2 * a2 ) * (     m1 / r    )   + \
        del * ( dx1x1 * a1 + dx2x1 * a2 ) *   pm2 ) ;
```

```
    a[ 2 ][ 4 ] =                              -dwx1 * pE              + \
         eps * psi *                                                    \
         ( del * ( dx1x1 * a1 + dx2x1 * a2 ) *    pE ) ;
/*-----------------------------------------------------*/

   return ; }
/*-----------------------------------------------------*/
```

Following the formation of the equations and Jacobians, the program can then determine the solution at the subsequent Newton iteration and / or time level. This is the program structure that has generated the essentially non-oscillatory results discussed in the following chapters.

Chapter 8

Non-Linearly Stable Implicit Runge-Kutta Time Integrations

The non-linear ordinary differential equations resulting from a spatial discretization of a set of partial differential equations, along with appropriate boundary conditions, can be abridged as the ODE system

$$\mathcal{M}\frac{dQ(t)}{dt} = \mathcal{F}(t, Q(t)) \tag{8.1}$$

where $\mathcal{M}\frac{dQ(t)}{dt}$ corresponds to a coupling of time derivatives and $F(t, Q(t))$ represents all the remaining terms in the spatial discretization and boundary-condition set. This system is cast as

$$\frac{dQ(t)}{dt} = F(t, Q(t)) \tag{8.2}$$

where $F \equiv \mathcal{M}^{-1}\mathcal{F}$.

The implicit algorithms in this chapter also sole the non-linear algebraic-differential system

$$\begin{cases} 0 & = F_1(t, Q(t)) \\ \dfrac{dQ_2}{dt} & = F_2(t, Q(t)) \end{cases} \tag{8.3}$$

where Q_2 denotes a subset of Q and F_1 and F_2 denote two subsets of F, with $F_1 \cap F_2 = \emptyset$ and $F_1 \cup F_2 = F$. This system may be viewed as a differential system for Q_2 subject to constraints F_1 to determine both Q_1 and Q_2. A CFD example of this algebraic-differential system is the set of incompressible Euler and Navier-Stokes equations, which feature no time derivatives in the continuity equation; in this example, Q_1 corresponds to pressure and Q_2 to velocity, with F_1 and F_2 respectively correlating with the continuity and linear-momentum equations.

For CFD simulations with local mesh refinements and resolution of viscous boundary layers, the spatial-discretization ODE system becomes extremely stiff. In this situation, the solution of such a system consists of components varying very rapidly in time that are superimposed to components that evolve more slowly. This is precisely the overall solution character that complicates numerical integration. The presence of disparate length and time scales in a flow then exacerbates this stiffness phenomenon. Hence, to achieve an accurate simulation of steady and especially time-dependent viscous flows this book presents a class of implicit Runge-Kutta (IRK) numerical time integration algorithms that are proven both high order accurate and non-linearly unconditionally stable for the stiff discrete Navier-Stokes equations.

8.1   Runge-Kutta Algorithm

Let "n" and "s" respectively denote a time level and the number of Runge- Kutta stages. A Runge-Kutta algorithm for system (8.1) can be cast as

$$Q_{n+1} - Q_n = \sum_{i=1}^{s} b_i K_i, \;\; \mathcal{M}K_i = \Delta t \cdot \mathcal{F}\left(t_n + c_i \Delta t, Q_n + \sum_{j=1}^{s} a_{ij} K_j\right) \tag{8.4}$$

the corresponding algorithm for systems (8.2), (8.3) becomes

$$Q_{n+1} - Q_n = \sum_{i=1}^{s} b_i K_i, \;\; PK_i = \Delta t \cdot F\left(t_n + c_i \Delta t, Q_n + \sum_{j=1}^{s} a_{ij} K_j\right) \tag{8.5}$$

where K_i's, $1 \leq i \leq s$, denote the Runge-Kutta arrays and P denotes either the identity matrix for (8.2), or a diagonal matrix for (8.3), with a "0" on the diagonal element corresponding to each equation in F_1 and a "1" on the diagonal element corresponding to each equation in F_2. The parameters b_i, c_i, a_{ij}, $1 \leq i, j \leq s$ are constant Runge-Kutta coefficients. The coefficients c_i and b_i satisfy the constraints

$$\sum_{i=1}^{s} b_i = 1, \;\; c_i = \sum_{j=s}^{s} a_{ij} \tag{8.6}$$

The coefficients b_i and a_{ij} are arranged in a column array and a square matrix as

$$b \equiv \{b_i\}, \;\; A \equiv \{a_{ij}\} \tag{8.7}$$

The coefficients a_{ij} determine whether the algorithm is explicit or implicit. The algorithm is explicit when $a_{ij} = 0$ for $j \geq i$; in this case the Runge-Kutta arrays K_i are sequentially computed without having to solve systems of equations. The algorithm is implicit when some $a_{ij} \neq 0$ for $j \geq i$; in this case all Runge-Kutta arrays are coupled with one another and are determined via the second expression in (8.4), (8.5), which are now systems of equations. A special kind of implicit process is the diagonally implicit algorithm, in which $a_{ij} = 0$ for $j > i$; this specification signifies that the Runge-Kutta arrays K_i are implicitly determined after one another.

As (8.4), (8.5) indicate this algorithm remains valid for non-autonomous system, given the explicit dependence upon time. This important feature allows using time dependent boundary conditions while maintaining accuracy and stability. As a result, this algorithm advantageously allows dependable simulations of transient and time- dependent flows for time dependent upstream and downstream boundary conditions.

This time integration algorithm can be made intrinsically stable independently of the stability of the spatial discretization. Furthermore, the stability analysis for this time integration remains separate from the stability analysis of the spatial discretization. Consequently, the stability of a CFD algorithm that employs (8.4), (8.5) only depends on the stability of the spatial discretization. This approach remarkably simplifies algorithm development since the associated stability analysis no longer requires simultaneous coupled analysis of the discretization and time-stepping algorithm.

This algorithm thus integrates the Euler and Navier-Stokes equations in a fully coupled fashion. Owing to its accuracy and stability properties, the implicit Runge-Kutta procedure allows dependable simulations of realistic steady and unsteady subsonic, transonic, and supersonic flows.

8.2 Non-Linear Stability

The stability of a numerical time integration procedure remains subtly different from the stability of a time-discrete solution. Since the numerical time integration procedure leads to an approximation of the solution of a time-continuum differential system, the resulting discrete solution should mirror the stability of the time-continuum solution. A numerical time integration procedure, therefore, is defined stable when the evolution of the generated discrete solution emulates the evolution of the time-continuum solution. In other words, a numerical time integration is stable when the discretization process does not lead to an increase in the solution energy. If the time-continuum solution decays, so should the discrete solution; if the time-continuum solution evolves like a packet of waves of different frequencies, so should the discrete solution, to the extent allowed by the accuracy of the algorithm; if the time-continuum solution grows, the discrete solution should also emulate such a trend. The goal of assessing the stability of the discrete solution can rapidly become as demanding as determining the stability of the time-continuum solution, which is a particularly challenging. On the other hand, it becomes comparatively easier as well as practical to assess the stability of a procedure.

The concept of non-linear stability fundamentally refers to the capability of a time-integration procedure, in this case the Runge-Kutta equations (8.4), (8.5), to generate a time-discrete solution Q_n that emulates the time evolution of the time-continuum solution $Q \equiv q^h$ for the infinite time interval and general non-linear rhs $F(\cdot, \cdot)$.

Algorithm (8.4), (8.5) satisfies a non-linear energy stability condition, stating that the time-discretization does not contribute an increase in an energy rate of change over an average energy rate of change of the time-continuum system. This condition can be expressed as

$$\frac{\|Q_{n+1}\|^2 - \|Q_n\|^2}{\Delta t} \leq \overline{\frac{d\|Q\|^2}{dt}} \tag{8.8}$$

where $\|.\|$ denotes an inner-product norm, and overbar signifies an average over the time interval $(t, t+\Delta t)$. An energy rate of change of the discrete system, therefore, does not exceed an associated average energy rate of change of the time continuum system. Algorithm (8.4), (8.5) can satisfy (8.8) regardless of the size of the time step Δt, which implies unconditional energy stability.

To describe how (8.4), (8.5) satisfy (8.8), express the energy associated with the time continuum solution Q of (8.2) as

$$\|Q\|^2 = (Q, \ Q) \tag{8.9}$$

The rate of change of this solution energy becomes a function of Q and the *rhs* of (8.2) as

$$\frac{d\|Q\|^2}{dt} = 2\left(Q, \ \frac{dQ}{dt}\right) = 2\,(Q, \ F) \tag{8.10}$$

An average energy rate can then be expressed as

$$\overline{\frac{d\|Q\|^2}{dt}} \equiv \sum_{i=1}^{s} b_i \frac{d\|Q\|^2}{dt}\bigg|_{Q=Q^i} = 2\sum_{i=1}^{s} b_i\left(Q^i, \ F(t^i, Q^i)\right) \tag{8.11}$$

Since (8.11) has to represent a meaningful average, the weights b_i, $1 \leq i \leq s$ must be non-negative, which leads to the constraints

$$b_i \geq 0, \quad 1 \leq i \leq s \tag{8.12}$$

Based on (8.4), (8.5), the array Q^i in (8.11) is defined as

$$Q^i \equiv Q_n + \sum_{j=1}^{s} a_{ij} K_j \tag{8.13}$$

which corresponds to a field at the intermediate time level $t^i \equiv t + c_i \Delta t \in (t, t+\Delta t)$. Together with (8.4), (8.5), this form for Q^i leads to the expressions

$$K_i = \Delta t F(t^i, Q^i) \equiv \Delta t F_i, \quad Q_n - Q^i = -\sum_{j=1}^{s} a_{ij} K_j = -\Delta t \sum_{j=1}^{s} a_{ij} F_j \tag{8.14}$$

The Runge-Kutta algorithm (8.4), (8.5) generates for the *lhs* of (8.8) the result

$$\frac{\|Q_{n+1}\|^2 - \|Q_n\|^2}{\Delta t} =$$
$$2\sum_{i=1}^{s} b_i\left(Q^i, \ F(t^i, Q^i)\right) - \Delta t \sum_{i,j=1}^{s} (b_i a_{ij} + b_j a_{ji} - b_i b_j)\left(F(t^i, Q^i), \ F(t^j, Q^j)\right) \tag{8.15}$$

which applies to completely general non-linear expressions $F(\cdot, \cdot)$. This result features a discrete energy rate at the *lhs* and an average time-continuum energy rate at the *rhs*. Employing this result, the energy stability condition (8.8) then translates into the quadratic-form inequality

$$-\Delta t \sum_{i,j=1}^{s} (b_i a_{ij} + b_j a_{ji} - b_i b_j)\left(F(t^i, Q^i), \ F(t^j, Q^j)\right) \leq 0 \tag{8.16}$$

With $\{\tilde{b}_{ik}\}$ denoting the diagonal matrix with diagonal entries $\{b_i\}$, this inequality can be unconditionally satisfied for any positive time step Δt when the symmetric energy matrix

$$E \equiv \left\{\tilde{b}_{ik} a_{kj} + \tilde{b}_{jk} a_{ki} - b_i b_j\right\} \tag{8.17}$$

is non-negative definite. Since this matrix E is symmetric, it will be non-negative definite when all of its principal minors remain non-negative. For an implicit algorithm it is possible to determine Runge-Kutta coefficients b_i and a_{ij} to satisfy this condition independently of Δt. Hence, an implicit Runge-Kutta algorithm can be unconditionally non-linearly energy stable. Specific Runge-Kutta coefficients that satisfy this unconditional stability constraint are presented in Section 8.4.4, 8.5. For an explicit algorithm, $a_{ij} = 0$ for $j \geq i$ and in this case the first principal minor of E equals $-b_1^2$. This minor is negative, hence an explicit Runge-Kutta algorithm cannot be unconditionally non-linearly energy stable. However, if the time-continuum system is dissipative, hence $2(Q, F) \leq 0$, then the condition $\|Q_{n+1}\|^2 - \|Q_n\|^2 \leq 0$ can still be met by limiting the size of Δt. On the other hand if the time-continuum system is not dissipative, then the condition $\|Q_{n+1}\|^2 - \|Q_n\|^2 \leq 0$ cannot be met. The discrete solution Q_n will grow, but so does the time continuum solution, hence an accuracy limit on the size of the time step Δt will generate a discrete solution that increases at a rate comparable to the rate of the time-continuum solution.

The energy result (8.15) is obtained by calculating the square of the norm $\|Q_{n+1}\|$ as

$$
\begin{aligned}
\|Q_{n+1}\|^2 &= \|Q_n + \sum_{i=1}^{s} b_i K_i\|^2 = \left(Q_n + \sum_{i=1}^{s} b_i K_i, Q_n + \sum_{j=1}^{s} b_j K_j\right) \\
&= (Q_n, Q_n) + \sum_{i=1}^{s} b_i (K_i, Q_n) + \sum_{j=1}^{s} b_j (Q_n, K_j) + \sum_{i,j=1}^{s} b_i b_j (K_i, K_j) \\
&= \|Q_n\|^2 + \Delta t \sum_{i=1}^{s} b_i \left(F_i, Q^i\right) + \Delta t \sum_{i=1}^{s} b_i \left(F_i, Q_n - Q^i\right) \\
&\quad + \Delta t \sum_{j=1}^{s} b_j \left(Q^j, F_j\right) + \Delta t \sum_{j=1}^{s} b_j \left(Q_n - Q^j, F_j\right) + \Delta t^2 \sum_{i,j=1}^{s} b_i b_j (F_i, F_j) \\
&= \|Q_n\|^2 + 2\Delta t \sum_{i=1}^{s} b_i \left(Q^i, F_i\right) \\
&\quad - \Delta t^2 \sum_{i,j=1}^{s} b_i a_{ij} (F_i, F_j) - \Delta t^2 \sum_{i,j=1}^{s} b_j a_{ji} (F_i, F_j) + \Delta t^2 \sum_{i,j=1}^{s} b_i b_j (F_i, F_j) \\
&= \|Q_n\|^2 + 2\Delta t \sum_{i=1}^{s} b_i \left(Q^i, F_i\right) - \Delta t^2 \sum_{i,j=1}^{s} (b_i a_{ij} + b_j a_{ji} - b_i b_j)(F_i, F_j) \quad (8.18)
\end{aligned}
$$

which leads to the energy result (8.15). In terms of the Runge-Kutta arrays, this energy expression leads to the normalized energy variation

$$
\frac{\|Q_{n+1}\|^2 - \|Q_n\|^2}{\|Q_n\|^2} = 2\frac{\sum_{i=1}^{s} b_i (Q^i, K_i)}{\|Q_n\|^2} - \frac{\sum_{i,j=1}^{s} (b_i a_{ij} + b_j a_{ji} - b_i b_j)(K_i, K_j)}{\|Q_n\|^2} \quad (8.19)
$$

This variation exposes the time-discretization expression

$$
D_D = -\frac{\sum_{i,j=1}^{s} (b_i a_{ij} + b_j a_{ji} - b_i b_j)(K_i, K_j)}{\|Q_n\|^2} \quad (8.20)
$$

which corresponds to dissipation for an unconditionally energy stable algorithm. The larger this dissipation becomes, the faster Q_n will converge to a steady state solution.

Any IRK algorithm that satisfies the energy-stability property is called algebraically stable. Its significance for the CFD implicit algorithms, and the subject IRK procedure, is investigated in Section 8.7. In the derivation of results (8.15), (8.18), crucially, no hypothesis was necessary on the structure of the residual $F(\cdot, \cdot)$ and/or the semi-discretization employed. Therefore, (8.4), (8.5) becomes unconditionally energy stable not just for a traditional linear model problem, but for the full semi-discrete non-linear Navier-Stokes equations on arbitrary structured/unstructured meshes.

8.3   Linear Stability

This type of stability investigates the stability of (8.4), (8.5) as applied to the linear system

$$\frac{dQ}{dt} = -\lambda Q \tag{8.21}$$

where λ denotes a positive eigenvalue. System (8.21) exemplifies a dissipative system, in that the associated energy rate of change remains negative

$$\frac{d\|Q\|^2}{dt} = 2(Q, F) = -2\lambda(Q, Q) \le 0 \tag{8.22}$$

The time continuum solution $Q = Q(t)$ then monotonically decreases and it is thus desirable that the discrete solution Q_n behaves similarly. Accordingly, the magnitude of the "amplification" ratio $\|Q_{n+1}\|/\|Q_n\|$ should remain less than one. Based on definition (8.13) of Q^i, this traditional amplification ratio can then be expressed as

$$\frac{Q_{n+1}}{Q_n} = 1 + \frac{\sum_{i=1}^{s} b_i K_i}{Q_n} \tag{8.23}$$

An energy stable integration is necessarily linearly stable, for in the linear ODE system case the *rhs* of (8.10) becomes negative for a dissipative F, which leads to an amplification factor less than one

$$\frac{\|Q_{n+1}\|^2 - \|Q_n\|^2}{\Delta t} \le 0 \;\Rightarrow\; \|Q_{n+1}\|^2 \le \|Q_n\|^2 \;\Rightarrow\; \frac{\|Q_{n+1}\|^2}{\|Q_n\|^2} \le 1 \tag{8.24}$$

Denote with I the $s \times s$ identity matrix and with u, for unity, a column array of size s with entries all equal to 1. The amplification ratio (8.23) may thus be expressed as

$$\frac{Q_{n+1}}{Q_n} = \frac{\det\left(I + \lambda\Delta t A - \lambda\Delta t \, u b^T\right)}{\det\left(I + \lambda\Delta t A\right)} \tag{8.25}$$

where A and b^T respectively denote the matrix of IRK coefficients and the transpose of the column array b in (8.7).

If the Runge-Kutta method is explicit, then $a_{ij} = 0$ for $j \ge i$ and A is a lower triangular matrix with zero's on its main diagonal. Accordingly, $\det(I + \lambda\Delta t A) = 1$ and (8.25) becomes a polynomial in $\lambda\Delta t$. In this case, linear stability for a dissipative system is conditional, with $\lambda\Delta t$ limited by the condition that (8.25) should be less than 1.

For an implicit Runge-Kutta algorithm, the amplification - ratio denominator $\det \left(I + \lambda \Delta t A \right)$ becomes a polynomial in $\lambda \Delta t$. In this case, the amplification ratio can remain less than 1 when its denominator does not vanish, which places linear-stability constraints on the coefficients a_{ij}. Naturally, a_{ij} satisfy these constraints if the algorithm is non-linearly energy stable, as shown in (8.24). For a diagonally implicit method, $a_{ij} = 0$ for $j > i$ and A is a lower triangular matrix with a_{ii}, $1 \leq i \leq s$, on its main diagonal. Hence, $\det \left(I + \lambda \Delta t A \right) = \prod_{i=1}^{s} \left(1 + \lambda \Delta t a_{ii} \right)$, which will not vanish for any $\Delta t \geq 0$ under the linear-stability condition

$$a_{ii} \geq 0, \quad 1 \leq i \leq s \tag{8.26}$$

With this condition, (8.25) will remain bounded.

8.4 IRK2: A Two-Stage Diagonally Implicit Algorithm

This section presents a two-stage diagonally implicit Runge-Kutta algorithm developed by this author. Specifically designed for the efficient numerical integration of stiff non-linear systems, this algorithm is second order accurate and non-linearly energy stable, with a linear amplification ratio that vanishes for both unbounded Jacobian eigenvalues and a finite value of the Courant number $C_{FL} = \lambda \Delta t$. For the numerical integration of the ODE system

$$\mathcal{M} \frac{dQ(t)}{dt} = \mathcal{F}\left(t, Q(t) \right) \tag{8.27}$$

this two-stage Runge-Kutta algorithm is

$$Q_{n+1} - Q_n = b_1 K_1 + b_2 K_2 \tag{8.28}$$
$$\mathcal{M} K_1 = \Delta t \cdot \mathcal{F}\left(t_n + c_1 \Delta t, Q_n + a_{11} K_1 \right) \tag{8.29}$$
$$\mathcal{M} K_2 = \Delta t \cdot \mathcal{F}\left(t_n + c_2 \Delta t, Q_n + a_{21} K_1 + a_{22} K_2 \right) \tag{8.30}$$

This algorithm is implicit because the entries in the arrays K_1 and K_2 remain coupled and are then computed by solving algebraic systems. Since the algorithm is diagonally implicit, K_1 is determined independently of K_2. Thus, given the solution Q_n at time t_n, K_1 is computed first, followed by K_2. The solution Q_{n+1} is then determined by way of (8.28).

Grossman et al. independently reported [73] that a linearized version of this diagonally implicit Runge-Kutta algorithm yielded accurate solutions for a thermo-chemical non-equilibrium shock tube flow analysis. They also found that the algorithm generated accurate solutions in approximately a factor of ten less of overall computational work in comparison to an explicit calculation with the same order of accuracy. The class of algorithms (8.28)-(8.30) has also generated accurate solutions for 2-D supersonic and equilibrium hypersonic flows. The following sections delineate the numerical calculation of the Runge-Kutta arrays K_i, $1 \leq i \leq 2$ and develop the equations for the determination of the Runge-Kutta coefficients b_1, b_2, a_{11}, a_{12}, a_{22}, for this algorithm.

8.4.1 Numerical Linear Algebra

The terminal numerical solution is then determined using Newton's method in linearized one step mode or non-linear form. For the implicit fully-coupled computation of the IRK arrays

K_i, Newton's method is cast as

$$\left[\mathcal{M} - a_{\underline{ii}}\Delta t \left(\frac{\partial \mathcal{F}}{\partial Q} \right)^p_{Q^p_i} \right] \left(K^{p+1}_i - K^p_i \right) = \Delta t \; \mathcal{F}\left(t_n + c_i \Delta t, Q^p_i\right) - \mathcal{M}K^p_i$$

$$Q^p_i \equiv Q_n + a_{i1}K^p_1 + a_{i2}K^p_2 \tag{8.31}$$

where $a_{ij} = 0$ for $j > i$, p is the iteration index, and $K^p_1 \equiv K_1$ for $i = 2$; for finite elements, the Jacobian

$$J_i(Q) \equiv \mathcal{M} - a_{\underline{ii}}\Delta t \left(\frac{\partial \mathcal{F}}{\partial Q} \right)^p_{Q^p_i} \tag{8.32}$$

then becomes a block sparse matrix. The initial estimate K^0_i can be conveniently set equal to the zero array, because, for $p = 0$, K^0_i may cancel out from (8.31). To show this result, let Q^{00}_i and $Q^{0\kappa}_i$ define the following arrays

$$Q^{00}_i \equiv Q_n + a_{21}\delta^2_i K_1, \;\; Q^{0\kappa}_i \equiv Q_n + (1 - \kappa)a_{21}\delta^2_i K_1 + \kappa \left(a_{i1}K^0_1 + a_{i2}K^0_2 \right) \tag{8.33}$$

where $0 \leq \kappa \leq 1$. For $p = 0$, the function $\mathcal{F}\left(t_n + c_i \Delta t, Q^0_i\right)$ in (8.31) may be cast as

$$\mathcal{F}\left(t_n + c_i \Delta t, Q^0_i\right) = \mathcal{F}\left(t_n + c_i \Delta t, Q^{00}_i\right) + a_{\underline{ii}}\left(\frac{\partial \mathcal{F}}{\partial Q} \right)^0_{Q^{0k}_i} K^0_i \tag{8.34}$$

Through this expansion, for $p = 0$, (8.31) becomes

$$\left[\mathcal{M} - a_{\underline{ii}}\Delta t \left(\frac{\partial \mathcal{F}}{\partial Q} \right)^0_{Q^{00}_i} \right] K^1_i = \Delta t \; \mathcal{F}\left(t_n + c_i \Delta t, Q^{00}_i\right)$$

$$+a_{\underline{ii}}\Delta t \left[\left(\frac{\partial \mathcal{F}}{\partial Q} \right)^0_{Q^{0k}_i} - \left(\frac{\partial \mathcal{F}}{\partial Q} \right)^0_{Q^0_i} \right] K^0_i + a_{\underline{ii}}\Delta t \left[\left(\frac{\partial \mathcal{F}}{\partial Q} \right)^0_{Q^0_i} - \left(\frac{\partial \mathcal{F}}{\partial Q} \right)^0_{Q^{00}_i} \right] K^1_i \tag{8.35}$$

For linear systems, the differences of Jacobians in this expression will vanish; for non-linear systems, these differences either approach zero towards steady state, when $K_i \rightarrow 0$, hence $Q^{0k}_i = Q^0_i = Q^{00}_i$, or, in any case, become higher-order terms that remain negligible in comparison to $\mathcal{F}\left(t_n + c_i \Delta t, Q^{00}_i\right)$, for a suitably constrained Δt. This expression, accordingly, reverts to (8.31) for $p = 0$ when $K^0_i \equiv 0$, which shows that this specification may be conveniently selected as an initial estimate. This estimate also beneficially provides the starting array for an iterative linear-algebra solver, which requires an admissible initial guess. When one Newton iteration only is executed for (8.31), within each time interval, Newton's method becomes akin to a classical direct linearized implicit solver. The linearized implicit "θ" schemes (e.g. Beam and Warming) are contained within (8.31) when only one Newton iteration is performed per time step, and the initial estimate for K_i is set to the zero array. For a convenient computer implementation, the terminal numerical solution of (8.31) for K_i can be obtained using available efficient linear algebra solvers, like a direct Gaussian solver and/or the Generalized Minimal Residual method (GMRES).

8.4.2   Stability Constraints

For algorithm (8.28)-(8.30), condition (8.12) becomes

$$b_1 \geq 0, \quad b_2 \geq 0 \tag{8.36}$$

The associated energy matrix E is

$$E = \begin{pmatrix} b_1(2a_{11} - b_1), & b_2(a_{21} - b_1) \\ b_2(a_{21} - b_1), & b_2(2a_{22} - b_2) \end{pmatrix} \tag{8.37}$$

and the stability condition that all of its principal minors must be non-negative translates into the constraints

$$
\begin{aligned}
&b_1(2a_{11} - b_1) \geq 0 \\
&b_1 b_2 (2a_{11} - b_1)(2a_{22} - b_2) - b_2^2(a_{21} - b_1)^2 \\
&= b_1 b_2 (4a_{22}a_{11} - 2a_{22}b_1 - 2a_{11}b_2 + b_1 b_2) - b_2^2(a_{21} - b_1)^2 \geq 0
\end{aligned}
\tag{8.38}
$$

The present IRK2 algorithm can also be made strongly linearly stable, which means that the amplification ratio for a linear system vanishes with increasing stiffness and/or solution component frequency. Consequently, the integration scheme linear dissipation increases with increasing stiffness. To determine the corresponding constraints on the Runge-Kutta coefficients, form the linear amplification ratio (8.25) for (8.28)-(8.30), which yields

$$\frac{Q_{n+1}}{Q_n} = \frac{\gamma(\lambda\Delta t)^2 + (a_{11} + a_{22} - b_1 - b_2)\lambda\Delta t + 1}{(1 + a_{11}\lambda\Delta t)(1 + a_{22}\lambda\Delta t)} \tag{8.39}$$

wherein the coefficient γ is

$$\gamma = a_{11}a_{22} - a_{22}b_1 - a_{11}b_2 + b_2 a_{21} \tag{8.40}$$

Upon setting $\gamma = 0$, this amplification ratio will approach zero as the denominator becomes larger and larger, hence

$$\lim_{\Delta t \to \infty} \frac{\|Q_{n+1}\|}{\|Q_n\|} = 0 \tag{8.41}$$

which signifies damping of high frequency modes. This amplification ratio also vanishes for a finite magnitude of $\lambda\Delta t$, as indicated in Section 8.4.5.

Because of the linear-stability constraints $a_{11} > 0$, $a_{22} > 0$, the roots

$$\lambda\Delta t_1 = -(a_{11})^{-1}, \qquad \lambda\Delta t_2 = -(a_{22})^{-1} \tag{8.42}$$

of the denominator in (8.39) are negative. Therefore the amplification ratio remains always bounded for any $\Delta t > 0$. Finally, $\|Q_{n+1}/Q_n\| \leq 1$, for any positive Δt.

8.4.3   Accuracy and B - Consistency / Convergence Constraints

The constraints on the IRK2 constants that guarantee second order accuracy in the classical sense are

$$b_1 + b_2 = 1, \quad b_1 a_{11} + b_2(a_{21} + a_{22}) = \frac{1}{2} \tag{8.43}$$

When equations (8.43) are satisfied, the following local time discretization error

$$\hat{l}_{n+1} = Q(t_{n+1}) - \hat{Q}_{n+1} \tag{8.44}$$

with

$$\hat{Q}_{n+1} = Q(t_n) + \Delta t(\ b_1 K_1(t_n, Q(t_n)\ ) + b_2 K_2(t_n, Q(t_n)\ )\ ) \tag{8.45}$$

is bounded by the inequality

$$\|\hat{l}_{n+1}\| \leq S\Delta t^3, \ \forall \Delta t \in (0, \Delta t_S] \tag{8.46}$$

where S and Δt_S in general depend upon the stiffness of the ODE system. For the stiff equations resulting from the spatial semi-discretization of the compressible viscous Navier-Stokes equations, inequality (8.46) is insufficient to ensure the expected accuracy for practical time steps Δt, since the constant S grows with increasing stiffness, while Δt_S may concurrently decrease.

The algorithm suitable for the numerical integration of the subject class of non-linear stiff ODE system should instead maintain a uniform order of accuracy irrespective of the stiffness magnitude. This is the essence of the B-consistency notion, which is mathematically characterized by the inequality

$$\|\hat{l}_{n+1}\| \leq D\Delta t^{p+1}, \ \forall \Delta t \in (0, \Delta t_1] \tag{8.47}$$

where the constants D and Δt_1 are totally independent of stiffness and exponent p is the order of consistency. For the IRK2 algorithm (8.28)-(8.30) this order is bounded by the inequality $1 \leq p \leq 2$ and the sufficient conditions for B-consistency of order one, at least, are the linear stability constraints (8.26)

$$a_{11} > 0, \quad a_{22} > 0 \tag{8.48}$$

The efficient numerical integration of system (8.2) demands that convergence rate not degrade with increasing stiffness. This is the allied concept of B-convergence which is formulated in terms of the global time discretization error

$$\hat{e}_{n+1} = Q(t_{n+1}) - Q_{n+1} \tag{8.49}$$

via the inequality

$$\|\hat{e}_{n+1}\| \leq C\Delta t^r, \ \forall \Delta t \in (0, \Delta t_2] \tag{8.50}$$

In estimate (8.50), the exponent r is the convergence order and satisfies the inequality $r \leq p+1$. Further, the constants C and Δt_2 do not depend on stiffness. For the present diagonal algorithm, inequality (8.50) is satisfied if, in addition to constraints (8.48), the algorithm is also non-linearly energy stable, which implies condition (8.12) on b_i, $1 \leq i \leq 2$.

8.4.4 Solution of the IRK Coefficient Equations

The derived expressions for the algorithm coefficients that guarantee non-linear energy stability, 2nd order accuracy, and B - consistency / convergence are

$$
\begin{cases}
b_1 + b_2 = 1 \\
b_1 a_{11} + b_2(a_{21} + a_{22}) = \frac{1}{2} \\
b_1 a_{22} + b_2(a_{11} - a_{21}) = a_{11} a_{22} \\
b_1 > 0, \quad b_2 > 0 \\
a_{11} > 0, \quad a_{22} > 0 \\
b_1(2a_{11} - b_1) \geq 0 \\
b_1 b_2(4a_{22}a_{11} - 2a_{22}b_1 - 2a_{11}b_2 + b_1 b_2) - b_2^2(a_{21} - b_1)^2 \geq 0
\end{cases}
\tag{8.51}
$$

In terms of a_{21} and a_{22}, the solution of the first three equations in this system

$$
b_1 = 1 - \frac{2a_{22}}{4(a_{21} + a_{22})(1 - a_{22}) - 2(1 - 2a_{22})}
$$

$$
b_2 = \frac{2a_{22}}{4(a_{21} + a_{22})(1 - a_{22}) - 2(1 - 2a_{22})}
$$

$$
a_{11} = \frac{1 - 2a_{22}}{2(1 - a_{22})}
\tag{8.52}
$$

In concert with the stability inequalities

$$
\begin{cases}
b_1 > 0, \quad b_2 > 0 \\
a_{11} > 0, \quad a_{22} > 0 \\
b_1(2a_1 - b_1) \geq 0 \\
b_1 b_2(4a_{22}a_{11} - 2a_{22}b_1 - 2a_{11}b_2 + b_1 b_2) - b_2^2(a_{21} - b_1)^2 \geq 0
\end{cases}
\tag{8.53}
$$

it is possible to find several numerical solutions for (8.52). Three such solutions are listed in Table 8.1 Note that for the present diagonally implicit IRK2 algorithm, the solutions K_1 and

Table 8.1: IRK2 Coefficients

	b_1	b_2	a_{11}	a_{21}	a_{22}
IRK21	$\dfrac{3 - \sqrt{3}}{4}$	$\dfrac{1 + \sqrt{3}}{4}$	$\dfrac{3 - \sqrt{3}}{6}$	$2 - \sqrt{3}$	$\dfrac{\sqrt{3} - 1}{2}$
IRK22	$\dfrac{3}{5}$	$\dfrac{2}{5}$	$\dfrac{1}{3}$	$\dfrac{1}{2}$	$\dfrac{1}{4}$
IRK23	$\dfrac{14}{29}$	$\dfrac{15}{29}$	$\dfrac{2}{7}$	$\dfrac{2}{5}$	$\dfrac{3}{10}$

K_2 for equations (8.29)-(8.30) do exist and are unique, since the coefficients b_1, b_2, a_{11}, and a_{22} are all positive. Of all two-stage diagonally implicit Runge-Kutta algorithms, IRK2 is the only algorithm that is simultaneously second-order accurate, non-linearly energy stable for stiff systems, and strongly linearly stable.

8.4.5   Accuracy and Stability Performance

The accuracy and stability of the IRK2 algorithms is assessed by comparing the variations of the time-continuum and discretization amplification ratio (8.23) and normalized energy variation (8.19). This comparison employs the model linear differential equation. As well known, the exact solution of the model linear differential statement

$$\frac{dQ}{dt} = -\lambda Q, \quad Q(0) = 1 \tag{8.54}$$

is

$$Q(t) = \exp(-\lambda t) \tag{8.55}$$

which leads to the time-continuum amplification ratio and normalized energy variation

$$\frac{Q(t + \Delta t)}{Q(t)} = \exp(-\lambda \Delta t), \quad \frac{Q^2(t + \Delta t) - Q^2(t)}{Q^2(t)} = \exp(-2\lambda \Delta t) - 1 \tag{8.56}$$

where the eigenvalue λ is positive. The IRK2 algorithms lead to the time-discretization amplification ratio and normalized energy variation

$$\frac{Q_{n+1}}{Q_n} = \frac{1 - (1 - a_{11} - a_{22})\lambda \Delta t}{(1 + a_{11}\lambda \Delta t)(1 + a_{22}\lambda \Delta t)} \tag{8.57}$$

$$\frac{\|Q_{n+1}\|^2 - \|Q_n\|^2}{\|Q_n\|^2} = 2\frac{\sum_{i=1}^{s} b_i (Q^i, K_i)}{\|Q_n\|^2} - \frac{\sum_{i,j=1}^{s} (b_i a_{ij} + b_j a_{ji} - b_i b_j) (K_i, K_j)}{\|Q_n\|^2} = D_S + D_D \tag{8.58}$$

where

$$D_S = \frac{2b_1 K_1}{Q_n}\left(1 + a_{11}\frac{K_1}{Q_n}\right) + \frac{2b_2 K_2}{Q_n}\left(1 + a_{21}\frac{K_1}{Q_n} + a_{22}\frac{K_2}{Q_n}\right)$$

$$D_D = -\left(2a_{11}b_1 - b_1^2\right)\frac{K_1^2}{Q_n^2} - 2\left(a_{21}b_2 - b_1 b_2\right)\frac{K_1 K_2}{Q_n^2} - \left(2a_{22}b_2 - b_2^2\right)\frac{K_2^2}{Q_n^2} \tag{8.59}$$

and

$$K_1 = -\frac{\lambda \Delta t Q_n}{1 + a_{11}\lambda \Delta t}, \quad K_2 = -\frac{\lambda \Delta t Q_n}{1 + a_{22}\lambda \Delta t} + \frac{a_{21}(\lambda \Delta t)^2 Q_n}{(1 + a_{11}\lambda \Delta t)(1 + a_{22}\lambda \Delta t)} \tag{8.60}$$

The closeness between the discrete and time-continuum amplification ratio and normalized energy variation measures the accuracy of the IRK2 algorithms. The magnitudes of the discretization amplification ratio and normalized energy variation then measure the stability degree of the IRK2 algorithms. Figure 8.1 compare the variations of the time-continuum and discretization amplification ratio and normalized energy variation for the IRK23 algorithm. The corresponding variations for IRK21 and IRK22 are similar, although, among these three algorithms, IRK23 is the one that most nearly maximizes the first stability expression in (8.38).

For moderate $\lambda \Delta t$, the time-continuum and discretization variations remain virtually indistinguishable from each other, which indicates accuracy. The departure between the

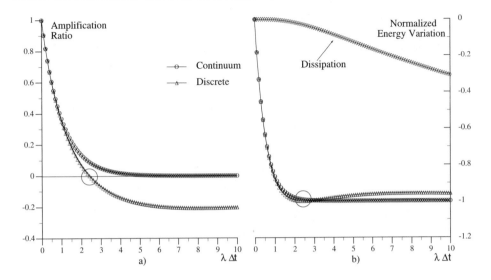

Figure 8.1: Stability Curves: a) Amplification Ratio, b) Normalized Energy Variation

time-continuum and discretization predictions begin at about $\lambda \Delta t = 1$, a well known reference, for $\lambda \Delta t = 1$ corresponds to a Courant number equal to one. Heed that the discrete amplification ratio vanishes for the zero-ratio values

$$\lambda \Delta t_0 = \frac{1}{1 - a_{11} - a_{22}} \quad \Rightarrow \quad \Delta t_0 = \frac{1}{\lambda(1 - a_{11} - a_{22})} \tag{8.61}$$

For the three IRK2 algorithms, the numerical values of the zero-ratio C_{FL} numbers are presented in Table 8.2 The significance of these results is that rapid convergence to steady

Table 8.2: Zero-Ratio CFL Numbers

	IRK21	IRK22	IRK23
$\lambda \Delta t_0$	$\dfrac{3 + \sqrt{3}}{2}$	$\dfrac{12}{5}$	$\dfrac{70}{29}$

state can be achieved when the C_{FL} number equals the zero-ratio magnitude.

Observe that for increasing $\lambda \Delta t$, the amplification ratio approaches nought. This variation corresponds to strong linear stability, which, of course, was enforced by design. The magnitude of the discretization dissipation D_D in (8.59) increases for growing $\lambda \Delta t$, as expected. The discretization normalized energy variation approaches the limit of -1, which coincides with the time-continuum limit. Note that this variation already nears this limit in the neighborhood of the zero-ratio value of $\lambda \Delta t$.

8.5 One- and Two-Stage Diagonally Implicit Algorithms

Various diagonally implicit algorithms emerge from specific choices of the Runge-Kutta coefficients. The following table lists sets of coefficients for several algorithms. The coefficients

Table 8.3: Runge-Kutta Coefficients

	b_1	b_2	a_{11}	a_{21}	a_{22}
"θ"	$1-\theta$	θ	0	$1-\theta$	θ
EU	1	0	1	0	0
TP	$\frac{1}{2}$	$\frac{1}{2}$	0	$\frac{1}{2}$	$\frac{1}{2}$
IRK1	$\frac{1}{2}$	$\frac{1}{2}$	$\frac{3-\sqrt{3}}{6}$	$2-\sqrt{3}$	$\frac{\sqrt{3}-1}{2}$
MP	1	0	$\frac{1}{2}$	0	0
IRK3	$\frac{1}{2}$	$\frac{1}{2}$	$\frac{3+\sqrt{3}}{6}$	$\frac{\sqrt{3}}{3}$	$\frac{3+\sqrt{3}}{6}$

in the first row correspond to the well-known "θ" implicit algorithm, which encompasses the backward Euler (EU) and Trapezoidal (TP) (or Crank-Nicolson) rules, of which the coefficients are listed in the second and third rows of the table. The fifth row corresponds to the one-stage, second-order, and non-linearly stable Middle-Point (MP) rule. IRK3 indicates a third-order algorithm discussed, while IRK1 corresponds to a first order algorithm whose coefficients can be used with those of IRK2 to implement the time step control strategy discussed in the following section.

8.6 Time Step Size Adjustment

For an explicit scheme, the step size is severely restricted by numerical stability rather than by accuracy considerations. The non-linearly absolutely stable IRK algorithm (8.4), (8.5) yields a bounded solution independent of the time step size. Therefore, the efficiency of these IRK schemes depend on a convenient method for adjusting the time step to meet a prescribed accuracy level. The following developments present a computationally convenient procedure for estimating the local truncation error and associated time step.

Denote with $\tilde{Q}_n$ and $\tilde{Q}_{n+1}$ the exact solution of the ODE system (8.2) at time stations n and $n+1$. Algorithm (8.28) can then be cast as

$$\tilde{Q}_{n+1} - \tilde{Q}_n = (b_1)_h K_1 + (b_2)_h K_2 + \mathcal{E} \tag{8.62}$$

$$\tilde{Q}_{n+1} - \tilde{Q}_n = (b_1)_l \, K_1 + (b_2)_l \, K_2 + \frac{c\mathcal{E}}{\Delta t} \tag{8.63}$$

where subscripts h and l denote Runge-Kutta coefficients corresponding to higher and lower order time-integration algorithms respectively, whereas $\mathcal{E}$ corresponds to an estimate of the local truncation error, with c a suitable scaling factor. Furthermore, it is possible to find sets of Runge Kutta coefficients c_i and a_{ij} so that the intermediate solutions K_i, $1 \le i \le 2$, are the same for expressions (8.62) and (8.63). Hence, no additional arrays must be computed to estimate Δt. Accordingly, subtracting (8.63) from (8.62) yields

$$0 = ((b_1)_h - (b_1)_l) \, K_1 + ((b_2)_h - (b_1)_l) \, K_2 - \mathcal{E} \left(\frac{c}{\Delta t} - 1 \right) \tag{8.64}$$

which leads to

$$\frac{\|c\| \cdot \|\mathcal{E}\|}{\Delta t} \le \| ((b_1)_h - (b_1)_l) \, K_1 + ((b_2)_h - (b_1)_l) \, K_2 \| + \|\mathcal{E}\| \tag{8.65}$$

Whence, a sufficient condition to estimate the time step Δt will be

$$\Delta t = \frac{\|c\| \cdot \|\mathcal{E}\|}{\| ((b_1)_h - (b_1)_l) \, K_1 + ((b_2)_h - (b_1)_l) \, K_2 \| + \|\mathcal{E}\|} \tag{8.66}$$

which depends on the time index n via K_i, and the now prescribed tolerance $\|\mathcal{E}\|$. Near steady state the denominator expression involving K_1 and K_2 becomes infinitesimal. Accordingly, Δt approaches $\|c\|$, which can be considered as an estimate of a terminal Δt. For a fixed $\|\mathcal{E}\|$ and away from steady state, the denominator in (8.66) will increase for rapidly varying solutions. In this case, (8.66) induces a corresponding decrease in the time step.

8.7 Theoretical Comparisons

Within the implicit Runge-Kutta framework, non-linear energy stability can be comparatively assessed for the CFD backwards Euler and trapezoidal (Crank-Nicolson) implicit algorithms. To derive these procedures from (8.28)-(8.30), consider the "θ" ODE algorithm family applied to (8.2) yielding

$$Q_{n+1} - Q_n = \Delta t(\ (1-\theta)F(t_n, Q_n) + \theta F(t_{n+1}, Q_{n+1}) \ ) \tag{8.67}$$

Equation (8.67) can be recast as

$$Q_{n+1} - Q_n = \Delta t(\ (1-\theta)K_1 + \theta K_2 \ ) \tag{8.68}$$
$$K_1 = F(t_n, Q_n) \tag{8.69}$$
$$K_2 = F(t_n + \Delta t, Q_n + (1-\theta)\Delta t K_1 + \theta \Delta t K_2) \tag{8.70}$$

and this form coincides with the structure of (8.28)-(8.30).

The corresponding coefficients b_1, b_2, and matrix E are

$$b_1 = 1 - \theta, \quad b_2 = \theta, \quad E = \begin{pmatrix} -(1-\theta)^2, & 0 \\ 0, & \theta^2 \end{pmatrix} \tag{8.71}$$

and the non-linear energy stability constraints become

$$(1 - \theta) \geq 0, \quad \theta \geq 0, \quad -(1 - \theta)^2 \geq 0, \quad -(1 - \theta)^2 \theta^2 \geq 0 \tag{8.72}$$

Accordingly, the backwards Euler ($\theta = 1$) algorithm is non-linearly energy stable, whereas the Crank-Nicolson rule ($\theta = \frac{1}{2}$) is not, even though the linear-stability constraints are identically satisfied by both algorithms. The dependence of numerical growth or decay upon non-linear energy stability can be inferred from the fundamental energy relation as applied to dissipative systems

$$\frac{\|Q_{n+1}\|^2 - \|Q_n\|^2}{\Delta t} =$$

$$2 \sum_{i=1}^{2} b_i \left(Q^i, \ F(t^i, Q^i) \right) - \Delta t \sum_{i,j=1}^{2} (b_i a_{ij} + b_j a_{ji} - b_i b_j) \left(F(t^i, Q^i), \ F(t^j, Q^j) \right) \tag{8.73}$$

A dissipative system is characterized by the inequality $d\|Q\|^2/dt \leq 0$. For such systems, then, the following inequality may hold

$$2 \sum_{i=1}^{2} b_i \left(Q^i, \ F(t^i, Q^i) \right) - \Delta t \sum_{i,j=1}^{2} (b_i a_{ij} + b_j a_{ji} - b_i b_j) \left(F(t^i, Q^i), \ F(t^j, Q^j) \right) \leq 0 \tag{8.74}$$

The backwards Euler and present IRK2 algorithms unconditionally satisfy this constraint for dissipative systems since $b_1 > 0$, $b_2 > 0$, and the energy matrix E is non-negative definite for both algorithms. Therefore both terms in (8.74) are always negative. Conversely, the Crank-Nicolson rule does not unconditionally satisfy this constraint, since its matrix E is negative definite. Hence, the second term in (8.74) can be negative and consequently the whole expression can be positive. Hence, non-linear energy stability can only be achieved in this case if the dissipation inner product $\sum_{i=1}^{2} b_i (Q^i, \ F(t^i, Q^i)) < 0$ reaches a sufficient magnitude such that (8.74) is satisfied. Therefore, the stability of this scheme critically depends on the dissipation property of the operator $F(\cdot, \cdot)$, via the magnitude of $\sum_{i=1}^{2} b_i (Q^i, \ F(t^i, Q^i)) < 0$.

Conversely, the backwards Euler and IRK2 algorithms may be dissipative, even for non dissipative systems, as long as the magnitude of the inner product $\sum_{i=1}^{2} b_i (Q^i, \ F(t^i, Q^i)) > 0$ is suitably moderate so that constraint (8.74) is still satisfied. However this phenomenon is less pronounced for the IRK2 algorithms by virtue of the relation

$$\sum_{i,j=1}^{2} (b_i a_{ij} + b_j a_{ji} - b_i b_j)_{EUL} \left(F(t^i, Q^i), \ F(t^j, Q^j) \right)$$

$$\geq \sum_{i,j=1}^{2} (b_i a_{ij} + b_j a_{ji} - b_i b_j)_{IRK2} \left(F(t^i, Q^i), \ F(t^j, Q^j) \right) \tag{8.75}$$

which derives from the ratio

$$\frac{(\lambda_E)_{EUL}}{(\lambda_E)_{IRK2}} > 1 \tag{8.76}$$

where (λ_E) is the largest eigenvalue of matrix E for each of the two algorithms. If some form of severe numerical instability develops in the implicit temporal integration with the subject IRK2 algorithm, applied to the semi-discrete CFD evolution equations, it is likely that the spatial semi-discretization employed has yielded an unstable operator $F(\cdot, \cdot)$, with $\sum_{i=1}^{2} b_i (Q^i, \ F(t^i, Q^i)) > 0$.

Chapter 9

One-Dimensional Non-Discrete Characteristics-Bias Resolution

This chapter introduces a non-discrete characteristics-bias representation of the Euler system at the partial differential equation level, in the continuum, before the spatial discretization. As one of its chief advantages, this continuum formulation yields an upstream-weighted system that leads to a set of stable discrete equations by way of a traditional and versatile centered Galerkin finite element procedure. Since the characteristics-bias development takes place in the continuum, it directly relies upon characteristics theory and the accompanying wealth of theoretical existence and convergence results in the field of hyperbolic and parabolic partial differential equations.

Characteristic lines and surfaces play a theoretically prominent role in the solution of first-order hyperbolic partial differential systems, for solutions of these systems remain constant along real characteristic lines and information propagates along these lines. Hyperbolic systems command a great deal of attention in fluid mechanics because the Euler equations form a hyperbolic system. This chapter highlights the theory of scalar non-linear equations to introduce the fundamental notions of characteristic lines, solutions in wave-like form, weak solutions, and shock formation from initially smooth initial conditions. The extension of solutions in wave-like form to systems then leads to multiple characteristic speeds and lines.

As summarized in this chapter, mathematics researchers of the caliber of Lax, Ladyzhenskaya, Olejnik, and Tadmor, among others, have established that entropy weak solutions of hyperbolic equations may be obtained as limiting solutions of parabolic-perturbation equations, as a perturbation parameter approaches nought. Based on these pivotal findings, the characteristics-bias system generalizes the notion of the parabolic-perturbation equation to systems and links the artificial-diffusion method with the characteristics-based upwind procedure.

The non-discrete, characteristics-bias representation of the Euler and Navier-Stokes equations is a partial differential system that encompasses these equations and also induces a non-linear solution-dependent upstream dissipation along characteristic directions. This upstream dissipation originates from a differential parabolic perturbation within the characteristics - bias system, a perturbation that emerges from a decomposition not of the Euler flux, but of the flux Jacobian.

As stated, a centered discretization of the characteristics-bias system automatically generates a consistent genuinely upstream-bias approximation of the original equations, without any need for additional numerical dissipation terms. As a result, this formulation, induces but a minimal amount of upstream dissipation that leads to essentially non-oscillatory solutions. The characteristics-bias system satisfies the same Galilean invariance principle satisfied by the original equation and, in integral form, also leads to a generalization of the Discontinuous Galerkin (DG) and Streamline Upwind Petrov Galerkin (SUPG) formulations.

9.1   Solutions of Non-Linear Hyperbolic Equations

For simplicity, consider the reference hyperbolic scalar conservation law without source

$$\frac{\partial q}{\partial t} + \frac{\partial f(q)}{\partial x} = 0 \tag{9.1}$$

where $f(q)$ is the flux with Jacobian $f'(q) = \frac{\partial f}{\partial q}$. For a non-linear problem, the second derivative $f''(q)$ does not vanish identically, which implies that in the flux gradient expression $\frac{\partial f(q)}{\partial x} = f'(q)\frac{\partial q}{\partial x}$ the convection speed $f'(q)$ is not constant. As a result, discontinuous solutions can arise when a rapidly propagated wave "overtakes" and breaks onto a wave that is slowly convected.

In the bounded Cartesian-product domain $D \equiv \Omega \times T$, with $\Omega \equiv (a, b)$, $T \equiv (t_0, t_{\text{fin}})$, a well posed initial boundary value (IBV) problem associated with (9.1) consists in determining the unique solution $q = q(x, t) \in C^1(D)$ of (9.1), subject to the following initial and boundary conditions

$$q(x, t_0) = q_0(x) \in C^1(\Omega) \tag{9.2}$$

$$q(a, t) = k_a(t) \ \text{ if } \ f'(q(a, t)) \geq 0, \ \text{ or } \ q(b, t) = k_b(t) \ \text{ if } \ f'(q(b, t)) \leq 0 \tag{9.3}$$

which collectively correspond to the auxiliary data. As a distinguishing non-linear feature, a solution of (9.1)-(9.3) may be non-unique. Furthermore, even when the initial condition is differentiable with $q_0 \in C^1(\Omega)$, a solution may remain smooth, but only for a limited time before becoming discontinuous, which implies that a classical solution $q \in C^1(\Omega)$ of (9.1)-(9.3) may not exist for all $t > 0$.

A generalized solution $q \in \mathcal{H}^0(D)$ of a differential problem equivalent to (9.1)-(9.3), however, may be identified as follows. Consider a smooth test function $v = v(x, t) \in C^1(D)$, vanishing both outside D and on the lines $t = \tau$, $x = a$, and $x = b$. Employing this function v, Green's theorem yields the equivalence

$$-\int_{t_0}^{\tau} \int_a^b v \left(\frac{\partial q}{\partial t} + \frac{\partial f(q)}{\partial x} \right) dx\, dt = \int_{t_0}^{\tau} \int_a^b \left(q\frac{\partial v}{\partial t} + f(q)\frac{\partial v}{\partial x} \right) dx\, dt + \int_{a(t=t_0)}^b vq_0(x)\, dx \tag{9.4}$$

A solution $q = q(x, t)$ of the equation

$$\int_{t_0}^{\tau} \int_a^b \left(q\frac{\partial v}{\partial t} + f(q)\frac{\partial v}{\partial x} \right) dx\, dt + \int_{a(t=t_0)}^b vq_0(x)\, dx = 0 \tag{9.5}$$

need not be either differentiable, or even continuous, since it suffices that both q and q_0 remain square integrable, hence $q_0(x) \in \mathcal{H}^0(\Omega)$ and $q(x, t) \in \mathcal{H}^0(D)$. With square integrable initial data q_0, a solution q of (9.5) is then termed a generalized or weak solution of the IBV problem (9.1)-(9.3) if q satisfies (9.5) for any test function v that vanishes on the lines $t = \tau$, $x = a$, and $x = b$

The following theorem, proven by other researchers, establishes the properties of the generalized solution. Let $q_0 \in \mathcal{H}^0(\Omega)$ and $f(q) \in C^2(\Re)$, with $\Re$ denoting the real-number field. Additionally, let the inequality $f''(q) > 0$ apply for any $q \in Q \equiv \{q : |q| \leq \kappa \|q_0\|_{\mathcal{H}^0(\Omega)}\}$, with κ a finite positive constant. Then:

I) a generalized solution $q(x,t)$ of (9.5) corresponding to q_0 exists, and remains bounded with

$$|q(x,t)| \leq \kappa \|q_0\|_{\mathcal{H}^0(\Omega)}, \quad \text{for all } (x,t) \in D \tag{9.6}$$

II) for a constant $E > 0$ depending on $\|q_0\|_{\mathcal{H}^0(\Omega)}$ and $\min_{q \in Q} \left[f''(q) \right]$, the solution of (9.5) as augmented by the entropy E-condition

$$\frac{q(x + \delta x, t) - q(x,t)}{\delta x} < \frac{E}{t} \tag{9.7}$$

is unique, for every $\delta x > 0$, $t > 0$, and $x \in \Omega$.

III) the solution q satisfying I and II is stable and continuously depend on q_0. Hence, for a perturbed initial condition $q_\delta = q_0 + \delta q_0 \in \mathcal{H}_0(\Omega)$, the deviation between the corresponding perturbation solution q_δ and q satisfies the stability bound

$$\int_{x_1}^{x_2} |q(x,t) - q_\delta(x,t)| \, dx \leq \int_{x_1 - ct}^{x_2 + ct} |\delta q_0| \, dx \tag{9.8}$$

for $c = \max_{q \in Q} \left[f'(q) \right]$, every x_1 and $x_2 \in \Re$, $x_1 < x_2$, and every $t > 0$.

The entropy E-condition (9.7), synthesized by Olejnik [25, 26, 27], establishes the time evolution of the discontinuities in the entropy solutions of (9.5). It shows that the magnitude of a positive discontinuous variation $q(x + \delta x, t) - q(x,t)$, e.g. an expansion shock, will decrease as time elapses, for this positive variation must remain less than the decreasing positive upper bound E/t. On the other hand, if q decreases across a discontinuity, then such a discontinuity can persist stable as time elapses, for the corresponding non-positive variation $q(x + \delta x, t) - q(x,t)$ always remains less than the positive upper bound E/t for all $t > 0$. The E-condition is thus called the entropy condition because a unique and stable solution discontinuity can take place in one direction only, in agreement with the second law of thermodynamics.

9.1.1 Solutions and Characteristic Lines

When the weak solution q of (9.5) is differentiable and q_0 is continuous, then the weak statement (9.5) implies the conservation law (9.1) for both positive and negative $f''(q)$; as a result, q becomes a classical solution almost everywhere (a.e.) in D. To see this, transform the domain integral in (9.5) by means of Green's theorem, which yields

$$\int_{t_0}^{\tau} \int_a^b v \left[\frac{\partial q}{\partial t} + \frac{\partial f(q)}{\partial x} \right] dx \, dt + \int_{a_{(t=t_0)}}^b v \left[q(x,t_0) - q_0(x) \right] dx = 0 \tag{9.9}$$

Since this statement must be satisfied for all $v \in C^1(D)$, the integrands between parentheses must consequently vanish, leading to

$$\frac{\partial q}{\partial t} + \frac{\partial f(q)}{\partial x} = 0, \quad q(x, t_0) = q_0(x_0) \tag{9.10}$$

which indicates that $q(x, t)$ satisfies equation (9.1) and the initial condition (9.2). In these results, q_0 denotes a prescribed initial-condition function and x_0 indicates the coordinate x at time $t = t_0 = 0$. A $C^1(D)$ solution is expressed as

$$q = q(x - \lambda(q)t), \quad \lambda(q) \equiv \frac{\partial f(q)}{\partial q} \quad \Rightarrow \quad q = q(\eta_1), \quad \eta_1 \equiv x - \lambda t \tag{9.11}$$

as is verified by substitution, according to the following developments. The hyperbolic equation in (9.10) can be cast as

$$\frac{\partial q}{\partial t} \frac{1}{\sqrt{1 + \lambda^2}} + \frac{\partial q}{\partial x} \frac{\lambda}{\sqrt{1 + \lambda^2}} = \frac{dq}{ds} = 0 \tag{9.12}$$

which defines a vanishing directional derivative of q along a line with direction cosines

$$\frac{dt}{ds} = \frac{1}{\sqrt{1 + \lambda^2}}, \quad \frac{dx}{ds} = \frac{\lambda}{\sqrt{1 + \lambda^2}} \quad \Rightarrow \quad \frac{dx}{dt} = f'(q) = \lambda(q) \tag{9.13}$$

The derivative λ of $f = f(q)$ provides the slope of the line along which the solution q remains constant. This line is known as a characteristic line and considering that dx/dt defines a speed, the derivative λ is called a characteristic speed. Since q is constant along this line, such a solution evolves by remaining equal to its initial value, at a point, along the characteristic line through the point. For a constant q along each characteristic line, the corresponding characteristic-line slope $\lambda = \lambda(q)$ is also constant, which means that for a scalar equation, the characteristic lines are straight lines. For a constant λ, the direction-cosine equations (9.13) are exactly integrated with $t(0) = 0$, $x(0) = x_0$ as

$$x = x_0 + \frac{\lambda}{\sqrt{1 + \lambda^2}} s, \quad t = \frac{1}{\sqrt{1 + \lambda^2}} s, \quad x = x_0 + \lambda t \tag{9.14}$$

These results define the coordinate transformation

$$t = t(s), \quad x = x(x_0, s) \tag{9.15}$$

which is valid whenever t and x remain differentiable functions of both s and x_0. With reference to Figure 9.1, these equations lead to

$$d^2 = (x - x_0)^2 + t^2 = s^2$$

which shows that the curvilinear coordinate "s" corresponds to the distance from $(0, x_0)$ along the line with equation

$$x = x_0 + \lambda(q)t, \quad \Rightarrow x_0 = x - \lambda(q)t \tag{9.16}$$

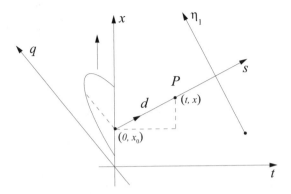

Figure 9.1: Solution and Characteristic Line

Since q remains constant along characteristic lines, as noted, its magnitude as a function of s remains equal to its magnitude for $s = 0$, which corresponds to the magnitude of the initial condition at $x = x_0$. This recognition justifies the following equalities

$$q(x(s), t(s)) = q(x(0), t(0)) = q(x_0, 0) = q_0(x_0) \tag{9.17}$$

and following (9.16) the terminal form of the solution becomes

$$q(x, t) = q_0(x_0) = q_0(x - \lambda(q)t) = q(x - \lambda(q)t) \tag{9.18}$$

where $q_0(x - \lambda(q)t)$ provides the solution corresponding to a prescribed initial condition $q_0(x_0)$ and $q(x - \lambda(q)t)$ expresses the general solution as a function of the coordinate $\eta_1 \equiv x - \lambda(q)t$. As time elapses, this solution describes a non-linear travelling-wave propagation, with speed λ, of the initial profile $q = q(x_0), t = 0$, along the x-axis. As a result, solution information arrives at point P from the direction upstream of P; accordingly, an upstream-weighted integral average of the solution and flux divergence around P remains consistent with this wave-propagation pattern, averages that are employed in Section 9.3. When λ remains independent of q, this propagation corresponds to a rigid translation, whereas when λ depends upon q, the propagation involves both translation and deformation. In either case, the initial profile $q = q(x_0)$ propagates as a wave. For this reason, solution (9.18) is termed a wave solution and its mathematical structure is a wave-like form. The line

$$x - \lambda t = C \tag{9.19}$$

with C a constant, is a characteristic line and λ denotes the corresponding characteristic speed.

Expression (9.18) implicitly defines the solution q corresponding to q_0 in the form

$$F\left(q(x, t)\right) \equiv q(x, t) - q_0\left(x - f'\left(q(x, t)\right)t\right) = 0 \tag{9.20}$$

which only in a few cases can yield an explicit solution. For an arbitrary initial condition $q_0 = q_0(x_0)$, a solution for q can be numerically obtained iteratively from Newton's method

$$\frac{dF}{dq}\bigg|_{q=q^p} \left(q^{p+1} - q^p\right) = -F\left(q^p\right) \tag{9.21}$$

where superscript p denotes the iteration index. For the function F defined by (9.20) and for fixed x and t, iteration (9.21) yields

$$\left(1 + q_0'(x - f'(q^p)t)f''(q^p)t\right)\left(q^{p+1} - q^p\right) = q_0\left(x - f'(q^p)t\right) - q^p \tag{9.22}$$

For a smooth q, (9.22) converges to a unique solution for a suitable initial estimate q^0. For a discontinuous q, however, (9.22) may converge to different solutions for fixed x and t depending on q^0. With reference to Figure 9.2, an initial estimate q^0 within D_ℓ, for example, may converge to q_ℓ, whereas an estimate q^0 within D_r may converge to q_r.

9.1.2   Breaking Time

Solution (9.18) also yields the time when it ceases to be continuous. In order to establish this breaking time, it suffices to determine the partial derivatives of q from (9.18). These derivatives are expressed as

$$\begin{aligned}
\frac{\partial q}{\partial t}\bigg|_x &= \frac{dq_0}{dx_0}\frac{\partial x_0}{\partial t}\bigg|_x = q_0'(x_0) \cdot \left[-f''(q)\frac{\partial q}{\partial t}t - f'(q)\right] \\
\frac{\partial q}{\partial x}\bigg|_t &= \frac{dq_0}{dx_0}\frac{\partial x_0}{\partial x}\bigg|_t = q_0'(x_0) \cdot \left[1 - f''(q)\frac{\partial q}{\partial x}t\right]
\end{aligned} \tag{9.23}$$

which lead to

$$\frac{\partial q}{\partial t} = -\frac{f'(q)q_0'(x_0)}{1 + tf''(q)q_0'(x_0)}, \quad \frac{\partial q}{\partial x} = \frac{q_0'(x_0)}{1 + tf''(q)q_0'(x_0)} \tag{9.24}$$

Since these expressions evolve from solution (9.18), they are seen identically to satisfy the hyperbolic equation in (9.10). With $f''(q) > 0$, these expressions imply uniqueness for a differentiable solution that evolves from an initial condition with $q_0'(x_0) > 0$. To achieve this conclusion, heed that if q is differentiable, its partial derivative with respect to x exists. The very definition of $\frac{\partial q}{\partial x}$ in (9.24) leads to the following result

$$\lim_{\delta x \to 0^+} \frac{q(x + \delta x, t) - q(x,t)}{\delta x} = \frac{q_0'(x_0)}{1 + tf''(q)q_0'(x_0)} = \frac{1}{1/q_0'(x_0) + tf''(q)} \leq \frac{1}{tf''(q)} \leq \frac{E}{t} \tag{9.25}$$

which shows that q is unique because it satisfies the uniqueness entropy condition (9.7).

The crucial element in this derivation is the first inequality, which remains valid for all $t > 0$ and $f''(q) > 0$ only if $q_0'(x_0) > 0$ for all x_0. This observation implies that solution non-uniqueness at a point may be triggered by $q_0'(x_0) < 0$, which corresponds to characteristics intersecting in the (x, t) plane, as illustrated in Figure 9.2. Analytically, the solution

q becomes non-unique when its partial derivatives from (9.24) become unbounded. Corresponding to a vanishing denominator, this condition leads to the first "wave breaking" time

$$t_B = \frac{1}{\max\limits_{x_0 \in (a,b)} (-q_0'(x_0) f''(q))} = \frac{1}{\max\limits_{x_0 \in (a,b)} (-q_0'(x_0) f''(q_0(x_0)))} \tag{9.26}$$

which is positive for $q_0'(x_0) < 0$. Let x_{0_B} denote that x_0 corresponding to the denominator in these expressions, that is $\max\limits_{x_0 \in (a,b)} (-q_0'(x_0) f''(q_0(x_0)))$. From (9.16), the location where wave breaking first takes place is

$$x_B = x_{0_B} + f'(q_0(x_{0_B})) t_B \tag{9.27}$$

These results indicates the crucial contribution of the initial condition to the formation of shocks.

9.1.3 Shocks

The slopes of the characteristic lines directly depend on the initial condition q_0 for q, as elaborated in this section, and for some initial conditions the characteristics may eventually intersect, which is the mechanism by which non unique and discontinuous solutions can arise. When the generalized solution q is discontinuous, the weak statement (9.5) yields the discontinuity propagation speed. To obtain this speed, consider a smooth curve Γ with equation $x_\Gamma = x_\Gamma(t)$, such that q is smooth and satisfies the hyperbolic equation in (9.10), on either side of Γ, but discontinuous at a point $P \in \Gamma$, see Figure 9.2. For $t > t_0$ and for a test function v with compact support on an open neighborhood D of P, the weak statement (9.5) becomes

$$\int \int_D \left(q \frac{\partial v}{\partial t} + f(q) \frac{\partial v}{\partial x} \right) dx \, dt =$$

$$= \int \int_{D_\ell} \left(q \frac{\partial v}{\partial t} + f(q) \frac{\partial v}{\partial x} \right) dx \, dt + \int \int_{D_r} \left(q \frac{\partial v}{\partial t} + f(q) \frac{\partial v}{\partial x} \right) dx \, dt = 0 \tag{9.28}$$

where D_ℓ and D_r are open subsets of D with $D_\ell \cup D_r = D - \Gamma$ and $D_\ell \cap D_r = \emptyset$. Since q is smooth on D_ℓ and D_r, the lhs of (9.28) can be transformed by way of Green's theorem, yielding

$$- \int \int_{D_\ell} v \left(\frac{\partial q}{\partial t} + \frac{\partial f(q)}{\partial x} \right) dx \, dt - \int \int_{D_r} v \left(\frac{\partial q}{\partial t} + \frac{\partial f(q)}{\partial x} \right) dx \, dt +$$

$$+ \oint_{\partial D_\ell} v(-q \, dx + f \, dt) + \oint_{\partial D_r} v(-q \, dx + f \, dt) = 0 \tag{9.29}$$

where the two domain integrals will vanish since q, as stated, satisfies the equation in (9.10) on both D_ℓ and D_r.

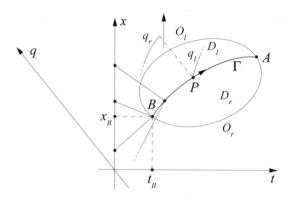

Figure 9.2: Shocked Solution and Characteristics Converging on Discontinuity Curve Γ

Furthermore, $v = 0$ along both AO_rB and $BO_\ell A$. Hence, (9.29) simplifies to

$$\int_{AP_\ell B} v(-q\,dx + f\,dt) + \int_{BP_r A} v(-q\,dx + f\,dt) = 0 \tag{9.30}$$

and

$$\int_{AP_\ell B} v(-q\,dx + f\,dt) - \int_{AP_r B} v(-q\,dx + f\,dt) = 0 \tag{9.31}$$

where P_ℓ and P_r indicate point P as approached from D_ℓ and D_r respectively. These line integrals reduce to

$$\int_{APB} v(-[q]\,dx + [f]\,dt) = 0, \quad [q] \equiv q_\ell - q_r, \quad [f] \equiv f_\ell - f_r \tag{9.32}$$

where $[q]$ and $[f]$ denote the jumps in q and $f(q)$ across Γ. Since the integral in (9.32) must be satisfied for all test functions v with compact support on D, the integrand between parentheses must consequently vanish

$$-[q]\,dx + [f]\,dt = 0 \tag{9.33}$$

which leads to the discontinuity speed

$$s \equiv \frac{dx_\Gamma}{dt} = \frac{[f]}{[q]} \tag{9.34}$$

It follows that an infinitesimal discontinuity propagates along a characteristic line, for (9.33)-(9.34) yield the following form for s

$$s = \lim_{q_\ell \to q_r} \frac{[f]}{[q]} = f'(q) \tag{9.35}$$

which coincides with the characteristic speed from (9.13), corresponding to the flux Jacobian.

The entropy condition (9.7) then leads to a condition on s that a unique stable discontinuity satisfies. As indicated previously, (9.7) implies that q can only decrease through a stable discontinuity, hence

$$q_\ell > q_r \tag{9.36}$$

Furthermore, for $f''(q) > 0$ the flux Jacobian can only increase as a function of q, hence

$$f'(q_\ell) > f'(q) > f'(q_r) \tag{9.37}$$

for any q such that $q_\ell > q > q_r$. A mean-value-theorem resolution of $[f]$ then yields

$$s(q_\ell - q_r) = f_\ell - f_r = f'(q_m) \cdot (q_\ell - q_r) \quad \Rightarrow \quad s = f'(q_m) \tag{9.38}$$

for a q_m such that $q_\ell > q_m > q_r$. Therefore, (9.37), (9.38), provide

$$f'(q_\ell) > s > f'(q_r) \tag{9.39}$$

which supplies a uniqueness entropy condition on s. Since the point where $q = q_\ell$ precedes that were $q = q_r$, and $q_\ell > q_r$, (9.39) geometrically signifies that the characteristic lines associated with a unique discontinuous solution will intersect on Γ, as depicted in Figure 9.2. For a positive $f''(q)$ a smooth initial condition with $q_0'(x_0) < 0$ propagates and deforms along characteristic lines with slopes (9.13). These characteristics converge towards Γ, which convergence satisfies the entropy condition (9.39) and implies that the evolving solution is unique. At a point P defined by (9.26)-(9.27) two or more characteristics will meet, triggering a solution discontinuity. From this point, with speed (9.34), the discontinuity then propagates along the curve Γ.

9.1.4 Hyperbolic Systems

Consider the non-linear system

$$\frac{\partial q}{\partial t} + \frac{\partial f(q)}{\partial x} = 0 \tag{9.40}$$

When the eigenvalues of the flux vector Jacobian $\frac{\partial f}{\partial q}$ are distinct and real the system is hyperbolic. Also for this system, is it possible to express a solution in non-linear wave-like form as

$$q = q(x - \lambda(q)t), \quad \eta_1 \equiv x - \lambda(q)t \tag{9.41}$$

with the characteristic speed $\lambda(q)$ to be determined by the system. Based on this form of the solution, Chapter 14 shows that the partial derivatives of the non-linear solution q with respect to t and x are expressed as

$$\frac{\partial q}{\partial t} = \frac{-\lambda(q)}{1 + t\dfrac{\partial \lambda}{\partial \eta_1}} M \frac{\partial q}{\partial \eta_1}$$

$$\frac{\partial q}{\partial x} = \frac{1}{1 + t\dfrac{\partial \lambda}{\partial \eta_1}} M \frac{\partial q}{\partial \eta_1} \tag{9.42}$$

where M denotes a non-singular matrix. The insertion of these derivatives into the hyperbolic system yields the statement

$$\frac{1}{1+t\dfrac{\partial\lambda}{\partial\eta_1}}\left(-\lambda(q)I+\frac{\partial f}{\partial q}\right)M\frac{\partial q}{\partial\eta_1}=0 \tag{9.43}$$

For non-vanishing vectors $M\frac{\partial q}{\partial\eta_1}$, this system thus defines the characteristic speeds as the eigenvalues of the flux vector Jacobian $\frac{\partial f}{\partial q}$. For a hyperbolic system with n equations, accordingly, there exist n characteristic speeds and associated characteristic lines. The algebraic signs of these eigenvalues determine whether the waves associated with q propagate in the same or opposing directions. When these eigenvalues are all positive or negative, the waves associated with q all propagate in the same direction.

9.2 Parabolic Perturbation Equations

The weak solution of (9.5) is generally non differentiable. The following considerations highlight, however, that the unique differentiable or weak solution of (9.5) satisfying the entropy condition may be obtained as the limit of a sequence of continuous differentiable functions. These developments essentially rely on singular-perturbation partial differential equations, equations that both contain a higher-order perturbation differential term and undergo a change of order and type when the associated perturbation parameter approaches zero.

9.2.1 Scalar Equations

In the late fifties, Lax [115, 116] studied the convergence of the solutions q^ε of the singular parabolic-perturbation equation

$$\frac{\partial q^\varepsilon}{\partial t}+\frac{\partial f(q^\varepsilon)}{\partial x}=\varepsilon\frac{\partial}{\partial x}\left(\frac{\partial q^\varepsilon}{\partial x}\right) \tag{9.44}$$

This is a parabolic-perturbation equation because the second order rhs differential perturbation transforms the original hyperbolic equation into a parabolic equation. Additionally, the superscript ε in this equation denotes solution dependence on the perturbation parameter ε. Lax's investigations proved that the unique weak solution satisfying an entropy condition for the weak statement (9.5) can be obtained as the limiting solution of the parabolic-perturbation equation (9.44) as the constant positive perturbation parameter ε approaches zero.

Ladyzhenskaya [109] and especially Olejnik [147, 148, 149, 150] generalized Lax's result using various forms of the parabolic perturbation. Significant theorems resulting from their studies apply to (9.5) and indicate that the unique weak solution of (9.5) that also satisfies an entropy condition is obtained as the limiting solution of the variable-coefficient parabolic-perturbation equation

$$\frac{\partial q^\varepsilon}{\partial t}+\frac{\partial f(q^\varepsilon)}{\partial x}=\frac{\partial}{\partial x}\left(\varepsilon\alpha(x,t)\frac{\partial q^\varepsilon}{\partial x}\right) \tag{9.45}$$

as the positive constant ε approaches zero, while $\alpha(x,t) > 0$, for all x, and $t > 0$. Furthermore, these studies prove that a unique entropy solution for (9.5) derives from the limiting solution of the non-linear parabolic-perturbation equation

$$\frac{\partial q^{\varepsilon}}{\partial t} + \frac{\partial f(q^{\varepsilon})}{\partial x} = \frac{\partial}{\partial x}\left(\varepsilon \xi \left(\frac{\partial q^{\varepsilon}}{\partial x} \right) \right) \tag{9.46}$$

as the positive constant ε approaches zero, where $\xi\left(\frac{\partial q^{\varepsilon}}{\partial x}\right)$ is a smooth function of $\frac{\partial q^{\varepsilon}}{\partial x}$.

As far as the initial-value Cauchy problem for (9.5) is concerned, the parabolic - perturbation approach provides a complete methodology for determining the unique hyperbolic -problem solution, since no boundary conditions are involved. Concerning the initial boundary value problem, the issue of convergence of a solution q^{ε} that has to satisfy two boundary conditions, one at each boundary point, to a limiting solution q of (9.5) that can only satisfy one boundary condition, for a positive flux Jacobian at both boundary points, requires delicate analytical considerations in addition to convergence for the Cauchy problem.

Significantly, Olejnik's contributions have completely resolved this issue. In a sequence of detailed theorems [147, 148, 149, 150], she has proved that the limiting solution is no longer connected to the additional boundary condition of the perturbation parabolic equation, but only depends on the single boundary condition for the original hyperbolic equation. These theorems reiterate that the solutions of the parabolic-perturbation equations are smooth, hence $q^{\varepsilon} \in C^2(D)$, and show that q^{ε} and its derivatives converge pointwise almost everywhere (a.e.) to the unique, classical or weak, solution q of (9.5). Furthermore, convergence is also proven in the following norm

$$\lim_{\varepsilon \to 0} \|q^{\varepsilon} - q\| = \lim_{\varepsilon \to 0} \int_a^b |q^{\varepsilon} - q| \, dx = 0 \tag{9.47}$$

The linear and non-linear parabolic perturbations in (9.44)-(9.46) thus direct the solutions of the respective equations to converge to entropy solutions of (9.5). These results, which also apply to classical solutions as smooth forms of entropy solutions, theoretically legitimize the use of a linear or even non-linear artificial-diffusion equation to determine the unique entropy solution of a hyperbolic conservation law.

Continuing the tradition of these fundamental investigations, Tadmor [181, 182] reported that the unique weak solution of (9.5) is also obtained as the limiting solution of the non-linear equation

$$\frac{\partial q^{\varepsilon}}{\partial t} + \frac{\partial f(q^{\varepsilon})}{\partial x} = \varepsilon \frac{\partial}{\partial x}\left(\frac{\partial \varphi(q^{\varepsilon})}{\partial x} \right) \tag{9.48}$$

as the positive parameter ε approaches zero, while $\varphi(q^{\varepsilon})$ can be any differentiable function of q^{ε} such that $\partial \varphi / \partial q > 0$. In particular, the author of this book recognizes the connection between (9.48) and an upwind approximation, as results via the selection

$$\varphi(q) = \psi \mathrm{sgn}(f')f(q) \tag{9.49}$$

With this specification, (9.48) becomes

$$\frac{\partial q^{\varepsilon}}{\partial t} + \left(I - \varepsilon \psi \mathrm{sgn}(f') \frac{\partial}{\partial x} \right) \frac{\partial f(q^{\varepsilon})}{\partial x} = 0 \tag{9.50}$$

which corresponds to a non-discrete upwind representation of $\frac{\partial f(q^\varepsilon)}{\partial x}$. An integral statement for this equation with a test function $\widehat{w}$ with compact support in the integration domain $\widehat{\Omega}$ becomes

$$\int_{\widehat{\Omega}} \widehat{w}\frac{\partial q^\varepsilon}{\partial t}d\Omega + \int_{\widehat{\Omega}}\left(\widehat{w} + \frac{\partial \widehat{w}}{\partial x}\varepsilon\psi\mathrm{sgn}(f')\right)\frac{\partial f(q^\varepsilon)}{\partial x}d\Omega = 0 \qquad (9.51)$$

which corresponds to an upstream-weighted integral average. These crucial non-discrete upwind representations are amplified in Section 9.3.

9.2.2 Perturbation Systems

The generalization of a parabolic-perturbation scalar equation to a system with n equations may be expressed as

$$\frac{\partial q}{\partial t} + \frac{\partial f(q)}{\partial x} - \frac{\partial}{\partial x}\left(\varepsilon\psi B(q)\frac{\partial q}{\partial x}\right) = 0 \qquad (9.52)$$

where $\varepsilon\psi$ denotes a positive perturbation parameter and $B = B(q)$ indicates a non-linear perturbation matrix. This matrix is determined under the condition that the corresponding differential operator should prevent at each point the unbounded growth of any perturbation δq to the solution. Since it is the growth of such a perturbation that B should prevent, it suffices to investigate the evolution of infinitesimal perturbations. If B is effective, the perturbation does not grow beyond an infinitesimal magnitude; if the initial perturbation is not infinitesimal, an infinitesimal portion of it may be regarded as the emerging perturbation and the remaining magnitude may be associated with the solution itself. In either case, the evolution of perturbations in q at each point is controlled by the linear counterpart of (9.52). The analysis of the time evolution of the energy in the solution of this linear counterpart indicates that (9.52) is stable in the sense indicated when B possesses n distinct positive eigenvalues, as shown in the following developments.

The solution q, flux vector $f(q)$, Jacobian $A \equiv \frac{\partial f(q)}{\partial q}$, and matrix $B(q)$ are expressed as

$$
\begin{aligned}
q &= q_0 + \delta q \\
f(q) &= f(q_0) + A(\widehat{q})\delta q \\
A(\widehat{q}) &= A(\overline{q_0}) + \frac{\partial A}{\partial x}\delta x \\
B(q) &= B(q_0) + \frac{\partial B}{\partial q}\delta q \\
B(q_0) &= B(\overline{q_0}) + \frac{\partial B}{\partial x}\delta x
\end{aligned}
\qquad (9.53)
$$

where δq denotes a perturbation to q_0, with both q_0 and q corresponding to solutions of (9.52). In the expressions after the first, the mean-value theorem justifies the strict equalities, with $\widehat{q}$ indicating an appropriate mean-value average, all the derivatives of A and B being evaluated at mean-value averages, and $\overline{q_0}$ denoting the constant value of q_0 at each point considered. With q_0 a solution of (9.52), the insertion of (9.53) into (9.52) and elimination of higher order terms yield the linear system

$$\frac{\partial(\delta q)}{\partial t} + A(\overline{q_0})\frac{\partial(\delta q)}{\partial x} - \frac{\partial}{\partial x}\left(\varepsilon\psi B(\overline{q_0})\frac{\partial(\delta q)}{\partial x}\right) = 0 \qquad (9.54)$$

which governs at each point the time evolution of the perturbation (δq). Let the matrix B possess n distinct eigenvalues for any $\overline{q}_0$, hence any q; under this condition, this matrix may be expressed via the similarity transformation

$$B(\overline{q}_0) = X\Lambda X^{-1} \tag{9.55}$$

where the diagonal matrix Λ contains all the eigenvalues of B and the constant matrix $X = X(\overline{q}_0)$ denotes the associated eigenvector matrix. Next, multiply system (9.54) by X^{-1}, which results in the associated linear system

$$\frac{\partial\left(X^{-1}\delta q\right)}{\partial t} + X^{-1}A(\overline{q}_0)X\frac{\partial\left(X^{-1}\delta q\right)}{\partial x} - \frac{\partial}{\partial x}\left(\varepsilon\psi\Lambda\frac{\partial\left(X^{-1}\delta q\right)}{\partial x}\right) = 0 \tag{9.56}$$

The energy $\mathcal{E}$ in the solution $(X^{-1}\delta q)$ of this system and the time rate of change of this energy may be expressed as

$$\mathcal{E} \equiv \frac{1}{2}\int_\Omega \left(X^{-1}\delta q\right)^T \left(X^{-1}\delta q\right) d\Omega \quad \Rightarrow \quad \frac{d\mathcal{E}}{dt} = \int_\Omega \left(X^{-1}\delta q\right)^T \frac{\partial\left(X^{-1}\delta q\right)}{\partial t} d\Omega \tag{9.57}$$

The time derivative of $(X^{-1}\delta q)$ in this integral may be replaced *à la* Lyapunov in terms of the system itself, which yields

$$\frac{d\mathcal{E}}{dt} = -\int_\Omega \left(X^{-1}\delta q\right)^T X^{-1}AX\frac{\partial\left(X^{-1}\delta q\right)}{\partial x}d\Omega + \oint_{\partial\Omega}\left(\varepsilon\psi\left(X^{-1}\delta q\right)^T\Lambda\frac{\partial\left(X^{-1}\delta q\right)}{\partial x}\right)n_x d\Gamma$$

$$-\int_\Omega \varepsilon\psi\frac{\partial\left(X^{-1}\delta q\right)^T}{\partial x}\Lambda\frac{\partial\left(X^{-1}\delta q\right)}{\partial x}d\Omega \tag{9.58}$$

In this expression, the first domain integral results from the hyperbolic part of the perturbation system, a part that always features in the time rate of change of $\mathcal{E}$, independently of the presence of the perturbation matrix B. This expression thus allows determining the contribution of this matrix to the rate of change of $\mathcal{E}$. The surface integral results from an integration by parts of the domain integral of the second-order term containing Λ; for a one-dimensional formulation this surface integral reduces to evaluations at two boundary points, evaluations that remain bounded because of enforced needed boundary conditions at these points. The crucial contribution of the Λ matrix, in comparison to the case when this matrix is absent, is the domain integral of the associated quadratic form. Due to the minus sign preceding the integral, it is when this form is positive that the evolution of $\mathcal{E}$ is limited, which prevents any unbounded growth of the perturbation δq; additionally, the limiting action of this quadratic form increases with a rise in the magnitude of $\varepsilon\psi$. This quadratic form remains positive when all the diagonal elements of Λ remain positive. As anticipated at the beginning of this section, accordingly, the perturbation δq will not grow when all the eigenvalues of B are distinct, so that (9.55) definitely exists, and positive, so that the quadratic form associated with Λ in (9.58) remains positive. For any solution $q = q(x, t)$, the parabolic perturbations developed in this book will all correspond to a matrix $B = B(q)$ with distinct and positive eigenvalues.

9.3 Characteristics-Bias Resolution

The characteristics-bias formulation induces an upwind bias at the partial differential equation level, in the continuum, before any spatial discretization. Following the wave-propagation and parabolic-perturbation results respectively in Sections 9.1.4, 9.2.2 and relying upon a Flux Jacobian Decomposition (FJD), the formulation generates a companion system that features a parabolic-perturbation matrix with distinct positive eigenvalues. A traditional centered discretization of the companion system then automatically yields a consistent upwind discretization of the original Euler equations. As its chief advantage, this formulation performs all the characteristics-bias developments in the continuum, which allows both to ensure stability of the resulting companion system and to employ a traditional Galerkin finite element spatial discretization.

The non-discrete characteristics-bias formulation derives from an upstream weighted integral statement associated with the governing system

$$\frac{\partial q}{\partial t} - \phi + \frac{\partial f(q)}{\partial x} = 0 \tag{9.59}$$

The reference integral statement may be expressed as

$$\int_{\widehat{\Omega}} \widehat{w} \left(\frac{\partial q}{\partial t} - \phi + \frac{\partial f(q)}{\partial x} \right) d\Omega = 0 \tag{9.60}$$

which is equivalent to the governing system when it holds for arbitrary subdomains $\widehat{\Omega} \subseteq \Omega$ and test functions $\widehat{w} \in \mathcal{H}^1(\widehat{\Omega}) \subseteq \mathcal{H}^1(\Omega)$. By comparison, the characteristics-bias integral is then defined as

$$\int_{\widehat{\Omega}} \widehat{w} \left(G^C - \phi^C + \frac{\partial f^C}{\partial x} \right) d\Omega = 0 \tag{9.61}$$

which must also hold for arbitrary subdomains $\widehat{\Omega} \subseteq \Omega$ and test functions $\widehat{w} \in \mathcal{H}^1(\widehat{\Omega}) \subseteq \mathcal{H}^1(\Omega)$. In this integral statement, superscript "C" denotes characteristics-bias terms. In particular, the characteristics-bias flux f^C automatically induces within (9.61) an upstream-bias approximation for the hyperbolic flux divergence $\frac{\partial f}{\partial x}$.

9.3.1 Upstream-Bias Time Derivative and Source

The expressions for G^C and ϕ^C are defined by the upstream weighted integral average

$$\int_{\widehat{\Omega}} \widehat{w} G^C \, d\Omega \equiv \int_{\widehat{\Omega}} (\widehat{w} + \psi \, \delta\widehat{w}) \frac{\partial q}{\partial t} \, d\Omega, \quad \int_{\widehat{\Omega}} \widehat{w} \phi^C \, d\Omega \equiv \int_{\widehat{\Omega}} (\widehat{w} + \psi \, \delta\widehat{w}) \phi \, d\Omega \tag{9.62}$$

the variation $\delta\widehat{w}$ is then expressed as

$$\delta\widehat{w} = \frac{\partial \widehat{w}}{\partial x} \delta x, \quad \delta x = \text{sgn}(u)\varepsilon \tag{9.63}$$

where ε indicates a length measure and the non-dimensional controller $\psi \in \mathcal{H}^0(\Omega)$, $0 < \psi \leq \mathcal{O}(1)$, determines the amount of upstream bias in this average. On the basis of this average, via integrating the variation of $\widehat{w}$ by parts, the characteristics-bias resolutions of the time-derivative of q and source term ϕ emerge as

$$G^C \equiv \frac{\partial q}{\partial t} - \frac{\partial}{\partial x}\left[\varepsilon\psi\mathrm{sgn}(u)\frac{\partial q}{\partial t}\right], \quad \phi^C \equiv \phi - \frac{\partial}{\partial x}\left[\varepsilon\psi\mathrm{sgn}(u)\,\phi\right] \qquad (9.64)$$

9.3.2 Flux Jacobian Decomposition and Upstream-Bias Integral Average

According to the derivations in Section 9.1.4, when the eigenvalues of the flux Jacobian matrix all display the same algebraic sign, all the associated solution waves propagate in the same direction. As a result, a weighted integral average of the flux Jacobian with greater weight in the upstream direction, from where waves are convected, remains consistent with wave propagation. Additionally, Section 9.2.2 has indicated that a stable parabolic perturbation originates when all the eigenvalues of the perturbation matrix are positive. Relying upon the results of these two sections, the characteristics-bias flux f^C is established by first developing the flux Jacobian decomposition (FJD) into L contributions

$$\frac{\partial f}{\partial q} = \sum_{\ell=1}^{L}\alpha_\ell A_\ell \quad \Rightarrow \quad \frac{\partial f}{\partial x} = \sum_{\ell=1}^{L}\alpha_\ell A_\ell\frac{\partial q}{\partial x} \qquad (9.65)$$

where A_ℓ corresponds to a flux-Jacobian matrix component with uniform-sign eigenvalues and α_ℓ denotes a linear-combination function, possibly depending upon q. An integral average of the Euler flux divergence $\frac{\partial f}{\partial x}$ as expressed through decomposition (9.65) becomes

$$\int_{\widehat{\Omega}}\widehat{w}\frac{\partial f}{\partial x}\,d\Omega = \int_{\widehat{\Omega}}\sum_{\ell=1}^{L}\widehat{w}\alpha_\ell A_\ell\frac{\partial q}{\partial x}\,d\Omega \qquad (9.66)$$

By comparison, the flux f^C is therefore defined by way of an upstream-bias integral average as

$$\int_{\widehat{\Omega}}\widehat{w}\frac{\partial f^C}{\partial x}\,d\Omega = \int_{\widehat{\Omega}}\sum_{\ell=1}^{L}\left(\widehat{w}+\psi\delta_\ell\widehat{w}\right)\alpha_\ell A_\ell\frac{\partial q}{\partial x}\,d\Omega \qquad (9.67)$$

where the rhs integral provides an upstream bias for each matrix component within the FJD in (9.65).

The variation $\delta_\ell\widehat{w}$ induces the appropriate upstream-bias for the test function $\widehat{w}$ for each "ℓ" component within (9.67). Depending on the physical significance, magnitude and algebraic sign of the eigenvalues of A_ℓ, the variation $\delta_\ell\widehat{w}$ can vanish or become algebraically positive or negative, which corresponds to an upstream bias respectively in the negative or positive sense of the x- axis.

9.3.3 Characteristics-Bias Flux

The variation $\delta_\ell\widehat{w}$ in (9.67) becomes

$$\delta_\ell\widehat{w} = \frac{\partial\widehat{w}}{\partial x}\delta_\ell x = \frac{\partial\widehat{w}}{\partial x}a_\ell\varepsilon, \quad a_\ell\varepsilon = \delta_\ell x \qquad (9.68)$$

where ε denotes a local positive length scale, while the direction cosine a_ℓ can equal 0 or $+1$, -1, possibly also depending upon q.

With these specifications, the upstream-bias integral average (9.67) becomes

$$\int_{\widehat{\Omega}} \widehat{w} \frac{\partial f^C}{\partial x} \, d\Omega = \int_{\widehat{\Omega}} \widehat{w} \frac{\partial f}{\partial x} \, d\Omega + \int_{\widehat{\Omega}} \varepsilon \psi \frac{\partial \widehat{w}}{\partial x} \sum_{\ell=1}^{L} a_\ell \alpha_\ell A_\ell \frac{\partial q}{\partial x} \, d\Omega \tag{9.69}$$

Considering that $\widehat{w}$ has compact support in $\widehat{\Omega}$, it vanishes on the boundary $\partial\widehat{\Omega}$ of $\widehat{\Omega}$. As a result, integrating (9.69) by parts generates

$$\int_{\widehat{\Omega}} \widehat{w} \left[\frac{\partial f^C}{\partial x} - \frac{\partial f}{\partial x} + \frac{\partial}{\partial x} \left(\varepsilon \psi \sum_{\ell=1}^{L} a_\ell \alpha_\ell A_\ell \frac{\partial q}{\partial x} \right) \right] d\Omega = 0 \tag{9.70}$$

which contains no boundary integrals. Since this integral must vanish for arbitrary test functions $\widehat{w}$ and domains $\widehat{\Omega}$, its integrand must identically equal zero, which generates the following expression for the divergence of the characteristics-bias flux f^C

$$\frac{\partial f^C}{\partial x} = \frac{\partial f}{\partial x} - \frac{\partial}{\partial x} \left(\varepsilon \psi \sum_{\ell=1}^{L} a_\ell \alpha_\ell A_\ell \frac{\partial q}{\partial x} \right) \tag{9.71}$$

and associated flux f^C

$$f^C = f - \varepsilon \psi \sum_{\ell=1}^{L} a_\ell \alpha_\ell A_\ell \frac{\partial q}{\partial x} \tag{9.72}$$

Expression (9.71) exhibits an upstream-bias artificial diffusion, in the form of a second-order differential expression with matrix

$$\mathcal{A} \equiv \sum_{\ell=1}^{L} a_\ell \alpha_\ell A_\ell \tag{9.73}$$

For physical consistency of the upstream bias in (9.67)- (9.71) and associated mathematical stability of the corresponding second-order differential expression, as shown in Section 9.2.2, all the eigenvalues of this upstream matrix must be distinct and positive. This requirement becomes a fundamental upstream-bias stability condition.

The continuum expression (9.71) for the divergence of the characteristics-bias flux constitutes a non-discrete generalization of the various numerical-flux formulae employed in several CFD upwind schemes. It encompasses, generalizes, and unifies flux-vector and flux-difference schemes as shown by the following representative examples.

9.3.4   Flux Vector Splitting

Consider a representative Flux Vector Splitting (FVS), e.g. van Leer's. In this formulation, the inviscid flux f is "split" as

$$f = f^+ + f^- \tag{9.74}$$

where the Jacobian matrices of f^+ and f^- respectively possess non-negative and non-positive eigenvalues.

The FJD expression (9.65) encompasses (9.74) with $L = 2$ as

$$\sum_{\ell=1}^{L} \alpha_\ell A_\ell = \frac{\partial f^+}{\partial q} + \frac{\partial f^-}{\partial q}, \quad \alpha_1 = 1, \quad \alpha_2 = 1 \tag{9.75}$$

The corresponding characteristics-bias flux divergence for FVS accrues from (9.71) with $\psi = 1$, $a_1 = 1$, $a_2 = -1$ as

$$\frac{\partial f^C}{\partial x} = \frac{\partial f}{\partial x} - \frac{\partial}{\partial x}\left(\varepsilon\left(\frac{\partial f^+}{\partial q} - \frac{\partial f^-}{\partial q}\right)\frac{\partial q}{\partial x}\right) = \frac{\partial f}{\partial x} - \frac{\partial}{\partial x}\left(\varepsilon\left(\frac{\partial f^+}{\partial x} - \frac{\partial f^-}{\partial x}\right)\right) \tag{9.76}$$

which generalizes in the continuum the traditional numerical flux formulae for FVS constructions.

The associated upstream matrix $\mathcal{A}$ is

$$\mathcal{A} = \left(\frac{\partial f^+}{\partial q}\right) + \left(-\frac{\partial f^-}{\partial q}\right) \tag{9.77}$$

The upstream-bias stability condition, however, is not automatically satisfied, even though each of the two matrices $\left(\frac{\partial f^+}{\partial q}\right)$ and $\left(-\frac{\partial f^-}{\partial q}\right)$ has positive eigenvalues. This stability condition is not unconditionally satisfied because the sum of two positive-eigenvalue matrices does not necessarily yield a matrix with positive eigenvalues. As an example consider the following matrix sum of two positive - eigenvalue matrices

$$\begin{pmatrix} 2 & , & \sigma \\ 3 & , & 6 \end{pmatrix} = \begin{pmatrix} 1 & , & 0 \\ 3 & , & 2 \end{pmatrix} + \begin{pmatrix} 1 & , & \sigma \\ 0 & , & 4 \end{pmatrix} \tag{9.78}$$

where σ is a real number. One of the eigenvalues of this matrix sum is negative for $\sigma > 4$. For instance, for $\sigma = 7$ the eigenvalues are $+9$ and -1.

Most likely, however, (9.77) satisfies the upstream-bias stability condition for most of the flow conditions considered in the technical literature, in view of the stable results reported. For subsonic flows, each of the two flux vector components in (9.74) remains unrelated to the physics of acoustics or convection. On the other hand, (9.74) is computationally advantageous, for it calls for the discretization of simple flux-vector components.

9.3.5 Flux Difference Splitting

Consider next a representative Flux Difference Splitting (FDS) development, e.g. Roe's. In this formulation, the inviscid flux Jacobian of f is "split" as

$$\frac{\partial f}{\partial q} = X\Lambda^+ X^{-1} + X\Lambda^- X^{-1} \tag{9.79}$$

where X and $\Lambda = \Lambda^+ + \Lambda^-$ denote the right eigenvector matrix and eigenvalue diagonal matrix of the Jacobian, all evaluated at special average values of q, with Λ^+ and Λ^- respectively containing non- negative and non-positive eigenvalues. The matrices at the rhs of (9.79), therefore, will respectively possess non-negative and non-positive eigenvalues.

The FJD expression (9.65) encompasses (9.79) with $L = 2$ as

$$\sum_{\ell=1}^{L} \alpha_\ell A_\ell = X\Lambda^+ X^{-1} + X\Lambda^- X^{-1}, \qquad \alpha_1 = 1, \qquad \alpha_2 = 1 \tag{9.80}$$

The corresponding characteristics-bias divergence for this formulation accrues from (9.71) with $\psi = 1$, $a_1 = 1$, $a_2 = -1$ as

$$\frac{\partial f^C}{\partial x} = \frac{\partial f}{\partial x} - \frac{\partial}{\partial x}\left(\varepsilon \left(X\Lambda^+ X^{-1} - X\Lambda^- X^{-1} \right) \frac{\partial q}{\partial x} \right)$$

$$= \frac{\partial f}{\partial x} - \frac{\partial}{\partial x}\left(\varepsilon X \left(\Lambda^+ - \Lambda^- \right) X^{-1} \frac{\partial q}{\partial x} \right) \tag{9.81}$$

which generalizes in the continuum the traditional numerical flux formulae for FDS constructions.

The associated upstream matrix $\mathcal{A}$ is

$$\mathcal{A} = X \left(\Lambda^+ - \Lambda^- \right) X^{-1} \tag{9.82}$$

which has non-negative eigenvalues and therefore automatically satisfies the upstream-bias stability condition for any flow state for which no eigenvalue vanishes. The discretization of (9.81) calls for more computational operations than (9.76), while each of the two rhs components in (9.79) lumps into one matrix the matrices representative of the distinct acoustics and convection wave propagation mechanisms. On the other hand, numerous numerical results bear out the accuracy of an FDS formulation.

9.3.6   Characteristics-Bias System

As noted in Section 9.2, the unique weak solution associated with the unperturbed system is also obtained as the limiting solution of the non-linear equation

$$\frac{\partial q^\varepsilon}{\partial t} + \frac{\partial f(q^\varepsilon)}{\partial x} = \varepsilon \frac{\partial}{\partial x} \left(\frac{\partial \varphi\left(q^\varepsilon\right)}{\partial x} \right) \tag{9.83}$$

as the positive parameter ε approaches zero. This equation is but a prototype of the characteristics-bias system. Since practical computations solve the initial-boundary value problem with a perturbation parameter $\varepsilon\psi$ that remains commensurate with a decreasing, but finite, mesh spacing, for efficiency's and accuracy's sake, system (9.83) should be solved by an exact solution of the unperturbed system for finite $\varepsilon\psi$, in the case of a Jacobian $\frac{\partial f}{\partial q}$ with equal-sign eigenvalues, and generally admit solutions q^ε that should rapidly converge to a weak-solution q for decreasing $\varepsilon\psi$. It is to promote these properties that the perturbation in the previous sections has been developed from a characteristics-bias representation of the original system. Dispensing with superscript ε, for simplicity, (9.61) and the developments in the previous sections lead to the characteristics-bias system

$$\frac{\partial q}{\partial t} - \frac{\partial}{\partial x}\left[\varepsilon\psi\text{sgn}(u)\frac{\partial q}{\partial t} \right] - \phi + \frac{\partial}{\partial x}\left[\varepsilon\psi\text{sgn}(u)\,\phi \right] + \frac{\partial f}{\partial x} - \frac{\partial}{\partial x}\left(\varepsilon\psi \sum_{\ell=1}^{L} a_\ell \alpha_\ell A_\ell \frac{\partial q}{\partial x} \right) = 0 \tag{9.84}$$

Since

$$\frac{\partial f}{\partial x} = \sum_{\ell=1}^{L} \alpha_\ell A_\ell \frac{\partial q}{\partial x} \qquad (9.85)$$

this system may be more lucidly expressed as

$$\left[I - \frac{\partial}{\partial x} \varepsilon \psi \mathrm{sgn}(u) \cdot \right] \left(\frac{\partial q}{\partial t} - \phi + \frac{\partial f(q)}{\partial x} \right) - \frac{\partial}{\partial x} \left(\varepsilon \psi \mathrm{sgn}(u) \sum_{\ell=1}^{L} (a_\ell - 1) \alpha_\ell A_\ell \frac{\partial q}{\partial x} \right) = 0 \quad (9.86)$$

This "companion" system contains the original system, its derivative with respect to x and a compensating second-order term that emerges from the upstream-bias stability condition on $\mathcal{A}$. This compensating term, however, is only present when the eigenvalues of the flux Jacobian $\frac{\partial f}{\partial q}$ do not all have the same algebraic sign; when they do, $a_\ell \equiv 1$, $1 \leq \ell \leq L$, and the characteristics-bias system becomes

$$\left[I - \frac{\partial}{\partial x} \varepsilon \psi \mathrm{sgn}(u) \cdot \right] \left(\frac{\partial q}{\partial t} - \phi + \frac{\partial f(q)}{\partial x} \right) = 0 \qquad (9.87)$$

which only contains the original system and its derivative with respect to x. In these conditions, significantly, an exact solution of the original system will also satisfy this companion system for any finite $\varepsilon \psi$. When the equations in this system are viewed as quasi-linear equations, independently of one another, and the fluid velocity u is considered constant, the equations are more clearly seen to remain hyperbolic, because of the presence of the mixed derivative with respect to both x and t. In a suitable neighborhood of a shock, where sufficient dissipation is needed to ensure essential non-oscillatory shock capturing, this mixed derivative will have to be eliminated, so that in the vicinity of a shock the characteristics-bias system becomes a more dissipative parabolic system. In this manner, $\mathcal{H}^1(\Omega)$ solutions of this system may converge to $\mathcal{H}^0(\Omega)$ solution of the unperturbed system.

9.4 Galilean Invariance and Characteristics-Bias Diffusion

The unperturbed Euler and Navier-Stokes systems are Galilean invariant. Only because of its characteristics-bias time derivative and source term can the characteristics-bias system (9.86) also satisfy this fundamental physical property. Satisfying this property is important because in this case the characteristics-bias system reduces the amount of induced characteristics-bias diffusion, as shown in this section.

The Galilean invariance property reflects the notion that the law of conservation of mass, the second law of mechanics, and the first law of thermodynamics remain independent from the state of motion of flow reference frames that move with constant velocity with respect to one another, called Galilean frames and illustrated in Figure 9.3. The corresponding continuity, linear-momentum, and energy equation thus remain formally invariant with respect to any co-ordinate transformation between Galilean frames.

With V_B the absolute velocity of the moving-reference origin B, and with V and u respectively the absolute and relative velocity of a fluid particle P, the co-ordinate sets

(X, Y) and (x, y) as well as velocity of P with respect to any two Galilean frames may be related as

$$X = X_B + x = V_B t + x \quad \Rightarrow \quad \left.\frac{\partial X}{\partial t}\right|_x = V_B, \quad u = V - V_B \qquad (9.88)$$

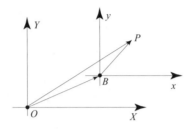

Figure 9.3: Galilean Reference Frames

By virtue of this co-ordinate transformation, a function of (X, t) is cast as $\rho(X, t) = \rho(X(x, t), t)$ and the derivatives with respect to (X, t) in the generalized Euler continuity equation, for instance

$$\left.\frac{\partial \rho A}{\partial t}\right|_x + \left.\frac{\partial \rho u A}{\partial x}\right|_t = \widetilde{m}_b \qquad (9.89)$$

correlate with the derivatives with respect to (x, t) as

$$\left.\frac{\partial \rho A}{\partial t}\right|_x = \left.\frac{\partial \rho A}{\partial X}\right|_t \left.\frac{\partial X}{\partial t}\right|_x + \left.\frac{\partial \rho A}{\partial t}\right|_X, \quad \left.\frac{\partial \rho u A}{\partial x}\right|_t = \left.\frac{\partial \rho u A}{\partial X}\right|_t \qquad (9.90)$$

Upon substituting these correlations for the corresponding derivatives, equation (9.89) becomes

$$V_B \left.\frac{\partial \rho A}{\partial X}\right|_t + \left.\frac{\partial \rho A}{\partial t}\right|_X + \left.\frac{\partial \rho (V - V_B) A}{\partial X}\right|_t = \widetilde{m}_b \quad \Rightarrow \quad \left.\frac{\partial \rho A}{\partial t}\right|_X + \left.\frac{\partial \rho V A}{\partial X}\right|_t = \widetilde{m}_b \qquad (9.91)$$

which coincides with (9.89) and thus proves Galilean invariance of the continuity equation for any magnitude of the mass transfer term $\widetilde{m}_b$, since this term remains unaltered by a Galilean co-ordinate transformation. Likewise, all the other source terms as well as several thermodynamic variables and the upstream-bias functions remain unaffected by this co-ordinate transformation because either they do not depend on the velocity of a reference frame or only involve relative velocities with respect to the flow channel, which velocities transform into one another, thereby retaining their form in the transformation. As a result, similar transformations of the other governing equations, also inserting the continuity equation into the linear-momentum equation and these two equations into the energy equation, show that both the linear-momentum and energy equations are Galilean invariant.

Owing to the characteristics-bias expressions for the source term and time derivative of q, system (9.86) is also Galilean invariant. To show this it suffices to consider the characteristics-bias generalized Euler continuity equation. Anticipating the developments of following chapters, this equation may be conveniently expressed for sgn(u)=1 as

$$
\left.\frac{\partial \rho A}{\partial t}\right|_x + \left.\frac{\partial \rho u A}{\partial x}\right|_t - A\left.\frac{\partial}{\partial x}\left(\varepsilon\psi c\alpha \left.\frac{\partial \rho}{\partial x}\right|_t\right)\right|_t - \widetilde{m}_b + \left.A\frac{\partial}{\partial x}\left(\varepsilon\psi\frac{\widetilde{m}_b}{A}\right)\right|_t
$$
$$
-\left.A\frac{\partial}{\partial x}\left(\frac{\varepsilon\psi}{A}\left(\left(\left.\frac{\partial \rho A}{\partial t}\right|_x + \rho u \left.\frac{\partial A}{\partial x}\right|_t\right) + \left(\left.A\frac{\partial \rho u}{\partial x}\right|_t\right)\right)\right)\right|_t = 0 \qquad (9.92)
$$

where the last term corresponds to the characteristics-bias expression for the continuity-equation component of the flux f and the three terms to the left of it correspond to the characteristics-bias expressions for the source term and time derivative of ρ. A similar expression is obtained for incompressible and free-surface flows. Transforming the derivatives in this equation following developments (9.89)-(9.91) yields

$$
\left.\frac{\partial \rho A}{\partial t}\right|_X + \left.\frac{\partial \rho V A}{\partial X}\right|_t - A\left.\frac{\partial}{\partial X}\left(\varepsilon\psi c\alpha \left.\frac{\partial \rho}{\partial X}\right|_t\right)\right|_t - \widetilde{m}_b + \left.A\frac{\partial}{\partial X}\left(\varepsilon\psi\frac{\widetilde{m}_b}{A}\right)\right|_t
$$
$$
-A\frac{\partial}{\partial X}\left(\frac{\varepsilon\psi}{A}\left(\left(V_B\left.\frac{\partial \rho A}{\partial X}\right|_t + \left.\frac{\partial \rho A}{\partial t}\right|_X + \rho V\left.\frac{\partial A}{\partial X}\right|_t - \rho V_B\left.\frac{\partial A}{\partial X}\right|_t\right)\right.\right.
$$
$$
\left.\left.\left.-A\left(V_B\left.\frac{\partial \rho}{\partial X}\right|_t - \left.\frac{\partial \rho V}{\partial X}\right|_t\right)\right)\right)\right|_t = 0 \qquad (9.93)
$$

with analogous results for the linear-momentum and energy equations. All the terms containing V_B cancel one another out, and as a result this equation coincides with (9.93). Not only is this result showing that the characteristics-bias system is Galilean invariant, but also that this system satisfies this fundamental physical property only when commensurate characteristics-bias expressions are also present for the source term and the time derivative of q.

The importance of this fundamental, but seldom, if ever, mentioned, property in CFD lies in its connection with artificial diffusion. In order to highlight this notion it suffices to consider a linear continuity equation with constant positive A, u, ε, ψ_t, ψ, c, α

$$
\frac{\partial \rho}{\partial t} + u\frac{\partial \rho}{\partial x} - \frac{\partial}{\partial x}\left(\varepsilon\psi_t\frac{\partial \rho}{\partial t} + \varepsilon\psi(c\alpha + u)\frac{\partial \rho}{\partial x}\right) = 0 \qquad (9.94)
$$

where $\psi_t = \psi$ for a Galilean invariant formulation. A generalized eigenfunction solution for this equation may be expressed as

$$
\rho(x,t) = \exp(\lambda t)\exp(\mathbf{j}\,\omega(x - ut)), \quad \mathbf{j} \equiv \sqrt{-1} \qquad (9.95)
$$

where $\lambda = 0$ corresponds to an exact solution of not only the equation with no bias, hence $\varepsilon\psi = 0$, but also the characteristics-bias equation with $\alpha = 0$ and $\psi_t = \psi$, as directly established by substitution. When this substitution is executed for any positive α, ψ_t, the resulting expressions for the dissipation eigenvalue λ is

$$
\lambda = -\frac{\omega c}{2}\lambda_{\omega\varepsilon} = -\frac{\varepsilon c\omega^2((M + \alpha)\psi - M\psi_t)(1 + \mathbf{j}\,\omega\varepsilon\psi_t)}{1 + \omega^2\varepsilon^2\psi_t^2} \qquad (9.96)
$$

Since the real part of this eigenvalue remains negative, the characteristics-bias induces some artificial diffusion, which reaches a minimum for $\psi_t = \psi$, that is for a Galilean invariant formulation, a minimum that vanishes for $\alpha = 0$. As discussed in Chapter 10, the semi-discrete formulation sets $\varepsilon = \Delta x/2$. Figure 9.4 thus displays the variation of the absolute value of the real part of $\lambda_{\omega\varepsilon}$ for $0 \leq 2\omega\varepsilon \leq \pi$, as obtained for two representative cases: a) $M = 0.3$ hence $\alpha = 0.4$, and b) $M = 1.2$ hence $\alpha \equiv 0$. As this chart shows, for both subsonic and supersonic flows, the induced diffusion rapidly decreases as the characteristics-bias system approaches the Galilean invariant form. For supersonic flows, in particular, this induced diffusion vanishes.

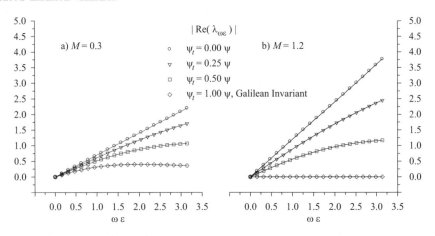

Figure 9.4: Characteristics-Bias Diffusion: a) Subsonic Flow, b) Supersonic Flow

A Galilean invariant characteristics-bias formulation, accordingly, provides a consistent up-winding, but with reduced levels of induced diffusions. As discussed in Chapter 13, the CFD procedure built upon a Galilean invariant characteristics-bias system generates minimally dissipative and essentially non-oscillatory solutions with a sharply captured contact discontinuity, which provides another justification for the particular form of upstream bias developed in this formulation for both the time derivative and source term.

9.5 Non-Discrete Discontinuous Galerkin (DG) Form

For both a weight function w with non-vanishing trace and a solution q that is discontinuous on the boundary $\partial\widehat{\Omega}$ of a subdomain $\widehat{\Omega} \subset \Omega$, an integration by parts of both the flux divergence and the characteristics-bias expression in (9.86) generates the weak statement

$$\int_{\widehat{\Omega}} w \left(\frac{\partial q}{\partial t} - \phi \right) d\Omega + \int_{\widehat{\Omega}} \varepsilon\psi \operatorname{sgn}(u) \frac{\partial w}{\partial x} \left(\frac{\partial q}{\partial t} - \phi + \frac{\partial f}{\partial x} + \sum_{\ell=1}^{L} (a_\ell - 1)\, \alpha_\ell A_\ell \frac{\partial q}{\partial x} \right) d\Omega$$

$$-\int_{\widehat{\Omega}} \frac{\partial w}{\partial x} f \, d\Omega + \oint_{\partial\widehat{\Omega}} w \left(f - \varepsilon\psi \, \mathrm{sgn}(u) \left(\frac{\partial q}{\partial t} - \phi + \frac{\partial f}{\partial x} + \sum_{\ell=1}^{L} (a_\ell - 1)\, \alpha_\ell A_\ell \frac{\partial q}{\partial x} \right) \right) d\,\partial\widehat{\Omega} = 0$$

$$(9.97)$$

This statement may be viewed as a non-discrete Galilean invariant DG formulation. In respect of the second domain integral, this expression presents a direct counterpart of the "shock-capturing" term employed in current DG algorithms; rather than being a Laplacian expression added to the formulation, this term naturally emerges from the characteristics-bias procedure and system. With reference to the boundary integral, this expression features not only the flux vector f, but also an expression that corresponds to the characteristics bias resolution of f; as a result, the entire boundary expression may be viewed as a non-discrete "numerical flux" for f. Rather than requiring the substitution of f with a numerical flux, as takes place in reported DG developments, the characteristics-bias system procedure directly leads to a surface integral with a specific non-discrete numerical flux, in this case the characteristics-bias resolution flux. A discrete DG algorithm may then result from a finite element discretization of (9.97).

9.6   Non-Discrete Streamline Upwind Petrov-Galerkin (SUPG) Form

For a weight function w with compact support in Ω, a continuous solution q, and I indicating the identity matrix of appropriate rank, an integration by parts of the characteristics-bias expression in (9.86) generates the weak statement

$$\int_{\Omega} \left(w + \varepsilon\psi\mathrm{sgn}(u)\frac{\partial w}{\partial x} \right) \left(\frac{\partial q}{\partial t} - \phi + \frac{\partial f}{\partial x} \right) d\Omega$$

$$+ \int_{\Omega} \varepsilon\psi \frac{\partial w}{\partial x} \left(\mathrm{sgn}(u) \sum_{\ell=1}^{L} (a_\ell - 1)\, \alpha_\ell A_\ell \right) \frac{\partial q}{\partial x} \, d\Omega = 0 \qquad (9.98)$$

The boundary integrals from the integration by parts do not feature in this formulation because either the weight function has vanishing trace on the boundary or the boundary conditions described in Chapter 13 are enforced. This statement is recognized as a Galilean invariant non-discrete SUPG integral statement for the generalized equations. As an alternative to the reported SUPG formulations, the upstream-bias term for the generalized equations in the first domain integral in (9.98) does not require any premultiplication by the transpose of the Euler flux Jacobian matrix, which reduces the number of required calculations; as the counterparts of the τ_{SUPG}, and δ_{SUPG} stability parameters, moreover, the characteristics-bias parameters $\varepsilon\psi$ and $\varepsilon\psi(a_\ell-1)$ will explicitly depend on the square or cube of a local mesh spacing, respectively for shocked and smooth flows, which dependence has contributed to the definite asymptotic convergence rates detailed in Chapter 13. The second integral in this statement corresponds to the stability and shock-capturing terms of reported SUPG formulations. Unlike the common ad-hoc terms, the stability and shock-capturing terms in this formulation naturally emerge from characteristic wave propagation, for they are directly generated by the characteristics-bias decomposition of the Euler flux divergence

$\partial f / \partial x$. Considering that $a_\ell = 1$, $1 \leq \ell \leq L$, in regions of supersonic flow, these terms even identically vanish in such regions. In these conditions, the exact solution of the generalized unperturbed Euler system can only satisfy this statement when the characteristics-bias formulation is extended to the source term ϕ, as displayed in (9.98). For quasi-one-dimensional flows, the spatially discrete equations in this book derive from a finite element discretization of this integral statement.

Chapter 10

Characteristics-Bias Controller and Length

This chapter introduces a controller of induced upstream-bias dissipation and a reference length for the characteristics-bias formulation. Denoted ψ, this controller locally varies the amount of upstream bias within the algorithm, as non-linearly determined by the evolving solution. In this algorithm, $\psi = 0$ corresponds to no upstream bias, which leads to a classical centered formulation, whereas $\psi = \psi_{\max} = \mathcal{O}(1)$ corresponds to maximum induced upstream dissipation, for essentially non-oscillatory capturing of shock waves. The reference length ε is established in terms of gradients of local coordinate transformations so that it can lead to a consistent upstream formulation along characteristic directions for multi-dimensional flows and arbitrarily shaped computational cells.

The specific objective of letting ψ vary locally as the solution evolves is to minimize induced upstream-bias dissipation for maximum accuracy within the prescribed computational stencil. As its distinguishing design feature, the characteristics-bias representation remains an authentic multi-dimensional upstream formulation for any ψ with $0 < \psi \leq \psi_{\max}$.

The magnitude of the controller ψ within each element correlates with a local measure of solution smoothness. In regions of smooth flow, the algorithm locally minimizes upstream-bias diffusion, hence ψ approaches $\psi_{\min}$, for accuracy. In regions of discontinuous solutions, the algorithm increases local upstream diffusion, hence ψ approaches $\psi_{\max}$, for stability.

The controller will thus correlate with a bounded local measure φ of solution smoothness and monotonicity. The following sections will detail this measure and then develop the correlation between this measure and the controller. The concluding sections will develop the multi-dimensional length measure ε and then establish the magnitude of $\varepsilon\psi$ with respect to a mesh measure, a magnitude that will influence the order of accuracy of the characteristics-bias equations.

10.1   Measure of Solution Smoothness

A practical measure of local solution smoothness relies upon a set of points where the slopes of the discrete solution are generally discontinuous. For a finite element approximation this set of points is the set of finite-element side nodes, edges, or faces, shared by distinct finite elements, for the continuous finite-element expansion changes basis function from element to element, which implies that solution slopes are generally discontinuous at element side nodes.

For a one-dimensional case, this slope discontinuity at element end nodes is depicted in Figures 10.1-10.2 via the local normal unit vectors n^L and n^R respectively to the left and right of the slope-discontinuity point P, where there exist two distinct normal unit vectors, one for each of two elements sharing the node.

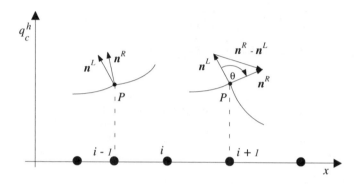

Figure 10.1: Slope Discontinuities and Local Unit Vectors

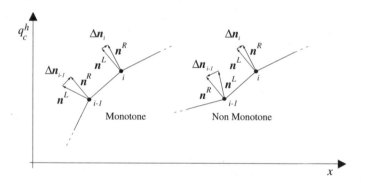

Figure 10.2: Monotone and Non Monotone Slope Distributions

With reference to these figures, the magnitude $\|n^R - n^L\|$ of the vector difference $n^R - n^L$ becomes proportional to a bounded measure of local slope discontinuity. If the graph slope becomes continuous at P, then n^L approaches n^R and $\|n^R - n^L\|$ approaches nought. On the other hand, when a slope discontinuity exists at P, as shown in the figure, $\|n^R - n^L\|$ increases depending on the magnitude of the slope jump. With θ the angle between the unit vectors in Figure 10.1, a positive measure φ of slope discontinuity that vanishes for continuous slopes, remains bounded, and varies between 0 and 1, for $0 \leq \theta \leq 90^0$, may be defined as

$$\varphi \equiv \frac{1}{\sqrt{2}}\|\Delta n\|_i = \frac{1}{\sqrt{2}}\|n^R - n^L\|_{x=x_i} = \sqrt{(1 - \cos\theta)}\Big|_{x=x_i} \tag{10.1}$$

where the final expression follows from the law of cosines.

This concept of unit vector difference also allows measuring local solution monotonicity. Solution monotonicity signifies that solution slopes are locally increasing or constant or decreasing in the neighborhood of a finite element. With reference to Figure 10.2, consider the magnitude of the difference of vector differences $\|\Delta n_i - \Delta n_{i-1}\|$. When the distribution is monotone, then this difference can approach zero because Δn_{i-1} and Δn_i point in virtually the same direction. On the other hand, when the distribution is not monotone, then this difference will exceed zero, because Δn_{i-1} and Δn_i point in opposite directions.

The unit-vector difference and difference of differences may also be employed as indicators of needed local mesh refinement. The following developments determine the local solution measure for a one-dimensional flow. On the basis of these results, the subsequent section generalizes this determination to multi-dimensional flows.

10.1.1   One-Dimensional Flows

The general expression of φ_i corresponding to a scalar measure q_c^h of the solution array q^h directly derives from a finite element expansion. Such an expansion can be expressed in synthetic implicit form as $F(q_c^h, x, t) \equiv q_c^h - q_c^h(x, t) = 0$. Hence, a normal unit vector n can be cast at each time level t as $n \equiv \text{grad } F(q_c^h, x, t)/\|\text{grad } F\|$, where the vector operator "grad" encompasses the dependent variable q_c^h. With e_0 and e_1 the unit vectors in the q- and x-directions, respectively, the expression for n is cast as

$$n(x, t) = \frac{e_0 - \dfrac{\partial q_c^h}{\partial x} e_1}{\sqrt{1 + \left(\dfrac{\partial q_c^h}{\partial x}\right)^2}} \tag{10.2}$$

Let L and R denote the two elements that share a node "i". In terms of the partial derivatives of q_c^h, the unit-vector difference Δn_i becomes

$$\begin{aligned}
\Delta n_i &= \left(\frac{1}{\sqrt{1 + \left(\dfrac{\partial q_c^h}{\partial x}\right)^2}}\Bigg|_R - \frac{1}{\sqrt{1 + \left(\dfrac{\partial q_c^h}{\partial x}\right)^2}}\Bigg|_L\right) e_0 \\
&+ \left(\frac{-\dfrac{\partial q_c^h}{\partial x}}{\sqrt{1 + \left(\dfrac{\partial q_c^h}{\partial x}\right)^2}}\Bigg|_R - \frac{-\dfrac{\partial q_c^h}{\partial x}}{\sqrt{1 + \left(\dfrac{\partial q_c^h}{\partial x}\right)^2}}\Bigg|_L\right) e_1
\end{aligned} \tag{10.3}$$

where the partial derivatives are determined through the finite element expansion. For a uniform grid, the form of expression (10.3) from the two elements sharing node "i" becomes

$$
\Delta n_i = \left(\frac{\Delta x}{\sqrt{\Delta x^2 + \left(q_{c_{i+1}} - q_{c_i} \right)^2}} - \frac{\Delta x}{\sqrt{\Delta x^2 + \left(q_{c_i} - q_{c_{i-1}} \right)^2}} \right) e_0
$$

$$
+ \left(-\frac{q_{c_{i+1}} - q_{c_i}}{\sqrt{\Delta x^2 + \left(q_{c_{i+1}} - q_{c_i} \right)^2}} + \frac{q_{c_i} - q_{c_{i-1}}}{\sqrt{\Delta x^2 + \left(q_{c_i} - q_{c_{i-1}} \right)^2}} \right) e_1 \qquad (10.4)
$$

where the denominator never vanishes. This expression, furthermore, remains bounded for arbitrary nodal values of q_c^h. These expressions are compactly cast as

$$
\Delta n_i = v_0 e_0 + v_1 e_1 \quad \Rightarrow \quad \varphi = \sqrt{v_0^2 + v_1^2} \qquad (10.5)
$$

where the coefficients v_0 and v_1 follow from inspection of (10.3). Accordingly, the square of the solution smoothness measure is calculated as

$$
\varphi_i^2 = \frac{1}{2} \left(\frac{1}{\sqrt{1 + \left(\frac{\partial q_c^h}{\partial x} \right)^2}} \Bigg|_R - \frac{1}{\sqrt{1 + \left(\frac{\partial q_c^h}{\partial x} \right)^2}} \Bigg|_L \right)^2
$$

$$
+ \frac{1}{2} \left(\frac{-\frac{\partial q_c^h}{\partial x}}{\sqrt{1 + \left(\frac{\partial q_c^h}{\partial x} \right)^2}} \Bigg|_R - \frac{-\frac{\partial q_c^h}{\partial x}}{\sqrt{1 + \left(\frac{\partial q_c^h}{\partial x} \right)^2}} \Bigg|_L \right)^2 \qquad (10.6)
$$

10.1.2 Multi-Dimensional Flows

For a multi-dimensional flow the finite element scalar measure $q_c^h = q_c^h(\boldsymbol{x}, t)$ of the solution array q^h is expressed in implicit form as $F(q_c^h, \boldsymbol{x}, t) \equiv q_c^h - q_c^h(\boldsymbol{x}, t) = 0$. With e_0 and e_ℓ the unit vectors in the q- and x_ℓ-directions, $1 \leq \ell \leq n$, respectively, the expression for the unit vector $\boldsymbol{n}$ is cast as

$$
\boldsymbol{n}(\boldsymbol{x}, t) \equiv \frac{\operatorname{grad} F(q_c^h, \boldsymbol{x}, t)}{\|\operatorname{grad} F\|} = \frac{e_0 - \sum_{\ell=1}^{n} \frac{\partial q_c^h}{\partial x_\ell} e_\ell}{\sqrt{1 + \sum_{\ell=1}^{n} \frac{\partial q_c^h}{\partial x_\ell} \frac{\partial q_c^h}{\partial x_\ell}}} \qquad (10.7)
$$

A systematic and unique procedure is next presented to determine the local unit vectors $\boldsymbol{n}_L$ and $\boldsymbol{n}_R$. This procedure generalizes the one-dimensional procedure, yet it still requires only two sets of unit vectors, as in the one-dimensional case. Consider the configuration in

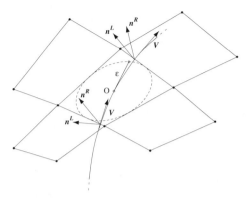

Figure 10.3: Multidimensional Determination of Unit Vectors

Figure 10.3, for a two-dimensional flow, for graphical clarity. The finite element associated with the two sets of unit vectors $\boldsymbol{n}_L$ and $\boldsymbol{n}_R$ has centroid O; let this element be denoted as element O. Consider the streamline through O and identify the two element edges intersected by the line through O that is tangent to this streamline. These two element edges are called the streamline edges. Besides element O, each of these two edges corresponds to another element. The solution over this other element is then used to determine one unit vector $\boldsymbol{n}$ and the solution over element O leads to the second unit vector in the set $\boldsymbol{n}_L$ and $\boldsymbol{n}_R$ at each streamline edge. The second set of unit vectors $\boldsymbol{n}_L$ and $\boldsymbol{n}_R$ corresponds to the other streamline edge. When the tangent to the centroidal streamline within each element happens to intersect not an element edge, but a node, the procedure simply selects any edge that shares both that node and another element. When the tangent to the centroidal streamline leads to a boundary edge or node, then $\Delta\boldsymbol{n} = 0$ at that location. This two-dimensional procedure naturally generalizes the one-dimensional formulation and directly extends to three-dimensional flows. In this case, the tangent to the centroidal streamline intersects two element faces, at which the differences of unit vectors $\Delta\boldsymbol{n}^-$ and $\Delta\boldsymbol{n}^+$ are calculated.

Employing this multi-dimensional procedure, in terms of the partial derivatives of q_c^h and with L and R denoting the two elements sharing the edge or face where $\Delta\boldsymbol{n}$ is calculated, the expression for this vector difference becomes

$$
\Delta\boldsymbol{n} = \left(\left. \frac{1}{\sqrt{1 + \sum_{m=1}^{n}\left(\frac{\partial q_c^h}{\partial x_m}\right)^2}} \right|_R - \left. \frac{1}{\sqrt{1 + \sum_{m=1}^{n}\left(\frac{\partial q_c^h}{\partial x_m}\right)^2}} \right|_L \right) \boldsymbol{e}_0
$$

$$
+ \sum_{\ell=1}^{n} \left(\left. \frac{-\frac{\partial q_c^h}{\partial x_\ell}}{\sqrt{1 + \sum_{m=1}^{n}\left(\frac{\partial q_c^h}{\partial x_m}\right)^2}} \right|_R - \left. \frac{-\frac{\partial q_c^h}{\partial x_\ell}}{\sqrt{1 + \sum_{m=1}^{n}\left(\frac{\partial q_c^h}{\partial x_m}\right)^2}} \right|_R \right) \boldsymbol{e}_\ell
$$

(10.8)

This expression is compactly cast as

$$\Delta \boldsymbol{n} = v_0 \boldsymbol{e}_0 + \sum_{\ell=1}^{n} v_\ell \boldsymbol{e}_\ell \tag{10.9}$$

where the coefficients v_0 and v_ℓ, $1 \leq \ell \leq n$ follow from inspection of (10.8). Accordingly, the solution smoothness measure φ is calculated as

$$\varphi = \sqrt{v_0^2 + \sum_{\ell=1}^{n} v_\ell^2} \tag{10.10}$$

In expression (10.8) each set of partial derivatives of q_c is evaluated within each element that shares a streamline edge or face.

10.2   Correlation between φ and ψ

The amount of induced upstream bias dissipation is determined by the upstream bias controller ψ. Within each element, independently of flow dimensionality, there will be two edge controllers $(\psi)_1$ and $(\psi)_2$; the element controller ψ may then be expressed as the average of these two magnitudes. The upstream dissipation will locally reduce to a minimum, hence $\psi = \psi_{\min}$, when the solution is both smooth and monotone; this dissipation, instead, will approach a maximum, hence $\psi = \psi_{\max}$, in the vicinity of a discontinuity, e.g. a shock.

Each edge controller ψ directly correlates with the solution smoothness measure φ. Given the definition of φ in (10.1), a magnitude for this measure that correlates with a locally continuous solution is $\varphi_c = 0$. The magnitude of φ that should correlate with a local discontinuity may be selected with reference to Figure 10.4. In the neighborhood of a shock,

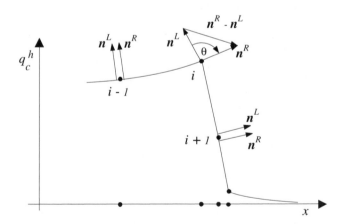

Figure 10.4: Local Unit Vectors at a Shock

the angle θ can become greater than $90°$, as shown in the figure for point i. The threshold $\theta = 90°$ is thus selected for maximum induced dissipation, which from (10.1) leads to $\varphi_D = 1$.

The controller ψ is then expressed as a cubic spline variation that depends upon φ, so that $\psi(\varphi_C) = \psi_{\min}$, $\psi(\varphi_D) = \psi_{\max}$ and both $\frac{\partial\psi}{\partial\varphi}(\varphi_C)$ and $\frac{\partial\psi}{\partial\varphi}(\varphi_D)$ vanish; in this fashion, ψ induces a gradual upstream-bias. The expression for ψ that satisfies these requirements is

$$
\psi \equiv \begin{cases}
\psi_{\min} & ,\ \varphi \leq \varphi_C \\[2ex]
\psi_{\min} + (\psi_{\max} - \psi_{\min}) \left(3 \left(\dfrac{\varphi - \varphi_C}{\varphi_D - \varphi_C} \right)^2 - 2 \left(\dfrac{\varphi - \varphi_C}{\varphi_D - \varphi_C} \right)^3 \right) & ,\ \varphi_C < \varphi < \varphi_D \\[2ex]
\psi_{\max} & ,\ \varphi_D \leq \varphi
\end{cases}
$$

$$(10.11)$$

as shown in Figure 10.5

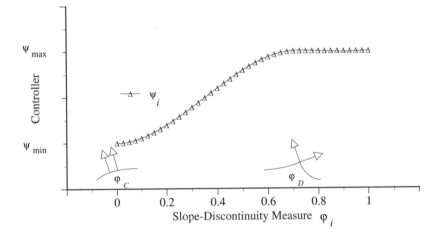

Figure 10.5: Variation of Controller ψ_i

In the neighborhood of a boundary geometric-discontinuity point x_0, the controller ψ decreases from 2.0 to its local magnitude as follows. Let ε, ε^{N_1}, and ε^{N_2} denote the characteristics-bias lengths, derived in the next section, as based on the mutually perpendicular unit vectors a, a^{N_1}, a^{N_2} and consider the ellipsoids Q_1 and Q_2 both centered at x_0 and respectively with semi-axes (2ε, $2\varepsilon^{N_1}$, $2\varepsilon^{N_2}$) and (4ε, $4\varepsilon^{N_1}$, $4\varepsilon^{N_2}$) along the a, a^{N_1}, a^{N_2} directions, with a parallel to velocity. With these specifications, ψ equals two within Q_1 and linearly decreases to its local magnitude towards the surface of Q_2. The characteristics-bias formulation also employs controllers ψ_ϕ and ψ_t to control the upstream bias on the source term and partial derivative with respect to time, with both of these controllers subject to the constraints $\psi_\phi \leq 2/3$, $\psi_t \leq 2/3$, obtained as follows, after determining the one-dimensional characteristics-bias length ε. As shown in Chapter 13, at each internal node of a uniform grid, for the simplifying case of constant ψ, ψ_t, and ψ_ϕ on Ω^h, the discrete system corresponding to the characteristics-bias formulation for the quasi-one-dimensional Euler system for a supersonic flow becomes

$$\left[\left(\frac{\partial q}{\partial t}\right)_{i-1} + 2\left(\frac{\partial q}{\partial t}\right)_i\right]\left(\frac{1}{6} + \frac{\varepsilon\psi_t}{2\Delta x}\right) + \left[2\left(\frac{\partial q}{\partial t}\right)_i + \left(\frac{\partial q}{\partial t}\right)_{i+1}\right]\left(\frac{1}{6} - \frac{\varepsilon\psi_t}{2\Delta x}\right)$$

$$+ (\phi_{i-1} + 2\phi_i)\left(\frac{1}{6} + \frac{\varepsilon\psi_\phi}{2\Delta x}\right) + (2\phi_i + \phi_{i+1})\left(\frac{1}{6} - \frac{\varepsilon\psi_\phi}{2\Delta x}\right)$$

$$+ (f_i - f_{i-1})\left(\frac{1}{2\Delta x} + \frac{\varepsilon\psi}{\Delta x^2}\right) + (f_{i+1} - f_i)\left(\frac{1}{2\Delta x} - \frac{\varepsilon\psi}{\Delta x^2}\right) = 0 \qquad (10.12)$$

This result corresponds to a compact three-node upwind scheme as a linear combination of upstream and downstream contributions, with greater weight on the upstream terms, for ψ_t, ψ_ϕ, ψ are positive. This equation becomes a consistent fully upwind formula when all downwind contributions to (10.12) vanish. In order to establish a rational choice for ε for this one-dimensional case, such a fully upwind formula for f is set to correspond to $\psi = 1$, which leads to the expression

$$\frac{1}{2\Delta x} - \frac{\varepsilon\psi}{\Delta x^2} = \frac{1}{2\Delta x} - \frac{\varepsilon}{\Delta x^2} = 0 \;\Rightarrow\; \varepsilon = \frac{\Delta x}{2} \qquad (10.13)$$

Since ψ controls the amount of induced diffusion, it is possible for $\psi_{\max} > 1$, yet $\psi_{\max} \leq \mathcal{O}(1)$, with $\psi_{\max} > 1$ simply inducing additional diffusion. The discrete upwinding of the source term does not involve any diffusion and the upwinding of the time derivatives induces antidiffusion. Accordingly, this kind of upwinding is constrained in this study, with a maximum set to full upwinding. This requirement implies the following constraints

$$\frac{1}{6} - \frac{\varepsilon\psi_t}{2\Delta x} \geq 0, \quad \frac{1}{6} - \frac{\varepsilon\psi_\phi}{2\Delta x} \geq 0 \qquad (10.14)$$

which with $\varepsilon = \Delta x/2$, leads to

$$0 < \psi_t \leq \frac{2}{3}, \quad 0 < \psi_\phi \leq \frac{2}{3} \qquad (10.15)$$

With reference to ψ_ϕ, when $\psi \leq 2/3$, ψ_ϕ equals ψ, otherwise $\psi_\phi = 2/3$, hence $\psi_\phi \equiv \min(\psi, 2/3)$. In respect of ψ_t, with $\psi_t \leq 2/3$, this controller must rapidly decrease in the neighborhood of a discontinuity, as mentioned in Section 9.3.6, so that the characteristics-bias system may become parabolic in a shock region and thereby locally induce sufficient dissipation, for essentially non-oscillatory shock capturing. This study has established this needed variation as a relationship between ψ_t and φ. In this function, ψ_t equals ψ up to the threshold $\psi_M = \min(\psi(\varphi_M), 2/3)$, with $\varphi_M = 3/4$ and a threshold derivative ψ'_M that equals either the derivative $\psi'(\varphi_M)$, if $\psi_M < 2/3$, or zero, otherwise. Beyond the threshold ψ_M, the controller ψ_t decreases via another cubic spline down to $\psi_t(\varphi_D) = 0$. The expression for ψ_t

that satisfies these requirements is

$$
\psi_t \equiv
\begin{cases}
\min(\psi(\varphi),2/3) & , \quad \varphi_{\min} \leq \varphi \leq \varphi_M \\[2ex]
\begin{aligned}
&\psi_M\left(1 - 3\left(\dfrac{\varphi - \varphi_M}{\varphi_D - \varphi_M}\right)^2 + 2\left(\dfrac{\varphi - \varphi_M}{\varphi_D - \varphi_M}\right)^3\right) + \\
&\psi'_M\left(\left(\dfrac{\varphi - \varphi_M}{\varphi_D - \varphi_M}\right) - 2\left(\dfrac{\varphi - \varphi_M}{\varphi_D - \varphi_M}\right)^2 + \left(\dfrac{\varphi - \varphi_M}{\varphi_D - \varphi_M}\right)^3\right)
\end{aligned} & , \quad \varphi_M < \varphi < \varphi_D \\[2ex]
0 & , \quad \varphi_D \leq \varphi
\end{cases}
\tag{10.16}
$$

The variations of ψ, ψ_ϕ, and ψ_t are displayed in Figure 10.6.

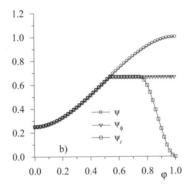

Figure 10.6: Controllers ψ, ψ_ϕ, ψ_t

Expressions (10.11) and (10.16) lead to $\psi = \psi_{\max}$ and $\psi_t = 0$ at a node where the computational solution experiences a discontinuity. These magnitudes are also sustained within a shock region where $\psi > 2/3$. With "n" denoting the flow dimensions, this region encompasses the base element where $\psi > 2/3$ and the $2n$ neighboring elements that share with the base element 2 nodes, 4 edges, or 6 faces, respectively for one-, two-, or three-dimensional flows. The specific individual magnitudes of these controllers within a shock region most directly contribute to non-oscillatory shock capturing. As the computational results have shown, only within a shock region may the three controllers differ from one another; in all other solution regions, instead, $\psi_t = \psi_\phi = \psi$, which, in this case, only requires one controller.

10.3 Multi-Dimensional Length ε

Since the streamline is a characteristic line, the multi-dimensional characteristics-bias length ε is calculated along the tangent to the centroid streamline, within each element, in such a way that this length smoothly varies as the centroid velocity changes orientation. The method presented in this section determines this length for multi-dimensional grids as a natural generalization of a one-dimensional procedure that establishes the result $\varepsilon = \Delta x/2$, obtained in the previous section. For a one-dimensional flow, express the position vector r of a point along the x-axis in terms of a unit vector i and the one-dimensional coordinate transformation

$$r = \frac{1}{2}(1-s)x_L i + \frac{1}{2}(1+s)x_R i \qquad (10.17)$$

where x_L and x_R denote the coordinates of the end nodes of a linear element. On the basis of this expression, the first-order variation of r with respect to s yields

$$\Delta r = \frac{\partial r}{\partial s}\Delta s = \left(-\frac{1}{2}x_L i + \frac{1}{2}x_R i\right)\Delta s = \frac{\Delta x}{2}i\Delta s = \varepsilon i, \quad \Delta s = 1 \qquad (10.18)$$

where Δs denotes a stretching factor such that, for a smooth directional variation, Δr always results from the semi-length of a transformed 1-D element, or a radius of a circle, in 2-D, or sphere, in 3-D, inscribed in the transformed element, a semi-length or radius that equals one unit for isoparametric elements. With these considerations, (10.18) leads to the following generalized expression for ε

$$\varepsilon = \|\Delta r\|_0 \qquad (10.19)$$

where subscript "0" denotes a centroidal evaluation. This expression yields the correct result $\varepsilon = \Delta x_\ell/2$, $1 \leq \ell \leq n$, with n the flow dimensions, for two- or three-dimensional flows over a uniform grid of rectangular elements, when the flow is unidirectionally parallel to the x_ℓ axis. This expression is expanded for the case of a general multi-dimensional flow on arbitrary grids of elements, of any admissible orientation and shape, as follows.

    With reference to Figure 10.7

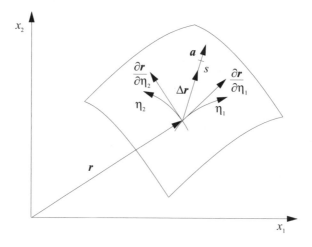

Figure 10.7: Multi-Dimensional Position Vector Variation Δr

the multi-dimensional expression for ε in terms of the local coordinate-transformation derivatives of the position vector r is elaborated as

$$\varepsilon = \|\Delta r\|_0, \quad \Delta r = \frac{\partial r}{\partial s}\Delta s = \left(\frac{\partial r}{\partial \eta_1}\frac{\partial \eta_1}{\partial s} + \frac{\partial r}{\partial \eta_2}\frac{\partial \eta_2}{\partial s} + \frac{\partial r}{\partial \eta_3}\frac{\partial \eta_3}{\partial s}\right)\Delta s \tag{10.20}$$

where s in this case denotes a local coordinate along the straight axis originating at the element centroid and remaining tangent to the centroidal streamline, like the unit vector a. With reference to Figure 10.8 the elementary variations δs and $\delta \eta_i$, $1 \le i \le n$, at the element centroid, are correlated as

$$\delta s \cos \theta_i = \delta \eta_i, \quad 1 \le i \le n \tag{10.21}$$

The direction cosines $\cos \theta_i$, $1 \le i \le n$, are then expressed in terms of inner products of unit vectors as

$$\cos \theta_i = a \cdot \frac{\dfrac{\partial r}{\partial \eta_i}}{\left\|\dfrac{\partial r}{\partial \eta_i}\right\|}, \quad 1 \le i \le n \tag{10.22}$$

where the derivatives are evaluated at the element centroid.

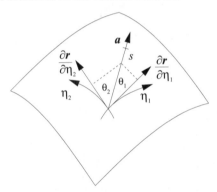

Figure 10.8: Direction Cosines of a

The partial derivatives $\frac{\partial \eta_i}{\partial s}$ and variations $\Delta \eta_i$, $1 \le i \le n$, are thus expressed in terms of readily computed terms as

$$\frac{\partial \eta_1}{\partial s} = a \cdot \frac{\dfrac{\partial r}{\partial \eta_1}}{\left\|\dfrac{\partial r}{\partial \eta_1}\right\|}, \quad \frac{\partial \eta_2}{\partial s} = a \cdot \frac{\dfrac{\partial r}{\partial \eta_2}}{\left\|\dfrac{\partial r}{\partial \eta_2}\right\|}, \quad \frac{\partial \eta_3}{\partial s} = a \cdot \frac{\dfrac{\partial r}{\partial \eta_3}}{\left\|\dfrac{\partial r}{\partial \eta_3}\right\|}, \quad \Delta \eta_i = \frac{\partial \eta_i}{\partial s}\Delta s \tag{10.23}$$

The condition that Δr should result from a unitary radius leads to the expression for Δs as

$$\Delta \eta_1^2 + \Delta \eta_2^2 + \Delta \eta_3^2 = 1 \quad \Rightarrow \quad \left(\frac{\partial \eta_1}{\partial s}\right)^2 + \left(\frac{\partial \eta_2}{\partial s}\right)^2 + \left(\frac{\partial \eta_3}{\partial s}\right)^2 = \frac{1}{(\Delta s)^2} \tag{10.24}$$

from which Δs always equals one when (η_1, η_2, η_3) correspond to orthogonal coordinates. Results (10.23) and (10.24) complete the multi-dimensional expression (10.20) for computing ε. This expression returns the correct result $\varepsilon = \Delta x_i / 2$, $1 \leq i \leq n$, for any flow direction, in a uniform grid of square elements. Since ε is measured along the velocity direction within each element, (10.20), for uniform rectangular grids, correctly computes $\varepsilon = \Delta x_1 / 2$ or $\varepsilon = \Delta x_2 / 2$, depending on whether the flow is parallel to either the x_1 or x_2 axis, in complete agreement with one-dimensional results.

When the flow is not locally parallel to any Cartesian axis and since each of the first three terms in (10.23) results from the product of a constant and the velocity direction vector $\boldsymbol{a}$, expression (10.20) always returns the radius of an ellipse associated with each element, as shown in Figure 10.9 for two-dimensional flows; for three-dimensional flows, (10.20) returns the radius of an ellipsoid ellipsoid associated with each element. When the element is either a rhomboid or a regular tetrahedron, this ellipse, in 2-D, or ellipsoid, in 3-D, is inscribed within the associated physical-space element.

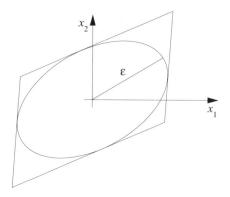

Figure 10.9: Characteristics-Bias Length as Ellipse Radius

Considering that the characteristics-bias length ε is a radius within a smooth closed surface, this mesh measure will always vary smoothly as a function of the velocity direction, within each element of an arbitrary grid. By virtue of (10.18), additionally, this length will always remain commensurate with a mesh spacing, which is beneficial because the upstream-bias coefficient $\varepsilon \psi$ influences the order of accuracy of the discrete equations.

10.4   Magnitude of φ, ψ, $\varepsilon \psi$

This section correlates the magnitudes of φ, ψ, and $\varepsilon \psi$ with a power of a mesh spacing Δx. For the sole purpose of determining the order of φ with respect to Δx, for a smooth solution over any two contiguous elements, the discrete solution values q_{c_j} in (10.4), $i - 1 \leq j \leq i + 1$, over the corresponding elements, can be considered as the nodal values of a single auxiliary continuous functions $q_c(x, t)$ that interpolates the q_{c_j}'s over the elements. With this specification, the Taylor's series expansion of (10.4) yields

$$\varphi_i = \frac{\left|q_c''(x_i,t)\right|}{1 + \left(q_c'(x_i,t)\right)^2} \frac{\Delta x}{2} + \mathcal{O}\left(\Delta x^2\right) = \mathcal{K}\left[1 + \left(q_c'(x_i,t)\right)^2\right]^{\frac{1}{2}} \frac{\Delta x}{2} + \mathcal{O}\left(\Delta x^2\right) \qquad (10.25)$$

where superscript prime indicates differentiation with respect to x and $\mathcal{K}$ denotes the local curvature. This expansion reveals that φ_i decreases for vanishing Δx. Even for large slopes, furthermore, φ_i remains of order Δx in regions of small curvature. Only when both curvature and slope drastically rise, e.g. at a shock, will φ_i increase, precisely as needed.

With reference to $\psi_{\min}$, the characteristics-bias system formulation allows this stability parameter to decrease in relation to a grid spacing measure as

$$\psi_{\min} = C_0 \left(\frac{\Delta x}{\Delta x_0}\right)^{m_\psi} = C_\psi \Delta x^{m_\psi} \qquad (10.26)$$

with C_0 and Δx_0 specified constants, e.g. $\Delta x_0 = 1/100$, $C_0 \leq \frac{1}{2}\left(\frac{\Delta x}{\Delta x_0}\right)^{-m_\psi}$, and with $m_\psi = 1$ for shocked flows, and $m_\psi = 2$ for continuous flows; expression (10.26) also applies to $\psi_{\max}$, with $m_\psi = 0$ for shocked flows. For small φ, significantly, the order of the variable ψ also coincides with Δx^{m_ψ}. To show this result, ψ may be expanded as

$$\psi(\varphi) = \psi(0) + \frac{\partial\psi}{\partial\varphi}(0)\varphi + \frac{\partial^2\psi}{\partial\varphi^2}(0)\frac{\varphi^2}{2} + \mathcal{O}(\varphi^3) = \psi_{\min} + \frac{\partial^2\psi}{\partial\varphi^2}(0)\frac{\varphi^2}{2} + \mathcal{O}(\varphi^3) \qquad (10.27)$$

as obtained with $\psi(0) = \psi_{\min}$ and $\frac{\partial\psi}{\partial\varphi}(0) = 0$. As shown in (10.25), the order of φ with respect to Δx is

$$\varphi = \mathcal{K}\left[1 + \left(q_c'(x_i,t)\right)^2\right]^{\frac{1}{2}} \frac{\Delta x}{2} + \mathcal{O}\left(\Delta x^2\right) \qquad (10.28)$$

As a result, ψ, ψ_ϕ, and ψ_t remain of order $(\Delta x)^{m_\psi}$, with $\varepsilon\psi$ of consequent order $(\Delta x)^{m_\psi+1}$. This order has contributed to convergent computational solutions, with and without shocks.

Chapter 11

Computational Analysis of Quasi-1-D Incompressible Flows

This chapter describes a characteristics-bias procedure for the computational analysis of incompressible flows. Within an incompressible-flow field, the propagation speeds of acoustic and convection waves become unbounded. It is however possible to investigate characteristic information propagation for the finite wave speeds associated with a slight-compressibility form of the continuity equation, which emerges from a direct polytropic correlation between pressure and density. This procedure naturally resembles an artificial-compressibility formulation for the incompressible-flow Navier-Stokes equations. In this book, however, this formulation is exclusively employed both to investigate information propagation and to develop a corresponding characteristics representation. The terminal continuity equation in this representation arises from a limiting form for an unbounded polytropic exponent and eventually features no time derivative of pressure. This result ensures an unbounded signal propagation velocity and thus leads to consistent numerical investigations of both steady and unsteady incompressible flows.

The slight-compressibility system matrices model coupled convection and acoustic propagation. The acoustic components correspond to the slight-compressibility acoustics equations, which physically meaningful building blocks for the non-discrete upstream-bias formulation detailed in Chapter 9. The characteristics-bias formulation for incompressible flows, rests on a decomposition of the Euler system matrix components that model the physics of acoustics and convection and this chapter documents performance of the Acoustics-Convection Upstream Resolution Algorithm for the computational solution of the quasi-one-dimensional incompressible Euler equations. Despite the absence of any time derivative of pressure within the continuity equation, the algorithm simultaneously solves the coupled continuity and linear - momentum equations and directly generates pressure and velocity fields for both steady and unsteady flows, without requiring any post processing of the pressure and velocity.

The discrete equations originate from a Galerkin finite element discretization of the characteristic-bias system, which employs the same order of elements for both pressure and speed. This finite element discretization naturally and automatically leads to consistent boundary differential equations and a new outlet pressure boundary condition that requires no algebraic extrapolation of variables and allows the direct calculation of pressure-driven flows. The resulting discrete equations correspond to an essentially centered discretization in the form of a non-linear combination of upstream diffusive and downstream anti-diffusive flux differences, with greater bias on the upstream diffusive flux difference. This formulation, furthermore, directly accommodates an implicit solver, for required Jacobian matrices are determined in a straightforward manner. The finite element equations form a system of algebraic and differential equations. Despite the absence of any pressure time derivative in this system, the equations are directly integrated in time within a compact block tri-diagonal matrix statement by way of the implicit non-linearly stable Runge-Kutta algorithm (8.5).

As documented by several computational results for both steady and unsteady flows, the acoustics-convection solver induces but minimal artificial diffusion and generates essentially non-oscillatory solutions that reflect available exact solutions. The calculations for steady flows rapidly converge to steady state and for steady and unsteady flows the algorithm conserves both mass and stagnation pressure.

11.1   Slight-Compressibility Continuity Equation

For the quasi-one-dimensional compressible-flow continuity equation

$$\frac{\partial \rho}{\partial t} + \frac{\partial \rho u}{\partial x} + \frac{\rho u}{A}\frac{dA}{dx} = 0 \tag{11.1}$$

consider the isentropic correlation between pressure p and density ρ

$$p = p(\rho) \tag{11.2}$$

For a perfect gas this result becomes

$$p = p_o \left(\frac{\rho}{\rho_o}\right)^{\gamma} \tag{11.3}$$

where $\gamma = c_p/c_v$ denotes the ratio of specific heats. A general polytropic equation of state emerges as

$$p = p_o \left(\frac{\rho}{\rho_o}\right)^{\kappa\gamma} \tag{11.4}$$

with $\kappa \geq 0$ denoting the polytropic exponent. With this result, the time derivative of density correlates with the time derivative of pressure as

$$\frac{\partial \rho}{\partial t} = \frac{\rho}{\kappa\gamma p}\frac{\partial p}{\partial t} \tag{11.5}$$

which leads to the continuity equation

$$\frac{\rho}{\kappa\gamma p}\frac{\partial p}{\partial t} + \frac{\partial \rho u}{\partial x} + \frac{\rho u}{A}\frac{dA}{dx} = 0 \tag{11.6}$$

For an incompressible flow, hence $\rho \simeq$ constant, this equation transforms into

$$\frac{1}{\kappa\gamma p}\frac{\partial p}{\partial t} + \frac{\partial u}{\partial x} + \frac{u}{A}\frac{dA}{dx} = 0 \tag{11.7}$$

For $\kappa = 1$, this equation becomes the slight-compressibility continuity equation, which governs the low-speed flow of a perfect gas. This is the equation employed in the characteristic analysis of incompressible flows and associated development of a characteristics-bias formulation. As noted in Section 4.2.3, $c_p = c_v$ for incompressible flows, hence $\gamma = 1$, which is the magnitude of γ used in this formulation. The corresponding formulation for incompressible flows then results by replacing $\frac{\partial p}{\partial t}$ with $\frac{1}{\kappa}\frac{\partial p}{\partial t}$, showing in Section 11.6 that the corresponding characteristics-bias perturbation in the new system remains stable for any $\kappa > 0$, and then taking the limit for $\kappa \to \infty$. With this limit, the resulting continuity equation achieves a form that does not contain any time derivative of pressure.

11.2 Incompressible Flow "Characteristics"

The unsteady quasi-one-dimensional Euler equations are cast as

$$\begin{cases} \dfrac{\partial u}{\partial x} + \dfrac{u}{A}\dfrac{dA}{dx} = 0 \\[2mm] \dfrac{\partial u}{\partial t} + \dfrac{\partial}{\partial x}\left(u^2 + \dfrac{p}{\rho}\right) + \dfrac{u^2}{A}\dfrac{dA}{dx} = 0 \end{cases} \tag{11.8}$$

with steady-state form

$$\begin{cases} \dfrac{\partial u}{\partial x} + \dfrac{u}{A}\dfrac{dA}{dx} = 0 \\[2mm] \dfrac{\partial}{\partial x}\left(u^2 + \dfrac{p}{\rho}\right) + \dfrac{u^2}{A}\dfrac{dA}{dx} = 0 \end{cases} \tag{11.9}$$

The solution of this time independent system also accrues as the steady-state solution of the slight-compressibility unsteady system

$$\begin{cases} \dfrac{1}{\kappa\gamma p}\dfrac{\partial p}{\partial t} + \dfrac{\partial u}{\partial x} + \dfrac{u}{A}\dfrac{dA}{dx} = 0 \\[2mm] \dfrac{\partial u}{\partial t} + \dfrac{\partial}{\partial x}\left(u^2 + \dfrac{p}{\rho}\right) + \dfrac{u^2}{A}\dfrac{dA}{dx} = 0 \end{cases} \tag{11.10}$$

which formally coincides with (11.8) for $\kappa \to \infty$.

This Chapter develops the characteristics-bias formulation for an equation system that contains the slight-compressibility continuity equation

$$\frac{1}{\gamma p}\frac{\partial p}{\partial t} + \frac{\partial u}{\partial x} + \frac{u}{A}\frac{dA}{dx} = 0 \tag{11.11}$$

The slight-compressibility system can be cast as

$$\frac{\partial q}{\partial t} - \phi + A(q)\frac{\partial q}{\partial x} = 0 \tag{11.12}$$

For non-trivial solutions q, the characteristic speeds of this system are the Euler eigenvalues of the system matrix

$$A(q) = \begin{pmatrix} 0 & , & \gamma p \\ \frac{1}{\rho} & , & 2u \end{pmatrix} \tag{11.13}$$

These eigenvalues have been exactly determined in closed form as

$$\lambda_1^E = u + \sqrt{\left(u^2 + \gamma\frac{p}{\rho}\right)}, \qquad \lambda_2^E = u - \sqrt{\left(u^2 + \gamma\frac{p}{\rho}\right)} \tag{11.14}$$

The expression $\gamma p/\rho$ formally coincides with the square of the compressible-flow speed of sound c. These eigenvalues are thus expressed as

$$\lambda_1^E = u + \sqrt{(u^2 + c^2)}, \qquad \lambda_2^E = u - \sqrt{(u^2 + c^2)} \tag{11.15}$$

with non-dimensional form

$$\lambda_1^\Im = \Im + \sqrt{1 + \Im^2}, \qquad \lambda_2^\Im = \Im - \sqrt{1 + \Im^2} \tag{11.16}$$

where $\Im$ denotes the incompressible-flow number

$$\Im \equiv \frac{u}{\sqrt{\gamma\frac{p}{\rho}}} \tag{11.17}$$

For a vanishing magnitude of $\Im$, these eigenvalues vary like $\Im \pm 1$; for any magnitude of $\Im$, one eigenvalues always keeps positive, the other always remains negative, as Figure 11.1 shows.

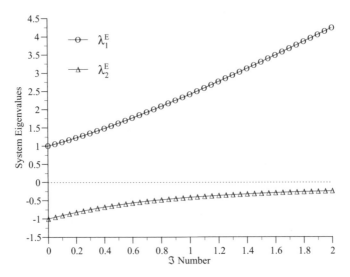

Figure 11.1: Slight-Compressibility Eigenvalues

This situation portray an incompressible flow. These eigenvalues correspond to the slopes of the characteristics, as portrayed in Figure 11.2

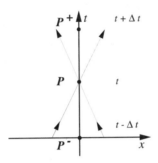

Figure 11.2: Slight-Compressibility Characteristics

This figure shows the characteristics in a suitable neighborhood of a flow field point P in a (t, x) plane. Wave propagation remains subsonic, which occurs by both convection and acoustics, bi-modally from both upstream and downstream toward and then away from P; for small $\Im$, wave propagation is essentially acoustic. The upstream CFD algorithm in this book mathematically models this coupled acoustic-convection wave propagation. The algorithm identifies the genuine convection and acoustics components within the incompressible-flow system matrix and then establishes a physically consistent upstream approximation for each of these components.

11.3   Slight-Compressibility Acoustics Equations

The quasi-one-dimensional slight-compressibility equations

$$\frac{\partial q}{\partial t} - \phi + A(q)\frac{\partial q}{\partial x} = 0 \tag{11.18}$$

where

$$A(q) = \begin{pmatrix} 0 & , & \gamma p \\ \frac{1}{\rho} & , & 2u \end{pmatrix} \tag{11.19}$$

contain the pseudo-acoustics equations for vanishing $\Im$ numbers. Identification of these equations yields the acoustics component of the system matrix for any $\Im$ number. Upon writing the velocity component u in terms of the $\Im$ number as $u = c\Im u/|u|$ the quasi-one dimensional system (11.18) becomes

$$\frac{\partial p}{\partial t} + \gamma p\frac{\partial u}{\partial x} + \gamma pc\Im\frac{u/|u|}{A}\frac{dA}{dx} = 0$$

$$\frac{\partial u}{\partial t} + \frac{\partial p/\rho}{\partial x} + c\Im \frac{u}{|u|}\left(2\frac{\partial u}{\partial x} + \frac{u/|u|}{A}\frac{dA}{dx}\right) = 0 \tag{11.20}$$

and for a vanishing $\Im$ number, these equations reduce to the non-linear acoustics system

$$\frac{\partial}{\partial t}\begin{pmatrix} p \\ u \end{pmatrix} + \begin{pmatrix} 0 & , & \gamma p \\ 1/\rho & , & 0 \end{pmatrix}\frac{\partial}{\partial x}\begin{pmatrix} p \\ u \end{pmatrix} = 0 \quad \Rightarrow \quad \frac{\partial q}{\partial t} + A^a\frac{\partial q}{\partial x} = 0 \tag{11.21}$$

The dependent variables in these equations correspond to those in a flow field that originates from slight perturbations to an otherwise quiescent field.

The matrix A^a

$$A^a = \begin{pmatrix} 0 & , & \gamma p \\ 1/\rho & , & 0 \end{pmatrix} \tag{11.22}$$

has the eigenvalues

$$\lambda^a_{1,2} = \pm\sqrt{\gamma\frac{p}{\rho}} = \pm c \tag{11.23}$$

Where c corresponds to the isentropic speed of sound. This matrix, accordingly, is the acoustic matrix. This matrix naturally features a complete set of eigenvectors and thus possesses the similarity form

$$A^a \equiv X\Lambda^a X^{-1} = X\Lambda^{a+}X^{-1} + X\Lambda^{a-}X^{-1}, \quad \Lambda^a = \Lambda^{a+} + \Lambda^{a-} \tag{11.24}$$

where Λ^{a+} and Λ^{a-} respectively contain non-negative and non-positive eigenvalues as follows

$$\Lambda^{a+} \equiv \begin{pmatrix} c & , & 0 \\ 0 & , & 0 \end{pmatrix} \quad , \quad \Lambda^{a-} \equiv -\begin{pmatrix} 0 & , & 0 \\ 0 & , & c \end{pmatrix} \tag{11.25}$$

The eigenvector matrices X and X^{-1} are

$$X = \begin{pmatrix} \gamma p & , & \gamma p \\ c & , & -c \end{pmatrix}, \quad X^{-1} = \frac{1}{2\gamma pc}\begin{pmatrix} c & , & \gamma p \\ c & , & -\gamma p \end{pmatrix} \tag{11.26}$$

This similarity transformation is employed in the following section to determine a decomposition of the system matrix in terms of acoustic components.

11.4 Incompressible-Flow Acoustics-Convection System-Matrix Decomposition

The characteristic-bias formulation for the quasi-one dimensional slight-compressibility system

$$\begin{cases} \dfrac{\partial p}{\partial t} + \gamma p\dfrac{\partial u}{\partial x} + \dfrac{\gamma pu}{A}\dfrac{dA}{dx} = 0 \\[2mm] \dfrac{\partial u}{\partial t} + \dfrac{\partial}{\partial x}\left(u^2 + \dfrac{p}{\rho}\right) + \dfrac{u^2}{A}\dfrac{dA}{dx} = 0 \end{cases} \tag{11.27}$$

emerges from a decomposition of the matrix $A = A(q)$ in this system into acoustic, convection, and pressure-Jacobian components. The following developments apply for arbitrary bounded as well as unbounded magnitudes of γ, which makes the formulation also applicable to authentic incompressible flows.

11.4.1   Convection and Pressure-Jacobian Components

The system matrix $A = A(q)$ can be decomposed into convection and pressure-Jacobian components as

$$A(q) = A^q(q) + A^p(q) \tag{11.28}$$

where the matrices $A^q(q)$ and $A^p(q)$ are defined as

$$A^q(q) \equiv \begin{pmatrix} 0 & , & \gamma p \\ 0 & , & 2u \end{pmatrix}, \quad A^p(q) \equiv \begin{pmatrix} 0 & , & 0 \\ 1/\rho & , & 0 \end{pmatrix} \tag{11.29}$$

With reference to Section 11.2, for any $\Im$ number, the eigenvalues of $A(q)$

$$\lambda_1 = \Im + \sqrt{1 + \Im^2}, \quad \lambda_2 = \Im - \sqrt{1 + \Im^2}, \tag{11.30}$$

have mixed algebraic sign and an upstream approximation of the matrix product $A\partial q/\partial x$ along one single direction remains inconsistent with the two-way propagation of acoustic waves. Without pressure in the momentum equation, however, the speed of sound c vanishes and the corresponding system eigenvalues remain all non-negative. The resulting pure-convection system matrix can then be upstream approximated along one single direction. The system matrix product is thus decomposed as

$$A\frac{\partial q}{\partial x} = A^q \frac{\partial q}{\partial x} + A^p \frac{\partial q}{\partial x} \tag{11.31}$$

For low and vanishing $\Im$ numbers, however, this decomposition is insufficient for an accurate upstream modeling of acoustic waves. For an $\Im$ number that approaches zero, the eigenvalues of the matrix components A^q and A^p are

$$\lambda_1^q = 0, \quad \lambda_2^q = 2u, \quad \lambda_{1,2}^p = 0 \tag{11.32}$$

which certainly all keep the same algebraic sign, but for vanishing $\Im$ number remain far less than the dominant speed of sound c. For low $\Im$ numbers, therefore, an upstream approximation for the components in (11.31) would inaccurately model the physics of acoustics. This difficulty is resolved by further decomposing the pressure matrix in (11.31) in terms of a genuine acoustic component, for accurate upstream modeling of acoustic waves.

11.4.2   Acoustic Components

Based on the acoustic equations (11.21), the system matrix product can be alternatively decomposed for arbitrary $\Im$ numbers and corresponding dependent variables p and u as

$$A(q)\frac{\partial q}{\partial x} = A^q(q)\frac{\partial q}{\partial x} + A^p(q)\frac{\partial q}{\partial x} = A^q(q)\frac{\partial q}{\partial x} + (A^a(q) + A^{nc}(q))\frac{\partial q}{\partial x} \tag{11.33}$$

where the matrices $A^a(q)$ and $A^{nc}(q)$ are defined as

$$A^a \equiv \begin{pmatrix} 0 & , & \gamma p \\ 1/\rho & , & 0 \end{pmatrix}, \quad A^{nc} \equiv \begin{pmatrix} 0 & , & -\gamma p \\ 0 & , & 0 \end{pmatrix} \tag{11.34}$$

The matrix A^{nc} can be termed a "non-linear coupling" matrix, for it completes the non-linear coupling between convection and acoustics within (11.33). All the eigenvalues of A^{nc} identically vanish. No need exists, therefore, to involve A^{nc} in the upstream-bias approximation of the system matrix $A = A(q)$. The eigenvalues of A^a are exactly determined in closed form as

$$\lambda_{1,2}^a = \pm \sqrt{\gamma \frac{p}{\rho}} = \pm c \tag{11.35}$$

This matrix, accordingly, can be used for an acoustic upstream bias formulation. With reference to the similarity transformation (11.24) for this matrix, decomposition (11.33) becomes

$$A(q)\frac{\partial q}{\partial x} = \left(X\Lambda^{a+}X^{-1} + X\Lambda^{a-}X^{-1} + A^a(q) + A^{nc}(q) \right) \frac{\partial q}{\partial x} \tag{11.36}$$

Since the two acoustics matrices at the rhs of (11.36) respectively possess non-negative and non-positive eigenvalues, a characteristics-bias resolution of these matrices involves an upstream approximation of the first matrix and a downstream approximation of the second matrix. These approximations naturally lead to the following absolute acoustics-matrix upstream expression

$$|A^a|\frac{\partial q}{\partial x} \equiv X\left(\Lambda^{a+} - \Lambda^{a-} \right) X^{-1}\frac{\partial q}{\partial x} = \begin{pmatrix} c & , & 0 \\ 0 & , & c \end{pmatrix} \frac{\partial q}{\partial x} \tag{11.37}$$

The beautifully simple result ensues

$$|A^a|\frac{\partial q}{\partial x} = c\frac{\partial q}{\partial x} = cI\frac{\partial q}{\partial x}, \qquad I \equiv \text{identity matrix} \tag{11.38}$$

which indicates for this matrix product the equivalence of replacing $|A^a|$ with the matrix cI, of which all eigenvalues equal $+c$, for all $\Im$ numbers.

11.4.3   Combination of Matrix Decompositions

The previous sections have shown that the system matrix product $A(q)\partial q/\partial x$ can be equivalently expressed as

$$A(q)\frac{\partial q}{\partial x} = \begin{cases} A^q(q)\dfrac{\partial q}{\partial x} + A^p\dfrac{\partial q}{\partial x}; \\[2mm] \left(X\Lambda^{a+}X^{-1} + X\Lambda^{a-}X^{-1} + A^q(q) + A^{nc} \right) \dfrac{\partial q}{\partial x} \end{cases} \tag{11.39}$$

where the first expression is convenient for a characteristics-bias approximation for high $\Im$ numbers and the second expression is needed for low $\Im$ numbers. A decomposition for all $\Im$ numbers is thus cast as a linear combination of these two decompositions, with linear combination parameter α, $0 \leq \alpha \leq 1$

$$A(q)\frac{\partial q}{\partial x} = (1-\alpha)\left(A^q(q) + A^p(q) \right) \frac{\partial q}{\partial x}$$

$$+\alpha\left(X\Lambda^{a+}X^{-1} + X\Lambda^{a-}X^{-1} + A^q(q) + A^{nc}\right)\frac{\partial q}{\partial x} \tag{11.40}$$

which leads to the following acoustics-convection decomposition, with $\delta \equiv 1 - \alpha$

$$A(q)\frac{\partial q}{\partial x} = \alpha\left(X\Lambda^{a+}X^{-1} + X\Lambda^{a-}X^{-1}\right)\frac{\partial q}{\partial x} + A^q(q)\frac{\partial q}{\partial x} + \delta A^p(q)\frac{\partial q}{\partial x} + \alpha A^{nc}\frac{\partial q}{\partial x} \tag{11.41}$$

The streamline acoustic expression $\alpha\left(X\Lambda^{a+}X^{-1} + X\Lambda^{a-}X^{-1}\right)\frac{\partial q}{\partial x}$ accounts for the bi-modal propagation of acoustic waves; this expression is thus employed for an acoustic upstream-bias approximation for low $\Im$ numbers. As $\Im$ increases a greater fraction of the pressure gradient $A^p(q)\frac{\partial q}{\partial x}$ may be involved in the upstream-bias approximation along with the acoustic matrix; accordingly, as $\Im$ increases, the upstream-bias function α may concurrently decrease. No need exists to involve the matrix A^{nc} in the upstream-bias formulation because this matrix is devoid of physical significance and all its eigenvalues vanish. For any magnitude of both pressure and pressure gradient, the convection field uniformly carries information along streamlines; hence, the entire convection term $A^q(q)\frac{\partial q}{\partial x}$ can receive an upstream bias along the single streamline direction.

11.5 Characteristics-Bias System Matrix

Given the algebraic sign of the eigenvalue set of each matrix term in (11.41), the associated direction cosines a_ℓ for the upstream-bias expression corresponding to (9.71) are

$$a_1 = +1, \quad a_2 = -1, \quad a_3 = a_4 = s = \text{sgn}(u), \quad a_5 = 0 \tag{11.42}$$

where $s = \text{sgn}(u)$ denotes the algebraic sign of u. With (11.38), (11.41), and (11.42) the general expression (9.71) leads to the acoustics-convection characteristics matrix resolution

$$F^C = A(q)\frac{\partial q}{\partial x} - \frac{\partial}{\partial x}\left[\varepsilon\psi\left(\alpha cI + sA^q(q) + s\delta A^p(q)\right)\frac{\partial q}{\partial x}\right] \tag{11.43}$$

where I denotes the identity matrix of appropriate size. The terms in this expression, furthermore, directly correspond to the physics of acoustics and convection. For low $\Im$ numbers, $\delta \simeq 0$ and (11.43) reduces to

$$F^C = A(q)\frac{\partial q}{\partial x} - \frac{\partial}{\partial x}\left[\varepsilon\psi\left(\alpha cI + sA^q(q)\right)\frac{\partial q}{\partial x}\right] \tag{11.44}$$

which essentially induces only an acoustics upstream. In the limit of a "vacuum" flow, hence large $\Im$, $\alpha \to 0$ and $\delta \to 1$. Expression (11.43) in this case becomes

$$F^C = A(q)\frac{\partial q}{\partial x} - \frac{\partial}{\partial x}\left[\varepsilon\psi\left(sA(q)\right)\frac{\partial q}{\partial x}\right] \tag{11.45}$$

which corresponds to an upstream approximation of the entire system matrix product.

11.5.1 Consistent Upstream-Bias

The acoustics-convection upstream functions α and δ depend on the $\Im$ number. They are determined by enforcing the stability condition on the perturbation matrix within the characteristics-bias resolution

$$F^C = A(q)\frac{\partial q}{\partial x} - \frac{\partial}{\partial x}\left[\varepsilon\psi\left(\alpha cI + sA^q(q) + s\delta A^p(q)\right)\frac{\partial q}{\partial x}\right] \tag{11.46}$$

The terms between parentheses collectively constitute the upstream-bias dissipation matrix

$$\mathcal{A} \equiv \alpha cI + sA^q(q) + s\delta A^p(q) \tag{11.47}$$

All the eigenvalues of the stability matrix $\mathcal{A}$ have been analytically determined exactly in closed form. Dividing through by the speed of sound c, the non-dimensional form of these eigenvalues is

$$\lambda_{1,2} = \alpha + \Im \pm \left(\Im^2 + \delta\right)^{1/2} \tag{11.48}$$

In order to ensure physical significance for the characteristics-bias matrix in (11.43), hence for the upstream-bias approximation to decomposition (11.41), the upstream bias functions α and δ are determined by forcing the upstream-bias eigenvalues (11.48) to remain positive for all $\Im$ numbers.

Eigenvalues (11.48) remain real for any positive α and δ and λ_1 remains positive for any positive α. The eigenvalue λ_2 remains positive when α and δ lead to the constraint

$$\alpha + \Im - \sqrt{\Im^2 + \delta} \geq 0 \quad \Rightarrow \alpha^2 + 2\alpha\Im \geq \delta \tag{11.49}$$

In particular, since $\delta = 1 - \alpha$, this constraint becomes the following constraint on α

$$\alpha \geq \sqrt{1 + (\Im + 1/2)^2} - (\Im + 1/2) \tag{11.50}$$

Any α and δ that satisfy (11.49), (11.50), therefore, will lead to non-negative characteristic-bias eigenvalues.

11.5.2 Upstream-Bias Eigenvalues and Functions α and δ

Rather than prescribing some expression for α or δ and accepting the resulting variations for these eigenvalues, the function α is determined by correlating the characteristic-bias eigenvalues (11.48) with the system matrix eigenvalues (11.16). From λ_1 and λ_2 in (11.48), the corresponding expression for $\alpha = \alpha(\Im)$ is

$$\alpha(\Im) = \frac{\lambda_1 + \lambda_2}{2} - \Im \tag{11.51}$$

As one way to correlate the characteristic-bias eigenvalues (11.48) with the system-matrix eigenvalues, consider the absolute non-dimensional system-matrix eigenvalues

$$\left|\lambda_1^\Im\right| = \Im + \sqrt{1 + \Im^2}, \quad \left|\lambda_2^\Im\right| = \sqrt{1 + \Im^2} - \Im \tag{11.52}$$

These expressions lead to the average absolute eigenvalue

$$\frac{\left|\lambda_1^\Im\right| + \left|\lambda_2^\Im\right|}{2} = \sqrt{1 + \Im^2} \tag{11.53}$$

The average characteristic-bias eigenvalue in (11.51) is thus set equal to this expression. With this connection, the expression for α from (11.51) becomes

$$\alpha(\Im) = \sqrt{1 + \Im^2} - \Im \tag{11.54}$$

which happens to coincide with $\left|\lambda_2^\Im\right|$, satisfies (11.49), and rapidly decreases for increasing $\Im$. With this result for α, the corresponding expression for δ becomes

$$\delta = 1 - \alpha = 1 + \Im - \sqrt{1 + \Im^2} \tag{11.55}$$

which increases for increasing $\Im$. Owing to expressions (11.54) and (11.55), the upstream bias eigenvalues remain positive, which corresponds to a physically consistent characteristic-bias formulation.

The variations of the upstream-bias functions $\alpha = \alpha(\Im)$ and $\delta = \delta(\Im)$ and the corresponding eigenvalues (11.48) are presented in Figures 11.3, 11.4, respectively. Figure 11.3 indicates that the upstream-bias functions as well as their slopes remain continuous for all $\Im$ numbers. As $\delta = \delta(\Im)$ rises, the upstream-bias contribution from the acoustics matrix correspondingly decreases.

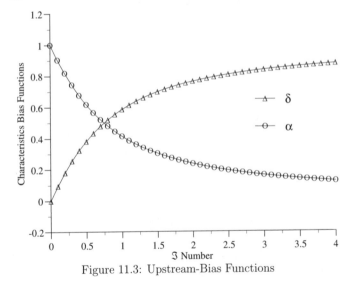

Figure 11.3: Upstream-Bias Functions

With reference to Figure 11.4, the characteristics-bias eigenvalues (11.48) remain positive. These eigenvalues and their slopes remain continuous for all $\Im$ numbers and smoothly approach 1 for vanishing $\Im$, indicating a physically consistent upstream-bias approximation of the acoustic equations embedded within the slight-compressibility equations. For increasing $\Im$, these eigenvalues approach the corresponding absolute slight-compressibility eigenvalues.

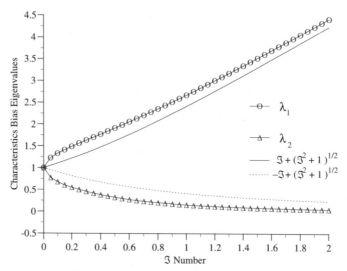

Figure 11.4: Upstream-Bias Eigenvalues

11.6   Incompressible-Flow Formulation

Following the developments in Chapter 9 and in the previous sections, the characteristics-bias formulation for the slight-compressibility continuity equation can be expressed as

$$\frac{\partial p}{\partial t} - \gamma p \phi_p + \gamma p \frac{\partial u}{\partial x} - \frac{\partial}{\partial x}\left[\varepsilon \psi s \left(\frac{\partial p}{\partial t} - \gamma p \phi_p + \gamma p \frac{\partial u}{\partial x} + \alpha c s \frac{\partial p}{\partial x} \right) \right] = 0 \qquad (11.56)$$

where ϕ_p denotes the complete source term in the continuity equation. As anticipated in Section 11.1, the corresponding characteristics-bias continuity equation for totally incompressible flows originates from this equation upon replacing $\frac{\partial p}{\partial t}$ with $\frac{1}{\kappa}\frac{\partial p}{\partial t}$. The complete characteristics-bias system becomes

$$\frac{\partial p}{\partial t} - \kappa \gamma p \phi_p + \kappa \gamma p \frac{\partial u}{\partial x} - \kappa \frac{\partial}{\partial x}\left[\varepsilon \psi s \left(\frac{1}{\kappa}\frac{\partial p}{\partial t} - \gamma p \phi_p + \gamma p \frac{\partial u}{\partial x} + \alpha c s \frac{\partial p}{\partial x} \right) \right] = 0$$

$$\frac{\partial u}{\partial t} - \phi_u + \frac{\partial}{\partial x}\left(u^2 + \frac{p}{\rho} \right) - \frac{\partial}{\partial x}\left[\varepsilon \psi s \left(\frac{\partial u}{\partial t} - \phi_u \frac{\partial}{\partial x}\left(u^2 + \frac{p}{\rho} \right) + \alpha c s \frac{\partial u}{\partial x} + (\delta - 1)\frac{\partial (p/\rho)}{\partial x} \right) \right] = 0$$
$$(11.57)$$

Since the polytropic exponent κ no longer equals one, the eigenvalues of the characteristics-bias parabolic expression within this system will now equal the eigenvalues of the matrix product $I_\kappa A$, where I_κ denotes a positive diagonal matrix with ones on the leading diagonal with the exception of the entry on the first row and column, which equals κ. The eigenvalues of this matrix have been exactly determined as

$$\lambda_{1,2}^\kappa = \alpha\left(\frac{1+\kappa}{2} \right) + \Im \pm \sqrt{\left(\alpha\left(\frac{1-\kappa}{2} \right) + \Im \right)^2 + \kappa\delta} \qquad (11.58)$$

which revert to (11.48) for $\kappa = 1$. These eigenvalues remain positive for any κ as results from the inequality

$$\alpha\left(\frac{1+\kappa}{2}\right) + \Im \geq \sqrt{\left(\alpha\left(\frac{1-\kappa}{2}\right) + \Im\right)^2 + \kappa\delta} \tag{11.59}$$

This inequality leads to

$$\alpha^2 + 2\alpha\Im \geq \delta \tag{11.60}$$

which no longer depends upon κ. This results mirrors (11.49) and is unconditionally satisfied by the expressions already selected for both α and δ. Eigenvalues (11.58) thus remain positive. Figure 11.5 portrays the variation of the eigenvalues (11.58) for $\kappa = 10$, with

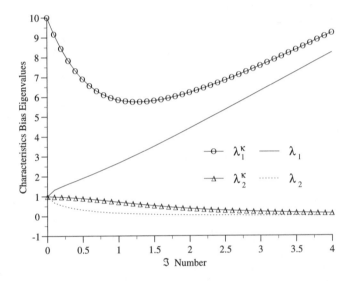

Figure 11.5: Asymptotic Upstream-Bias Eigenvalues

topologically similar curves arising for any other $\kappa > 1$. These curves show these eigenvalues remain positive and approach eigenvalues (11.48) in the limit of a vacuum flow, that is $\Im \to \infty$.

As shown, these eigenvalues remain positive for any positive κ including unbounded magnitudes. The characteristics-bias expression within (11.57), therefore, remains stable for unbounded κ. This special case corresponds to the incompressible-flow equations, which are formally obtained by dividing the continuity equation in system (11.57) by $\kappa\gamma p$, setting $\gamma = 1$, and taking the limit for $\kappa \to \infty$. The limiting system becomes

$$\frac{\partial u}{\partial x} - \phi_p - \frac{1}{p}\frac{\partial}{\partial x}\left[\varepsilon\psi s\left(p\frac{\partial u}{\partial x} - p\phi_p + \alpha cs\frac{\partial p}{\partial x}\right)\right] = 0$$

$$\frac{\partial u}{\partial t} - \phi_u + \frac{\partial}{\partial x}\left(u^2 + \frac{p}{\rho}\right)$$

$$-\frac{\partial}{\partial x}\left[\varepsilon\psi s\left(\frac{\partial u}{\partial t} - \phi_u + \frac{\partial}{\partial x}\left(u^2 + \frac{p}{\rho}\right) + \alpha cs\frac{\partial u}{\partial x} + (\delta - 1)\frac{\partial(p/\rho)}{\partial x}\right)\right] = 0 \qquad (11.61)$$

This system supplies the characteristics-bias formulation for incompressible flows. It does not contain any time pressure derivative in the continuity equation, which thus allows consistent computational simulations of not only steady, but also unsteady flows.

11.7   Variational Statement and Boundary Conditions

In terms of the arrays q_p, q_u, f_p, and f_u^p, which follow from inspection, and for $w \in \mathcal{H}^1(\Omega)$, an integral statement associated with (11.61) becomes

$$\int_\Omega w\left\{\frac{\partial f_p}{\partial x} - \phi_p - \frac{1}{p}\frac{\partial}{\partial x}\left[\varepsilon\psi s\left(p\frac{\partial f_p}{\partial x} - p\phi_p + \alpha cs\frac{\partial q_p}{\partial x}\right)\right]\right\} dx = 0$$

$$\int_\Omega w\left\{\frac{\partial q_u}{\partial t} - \phi_u + \frac{\partial f_u}{\partial x}\right\} dx$$

$$-\int_\Omega w\frac{\partial}{\partial x}\left\{\varepsilon\psi s\left(\frac{\partial q_u}{\partial t} - \phi_u + \frac{\partial f_u}{\partial x} + \alpha cs\frac{\partial q_u}{\partial x} + (\delta - 1)\frac{\partial f_u^p}{\partial x}\right)\right\} dx = 0 \qquad (11.62)$$

An integration by parts of the characteristics-bias perturbation in this statement generates the following weak statement

$$\int_\Omega\left(w + \varepsilon\psi sp\frac{\partial}{\partial x}\left(\frac{w}{p}\right)\right)\left(\frac{\partial f_p}{\partial x} - \phi_p\right) dx + \int_\Omega\frac{\partial}{\partial x}\left(\frac{w}{p}\right)\varepsilon\psi\left(\alpha cs\frac{\partial q_p}{\partial x}\right) dx$$

$$-\left(ws\varepsilon\psi\left(\frac{\partial f_p}{\partial x} - \phi_p + \frac{\alpha cs}{p}\frac{\partial q_p}{\partial x}\right)\right)\Big|_{x=x_{\text{in}}}^{x=x_{\text{out}}} = 0$$

$$\int_\Omega\left(w + \varepsilon\psi s\frac{\partial w}{\partial x}\right)\left(\frac{\partial q_u}{\partial t} - \phi_u + \frac{\partial f_u}{\partial x}\right) dx + \int_\Omega\frac{\partial w}{\partial x}\varepsilon\psi s\left(\alpha cs\frac{\partial q_u}{\partial x} + (\delta - 1)\frac{\partial f_u^p}{\partial x}\right) dx$$

$$-\left(ws\varepsilon\psi\left(\frac{\partial q_u}{\partial t} - \phi_u + \frac{\partial f_u}{\partial x} + \alpha cs\frac{\partial q_u}{\partial x} + (\delta - 1)\frac{\partial f_u^p}{\partial x}\right)\right)\Big|_{x=x_{\text{in}}}^{x=x_{\text{out}}} = 0 \qquad (11.63)$$

The $\varepsilon\psi$ boundary expression is eliminated by enforcing a weak Neumann-type boundary condition for the hyperbolic-parabolic characteristics-bias perturbation. One part of this condition imposes that the original system should be satisfied at the boundary; the remaining terms are also set to zero as part of this weak boundary condition. Clearly, the single Neumann-type condition corresponds to the entire boundary expression set to nought. For increasing $\Im$ number, α decreases and δ increases; as a result, this boundary condition approaches the enforcement of the governing system on the boundary. The resulting weak statement is expressed as

$$\int_\Omega\left(w + \varepsilon\psi sp\frac{\partial}{\partial x}\left(\frac{w}{p}\right)\right)\left(\frac{\partial f_p}{\partial x} - \phi_p\right) dx + \int_\Omega\frac{\partial}{\partial x}\left(\frac{w}{p}\right)\varepsilon\psi s\left(\alpha cs\frac{\partial q_p}{\partial x}\right) dx = 0$$

$$\int_\Omega \left(w + \varepsilon \psi s \frac{\partial w}{\partial x} \right) \left(\frac{\partial q_u}{\partial t} - \phi_u + \frac{\partial f_u}{\partial x} \right) dx + \int_\Omega \frac{\partial w}{\partial x} \varepsilon \psi s \left(\alpha c s \frac{\partial q_u}{\partial x} + (\delta - 1) \frac{\partial f_u^p}{\partial x} \right) dx = 0$$

$$(11.64)$$

For all test functions $w \in \mathcal{H}^1(\Omega)$, the variational formulation seeks a solution $q \in \mathcal{H}^1(\Omega)$ that satisfies this weak statement.

This statement is subject to prescribed initial conditions $q(x,0) = q_0(x)$ and boundary conditions on $\partial\Omega \equiv \overline{\Omega}\backslash\Omega$. Synthetically, these boundary conditions are expressed as

$$B(x_{\partial\Omega}) q(x_{\partial\Omega}, t) = G_v(x_{\partial\Omega}, t) \qquad (11.65)$$

where $G_v(x_{\partial\Omega}, t)$ corresponds to the array of prescribed Dirichlet boundary conditions, with a zero entry for each corresponding unconstrained component of q, and $B(x_{\partial\Omega})$ denotes a square diagonal matrix, with a 1 for each diagonal entry, but replaced by zero for each corresponding unconstrained component of q.

The number of boundary conditions at inlet and outlet depends on the number of negative eigenvalues of $A(q)n_x$, where n_x denotes the x-component of the outward pointing normal unit vector $\mathbf{n}$. A negative eigenvalue signifies a wave propagation in the direction opposite $\mathbf{n}$, hence toward the flow domain, a propagation that carries information from the flow outside the domain, which information is embodied in a corresponding boundary condition.

At an inlet, $A(q)n_x$ always displays one negative eigenvalue; pressure is thus constrained by the Dirichlet boundary condition

$$p(x_{\text{in}}, t) = q_p(t) \qquad (11.66)$$

where $q_p(t)$ denotes a prescribed bounded function.

Similarly, at an outlet, $A(q)n_x$ always displays one negative eigenvalue, which requires one boundary condition. This study has selected a physically meaningful boundary condition on the pressure p, enforced via the linear-momentum integral statement as follows. With $f_u^q \equiv f_u - f_u^p$, $w(x_{\text{in}}) = 0$, and $w(x_{\text{out}}) = 1$ in this statement, an integration by parts of not only the characteristics-bias expression, but also the pressure gradient yields

$$\int_\Omega \left(w + \varepsilon \psi s \frac{\partial w}{\partial x} \right) \left(\frac{\partial q_u}{\partial t} - \phi_u + \frac{\partial f_u^q}{\partial x} \right) dx + \int_\Omega \frac{\partial w}{\partial x} \varepsilon \psi s \left(\alpha c s \frac{\partial q_u}{\partial x} + \delta \frac{\partial f_u^p}{\partial x} \right) dx$$

$$- \int_\Omega \frac{\partial w}{\partial x} \frac{p}{\rho} dx + \left(\frac{p}{\rho} - w s \varepsilon \psi \left(\frac{\partial q_u}{\partial t} - \phi_u + \frac{\partial f_u}{\partial x} + \alpha c s \frac{\partial q_u}{\partial x} + (\delta - 1) \frac{\partial f_u^p}{\partial x} \right) \right)\bigg|_{x=x_{\text{out}}} = 0 \quad (11.67)$$

The entire boundary expression is then set to a prescribed outlet external pressure p_{ext}/ρ; this boundary condition selection serves the double duty of both providing a boundary condition for the parabolic perturbation in the characteristics-bias system and forcing the outlet pressure to approach p_{ext} since ε is set equal to a mesh spacing. The stability, accuracy, and numerical performance of this pressure boundary condition are discussed in Section 11.11.

As elaborated in Section 11.10.3, this set of boundary conditions remains consistent with the physics and mathematics of a quasi-one-dimensional incompressible flow. Mechanically, an incompressible flow within a duct can be induced by lowering the outlet pressure in comparison to the inlet pressure, which generates a force that promotes the flow. The outlet pressure boundary condition is thus naturally enforced as shown. The inlet pressure boundary condition cannot be enforced within the inlet linear-momentum equation. If this were done, the continuity equation would receive no boundary equation and the resulting discrete continuity equations could not contribute to the determination of pressure and speed because these equations form a linearly-dependent system. Any other pressure boundary-condition enforcement strategy that does not involve at least the inlet discrete continuity equation has not generated a stable numerical solution. Instead, the first discrete continuity equation at node $i = 1$ is replaced by the pressure boundary condition shown.

11.8   Finite Element Weak Statement

The finite element solution exists on a partition Ω^h, $\Omega^h \subseteq \Omega$, of Ω, where superscript "h" signifies a spatial discrete approximation. This partition Ω^h has its boundary nodes in $\partial\Omega^h$ on the boundary $\partial\Omega$ of Ω and results from the union of N_e non-overlapping elements Ω_e, $\Omega^h = \bigcup_{e=1}^{N_e} \Omega_e$. With $\mathcal{P}_1(\Omega_e)$ and $\mathcal{P}_{1v}(\Omega_e)$ the finite-dimensional function spaces of respectively diagonal square matrix-valued and vector-valued linear polynomials within each Ω_e, for each "t", the corresponding diagonal square matrix-valued and vector-valued finite element discretization spaces employed in this study are defined as

$$\mathcal{S}^1(\Omega^h) \equiv \left\{ w^h \in \mathcal{H}^1(\Omega^h) : w^h\big|_{\Omega_e} \in \mathcal{P}_1(\Omega_e), \forall \Omega_e \in \Omega^h, B(x_{\partial\Omega^h}) w^h(x_{\partial\Omega^h}) = 0 \right\}$$

$$\mathcal{S}_v^1(\Omega^h) \equiv \left\{ w_v^h \in \mathcal{H}_v^1(\Omega^h) : w_v^h\big|_{\Omega_e} \in \mathcal{P}_{1v}(\Omega_e), \forall \Omega_e \in \Omega^h, B(x_{\partial\Omega^h}) w_v^h(x_{\partial\Omega^h}, t) = G_v(x_{\partial\Omega^h}, t) \right\}$$

$$\tag{11.68}$$

Based on these spaces, the finite element approximation for both speed u and pressure p is sought as $q^h \in \mathcal{S}_v^\eta$. Significantly, the characteristics-bias formulation does not seem to be constrained by the LBB condition and thus permits use of the same approximation space for both u and p. As a result, the spatially discrete solution q^h is determined for each "t" as the solution of the finite element weak statement associated with (11.64)

$$\int_{\Omega^h} \left(w^h + \varepsilon^h \psi^h s^h p^h \frac{\partial}{\partial x}\left(\frac{w^h}{p^h}\right) \right) \left(\frac{\partial f_p^h}{\partial x} - \phi_p^h \right) dx + \int_{\Omega^h} \varepsilon^h \psi^h \frac{\partial}{\partial x}\left(\frac{w^h}{p^h}\right) \left(\alpha^h c^h \frac{\partial q_p^h}{\partial x} \right) dx = 0,$$

$$\int_{\Omega^h} \left(w^h + \varepsilon^h \psi^h s^h \frac{\partial w^h}{\partial x} \right) \left(\frac{\partial q_u^h}{\partial t} - \phi_u^h + \frac{\partial f_u^h}{\partial x} \right) dx$$

$$+ \int_{\Omega^h} \frac{\partial w^h}{\partial x} \varepsilon^h \psi^h s^h \left(\alpha^h c^h s^h \frac{\partial q_u^h}{\partial x} + (\delta^h - 1)\frac{\partial f_u^{p^h}}{\partial x} \right) dx = 0 \tag{11.69}$$

where $s^h \equiv \text{sgn}(u^h)$ denotes the sign of u^h at the centroid of each element; the linear-momentum discrete weak statement for an outlet node then results from a finite element discretization of (11.67) when an outlet pressure boundary condition is enforced. Since every member w^h of $\mathcal{S}^1$ results from a linear combination of the linearly independent basis functions of this finite dimensional space, statement (11.69) is satisfied for N independent basis functions of the space, where N denotes the dimension of the space. For N mesh nodes within Ω^h, there exist clusters of "master" elements Ω_i^m, each comprising only those adjacent elements that share a mesh node x_i, which implies existence of exactly N master elements. Note that each master element represents a "finite volume" as used in finite volume schemes, which, however, do not employ the following finite element "pyramid" test functions. On each master element Ω_i^m, the discrete test function $w^h \equiv w_i = w_i(x)$, $1 \leq i \leq N$, will coincide with the pyramid basis function with compact support on Ω_i^m.

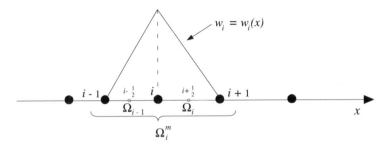

Figure 11.6: Adjacent Finite Elements, Master Element, Pyramid Test Function $w_i(x)$

Such a function equals one at node x_i, zero at all other mesh nodes, and also identically vanishes both on the boundary segments of Ω_i^m not containing x_i, and elsewhere within the computational domain outside Ω_i^m. Statement (11.69) yields a non-linear SUPG formulation, because the controller ψ^h does not remain constant, but varies as a function of a local measure of the discrete solution q^h, as detailed in Chapter 10. As supported by several computations, with subscript "i" denoting numerical value at node i, this measure is established as

$$q_{\text{meas}}^h\Big|_i = \frac{p_{\text{dim}}^h}{p_{\text{ref}}}\Big|_i + \frac{u_{\text{dim}}^h}{u_{\text{ref}}}\Big|_i = p^h\Big|_i + u^h\Big|_i \tag{11.70}$$

and when the nodal u_i^h is positive, this measure becomes the nodal ℓ_1 norm of q_i.

11.8.1 Galerkin Finite Element Expansions and Integrals

The discrete solution $q^h \in \mathcal{S}_v^1$ at each time t assumes the form of the following linear combination

$$q^h(x,t) \equiv \sum_{j=1}^{N} w_j(x) \cdot q^h(x_j, t) \tag{11.71}$$

of nodal solution values and trial basis functions that coincide with the test functions $w_j(x)$ for a Galerkin formulation. Similarly, the source $\phi = \phi(x, q(x,t))$ and fluxes $f = f(q(x,t))$, $f^q = f^q(q(x,t))$ and $f^p = f^p(q(x,t))$ are discretized through the group expressions

$$\phi^h(x,t) \equiv \sum_{j=1}^{N} w_j(x) \cdot \phi\left(x_j, q^h(x_j,t)\right), \qquad f^h(x,t) \equiv \sum_{j=1}^{N} w_j(x) \cdot f\left(q^h(x_j,t)\right)$$

$$f^{q^h}(x,t) \equiv \sum_{j=1}^{N} w_j(x) \cdot f^q\left(q^h(x_j,t)\right), \qquad f^{p^h}(x,t) \equiv \sum_{j=1}^{N} w_j(x) \cdot f^p\left(q^h(x_j,t)\right) \tag{11.72}$$

The notation for the discrete nodal variable and fluxes is then simplified as $q_j(t) \equiv q^h(x_j, t)$, $\phi_j(t) \equiv \phi^h(x_j, t)$, $f_j \equiv f^h(x_j, t)$, $f_j^q \equiv f^{q^h}(x_j, t)$, $f_j^p \equiv f^{p^h}(x_j, t)$. With implied summation on the repeated subscript index j, expansions (11.71)-(11.72) are then inserted into (11.69), which yields the discrete finite element weak statement

$$\sum_{e=1}^{N_e} \int_{\Omega_e} \left(w_i + \varepsilon^h \psi^h s^h p^h \frac{\partial}{\partial x}\left(\frac{w_i}{p^h}\right)\right)\left(\frac{\partial w_j}{\partial x} f_{p_j} - w_j \phi_{p_j}\right) d\Omega$$

$$+ \sum_{e=1}^{N_e} \int_{\Omega_e} \frac{\partial}{\partial x}\left(\frac{w_i}{p^h}\right)\varepsilon^h \psi^h \left(\alpha^h c^h \frac{\partial w_j}{\partial x} q_{p_j}\right) d\Omega = 0,$$

$$\sum_{e=1}^{N_e} \int_{\Omega_e} \left(w_i + \varepsilon^h \psi^h s^h \frac{\partial w_i}{\partial x}\right)\left(w_j \frac{dq_{u_j}}{dt} - w_j \phi_{u_j} + \frac{\partial w_j}{\partial x} f_{u_j}\right) d\Omega$$

$$+ \sum_{e=1}^{N_e} \int_{\Omega_e} \frac{\partial w_i}{\partial x}\frac{\partial w_j}{\partial x}\varepsilon^h \psi^h s^h \left(\alpha^h c^h s^h q_{u_j} + (\delta^h - 1)f_{u_j}^p\right) d\Omega = 0 \qquad (11.73)$$

where $1 \leq i, j \leq N$ and ε^h is set equal to a reference length within each element Ω_e, as shown in Chapter 10. While an expression like (11.71) for ψ^h, α^h, c^h, s^h, and δ^h can be directly accommodated within (11.73), each of these variables in this study has been set equal to a piece wise constant for computational simplicity, one centroidal constant value per element Ω_e, hence $\psi^h = \psi_e$, $\alpha^h = \alpha_e$, $c^h = c_e$, $s^h = s_e$, $\delta^h = \delta_e$, where ψ_e, α_e, c_e, s_e, δ_e denote centroidal constants. By the same token, the remaining semi-discrete pressure term p^h in the characteristics-bias expression within the continuity equation is also set to one centroidal constant value p_e within each element Ω_e. This simplification is only and exclusively employed in the characteristics-bias expression within the continuity equation; the discrete pressure within the entire momentum equation, including the characteristics-bias expression within such an equation, is precisely expressed through expansion (11.71). With this specification, the discrete equations simplify as

$$\sum_{e=1}^{N_e} \int_{\Omega_e} \left(w_i + \varepsilon_e \psi_e s_e \frac{\partial w_i}{\partial x}\right)\left(\frac{\partial w_j}{\partial x} f_{p_j} - w_j \phi_{p_j}\right) d\Omega$$

$$+ \sum_{e=1}^{N_e} \int_{\Omega_e} \frac{\partial w_i}{\partial x}\varepsilon_e \psi_e \left(\frac{\alpha_e c_e}{p_e}\frac{\partial w_j}{\partial x} q_{p_j}\right) d\Omega = 0,$$

$$\sum_{e=1}^{N_e} \int_{\Omega_e} \left(w_i + \varepsilon_e \psi_e s_e \frac{\partial w_i}{\partial x}\right)\left(w_j \frac{dq_{u_j}}{dt} - w_j \phi_{u_j} + \frac{\partial w_j}{\partial x} f_{u_j}\right) d\Omega$$

$$+ \sum_{e=1}^{N_e} \int_{\Omega_e} \frac{\partial w_i}{\partial x}\frac{\partial w_j}{\partial x}\varepsilon_e \psi_e s_e \left(\alpha_e c_e s_e q_{u_j} + (\delta_e - 1)f_{u_j}^p\right) d\Omega = 0 \qquad (11.74)$$

Since the test and trial functions w_i are prescribed functions of x, the spatial integrations in (11.74) are then exactly carried out, which transforms (11.74) into a system of ordinary algebraic-differential equations (OADE) in continuum time for determining at each time level t the unknown nodal values $q^h(x_j, t)$.

11.8.2   Discrete Upstream Bias

For a clear comparison between traditional finite difference/volume schemes and the acoustics-convection finite element algorithm (11.74), at any interior node "i" of the representative grid in Figure 11.6, equations (11.74) with $\varepsilon_e \equiv (\Delta x_{i+\frac{1}{2}})/2$ can be equivalently recast in difference notation as

$$-\left(\frac{\Delta x}{2}\left(\frac{1}{3}+\frac{s\psi}{2}\right)\right)_{i-\frac{1}{2}}\left(\phi_{p_{i-1}}+2\phi_{p_i}\right) - \left(\frac{\Delta x}{2}\left(\frac{1}{3}-\frac{s\psi}{2}\right)\right)_{i+\frac{1}{2}}\left(2\phi_{p_i}+\phi_{p_{i+1}}\right)$$

$$= -\left(\frac{\psi ac}{p}\right)_{i-\frac{1}{2}}(p_i - p_{i-1}) + \left(\frac{\psi ac}{p}\right)_{i+\frac{1}{2}}(p_{i+1}-p_i)$$

$$-\left((u_i - u_{i-1})\cdot\left(\frac{1+(s\psi)_{i-\frac{1}{2}}}{2}\right) + (u_{i+1}-u_i)\cdot\left(\frac{1-(s\psi)_{i+\frac{1}{2}}}{2}\right)\right) \quad (11.75)$$

for the continuity equation and

$$\left(\frac{\Delta x}{2}\left(\frac{1}{3}+\frac{s\psi}{2}\right)\right)_{i-\frac{1}{2}}\left(\frac{du_{i-1}}{dt}+2\frac{du_i}{dt}\right) + \left(\frac{\Delta x}{2}\left(\frac{1}{3}-\frac{s\psi}{2}\right)\right)_{i+\frac{1}{2}}\left(2\frac{du_i}{dt}+\frac{du_{i+1}}{dt}\right)$$

$$-\left(\frac{\Delta x}{2}\left(\frac{1}{3}+\frac{s\psi}{2}\right)\right)_{i-\frac{1}{2}}\left(\phi_{u_{i-1}}+2\phi_{u_i}\right) - \left(\frac{\Delta x}{2}\left(\frac{1}{3}-\frac{s\psi}{2}\right)\right)_{i+\frac{1}{2}}\left(2\phi_{u_i}+\phi_{u_{i+1}}\right)$$

$$= -(\psi ac)_{i-\frac{1}{2}}(u_i - u_{i-1}) + (\psi ac)_{i+\frac{1}{2}}(u_{i+1}-u_i)$$

$$-\left((u_i^2 - u_{i-1}^2)\cdot\left(\frac{1+(s\psi)_{i-\frac{1}{2}}}{2}\right) + (u_{i+1}^2-u_i^2)\cdot\left(\frac{1-(s\psi)_{i+\frac{1}{2}}}{2}\right)\right)$$

$$-\left((p_i - p_{i-1})\cdot\left(\frac{1+(s\psi\delta)_{i-\frac{1}{2}}}{2}\right) + (p_{i+1}-p_i)\cdot\left(\frac{1-(s\psi\delta)_{i+\frac{1}{2}}}{2}\right)\right) \quad (11.76)$$

for the linear-momentum equation. The set of equations (11.75) and (11.76) provide an algebraic-differential system for the determination of both p_i and u_i at each and every node "i" within Ω^h. Equations (11.75), which are the discrete characteristics-bias counterpart of the incompressible-flow continuity equation, are the algebraic equations, for they do not feature any time derivative of pressure. Equations (11.75), instead, are the differential equations in this system, for they are the discrete characteristics-bias counterpart of the incompressible-flow linear-momentum equations. As their structure shows, equations (11.75) and (11.76) feature a linear combination of two-point upstream and downstream flux differences. In these finite element equations, the values of the controller ψ^h determines the combination weights of the downstream and upstream expressions, and since ψ^h remains non-negative, these equations induce the appropriate upstream bias, since the upstream weight $1 + \psi_{i-\frac{1}{2}}$ always exceeds the downstream weight $1 - \psi_{i+\frac{1}{2}}$. As a result, the finite element weak statement (11.74) generates consistent variable-upstream-bias discrete equations that correspond to an upstream-bias discretizations for the original incompressible-flow quasi-one-dimensional Euler system within a compact block tri-diagonal matrix statement.

11.8.3   Boundary Equations and Pressure Boundary Condition

The integral statement (11.74) directly yields a set of boundary equations, for both inlet and outlet pressure and speed. Unlike algebraic extrapolations of variables, these boundary equations consistently couple boundary- and interior-node variables, including time derivatives of nodal variables when the corresponding differential equation feature a derivative with respect to time.

For the linear elements in this study, let N and $N-1$ denote the nodes within the outlet boundary element, with N corresponding to the outlet node. For the discrete finite element equation associated with boundary node x_N, the test function w satisfies the conditions $w(x_{N-1}) = 0$, $w(x_N) = 1$. From (11.74), the outlet-node algebraic-differential equations are

$$-\left(\Delta x \left(\frac{1}{3} + \frac{s\psi}{2}\right)\right)_{N-\frac{1}{2}} \left(\phi_{p_{N-1}} + 2\phi_{p_N}\right)$$

$$= -\left(\frac{2\psi ac}{p}\right)_{N-\frac{1}{2}} (p_N - p_{N-1}) - \left(1 + (s\psi)_{N-\frac{1}{2}}\right)(u_N - u_{N-1}) \tag{11.77}$$

for the continuity equation and

$$\left(\Delta x \left(\frac{1}{3} + \frac{s\psi}{2}\right)\right)_{N-\frac{1}{2}} \left(\frac{du_{N-1}}{dt} + 2\frac{du_N}{dt}\right)$$

$$-\left(\Delta x \left(\frac{1}{3} + \frac{s\psi}{2}\right)\right)_{N-\frac{1}{2}} \left(\phi_{u_{N-1}} + 2\phi_{u_N}\right) = -(2\psi ac)_{N-\frac{1}{2}}(u_N - u_{N-1})$$

$$-\left(1 + (s\psi)_{N-\frac{1}{2}}\right)\left(u_N^2 - u_{N-1}^2\right) - \left(1 + (s\psi\delta)_{N-\frac{1}{2}}\right)(p_N - p_{N-1}) \tag{11.78}$$

for the linear-momentum equation. These equations directly couple the time derivatives of the speed u at the adjacent boundary and interior nodes x_N and x_{N-1}. A similar equation is then obtained at an inlet, *mutatis mutandis*.

Concerning the pressure outlet boundary condition, this is naturally enforced within the surface integral that emerges in the momentum-equation weak statement (11.67), as shown in the previous section. The linear finite element discretization of such an equation leads to

$$\left(\Delta x \left(\frac{1}{3} + \frac{s\psi}{2}\right)\right)_{N-\frac{1}{2}} \left(\frac{du_{N-1}}{dt} + 2\frac{du_N}{dt}\right)$$

$$-\left(\Delta x \left(\frac{1}{3} + \frac{s\psi}{2}\right)\right)_{N-\frac{1}{2}} \left(\phi_{u_{N-1}} + 2\phi_{u_N}\right) = -(2\psi ac)_{N-\frac{1}{2}}(u_N - u_{N-1})$$

$$-\left(1 + (s\psi)_{N-\frac{1}{2}}\right)\left(u_N^2 - u_{N-1}^2\right) - (s\psi\delta)_{N-\frac{1}{2}}(p_N - p_{N-1}) - ((2p_{\text{out}} - p_N) - p_{N-1}) \tag{11.79}$$

In this equation, p_N denotes the outlet-node pressure, as calculated through the discrete equations, whereas, quite significantly, p_{out} corresponds to the specified outlet pressure boundary condition. With reference to the inlet pressure boundary condition, this is directly imposed as $p_1 = p_{\text{in}}$, which replaces the discrete continuity equation at the inlet node.

11.9   Implicit Runge Kutta Time Integration

The finite element equations (11.75)-(11.76) along with appropriate boundary equations and conditions form a system of algebraic and differential equations. The discrete continuity equations provide the algebraic equations and the discrete linear-momentum equations supply the differential equations in the time variable. Although these equations feature no time derivatives of pressure, the implicit Runge-Kutta time integration procedure (8.5) succeeds in simultaneously solving the discrete continuity and linear-momentum equations and directly yielding the solution for both pressure and speed.

The complete discrete characteristics-bias algebraic-differential system, inclusive of inlet and outlet pressure boundary conditions is cast as

$$0 = p_1 - p_{\text{in}},$$

$$\left(\Delta x \left(\frac{1}{3} + \frac{s\psi}{2} \right) \right)_{1+\frac{1}{2}} \left(2\frac{du_1}{dt} + \frac{du_2}{dt} \right)$$

$$- \left(\Delta x \left(\frac{1}{3} + \frac{s\psi}{2} \right) \right)_{1+\frac{1}{2}} (2\phi_{u_1} + \phi_{u_2}) = -(2\psi ac)_{1+\frac{1}{2}} (u_2 - u_1)$$

$$- \left(1 + (s\psi)_{1+\frac{1}{2}} \right) \left(u_2^2 - u_1^2 \right) - \left(1 + (s\psi\delta)_{1+\frac{1}{2}} \right) (p_2 - p_1) \tag{11.80}$$

for the inlet node, $i = 1$,

$$- \left(\frac{\Delta x}{2} \left(\frac{1}{3} + \frac{s\psi}{2} \right) \right)_{i-\frac{1}{2}} \left(\phi_{p_{i-1}} + 2\phi_{p_i} \right) - \left(\frac{\Delta x}{2} \left(\frac{1}{3} - \frac{s\psi}{2} \right) \right)_{i+\frac{1}{2}} \left(2\phi_{p_i} + \phi_{p_{i+1}} \right)$$

$$= - \left(\frac{\psi ac}{p} \right)_{i-\frac{1}{2}} (p_i - p_{i-1}) + \left(\frac{\psi ac}{p} \right)_{i+\frac{1}{2}} (p_{i+1} - p_i)$$

$$- \left((u_i - u_{i-1}) \cdot \left(\frac{1 + (s\psi)_{i-\frac{1}{2}}}{2} \right) + (u_{i+1} - u_i) \cdot \left(\frac{1 - (s\psi)_{i+\frac{1}{2}}}{2} \right) \right),$$

$$\left(\frac{\Delta x}{2} \left(\frac{1}{3} + \frac{s\psi}{2} \right) \right)_{i-\frac{1}{2}} \left(\frac{du_{i-1}}{dt} + 2\frac{du_i}{dt} \right) + \left(\frac{\Delta x}{2} \left(\frac{1}{3} - \frac{s\psi}{2} \right) \right)_{i+\frac{1}{2}} \left(2\frac{du_i}{dt} + \frac{du_{i+1}}{dt} \right)$$

$$- \left(\frac{\Delta x}{2} \left(\frac{1}{3} + \frac{s\psi}{2} \right) \right)_{i-\frac{1}{2}} \left(\phi_{u_{i-1}} + 2\phi_{u_i} \right) - \left(\frac{\Delta x}{2} \left(\frac{1}{3} - \frac{s\psi}{2} \right) \right)_{i+\frac{1}{2}} \left(2\phi_{u_i} + \phi_{u_{i+1}} \right)$$

$$= -(\psi ac)_{i-\frac{1}{2}} (u_i - u_{i-1}) + (\psi ac)_{i+\frac{1}{2}} (u_{i+1} - u_i)$$

$$- \left(\left(u_i^2 - u_{i-1}^2 \right) \cdot \left(\frac{1 + (s\psi)_{i-\frac{1}{2}}}{2} \right) + \left(u_{i+1}^2 - u_i^2 \right) \cdot \left(\frac{1 - (s\psi)_{i+\frac{1}{2}}}{2} \right) \right)$$

$$- \left((p_i - p_{i-1}) \cdot \left(\frac{1 + (s\psi\delta)_{i-\frac{1}{2}}}{2} \right) + (p_{i+1} - p_i) \cdot \left(\frac{1 - (s\psi\delta)_{i+\frac{1}{2}}}{2} \right) \right) \tag{11.81}$$

for the internal nodes, $2 \leq i \leq N - 1$, and

$$-\left(\Delta x \left(\frac{1}{3} + \frac{s\psi}{2}\right)\right)_{N-\frac{1}{2}} \left(\phi_{p_{N-1}} + 2\phi_{p_N}\right)$$

$$= -\left(\frac{2\psi ac}{p}\right)_{N-\frac{1}{2}} (p_N - p_{N-1}) - \left(1 + (s\psi)_{N-\frac{1}{2}}\right)(u_N - u_{N-1}),$$

$$\left(\Delta x \left(\frac{1}{3} + \frac{s\psi}{2}\right)\right)_{N-\frac{1}{2}} \left(\frac{du_{N-1}}{dt} + 2\frac{du_N}{dt}\right)$$

$$-\left(\Delta x \left(\frac{1}{3} + \frac{s\psi}{2}\right)\right)_{N-\frac{1}{2}} \left(\phi_{u_{N-1}} + 2\phi_{u_N}\right) = -(2\psi ac)_{N-\frac{1}{2}} (u_N - u_{N-1})$$

$$-\left(1 + (s\psi)_{N-\frac{1}{2}}\right)\left(u_N^2 - u_{N-1}^2\right) - (s\psi\delta)_{N-\frac{1}{2}} (p_N - p_{N-1}) - ((2p_{\text{out}} - p_N) - p_{N-1}) \quad (11.82)$$

for the outlet node, $i = N$. With Q denoting the array of nodal pressure and speed

$$Q \equiv \left\{ \begin{array}{c} p_1 \\ u_1 \\ p_2 \\ u_2 \\ \cdot \\ \cdot \\ p_i \\ u_i \\ \cdot \\ \cdot \\ p_N \\ u_N \end{array} \right\} \quad (11.83)$$

and $\mathcal{F}$ denoting an array of fluxes, source terms and characteristics-bias expressions, equations (11.80)-(11.82) are abridged as the non-linear algebraic-differential system

$$\mathcal{M}\frac{dQ(t)}{dt} = \mathcal{F}(t, Q(t)) \quad (11.84)$$

Since the discrete continuity equations within (11.80)-(11.82) do not contain any time derivatives of pressure, the "mass" matrix $\mathcal{M}$ contains N zero rows in addition to rows expressing the coupling of time derivatives of nodal speeds. This matrix can be expressed in the form

$$\mathcal{M} \equiv \left(\begin{array}{ccccccc} 0 & & & & & & \\ m_{21} & m_{22} & m_{23} & & & & \\ & & 0 & & & & \\ & & m_{43} & m_{44} & m_{45} & & \\ & & & & \cdot & & \\ & & & & & 0 & \\ & & & & & m_{NN-1} & m_{NN} \end{array} \right) \quad (11.85)$$

A set of representative equations for pressure and speed at an internal node "i" are thus cast as

$$0 = \mathcal{F}_{p_i}(t, Q)$$
$$m_{i,i-1}\frac{du_{i-1}}{dt} + m_{i,i}\frac{du_i}{dt} + m_{i,i+1}\frac{du_{i+1}}{dt} = \mathcal{F}_{u_i}(t, Q) \qquad (11.86)$$

where $\mathcal{F}_{p_i}(t, Q)$ and $\mathcal{F}_{p_i}(t, Q)$ correspond to the right-hand sides in (11.80)-(11.82).

The algebraic-differential system (11.84) is integrated through the two-stage implicit Runge-Kutta procedure

$$
\begin{aligned}
Q_{n+1} - Q_n &= b_1 K_1 + b_2 K_2 \\
\mathcal{M}K_1 &= \Delta t \cdot \mathcal{F}\left(t_n + c_1\Delta t, Q_n + a_{11}K_1\right) \\
\mathcal{M}K_2 &= \Delta t \cdot \mathcal{F}\left(t_n + c_2\Delta t, Q_n + a_{21}K_1 + a_{22}K_2\right)
\end{aligned}
\qquad (11.87)
$$

The Runge-Kutta arrays K_1 and K_2 correspond to variations ΔQ_1 and ΔQ_2 of the solution array Q. The first expression in (11.87) thus corresponds to the linear combination

$$Q_{n+1} - Q_n = b_1\Delta Q_1 + b_2\Delta Q_2 \qquad (11.88)$$

The terminal numerical solution for the second and third expression in (11.87) is then determined using Newton's method in linearized one step mode or non-linear form. For the implicit fully-coupled computation of the IRK arrays K_i, Newton's method is cast as

$$\left[\mathcal{M} - a_{\underline{ii}}\Delta t \left(\frac{\partial \mathcal{F}}{\partial Q}\right)^\ell_{Q^\ell_i}\right]\left(K^{\ell+1}_i - K^\ell_i\right) = \Delta t\, \mathcal{F}\left(t_n + c_i\Delta t, Q^\ell_i\right) - \mathcal{M}K^\ell_i$$

$$Q^\ell_i \equiv Q_n + a_{i1}K^1_1 + a_{i2}K^\ell_2 \qquad (11.89)$$

where $a_{ij} = 0$ for $j > i$, superscript ℓ denotes the iteration index, and $K^\ell_1 \equiv K_1$ for $i = 2$; in direct solution mode, $\ell = 1$ and $K^1_i = 0$. For a representative set of discrete continuity and linear-momentum equations, expression (11.89) becomes

$$\left[-a_{\underline{ii}}\left(\frac{\partial \mathcal{F}_p}{\partial Q}\right)^\ell_{Q^\ell_i}\right]\left(K^{\ell+1}_i - K^\ell_i\right) = \mathcal{F}_p\left(t_n + c_i\Delta t, Q^\ell_i\right)$$

$$\left[\mathcal{M}_u - a_{\underline{ii}}\Delta t\left(\frac{\partial \mathcal{F}_u}{\partial Q}\right)^\ell_{Q^\ell_i}\right]\left(K^{\ell+1}_i - K^\ell_i\right) = \Delta t\, \mathcal{F}_u\left(t_n + c_i\Delta t, Q^\ell_i\right) - \mathcal{M}_u K^\ell_i \qquad (11.90)$$

Despite the absence of any time derivatives of pressure within (11.84), such equations simultaneously couple the discrete continuity and linear-momentum equations and yield the solution for Δu and Δp. The distributions of p and u within Q then directly evolve from the first expression in (11.87), hence

$$
\begin{aligned}
p_{n+1} - p_n &= b_1\Delta p_1 + b_2\Delta p_2 \\
u_{n+1} - u_n &= b_1\Delta u_1 + b_2\Delta u_2
\end{aligned}
\qquad (11.91)
$$

As the results in the following section bear out, the simultaneous solution of the coupled discrete characteristics-bias continuity and linear-momentum equations directly generates a stable solution for both pressure and speed without requiring any further post-processing calculations for pressure or speed.

11.10   Reference Exact Solutions

The incompressible quasi-one-dimensional flow within a converging - diverging nozzle is uniquely defined by the nozzle cross-sectional area distribution and the inlet and outlet pressure. The cross-sectional area and one pressure, inlet or outlet, can be freely prescribed. The remaining pressure has to satisfy a constraint that prevents the manifestation of a vacuum flow at the nozzle throat. This section details the reference exact solution of the quasi-one-dimensional incompressible-flow equations

$$\begin{cases} \dfrac{\partial u}{\partial x} + \dfrac{u}{A}\dfrac{dA}{dx} = 0 \\[2ex] \dfrac{\partial u}{\partial t} + \dfrac{\partial}{\partial x}\left(u^2 + p\right) + \dfrac{u^2}{A}\dfrac{dA}{dx} = 0 \end{cases} \tag{11.92}$$

The absence of any time derivative of pressure in the continuity equations, as shown, leads to unbounded wave propagation speeds. As a direct result, inlet and outlet pressure boundary conditions lead to physically meaningful pressure and speed that do depend upon time, but do not exhibit a finite-speed wave propagation structure and remain independent of the initial spatial distributions of speed and pressure. A theoretically instantaneous adjustment of speed and pressure throughout the nozzle occurs from an initial distribution to the distributions induced by both the nozzle shape and pressure boundary conditions.

11.10.1   General Solution

System 11.92 can be cast as

$$\begin{cases} \dfrac{\partial u}{\partial x} = -\dfrac{u}{A}\dfrac{dA}{dx} \\[2ex] \dfrac{\partial u}{\partial t} = -\dfrac{\partial}{\partial x}\left(\dfrac{u^2}{2} + p\right) \end{cases} \tag{11.93}$$

The continuity equation in this system yields

$$u(x,t) = u_i(t)\dfrac{A_i}{A(x)} \quad \Rightarrow u_o(t) = u_i(t)\dfrac{A_i}{A_o} \tag{11.94}$$

where u_i and u_o respectively denote the inlet and outlet nozzle speeds. This solution leads to the following time derivative of the local speed $u = u(x,t)$

$$\dfrac{\partial u}{\partial t} = \dfrac{du_i}{dt}\dfrac{A_i}{A(x)} \tag{11.95}$$

Insertion of this derivative into the linear-momentum equation in system (11.93) yields the differential equation

$$\dfrac{du_i}{dt}\dfrac{A_i}{A(x)} = -\dfrac{\partial}{\partial x}\left(\dfrac{u^2}{2} + p\right) \tag{11.96}$$

which can be integrated in space to yield

$$-\frac{du_i}{dt}\int_{x_i}^{x}\frac{A_i}{A(X)}\,dX = \frac{u^2(x,t)}{2} + p(x,t) - \frac{u_i^2(t)}{2} - p_i(t) \tag{11.97}$$

11.10.2 Particular Solution

The particular solution corresponding to prescribed inlet and outlet pressure boundary conditions evolves from the general solution as follows. Let x_o denote the nozzle outlet coordinate. Equation (11.97) at $x = x_o$ thus yields the ordinary differential equation

$$-\frac{du_i}{dt}\int_{x_i}^{x_o}\frac{A_i}{A(x)}\,dx = \frac{u_o^2(t)}{2} + p_o(t) - \frac{u_i^2(t)}{2} - p_i(t) \tag{11.98}$$

Through (11.94), this result becomes

$$\frac{du_i}{dt}\int_{x_i}^{x_o}\frac{A_i}{A(x)}\,dx + \frac{u_i^2(t)}{2}\left[\left(\frac{A_i}{A_o}\right)^2 - 1\right] = p_i(t) - p_o(t) \tag{11.99}$$

and thus

$$\frac{du_i}{dt} + au_i^2 = b\left(p_i(t) - p_o(t)\right)$$

$$a = \frac{b}{2}\left[\left(\frac{A_i}{A_o}\right)^2 - 1\right], \quad b = \frac{1}{\displaystyle\int_{x_i}^{x_o}\frac{A_i}{A(x)}\,dx} \tag{11.100}$$

For prescribed $p_i = p_i(t)$ and $p_o = p_o(t)$, therefore, this differential equation yields both the inlet speed $u_i = u_i(t)$ and its time derivative, consistent to a prescribed initial inlet speed $u_i(0) = u_{i0}$. The distribution of nozzle speed $u = u(x,t)$ follows as

$$u(x,t) = u_i\frac{A_i}{A(x)} \tag{11.101}$$

and with this solution, equation (11.97) fully determines the distribution of nozzle pressure $p = p(x,t)$.

$$p(x,t) = p_i(t) + \frac{u_i^2(t)}{2}\left[1 - \left(\frac{A_i}{A(x)}\right)^2\right] - \frac{du_i}{dt}\int_{x_i}^{x}\frac{A_i}{A(X)}\,dX \tag{11.102}$$

11.10.3 Solution Properties

The particular solution can be expressed as

$$\begin{cases} \dfrac{du_i}{dt} + au_i^2 = b\left(p_i(t) - p_o(t)\right), \quad u_i(0) = u_{i0} \\[2mm] u(x,t) = u_i(t)\dfrac{A_i}{A(x)} \\[2mm] p(x,t) = p_i(t) + \dfrac{u_i^2(t)}{2}\left[1 - \left(\dfrac{A_i}{A(x)}\right)^2\right] - \dfrac{du_i}{dt}\int_{x_i}^{x}\dfrac{A_i}{A(X)}\,dX \end{cases} \tag{11.103}$$

A visual inspection of this solution reveals several unique properties. Despite the absence of any time-derivative of pressure in the continuity equation, both speed u and pressure p do depend upon time. However, the solution does not posses a finite-speed wave structure for either speed u or pressure p. Furthermore, no dependence exists on any initial spatial distribution of either u or p. The spatial distributions of both u and p throughout the nozzle, therefore, instantaneously follow the inlet and outlet time variations of pressure. The second equation in (11.103) shows the volumetric flow rate $Q(t) = u(x,t)A(x) = u_i A_i$ does not depend upon x but only on t. At each time level, therefore, Q remains constant at every location along a duct axis.

The discussion on the boundary conditions that promote a physically meaningful solution remains delicate. The characteristic analysis for the incompressible quasi-one-dimensional equations has revealed that at both inlet and outlet one eigenvalue remains negative and the other remains positive. One boundary condition only is thus required at both the inlet and outlet

Suppose the inlet speed were prescribed. This selection would immediately constrain the outlet speed, following the second equation in (11.103). A constrained outlet speed corresponds to an outlet boundary condition on speed. No further boundary conditions could then be prescribed. This selection of boundary constraints, however, would not completely determine pressure, for the first equation in (11.103) would only determine the pressure difference $p_i(t) - p_o(t)$, hence pressure would remain defined up to an arbitrary constant only. Similar conclusions follow when the outlet speed is initially specified. The specification of any boundary speed, therefore, yields no complete solution for pressure.

Even though it appears system (11.103) algebraically allows the specification of one speed and one pressure, effectively such specification corresponds to 3 boundary conditions, hence an over specification. Likewise, it even appears (11.103) allows the specification of pressure and speed both at the inlet or the outlet, but such exclusive specification contradicts the two-way unbounded-speed information propagation of an incompressible flow, which requires one boundary condition at the inlet and one at the outlet.

The resolution of this seemingly paradoxical situation remains simple. From whatever combination of boundary specifications that include one boundary pressure, system (11.103) provides the corresponding inlet and outlet pressure magnitudes. Use of these magnitudes as pressure boundary conditions in (11.103) or in a numerical solution of (11.92) will then yield a physical flow that provides the initial combination of boundary specifications.

11.10.4   Transient Solution

System (11.103) generates a transient solution that asymptotically attains a steady state, subject to the simple specifications

$$\Delta p \equiv p_i(t) - p_o(t) = \text{constant}, \quad u_i(0) = 0 \qquad (11.104)$$

With these specifications, the ordinary differential equation in (11.103) is exactly integrated, which yields the following solution for the inlet speed $u_i = u_i(t)$

$$u_i(t) = \frac{1 - \exp\left(-2t\sqrt{ab\Delta p}\right)}{1 + \exp\left(-2t\sqrt{ab\Delta p}\right)}\sqrt{\frac{b\Delta p}{a}} \tag{11.105}$$

This solution for u_i, in turn, leads to the following distributions for the nozzle speed and pressure

$$\begin{cases} u(x,t) &= u_i(t)\dfrac{A_i}{A(x)} \\[2mm] p(x,t) &= p_i(t) + \dfrac{u_i^2(t)}{2}\left[1 - \left(\dfrac{A_i}{A(x)}\right)^2\right] - \dfrac{du_i}{dt}\displaystyle\int_{x_i}^{x}\dfrac{A_i}{A(X)}dX \end{cases} \tag{11.106}$$

11.10.5 Steady State Solution

For unbounded time, the transient solution

$$u_i(t) = \frac{1 - \exp\left(-2t\sqrt{ab\Delta p}\right)}{1 + \exp\left(-2t\sqrt{ab\Delta p}\right)}\sqrt{\frac{2\Delta p}{\left(\dfrac{A_i}{A_o}\right)^2 - 1}} \tag{11.107}$$

and (11.106) lead to the steady state solution

$$u_i = \sqrt{\frac{2\Delta p}{\left(\dfrac{A_i}{A_o}\right)^2 - 1}}, \quad u_o = \frac{A_i}{A_o}\sqrt{\frac{2\Delta p}{\left(\dfrac{A_i}{A_o}\right)^2 - 1}}, \quad u(x) = \frac{A_i}{A(x)}\sqrt{\frac{2\Delta p}{\left(\dfrac{A_i}{A_o}\right)^2 - 1}} \tag{11.108}$$

$$p(x,t) + \frac{u^2(x,t)}{2} = p_i(t) + \frac{u_i^2(t)}{2} \equiv p_{st}(t), \quad p(x) = p_i - \Delta p\frac{\left(\dfrac{A_i}{A(x)}\right)^2 - 1}{\left(\dfrac{A_i}{A_o}\right)^2 - 1} \tag{11.109}$$

$$p_{st} = p_i + \frac{p_i - p_o}{\left(\dfrac{A_i}{A_o}\right)^2 - 1}, \quad p_T = p_i - \Delta p\frac{\left(\dfrac{A_i}{A_T}\right)^2 - 1}{\left(\dfrac{A_i}{A_o}\right)^2 - 1} \tag{11.110}$$

where p_{st} and p_T respectively denote the stagnation and throat pressure. The first equation in (11.109) corresponds to a non-dimensional Bernoulli equation and shows the stagnation pressure p_{st} results from a theoretically vanishing speed, which, in turn, correlates with a theoretically unbounded "reservoir" cross section A_S. The equation additionally shows p_{st} does not depend upon x, but only on t. At each time level, therefore, the sum of local pressure and half of the speed squared remains constant at every location along a duct axis. Accordingly, as the duct speed increases the local pressure p will decrease.

The distribution of $u = u(x)$, in particular, shows that the speed increases as the cross section decreases and thus the maximum speed takes place at the nozzle throat. For a physical flow, however, this maximum speed cannot increase unboundedly, but must remain constrained by the condition that pressure must remain positive and cannot vanish. The minimum pressure, that is the throat pressure p_T, must remain non-negative, which leads to the constraints

$$p_T \geq 0 \quad \Rightarrow \quad p_o \geq p_i \frac{\left(\dfrac{A_i}{A_T}\right)^2 - \left(\dfrac{A_i}{A_o}\right)^2}{\left(\dfrac{A_i}{A_T}\right)^2 - 1} \tag{11.111}$$

Furthermore, for a mechanically realistic flow from duct inlet to outlet the inlet pressure p_i must exceed the outlet pressure p_o, hence

$$p_i \geq p_o \quad \Rightarrow \quad \frac{\left(\dfrac{A_i}{A_T}\right)^2 - \left(\dfrac{A_i}{A_o}\right)^2}{\left(\dfrac{A_i}{A_T}\right)^2 - 1} \leq 1 \quad \Rightarrow \quad A_o \leq A_i \tag{11.112}$$

These pressure specifications lead to the following constraint on the ratio of outlet pressure over inlet pressure

$$\frac{\left(\dfrac{A_i}{A_T}\right)^2 - \left(\dfrac{A_i}{A_o}\right)^2}{\left(\dfrac{A_i}{A_T}\right)^2 - 1} \leq \frac{p_o}{p_i} \leq 1 \tag{11.113}$$

The condition of a "vacuum" corresponds to a vanishing throat pressure p_T, which theoretically can exist at the "vacuum" throat A_v. This reference cross-section area is defined by

$$0 = p_i - \Delta p \frac{\left(\dfrac{A_i}{A_v}\right)^2 - 1}{\left(\dfrac{A_i}{A_o}\right)^2 - 1} \quad \Rightarrow \quad A_v = \frac{A_i}{\sqrt{1 + \left[\left(\dfrac{A_i}{A_o}\right)^2 - 1\right]\dfrac{p_i}{\Delta p}}} \tag{11.114}$$

The corresponding vacuum speed evolves as

$$u_v = u_i \frac{A_i}{A_v} = \sqrt{\frac{2\Delta p}{\left(\dfrac{A_i}{A_o}\right)^2 - 1}} \sqrt{1 + \left[\left(\dfrac{A_i}{A_o}\right)^2 - 1\right]\dfrac{p_i}{\Delta p}} \tag{11.115}$$

hence

$$u_v = \sqrt{2p_i + \frac{2\Delta p}{\left(\dfrac{A_i}{A_o}\right)^2 - 1}} = \sqrt{2p_{st}} \tag{11.116}$$

11.10.6 Similarity Solution

The defining flow information can be expressed as

$$A = A(x), \quad p_i, \quad p_o \quad \text{with} \quad \frac{\left(\dfrac{A_i}{A_T}\right)^2 - \left(\dfrac{A_i}{A_o}\right)^2}{\left(\dfrac{A_i}{A_T}\right)^2 - 1} \leq \frac{p_o}{p_i} \leq 1$$

which uniquely establishes a quasi-one dimensional isentropic incompressible flow within a converging-diverging nozzle. The corresponding vacuum cross sectional area and stagnation pressure follow as

$$p_{st} = p_i + \frac{p_i - p_o}{\left(\dfrac{A_i}{A_o}\right)^2 - 1}, \quad A_v = \frac{A_i}{\sqrt{1 + \left[\left(\dfrac{A_i}{A_o}\right)^2 - 1\right]\dfrac{p_i}{p_i - p_o}}}$$

Likewise, the vacuum and inlet speeds become

$$u_v = \sqrt{2p_{st}}, \quad u_i = \sqrt{\frac{2(p_i - p_o)}{\left(\dfrac{A_i}{A_o}\right)^2 - 1}}$$

Stagnation- and Vacuum-State Reference

By definition, the incompressible-flow number is expressed as

$$\Im \equiv \frac{u}{\sqrt{p}} \tag{11.117}$$

The ratio of local over stagnation pressure then depends on $\Im$ as

$$p_{st} = p + \frac{u^2}{2} = p + p\frac{\Im^2}{2} = p\left(1 + \frac{\Im^2}{2}\right) \Rightarrow \frac{p}{p_{st}} = \frac{2}{2 + \Im^2} \tag{11.118}$$

The vacuum state corresponds to a vanishing pressure. Result (11.118) shows a vanishing pressure obtains for an unbounded $\Im$ number

$$\lim_{\Im \to \infty} \frac{2}{2 + \Im^2} = 0 \tag{11.119}$$

The vacuum speed u_v, therefore, corresponds to an unbounded $\Im$ number. Independently of the vacuum condition, let u_T denote the throat speed. The definition of the $\Im$ number leads to the ratio

$$\frac{u^2}{u_T^2} = \frac{p}{p_T}\frac{\Im^2}{\Im_T^2} = \frac{p}{p_{st}}\frac{p_{st}}{p_T}\frac{\Im^2}{\Im_T^2} = \frac{\Im^2}{2 + \Im^2}\frac{2 + \Im_T^2}{\Im_T^2} \tag{11.120}$$

The corresponding ratio based on the vacuum speed follows as

$$\frac{u^2}{u_v^2} = \lim_{\Im_T \to \infty} \frac{u^2}{u_T^2} = \lim_{\Im_T \to \infty} \frac{\Im^2}{2 + \Im^2} \frac{2 + \Im_T^2}{\Im_T^2} = \frac{\Im^2}{2 + \Im^2} \tag{11.121}$$

which leads to

$$\frac{u}{u_v} = \frac{\Im}{\sqrt{2 + \Im^2}} \tag{11.122}$$

The corresponding area ratio and $\Im$ number in terms of area ratio become

$$\frac{A}{A_v} = \frac{\sqrt{2 + \Im^2}}{\Im} \quad \Rightarrow \quad \Im = \sqrt{\frac{2}{\left(\dfrac{A}{A_v}\right)^2 - 1}} \tag{11.123}$$

Inlet-State Reference

With these results, the inlet and local $\Im$ number evolve as

$$\Im_i = \frac{u_i}{\sqrt{p_i}} = \sqrt{\frac{2(p_i - p_o)}{p_i\left(\left(\dfrac{A_i}{A_o}\right)^2 - 1\right)}}, \quad \Im = \sqrt{\frac{2}{\left(\dfrac{A}{A_v}\right)^2 - 1}}$$

The local flow pressure and speed can then result from the ratios

$$\frac{p}{p_i} = \frac{p}{p_{st}}\frac{p_{st}}{p_i} = \frac{2 + \Im_i^2}{2 + \Im^2}, \quad \frac{u}{\sqrt{p_i}} = \frac{u}{\sqrt{p}}\frac{\sqrt{p}}{\sqrt{p_i}} = \Im\sqrt{\frac{2 + \Im_i^2}{2 + \Im^2}}$$

which provide a complete solution for the steady quasi-one-dimensional flow within a converging-diverging nozzle.

11.11   Computational Results

The computational results have validated the accuracy and essential-monotonicity performance of the acoustics-convection upstream resolution algorithm for unsteady, transient, and steady incompressible flows. The benchmarks in this section cover a total of 3 different flows within a convergent-divergent, Venturi, tube.

The non-dimensional cross-sectional area distribution for this Venturi tube is

$$A(x) = \begin{cases} \frac{7}{4} - \frac{3}{4}\cos\left(\pi(2x - 1)\right), & 0 \le x \le \frac{1}{2} \\ \frac{5}{4} - \frac{1}{4}\cos\left(\pi(2x - 1)\right), & \frac{1}{2} \le x \le 1 \end{cases}, \quad A_{in} = \frac{5}{2}, \quad A_{out} = \frac{3}{2} \tag{11.124}$$

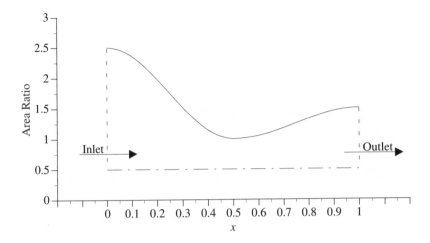

Figure 11.7: Variation of Non-dimensional Cross-sectional Area $A = A(x)$

The nozzle cross-section area distribution remains continuous with continuous slopes, but contains a discontinuous throat curvature, as shown in Figure 11.7. This feature induces a nozzle-throat discontinuous curvature in the flow variables. For this cross-sectional area, the exact-solution integrals exactly evaluate as follows

$$\int_{x_i}^{x_o} \frac{A_i}{A(x)}\, dx = \int_0^1 \frac{A_i}{A(x)}\, dx = \frac{3\sqrt{10} + 5\sqrt{6}}{12} \tag{11.125}$$

$$\int_{x_i}^{x} \frac{A_i}{A(X)}\, dX = \left\{ \begin{array}{ll} \frac{\sqrt{10}}{2\pi} \left\{ \frac{\pi}{2} + \tan^{-1}\left[\frac{\sqrt{10}}{2} \tan\left((2x-1)\frac{\pi}{2} \right) \right] \right\}, & 0 \leq x \leq \frac{1}{2} \\[3mm] \frac{5\sqrt{6}}{6\pi} \left\{ \frac{\pi\sqrt{15}}{10} + \tan^{-1}\left[\frac{\sqrt{6}}{2} \tan\left((2x-1)\frac{\pi}{2} \right) \right] \right\}, & \frac{1}{2} \leq x \leq 1 \end{array} \right. \tag{11.126}$$

The corresponding spatial computational domain Ω for all the results presented in this section is defined as: $\Omega \equiv [a, b] = [0, 1]$, uniformly discretized into 100 linear finite elements, hence $\Delta x = 0.01$. For each benchmark, the calculations proceeded with a prescribed constant maximum Courant number $C_{\max}$ defined as

$$C_{\max} \equiv \max\{|u + \sqrt{u^2 + c}|, |u - \sqrt{u^2 + c}|\} \frac{\Delta t}{\Delta x} \tag{11.127}$$

Given Δx and $C_{\max}$ for each benchmark, the corresponding time step Δt is thus determined as

$$\Delta t = \frac{C_{\max} \Delta x}{\max\{|u + \sqrt{u^2 + c}|, |u - \sqrt{u^2 + c}|\}} \tag{11.128}$$

As detailed in Chapter 10, the upstream-bias controller uses one scalar component of the dependent variable q. In this study, the algorithm has employed static pressure p to calculate ψ.

All the solutions in these validations are presented in non dimensional form, with pressure p made dimensionless through inlet pressure p_i. The non-dimensional speed is obtained by the reference speed $u_r = \sqrt{p_i/\rho}$. The results in this section show the characteristics-bias procedure has consistently and directly generated non-oscillatory results for both pressure p and speed u that reflect available exact solutions.

11.11.1 Unsteady Flow: Periodic Outlet Pressure

This benchmark determines the genuinely unsteady incompressible flow that results from a periodic time variation of the outlet pressure. With a constant inlet pressure, the flow within the Venturi tube is characterized by a pressure distribution that approaches the time variation of the outlet pressure, at locations near the tube outlet. The associated distribution of speed remains unsteady throughout the tube.

For the given nozzle, the pressure-ratio constraint (11.113) yields the minimum outlet-inlet pressure ratio

$$\left(\frac{p_o}{p_i}\right)_{\min} = \frac{125}{189} \simeq 0.6614 \tag{11.129}$$

For a genuinely unsteady incompressible flow, consider then the following pressure magnitudes

$$p_i = 1, \quad p_{o_{\min}} = \frac{2}{3} = 0.\overline{6} \tag{11.130}$$

and impose a cosine-wave variation of the outlet pressure. These specifications yield the time-dependent variations

$$p_o(t) = \frac{1}{120}\left(97 + 17\cos(t)\right), \quad p_i - p_o(t) = \frac{1}{120}\left(23 - 17\cos(t)\right) \tag{11.131}$$

The general ordinary differential equation for the inlet speed u_i

$$\frac{du_i}{dt} + au_i^2 = b\left(p_i(t) - p_o(t)\right)$$

specifically corresponds to the prescribed nozzle and pressure boundary conditions through the following magnitudes of the constants a and b

$$b = \frac{1}{\displaystyle\int_{x_i}^{x_o} \frac{A_i}{A(x)}\,dx} = \frac{12}{3\sqrt{10}+5\sqrt{6}}, \quad a = \frac{b}{2}\left[\left(\frac{A_i}{A_o}\right)^2 - 1\right] = \frac{32}{3(3\sqrt{10}+5\sqrt{6})} \tag{11.132}$$

The terminal non-linear differential equation for u_i thus becomes

$$\frac{du_i}{dt} = \frac{1}{3\sqrt{10}+5\sqrt{6}}\left[-\frac{32}{3}u_i^2 + \frac{1}{10}\left(23 - 17\cos(t)\right)\right] \tag{11.133}$$

With this equation, both u_i and $\frac{du_i}{dt}$ have been determined numerically, subject to the initial condition $u_i(0) = 0$. Since at every instant t the inlet pressure remains greater than the outlet pressure, the flow will continuously proceed from the inlet to the outlet. At $t = 0$, the outlet pressure begins to decrease from its maximum and the flow starts developing. After the initial transient, the distributions of pressure and speed at each location within the tube become periodic functions of time.

Figure 11.8-a) shows the variation of outlet pressure p_o and the calculated variations of both inlet speed u_i and its time derivative.

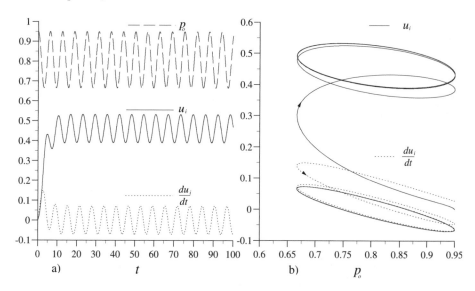

Figure 11.8: Unsteady Flow, $0 \leq t \leq 100$: a) p_o, u_i, $\frac{du_i}{dt}$ vs. t; b) u_i, $\frac{du_i}{dt}$ vs. p_o

This variation for u_i results from a numerical solution of equation (11.133) and reflects the corresponding variation from the characteristics-bias discrete equations. This figure presents these variations for a total of 16 complete periods of outlet pressure, corresponding to the time range $0 \leq t \leq 100$. As the figure shows, the transient for both u_i and $\frac{du_i}{dt}$ completes in about 1.5 periods of outlet pressure. Thereafter, u_i and $\frac{du_i}{dt}$ become periodic functions of t. Figure 11.8-b) also shows the evolution of u_i and $\frac{du_i}{dt}$ toward their periodic states, but in a different format. This figure displays u_i and $\frac{du_i}{dt}$ plotted versus p_o and shows that the periodic variations of these two variables versus p_o correspond to variations within closed elliptical orbits.

Figures 11.9-a)-b) present similar variations for u_i and $\frac{du_i}{dt}$. Figure 11.9-a) presents the variations for a reduced time range, $0 \leq t \leq 30$ and Figure 11.9-b) shows $\frac{du_i}{dt}$ versus u_i.

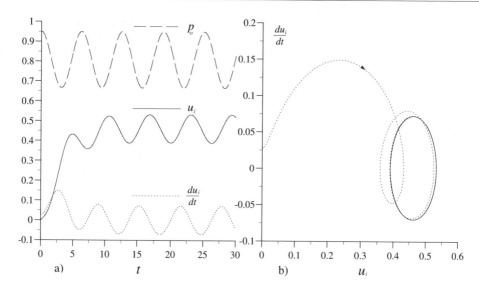

Figure 11.9: Unsteady Flow, $0 \leq t \leq 30$: a) p_o, u_i, $\frac{du_i}{dt}$ vs. t; b) $\frac{du_i}{dt}$ vs. u_i

The characteristics-bias algorithm has generated the spatial distributions for pressure p and speed u at a $C_{\max} = 1$, as shown in Figures 11.10-a)-b) and Figures 11.11-a)-b). The controller minimum and maximum were set to $\psi_{\min} = 0.3$ and $\psi_{\max} = 0.5$; nevertheless, the calculated ψ remained nearly equal to its minimum without ever rising to its maximum. It was observed the calculated solution for both p and u remained close to the exact solution for $0.3 \leq \psi$. These calculated distributions for p and u reflect the corresponding exact solutions indicated with a solid line. In particular, the CFD solutions remain smooth and undistorted.

The figures present these distributions for 5 time levels. The first time level correspond to $t = 2.69$, which virtually lies in the middle of the transient phase. The corresponding distribution of pressure at this stage does not yet reflect the variation of the tube cross-sectional area, unlike the distribution of u, which must always correlate with the distribution of cross-sectional area, because of the absence of any time derivative of pressure within the continuity equation. The other 4 time levels correspond to instants after the first 9 periods of outlet pressure.

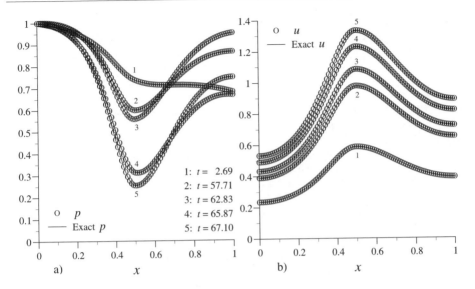

Figure 11.10: Unsteady-Flow Distributions: a) Pressure p; b) Speed u

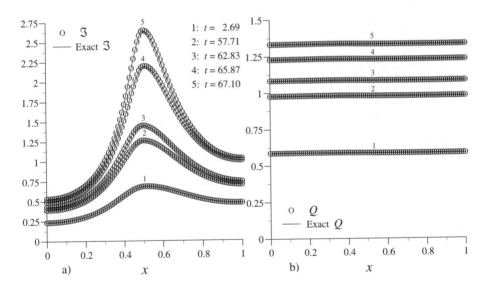

Figure 11.11: Unsteady-Flow Distributions: a) $\Im$; b) Volumetric Flow Rate Q

These time levels respectively correspond to minimum inlet speed, maximum outlet pressure, minimum outlet pressure and maximum inlet speed. For all time levels considered, the characteristics-bias numerical solutions for p and u remain indistinguishable from the exact solutions. In particular, the outlet boundary-condition enforcement strategy within the linear-momentum equation leads to a smooth pressure distribution throughout the tube.

The numerical solution in figures 11.11-a)-b) correspond to the incompressible-flow number $\Im$ and the volumetric flow rate $Q = u(x,t)A(x)$. For all time levels, the distribution of $\Im$ closely reflect the exact distribution. The calculated distribution of volumetric mass low rate Q conveys the ability of the characteristics-bias algorithm to conserve mass, a critical requirement of any incompressible-flow solver. For any tube cross-sectional area distribution, $Q = u(x,t)A(x)$ must not vary with x, as implied by the continuity equation. However, Q can vary with time, as a result of a time variation of the inlet speed u_i. Although the characteristics-bias continuity equation features a reference acoustic term that depends on pressure, the calculated distribution of Q remains constant at all positions within the tube and it increases following the rise in inlet speed. These distributions, furthermore, reflect the exact volumetric flow rate. Despite the absence of any time-derivative of pressure, the characteristics-bias algorithm, therefore, has generated distributions for p and u that follow the exact solution at all time levels. These results have been generated by a direct numerical solution the incompressible-flow continuity and linear-momentum equations subject to physical pressure boundary conditions.

11.11.2　Transient and Steady State Flow

This benchmark tests the capability of the algorithm to calculate transient and steady distributions of pressure and speed. The flow evolves from an initially quiescent field with uniform pressure that is the subject to a constant inlet pressure, which exceeds a constant outlet pressure.

The specified distribution of tube cross-sectional areas lead to the following area magnitudes

$$A_i = \frac{5}{2}, \quad A_T = 1, \quad A_o = \frac{3}{2}, \quad \frac{A_i}{A_o} = \frac{5}{3} \tag{11.134}$$

From result (11.113), the outlet-inlet pressure ratio has to satisfy the constraint

$$0.6614 \simeq \frac{125}{189} \leq \frac{p_o}{p_i} \leq 1 \tag{11.135}$$

A realistic flow can thus evolve from the pressure specifications

$$p_i = 1, \quad p_o = \frac{3}{4} \simeq 0.77, \quad p_i - p_o = \frac{1}{4} \tag{11.136}$$

This flow is numerically calculated by enforcing a constant step decrease in outlet pressure, from the initial $p_o = 1$ to the terminal $p_o = \frac{3}{4}$.

Transient Flow

Figures 11.12-11.13 present the solution generated with $C_{\max} = 1.0$. The controller minimum and maximum were again to $\psi_{\min} = 0.3$ and $\psi_{\max} = 0.5$; nevertheless, the calculated ψ remained nearly equal to its minimum without ever rising to its maximum. It was observed the calculated solution for both p and u reflect the exact solution for $0.3 \leq \psi$. With these settings for ψ, the calculated solution remains smooth throughout the computational domain.

Figure 11.12-a)-b) present the transient distributions for p and u for 4 time levels. At the first time level $t = 0.5$ the corresponding distributions for p and u lie close to the initial state. The distribution of pressure at this stage does not yet reflect the variation of the tube cross-sectional area, yet the sudden change from the initial condition remains quite evident in the figure.

The distributions for p and u at the other time levels show the transient progression toward the terminal steady state. In particular these distributions display a curvature discontinuity at the tube throat, as induced by the curvature discontinuity in the distribution of tube cross-sectional area. The smooth variations of pressure reflect the exact solution at all time levels, and show that the calculated outlet pressure remains equal to the prescribed pressure, which confirms the reliability of the boundary-condition enforcement strategy through the linear-momentum equation. The variations of speed u also reflect the exact solution at all time levels. These variations remain smooth and display the expected increase in u at the tube throat as the solution approaches its steady state

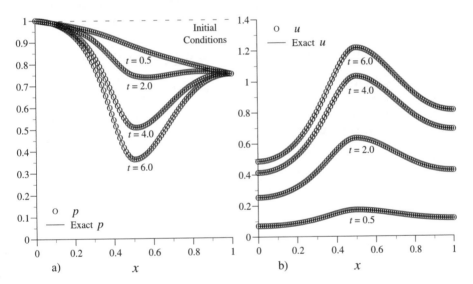

Figure 11.12: Transient-Flow Distributions: a) Pressure p; b) Speed u

Figure 11.13-a)-b) show the calculated incompressible-flow number $\Im$ and volumetric flow rate $Q = u(x,t)A(x)$.

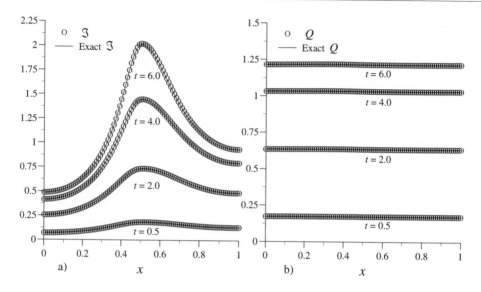

Figure 11.13: Transient-Flow Distributions: a) $\Im$; b) Volumetric Flow Rate Q

For all time levels, the distribution of $\Im$ remain indistinguishable from the exact variation. The distribution of Q shown in Figure 11.13-b) remains constant at all positions within the tube and it increases following the rise in inlet speed. These distributions, furthermore, reflect the exact volumetric flow rate.

Steady Flow

The pressure specifications (11.136) induce the following reference quantities

$$A_v = \frac{A_i}{\sqrt{1 + \left[\left(\frac{A_i}{A_o}\right)^2 - 1\right]\frac{p_i}{\Delta p}}} = \frac{15}{2\sqrt{73}} \simeq 0.8778 \tag{11.137}$$

$$u_i = \sqrt{\frac{2\Delta p}{\left(\frac{A_i}{A_o}\right)^2 - 1}} = \frac{3}{4\sqrt{2}} \simeq 0.530033, \quad u_T = u_i\frac{A_i}{A_T} = \frac{15}{8\sqrt{2}} \simeq 1.32583 \tag{11.138}$$

$$u_v = u_i\frac{A_i}{A_v} = \frac{\sqrt{73}}{4\sqrt{2}} \simeq 1.51038, \quad u_o = u_i\frac{A_i}{A_o} = \frac{5}{4\sqrt{2}} \simeq 0.88388 \tag{11.139}$$

$$p_{st} = p_i + \frac{u_i^2}{2} = \frac{73}{64} \simeq 1.141, \quad p_T = p_{st} - \frac{u_T^2}{2} = \frac{67}{256} \simeq 0.26172 \tag{11.140}$$

Figures 11.14-11.16 present the steady solution generated with $C_{\max} = 200$.

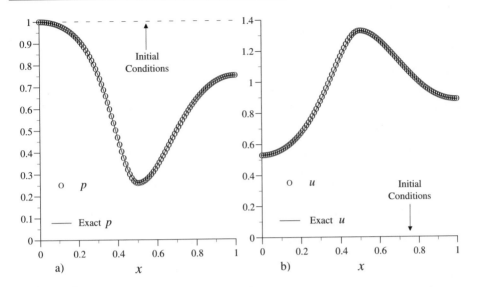

Figure 11.14: Steady-Flow Distributions: a) Pressure p; b) Speed u

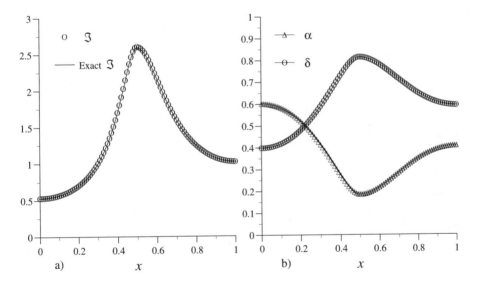

Figure 11.15: Steady-Flow Distributions: a) $\Im$; b) Characteristics-Bias Functions α, δ

As in the calculation of the transient flow, the controller minimum and maximum were again set to $\psi_{\min} = 0.3$ and $\psi_{\max} = 0.5$. The calculated ψ remained nearly equal to its minimum without ever rising to its maximum. Again, the calculated solution for both p and u remained close to the exact solution for $0.3 \leq \psi$. With these settings for ψ, the calculated solution remains smooth throughout the computational domain.

The distributions of pressure and speed in Figure 11.14 a)-b) follow the corresponding exact solutions, with the curvature discontinuity at the tube throat clearly resolved. These distributions of pressure and speed also reflect the reference magnitudes (11.137)-(11.140).

Figure 11.15 presents the distribution of the incompressible-flow number $\Im$, which also reflect the exact solution, and the corresponding distributions of the acoustics and pressure-gradient characteristics-bias functions α and δ. According to these distributions, the acoustics upstream-bias α controls rapidly decreases for increasing $\Im$. For these $\Im$ magnitudes, therefore, the effect of the characteristics-bias spatial derivative of pressure term within the continuity equation quickly diminishes. Regardless of this pressure term, however, the calculated solution is seen to mirror the exact distributions. For increasing $\Im$, the pressure-gradient controller δ briskly rises, which increases the upstream-bias in the approximation of the pressure gradient.

The characteristics-bias approximation ensures stability in the direct numerical solution of the continuity and linear-momentum equations. It also leads to distributions that conserve both mass and stagnation pressure, as Figure 11.16-a) illustrates in terms of the calculated $\Im$, Q, and p_{st}. Despite the significant variation of $\Im$, hence p and u, the calculated volumetric flow rate Q and stagnation pressure p_{st} remain essentially constant throughout the Venturi tube and reflect the exact magnitudes in (11.137)-(11.140).

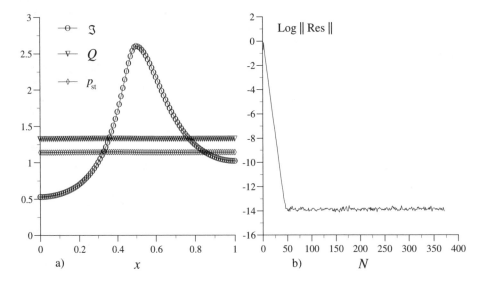

Figure 11.16: Steady-Flow Distributions: a) $\Im$, Q, p_{st}; b) Convergence Rate

Figure 11.16-b) presents the convergence rate toward steady state, measured through the maximum residual at each time level. The steady state was achieved in about 50 time steps, with $C_{max} = 200$ and a reduction of the maximum residual by 14 orders of magnitude, down to machine zero. A reduction of the maximum residual by over 8 orders of magnitude was achieved within 25 time steps. Not only does the characteristics-bias formulation generate smooth solutions that mirror exact distributions, but it also provides an essentially monotone rapid convergence to a steady state that remains genuinely steady.

11.12   Computational Performance

The acoustics-convection upstream resolution algorithm rests on the physics and mathematics of acoustics and convection. It introduces a decomposition of the flux vector Jacobian into acoustics and convection matrix components and generates the upstream bias at the differential equation level before any discrete approximation. This characteristics-bias algorithm generates a characteristics-bias flux that generalizes in the continuum the traditional upwind-scheme numerical fluxes. Such a formulation also admits a straightforward implicit implementation, features a computational simplicity that parallels a traditional centered discretization, and rationally decreases superfluous artificial diffusion. Despite the absence of any time derivative of pressure within the governing equations, the algorithm simultaneously solves the coupled continuity and linear - momentum equations. Without employing any artificial time derivative of pressure, the procedure directly generates pressure and velocity fields for both steady and unsteady flows. This formulation directly enforces the outlet pressure within the surface integral that emerges in the momentum-equation weak statement. The computed solutions at nozzle outlets remain smooth and undistorted and mirror the exact reference solutions, which bears out the reliability of this pressure boundary condition.

A natural finite element discretization of the characteristics-bias system, with equal-order elements for pressure and speed, generates an essentially centered approximation of the incompressible-flow Euler equations, in the form of a non-linear combination of upstream diffusive and downstream anti-diffusive flux differences, with greater bias on the upstream diffusive flux difference. The study in this chapter has implemented the algorithm using a linear approximation of fluxes within two-noded cells, without any MUSCL-type local extrapolation of variables. According to the solution-driven numerical values of the upstream-bias controller, the computed solutions resulted from a mostly centered discretization. This finding indicates that a uniformly fully upwind formulation is not strictly necessary within a characteristics-bias algorithm to generate essentially crisp and non-oscillatory solutions for genuinely unsteady, transient, and steady incompressible flows. The algorithm reduces artificial dissipation and generates asymptotically converging solutions that agree with the exact reference solutions. The calculations for steady flows rapidly converge to steady state and for steady and unsteady flows the algorithm conserves both mass and stagnation pressure.

Chapter 12

Numerical Study of Generalized Quasi-1-D Free Surface Flows

This chapter documents performance of the Gravity-Wave-Convection Upstream Resolution Algorithm for the numerical investigation of adiabatic and non-adiabatic generalized free-surface flows also with embedded hydraulic jumps. The algorithm employs a gravity-wave-convection decomposition of the system flux Jacobian and induces the upstream bias at the differential - equation level, before any discretization, within a characteristics-bias system associated with the governing equations. This characteristics-bias system essentially combines the governing system with a regularizing hyperbolic-parabolic; the perturbation parameter is then linked to the product of a mesh measure and solution smoothness measure, so that the regularized solution approaches the corresponding weak solution of the governing equations as the mesh is refined.

This formulation becomes Galilean invariant and allows the exact solution of governing equations also to satisfy the characteristics-bias system in regions of supercritical flow. Directly resulting from the upwind-bias representation of the source term and time derivative, this property reduces upwind diffusion and practically leads to converging essentially non-oscillatory or monotone solutions for both steady and unsteady problems, with crisp capturing of not only hydraulic jumps, but also energy contact discontinuities. For low subcritical Froude numbers, this formulation returns a consistent upstream-bias approximation for the wave celerity equations. For supercritical Froude numbers, the formulation smoothly becomes an upstream-bias approximation of the entire inviscid flux. With the objective of minimizing induced artificial diffusion, the formulation non-linearly induces upstream-bias essentially locally in regions of solution discontinuities, whereas it decreases the upstream-bias in regions of solution smoothness.

The gravity-wave-convection flux Jacobian decomposition consists of components that genuinely model the physics of gravity-wave propagation and convection. These components combine the computational simplicity of FVS with the accuracy and stability of FDS and also feature eigenvalues with uniform algebraic sign. This formulation eliminates the unstable linear-dependence problem in steady low-Froude-number flows and satisfies by design the upstream-bias stability condition.

The discrete equations originate from a Galerkin finite element discretization of the characteristic-bias system. This finite element discretization naturally and automatically leads to consistent boundary differential equations and a new outlet pressure boundary condition that does not require any algebraic extrapolation of variables. The resulting discrete equations correspond to an essentially centered discretization in the form of a non-linear combination of upstream diffusive and downstream anti-diffusive flux differences, with greater bias on the upstream diffusive flux difference. This formulation also directly accommodates an implicit solver, for the required Jacobian matrices are analytically determined in a straightforward manner. The finite element equations are then integrated in time within a compact block tri-diagonal matrix statement by way of an implicit non-linearly stable Runge-Kutta algorithm for stiff systems.

The operation count for this algorithm is comparable to that of a simple flux vector splitting algorithm. The developments in this study have employed basic two-noded cells, which has thus led to a block tridiagonal matrix system, for the implicit formulation. To determine the ultimate accuracy of linear approximations of fluxes within two-noded cells, for a computationally efficient implementation, this study employs no MUSCL-type local extrapolation of dependent variables. Several numerical investigations document the theoretical accuracy of the formulation and show that for the linear-basis implementation, the corresponding solutions asymptotically converge in the $\mathcal{H}^0$ and $\mathcal{H}^1$ norms, at a rate that equals up to 3, in the $\mathcal{H}^0$ norm, and exponentially converge to steady state, with Courant numbers up to 200. Displaying but minimal upwind artificial diffusion, the computational results for generalized free-surface flows reflect available exact solutions for flows within straight and converging-diverging channels with shaft work, mass as well as heat transfer, wall friction and variable bathymetry. These results keep mass flow and enthalpy constant across hydraulic jumps and show the direct mass-transfer and bed-slope effect on the location and strength of a hydraulic jump, with sufficient mass transfer and upward bed slope capable of leading to complete hydraulic jump elimination.

12.1 One-Dimensional Free-Surface Flow Characteristics

Following the developments in previous chapters, the generalized non-dimensional depth-averaged equations, with respect to an inertial reference frame, are expressed as

$$
\begin{cases}
\dfrac{\partial h}{\partial t} + \dfrac{\partial m}{\partial x} + \dfrac{m}{W}\dfrac{\partial W}{\partial x} = \dfrac{\widetilde{m}_b}{W} \\[2ex]
\dfrac{\partial m}{\partial t} + \dfrac{\partial}{\partial x}\left(\dfrac{m^2}{h} + \dfrac{h^2}{2}\right) + \dfrac{m^2}{hW}\dfrac{\partial W}{\partial x} = -\left(1 + \dfrac{2h}{W(W_{\mathrm{ref}}/h_{\mathrm{ref}})}\right)f_D\dfrac{m|m|}{h^2} \\[2ex]
\qquad - h\dfrac{\partial b}{\partial x} - s\chi\, hw_m + \dfrac{\dot{m}_b\widetilde{m}_b(W/W_{\mathrm{ref}})\sin\theta}{h_s W} \\[2ex]
\dfrac{\partial E}{\partial t} + \dfrac{\partial}{\partial x}\left(\dfrac{m}{h}\left(E + \dfrac{h^2}{2}\right)\right) + \dfrac{m(E + h^2/2)}{hW}\dfrac{\partial W}{\partial x} = \\[2ex]
\qquad |m|\,(g_{\mathrm{T}} - w_m) + \dfrac{f_{\mathrm{air}}}{8}\dfrac{\rho_{\mathrm{air}}}{\rho}u(u_{\mathrm{wind}} - u)|u_{\mathrm{wind}} - u| + \dfrac{(E + h^2/2)_b\,\widetilde{m}_b}{h_b W}
\end{cases}
\tag{12.1}
$$

where $s \equiv \text{sgn}(m)$ and $\widetilde{m}_b$ is defined as

$$\widetilde{m}_b \equiv \dot{m}_b \left(\cos\theta + \sin\theta \frac{\partial W}{\partial x} \frac{W_{\text{ref}}/L}{2} \right) \tag{12.2}$$

This quasi-one-dimensional system can be expressed as

$$\frac{\partial q}{\partial t} + \frac{\partial f(q)}{\partial x} = \phi \tag{12.3}$$

where the independent variable (x, t) varies in the domain $D \equiv \Omega_1 \times [t_o, T]$, $\Omega_1 \equiv [x_{\text{in}}, x_{\text{out}}]$ and the system consists of the continuity, linear-momentum, and energy equations. This system consists of the continuity, linear-momentum and total-energy equations, with arrays of dependent variables $q = q(x, t)$, $f = f(q)$ and $\phi = \phi(x, q)$ defined as

$$q \equiv \left\{ \begin{array}{c} h \\ m \\ E \end{array} \right\}, \quad f(q) \equiv \left\{ \begin{array}{c} m \\ \dfrac{m^2}{h} + \dfrac{h^2}{2} \\ \dfrac{m}{h}\left(E + \dfrac{h^2}{2}\right) \end{array} \right\},$$

$$\phi \equiv - \left\{ \begin{array}{c} 0 \\ \left(1 + \dfrac{2h}{W(W_{\text{ref}}/h_{\text{ref}})}\right) f_D \dfrac{m|m|}{h^2} + h\dfrac{\partial b}{\partial x} + s\chi\, hw_m \\ |m|(w_m - g_{\text{T}}) - \dfrac{f_{\text{air}}}{8}\dfrac{\rho_{\text{air}}}{\rho} u(u_{\text{wind}} - u)|u_{\text{wind}} - u| \end{array} \right\}$$

$$- \frac{m}{hW}\frac{\partial W}{\partial x} \left\{ \begin{array}{c} h \\ m \\ E + \dfrac{h^2}{2} \end{array} \right\} + \frac{\widetilde{m}_b}{W} \left\{ \begin{array}{c} 1 \\ \dfrac{\dot{m}_b}{h_b}(W_{\text{ref}}/L)\sin\theta \\ \left.\dfrac{E + h^2/2}{h}\right|_b \end{array} \right\} \tag{12.4}$$

The characteristic speeds associated with system (12.3) are the eigenvalues of the flux vector Jacobian

$$\frac{\partial f(q)}{\partial q} = \left(\begin{array}{ccc} 0 & , & 1 & , & 0 \\ -\dfrac{m^2}{h^2} + h & , & \dfrac{2m}{h} & , & 0 \\ -\dfrac{m}{h^2}\left(E + \dfrac{h^2}{2}\right) + m & , & \dfrac{1}{h}\left(E + \dfrac{h^2}{2}\right) & , & \dfrac{m}{h} \end{array} \right) \tag{12.5}$$

These eigenvalues have been exactly determined in closed form as

$$\lambda_1^{f_s} = u, \quad \lambda_{2,3}^{f_s} = u \pm c, \quad c \equiv \sqrt{h} \tag{12.6}$$

where superscript "f_s" indicates a free-surface eigenvalue and "c" denotes the gravity-wave celerity. Dividing by c, the corresponding form of these eigenvalues is

$$\lambda_1 = Fr, \qquad \lambda_{2,3} = Fr \pm 1, \qquad Fr \equiv \frac{|u|}{c} \tag{12.7}$$

where Fr denotes the Froude number. Figure 12.1

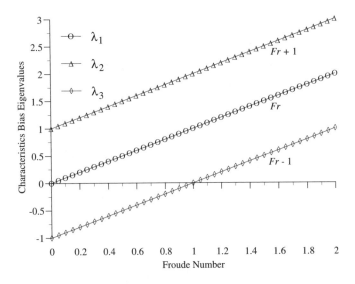

Figure 12.1: Free-Surface Flow Eigenvalues

portrays the variation of these eigenvalues and shows λ_3 remains negative for subcritical flows, $Fr < 1$, vanishes for critical flows, $Fr = 1$, and becomes positive for supercritical flows, $Fr > 1$. This change in sign affects the direction of wave propagation depending on Fr.

These eigenvalues correspond to the slopes of the characteristics, as portrayed in Figure 12.2 for representative hypercritical, supercritical, critical, and subcritical flows. This figure shows the characteristics in a neighborhood of a flow field point P in a (t, x) plane. As an interesting geometric difference among supercritical, critical, and subcritical flows, a time axis through P is respectively outside, on the boundary, and inside the domain of dependence and range of influence of point P. Wave propagation for supercritical flows essentially occurs by convection, mono-axially from upstream to downstream of P; the critical case becomes a limiting case; for subcritical flows, instead, wave propagation occurs by both convection and gravitation, bi-modally from both upstream and downstream toward and then away from P; for vanishing Froude number, wave propagation is essentially gravitational.

Since free-surface wave propagation physically occurs by gravitation and convection, the characteristics-bias formulation in the following sections is mathematically based on this coupled gravitational-convection wave propagation. The formulation identifies the genuine

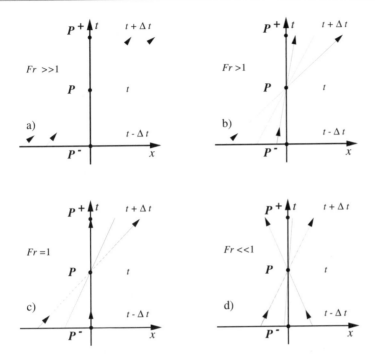

Figure 12.2: Characteristics: a) Hypercritical, b) Supercritical, c) Critical, d) Subcritical

convection and gravitation components within the flux Jacobian and then establishes a physically consistent upstream representation for each of these components. The next section identifies and discusses the gravity-wave components.

12.2 Gravity-Wave Equations

The free-surface system

$$\frac{\partial q}{\partial t} - \phi + \frac{\partial f(q)}{\partial x} = 0 \tag{12.8}$$

contains the gravity-wave equations for vanishing Froude numbers. Identification of these equations yields the gravity-wave component of the free-surface flux Jacobian for any Froude number. Upon restricting the source term ϕ to the variable-width terms, for simplicity, and expressing momentum m in terms of the Froude number Fr as $m = hcFr\, u/|u|$, $c \equiv \sqrt{gh}$, the free-surface system (12.8) becomes

$$\frac{\partial h}{\partial t} + \frac{\partial m}{\partial x} + cFr\frac{u}{|u|}\frac{h}{W}\frac{dW}{dx} = 0$$

$$\frac{\partial m}{\partial t} + c^2\frac{\partial h}{\partial x} + cFr\frac{u}{|u|}\left[2\frac{\partial m}{\partial x} - \frac{m}{h}\frac{\partial h}{\partial x} + \frac{m}{W}\frac{dW}{dx}\right] = 0$$

$$\frac{\partial E}{\partial t} + \frac{1}{h}\left(E + \frac{h^2}{2}\right)\frac{\partial m}{\partial x} + cFr\frac{u}{|u|}\left[h\frac{\partial}{\partial x}\left(\frac{1}{h}\left(E + \frac{h^2}{2}\right)\right) + \frac{(E + h^2/2)}{W}\frac{dW}{dx}\right] = 0 \quad (12.9)$$

and for a vanishing Froude number, these equations reduce to the gravity-wave system

$$\frac{\partial}{\partial t}\begin{pmatrix} h \\ m \\ E \end{pmatrix} + \begin{pmatrix} 0 &,& 1 &,& 0 \\ c^2 &,& 0 &,& 0 \\ 0 &,& \frac{1}{h}\left(E + \frac{h^2}{2}\right) &,& 0 \end{pmatrix}\frac{\partial}{\partial x}\begin{pmatrix} h \\ m \\ E \end{pmatrix} = 0 \quad\Rightarrow\quad \frac{\partial q}{\partial t} + A^{gw}\frac{\partial q}{\partial x} = 0 \quad (12.10)$$

Heed that the energy equation toward steady state, in this case, is no longer linearly independent from the continuity equation. The dependent variables in these equations correspond to those in a flow field that originates from slight perturbations to an otherwise quiescent field. The matrix A^{gw}

$$A^{gw} = \begin{pmatrix} 0 &,& 1 &,& 0 \\ c^2 &,& 0 &,& 0 \\ 0 &,& \frac{1}{h}\left(E + \frac{h^2}{2}\right) &,& 0 \end{pmatrix} \quad (12.11)$$

has the eigenvalues

$$\lambda_1^{gw} = 0, \quad \lambda_{2,3}^{gw} = \pm c \quad (12.12)$$

This matrix, accordingly, can be termed the "gravity-wave" matrix, for its eigenvalues equal the gravity-wave celerity c for any Froude number. Despite its zero eigenvalue, A^{gw} features a complete set of eigenvectors and thus possesses the similarity form

$$A^{gw} \equiv X\Lambda^{gw}X^{-1} = X\Lambda^{gw+}X^{-1} + X\Lambda^{gw-}X^{-1}, \quad \Lambda^{gw} = \Lambda^{gw+} + \Lambda^{gw-} \quad (12.13)$$

where the diagonal matrix Λ^{gw} equals

$$\Lambda^{gw} \equiv \begin{pmatrix} c &,& 0 &,& 0 \\ 0 &,& -c &,& 0 \\ 0 &,& 0 &,& 0 \end{pmatrix} \quad (12.14)$$

The corresponding eigenvector matrices X and X^{-1} are

$$X = \begin{pmatrix} 1 &,& 1 &,& 0 \\ c &,& -c &,& 0 \\ \frac{E + h^2/2}{h} &,& \frac{E + h^2/2}{h} &,& 1 \end{pmatrix}, \quad X^{-1} = \frac{1}{2c}\begin{pmatrix} c &,& 1 &,& 0 \\ c &,& -1 &,& 0 \\ -\frac{2c(E + h^2/2)}{h} &,& 0 &,& 2c \end{pmatrix}$$

$$(12.15)$$

This similarity transformation is employed in the gravity-wave flux-Jacobian decomposition, as shown in the next section.

12.3   Free-Surface Gravity-Wave-Convection Flux Jacobian Decomposition

The characteristics-bias formulation for the free-surface system

$$\frac{\partial q}{\partial t} - \phi + \frac{\partial f(q)}{\partial x} = 0 \tag{12.16}$$

emerges from a decomposition of the flux-vector Jacobian $\partial f/\partial q$ into gravity-wave and convection components. The following section determines these components.

12.3.1   Convection and Pressure-Gradient Components

The flux divergence $\frac{\partial f}{\partial x}$ can be decomposed into convection and pressure-gradient components as

$$\frac{\partial f}{\partial x} = \frac{\partial f^q}{\partial x} + \frac{\partial f^p}{\partial x} \tag{12.17}$$

where f^q and f^p respectively denote the convection and pressure fluxes, defined as

$$f^q(q) \equiv \left\{ \begin{array}{c} m \\[2mm] \dfrac{m^2}{h} \\[3mm] \dfrac{m}{h}\left(E + \dfrac{h^2}{2}\right) \end{array} \right\} = \frac{m}{h} \cdot \left\{ \begin{array}{c} h \\[2mm] m \\[3mm] \left(E + \dfrac{h^2}{2}\right) \end{array} \right\}, \quad f^p \equiv \left\{ \begin{array}{c} 0 \\[2mm] \dfrac{h^2}{2} \\[3mm] 0 \end{array} \right\}, \quad p \equiv \frac{h^2}{2} \tag{12.18}$$

The non-dimensional eigenvalues of the flux vector Jacobian $\partial f/\partial q = \partial f^q/\partial q + \partial f^p/\partial q$ are

$$\lambda_1 = Fr, \qquad \lambda_{2,3} = Fr \pm 1 \tag{12.19}$$

For supercritical flows, the eigenvalues (12.19) all have the same algebraic sign and the entire flux divergence can be upstream approximated along one single direction. For subcritical flows these eigenvalues have mixed algebraic sign and an upstream approximation for the flux divergence along one single direction remains inconsistent with the two-way propagation of gravity waves. Without the pressure gradient in the momentum equation, however, the corresponding flux-Jacobian eigenvalues all have the same algebraic sign and the resulting convection flux divergence can then be upstream approximated along one single direction. The flux divergence can thus be decomposed as the linear combination

$$\frac{\partial f}{\partial x} = \left[\frac{\partial f^q}{\partial x} + \beta \frac{\partial f^p}{\partial x}\right] + \left[(1 - \beta)\frac{\partial f^p}{\partial x}\right], \quad 0 \le \beta \le 1 \tag{12.20}$$

where the positive pressure-gradient partition function β can be chosen in such a way that all the eigenvalues of each of the two components between brackets in (12.20) keep the same algebraic sign for all Froude numbers. In this manner, these entire components can be upstream approximated along single directions. This choice for β is possible because the eigenvalues of a matrix are continuous functions of the matrix entries and hence all the eigenvalues for the components in (12.20) will continuously depend upon β. The function β will gradually increase toward 1 for increasing Froude number, so that an upstream approximation for the components in (12.20) smoothly approaches and then becomes an upstream approximation for the entire $\frac{\partial f}{\partial x}$ along one single direction. Decomposition (12.20) is thus used for an upstream approximation of the flux divergence for subcritical and supercritical flows.

For low and vanishing Froude numbers, decomposition (12.20), however, is insufficient for an accurate upstream modeling of gravity waves. For a Froude number that approaches zero, the free-surface eigenvalues (12.19) can all keep the same algebraic sign only if the gravity-wave celerity contribution vanishes, which corresponds to a vanishing pressure gradient contribution and hence β approaching zero. But for β approaching zero, the eigenvalues associated with the components in (12.20) approach the eigenvalues of the Jacobians

$$
\frac{\partial f^q(q)}{\partial q} = \begin{pmatrix} 0 & , & 1 & , & 0 \\ -\dfrac{m^2}{h^2} & , & \dfrac{2m}{h} & , & 0 \\ -\dfrac{m}{h^2}\left(E+\dfrac{h^2}{2}\right)+m & , & \dfrac{1}{h}\left(E+\dfrac{h^2}{2}\right) & , & \dfrac{m}{h} \end{pmatrix} \tag{12.21}
$$

and

$$
\frac{\partial f^p}{\partial q} = \begin{pmatrix} 0 & , & 0 & , & 0 \\ c^2 & , & 0 & , & 0 \\ 0 & , & 0 & , & 0 \end{pmatrix} \tag{12.22}
$$

The eigenvalues of these Jacobians respectively are

$$
\lambda_{1,3}^q = \frac{m}{h} \tag{12.23}
$$

and

$$
\lambda_{1,3}^p = 0 \tag{12.24}
$$

which certainly all keep the same algebraic sign, but for vanishing Froude number remain far less than the dominant gravity-wave celerity c. For low Froude numbers, therefore, an upstream approximation for the components in (12.20) would inaccurately model the physics of gravity-wave propagation. This difficulty is resolved by further decomposing the pressure gradient in (12.17) in terms of a genuine gravity-wave component, for accurate upstream modeling of gravity waves.

12.3.2   Gravity-Wave Components

Based on the gravity-wave equations (12.10), the flux divergence $\frac{\partial f}{\partial x}$ can be alternatively decomposed for arbitrary Froude numbers and corresponding dependent variables h, m, and E as

$$
\frac{\partial f}{\partial x} = \frac{\partial f^q}{\partial x} + \frac{\partial f^p}{\partial x} = \frac{\partial f^q}{\partial x} + (A^{gw} + A^{nc})\frac{\partial q}{\partial x} \tag{12.25}
$$

In this decomposition, the matrices A^{gw} and A^{nc} are defined as

$$A^{gw} \equiv \begin{pmatrix} 0 & , & 1 & , & 0 \\ c^2 & , & 0 & , & 0 \\ 0 & , & \frac{1}{h}\left(E + \frac{h^2}{2}\right) & , & 0 \end{pmatrix}, \quad A^{nc} \equiv \begin{pmatrix} 0 & , & -1 & , & 0 \\ 0 & , & 0 & , & 0 \\ 0 & , & -\frac{1}{h}\left(E + \frac{h^2}{2}\right) & , & 0 \end{pmatrix} \quad (12.26)$$

Heed, in particular, that no flux component of $f(q)$ exists, of which the Jacobian equals A^{gw}. The eigenvalues of the matrix A^{nc} have been determined in closed form as

$$\lambda^{nc}_{1,3} = 0 \quad (12.27)$$

which identically vanish for any Fr. The matrix A^{nc} can be termed a "non-linear coupling" matrix, for it completes the non-linear coupling between convection and gravity-wave propagation within (12.25) so that the two free-surface eigenvalues $\lambda^{fs}_{2,3}$ in (12.6) do correspond to the sum of a convection speed and gravity-wave celerity. Since decomposition (12.25) will be used in the upstream- bias formulation for small Froude numbers only and considering that the eigenvalues in (12.27) identically vanish, no need exists to involve A^{nc} in the upstream-bias approximation of the flux Jacobian in (12.25).

The eigenvalues of A^{gw} are exactly determined in closed form as

$$\lambda^{gw}_1 = 0, \qquad \lambda^{gw}_{2,3} = \pm c \quad (12.28)$$

This matrix is thus used for the a gravity-wave upstream bias. With reference to the similarity transformation (12.13) for this matrix, the free-surface flux divergence decomposition (12.25) becomes

$$\frac{\partial f}{\partial x} = \left(X\Lambda^{gw+}X^{-1} + X\Lambda^{gw-}X^{-1} + \frac{\partial f^q}{\partial q} + A^{nc} \right)\frac{\partial q}{\partial x} \quad (12.29)$$

which exposes the gravity-wave and convection components. Since the two gravity-wave matrices at the "rhs" of expression (12.29) respectively possess non-negative and non-positive eigenvalues, a characteristics-bias representation of these matrices involves an upstream representation of the first matrix and a downstream representation of the second matrix. The combination of these two representations directly leads to the following absolute gravity-wave matrix

$$|A^{gw}| \equiv X\left(\Lambda^{gw+} - \Lambda^{gw-}\right)X^{-1} \quad (12.30)$$

which depends on the diagonal matrices Λ^{gw+}, Λ^{gw-}, with different selections inducing distinct magnitudes of upwind diffusion. The following matrices

$$\Lambda^{gw+} \equiv \frac{1}{2}\begin{pmatrix} 2c & , & 0 & , & 0 \\ 0 & , & 0 & , & 0 \\ 0 & , & 0 & , & c \end{pmatrix}, \quad \Lambda^{gw-} \equiv -\frac{1}{2}\begin{pmatrix} 0 & , & 0 & , & 0 \\ 0 & , & 2c & , & 0 \\ 0 & , & 0 & , & c \end{pmatrix} \quad (12.31)$$

lead to the following absolute gravity-wave matrix statement

$$|A^{gw}|\frac{\partial q}{\partial x} \equiv X\left(\Lambda^{gw+} - \Lambda^{gw-}\right)X^{-1}\frac{\partial q}{\partial x} = cI\frac{\partial q}{\partial x} \quad (12.32)$$

which indicates for this matrix product the equivalence of replacing $|A^{gw}|$ with the matrix cI, of which all eigenvalues approach $+c$. This computationally advantageous form of the gravity-wave upstream-bias is used essentially in the low subcritical-flow Froude number regime.

12.3.3   Combination of Flux-Jacobian Decompositions

The previous sections have shown that the flux Jacobian can be equivalently expressed as

$$
\frac{\partial f}{\partial x} =
\begin{cases}
\left[\dfrac{\partial f^q}{\partial x} + \beta\dfrac{\partial f^p}{\partial x}\right] + \left[(1-\beta)\dfrac{\partial f^p}{\partial x}\right]; \\[3mm]
\left(X\Lambda^{gw+}X^{-1} + X\Lambda^{gw-}X^{-1} + \dfrac{\partial f^q}{\partial q} + A^{nc}\right)\dfrac{\partial q}{\partial x}
\end{cases}
\tag{12.33}
$$

where the first expression is convenient for a characteristics-bias approximation for high-subcritical and supercritical Froude numbers and the second expression is needed for low-subcritical Froude numbers.

A decomposition for all Froude numbers may thus be cast as a linear combination of these two system decompositions, with linear combination function α, $0 \leq \alpha \leq 1$

$$
\begin{aligned}
\frac{\partial f}{\partial x} = {} & (1-\alpha)\left\{\left[\frac{\partial f^q}{\partial x} + \beta\frac{\partial f^p}{\partial x}\right] + \left[(1-\beta)\frac{\partial f^p}{\partial x}\right]\right\} \\[2mm]
& + \alpha\left\{X\Lambda^{gw+}X^{-1} + X\Lambda^{gw-}X^{-1} + \frac{\partial f^q}{\partial q} + A^{nc}\right\}\frac{\partial q}{\partial x}
\end{aligned}
\tag{12.34}
$$

Relying on the simplifying function $\delta \equiv \beta(1-\alpha)$, the final form of the gravity-wave-convection flux-divergence decomposition becomes

$$
\frac{\partial f}{\partial q}\frac{\partial q}{\partial x} = \alpha\left(X\Lambda^{gw+}X^{-1} + X\Lambda^{gw-}X^{-1}\right)\frac{\partial q}{\partial x}
$$

$$
+ \left[\frac{\partial f^q}{\partial q} + \delta\frac{\partial f^p}{\partial q}\right]\frac{\partial q}{\partial x} + (1-\alpha-\delta)\frac{\partial f^p}{\partial x} + \alpha A^{nc}\frac{\partial q}{\partial x}
\tag{12.35}
$$

This decomposition leads to a consistent upstream-bias representation due to the physical significance of its terms. The streamline gravity-wave term $\alpha\left(X\Lambda^{gw+}X^{-1} + X\Lambda^{gw-}X^{-1}\right)\frac{\partial q}{\partial x}$ accounts for the bi-modal streamline propagation of gravity waves; this term is thus employed for a gravity-wave upstream - bias approximation along the streamline, for low Froude numbers. Since this term already models free-surface-height induced gravity-wave propagation along each streamline, no upstream-bias is needed for $\alpha A^{nc}\frac{\partial q}{\partial x}$, for an upstream of this term would essentially add only algebraic complexity to the formulation, since its coefficient α rapidly decreases and then vanishes for supercritical and high subcritical Froude numbers and all the eigenvalues of the Jacobian of this term vanish. By the same token, there is no need to develop an additional upstream bias for the $(1 - \alpha - \delta)\frac{\partial f^p}{\partial x}$, also considering that the coefficient $(1 - \alpha - \delta)$ term vanishes for gravity-wave and supercritical flows and all the eigenvalues of the Jacobian of this term identically vanish. Besides, as the Mach number rises, an increasing fraction of the pressure gradient receives an upstream bias in the term $\left[\frac{\partial f^q}{\partial x_j} + \delta\frac{\partial f^p}{\partial x_j}\right]$. This expression is counted as one term because the eigenvalues associated with this term will all keep the same sign, since $\delta = (1 - \alpha)\beta \leq \beta$. For any magnitude of both pressure and pressure gradient, the convection field uniformly carries information along the streamline; accordingly, this entire term can receive an upstream bias along the streamline direction.

12.4 Characteristics-Bias Flux Divergence

Given the algebraic sign of the eigenvalue set of each matrix term in (12.35), the associated direction cosines a_ℓ for the general upstream-bias expression (9.71) are

$$a_1 = +1, \quad a_2 = -1, \quad a_3 = s = \text{sgn}(u), \quad a_4 = a_5 = 0 \tag{12.36}$$

where $s = \text{sgn}(u)$ denotes the algebraic sign of u. With (12.35) and (12.36) the upstream-bias expression (9.71) leads to the gravity-wave-convection characteristics flux divergence

$$\frac{\partial f^C}{\partial x} = \frac{\partial f}{\partial x} - \frac{\partial}{\partial x}\left[\varepsilon\psi\left(\alpha c\frac{\partial q}{\partial x} + s\frac{\partial f^q}{\partial x} + s\delta\frac{\partial f^p}{\partial x}\right)\right] \tag{12.37}$$

The operation count for expression (12.37) is then comparable to that of an FVS formulation. The terms in this expression, furthermore, directly correspond to the physics of gravity waves and convection. Up to an arbitrary constant, this expression leads to the characteristics-bias flux

$$f^C = f - \varepsilon\psi\left(\alpha c\frac{\partial q}{\partial x} + s\frac{\partial f^q}{\partial x} + s\delta\frac{\partial f^p}{\partial x}\right) \tag{12.38}$$

Heed that the components within f^C remain linearly independent of one another, which avoids the linear-dependence instability in the steady low-Froude-number free-surface equations. For low Froude numbers, $\delta = 0$ and (12.37) reduces to

$$\frac{\partial f^C}{\partial x} = \frac{\partial f}{\partial x} - \frac{\partial}{\partial x}\left[\varepsilon\psi\left(\alpha c\frac{\partial q}{\partial x} + s\frac{\partial f^q}{\partial x}\right)\right] \tag{12.39}$$

which essentially induces only a gravity-wave upstream. For supercritical flow, $\alpha = 0$ and $\delta = 1$. Expression (12.37) in this case becomes

$$\frac{\partial f^C}{\partial x} = \frac{\partial f}{\partial x} - \frac{\partial}{\partial x}\left[\varepsilon\psi\left(s\frac{\partial f}{\partial x}\right)\right] \tag{12.40}$$

which corresponds to an upstream approximation of the entire free-surface flux divergence.

12.4.1 Consistent Upstream-Bias

The gravity-wave-convection characteristics-bias functions α and δ depend on the Froude number. They are determined by enforcing the upstream stability condition on the upstream matrix for (12.37). The divergence of the characteristics flux f^C in (12.37) becomes

$$\frac{\partial f^C}{\partial x} = \frac{\partial f}{\partial x} + \frac{\partial}{\partial x}\left[\varepsilon\psi\left(\alpha cI + s\frac{\partial f^q}{\partial q} + s\delta\frac{\partial f^p}{\partial q}\right)\frac{\partial q}{\partial x}\right] \tag{12.41}$$

where I denotes the identity matrix of appropriate size. The terms between parentheses collectively constitute the upstream- bias dissipation matrix

$$\mathcal{A} \equiv \alpha cI + s\frac{\partial f^q}{\partial q} + s\delta\frac{\partial f^p}{\partial q} \tag{12.42}$$

Despite the formidable algebraic complexity of $\mathcal{A}$, all of its eigenvalues have been analytically determined exactly in closed form. Dividing through by the gravity-wave celerity c, the non-dimensional form of these eigenvalues is

$$\lambda_1 = \alpha + Fr, \qquad \lambda_{2,3} = \alpha + Fr \pm \sqrt{\delta} \tag{12.43}$$

In order to ensure physical significance for the characteristics- bias flux within (12.37), hence for the upstream-bias approximation to decomposition (12.35), the upstream bias functions α and δ are determined by forcing the upstream-bias eigenvalues (12.43) to remain positive for all Froude numbers. Rather than prescribing some expressions for α and δ and accepting the resulting variations for these eigenvalues, physically reasonable expressions for these eigenvalues are instead prescribed and the corresponding functions for α and δ determined.

12.4.2   Upstream-Bias Eigenvalues and Functions α and δ

The prescribed eigenvalues λ_1 and λ_3 respectively correlate with Fr and $|Fr - 1|$. These eigenvalues directly lead to α and δ.

The eigenvalue λ_1 correlates with Fr, but must remain distinct from $|Fr - 1|$, for eigenvalue separation, and approach 1 for vanishing Fr, because all the non-dimensional eigenvalues of the absolute gravity-wave matrix equal 1. With $\varepsilon_{Fr} = \frac{1}{4}$, in this work, one expression for λ_1 that meets this requirements is the composite spline

$$\lambda_1(Fr) \equiv \begin{cases} 1 - Fr + \dfrac{\varepsilon_{Fr}}{2}(2Fr)^{1/\varepsilon_{Fr}} &, \quad 0 \;\; \le \;\; Fr \;\; < \;\; \frac{1}{2} \\[2ex] \dfrac{(Fr - \frac{1}{2})^2}{2\varepsilon_{Fr}} + \dfrac{1 + \varepsilon_{Fr}}{2} &, \quad \frac{1}{2} \;\; \le \;\; Fr \;\; < \;\; \frac{1}{2} + \varepsilon_{Fr} \\[2ex] Fr &, \quad \frac{1}{2} + \varepsilon_{Fr} \;\; \le \;\; Fr \end{cases} \tag{12.44}$$

The eigenvalue λ_3 correlates with the absolute-value free-surface eigenvalue $|Fr - 1|$. As a consequence, λ_3 will vary between 1 and $1 - Fr$ for $0 \le Fr \le 1 - \varepsilon_\lambda$ and smoothly shift from $1 - Fr$ to $Fr - 1$ within the critical transition layer $1 - \varepsilon_{Fr} \le Fr \le 1 + \varepsilon_{Fr}$, where ε_{Fr} denotes a transition-layer parameter; in this work $\varepsilon_{Fr} = \frac{1}{4}$. One expression for λ_3 that remains smooth and meets these requirements is the composite spline

$$\lambda_2(Fr) \equiv \begin{cases} 1 - Fr &, \quad 0 \;\; \le \;\; Fr \;\; \le \;\; 1 - \varepsilon_{Fr} \\[2ex] \dfrac{(Fr - 1)^2}{2\varepsilon_{Fr}} + \dfrac{\varepsilon_{Fr}}{2} &, \quad 1 - \varepsilon_{Fr} \;\; < \;\; Fr \;\; < \;\; 1 + \varepsilon_{Fr} \\[2ex] Fr - 1 &, \quad 1 + \varepsilon_{Fr} \;\; \le \;\; Fr \end{cases} \tag{12.45}$$

which is graphed in Figure 12.3. Observe in this curve the smooth transition in the critical region within the transition layer, where this λ_3 does not vanish, but remains not less than $\varepsilon_{Fr}/2$.

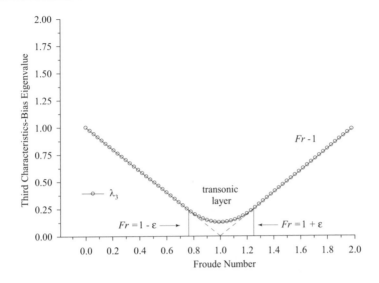

Figure 12.3: Characteristics-Bias Eigenvalue λ_3

With these λ_1 and λ_3, the corresponding expressions for both $\alpha = \alpha(Fr)$ and $\delta = \delta(Fr)$ from (12.43) have been exactly determined as

$$\alpha(Fr) = \lambda_1(Fr) - Fr, \quad \delta(Fr) = (\lambda_1(Fr) - \lambda_3(Fr))^2 \tag{12.46}$$

The variations of the upstream-bias functions $\alpha = \alpha(Fr)$ and $\delta = \delta(Fr)$ and the corresponding eigenvalues from (12.43) are presented in Figures 12.4-12.6. Figure 12.4 indicates that the upstream-bias functions as well as their slopes remain continuous for all Froude numbers, with $0 \leq \alpha, \delta \leq 1$ and $\alpha \equiv 0$ as well as $\delta \equiv 1$ for $Fr > \frac{1}{2} + \varepsilon_\lambda$. As $\delta = \delta(Fr)$ rises, the upstream-bias contribution from the gravity-wave matrix decreases rapidly, reducing to less than 25% of its maximum at $Fr = 0.39$. The variation of $\delta = \delta(Fr)$ shows that the pressure-gradient contribution to this upstream-bias formulation increases monotonically, while remaining less than 25% of its maximum, for $0 \leq Fr \leq 0.7$. When $\delta(Fr) \equiv 1$ for supercritical Froude numbers, the entire pressure-gradient is upstreamed with the same weight as in the convection flux, in complete agreement with the physical mono-axial wave propagation within supercritical flows.

Figure 12.5 presents the eigenvalues $\lambda_{1,3}^{cp}$ of the convection / pressure-gradient Jacobian $\left[\frac{\partial f^q}{\partial q} + \delta \frac{\partial f^p}{\partial q}\right]$. Obtained from (12.43) as $\lambda_{1,3}^{cp} = \lambda_{1,3} - \alpha$, these eigenvalues, as anticipated, remain positive, which justifies treating this Jacobian as one single term in the characteristics-bias divergence.

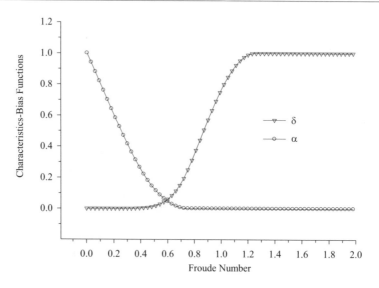

Figure 12.4: Upstream-Bias Functions

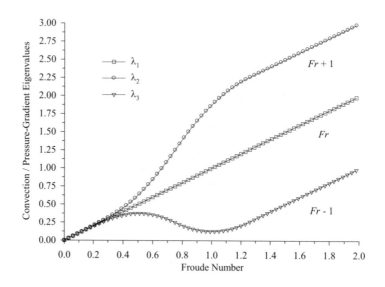

Figure 12.5: Convection / Pressure-Gradient Eigenvalues

For vanishing Froude numbers, δ approaches zero and the contribution to this Jacobian from pressure is virtually eliminated. As the Froude number, hence δ, increases these eigenvalues separate and smoothly approach the eigenvalues Fr, $Fr \pm 1$ for supercritical flows. These eigenvalues do not approach $|Fr \pm 1|$ for vanishing Fr, which is the reason why the gravity-wave upstream-bias, hence the α term, is developed. The complete characteristics-bias streamline eigenvalues (12.43) are shown in Figure 12.6. As shown in Figure 12.6, these eigenvalues remain positive and their variations and slopes remain continuous for all Froude numbers. These eigenvalues smoothly approach 1 for vanishing Fr, indicating a physically consistent upstream-bias approximation of the gravity-wave equations embedded within the full system. For subcritical flows, λ_1 remains greater than Fr, a feature that is compensated by a λ_2 that is less than $Fr + 1$ in the same flow regime. These eigenvalues remain positive as well as smooth, and their slopes remain continuous for all Mach numbers. For $Fr > 1 + \varepsilon_{Fr}$, these eigenvalues respectively coincide with the system flux Jacobian eigenvalues Fr, $Fr + 1$, $Fr - 1$, which corresponds to an upstream-bias approximation of the entire flux vector, for supercritical flows.

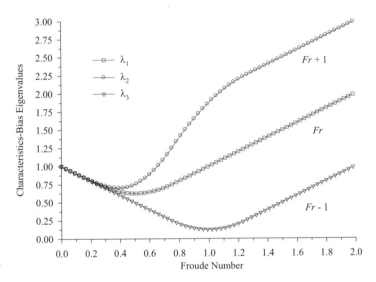

Figure 12.6: Upstream-Bias Eigenvalues

12.5   Variational Statement and Boundary Conditions

With ϕ_h, ϕ_m, ϕ_E denoting the components of the source term ϕ, respectively in the continuity, linear-momentum, and energy equation, the developments in Chapter 9 and the preceding sections lead to the following characteristics-bias integral equations for generalized quasi-one-dimensional depth-averaged free-surface flows

$$\int_\Omega w \left(\frac{\partial h}{\partial t} - \phi_h + \frac{\partial m}{\partial x} \right) d\Omega - \int_\Omega w \frac{\partial}{\partial x} \left(\varepsilon \psi s \left(\frac{\partial h}{\partial t} - \phi_h + \frac{\partial m}{\partial x} + \alpha c s \frac{\partial h}{\partial x} \right) \right) d\Omega = 0 \quad (12.47)$$

$$\int_\Omega w \left(\frac{\partial m}{\partial t} - \phi_m + \frac{\partial}{\partial x} \left(\frac{m^2}{h} + \frac{h^2}{2} \right) \right) d\Omega$$

$$- \int_\Omega w \frac{\partial}{\partial x} \left(\varepsilon \psi s \left(\frac{\partial m}{\partial t} - \phi_m + \frac{\partial}{\partial x} \left(\frac{m^2}{h} + \frac{h^2}{2} \right) + \alpha c s \frac{\partial m}{\partial x} + (\delta - 1) \frac{\partial h^2/2}{\partial x} \right) \right) d\Omega = 0$$

$$(12.48)$$

$$\int_\Omega w \left(\frac{\partial E}{\partial t} - \phi_E + \frac{\partial}{\partial x} \left(\frac{m}{h} \left(E + \frac{h^2}{2} \right) \right) \right) d\Omega$$

$$- \int_\Omega w \frac{\partial}{\partial x} \left(\varepsilon \psi s \left(\frac{\partial E}{\partial t} - \phi_E + \frac{\partial}{\partial x} \left(\frac{m}{h} \left(E + \frac{h^2}{2} \right) \right) + \alpha c s \frac{\partial E}{\partial x} \right) \right) d\Omega = 0 \qquad (12.49)$$

These equations are expressed as the single system

$$\int_\Omega w \left\{ \left(\frac{\partial q}{\partial t} - \phi + \frac{\partial f}{\partial x} \right) - \frac{\partial}{\partial x} \left[\varepsilon \psi s \left(\frac{\partial q}{\partial t} - \phi + \frac{\partial f}{\partial x} \right) \right] \right\} d\Omega$$

$$- \int_\Omega w \frac{\partial}{\partial x} \left[\varepsilon \psi s \left(\alpha c s \frac{\partial q}{\partial x} + (\delta - 1) \frac{\partial f^p}{\partial x} \right) \right] d\Omega = 0 \qquad (12.50)$$

An integration by parts of the characteristics-bias perturbation in this statement generates the following weak statement

$$\int_\Omega \left(w + \varepsilon \psi s \frac{\partial w}{\partial x} \right) \left(\frac{\partial q}{\partial t} - \phi + \frac{\partial f}{\partial x} \right) d\Omega + \int_\Omega \varepsilon \psi s \frac{\partial w}{\partial x} \left(\alpha c s \frac{\partial q}{\partial x} + (\delta - 1) \frac{\partial f^p}{\partial x} \right) d\Omega$$

$$- \left(w s \varepsilon \psi \left(\frac{\partial q}{\partial t} - \phi + \frac{\partial f}{\partial x} + \alpha c s \frac{\partial q}{\partial x} + (\delta - 1) \frac{\partial f^p}{\partial x} \right) \right) \Big|_{x=x_{\text{in}}}^{x=x_{\text{out}}} = 0 \qquad (12.51)$$

The $\varepsilon \psi$ boundary expression is eliminated by enforcing a weak Neumann-type boundary condition for the hyperbolic-parabolic characteristics-bias perturbation. One part of this condition imposes that the original system should be satisfied at the boundary; the remaining terms are also set to zero as part of this weak boundary condition. Clearly, the single Neumann-type condition corresponds to the entire boundary expression set to nought. For supercritical flows with $Fr \geq 1 + \varepsilon_{Fr}$, $\alpha = 0$ and $\delta = 1$ and this boundary condition, accordingly, enforces the governing system on the boundary. The resulting weak statement is expressed as

$$\int_\Omega \left(w + \varepsilon \psi s \frac{\partial w}{\partial x} \right) \left(\frac{\partial q}{\partial t} - \phi + \frac{\partial f}{\partial x} \right) d\Omega + \int_\Omega \varepsilon \psi s \frac{\partial w}{\partial x} \left(\alpha c s \frac{\partial q}{\partial x} + (\delta - 1) \frac{\partial f^p}{\partial x} \right) d\Omega = 0 \quad (12.52)$$

For all test functions $w \in \mathcal{H}^1(\Omega)$, the variational formulation seeks a solution $q \in \mathcal{H}^\eta(\Omega)$ with $\eta = 0$ or $\eta = 1$ respectively for shocked or smooth flows, that satisfies this weak statement.

This statement is subject to prescribed initial conditions $q(x, 0) = q_0(x)$ and boundary conditions on $\partial\Omega \equiv \overline{\Omega} \backslash \Omega$. Synthetically, these boundary conditions are expressed as

$$B(x_{\partial\Omega}) q(x_{\partial\Omega}, t) = G_v(x_{\partial\Omega}, t) \qquad (12.53)$$

where $G_v(x_{\partial\Omega}, t)$ corresponds to the array of prescribed Dirichlet boundary conditions, with a zero entry for each corresponding unconstrained component of q, and $B(x_{\partial\Omega})$ denotes a square diagonal matrix, with a 1 for each diagonal entry, but replaced by zero for each corresponding unconstrained component of q.

The number of boundary conditions at inlet and outlet depends on the number of negative eigenvalues of $\frac{\partial f}{\partial q} n_x$, where n_x denotes the x-component of the outward pointing normal unit vector $\mathbf{n}$. A negative eigenvalue signifies a wave propagation in the direction opposite $\mathbf{n}$, hence toward the flow domain, a propagation that carries information from the flow outside the domain, which information in embodied in a corresponding boundary condition.

At an inlet, $\frac{\partial f}{\partial q} n_x$ displays two negative eigenvalues; free-surface height or mass flow and total energy are thus constrained by the Dirichlet boundary conditions

$$\frac{\partial h}{\partial t}(x_{\text{in}}, t) = q_h^{'}(t), \quad \frac{\partial E}{\partial t}(x_{\text{in}}, t) = q_E^{'}(t) \tag{12.54}$$

where $q_h^{'}(t)$ and $q_E^{'}(t)$ denote prescribed bounded functions. If the inlet is supercritical, the third eigenvalue of $\frac{\partial f}{\partial q} n_x$ is negative and the similar Dirichlet constraint

$$\frac{\partial m}{\partial t}(x_{\text{in}}, t) = q_m^{'}(t) \tag{12.55}$$

is enforced on mass flow, with $q_m^{'}(t)$ a prescribed condition.

No Dirichlet constraint is enforced on either m or h at a subcritical inlet because the third eigenvalue of $\frac{\partial f}{\partial q} n_x$ is positive, implying that the boundary magnitude of either m or h is determined by the downstream flow conditions. In this case, the integral statement (12.52) for the corresponding linear-momentum or continuity equation is employed, using a weight function w with a non-vanishing trace at the inlet.

Similarly, at a supercritical outlet no Dirichlet constraints are imposed on the dependent variable "q" because all the eigenvalues of $\frac{\partial f}{\partial q} n_x$ are positive, implying that the boundary magnitude of q is determined by the upstream flow conditions. In this case, the integral statement (12.52) is employed, using a weight function w with a non-vanishing trace at the outlet.

At a subcritical outlet, two eigenvalues of $\frac{\partial f}{\partial q} n_x$ are positive, but one remains negative, implying that one exterior condition is required to determine the upstream flow. This study has selected a physically meaningful boundary condition on the pressure $p = h^2/2$, enforced via the linear-momentum integral statement as follows. Considering that $w(x_{\text{in}}) = 0$ and $w(x_{\text{out}}) = 1$ in this statement, an integration by parts of not only the characteristics-bias expression, but also the pressure gradient yields

$$\int_\Omega \left(w + \varepsilon\psi s \frac{\partial w}{\partial x}\right)\left(\frac{\partial m}{\partial t} - \phi_m + \frac{\partial f_m^q}{\partial x}\right) d\Omega + \int_\Omega \varepsilon\psi s \frac{\partial w}{\partial x}\left(\alpha cs \frac{\partial m}{\partial x} + \delta\frac{\partial f_m^p}{\partial x}\right) d\Omega$$

$$- \int_\Omega \frac{\partial w}{\partial x}\frac{h^2}{2} d\Omega + \left.\left(\frac{h^2}{2} - \varepsilon\psi s\left(\frac{\partial m}{\partial t} - \phi_m + \frac{\partial f_m}{\partial x} + \alpha cs\frac{\partial m}{\partial x} + (\delta - 1)\frac{\partial f_m^p}{\partial x}\right)\right)\right|_{x=x_{\text{out}}} = 0 \tag{12.56}$$

The entire boundary expression is then set to a prescribed outlet external pressure $p_{\text{ext}} = (h^2/2)_{\text{ext}}$; this boundary condition selection serves the double duty of both providing a boundary condition for the hyperbolic-parabolic perturbation in the characteristics-bias system and forcing the outlet pressure to approach p_{ext} since ε is set equal to a mesh spacing. The stability, accuracy, and numerical performance of this pressure boundary condition are discussed in Sections 12.6.3, 12.9.

12.6   Finite Element Weak Statement

The finite element solution exists on a partition Ω^h, $\Omega^h \subseteq \Omega$, of Ω, where "h" as a superscript in this case, signifies not free-surface height, but rather spatial discrete approximation. This partition Ω^h has its boundary nodes in $\partial\Omega^h$ on the boundary $\partial\Omega$ of Ω and results from the union of N_e non-overlapping elements Ω_e, $\Omega^h = \bigcup_{e=1}^{N_e} \Omega_e$. With $\eta = 0$ or $\eta = 1$, respectively for shocked or smooth flows, and with $\mathcal{P}_1(\Omega_e)$ and $\mathcal{P}_{1v}(\Omega_e)$ the finite-dimensional function spaces of respectively diagonal square matrix-valued and vector-valued linear polynomials within each Ω_e, for each "t", the corresponding diagonal square matrix-valued and vector-valued finite element discretization spaces employed in this study are defined as

$$\mathcal{S}^1(\Omega^h) \equiv \left\{ w^h \in \mathcal{H}^1(\Omega^h) : w^h\big|_{\Omega_e} \in \mathcal{P}_1(\Omega_e), \forall\Omega_e \in \Omega^h, B(x_{\partial\Omega^h}) w^h(x_{\partial\Omega^h}) = 0 \right\}$$

$$\mathcal{S}_v^\eta(\Omega^h) \equiv \left\{ w_v^h \in \mathcal{H}_v^\eta(\Omega^h) : w_v^h\big|_{\Omega_e} \in \mathcal{P}_{1v}(\Omega_e), \forall\Omega_e \in \Omega^h, B(x_{\partial\Omega^h}) w_v^h(x_{\partial\Omega^h}, t) = G_v(x_{\partial\Omega^h}, t) \right\}$$

$$(12.57)$$

Based on these spaces, the finite element approximation $q^h \in \mathcal{S}_v^\eta$, is determined for each "t" as the solution of the finite element weak statement associated with (16.86)

$$\int_{\Omega^h} \left(w^h + \varepsilon^h \psi^h s^h \frac{\partial w^h}{\partial x} \right) \left(\frac{\partial q^h}{\partial t} - \phi^h + \frac{\partial f^h}{\partial x} \right) d\Omega$$

$$+ \int_{\Omega^h} \varepsilon^h \psi^h \frac{\partial w^h}{\partial x} \left(\alpha^h c^h \frac{\partial q^h}{\partial x} + s^h (\delta^h - 1) \frac{\partial f^{p^h}}{\partial x} \right) d\Omega = 0, \quad \forall w^h \in \mathcal{S}^1 \qquad (12.58)$$

where $s^h \equiv \mathrm{sgn}(u^h)$ denotes the sign of u^h at the centroid of each element; the linear-momentum discrete weak statement for an outlet node then results from a finite element discretization of (12.56) when an outlet pressure boundary condition is enforced. Since every member w^h of $\mathcal{S}^1$ results from a linear combination of the linearly independent basis functions of this finite dimensional space, statement (12.58) is satisfied for N independent basis functions of the space, where N denotes the dimension of the space. For N mesh nodes within Ω^h, there exist clusters of "master" elements Ω_i^m, each comprising only those adjacent elements that share a mesh node x_i, which implies existence of exactly N master elements. Note that each master element represents a "finite volume" as used in finite volume schemes, which, however, do not employ the following finite element "pyramid" test functions. On each master element Ω_i^m, the discrete test function $w^h \equiv w_i = w_i(x)$, $1 \leq i \leq N$, will coincide with the pyramid basis function with compact support on Ω_i^m.

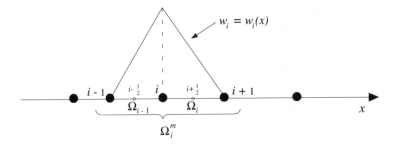

Figure 12.7: Adjacent Finite Elements, Master Element, Pyramid Test Function $w_i(x)$

Such a function equals one at node x_i, zero at all other mesh nodes, and also identically vanishes both on the boundary segments of Ω_i^m not containing x_i, and elsewhere within the computational domain outside Ω_i^m. Statement (12.58) yields a non-linear SUPG formulation, because the controller ψ^h does not remain constant, but varies as a function of a local measure of the discrete solution q^h, as detailed in Chapter 10. As supported by several computations, with subscript "i" denoting numerical value at node i, this measure is established as

$$q_{\text{meas}}^h \Big|_i = \frac{h_{\text{dim}}^h}{h_{\text{ref}}}\Big|_i + \frac{m_{\text{dim}}^h}{m_{\text{ref}}}\Big|_i + \frac{E_{\text{dim}}^h}{E_{\text{ref}}}\Big|_i = h^h\Big|_i + m^h\Big|_i + E^h\Big|_i \qquad (12.59)$$

and when the nodal m_i^h is positive, this measure becomes the nodal ℓ_1 norm of q_i.

12.6.1 Galerkin Finite Element Expansions and Integrals

The discrete solution $q^h \in \mathcal{S}_v^\eta$ at each time t assumes the form of the following linear combination

$$q^h(x,t) \equiv \sum_{j=1}^{N} w_j(x) \cdot q^h(x_j,t) \qquad (12.60)$$

of nodal solution values and trial basis functions that coincide with the test functions $w_j(x)$ for a Galerkin formulation. Similarly, the source $\phi = \phi(x,q(x,t))$ and fluxes $f = f(q(x,t))$, $f^q = f^q(q(x,t))$ and $f^p = f^p(q(x,t))$ are discretized through the group expressions

$$\phi^h(x,t) \equiv \sum_{j=1}^{N} w_j(x) \cdot \phi\left(x_j, q^h(x_j,t)\right), \qquad f^h(x,t) \equiv \sum_{j=1}^{N} w_j(x) \cdot f\left(q^h(x_j,t)\right)$$

$$f^{q^h}(x,t) \equiv \sum_{j=1}^{N} w_j(x) \cdot f^q\left(q^h(x_j,t)\right), \qquad f^{p^h}(x,t) \equiv \sum_{j=1}^{N} w_j(x) \cdot f^p\left(q^h(x_j,t)\right) \qquad (12.61)$$

The notation for the discrete nodal variable and fluxes is then simplified as $q_j(t) \equiv q^h(x_j,t)$, $\phi_j(t) \equiv \phi^h(x_j,t)$, $f_j \equiv f^h(x_j,t)$, $f_j^q \equiv f^{q^h}(x_j,t)$, $f_j^p \equiv f^{p^h}(x_j,t)$. With implied summation on the repeated subscript index j, expansions (12.60)-(12.61) are then inserted into (12.58), which yields the discrete finite element weak statement

$$\int_{\Omega^h} \left(w_i + \varepsilon^h \psi_t^h s^h \frac{\partial w_i}{\partial x} \right) w_j \frac{\partial q^h}{\partial t} \bigg|_j \, d\Omega - \int_{\Omega^h} \left(w_i + \varepsilon^h \psi_\phi^h s^h \frac{\partial w_i}{\partial x} \right) w_j \phi_j \, d\Omega$$

$$+ \int_{\Omega^h} \left(w_i + \varepsilon^h \psi^h s^h \frac{\partial w_i}{\partial x} \right) \left(\frac{\partial w_j}{\partial x} f_j \right) \, d\Omega + \int_{\Omega^h} \varepsilon^h \psi^h \frac{\partial w_i}{\partial x} \frac{\partial w_j}{\partial x} \left(\alpha^h c^h q_j + s^h (\delta^h - 1) f_j^p \right) \, d\Omega = 0$$

$$(12.62)$$

for $1 \le i \le N$, with $\psi_t^h = \psi_\phi^h = \psi^h$ for smooth flows, whereas $\psi_t^h \ne \psi_\phi^h \ne \psi^h$ in a three-element hydraulic jump region; in respect of ε, this term is set equal to a reference length within each element, typically a measure of the element size, equal to $\Delta x / 2$ within each element Ω_e of length Δx, as described in Chapter 10.

Within each element Ω_e, in the upstream-bias integrals

$$\int_{\Omega_e} \psi_{t,\phi}^h \frac{\partial w_i}{\partial x} w_j \, dx, \quad \int_{\Omega_e} \psi^h \frac{\partial w_i}{\partial x} \frac{\partial w_j}{\partial x} \, dx, \quad \int_{\Omega_e} \psi^h \alpha^h c^h \frac{\partial w_i}{\partial x} \frac{\partial w_j}{\partial x} \, dx, \quad \int_{\Omega_e} \psi^h (\delta^h - 1) \frac{\partial w_i}{\partial x} \frac{\partial w_j}{\partial x} \, dx$$

$$(12.63)$$

the functions α^h, δ^h, c^h, ψ^h, and $\psi_{t,\phi}^h$, standing for both ψ_t^h and ψ_ϕ^h, all depend on the variable (x,t), ψ^h and $\psi_{t,\phi}^h$ through an expression like (12.60) and α^h, δ^h, and c^h through

$$\alpha^h = \alpha \left(Fr \left(q^h(x,t) \right) \right), \quad \delta^h = \delta \left(Fr \left(q^h(x,t) \right) \right), \quad c^h = c \left(q^h(x,t) \right) \qquad (12.64)$$

For the linear basis functions in this paper, each derivative $\partial w_i / \partial x$ becomes a constant, which simplifies (12.63). These integrals directly influence non only the stability, but also the accuracy of the computational solution, especially in the neighbourhood of a discontinuity. Accordingly, within each element where the flow is smooth, these integrals are evaluated by way of a 3-point Gaussian quadrature, with abscissae $\{ -\sqrt{3/5}, 0, \sqrt{3/5} \}$ and weights $\{ 5/9, 8/9, 5/9 \}$, a rule that supplies a 5th-order accurate integration. A 5th-degree continuous polynomial representation of a hydraulic jump, however, contrasts with its discontinuous nature. As a result, within a three-element hydraulic jump region, integrals (12.63) are evaluated via a 1-point quadrature, with abscissa $\{ 0 \}$ and weight $\{ 2 \}$, a rule that returns a simple centroidal average for each of these integrals. According to the computational results, in comparison to a uniform use of either the 3-point or 1-point quadrature throughout the computational domain, the use of the separate-quadrature strategy has contributed to non-oscillatory hydraulic jumps and accurate computation of the jumps in the discontinuous dependent variables.

Since the test and trial functions w_i are prescribed functions of x, linear functions in this paper, the remaining spatial integrations in (12.62) are then exactly carried out. The completion of each indicated spatial integration transforms (12.62) into a system of ordinary differential equations (ODE) in continuum time for determining at each time level t the unknown nodal values $q^h (x_j, t)$.

12.6.2   Discrete Upstream Bias

For a clear comparison between traditional finite difference/volume schemes and the gravity-wave-convection finite element algorithm (12.62), at any interior node "i" of the representative grid in Figure 12.7, equations (12.62) with $\varepsilon_{i+\frac{1}{2}} \equiv (\Delta x_{i+\frac{1}{2}})/2$ can be equivalently recast in difference notation as

$$\left(\frac{\Delta x}{2}\left(\frac{1}{3}+\frac{s\psi_t}{2}\right)\right)_{i-\frac{1}{2}}\left(\frac{dq_{i-1}}{dt}+2\frac{dq_i}{dt}\right)+\left(\frac{\Delta x}{2}\left(\frac{1}{3}-\frac{s\psi_t}{2}\right)\right)_{i+\frac{1}{2}}\left(2\frac{dq_i}{dt}+\frac{dq_{i+1}}{dt}\right)$$

$$-\left(\frac{\Delta x}{2}\left(\frac{1}{3}+\frac{s\psi_\phi}{2}\right)\right)_{i-\frac{1}{2}}(\phi_{i-1}+2\phi_i)-\left(\frac{\Delta x}{2}\left(\frac{1}{3}-\frac{s\psi_\phi}{2}\right)\right)_{i+\frac{1}{2}}(2\phi_i+\phi_{i+1})$$

$$=-(\psi\alpha c)_{i-\frac{1}{2}}(q_i-q_{i-1})+(\psi\alpha c)_{i+\frac{1}{2}}(q_{i+1}-q_i)$$

$$-\frac{1}{2}\left(\left(1+(s\psi)_{i-\frac{1}{2}}\right)(f_i^q-f_{i-1}^q)+\left(1-(s\psi)_{i+\frac{1}{2}}\right)(f_{i+1}^q-f_i^q)\right)$$

$$-\frac{1}{2}\left(\left(1+(s\psi\delta)_{i-\frac{1}{2}}\right)(f_i^p-f_{i-1}^p)+\left(1-(s\psi\delta)_{i+\frac{1}{2}}\right)(f_{i+1}^p-f_i^p)\right) \qquad (12.65)$$

which couples several time derivatives at each node "i" and features a linear combination of two-point upstream and downstream flux differences. In these finite element equations, the values of the controller ψ^h determines the combination weights of the downstream and upstream expressions, and since ψ^h remains non-negative, these equations induce the appropriate upstream bias, since the upstream weight $1+\psi_{i-\frac{1}{2}}$ always exceeds the downstream weight $1-\psi_{i+\frac{1}{2}}$. As a result, the finite element weak statement (12.62) generates consistent variable-upstream-bias discrete equations that correspond to an upstream-bias discretization for the original system (12.1), within a compact block tri-diagonal matrix statement.

For smooth solutions, these equations will still couple upstream and downstream points even for supercritical flows. Such an algorithm remains mathematically consistent with the physics of mono-directional wave propagation for supercritical flows on the basis of Courant's and Hilbert's classical developments for non-linear hyperbolic systems. These developments show that while waves do propagate along characteristics, smooth solutions can be expanded in Taylor's series within arbitrary regions encircling any given point and along any direction, radiating upstream or downstream from the point.

For a closer comparison with upwind finite-volume schemes, the finite element equations (12.65) can be rearranged to generate the "numerical flux"

$$\mathrm{F}_{i+\frac{1}{2}}\equiv\frac{f_i+f_{i+1}}{2}-\psi_{i+\frac{1}{2}}\left[(\alpha c)_{i+\frac{1}{2}}(q_{i+1}-q_i)+\frac{s_{i+\frac{1}{2}}}{2}(f_{i+1}^q-f_i^q)+\frac{(s\delta)_{i+\frac{1}{2}}}{2}(f_{i+1}^p-f_i^p)\right] \tag{12.66}$$

which corresponds to the discrete counterpart of the characteristics-bias flux (12.38). By virtue of this numerical flux, equations (12.65) are recast as

$$\left(\frac{\Delta x}{2}\left(\frac{1}{3}+\frac{s\psi_t}{2}\right)\right)_{i-\frac{1}{2}}\left(\frac{dq_{i-1}}{dt}+2\frac{dq_i}{dt}\right)+\left(\frac{\Delta x}{2}\left(\frac{1}{3}-\frac{s\psi_t}{2}\right)\right)_{i+\frac{1}{2}}\left(2\frac{dq_i}{dt}+\frac{dq_{i+1}}{dt}\right)$$

$$-\left(\frac{\Delta x}{2}\left(\frac{1}{3}+\frac{s\psi_\phi}{2}\right)\right)_{i-\frac{1}{2}}(\phi_{i-1}+2\phi_i)-\left(\frac{\Delta x}{2}\left(\frac{1}{3}-\frac{s\psi_\phi}{2}\right)\right)_{i+\frac{1}{2}}(2\phi_i+\phi_{i+1})$$

$$=-\left(\mathrm{F}_{i+\frac{1}{2}}-\mathrm{F}_{i-\frac{1}{2}}\right) \tag{12.67}$$

which shows that the finite element weak statement (12.62) naturally leads to a conservative discrete algorithm.

12.6.3   Boundary Equations and Pressure Boundary Conditions

The integral statement (12.62) directly yields a set of consistent boundary differential equations, for both unconstrained boundary variables and for pressure, to enforce a pressure boundary condition at a subcritical outlet. These equations do not require any algebraic extrapolation of variables, but rather couple the time derivatives of boundary- and interior-node variables within the boundary cell.

For the linear elements in this study, let N and $N-1$ denote the nodes within the outlet boundary element, with N corresponding to the outlet node. For the discrete finite element equation associated with boundary node x_N, the test function w satisfies the conditions $w(x_{N-1}) = 0$, $w(x_N) = 1$.

The boundary differential equation from (12.62) corresponding to an outlet node becomes

$$\left(\Delta x \left(\frac{1}{3} + \frac{s\psi_t}{2}\right)\right)_{N-\frac{1}{2}} \left(\frac{dq_{N-1}}{dt} + 2\frac{dq_N}{dt}\right)$$

$$-\left(\Delta x \left(\frac{1}{3} + \frac{s\psi_\phi}{2}\right)\right)_{N-\frac{1}{2}} (\phi_{N-1} + 2\phi_N) = -2(\psi ac)_{N-\frac{1}{2}} (q_N - q_{N-1})$$

$$-\left(1 + (s\psi)_{N-\frac{1}{2}}\right) \left(f_N^q - f_{N-1}^q\right) - \left(1 + (s\psi\delta)_{N-\frac{1}{2}}\right) \left(f_N^p - f_{N-1}^p\right) \tag{12.68}$$

This equation directly couples the time derivatives of the solution q at the adjacent boundary and interior nodes x_N and x_{N-1}. A similar equation is then obtained at an inlet, *mutatis mutandis*.

Mechanically, a quasi-one-dimensional flow within a channel is induced by lowering the outlet pressure in comparison to the inlet pressure. Such an outlet pressure boundary condition is naturally enforced within the surface integral that emerges in the momentum-equation weak statement, as shown in the previous section. The linear finite element discretization of (12.56) yields

$$\left(\Delta x \left(\frac{1}{3} + \frac{s\psi_t}{2}\right)\right)_{N-\frac{1}{2}} \left(\frac{dq_{N-1}}{dt} + 2\frac{dq_N}{dt}\right)$$

$$-\left(\Delta x \left(\frac{1}{3} + \frac{s\psi_\phi}{2}\right)\right)_{N-\frac{1}{2}} (\phi_{N-1} + 2\phi_N) = -2(\psi ac)_{N-\frac{1}{2}} (q_N - q_{N-1})$$

$$-\left(1 + (s\psi)_{N-\frac{1}{2}}\right) \left(f_N^q - f_{N-1}^q\right) - (s\psi\delta)_{N-\frac{1}{2}} \left(f_N^p - f_{N-1}^p\right) - \left((2f_{\text{out}}^p - f_N^p) - f_{N-1}^p\right) \tag{12.69}$$

In this equation, f_N^p denotes the outlet-node pressure, whereas, quite significantly, f_{out}^p can correspond to the specified outlet pressure boundary condition. This strategy for imposing an outlet pressure boundary condition remains intrinsically stable. Suppose, for instance, that some numerical perturbation forces f_N^p to decrease below the imposed f_{out}^p. In this case the outlet boundary equation (12.69) induces a negative time rate of change for m_N, which leads to a corresponding reduction in m_N and consequent negative space gradient of momentum at the outlet. From the continuity equation in (12.62), such a negative gradient leads to an increase in h, which corresponds to a stable restoration of the outlet height toward the imposed outlet boundary condition. A similar conclusion on the stability of (12.69) is achieved by considering a perturbation increase in f_N^p. The results in Section 12.9 confirm the accuracy and stability of this pressure boundary condition procedure.

12.7   Implicit Runge Kutta Time Integration

The finite element equations (12.65) along with appropriate boundary equations and conditions form a system of non-linear ordinary differential equations in the time variable. The discrete characteristics-bias differential system, inclusive of inlet and outlet pressure boundary condition is cast as

$$
\left(\Delta x \left(\frac{1}{3} - \frac{s\psi_t}{2} \right) \right)_{1+\frac{1}{2}} \left(2\frac{dq_1}{dt} + \frac{dq_2}{dt} \right)
$$

$$
- \left(\Delta x \left(\frac{1}{3} - \frac{s\psi_\phi}{2} \right) \right)_{1+\frac{1}{2}} (2\phi_1 + \phi_2) = 2(\psi\alpha c)_{1+\frac{1}{2}} (q_2 - q_1)
$$

$$
- \left(1 - (s\psi)_{1+\frac{1}{2}} \right) (f_2^q - f_1^q) - \left(1 - (s\psi\delta)_{1+\frac{1}{2}} \right) (f_2^p - f_1^p)
\tag{12.70}
$$

for the inlet node, $i = 1$, with appropriate components replaced by (12.54),

$$
\left(\frac{\Delta x}{2} \left(\frac{1}{3} + \frac{s\psi_t}{2} \right) \right)_{i-\frac{1}{2}} \left(\frac{dq_{i-1}}{dt} + 2\frac{dq_i}{dt} \right) + \left(\frac{\Delta x}{2} \left(\frac{1}{3} - \frac{s\psi_t}{2} \right) \right)_{i+\frac{1}{2}} \left(2\frac{dq_i}{dt} + \frac{dq_{i+1}}{dt} \right)
$$

$$
- \left(\frac{\Delta x}{2} \left(\frac{1}{3} + \frac{s\psi_\phi}{2} \right) \right)_{i-\frac{1}{2}} (\phi_{i-1} + 2\phi_i) - \left(\frac{\Delta x}{2} \left(\frac{1}{3} - \frac{s\psi_\phi}{2} \right) \right)_{i+\frac{1}{2}} (2\phi_i + \phi_{i+1})
$$

$$
= -(\psi\alpha c)_{i-\frac{1}{2}} (q_i - q_{i-1}) + (\psi\alpha c)_{i+\frac{1}{2}} (q_{i+1} - q_i)
$$

$$
- \frac{1}{2} \left(\left(1 + (s\psi)_{i-\frac{1}{2}} \right) (f_i^q - f_{i-1}^q) + \left(1 - (s\psi)_{i+\frac{1}{2}} \right) (f_{i+1}^q - f_i^q) \right)
$$

$$
- \frac{1}{2} \left(\left(1 + (s\psi\delta)_{i-\frac{1}{2}} \right) (f_i^p - f_{i-1}^p) + \left(1 - (s\psi\delta)_{i+\frac{1}{2}} \right) (f_{i+1}^p - f_i^p) \right)
\tag{12.71}
$$

for the internal nodes, $2 \le i \le N - 1$, and

$$
\left(\Delta x \left(\frac{1}{3} + \frac{s\psi_t}{2} \right) \right)_{N-\frac{1}{2}} \left(\frac{dq_{N-1}}{dt} + 2\frac{dq_N}{dt} \right)
$$

$$
- \left(\Delta x \left(\frac{1}{3} + \frac{s\psi_\phi}{2} \right) \right)_{N-\frac{1}{2}} (\phi_{N-1} + 2\phi_N) = -2(\psi\alpha c)_{N-\frac{1}{2}} (q_N - q_{N-1})
$$

$$
- \left(1 + (s\psi)_{N-\frac{1}{2}} \right) (f_N^q - f_{N-1}^q) - (s\psi\delta)_{N-\frac{1}{2}} \left(f_N^p - f_{N-1}^p \right)
$$

$$
- \left((2f_{\text{out}}^p - f_N^p) - f_{N-1}^p \right)
\tag{12.72}
$$

for the outlet node, $i = N$.

With Q denoting the array of nodal free-surface height, momentum, and total energy

$$
Q \equiv \left\{
\begin{array}{c}
h_1 \\
m_1 \\
E_1 \\
h_2 \\
m_2 \\
E_2 \\
. \\
. \\
h_i \\
m_i \\
E_i \\
. \\
. \\
h_N \\
m_N \\
E_N
\end{array}
\right\}
\tag{12.73}
$$

and $\mathcal{F}$ denoting an array of fluxes, source terms and characteristics-bias expressions, and boundary conditions, equations (12.70)-(12.72) are abridged as the non-linear differential system

$$
M\frac{dQ(t)}{dt} = \mathcal{F}\left(t, Q(t)\right)
\tag{12.74}
$$

where the "mass" matrix M couple the discretization nodal time derivatives. A representative equation at an internal node "i" is thus cast as

$$
m_{i,i-1}\frac{dq_{i-1}}{dt} + m_{i,i}\frac{dq_i}{dt} + m_{i,i+1}\frac{dq_{i+1}}{dt} = \mathcal{F}_i(t, Q)
\tag{12.75}
$$

where $\mathcal{F}_i(t, Q)$ corresponds to the right-hand sides in (12.70)-(12.72) at node "i".

The differential system (12.74) is integrated through the two-stage implicit Runge-Kutta procedure

$$
\begin{aligned}
Q_{n+1} - Q_n &= b_1 K_1 + b_2 K_2 \\
MK_1 &= \Delta t \cdot \mathcal{F}\left(t_n + c_1\Delta t, Q_n + a_{11}K_1\right) \\
MK_2 &= \Delta t \cdot \mathcal{F}\left(t_n + c_2\Delta t, Q_n + a_{21}K_1 + a_{22}K_2\right)
\end{aligned}
\tag{12.76}
$$

The Runge-Kutta arrays K_1 and K_2 correspond to variations ΔQ_1 and ΔQ_2 of the solution array Q. The first expression in (12.76) thus corresponds to the linear combination

$$
Q_{n+1} - Q_n = b_1\Delta Q_1 + b_2\Delta Q_2
\tag{12.77}
$$

The terminal numerical solution for the second and third expression in (12.76) is established using Newton's method, which for the implicit fully-coupled computation of the IRK2 arrays K_i, $1 \le i \le 2$, is cast as

$$
\left[M - a_{\underline{ii}}\Delta t\left(\frac{\partial \mathcal{F}}{\partial Q}\right)_{Q_i^p}^p\right]\left(K_i^{p+1} - K_i^p\right) = \Delta t\,\mathcal{F}\left(t_n + c_i\Delta t, Q_i^p\right) - MK_i^p
$$

$$Q_i^p \equiv Q_n + a_{i1}K_1^p + a_{i2}K_2^p \tag{12.78}$$

where $a_{ij} = 0$ for $j > i$, p is the iteration index, and $K_1^p \equiv K_1$ for $i = 2$; for linear finite elements, the Jacobian

$$J_i(Q) \equiv \mathcal{M} - a_{\underline{ii}}\Delta t \left(\frac{\partial \mathcal{F}}{\partial Q}\right)^p_{Q_i^p} \tag{12.79}$$

becomes a block tri-diagonal matrix. Since the Jacobians of the controllers ψ, ψ_t and ψ_ϕ and functions α and δ are computationally expensive, these parameters are kept equal to their magnitudes at each time level n, during the determination of the Runge-Kutta arrays K_1 and K_2, a specification that does not require their Jacobians; all the other Jacobians have been exactly determined analytically. For all the results documented in the next section, the initial estimate K_i^0 is set equal to the zero array. The practical implementation of Newton's method for (12.78) has specified a maximum number of Newton iterations for the non-linear determination of each IRK array K_i. This number was set to 4, yet the algorithm only employed 4 Newton iterations when the absolute value of the maximum component of the residual $\mathcal{F}$ in (12.78) exceeded the prescribed tolerance of 5.0×10^{-6}, otherwise the algorithm employed fewer iterations, mostly 1. As detailed in Section 12.9, this implementation has led to a rapid convergence to steady states, with Courant number reaching 200.

12.8 Reference Exact Solutions

In terms of the primitive variables u and h, the steady-flow continuity and linear-momentum free-surface equations become

$$\begin{cases} \dfrac{\partial}{\partial x}(huW) = 0 \\[2mm] \dfrac{\partial}{\partial x}\left(u^2 hW\right) + W\dfrac{\partial}{\partial x}\left(\dfrac{h^2}{2}\right) = 0 \end{cases} \tag{12.80}$$

For smooth flows, for which the derivatives of the primitive variables exist everywhere within the flow, the continuity and linear-momentum equations yield

$$\frac{\partial}{\partial x}\left(\frac{u^2}{2} + h\right) = 0 \tag{12.81}$$

which expresses conservation of mechanical energy.

12.8.1 Channel-Width Rule

From the equation expressing conservation of mechanical energy, the gradient of the free-surface height is expressed in terms of the gradient of speed as

$$\frac{\partial h}{\partial x} = -u\frac{\partial u}{\partial x} \tag{12.82}$$

From the continuity equation, the same free-surface height gradient depends upon the speed and channel width gradients in the form

$$\frac{\partial h}{\partial x} = -\frac{h}{u}\frac{\partial u}{\partial x} - \frac{h}{W}\frac{\partial W}{\partial x} \tag{12.83}$$

The combination of these two expressions for the gradient of the free-surface height yields the expression for the channel-width rule

$$\frac{1}{W}\frac{\partial W}{\partial x} = -\frac{1}{u}\frac{\partial u}{\partial x}\left(1 - \frac{u^2}{h}\right) - \frac{1}{u}\frac{\partial u}{\partial x}\left(1 - Fr^2\right) \tag{12.84}$$

This result shows that besides the speed gradient, the sign of the area gradient depends on the Froude number Fr. This number remains less than one as the speed increases from a negligible magnitude. In these conditions the channel-width rule predicts a speed increase along a channel with decreasing width. As the speed keeps on increasing, Fr will eventually equal one. The channel-width rule then shows the channel width no longer decreases for the width gradient vanishes. Hence, the channel width where Fr equals one corresponds to the channel reference critical throat. When Fr exceeds one, the speed keeps on increasing along a channel with increasing width, since in these conditions the channel-width rule provides a positive width gradient for a positive speed gradient. These basic considerations show that Fr critically impacts the correlation between speed and width gradients.

12.8.2   Exact Integrals

The continuity equation in this system yields

$$u(x)h(x)W(x) = u_i h_i W_i \quad \Rightarrow u_o = u_i\frac{h_i W_i}{h_o W_o} \tag{12.85}$$

where u_i, h_i, W_i and u_o, h_o, and W_o respectively denote the inlet and outlet speed, free-surface height and channel width. The integrated form of the mechanical-energy equation becomes

$$\frac{u^2(x)}{2} + h(x) = \frac{u_i^2}{2} + h_i \tag{12.86}$$

Insertion of the expression for u_o into this result leads to u_i, u_o, and $u = u(x)$ in terms of prescribed inlet and outlet free-surface height and channel width

$$u_i = \sqrt{\frac{2\Delta h}{\left(\dfrac{h_i W_i}{h_o W_o}\right)^2 - 1}}, \quad u_o = \frac{h_i W_i}{h_o W_o}\sqrt{\frac{2\Delta h}{\left(\dfrac{h_i W_i}{h_o W_o}\right)^2 - 1}}$$

$$u(x) = \frac{h_i W_i}{h(x)W(x)}\sqrt{\frac{2\Delta h}{\left(\dfrac{h_i W_i}{h_o W_o}\right)^2 - 1}} \tag{12.87}$$

Substitution in the mechanical-energy equation of the expression for $u(x)$ yields a cubic equation for $h = h(x)$. The complete solution for $h = h(x)$ and $u = u(x)$ can thus be cast as

$$\begin{cases} h^3(x) - \left(\dfrac{u_i^2}{2} + h_i \right) h(x) + \dfrac{u_i^2 h_i^2 W_i^2}{2W(x)} = 0 \\[4mm] u(x) = \dfrac{h_i W_i}{h(x)W(x)} \sqrt{\dfrac{2\Delta h}{\left(\dfrac{h_i W_i}{h_o W_o} \right)^2 - 1}} \end{cases} \qquad (12.88)$$

The first equation can be solved for $h = h(x)$. With this h, the second equation in this system supplies u. Even though this solution procedure is theoretically complete, an elegant alternative procedure results from formulating the solution in similarity form.

12.8.3 Similarity Solution

The defining flow information is

$$W = W(x), \quad h_i, \quad h_o$$

which uniquely establishes a quasi-one dimensional free-surface flow within a converging-diverging channel. On the basis of this information, a unique steady similarity solution emerges solution emerges from (12.88).

Reference Variables

The mechanical-energy equation can be cast as

$$\frac{u^2(x)}{2} + h(x) = \frac{u_i^2}{2} + h_i \equiv h_{st} \qquad (12.89)$$

where h_{st} denotes the "stagnation" free-surface height. With reference to the expression for u_i, the free-surface height is expressed as

$$h_{st} = h_i + \frac{h_i - h_o}{\left(\dfrac{h_i W_i}{h_o W_o} \right)^2 - 1} \qquad (12.90)$$

The maximum speed $u_{\max}$ obtains in the limit of a vanishing height h. Correspondingly, the mechanical-energy equation leads to

$$\frac{u_{\max}^2}{2} = h_{st} \quad \Rightarrow u_{\max} = \sqrt{2h_{st}} \qquad (12.91)$$

The associated "maximum" channel width becomes unbounded as

$$W_{\max} = \lim_{h \to 0} u_i \frac{h_i W_i}{u_{max} h} = \infty \qquad (12.92)$$

The channel-throat Froude number equals one, hence

$$Fr_T = 1 \quad \Rightarrow u_T^2 = h_T \tag{12.93}$$

With the mechanical-energy and continuity equation, this equality leads to expressions for the throat free-surface height, speed, and channel width

$$h_T = \frac{2}{3}h_{st}, \quad u_T = \sqrt{\frac{2}{3}h_{st}}, \quad W_T = \frac{u_i h_i W_i}{\left(\frac{2}{3}h_{st}\right)^{3/2}} \tag{12.94}$$

Non-Dimensional Ratios

The ratio of local over stagnation pressure depends on the Froude number Fr as

$$h_{st} = h + \frac{u^2}{2} = h + h\frac{Fr^2}{2} = h\left(1 + \frac{Fr^2}{2}\right) \quad \Rightarrow \quad \frac{h}{h_{st}} = \frac{2}{2 + Fr^2} \tag{12.95}$$

Let $u_T = \sqrt{h_T}$ denote the throat speed. The definition of Fr leads to the ratio

$$\frac{u^2}{u_T^2} = Fr^2\frac{h}{h_T} = Fr^2\frac{h}{h_{st}}\frac{h_{st}}{h_T} = \frac{3Fr^2}{2 + Fr^2} \quad \Rightarrow \quad \frac{u}{u_T} = Fr\sqrt{\frac{3}{2 + Fr^2}} \tag{12.96}$$

From the continuity equation, the ratio W/W_T follows as

$$\frac{W}{W_T} = \frac{u_T}{u}\frac{h_T}{h} = \frac{u_T}{u}\frac{h_T}{h_{st}}\frac{h_{st}}{h} = \frac{1}{Fr}\left(\frac{2 + Fr^2}{3}\right)^{3/2} \tag{12.97}$$

The distribution of channel width $W = W(x)$ is given. The expression for the ratio W/W_T can thus be viewed as the following equation for Fr itself

$$Fr^2 - 3\left(\frac{W(x)}{W_T}\right)^{2/3}Fr^{2/3} + 2 = 0 \tag{12.98}$$

which can be solved numerically. With the solutions of this equation, the corresponding free-surface height and speed directly follow from (12.95) and (12.96).

12.8.4   Hydraulic Jump Calculation

Five quantities are necessary to determine a flow with embedded hydraulic jump. They consist of the inlet and outlet free-surface height h_i and h_o along with the inlet, throat, and outlet channel widths W_i, W_T, W_o. With reference to Figure 12.8, the complete determination of the flow with embedded hydraulic jump involves the calculation of 9 quantities. They are the inlet and outlet speeds u_i u_o, the pre- and post-jump free surface heights and speeds h_1, h_2 as well as u_1, u_2, the hydraulic-jump channel width $W_1 = W_J = W_2$ and the channel-throat free-surface height and speed h_T, u_T. This determination, therefore, requires 9 equations.

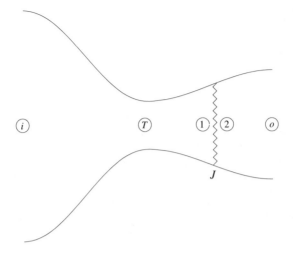

Figure 12.8: Hydraulic Jump

These equations involve $Fr_T = 1$ at the channel throat, the conservation of mass and linear momentum across the hydraulic jump, and the integrated continuity and mechanical-energy equations expressed in terms of the inlet and throat conditions, inlet and pre-jump conditions, and post-jump and outlet conditions. Grouped in sets, the forms of these 9 equations are

$$
a) \begin{cases} u_i h_i W_i = u_T h_T W_T \\ \dfrac{u_i^2}{2} + h_i = \dfrac{u_T^2}{2} + h_T \\ u_T^2 = h_T \end{cases} , \quad b) \begin{cases} u_i h_i W_i = u_1 h_1 W_J \\ \dfrac{u_i^2}{2} + h_i = \dfrac{u_1^2}{2} + h_1 \end{cases}
$$

$$
c) \begin{cases} u_1 h_1 = u_2 h_2 \\ u_1^2 h_1 + \dfrac{h_1^2}{2} = u_2^2 h_2 + \dfrac{h_2^2}{2} \end{cases} , \quad d) \begin{cases} u_2 h_2 W_J = u_o h_o W_o \\ \dfrac{u_2^2}{2} + h_2 = \dfrac{u_o^2}{2} + h_o \end{cases} \tag{12.99}
$$

The first system in this set directly yields the following equation for Fr_i

$$
Fr_i^2 - 3 \left(\frac{W_i}{W_T} \right)^{2/3} Fr_i^{2/3} + 2 = 0 \tag{12.100}
$$

which can be solved numerically for Fr_i. The difference between this equation and the formally identical equation (12.98) lies in W_T. In (12.98), W_T represents a reference channel throat width, which may or may not coincide with the actual throat width. In (12.100), instead, W_T equals the actual channel throat width, because the flow must be critical at the throat for a hydraulic jump to develop. The determination of the outlet Froude number Fr_o employs the following equation, which results from the continuity equation from systems

(12.99) b), c), d)

$$Fr_o = Fr_i \left(\frac{W_i}{W_o}\right) \left(\frac{h_i}{h_o}\right)^{3/2} \tag{12.101}$$

The corresponding inlet and outlet speeds result from the definition of the Froude number as

$$u_i = Fr_i \sqrt{h_i}, \quad u_o = Fr_o \sqrt{h_o} \tag{12.102}$$

Heed that the determination of the inlet and outlet Froude numbers and speeds do not require the hydraulic-jump Froude numbers.

The pre- and post-jump Froude numbers Fr_1 and Fr_2 satisfy an equation that results from system (12.99) c). The linear-momentum equation therein is cast with the difference $h_2^2 - h_1^2$ at the right-hand side. The square of u_2 in the resulting expression is then eliminated through the continuity equation. As a result of these substitutions a quadratic equation for h_2/h_1 emerges with physically meaningful solution

$$\frac{h_2}{h_1} = \frac{\sqrt{1 + 8Fr_1^2} - 1}{2} \tag{12.103}$$

The continuity equation is then expressed in terms of Fr_1, Fr_2, and h_1/h_2 as

$$Fr_2 = Fr_1 \left(\frac{h_1}{h_2}\right)^{3/2} \tag{12.104}$$

From these two results, the expression that supplies Fr_2 in terms of Fr_1 becomes

$$Fr_2 = \frac{2^{3/2} Fr_1}{\left(\sqrt{1 + 8Fr_1^2} - 1\right)^{3/2}} \tag{12.105}$$

The ratio h_{st_2}/h_{st_1} of the post- and pre-jump stagnation pressures h_{st_2} and h_{st_1} is expressed in terms of inlet and outlet free-surface height and Froude number using the mechanical - energy equations in systems (12.99) b), d). These equations are expressed as

$$\frac{u_i^2}{2} + h_i = \frac{u_1^2}{2} + h_1 = h_{st_1}, \quad \frac{u_2^2}{2} + h_2 = \frac{u_o^2}{2} + h_o = h_{st_2} \tag{12.106}$$

From these equations, the expressions for h_{st_2}, h_{st_1}, and h_{st_2}/h_{st_1} become

$$h_{st_1} = h_i \left(\frac{Fr_i^2}{2} + 1\right), \quad h_{st_2} = h_o \left(\frac{Fr_o^2}{2} + 1\right) \Rightarrow \frac{h_{st_2}}{h_{st_1}} = \frac{2 + Fr_o^2}{2 + Fr_i^2} \left(\frac{h_o}{h_i}\right) \tag{12.107}$$

The inlet and outlet free-surface heights are prescribed and the corresponding Froude numbers are determined through equations (12.100) and (12.101). The corresponding stagnation heights h_{st_1} and h_{st_2} as well as their ratio h_{st_2}/h_{st_1} are calculated independently of the pre- and post-jump Froude numbers.

In terms of h_{st_1}, h_{st_2}, and h_{st_2}/h_{st_1}, the ratio of post- and pre-jump free-surface heights h_2, h_1 and their ratio h_2/h_1 follow from (12.106) as

$$h_1 = \frac{2h_{st_1}}{2 + Fr_1^2}, \quad h_2 = \frac{2h_{st_2}}{2 + Fr_2^2} \Rightarrow \frac{h_2}{h_1} = \frac{2 + Fr_1^2}{2 + Fr_2^2} \left(\frac{h_{st_2}}{h_{st_1}}\right) \tag{12.108}$$

The substitution in this equation of results (12.103) and (12.105) respectively for h_2/h_1 and Fr_2 yields the single equation for the pre-jump Froude number Fr_1

$$2\left(\frac{h_{st_2}}{h_{st_1}}\right)\left(2+Fr_1^2\right)\left(\sqrt{1+8Fr_1^2}-1\right)^2 - 2\left(\sqrt{1+8Fr_1^2}-1\right)^3 - 8Fr_1^2 = 0 \qquad (12.109)$$

which can be solved numerically for Fr_1. With this solution, the post-jump Froude number Fr_2 follows from (12.105).

The pre- and post-jump free-surface heights are then obtained from (12.108) and the corresponding speeds follow from the definition of Froude number as

$$u_1 = Fr_1\sqrt{h_1}, \quad u_2 = Fr_1\sqrt{h_2} \qquad (12.110)$$

The channel width W_J, where the jump takes place, results from the continuity equation in system (12.99) b) as

$$W_J = W_i\frac{h_i u_i}{h_1 u_1} = W_i\frac{Fr_i}{Fr_1}\left(\frac{h_i}{h_1}\right)^{3/2}\frac{Fr_i}{Fr_1} = W_i\frac{Fr_i}{Fr_1}\left(\frac{2+Fr_1^2}{2+Fr_i^2}\right)^{3/2} \qquad (12.111)$$

The reference post-jump throat speed u_{T_2} and channel width W_{T_2} then follow from (12.96) and (12.97) as

$$u_{T_2} = \frac{u_o}{Fr_o}\left(\frac{2+Fr_o^2}{3}\right)^{1/2}, \quad W_{T_2} = W_o Fr_o\left(\frac{3}{2+Fr_o^2}\right)^{3/2} \qquad (12.112)$$

From the first and second expressions in (12.94), the pre- and post-jump throat speed and channel width ratios W_{T_2}/W_{T_1} and u_{T_1}/u_{T_2} depend upon h_{st_1}/h_{st_2} as

$$\frac{W_{T_2}}{W_{T_1}} = \frac{u_{T_1}}{u_{T_2}}\frac{h_{T_1}}{h_{T_2}} = \left(\frac{h_{st_1}}{h_{st_2}}\right)^{3/2} \qquad (12.113)$$

All of the results in this section allow a complete calculation of a free-surface flow with embedded hydraulic jump, based solely on the prescribed throat channel width and inlet and outlet channel widths and free-surface heights.

12.9 Computational Results

The characteristics-bias system formulation with continuum gravity-wave-convection upstream bias has generated physically consistent computational solutions for smooth and shocked mixed subcritical / supercritical flows with bed-shape and channel-width variations, shaft work, mass as well as heat transfer, and wall friction. The 10 different benchmarks in this section encompass one dam-break flow and several flows with shaft work as well as heat transfer, mass transfer, wall friction with bed slope, bed-shape and channel-width variations without and with hydraulic jumps, wall friction, and mass as well as heat transfer.

The spatial computational domain Ω for these benchmarks is defined as: $\Omega \equiv [x_{\text{in}}, x_{\text{out}}] = [0, 1]$, uniformly discretized through linear finite elements, for five progressively denser grids, as employed in the asymptotic convergence rate studies. These grids respectively contain 50, 100, 200, 400, and 800 linear elements. For each of these grids, every computational solution that admits a steady state rapidly converged, with the norm of the residual driven to the machine zero of 5.0×10^{-14}; the figures documenting the computational solutions in this section correspond to the 200-element grid. For each benchmark, the calculations proceeded with a prescribed constant maximum Courant number C_{max} defined as

$$C_{max} \equiv \max\{|u+c|, |u-c|, c\} \frac{\Delta t}{\Delta x} \tag{12.114}$$

Given Δx and C_{max} for each benchmark, the algorithm automatically calculated the corresponding Δt as

$$\Delta t = \frac{C_{max}\Delta x}{\max\{|u+c|, |u-c|, c\}} \tag{12.115}$$

with variable controller ψ determined as detailed in Chapter 10. Unless otherwise stated, all the computational solutions were obtained with a Courant number set equal to 200 and $\psi_{min} = 0.25$. Although ψ_{max} was set equal to 1.00, the formulation automatically kept ψ essentially equal to ψ_{min} in regions of solution smoothness and only increased it in the vicinity of a discontinuity. These computational solutions exhibit rapid convergence rate to a steady state and definite asymptotic convergence rate in both the $\mathcal{H}^0$ and $\mathcal{H}^1$ norms, for continuous solutions, and in the $\mathcal{H}^0$ norm for shocked solutions. The formulation has produced monotone and essentially non-oscillatory results that also reflect available exact solutions, with correctly predicted continuous mass flow and total enthalpy across hydraulic jumps.

12.9.1   Work Driven Flows

Although only summarily cited in some books on open-channel flows, shaft work affects these flows in the vicinity of pumps and turbines. Both shaft work and heat transfer feature in the same term in the generalized energy equation. The effect of this work on a flow, however, remains markedly different from the effect of heat because, unlike heat, mechanical work also induces a corresponding force in the linear momentum equation.

For a stage efficiency equal to 1, hence $\chi = 1$, it is possible to generate a basic steady-state exact solution, which has been determined for this study. On the basis of such a solution, the computational predictions for subcritical and supercritical flows in a straight rectangular channel are found in complete agreement with the mechanics of these flows. For the benchmarks for these flows, the non-dimensional specifications as well as boundary conditions for the subcritical case are: $h_{in} = 2.924$, $p_{out} \equiv (h^2/2)_{out} = 0.813$, with $w_m = 1.400$ and $E_{in} = 7.370$ without heat transfer, hence $g_T = 0.000$, or $E_{in} = 1.177$ with heat transfer and $g_T = 2.000$, corresponding from the exact solution to $Fr_{in} = 0.200$, $m_{in} = 1.000$ and $Fr_{out} = 0.695$. For the supercritical case, the corresponding specifications and boundary conditions are: $h_{in} = 0.481$, $m_{in} = 1.000$, with $w_m = 1.100$ and $E_{in} = 3.560$ without heat transfer, hence $g_T = 0.000$, or $E_{in} = 2.560$ with heat transfer and $g_T = 2.000$ corresponding to $Fr_{in} = 3.000$, $p_{out} = 0.358$ and $Fr_{out} = 1.283$ for the supercritical flow. The initial conditions throughout the duct correspond to a linear interpolation between the inlet and outlet states, resulting in an initial flow that is then advanced in time to a steady state.

According to basic thermodynamic considerations, extracting work from a subcritical flow decreases free-surface height, as expected, but consequently increases the Froude number; conversely when work is extracted from a supercritical flow, it is free-surface height that increases, while the Froude number decreases. These precise trends are reflected in the computational solutions shown in Figure 12.9, which mirror the exact solutions for both subcritical and supercritical flows.

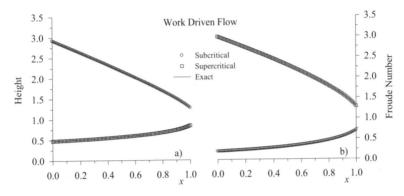

Figure 12.9: Work-Driven Flow: a) Free-Surface Height, b) Froude Number

While keeping mass flow unchanged, these flows can only produce work by expending total enthalpy. Such a canonical thermodynamic balance is reflected in the computational solutions graphed in Figure 12.10. Obtained for a constant shaft work, these solutions correctly display a constant mass flow coupled with a linearly decreasing total enthalpy.

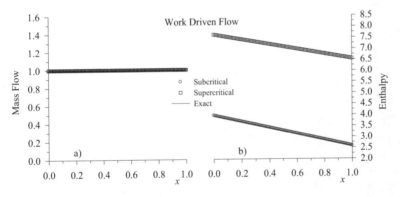

Figure 12.10: Work-Driven Flow: a) Mass Flow, b) Enthalpy

Since the total-energy equation ordinarily remains decoupled from the continuity and linear momentum equations, any heat transfer would only affect the energy distributions.

For both subcritical and supercritical flows, the results in Figure 12.11 indicate that a channel flow with significantly less inlet energy is still capable to evolve to an outlet-energy magnitude comparable to the adiabatic-flow level, when heat is transferred to the flow.

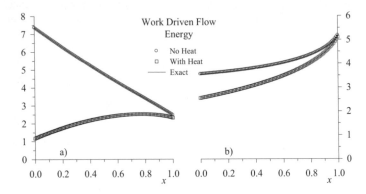

Figure 12.11: Energy in Work-Driven Flow: a) Subcritical Flow, b) Supercritical Flow

The asymptotic convergence rate of these solutions is graphed in Figure 12.12.

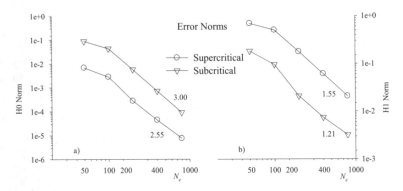

Figure 12.12: Error Norms: a) $\mathcal{H}^0$ Norm, b) $\mathcal{H}^1$ Norm

This rate achieves an order as high as 3.00, for the $\mathcal{H}^0$ norm, for the subcritical flow, and an order as high as 1.55 for the $\mathcal{H}^1$ norm, for the supercritical flow. These norms reflect convergence of not only the solution itself, but also its gradients.

12.9.2 Mass-Transfer Flows

Infrequently covered in reference books, mass-transfer flows remain of practical importance, for they model flows in the vicinity of tributaries and distributaries. The considerations and developments in these books have again led to an exact solution for the case of constant mass transfer in a direction orthogonal to the channel axis, a solution that depends the mass-transfer magnitude $\dot{m}_b$. The benchmarks for these flows encompass a subcritical and a supercritical flow in a straight rectangular duct, both flows subject to a uniform mass addition. For these benchmarks, the non-dimensional specifications as well as boundary conditions are: $h_{in} = 1.667$, $E_{in} = 1.111$, $p_{out} = 0.646$, with $\dot{m}_b = 0.555$, corresponding from the available exact solution to $Fr_{in} = 0.2000$, $m_{in} = 0.430$ and $Fr_{out} = 0.814$, for the subcritical flow, and $h_{in} = 0.397$, $m_{in} = 0.751$, $E_{in} = 2.107$, with $\dot{m}_b = 0.240$, corresponding to $Fr_{in} = 3.000$, $p_{out} = 0.397$ and $Fr_{out} = 1.178$ for the supercritical flow. The initial conditions throughout the duct correspond to a linear interpolation between the inlet and outlet states, resulting in an initial flow that is then advanced in time to a steady state.

Presented in Figure 12.13, the computational results are seen to reflet the exact solution, with mass addition inducing an increase or decrease in Froude number, respectively for a subcritical or a supercritical flow.

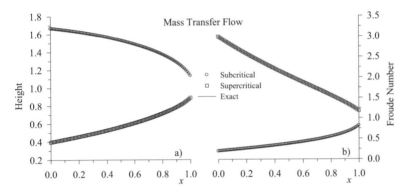

Figure 12.13: Mass-Transfer Flow: a) Free-Surface Height, b) Froude Number

Since no heat transfer affects this flow, the total enthalpy remains constant, yet the mass flow rate theoretically increases proportionately to the added mass. This feature is reflected in the computational results in Figure 12.14, which correctly presents a constant enthalpy and a linearly increasing mass flow rate.

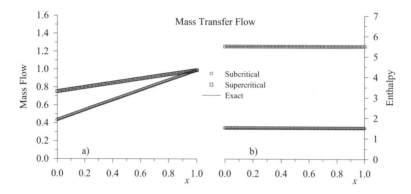

Figure 12.14: Mass-Transfer Flow: a) Mass Flow, b) Total Enthalpy

The available exact solution allowed the calculation of the $\mathcal{H}^0$ and $\mathcal{H}^1$ norms of the error associated with solutions for increasingly denser grids. Indicating solution convergence, these error norms are displayed in Figure 12.15

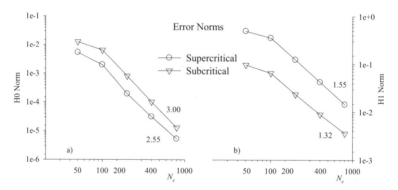

Figure 12.15: Error Norms: a) $\mathcal{H}^0$ Norm, b) $\mathcal{H}^1$ Norm

The asymptotic convergence rate achieves an order as high as 3.00, for the $\mathcal{H}^0$ norm, for the subcritical flow, and as high as 1.55, for the $\mathcal{H}^1$ norm, for the supercritical flow. These norms reflect convergence of not only the solution itself, but also its gradients.

Any amount of mass added in a direction parallel to the normal to the channel axis does not alter the linear momentum along the channel axis, for $\sin\theta = 0$ in the linear momentum equation in (5.124). The effects of mass addition along a direction at an angle with this normal depend on the ratio W_{ref}/L of channel reference width over length; for $W_{\mathrm{ref}}/L = 0.200$, the corresponding results are displayed in Figure 12.16

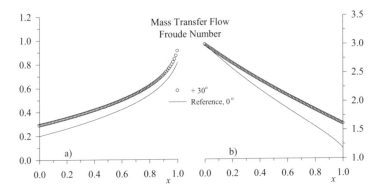

Figure 12.16: Inclined Mass Transfer Flow: a) Subcritical Flow, b) Supercritical Flow

In this case, mass addition increases the channel-flow linear momentum. Accordingly, these results correctly show an increase in the overall Froude-number distribution, for both subcritical and supercritical flows.

12.9.3 Adiabatic Smooth and Shocked Flows with Friction

Bed and wall friction affect any channel flow. Directly altering flow momentum, wall friction leads to not only smooth, but also shocked steady flows. This section details computational solutions of frictional subcritical and supercritical flows, without and with hydraulic jumps, within a straight duct with rectangular cross section. It is possible to develop an exact closed-form solution that depends upon not only the friction factor f_D, but also the non-dimensional ratio W_{ref}/h_{ref} of reference channel width over free-surface height; this solution has been employed to asses the accuracy of the computational solutions.

Employing the critical state as the reference state, the non-dimensional specifications as well as boundary conditions for the subcritical case are: $h_{in} = 1.406$, $E_{in} = 2.526$, $p_{out} = 0.604$, with $W_{ref}/h_{ref} = 10.000$, $f_D = 0.270$, corresponding to $Fr_{in} = 0.600$, $m_{in} = 1.000$ and $Fr_{out} = 0.868$, for the subcritical flow. For the supercritical case, the specifications as well as boundary conditions are $h_{in} = 0.481$, $m_{in} = 1.000$, $E_{in} = 2.529$, with $W_{ref}/h_{ref} = 10.000$, $f_D = 0.250$, $Fr_{in} = 3.000$ corresponding to $p_{out} = 0.398$ and $Fr_{out} = 1.187$ for the supercritical flow. The initial conditions throughout the duct correspond to a linear interpolation between the inlet and outlet states, resulting in an initial flow that is then advanced in time to a steady state.

Analogously to the Gas Dynamic Fanno flow, friction in a channel flow decreases free-surface height yet increases the Froude number, in a subcritical flow, but increases free-surface height, yet decreases the Froude number in a supercritical flow. Such trends resonate with those in flows with work extraction because both friction and work extraction induce a force that opposes the flow. These trends are also reflected in the computational solutions in Figure 12.17 that are seen to mirror the exact solutions for both subcritical and supercritical flows.

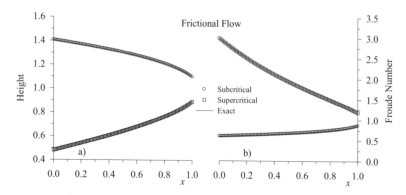

Figure 12.17: Frictional Flow: a) Free-Surface Height, b) Froude Number

The asymptotic convergence-rate trends of these solutions are displayed in Figure 12.18.

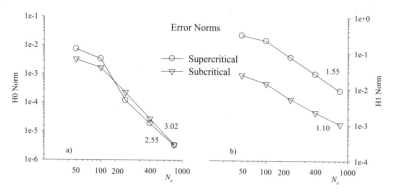

Figure 12.18: Error Norms: a) $\mathcal{H}^0$ Norm, b) $\mathcal{H}^1$ Norm

The asymptotic convergence rate achieves an order as high as 3.02, for the $\mathcal{H}^0$ norm, for the subcritical flow, and as high as 1.55 for the $\mathcal{H}^1$ norm, for supercritical flow. As noted for the previous results, these norms reflect convergence of not only the solution itself, but also its gradients.

The next benchmark involves a steady frictional channel flow incorporating a hydraulic jump that remains localized in the channel. This benchmark allows testing the capability of the characteristics-bias system formulation to generate an essentially non-oscillatory shocked frictional flow. The accuracy of the computational prediction is then assessed against the exact solution. For this benchmark with $W_{\text{ref}}/h_{\text{ref}} = 5.000$ and $f_D = 0.25$, the non-dimensional specifications as well as boundary conditions are: $h_{\text{in}} = 0.481$, $m_{\text{in}} = 1.000$, $E_{\text{in}} = 2.529$ at the supercritical inlet, and $p_{\text{out}} = 0.717$ at the subcritical outlet. These specifications correspond to $Fr_{\text{in}} = 3.000$ and $Fr_{\text{out}} = 0.763$ with a steady hydraulic jump that develops at $x = 0.75$

with upstream and downstream shock Froude numbers $Fr_u = 1.503$ and $Fr_d = 0.688$. Also for this case, the initial conditions result from a linear interpolation between the inlet supercritical state and outlet subcritical state. The resulting flow is then advanced in time to a steady state.

Achieved via a Courant number equal to 100, the computational solution is summarized in Figures 12.19-12.20 in terms of free-surface height, Froude number, speed, mass flow, energy and enthalpy. These variations remain monotone with a crisp hydraulic jump captured in at most one node; significantly, this hydraulic jump is captured at the theoretically exact location of $x = 0.75$, with accurately calculated free-surface height ratio and Froude numbers across the hydraulic jump. In the neighborhood of the outlet, the results remain devoid of any spurious oscillations, which reflects favorably on the enforcement of the pressure boundary condition indicated in Section 12.6.3. With accurately calculated hydraulic jump and outflow state, this solution reflects the reference exact solution.

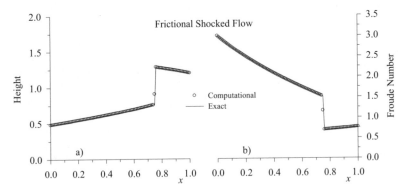

Figure 12.19: Frictional Flow: a) Free-Surface Height, b) Froude Number

Despite the hydraulic jump, mass flow remains unchanged and total enthalpy does not vary in an adiabatic flow. Reflecting the physical consistency of the gravity-wave-convection characteristics-bias formulation, the computational solution is consistent with these features, as shown in Figure 12.20. This solution correctly predicts constant mass flow and enthalpy that remain undistorted by the hydraulic jump in the other variables.

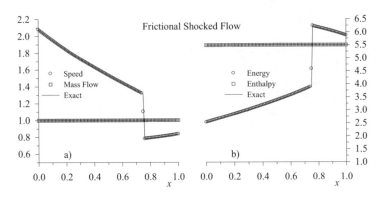

Figure 12.20: Frictional Flow: a) Speed and Mass Flow, b) Energy and Enthalpy

The characteristics-bias formulation with gravity-wave-convection flux-Jacobian decomposition depends on the upstream-bias functions α, δ, ψ. Presented in Figure 12.21 for this shocked frictional flow, the variations of these functions and the nodal ℓ_1 norm of the q^h solution show that α vanishes for $Fr > 0.6$, while δ, as expected, experiences a sharp decrease as the flow switches from supercritical to subcritical. Significantly, following the distribution of the nodal norm, the formulation leads to $\psi = \psi_{min}$, for most of the flow, and $\psi = \psi_{max}$ at the hydraulic jump. Accordingly, the formulation automatically manages the level of induced upstream-bias, keeping it to a minimum in regions of smooth flow, and focusing an increase only at hydraulic jumps, in order to generate a solution that remains both accurate and stable.

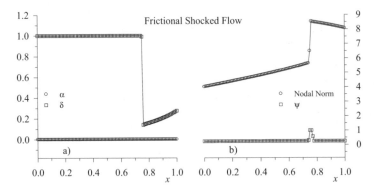

Figure 12.21: Frictional Flow: a) α and δ, b) Nodal Norm and ψ

Also for this shocked flow has the formulation generated an asymptotically convergent solution in the $\mathcal{H}^0$ norm, with corresponding convergence-rate curves presented in Figures 12.22 a)-b).

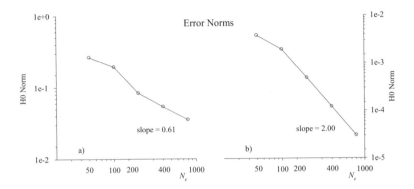

Figure 12.22: Error Norms: a) Shocked Solution, b) Continuous Solution Branches

Figure 12.22 a) displays the error norm for the entire solution, including the hydraulic jump, while Figure 12.22 b) displays the error norm for the union of the continuous subcritical and supercritical solution branches, a union obtained by only considering the computational and exact solutions in the interval $[0.00, 0.66] \cup [0.86, 1.0]$. The shocked solution converges at a rate of 0.61, yet in respect of the continuous solution branches their union is seen to converge at a rate of 2.0. This expected difference in convergence rates between these two solution sets indicates the normal hydraulic jump dominates the overall convergence rate. The formulation thus generates a computational solution that converges also for a discontinuous flow.

12.9.4   Shocked Flows with Friction, Mass Transfer and Bed Slope

Difficult to solve analytically in closed form, this problem involves the shocked flow described in the previous section as altered by uniform mass transfer and bed slope along the entire duct length. Numerically, this problem tests the performance of the algorithm accurately to compute a shocked flow simultaneously subject to wall friction, mass transfer and bed slope; physically, this problem illustrates the quantitative measurable effect of mass transfer and bed slope on a channel flow and the position as well as strength of a hydraulic jump within the flow.

The solutions generated by the characteristics-bias formulation for this investigation correspond to the same initial conditions as well as inlet and outlet boundary conditions discussed in the previous section for the shocked frictional flow, coupled with different magnitudes of the mass-transfer and bed-slope source terms in the generalized hydrodynamics equations. The computed results correspond to $\psi_{min} = 0.40$, for mass transfer alone, and $\psi_{min} = 0.60$ for mass transfer and bed slope; the non-dimensional magnitudes of bed slope and mass transferred to the flow are: $\dot{m}_b = \pm 5.00 \times 10^{-3}$, $\partial b / \partial x = \pm 5.00 \times 10^{-2}$, where the positive determination corresponds to an upward bed slope and mass addition, while the negative determination corresponds to a downward bed slope and mass removal.

Corresponding to a Courant number equal to 100, the computational solutions for mass addition and removal, upward and downward bed slope are summarized in Figures 12.23-12.24 in terms of free-surface height and Froude number variations. As in the previous investigations, these variations remain essentially non-oscillatory with crisp hydraulic jumps captured in at most two nodes. For all the cases considered in this frictional flow, no outflow distortion emerges for any position of the hydraulic jump within the duct, again reflecting favorably on the pressure boundary-condition enforcement procedure in Section 12.6.3.

For the prescribed single constant outlet-pressure boundary condition, the position of the hydraulic jump is entirely determined by the magnitude of mass transfer and bed slope. As the figures show, the hydraulic jump propagates upstream under mass addition and upward bed slope, but downstream under mass removal and downward bed slope. As noted before, mass addition tends to increase the Froude number for subcritical flow, but to decrease it for supercritical flow.

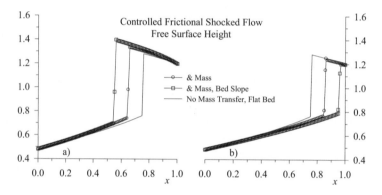

Figure 12.23: Free-Surface Height in Frictional Flow with Hydraulic Jump:
a) Mass Added, Upward Bed Slope, b) Mass Removed, Downward Bed Slope

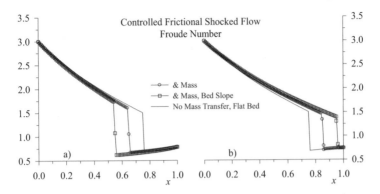

Figure 12.24: Froude Number in Frictional Flow with Hydraulic Jump:
a) Mass Added, Upward Bed Slope, b) Mass Removed, Downward Bed Slope

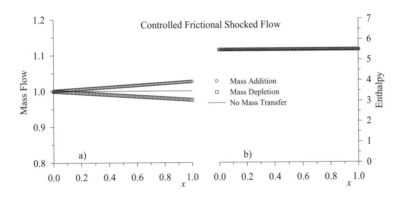

Figure 12.25: Frictional Flow with Hydraulic Jump and Mass Transfer:
a) Mass Flow, b) Enthalpy

In comparison to the baseline case of no mass transfer and flat bed, indicated by the unmarked solid line, an upward bed slope and mass addition move the hydraulic jump upstream; these perturbations force a greater drop in Froude number on the supercritical side of the flow, which in turn allows a greater increase in Froude number on the subcritical side of the flow. The addition of mass further disturbs a supercritical flow into an upstream-shifted stronger hydraulic jump, but further energizes a subcritical flow both to negotiate the adverse friction force and meet the imposed pressure boundary condition.

Conversely, a downward bed slope and mass removal move the hydraulic jump downstream in comparison to the baseline case. As it extracts energy from the flow, mass removal forces a smaller decrease in Froude number on the supercritical side of the flow, which in turn correlates with a smaller increase in Froude number on the subcritical side. The removal of mass lessens the disturbance of a supercritical flow and thereby sustains it for a longer distance, but also reduces energy of a subcritical flow and thereby decreases its spatial extent. For this reason the hydraulic jump moves downstream, so as to accommodate a shorter region of subcritical flow that can only negotiate a smaller amount of adverse friction force, but can still meet the prescribed outlet pressure boundary condition.

These are the only computations known to this author that employ CFD to investigate a channel flow with hydraulic jump, wall friction, bed slope and mass transfer. These results quantitatively indicate the effect of bed slope and mass transfer on the eventual position and strength of the normal hydraulic jump. These mechanism may be employed to control a hydraulic jump.

12.9.5   Smooth and Shocked Flows over a Bump

This section details the computational solutions for subcritical, supercritical and shocked adiabatic flows over a channel bed shaped like a bump. The accuracy of these solutions is then assessed against a corresponding exact solution of the governing equations. For these flows, the non-dimensional bed-shape expression is: $b(x) = (1 + \cos((2x - 1)\pi))/8$. The first flow involves a completely subcritical flow over the bump. The reference inlet Froude number is $Fr = 0.300$, with non-dimensional inlet mass flow and total energy $m_{in} = 1.000$, $E_{in} = 3.089$ as the two needed boundary conditions at the subcritical inlet; from the exact solution the corresponding free-surface height is $h_{in} = 2.231$. At the subcritical outlet, the pressure boundary condition is $p_{out} = 2.490$, which corresponds to $Fr_{out} = 0.300$. The initial conditions throughout the duct correspond to a uniform flow from the reference inlet Froude number, a flow that is then advanced in time to a steady state.

The computational solution, shown in Figure 12.26, remains smooth, correctly matches the imposed outlet pressure boundary condition, and reflects the corresponding exact solution.

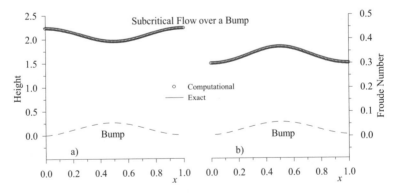

Figure 12.26: Subcritical Channel Flow: a) Free-Surface Height, b) Froude Number

As shown in Figure 12.27, the computational solutions for denser grids asymptotically converge in both the $\mathcal{H}^0$ and $\mathcal{H}^1$ norms with a converge rate as high as 3.0 in the $\mathcal{H}^0$ norm.

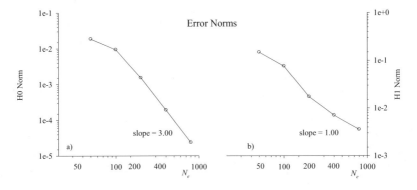

Figure 12.27: Error Norms: a) $\mathcal{H}^0$ Norm, b) $\mathcal{H}^1$ Norm

The next computational solution corresponds to a supercritical flow. The reference inlet Froude number is $Fr = 3.0$, with non-dimensional inlet free-surface height, mass flow, and total energy $h_{\text{in}} = 0.481$, $m_{\text{in}} = 1.000$ $E_{\text{in}} = 1.086$ as the three needed boundary conditions at the supercritical inlet. No boundary conditions are required at the outlet, for at this region of the channel, the flow is supercritical. The initial conditions throughout the duct correspond to a uniform flow from the reference inlet Froude number, a flow that is then advanced in time to a steady state.

The eventual steady-state computational solution in Figure 12.28 remains smooth and correctly predicts at the top of the bump a relative maximum for the free-surface height and a relative minimum for the Froude number.

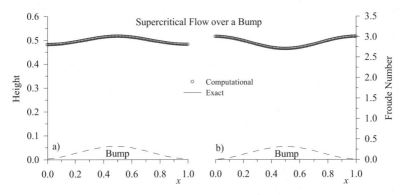

Figure 12.28: Supercritical Channel Flow: a) Free-Surface Height, b) Froude Number

This non-uniform bed shape, hence slope, affects the asymptotic convergence rates displayed in Figure 12.29.

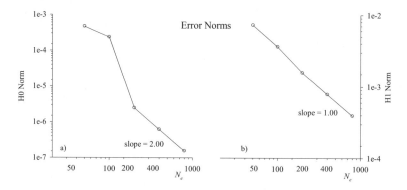

Figure 12.29: Error Norms: a) $\mathcal{H}^0$ Norm, b) $\mathcal{H}^1$ Norm

For this supercritical flow, the convergence rate equals 1.0 for the $\mathcal{H}^1$ norm and 2.00 for the $\mathcal{H}^0$ norm, rates that both indicate the convergence of this solution as well as its gradients.

For an appropriate back pressure, an isentropic flow throughout the channel may emerge that is subcritical upstream of the top of the bump and supercritical downstream of it. This is the next flow calculated, subject to the previous subcritical-flow inlet boundary conditions, but without any outlet flow pressure boundary condition, since the outlet is supercritical; the initial conditions have resulted from a linear variation between the inlet subcritical state and isentropic outlet supercritical state with $Fr_{\text{out}} = 1.763$ and $p_{\text{out}} = 0.235$. Rapidly achieved, the computational solution shown in Figure 12.30 remains indistinguishable from the exact solution, with computed outlet pressure and Froude number that reflect the corresponding isentropic magnitudes and a continuous expansion from inlet to outlet that remains smooth.

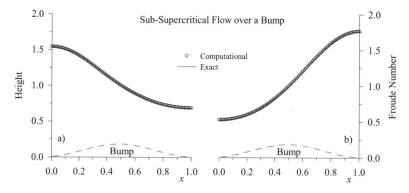

Figure 12.30: Sub-Supercritical Channel Flow: a) Free-Surface Height, b) Froude Number

As the grid is refined, the computational solution for this isentropic flow asymptotically converges, as illustrated in Figure 12.31.

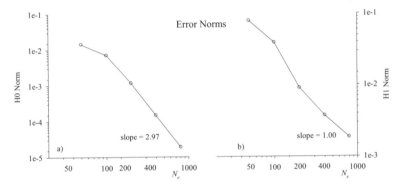

Figure 12.31: Error Norms: a) $\mathcal{H}^0$ Norm, b) $\mathcal{H}^1$ Norm

The convergence rate reaches 1.0 in the $\mathcal{H}^1$ norm and 2.97 in the $\mathcal{H}^1$ norm, indicating convergence of both the solution and solution gradients.

The final set of benchmarks for these channel flows over a bump involve the calculation of a shocked steady flow, with computational-solution accuracy assessed against the available exact solution. For this benchmark, the non-dimensional specifications as well as boundary conditions are: $m_{\text{in}} = 0.274$, $E_{\text{in}} = 2.663$ at the subcritical inlet, and $p_{\text{out}} = 1.105$ at the subcritical outlet. From the exact solution, these specifications correspond to $Fr_{\text{in}} = 0.524$, $h_{\text{in}} = 1.539$, and $Fr_{\text{out}} = 0.552$ with a steady hydraulic jump that develops at $x = 0.750$, with upstream and downstream shock Froude numbers $Fr_u = 1.508$ and $Fr_d = 0.686$. The initial conditions throughout the channel correspond to a linear interpolation between the inlet and outlet subcritical states.

Achieved with a Courant number equal to 100, the resulting solution is summarized in Figures 12.32-12.34 in terms of free-surface height, Froude number, speed, mass flow, energy and enthalpy. These variations remain monotone with a crisp hydraulic jump captured in at most one node; also in this case is this calculated jump predicted at the theoretically exact $x = 0.750$ location, with accurately calculated free-surface height ratio and Froude numbers across the jump. In the neighborhood of the outlet, the results remain devoid of any spurious oscillations, which again reflects favorably on the enforcement of the pressure boundary condition. With accurately calculated hydraulic jump and outflow states, this solution reflects the available exact solution.

Despite the hydraulic jump, the corresponding mass flow remains constant and the total enthalpy does not change in an adiabatic flow. These important features are correctly predicted by the computational solution, displayed in Figure 12.33, which shows a constant mass flow and enthalpy that remain undistorted by the hydraulic jump in the other variables.

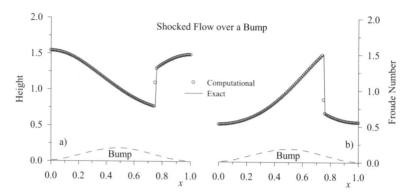

Figure 12.32: Channel Flow with Hydraulic Jump: a) Free-Surface Height,
b) Froude Number

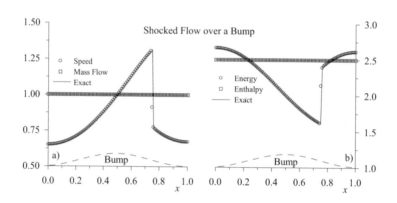

Figure 12.33: Channel Flow with Hydraulic Jump: a) Speed and Mass Flow,
b) Energy and Enthalpy

Presented in Figure 12.34 for this shocked channel flow are the variations of the upstream-bias functions "α", "δ" and "ψ" along with the nodal ℓ_1 norm of the q^h solution, variations that are directly determined by the distribution of the Froude number. Following the distribution of the nodal norm, also for this shocked channel flow can the formulation lead to $\psi = \psi_{min}$ for most of the flow and focus $\psi = \psi_{max}$ at the hydraulic jump. Accordingly, the formulation again keeps the upstream bias to a minimum in regions of smooth flow, and focuses an increase of this bias only at hydraulic jumps, in order to produce a solution that remains accurate and stable.

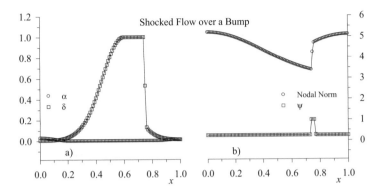

Figure 12.34: Channel Flow with Hydraulic Jump: a) α and δ, b) Nodal Norm and ψ

The formulation has generated an asymptotically convergent solution in the $\mathcal{H}^0$ norm for this flow with hydraulic jump, with corresponding convergence rates presented in Figures 12.35 a)-b).

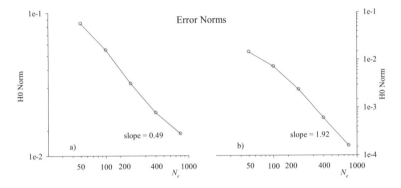

Figure 12.35: Error Norms: a) Shocked-Solution Norm, b) Continuous-Solution Norm

Figure 12.35 a) displays the error norm for the entire solution, including the hydraulic jump, while Figure 12.35 b) displays the error norm for the union of the continuous subcritical and supercritical solution branches, a union obtained by only considering the computational and exact solutions in the interval $[0.00, 0.66] \cup [0.86, 1.0]$. While the shocked solution containing the hydraulic jump converges at a rate of 0.49, the continuous solution set is seen to converge at a rate of 1.92, This expected difference in convergence rates between these two solution sets signals the normal hydraulic jump dominates the overall convergence rate. The formulation has thus again produced a computational solution that converges also for a discontinuous flow.

As noted previously, friction affects all channel flows. For $W_{\text{ref}}/h_{\text{ref}} = 5.000$ and $f_D = 0.25$, the effect of friction on this flow with hydraulic jump is presented in Figure 12.36

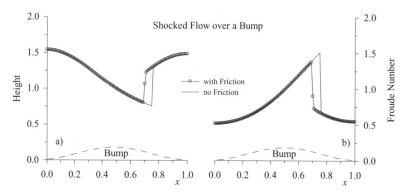

Figure 12.36: Channel Flow with Hydraulic Jump and Friction: a) Free-Surface Height, b) Froude-Number

Achieved with a Courant number equal to 100, this solution remains monotone with a hydraulic jump captured in at most one node. Inducing a force that opposes the flow, friction is seen to force the hydraulic jump upstream, with reduced free-surface height jump and peak Froude Number.

12.9.6   Dam-Break Flow

This benchmark has allowed the identification of a contact discontinuity in the distribution of unsteady total energy. Since an open channel flow involves an incompressible fluid with no contact discontinuity in the flow free-surface height and speed, the occurrence of this phenomenon in the total energy distribution signals that it is the temperature distribution that experiences such a discontinuity. As documented in this section, the developed characteristics-bias formulation has led to a crisp resolution of this discontinuity not only for the traditionally reported case of a flow in a frictionless flat-bed channel, but also for a flow affected by both friction and bed slope. In the dam-break flow in this section, the flow channel is initially divided in two regions separated by a dam located in the middle of the channel. In these regions, the non-dimensional initial conditions are

$$
\begin{aligned}
h = 1.0, \qquad & m = 0.0, \quad E = 1.0, \qquad 0.0 \le x \le 0.5 \\
h = 0.13827, \quad & m = 0.0, \quad E = 0.13827, \quad 0.5 < x \le 1.0
\end{aligned}
\tag{12.116}
$$

The exact solution for not only free-surface height h and speed u, but also total energy E has been developed in closed form. With $\eta \equiv (2 - (2x-1)/2t)/3$, $h_d = 0.13827$, $E_m = 0.66331$, $E_d = h_d$, $x_c = \frac{1}{2} + \frac{2}{3}t$, $x_s = \frac{1}{2} + 0.96774\,t$, this solution is expressed as

$$
\begin{array}{cccc}
h(x,t) & u(x,t) & E(x,t) & x\text{-range} \\
\hline
1 & 0 & 1 & x \leq \frac{1}{2} - t \\
\eta^2 & 2(1-\eta) & \frac{5}{2}\eta^2 - 4\eta^3 + \frac{5}{2}\eta^4 & \frac{1}{2} - t < x \leq \frac{1}{2} \\
\frac{4}{9} & \frac{2}{3} & \frac{34}{81} & \frac{1}{2} < x \leq x_c \\
\frac{4}{9} & \frac{2}{3} & E_m & x_c < x \leq x_s \\
h_d & 0 & E_d & x_s < x
\end{array}
\tag{12.117}
$$

The dam ruptures at $t = 0$ and the computational investigation with $C_{\max} = 1.0$ seeks the solution at $t = 0.155$ and $t = 0.310$. At these time stations, the exact solution features a hydraulic jump centered at respectively $x = 0.65$ and $x = 0.80$, for each of the components of the dependent variables in q, and a contact discontinuity centered at respectively $x = 0.603$ and $x = 0.707$, for the distribution of the total energy. Figures 12.37 a)-d) present the distributions of total energy at this time level, for several formulations of the mass matrix and magnitudes of the upstream-bias controller ψ. These distributions show a noticeable contact discontinuity in the distribution of this variable.

Figure 12.37 a) corresponds to a lumped mass matrix. This solution remains monotone, but diffused, with contact discontinuity and hydraulic jump spread over several elements; significantly, the lumped-mass formulation for this benchmark has been found to remain devoid of unphysical oscillations only when $\psi \geq 1$, uniformly throughout the computational domain. Figure 12.37 b) is also obtained for a uniform $\psi = 1$, but with a consistent mass matrix, which corresponds to no upstream bias on the time derivative. This distribution remains virtually indistinguishable from the previous one, which shows that a consistent mass-matrix formulation can generate unsteady solutions devoid of unphysical oscillations. The solution in Figure 12.37 c) also corresponds to a consistent mass matrix, but results from a variable ψ, with $\psi_{\min} = 0.25$ and $\psi_{\max} = 0.75$. The upstream bias in this case is no longer uniform, but is non-linearly applied in relation to the variation of the local solution slopes. Accordingly, the resolution of this solution has increased with a somewhat sharper, but still diffused contact discontinuity. It is the Galilean invariant formulation with a variable ψ and upstream bias on the time derivative that noticeably increases the resolution of the contact discontinuity, which thus provides another justification for the use of this formulation.

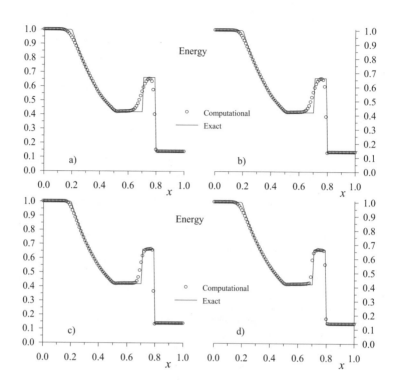

Figure 12.37: Energy: a) Linear, Lumped Mass Matrix, b) Linear, Consistent Mass Matrix c) Non-linear, Consistent Mass Matrix, d) Non-linear, Galilean Invariant

The formulation leads to essentially non-oscillatory shocked solutions not only for the total energy, but also for the Froude number, mass flow and free-surface height, as shown in Figures 12.38- 12.40, with practically horizontal plateau upstream of the hydraulic jump. This dam-break problem is particularly challenging because the flow remains critical along this plateau. With $Fr \simeq 1$, in this condition, the level of upstream-bias dissipation corresponding to one eigenvalue reaches a minimum, yet the formulation generates a converging solution.

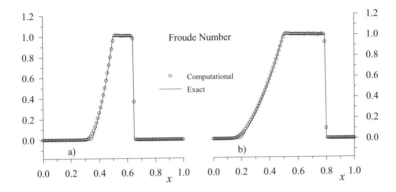

Figure 12.38: Froude Number: a) $t=0.07076$, b) $t=0.14152$

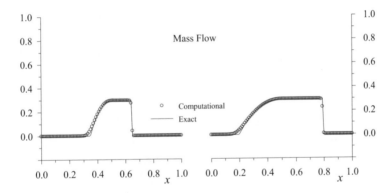

Figure 12.39: Mass Flow: a) $t=0.07076$, b) $t=0.14152$

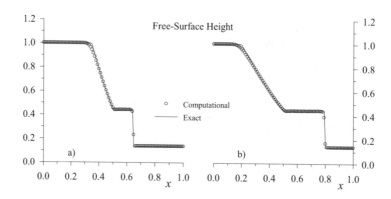

Figure 12.40: Free-Surface Height: a) t=0.07076, b) t=0.14152

These three figures display solutions at two distinct time stations. In particular, they are representative of the essentially non-oscillatory character of this transient solution.

Unavailable in closed form, the solution for this dam-break problem simultaneously subject to both friction and downward bed slope can only be generated computationally. Presented in Figures 12.41-12.44, the computational results for this realistic condition, for $W_{ref}/h_{ref} = 5.000$, $f_D = 0.50$, $\partial b/\partial x = -0.222$, remain essentially non-oscillatory with a crisp contact discontinuity and a definite hydraulic jump captured within at most two nodes.

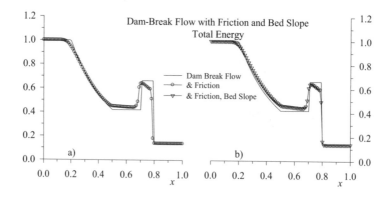

Figure 12.41: Energy: a) Effect of Friction, b) Effect of Friction and Bed Slope

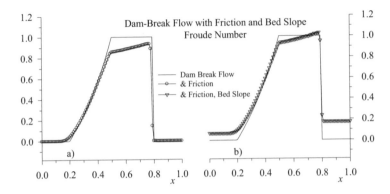

Figure 12.42: Froude Number: a) Effect of Friction, b) Effect of Friction and Bed Slope

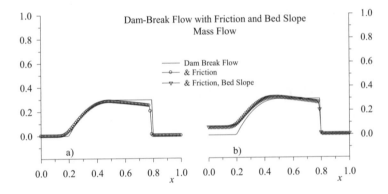

Figure 12.43: Mass Flow: a) Effect of Friction, b) Effect of Friction and Bed Slope

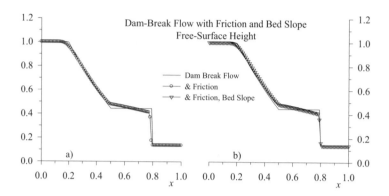

Figure 12.44: Free-Surface Height: a) Effect of Friction, b) Effect of Friction and Bed Slope

Friction in a channel with flat bed retards the flow. This effect leads to a decrease in the distribution of Froude number, as confirmed in Figure 12.42, and also marginally affects the location of the hydraulic jump. The further addition of a downward bed slope accelerates the flow, which increases the Froude number and mass flow, a consequence that is also reflected in the computational solutions. Also for this dam-break flow with friction and downward bed slope for which no closed form is available, the formulation has generated quantitative solutions that conform to the mechanics of the flow

12.9.7   Flow with Hydraulic Jumps in a Variable-Width Channel

This section details the computational solutions for subcritical, critical, supercritical and shocked adiabatic flows in a converging diverging channel, employing the critical state as a reference. The accuracy of these solutions is then assessed against the corresponding available exact solution. For these flows, the non-dimensional channel width distribution is

$$
W(x) = \begin{cases} 1.75 - 0.75\cos\left(\pi(2x-1)\right) & , \ 0 \leq x \leq \tfrac{1}{2} \\[2mm] 1.25 - 0.25\cos\left(\pi(2x-1)\right) & , \ \tfrac{1}{2} \leq x \leq 1 \end{cases} , \quad W(0) = 2.5, \ W(1) = 1.5
$$

(12.118)

Although continuous for every "x", this distribution exhibits a discontinuous curvature at the channel throat. At this location, such a geometric characteristic theoretically induces a slope discontinuity in the flow variables, a feature that provides another mechanism for assessing the resolution of the computational solutions.

The first flow involves a completely subcritical flow throughout the channel. The reference inlet Froude number is $Fr = 0.200$, with non-dimensional inlet free-surface height and total energy $h_{in} = 1.471$, $E_{in} = 1.125$ as the two needed boundary conditions at the subcritical inlet; from the exact solution the corresponding mass flow is $m_{\text{in}} = 0.357$. At the subcritical outlet, the pressure boundary condition is $p_{\text{out}} = 0.996$, which corresponds to $Fr_{\text{out}} = 0.355$. The initial conditions throughout the channel correspond to a linear interpolation between the inlet and outlet subcritical states.

The computational solution, shown in Figure 12.45, remains smooth and correctly matches the imposed outlet-pressure boundary condition. The flow expansion towards the throat and subsequent compression downstream of the throat are clearly resolved in these computational distributions that reflect the corresponding exact solution.

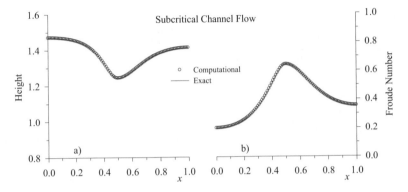

Figure 12.45: Subcritical Channel Flow: a) Free-Surface Height, b) Froude Number

As shown in Figure 12.46, the computational solutions for denser grids asymptotically converge in both the $\mathcal{H}^0$ and $\mathcal{H}^1$ norms with a converge rate as high as 3.00 in the $\mathcal{H}^0$ norm.

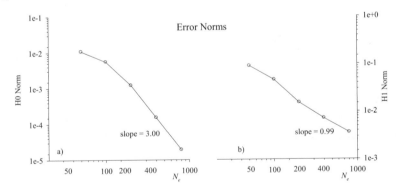

Figure 12.46: Error Norms: a) $\mathcal{H}^0$ Norm, b) $\mathcal{H}^1$ Norm

The next computational solution corresponds to a critical flow at the channel throat. The reference inlet Froude number is $Fr = 0.226$, with non-dimensional inlet free-surface height and total energy $h_{in} = 1.463$, $E_{in} = 1.124$ as the two needed boundary conditions at the subcritical inlet; from the exact solution the corresponding mass flow is $m_{in} = 0.400$. At the subcritical outlet, the pressure boundary condition is $p_{out} = 0.958$, which corresponds to $Fr_{out} = 0.409$. The initial conditions throughout the duct correspond to a uniform flow from the reference inlet Froude number, a flow that is then advanced in time to a steady state.

The eventual steady-state computational solution in Figure 12.47 remains smooth and subcritical. Reflecting the exact solution, this computational solution sharply resolves the slope discontinuity at the throat, which corresponds to a hydraulic jump of vanishing strength.

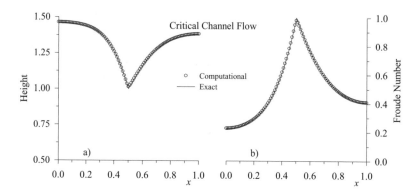

Figure 12.47: Critical Channel Flow: a) Free-Surface Height, b) Froude Number

This physical slope discontinuity affects the asymptotic convergence rates displayed in Figure 12.48.

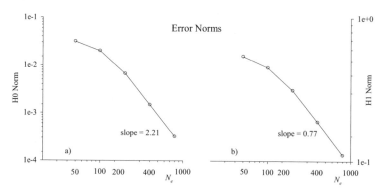

Figure 12.48: Error Norms: a) $\mathcal{H}^0$ Norm, b) $\mathcal{H}^1$ Norm

For this critical flow, the convergence rates equal 0.77 for the $\mathcal{H}^1$ norm and 2.21 for the $\mathcal{H}^0$ norm, rates that both indicate the convergence of this solution as well as its gradients.

For an appropriate back pressure, the emerging isentropic flow throughout the channel is subcritical upstream of the throat and supercritical downstream. This is the next flow calculated, subject to the previous inlet boundary conditions, but without any outlet flow pressure boundary condition, since the outlet is supercritical; the initial conditions have resulted from a linear variation between the inlet subcritical state and isentropic outlet supercritical state with $Fr_{out} = 2.117$ and $p_{out} = 0.107$. Rapidly achieved, the computational solution in Figure 12.49 remains indistinguishable from the exact solution, with computed outlet pressure and Froude number that reflect the córresponding isentropic magnitudes. The continuous expansion from inlet to outlet remains smooth and the mild slope discontinuity at the throat, induced by the throat discontinuous curvature, is sharply resolved.

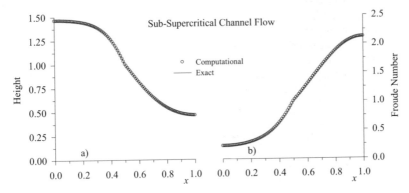

Figure 12.49: Sub-Supercritical Channel Flow: a) Free-Surface Height, b) Froude Number

As the grid is refined, the computational solution for this isentropic flow is asymptotically convergent, as illustrated in Figure 12.50.

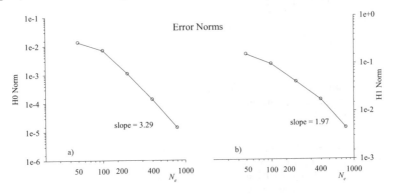

Figure 12.50: Error Norms: a) $\mathcal{H}^0$ Norm, b) $\mathcal{H}^1$ Norm

The convergence rate reaches 1.97 in the $\mathcal{H}^1$ norm and 3.20 in the $\mathcal{H}^0$ norm, indicating convergence of both the solution and solution gradients.

The final benchmark of this set of adiabatic channel flows involves the calculation of a shocked steady flow, with computational-solution accuracy assessed against the available exact solution. For this benchmark, the non-dimensional specifications as well as boundary conditions are: $h_{in} = 1.463$, $E_{in} = 2.587$ at the subcritical inlet, and $p_{out} = 0.830$ at the subcritical outlet. From the exact solution, these specifications correspond to $Fr_{in} = 0.226$, $m_{in} = 0.400$, and $Fr_{out} = 0.456$ with a steady hydraulic jump that develops at $x = 0.750$, with upstream and downstream shock Froude numbers $Fr_u = 1.745$ and $Fr_d = 0.609$. The initial conditions throughout the channel correspond to a linear interpolation between the inlet and outlet subcritical states.

Achieved with a Courant number equal to 150, the resulting solution is summarized in Figures 12.51-12.52 in terms of free-surface height, Froude number, speed, mass flow, energy and enthalpy. These variations remain essentially non-oscillatory with a crisp hydraulic jump captured in at most one node; also in this case have the computational results predicted this hydraulic jump at the theoretically exact $x = 0.750$ location, with accurately calculated free-surface height ratio and Froude numbers across the hydraulic jump.

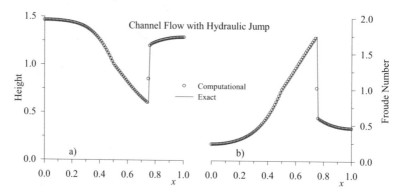

Figure 12.51: Channel Flow with Hydraulic Jump: a) Free-Surface Height, b) Froude Number

In the neighborhood of the outlet, the results remain devoid of any spurious oscillations, which again reflects favorably on the enforcement of the pressure boundary condition. With accurately calculated hydraulic jump and outflow states, this solution reflects the available exact solution.

Despite the hydraulic jump, the corresponding mass flow remains constant and the total enthalpy does not change in an adiabatic flow. These important features are correctly predicted by the computational solution, displayed in Figure 12.52, which shows a constant mass flow and enthalpy that remain undistorted by the hydraulic jump in the other variables.

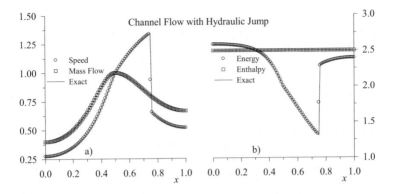

Figure 12.52: Channel Flow with Hydraulic Jump: a) Speed and Mass Flow, b) Energy and Enthalpy

Presented in Figure 12.53 for this shocked channel flow are the variations of the upstream-bias functions "α", "δ" and "ψ" along with the nodal ℓ_1 norm of the q^h solution, variations that are directly determined by the distribution of the Froude number. Following the distribution of the nodal norm, also for this shocked channel flow can the formulation lead to $\psi = \psi_{\min}$ for most of the flow and focus $\psi = \psi_{\max}$ at the hydraulic jump. Accordingly, the formulation again keeps the upstream bias to a minimum in regions of smooth flow, and focuses an increase of this bias only at hydraulic jumps, in order to produce a solution that remains accurate and stable.

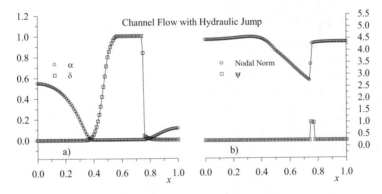

Figure 12.53: Channel Flow with Hydraulic Jump: a) α and δ, b) Nodal Norm and ψ

Also for a shocked channel flow has the formulation generated an asymptotically convergent solution in the $\mathcal{H}^0$ norm, with corresponding convergence rates presented in Figures 12.54 a)-b).

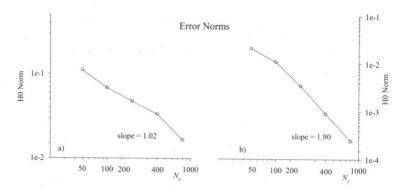

Figure 12.54: Error Norms: a) Shocked-Solution Norm, b) Continuous-Solution Norm

Figure 12.54 a) displays the error norm for the entire solution, including the hydraulic jump, while Figure 12.54 b) displays the error norm for the union of the continuous subcritical and supercritical solution branches, a union obtained by only considering the computational and exact solutions in the interval $[0.00, 0.66] \cup [0.86, 1.0]$. While the shocked solution converges at a maximum rate of 1.02, the continuous solution set is seen to converge at a rate of 1.90. This expected difference in convergence rates between these two solution sets again signals the hydraulic jump dominates the overall convergence rate. The formulation has thus produced a computational solution that converges also for a discontinuous flow.

12.9.8   Flow with Hydraulic Jumps in a Variable-Width Channel with Friction, Sloping Bed, Mass as well as Heat Transfer

Unavailable in analytical closed form, the solution of a hydrodynamic problem of a channel flow subject to variable channel width, wall friction, bed slope, and mass as well as heat transfer may only be obtained computationally. This section presents the computational solution for a channel flow subject to all of these effects. The shocked frictionless adiabatic flow discussed in the previous section provides one reference baseline channel flow. Obtained from the same boundary and initial conditions of this reference flow, the computed results in this section correspond to wall friction from channel inlet to outlet, with $W_{\mathrm{ref}}/h_{\mathrm{ref}} = 10.0$, $f_D = 0.05$, but bed slope, mass, and heat transfer, however, only imposed along the diverging part of the channel, in order to investigate the impact of these effects upon the hydraulic jump. The magnitudes of bed slopes and mass as well as heat transfer are: $\partial b/\partial x = \pm 0.075$, $\dot{m}_b = \pm 0.05$, with $W_{\mathrm{ref}}/L = 0.2$, $g_T = \pm 1.00$, where the positive determination corresponds to an upward slope and mass addition as well as heating while the negative determination corresponds to a downward bed slope and mass removal as well as cooling.

Achieved with a Courant number equal to 50, the computational results illustrated in Figures 12.55-12.56 remain essentially non-oscillatory, with crisp hydraulic jumps captured in at most one node, without any outflow distortion emerging for any position of the hydraulic jump within the duct. Far from being a numerical artefact, the clearly visible increasing slope discontinuity at the channel throat corresponds to both the curvature discontinuity in the channel, at this location, and the abrupt rise in local bed slope and mass as well as heat transfer; the computational results thus succeed in resolving this expected slope discontinuity without spurious oscillations.

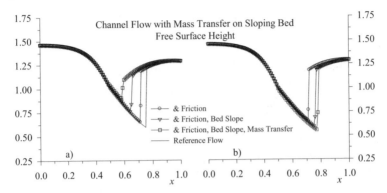

Figure 12.55: Free-Surface Height in Frictional Shocked Channel Flow:
a) Mass Added, Upward Sloping Bed, b) Mass Removed, Downward Sloping Bed

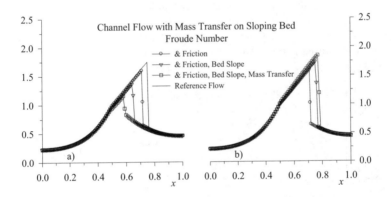

Figure 12.56: Froude Number in Frictional Shocked Channel Flow:
a) Mass Added, Upward Sloping Bed, b) Mass Removed, Downward Sloping Bed

Opposing the flow, the effect of friction weakens the hydraulic jump and shifts it upstream, as shown in the figures. An upward bed slope and mass addition also oppose the flow, which further shifts the hydraulic jump upstream. Sufficient upward bed slopes in combination with mass addition sizably weaken the jump and may even eliminate it totally. Conversely, a downward bed slope and mass depletion move the hydraulic jump downstream with consequent increase of the peak supercritical Froude number.

As noted, the energy equation ordinarily remains decoupled from the continuity and linear momentum equations. Accordingly, the exchange of heat with the flow only affects the total energy, hence temperature of the flow. The distribution of total energy resulting for $\psi_{min} = 0.35$ from heat exchange with the flow is presented in Figure 12.57.

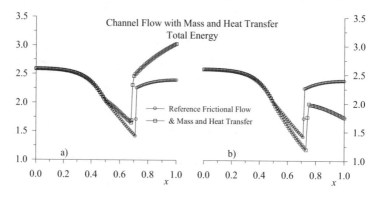

Figure 12.57: Energy: a) Mass and Heat Added, b) Mass and Heat Removed

As expected, heating increases the magnitude of total energy, whereas cooling decreases it. The exchange of both mass and heat with a flow alters the distributions of mass flow and total enthalpy. Such fundamental features have been correctly predicted by this solution, as illustrated in Figure 12.58. The results in the figure show a distribution of mass flow that reflects the mass transfer and a linear variation of total enthalpy from the point of initiation of constant heat transfer.

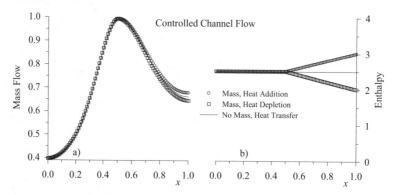

Figure 12.58: Shocked Flow with Mass and Heat Transfer: a) Mass Flow, b) Enthalpy

Again, these are the only CFD computations known to this author of shocked channel flows subject to wall friction, bed slope, mass and heat transfer. According to these results, also in a channel flow will these effects exercise significant control on the location and strength of a hydraulic jump.

12.9.9   Convergence to Steady State

The computational solutions in the previous sections rapidly converged to corresponding steady states for Courant numbers in excess of 50. The maximum Courant number for convergence has depended on flow features and the magnitude of the source terms. In the presence of heat, mass and work transfer in a shocked channel flow this number could reach 50; in the absence of all these transfers this number could rise to 150, for shocked adiabatic flows, and 200 for isentropic subcritical and supercritical flows. These magnitudes of Courant number, hence swift convergence, became available because the formulation allowed a specified maximum number of Newton iterations for the non-linear determination of each IRK array K_i, as described in Section 12.7. Although this number was set to 4, the algorithm employed fewer iterations throughout the pseudo-transient to convergence, at each time station, when the Newton-iteration residual fell below the prescribed tolerance of 5.0×10^{-6}.

Representative of the convergence histories of all the computed steady states in this study, Figure 12.59 a) illustrates this process for the shocked channel flow with wall friction, bed slope, and heat as well as mass transfer, discussed in the previous section, obtained for a Courant number equal to 50, and Figure 12.59 b) presents the corresponding process for the adiabatic shocked channel flow presented in Section 12.9.7, obtained for a Courant number equal to 150. In these figures, N_t denotes the number of time cycles and the integers below the convergence curves indicate the number of Newton-iterations employed in the corresponding region of the curves. In both cases, 3 or more Newton iterations were only required when the total residual exceeded about 5.0×10^{-6}. When the total residual fell below this threshold, for both flows the algorithm experienced a definite rapid convergence that only required 1 Newton iteration to determine each K_i at every time station.

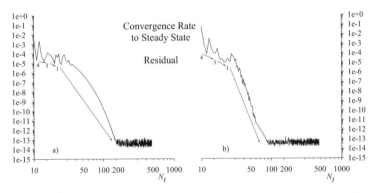

Figure 12.59: Convergence Rate Curves: a) Non-Adiabatic Flow, b) Adiabatic Flow

When, instead, the maximum number of Newton iterations per cycle was limited to 1, the calculations could only converge for a reduced Courant number that required more time cycles N_t to converge than in the multiple Newton-iteration mode. Whether in single or multiple Newton-iteration mode, for all the steady states determined in this study, the formulation is capable of driving the total residual to machine zero.

12.10 Computational Performance

This chapter has developed a characteristics-bias system for the generalized depth-averaged shallow-water equations and solved this system through a finite element spatial discretization and an implicit Runge-Kutta time integration, implemented in Newton-iteration mode. The generalized equations feature a source term that models the effects of not only channel width variations, but also wall friction, variable channel bed, shaft work and heat as well as mass transfer. Since no exact solutions of these equations are available for flows simultaneously influenced by all of these effects, these flows can only be investigated computationally. The characteristics bias system consists of the generalized equations as augmented with a continuum upstream-bias expression that evolves from a decomposition of the inviscid flux Jacobian into physically significant gravity-wave and convection components. This system induces an upwind representation directly at the differential equation level before the spatial discretization. Owing to this feature, a conventional centered or Galerkin spatial discretization of this system automatically generates an upwind stable discrete system that does not require any further stabilizing dissipation. Such a continuum upstream bias has been extended to both the source term and the time partial derivatives, which has resulted in a characteristics-bias system that becomes Galilean invariant, a fundamental physical principle satisfied by the original shallow water equations. According to the computational results, the extension of the continuum upstream bias to both the source term and time partial derivatives sizably reduces induced upwind dissipation, noticeably increases contact-discontinuity resolution, and beneficially leads to rapid convergence.

The characteristics-bias system has then been cast as an integral weak statement, which has also led to a continuum form of the SUPG procedure. Unlike the customary SUPG algorithm, however, the characteristics-bias SUPG requires no premultiplication by the transpose of the Euler flux Jacobian matrix. As counterparts of the usual SUPG "τ_{SUPG}" and "δ_{SUPG}" terms, moreover, the parameter products $\varepsilon\psi\alpha$ and $\varepsilon\psi\delta$ in this formulation explicitly depend on the square or cube of the local mesh spacing, which has also contributed to the recorded rapid convergence. The formulation keeps the upstream bias and associated dissipation to a minimum, in regions of smooth flow, and focuses an increase of this bias only at hydraulic jumps, in order to generate solutions that remain both accurate and stable. Accordingly, the formulation provides a physics based upwind algorithm, but with reduced levels of induced dissipation.

The mechanically significant computational solutions for smooth and shocked flows with channel width variations, shaft work, variable channel bed, wall friction, and heat as well as mass transfer reflect all the available corresponding exact solutions for flows influenced by only one effect, with computed mass flow and total enthalpy correctly remaining constant across hydraulic jumps. The numerical results remain essentially non-oscillatory or monotone, with crisp hydraulic jumps captured in at most one or two nodes. With reference to the available exact solutions, the corresponding computational solutions have converged asymptotically in both the $\mathcal{H}^0$ and $\mathcal{H}^1$ norms, for continuous flows, and in the $\mathcal{H}^0$ norm for shocked flows, with an order as high as 3 in the $\mathcal{H}^0$ norm for continuous solutions. Rapid convergence to steady state has also been recorded for booth smooth and shocked flows with Courant numbers ranging between 50 and 200, for the multiple Newton-iteration implementation.

From a physical viewpoint, mass addition and upward bed slope shift the hydraulic jump upstream, whereas mass depletion and downward bed slope shift the hydraulic jump downstream. From a fluid mechanics standpoint and respectively for frictional and diverging-channel flows, the upstream shift can take place in regions of both higher and lower Froude number, and the downstream shift can take place in regions of both lower and higher Froude number. According to the theoretical and computational findings discussed in this chapter, this formulation allows comprehensive, efficient and exhaustive investigations of physically realistic and mechanically relevant generalized open-channel flows.

Chapter 13

CFD Investigation of Generalized Quasi-1-D Compressible Flows

This chapter documents the performance of the Acoustics-Convection Upstream Resolution Algorithm for the computational analysis of adiabatic and non-adiabatic, smooth and shocked generalized compressible flows. The algorithm uses an acoustics-convection decomposition of the Euler flux Jacobian and induces the upstream bias at the differential - equation level, before any discretization, within a characteristics-bias system associated with the Euler equations with general equilibrium equations of state. The characteristics-bias system essentially adds to the Euler system a regularizing hyperbolic-parabolic perturbation; the perturbation parameter is then linked to the product of a mesh measure and solution smoothness measure, so that the regularized solution approaches the corresponding weak solution of the Euler equations as the mesh is refined. This formulation becomes Galilean invariant, which allows the exact solution of the Euler equations also to satisfy the characteristics-bias system in regions of supersonic flow. Directly resulting from the upwind-bias representation of the source term and time derivative, this property sizably reduces upwind diffusion and practically leads to converging essentially non-oscillatory solutions for both steady and unsteady problems, with crisp capturing of not only normal shocks, but also contact discontinuities. For low subsonic Mach numbers, this formulation returns a consistent upstream-bias approximation of the acoustics equations. For supersonic Mach numbers, the formulation smoothly becomes an upstream-bias approximation of the entire Euler flux. With the objective of minimizing induced artificial diffusion, the formulation non-linearly induces upstream-bias essentially locally in regions of solution discontinuities, whereas it decreases the upstream-bias in regions of solution smoothness. The acoustics-convection flux Jacobian decomposition consists of components that genuinely model the physics of acoustics and convection. This formulation eliminates the unstable linear-dependence problem in steady low-Mach-number flows and satisfies by design the upstream-bias stability condition.

The discrete equations originate from a Galerkin finite element discretization of the characteristic-bias system. This finite element discretization naturally and automatically leads to consistent boundary differential equations and a new outlet pressure boundary condition that does not require any algebraic extrapolation of variables. The resulting discrete equations correspond to an essentially centered discretization in the form of a non-linear combination of upstream diffusive and downstream anti-diffusive flux differences, with greater bias on the upstream diffusive flux difference. This formulation directly accommodates an implicit solver, for the required Jacobian matrices are determined in a straightforward manner. The finite element equations are then integrated in time within a compact block tri-diagonal matrix statement by way of an implicit non-linearly stable Runge-Kutta algorithm for stiff systems.

The operation count for this algorithm is comparable to that of a simple flux vector splitting algorithm. The developments in this study have employed basic two-noded cells, which has thus led to a block tridiagonal matrix system, for the implicit formulation. To determine the ultimate accuracy of linear approximations of fluxes within two-noded cells, for a computationally efficient implementation, this study employs no MUSCL-type local extrapolation of dependent variables. Several numerical investigations document theoretical accuracy of the formulation and shown that for the linear-basis implementation, the corresponding solutions exponentially achieve a steady state, with Courant numbers up to 200, and asymptotically converge in the $\mathcal{H}^0$ and $\mathcal{H}^1$ norms, at a rate that equals up to 3, in the $\mathcal{H}^0$ norm. The computational results for generalized adiabatic and non-adiabatic flows reflect available exact solutions for flows within straight and converging-diverging ducts with wall friction as well as heat, mass and work transfer. These results keep mass flow and enthalpy continuous across normal shocks and indicate that heating and cooling, as well as mass and work transfer, directly control the location and strength of a normal shock, with sufficient heating leading to complete shock elimination.

13.1   Characteristics Analysis

With respect to an inertial reference frame, the generalized quasi-1D Euler conservation law system

$$
\left\{
\begin{array}{l}
\dfrac{\partial \rho}{\partial t} + \dfrac{\partial m}{\partial x} + \dfrac{m}{A}\dfrac{\partial A}{\partial x} = \dfrac{\widetilde{m}_b}{A} \\[2ex]
\dfrac{\partial m}{\partial t} + \dfrac{\partial}{\partial x}\left(\dfrac{m^2}{\rho} + p\right) + \dfrac{m^2}{\rho A}\dfrac{\partial A}{\partial x} = \\[2ex]
\qquad - f_D \dfrac{m|m|}{2\rho D\sqrt{A_*/L}} - s\chi\,\rho w_m + \dfrac{b(\Delta p)_{b_x}}{A} + \dfrac{\dot{m}_b \widetilde{m}_b}{b\rho_b A}\sin\theta \\[2ex]
\dfrac{\partial E}{\partial t} + \dfrac{\partial}{\partial x}\left(m\dfrac{(E+p)}{\rho}\right) + \dfrac{m(E+p)}{\rho A}\dfrac{\partial A}{\partial x} = \\[2ex]
\qquad + |m|\,(g_{\mathrm{T}} - w_m) + mg_x - \dfrac{|\dot{m}_b|}{b}w_\tau + \left.\dfrac{E+p}{\rho}\right|_b \dfrac{\widetilde{m}_b}{A}
\end{array}
\right.
\tag{13.1}
$$

with $s \equiv \mathrm{sgn}(m)$ and

$$
\widetilde{m}_b \equiv \dot{m}_b\left(\cos\theta + \frac{\partial A}{\partial x}\frac{(\sqrt{A_*}/L)\sin\theta}{2\sqrt{\pi A}}\right)
\tag{13.2}
$$

is abridged as

$$
\frac{\partial q}{\partial t} + \frac{\partial f(q)}{\partial x} = \phi
\tag{13.3}
$$

where the independent variable (x, t) varies in the domain $D \equiv \Omega \times [t_o, T]$, $\Omega \equiv [x_{\text{in}}, x_{\text{out}}]$. This system consists of the continuity, linear-momentum and total-energy equations, with arrays of dependent variables $q = q(x, t)$, $f = f(q)$ and $\phi = \phi(x, q)$ defined as

$$
q \equiv \left\{ \begin{array}{c} \rho \\ m \\ E \end{array} \right\}, \quad
f(q) \equiv \left\{ \begin{array}{c} m \\ \dfrac{m^2}{\rho} + p \\ \dfrac{m}{\rho}(E + p) \end{array} \right\}
$$

$$
\phi \equiv -\dfrac{m}{\rho A}\dfrac{\partial A}{\partial x} \left\{ \begin{array}{c} \rho \\ m \\ E + p \end{array} \right\} + \dfrac{\widetilde{\dot{m}}_b}{A} \left\{ \begin{array}{c} 1 \\ \dfrac{\dot{m}_b}{b\rho_b}\sin\theta \\ \left.\dfrac{E + p}{\rho}\right|_b \end{array} \right\}
$$

$$
- \left\{ \begin{array}{c} 0 \\ f_D \dfrac{m|m|}{2\rho D \sqrt{A_*/L}} - \rho g_x + s\chi\, \rho w_m - \dfrac{b(\Delta p)_{b_x}}{A} \\ |m|(w_m - g_{\text{T}}) - mg_x + |\dot{m}_b|w_\tau/b \end{array} \right\} \tag{13.4}
$$

For non-trivial solutions "q" and general equilibrium pressure equations of state, the characteristic speeds associated with the Euler equations are the eigenvalues of the flux vector Jacobian

$$
\dfrac{\partial f(q)}{\partial q} = \left(\begin{array}{ccc} 0 & , \quad 1 & , \quad 0 \\ -\dfrac{m^2}{\rho^2} + p_\rho & , \quad \dfrac{2m}{\rho} + p_m & , \quad p_E \\ -\dfrac{m}{\rho^2}(E + p) + \dfrac{m}{\rho}p_\rho & , \quad \dfrac{E + p}{\rho} + \dfrac{m}{\rho}p_m & , \quad \dfrac{m}{\rho}(1 + p_E) \end{array} \right) \tag{13.5}
$$

These eigenvalues have been exactly determined in closed form as

$$
\lambda_1 = u, \qquad \lambda_{2,3} = u \pm \left(p_\rho + p_E \left(\dfrac{E + p}{\rho} - \dfrac{m^2}{\rho^2} \right) \right)^{1/2} \tag{13.6}
$$

In the absence of viscosity, shocks, and heat, mass, and work transfer, the flow remains isentropic, the governing equations become the Euler equations, and pressure only depends upon density as $p = p(\rho)$. For this reason, the eigenvalues $\lambda_{2,3}$ directly incorporate a sound speed expression that coincides with the isentropic partial derivative of pressure. These equilibrium-gas eigenvalues thus become

$$
\lambda_1^E = u, \qquad \lambda_{2,3}^E = u \pm c \tag{13.7}
$$

which have the same familiar form as the perfect-gas eigenvalues. The corresponding non-dimensional form is

$$
\lambda_1^{EM} = M, \qquad \lambda_{2,3}^{EM} = M \pm 1, \qquad M \equiv \dfrac{|u|}{c} \tag{13.8}
$$

where M denotes the Mach number. Figure 13.1 portrays the variation of these eigenvalues and shows $\lambda_3^{E_M}$ remains negative for subsonic flows, $M < 1$, vanishes for sonic flows, $M = 1$, and becomes positive for supersonic flows, $M > 1$. This change in sign affects the direction of wave propagation depending on M.

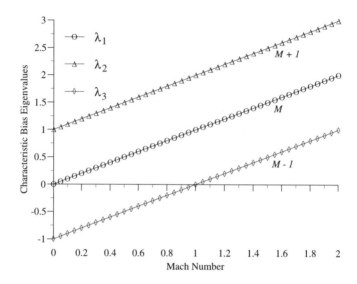

Figure 13.1: Compressible-Flow Euler Eigenvalues

These eigenvalues correspond to the slopes of the characteristics, as portrayed in Figure 13.2 for representative hypersonic, supersonic, sonic, and subsonic flows. Similar to the corresponding figure for free-surface flows, this figure shows the characteristics in a neighborhood of a flow field point P in a (t, x) plane. Also for compressible flows, an interesting geometric difference among supersonic, sonic, and subsonic flows is that a time axis through P is respectively outside, on the boundary, and inside the domain of dependence and range of influence of point P. Wave propagation for supersonic flows essentially occurs by convection, mono-axially from upstream to downstream of P; the sonic case becomes a limiting case; for subsonic flows, instead, wave propagation occurs by both convection and acoustics, bi- modally from both upstream and downstream toward and then away from P; for vanishing Mach number, wave propagation is essentially acoustic. Since gas dynamic wave propagation physically occurs by acoustics and convection, the characteristics-bias formulation in the following sections models this coupled acoustic-convection wave propagation. The formulation identifies the genuine convection and acoustics components within the flux Jacobian and then establishes a physically consistent upstream resolution for each of these components. The following section identifies and analyzes the acoustics equations within the Euler system.

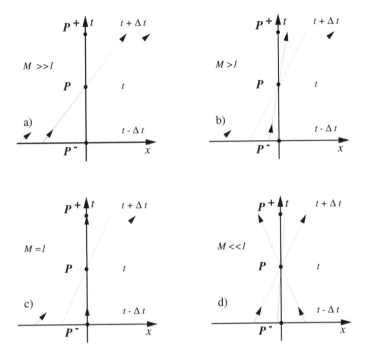

Figure 13.2: Characteristics: a) Hypersonic, b) Supersonic, c) Sonic, d) Subsonic

13.2   Acoustics Equations

The Euler system

$$\frac{\partial q}{\partial t} - \phi + \frac{\partial f(q)}{\partial x} = 0 \tag{13.9}$$

contains the acoustics equations for vanishing Mach numbers. Identification of these equations yields the acoustics component of the Euler flux Jacobian for any Mach number. Upon writing momentum m in terms of the Mach number M as $m = \rho c M u/|u|$ and using the pressure-derivative identities (2.22), the Euler system becomes

$$\frac{\partial \rho}{\partial t} + \frac{\partial m}{\partial x} + cM \frac{u}{|u|} \frac{\rho}{A} \frac{dA}{dx} = 0$$

$$\frac{\partial m}{\partial t} + p_\rho \frac{\partial \rho}{\partial x} + p_E \frac{\partial E}{\partial x} + cM \frac{u}{|u|} \left[(2 - p_E) \frac{\partial m}{\partial x} - \frac{m}{\rho} \frac{\partial \rho}{\partial x} + \frac{m}{A} \frac{dA}{dx} \right] = 0$$

$$\frac{\partial E}{\partial t} + \frac{E+p}{\rho} \frac{\partial m}{\partial x} + cM \frac{u}{|u|} \left[\rho \frac{\partial}{\partial x} \left(\frac{E+p}{\rho} \right) + \frac{E+p}{A} \frac{dA}{dx} \right] = 0 \tag{13.10}$$

and for a vanishing Mach number, these equations reduce to the acoustics system

$$
\frac{\partial}{\partial t}\begin{pmatrix}\rho\\m\\E\end{pmatrix}+\begin{pmatrix}0 & , & 1 & , & 0\\p_\rho & , & 0 & , & p_E\\0 & , & \dfrac{E+p}{\rho} & , & 0\end{pmatrix}\frac{\partial}{\partial x}\begin{pmatrix}\rho\\m\\E\end{pmatrix}=0 \quad\Rightarrow\quad \frac{\partial q}{\partial t}+A^a\frac{\partial q}{\partial x}=0 \tag{13.11}
$$

Heed that the energy equation toward steady state, in this case, is no longer linearly independent from the continuity equation. This phenomenon provides another explanation for the widely reported convergence difficulties experienced in the CFD simulation of incompressible, i.e. low-Mach-number, flows with a compressible flow formulation. The dependent variables in these equations correspond to those in a flow field that originates from slight perturbations to an otherwise quiescent field.

By virtue of the total enthalpy expression (2.31), the matrix A^a becomes

$$
A^{a0}=\begin{pmatrix}0 & , & 1 & , & 0\\p_\rho & , & 0 & , & p_E\\0 & , & \dfrac{c^2-p_\rho}{p_E} & , & 0\end{pmatrix} \tag{13.12}
$$

with eigenvalues

$$
\lambda_1^a=0,\qquad \lambda_{2,3}^a=\pm c \tag{13.13}
$$

where c corresponds to the isentropic speed of sound (2.30). This matrix, accordingly, can be termed the "acoustics" matrix.

Despite its zero eigenvalue, A^a features a complete set of eigenvectors and thus possesses the similarity form

$$
A^a\equiv X\Lambda^a X^{-1}=X\Lambda^{a+}X^{-1}+X\Lambda^{a-}X^{-1},\quad \Lambda^a=\Lambda^{a+}+\Lambda^{a-} \tag{13.14}
$$

where the eigenvalue matrix Λ^a is expressed as

$$
\Lambda^a\equiv\begin{pmatrix}c & , & 0 & , & 0\\0 & , & -c & , & 0\\0 & , & 0 & , & 0\end{pmatrix} \tag{13.15}
$$

and the diagonal matrices Λ^{a+} and Λ^{a-} respectively contain non-negative and non-positive eigenvalues. The eigenvector matrices X and X^{-1} are expressed as

$$
X=\begin{pmatrix}p_E & , & p_E & , & p_E\\cp_E & , & -cp_E & , & 0\\c^2-p_\rho & , & c^2-p_\rho & , & -p_\rho\end{pmatrix},\quad X^{-1}=\frac{1}{2c^2 p_E}\begin{pmatrix}p_\rho & , & c & , & p_E\\p_\rho & , & -c & , & p_E\\2\left(c^2-p_\rho\right) & , & 0 & , & -2p_E\end{pmatrix} \tag{13.16}
$$

This similarity transformation is employed in the acoustics flux-Jacobian decomposition, as shown in the next section.

13.3 Acoustics-Convection Flux Jacobian Decomposition

The characteristics-bias formulation for the Euler system

$$\frac{\partial q}{\partial t} - \phi + \frac{\partial f(q)}{\partial x} = 0 \tag{13.17}$$

emerges from a decomposition of the flux-vector Jacobian $\partial f/\partial q$ into acoustic and convection components. The following sections detail the formation of these components.

13.3.1 Convection and Pressure-Gradient Components

The flux divergence $\frac{\partial f}{\partial x}$ can be decomposed into convection and pressure-gradient components as

$$\frac{\partial f}{\partial x} = \frac{\partial f^q}{\partial x} + \frac{\partial f^p}{\partial x} \tag{13.18}$$

where f^q and f^p respectively denote the convection and pressure fluxes, defined as

$$f^q(q) \equiv \left\{ \begin{array}{c} m \\ \dfrac{m^2}{\rho} \\ \dfrac{m}{\rho}(E+p) \end{array} \right\} = \frac{m}{\rho} \cdot \left\{ \begin{array}{c} \rho \\ m \\ E+p \end{array} \right\}, \quad f^p \equiv \left\{ \begin{array}{c} 0 \\ p \\ 0 \end{array} \right\} \tag{13.19}$$

The non-dimensional eigenvalues of the flux vector Jacobian $\partial f/\partial q = \partial f^q/\partial q + \partial f^p/\partial q$ are

$$\lambda_1 = M, \quad \lambda_{2,3} = M \pm 1 \tag{13.20}$$

For supersonic flows, the eigenvalues (13.20) all have the same algebraic sign and the entire flux divergence can be upstream approximated along one single direction. For subsonic flows, these eigenvalues have mixed algebraic sign and an upstream approximation for the flux divergence along one single direction remains inconsistent with the two-way propagation of acoustic waves. Without the pressure gradient in the momentum equation, however, the corresponding flux-Jacobian eigenvalues all have the same algebraic sign and the resulting convection flux divergence can then be upstream approximated along one single direction. The flux divergence can thus be decomposed as the linear combination

$$\frac{\partial f}{\partial x} = \left[\frac{\partial f^q}{\partial x} + \beta \frac{\partial f^p}{\partial x} \right] + \left[(1-\beta) \frac{\partial f^p}{\partial x} \right], \quad 0 \le \beta \le 1 \tag{13.21}$$

where the positive pressure-gradient partition function β can be chosen in such a way that all the eigenvalues of each of the two components between brackets in (13.21) keep the same algebraic sign for all Mach numbers. In this manner, these entire components can be upstream approximated along single directions. This choice for β is possible because the eigenvalues of a matrix are continuous functions of the matrix entries and hence all the eigenvalues for the components in (13.21) will continuously depend upon β. The function β will gradually increase toward 1 for increasing Mach number, so that an upstream approximation for the components in (13.21) smoothly approaches and then becomes an upstream approximation for the entire $\frac{\partial f}{\partial x}$ along one single direction. Decomposition (13.21) is thus used for an upstream approximation of the flux divergence for subsonic and supersonic flows.

For low and vanishing Mach numbers, however, decomposition (13.21) is insufficient for an accurate upstream modeling of acoustic waves. For a Mach number that approaches zero, the Euler eigenvalues (13.20) can all keep the same algebraic sign only if the sound speed contribution vanishes, which corresponds to a vanishing pressure gradient contribution and hence β approaching zero. But for β approaching zero, the eigenvalues associated with the components in (13.21) approach the eigenvalues of the Jacobians

$$\frac{\partial f^q(q)}{\partial q} = \begin{pmatrix} 0 & , & 1 & , & 0 \\ -\dfrac{m^2}{\rho^2} & , & \dfrac{2m}{\rho} & , & 0 \\ -\dfrac{m}{\rho^2}(E+p) + \dfrac{m}{\rho}p_\rho & , & \dfrac{E+p}{\rho} + \dfrac{m}{\rho}p_m & , & \dfrac{m}{\rho}(1+p_E) \end{pmatrix} \tag{13.22}$$

and

$$\frac{\partial f^p}{\partial q} = \begin{pmatrix} 0 & , & 0 & , & 0 \\ p_\rho & , & p_m & , & p_E \\ 0 & , & 0 & , & 0 \end{pmatrix} \tag{13.23}$$

Using the pressure-derivative identity (2.22) the eigenvalues of these Jacobians respectively are

$$\lambda^q_{1,2} = \frac{m}{\rho}, \quad \lambda^q_3 = \frac{m}{\rho}(1+p_E) \tag{13.24}$$

and

$$\lambda^p_{1,2} = 0, \quad \lambda^p_3 = p_m = -\frac{m}{\rho}p_E \tag{13.25}$$

which certainly all keep the same algebraic sign, but for vanishing Mach number remain far less than the dominant speed of sound c. For low Mach numbers, therefore, an upstream approximation for the components in (13.21) would inaccurately model the physics of acoustics. This difficulty is resolved by further decomposing the pressure gradient in (13.21) in terms of a genuine acoustic component, for accurate upstream modeling of acoustic waves.

13.3.2 Acoustic Components

Based on the acoustics equations (13.11), the flux divergence $\frac{\partial f}{\partial x}$ can be alternatively decomposed for arbitrary Mach numbers and corresponding dependent variables ρ, m, and E as

$$\frac{\partial f}{\partial x} = \frac{\partial f^q}{\partial x} + \frac{\partial f^p}{\partial x} = \frac{\partial f^q}{\partial x} + (A^a + A^{nc})\frac{\partial q}{\partial x} \tag{13.26}$$

In this decomposition, the matrices A^a and A^{nc} are defined as

$$A^a \equiv \begin{pmatrix} 0 & , & 1 & , & 0 \\ p_\rho & , & 0 & , & p_E \\ 0 & , & \dfrac{c^2 - p_\rho}{p_E} & , & 0 \end{pmatrix}, \quad A^{nc} \equiv \begin{pmatrix} 0 & , & -1 & , & 0 \\ 0 & , & p_m & , & 0 \\ 0 & , & -\dfrac{c^2 - p_\rho}{p_E} & , & 0 \end{pmatrix} \tag{13.27}$$

Heed, in particular, that no flux component of $f(q)$ exists, of which the Jacobian equals A^a. The eigenvalues of the matrix A^{nc} have been determined in closed form as

$$\lambda^{nc}_{1,2} = 0, \qquad \lambda^{nc}_3 = -cMp_E u/|u| \tag{13.28}$$

which become infinitesimal for vanishing M. The matrix A^{nc} can be termed a "non-linear coupling" matrix, for it completes the non-linear coupling between convection and acoustics within (13.26) so that the two Euler eigenvalues $\lambda^E_{2,3}$ in (13.7) do correspond to the sum of convection and acoustic speeds. Since decomposition (13.26) will be used in the upstream-bias formulation for small Mach numbers only and considering that the eigenvalues in (13.28) vanish for these Mach numbers, no need exists to involve A^{nc} in the upstream-bias approximation of the Euler flux Jacobian. The eigenvalues of A^a are exactly determined in closed form as

$$\lambda^a_1 = 0, \qquad \lambda^a_{2,3} = \pm c \tag{13.29}$$

This matrix can thus be used to develop an acoustics upstream bias. With reference to the similarity transformation (13.14) for this matrix, the Euler flux divergence decomposition (13.26) becomes

$$\frac{\partial f}{\partial q} = X\Lambda^{a+}X^{-1} + X\Lambda^{a-}X^{-1} + \frac{\partial f^q}{\partial q} + A^{nc} \tag{13.30}$$

which exposes the acoustics and convection components. Since the two acoustics matrices at the rhs of this expression respectively possess non-negative and non-positive eigenvalues, a characteristics-bias representation of these matrices involves an upstream resolution of the first matrix and a downstream resolution of the second. These representations naturally lead to the following absolute acoustics-matrix expression

$$|A^a|\frac{\partial q}{\partial x} \equiv X\left(\Lambda^{a+} - \Lambda^{a-}\right)X^{-1}\frac{\partial q}{\partial x} \tag{13.31}$$

which depends on the selection of the diagonal matrices Λ^{a+} and Λ^{a-}, with different forms inducing distinct levels of upstream diffusion. With the following choices

$$\Lambda^{a+} \equiv \frac{1}{2}\begin{pmatrix} 2c & , & 0 & , & 0 \\ 0 & , & 0 & , & 0 \\ 0 & , & 0 & , & c \end{pmatrix}, \qquad \Lambda^{a-} \equiv -\frac{1}{2}\begin{pmatrix} 0 & , & 0 & , & 0 \\ 0 & , & 2c & , & 0 \\ 0 & , & 0 & , & c \end{pmatrix} \tag{13.32}$$

the associated absolute acoustic-matrix statement becomes

$$|A^a|\frac{\partial q}{\partial x} \equiv \left(X\Lambda^{a+}X^{-1} - X\Lambda^{a-}X^{-1}\right)\frac{\partial q}{\partial x} = X\left(\Lambda^{a+} - \Lambda^{a-}\right)X^{-1}\frac{\partial q}{\partial x} = cI\frac{\partial q}{\partial x} \tag{13.33}$$

which shows that $|A^a|$ equals cI, of which all eigenvalues approach $+c$. This computationally advantageous form of the acoustics upstream-bias is used essentially in the low subsonic-flow Mach number regime.

13.3.3   Combination of Flux-Jacobian Decompositions

The previous sections have shown that the Euler flux Jacobian can be equivalently expressed as

$$
\frac{\partial f}{\partial q} =
\begin{cases}
\left[\dfrac{\partial f^q}{\partial q} + \beta \dfrac{\partial f^p}{\partial q} \right] + \left[(1 - \beta) \dfrac{\partial f^p}{\partial q} \right]; \\[2ex]
X \Lambda^{a+} X^{-1} + X \Lambda^{a-} X^{-1} + \dfrac{\partial f^q}{\partial q} + A^{nc}
\end{cases}
\tag{13.34}
$$

where the first expression is convenient for a characteristics-bias approximation for high-subsonic and supersonic Mach numbers and the second expression is needed for low-subsonic Mach numbers.

A decomposition for all Mach numbers may thus be cast as a linear combination of these two Euler decompositions, with linear combination function α, $0 \le \alpha \le 1$

$$
\frac{\partial f}{\partial q} = (1 - \alpha) \left\{ \left[\frac{\partial f^q}{\partial q} + \beta \frac{\partial f^p}{\partial q} \right] + \left[(1 - \beta) \frac{\partial f^p}{\partial q} \right] \right\}
$$

$$
+ \alpha \left\{ X \Lambda^{a+} X^{-1} + X \Lambda^{a-} X^{-1} + \frac{\partial f^q}{\partial q} + A^{nc} \right\}
\tag{13.35}
$$

Owing to the simplifying function $\delta \equiv \beta(1 - \alpha)$, the final form of the acoustics-convection Euler flux-divergence decomposition becomes

$$
\frac{\partial f}{\partial x} = \alpha \left(X \Lambda^{a+} X^{-1} + X \Lambda^{a-} X^{-1} \right) \frac{\partial q}{\partial x} + \left[\frac{\partial f^q}{\partial q} + \delta \frac{\partial f^p}{\partial x} \right] + (1 - \alpha - \delta) \frac{\partial f^p}{\partial x} + \alpha A^{nc} \frac{\partial q}{\partial x}
\tag{13.36}
$$

This decomposition leads to a consistent upstream-bias representation due to the physical significance of its terms. The streamline acoustics term $\alpha \left(X \Lambda^{a+} X^{-1} + X \Lambda^{a-} X^{-1} \right) \frac{\partial q}{\partial x}$ accounts for the bi-modal streamline propagation of acoustic waves; this term is thus employed for an acoustics upstream - bias approximation along the streamline, for low Mach numbers. Since this term already models pressure-induced acoustic-wave propagation along each streamline, no upstream-bias is needed for $\alpha A^{nc} \frac{\partial q}{\partial x}$, for an upstream of this term would essentially add only algebraic complexity to the formulation, since its coefficient α rapidly decreases and then vanishes for supersonic and high subsonic Mach numbers and two eigenvalues of the Jacobian of this term vanish, with the third becoming negligible when α is significant. By the same token, there is no need to develop an additional upstream bias for the $(1 - \alpha - \delta) \frac{\partial f^p}{\partial x}$, term also considering that the coefficient $(1 - \alpha - \delta)$ vanishes for acoustic and supersonic flows, two eigenvalues of the Jacobian of this term identically vanish, and the third becomes negligible for vanishing Mach number. Besides, as the Mach number rises, an increasing fraction of the pressure gradient receives an upstream bias in the term $\left[\frac{\partial f^q}{\partial x_j} + \delta \frac{\partial f^p}{\partial x_j} \right]$. This expression is counted as one term because the eigenvalues associated with this term will all keep the same sign, since $\delta = (1 - \alpha)\beta \le \beta$. For any magnitude of both pressure and pressure gradient, the convection field uniformly carries information along the streamline; accordingly, this entire term can receive an upstream bias along the streamline direction.

13.4 Characteristics-Bias Flux-Divergence

Given the algebraic sign of the eigenvalue set of each matrix term in (13.36), the associated direction cosines a_ℓ for the upstream-bias expression (9.71) are

$$a_1 = +1, \quad a_2 = -1, \quad a_3 = s = \operatorname{sgn}(u), \quad a_4 = a_5 = 0 \tag{13.37}$$

where $s = \operatorname{sgn}(u)$ denotes the algebraic sign of u. With (13.36) and (13.37) the general upstream-bias expression (9.71) leads to the acoustics-convection characteristics-bias flux divergence

$$\frac{\partial f^C}{\partial x} = \frac{\partial f}{\partial x} - \frac{\partial}{\partial x}\left[\varepsilon\psi\left(\alpha c\frac{\partial q}{\partial x} + s\frac{\partial f^q}{\partial x} + s\delta\frac{\partial f^p}{\partial x}\right)\right] \tag{13.38}$$

The operation count for expression (13.38) is then comparable to that of an FVS formulation. The terms in this expression, furthermore, directly correspond to the physics of acoustics and convection. This expression also determines, up to an arbitrary constant, the non-discrete characteristics-bias flux

$$f^C = f - \varepsilon\psi\left(\alpha c\frac{\partial q}{\partial x} + s\frac{\partial f^q}{\partial x} + s\delta\frac{\partial f^p}{\partial x}\right) \tag{13.39}$$

Heed that the components within f^C remain linearly independent of one another, which avoids the linear-dependence instability in the steady low-Mach-number Euler equations.

For low Mach numbers, $\delta = 0$ and (13.38) reduces to

$$\frac{\partial f^C}{\partial x} = \frac{\partial f}{\partial x} - \frac{\partial}{\partial x}\left[\varepsilon\psi\left(\alpha c\frac{\partial q}{\partial x} + s\frac{\partial f^q}{\partial x}\right)\right] \tag{13.40}$$

which essentially induces only an acoustics upstream. For supersonic flow, $\alpha = 0$ and $\delta = 1$. Expression (13.38) in this case becomes

$$\frac{\partial f^C}{\partial x} = \frac{\partial f}{\partial x} - \frac{\partial}{\partial x}\left[\varepsilon\psi\left(s\frac{\partial f}{\partial x}\right)\right] \tag{13.41}$$

which corresponds to an upstream approximation of the entire Euler flux divergence.

13.4.1 Consistent Upstream-Bias

The acoustics-convection characteristics-bias functions α and δ depend on the Mach number. They are determined by enforcing the upstream stability condition on the characteristics-bias perturbation matrix for (13.38). The divergence of the characteristics flux f^C in (13.38) becomes

$$\frac{\partial f^C}{\partial x} = \frac{\partial f}{\partial x} + \frac{\partial}{\partial x}\left[\varepsilon\psi\left(\alpha c I + s\frac{\partial f^q}{\partial q} + s\delta\frac{\partial f^p}{\partial q}\right)\frac{\partial q}{\partial x}\right] \tag{13.42}$$

where I denotes the identity matrix of appropriate size. The terms between parentheses collectively constitute the upstream- bias dissipation matrix

$$\mathcal{A} \equiv \alpha c I + s\frac{\partial f^q}{\partial q} + s\delta\frac{\partial f^p}{\partial q} \tag{13.43}$$

Despite the formidable algebraic complexity of $\mathcal{A}$, all of its eigenvalues have been analytically determined exactly in closed form, not only for perfect, but also equilibrium gases. Dividing through by the speed of sound c, the non-dimensional form of these eigenvalues is

$$\lambda_1 = \alpha + M, \quad \lambda_{2,3} = \alpha + \left(1 + \frac{1-\delta}{2}p_E\right)M \pm \left(\left(\frac{1-\delta}{2}p_E M\right)^2 + \delta\right)^{1/2} \tag{13.44}$$

In order to ensure physical significance for the characteristics- bias flux within (13.38), hence for the upstream-bias approximation to decomposition (13.36), the upstream-bias functions α and δ are determined by forcing these eigenvalues to remain positive for all Mach numbers. Rather than prescribing some expressions for α and δ and accepting the resulting variations for these eigenvalues, physically reasonable expressions for these eigenvalues are instead prescribed and the corresponding functions for α and δ determined.

13.4.2   Upstream-Bias Eigenvalues and Functions α and δ

The prescribed eigenvalues λ_1 and λ_3 respectively correlate with M and $|M-1|$. These eigenvalues directly lead to α and δ.

The eigenvalue λ_1 correlates with M, but must remain distinct from $|M-1|$, for eigenvalue separation, and approach 1 for vanishing M, because all the non-dimensional eigenvalues of the absolute acoustics matrix equal 1. With $\varepsilon_M = \frac{1}{4}$, in this work, one expression for λ_1 that meets this requirements is the composite spline

$$\lambda_1(M) \equiv \begin{cases} 1 - M + \dfrac{\varepsilon_M}{2}(2M)^{1/\varepsilon_M} & , \quad 0 \;\le\; M \;<\; \dfrac{1}{2} \\[2ex] \dfrac{(M-\frac{1}{2})^2}{2\varepsilon_M} + \dfrac{1+\varepsilon_M}{2} & , \quad \dfrac{1}{2} \;\le\; M \;<\; \dfrac{1}{2}+\varepsilon_M \\[2ex] M & , \quad \dfrac{1}{2}+\varepsilon_M \;\le\; M \end{cases} \tag{13.45}$$

The eigenvalue λ_3 correlates with the absolute-value Euler eigenvalue $|M-1|$. As a consequence, λ_3 will vary between 1 and $1-M$ for $0 \le M \le 1-\varepsilon_M$ and smoothly shift from $1-M$ to $M-1$ within the sonic transition layer $1-\varepsilon_M \le M \le 1+\varepsilon_M$, where ε_M denotes a transition-layer parameter; again, $\varepsilon_M = \frac{1}{4}$ in this work. One expression for λ_3 that remains smooth and meets these requirements is the composite spline

$$\lambda_2(M) \equiv \begin{cases} 1 - M & , \quad 0 \;\le\; M \;\le\; 1-\varepsilon_M \\[2ex] \dfrac{(M-1)^2}{2\varepsilon_M} + \dfrac{\varepsilon_M}{2} & , \quad 1-\varepsilon_M \;<\; M \;<\; 1+\varepsilon_M \\[2ex] M - 1 & , \quad 1+\varepsilon_M \;\le\; M \end{cases} \tag{13.46}$$

which is graphed in Figure 13.3. Observe in particular the smooth transition in the critical region within the transition layer, where this λ_3 does not vanish, but remains not less than $\varepsilon_M/2$.

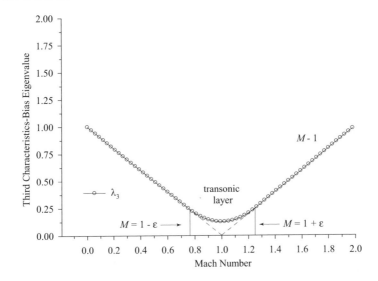

Figure 13.3: Characteristics-Bias Eigenvalue λ_3

With these λ_1 and λ_3, the corresponding expressions for both $\alpha = \alpha(M)$ and $\delta = \delta(M)$ from (13.44) have been exactly determined as

$$\alpha(M) = \lambda_1(M) - M, \quad \delta(M) = \frac{(\lambda_1(M) - \lambda_2(M))(\lambda_1(M) - \lambda_2(M) + p_E M)}{1 + p_E M (\lambda_1(M) - \lambda_2(M))} \quad (13.47)$$

where subscript "E" denotes partial differentiation with respect to this variable. The variations of the upstream-bias functions $\alpha = \alpha(M)$ and $\delta = \delta(M)$ and the corresponding eigenvalues from (13.44) are respectively presented in Figures 13.4-13.6.

Figure 13.4 indicates that the upstream-bias functions as well as their slopes remain continuous for all Mach numbers, with $0 \leq \alpha, \delta \leq 1$ and $\alpha \equiv 0$, for $M > \frac{1}{2} + \varepsilon_M$, as well as $\delta \equiv 1$ for $M > 1 + \varepsilon_M$. As $\delta = \delta(M)$ rises, the upstream-bias contribution from the acoustics matrix decreases rapidly, reducing to less than 25% of its maximum at $M = 0.39$. The variation of $\delta = \delta(M)$ shows that the pressure-gradient contribution to this upstream-bias formulation increases monotonically, while remaining less than 25% of its maximum, for $0 \leq M \leq 0.7$. When $\delta(M) \equiv 1$ for supersonic Mach numbers, the entire pressure-gradient is upstreamed with the same weight as in the convection flux, in complete agreement with the physical mono-axial wave propagation within supersonic flows. Figure 13.5 presents the eigenvalues $\lambda_{1,3}^{cp}$ of the convection / pressure-gradient Jacobian $\left[\frac{\partial f^q}{\partial q} + \delta \frac{\partial f^p}{\partial q}\right]$. Obtained from (13.44) as $\lambda_{1,3}^{cp} = \lambda_{1,3} - \alpha$, these eigenvalues, as anticipated, remain positive, which justifies treating this Jacobian as one single term in the characteristics-bias divergence. For vanishing Mach numbers, δ approaches zero and the contribution to this Jacobian from pressure is virtually eliminated. As the Mach number, hence δ, increases these eigenvalues separate and smoothly approach the Euler eigenvalues M, $M \pm 1$ for supersonic flows.

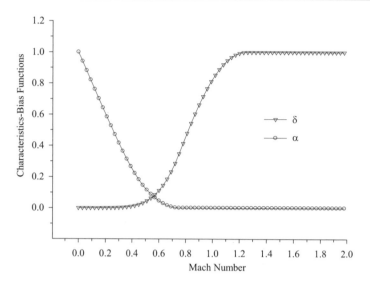

Figure 13.4: Upstream-Bias Functions

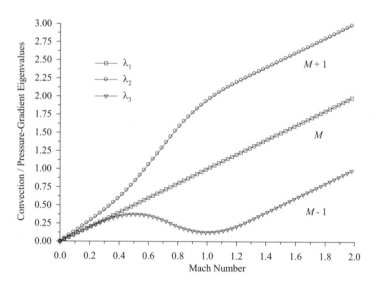

Figure 13.5: Convection / Pressure-Gradient Eigenvalues

These eigenvalues do not approach $|M \pm 1|$ for vanishing M, which is the reason why the acoustics upstream-bias, hence the α term, is developed. The complete characteristics-

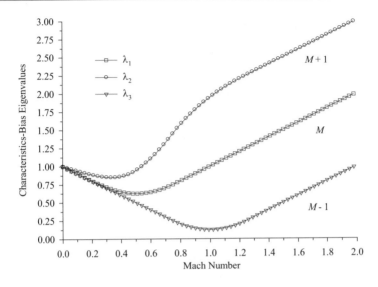

Figure 13.6: Characteristics-Bias Eigenvalues

bias streamline eigenvalues (13.44) are shown in Figure 13.6. These eigenvalues smoothly approach 1 for vanishing M, indicating a physically consistent upstream-bias approximation of the acoustic equations embedded within the Euler equations. For subsonic flows, λ_1 remains greater than M, a feature that is compensated by a λ_2 that is less than $M + 1$ in the same flow regime. These eigenvalues remain positive as well as smooth, and their slopes remain continuous for all Mach numbers. For $M > 1+\varepsilon_M$, the eigenvalues (13.44) respectively coincide with the Euler flux Jacobian eigenvalues M, $M + 1$, $M - 1$, which corresponds to an upstream-bias approximation of the entire flux vector, for supersonic flows.

13.5 Variational Statement and Boundary Conditions

With ϕ_ρ, ϕ_m, ϕ_E denoting the components of the source term ϕ, respectively in the continuity, linear-momentum, and energy equation, the developments in Chapter 9 and in the previous sections lead to the following characteristics-bias integral equations for modeling generalized compressible flows

$$\int_\Omega w \left(\frac{\partial \rho}{\partial t} - \phi_\rho + \frac{\partial m}{\partial x} \right) d\Omega - \int_\Omega w \frac{\partial}{\partial x} \left(\varepsilon \psi s \left(\frac{\partial \rho}{\partial t} - \phi_\rho + \frac{\partial m}{\partial x} + \alpha c s \frac{\partial \rho}{\partial x} \right) \right) d\Omega = 0 \quad (13.48)$$

$$\int_\Omega w \left(\frac{\partial m}{\partial t} - \phi_m + \frac{\partial}{\partial x} \left(\frac{m^2}{\rho} + p \right) \right) d\Omega$$

$$- \int_\Omega w \frac{\partial}{\partial x} \left(\varepsilon \psi s \left(\frac{\partial m}{\partial t} - \phi_m + \frac{\partial}{\partial x} \left(\frac{m^2}{\rho} + p \right) + \alpha c s \frac{\partial m}{\partial x} + (\delta - 1) \frac{\partial p}{\partial x} \right) \right) d\Omega = 0 \quad (13.49)$$

$$\int_\Omega w \left(\frac{\partial E}{\partial t} - \phi_E + \frac{\partial}{\partial x} \left(\frac{m}{\rho} (E + p) \right) \right) d\Omega$$

$$- \int_\Omega w \frac{\partial}{\partial x} \left(\varepsilon \psi s \left(\frac{\partial E}{\partial t} - \phi_E + \frac{\partial}{\partial x} \left(\frac{m}{\rho} (E + p) \right) + \alpha c s \frac{\partial E}{\partial x} \right) \right) d\Omega = 0 \tag{13.50}$$

These equations are expressed as the single system

$$\int_\Omega w \left\{ \left(\frac{\partial q}{\partial t} - \phi + \frac{\partial f}{\partial x} \right) - \frac{\partial}{\partial x} \left[\varepsilon \psi s \left(\frac{\partial q}{\partial t} - \phi + \frac{\partial f}{\partial x} \right) \right] \right\} d\Omega$$

$$- \int_\Omega w \frac{\partial}{\partial x} \left[\varepsilon \psi s \left(\alpha c s \frac{\partial q}{\partial x} + (\delta - 1) \frac{\partial f^p}{\partial x} \right) \right] d\Omega = 0 \tag{13.51}$$

An integration by parts of the characteristics-bias perturbation in this statement generates the following weak statement

$$\int_\Omega \left(w + \varepsilon \psi s \frac{\partial w}{\partial x} \right) \left(\frac{\partial q}{\partial t} - \phi + \frac{\partial f}{\partial x} \right) d\Omega + \int_\Omega \varepsilon \psi s \frac{\partial w}{\partial x} \left(\alpha c s \frac{\partial q}{\partial x} + (\delta - 1) \frac{\partial f^p}{\partial x} \right) d\Omega$$

$$- \left(w s \varepsilon \psi \left(\frac{\partial q}{\partial t} - \phi + \frac{\partial f}{\partial x} + \alpha c s \frac{\partial q}{\partial x} + (\delta - 1) \frac{\partial f^p}{\partial x} \right) \right) \Big|_{x=x_{\text{in}}}^{x=x_{\text{out}}} = 0 \tag{13.52}$$

The $\varepsilon \psi$ boundary expression is eliminated by enforcing a weak Neumann-type boundary condition for the hyperbolic-parabolic characteristics-bias perturbation. One part of this condition imposes that the original Euler system should be satisfied at the boundary; the remaining terms are also set to zero as part of this weak boundary condition. Clearly, the single Neumann-type condition corresponds to the entire boundary expression set to nought. For supersonic flows with $M \geq 1 + \varepsilon_M$, $\alpha = 0$ and $\delta = 1$ and this boundary condition, accordingly, enforces the governing system on the boundary. The resulting weak statement is expressed as

$$\int_\Omega \left(w + \varepsilon \psi s \frac{\partial w}{\partial x} \right) \left(\frac{\partial q}{\partial t} - \phi + \frac{\partial f}{\partial x} \right) d\Omega + \int_\Omega \varepsilon \psi \frac{\partial w}{\partial x} \left(c \alpha \frac{\partial q}{\partial x} + s (\delta - 1) \frac{\partial f^p}{\partial x} \right) d\Omega = 0 \tag{13.53}$$

which corresponds to a generalized non-discrete SUPG statement. Unlike the common SUPG formulation, however, this statement does not require a premultiplication of the governing system by the transpose of the flux-vector Jacobian $\frac{\partial f}{\partial q}$; additionally, the upstream-bias functions $\varepsilon \psi \alpha$ and $\varepsilon \psi \delta$ require no matrix norms, but directly originate from the upstream stability condition.

For all test functions $w \in \mathcal{H}^1(\Omega)$, the variational formulation seeks a solution $q \in \mathcal{H}^\eta(\Omega)$ with $\eta = 0$ or $\eta = 1$ respectively for shocked or smooth flows, that satisfies this weak statement. This statement is subject to prescribed initial conditions $q(x, 0) = q_0(x)$ and boundary conditions on $\partial \Omega \equiv \overline{\Omega} \backslash \Omega$. Synthetically, these boundary conditions are expressed as

$$B(x_{\partial \Omega}) q(x_{\partial \Omega}, t) = G_v(x_{\partial \Omega}, t) \tag{13.54}$$

where $G_v(x_{\partial \Omega}, t)$ corresponds to the array of prescribed Dirichlet boundary conditions, with a zero entry for each corresponding unconstrained component of q, and $B(x_{\partial \Omega})$ denotes a square diagonal matrix, with a 1 for each diagonal entry, but replaced by zero for each corresponding unconstrained component of q.

The number of boundary conditions at inlet and outlet depends on the number of negative eigenvalues of $\frac{\partial f}{\partial q} n_x$, where n_x denotes the x-component of the outward pointing normal unit vector $\mathbf{n}$. Again, a negative eigenvalue signifies a wave propagation in the direction opposite $\mathbf{n}$, hence toward the flow domain, a propagation that carries information from the flow outside the domain, which information in embodied in a corresponding boundary condition.

At an inlet, $\frac{\partial f}{\partial q} n_x$ displays two negative eigenvalues; density and total energy are thus constrained by the Dirichlet boundary conditions

$$\frac{\partial \rho}{\partial t}(x_{\text{in}}, t) = q'_\rho(t), \quad \frac{\partial E}{\partial t}(x_{\text{in}}, t) = q'_E(t) \tag{13.55}$$

where $q'_\rho(t)$ and $q'_E(t)$ denote prescribed bounded functions. If the inlet is supersonic, the third eigenvalue of $\frac{\partial f}{\partial q} n_x$ is negative and the similar Dirichlet constraint

$$\frac{\partial m}{\partial t}(x_{\text{in}}, t) = q'_m(t) \tag{13.56}$$

is enforced on mass flow, with $q'_m(t)$ a prescribed condition.

No Dirichlet constraint is enforced on m at a subsonic inlet because the third eigenvalue of $\frac{\partial f}{\partial q} n_x$ is positive, implying that the boundary magnitude of m is determined by the downstream flow conditions. In this case, the integral statement (13.53) for the corresponding linear-momentum is employed, using a weight function w with a non-vanishing trace at the inlet.

Similarly, at a supersonic outlet no Dirichlet constraints are imposed on the dependent variable "q" because all the eigenvalues of $\frac{\partial f}{\partial q} n_x$ are positive, implying that the boundary magnitude of q is determined by the upstream flow conditions. In this case, the integral statement (13.53) is employed, using a weight function w with a non-vanishing trace at the outlet.

At a subsonic outlet, two eigenvalues of $\frac{\partial f}{\partial q} n_x$ are positive, but one remains negative, implying that one exterior condition is required to determine the upstream flow. This study has selected a physically meaningful boundary condition on pressure p, enforced via the linear-momentum integral statement as follows. Considering that $w(x_{\text{in}}) = 0$ and $w(x_{\text{out}}) = 1$ in this statement, an integration by parts of not only the characteristics-bias expression, but also the pressure gradient yields

$$\int_\Omega \left(w + \varepsilon \psi s \frac{\partial w}{\partial x} \right) \left(\frac{\partial m}{\partial t} - \phi_m + \frac{\partial f_m^q}{\partial x} \right) d\Omega + \int_\Omega \varepsilon \psi s \frac{\partial w}{\partial x} \left(\alpha c s \frac{\partial m}{\partial x} + \delta \frac{\partial f_m^p}{\partial x} \right) d\Omega$$
$$- \int_\Omega \frac{\partial w}{\partial x} p \, d\Omega + \left(p - \varepsilon \psi s \left(\frac{\partial m}{\partial t} - \phi_m + \frac{\partial f_m}{\partial x} + \alpha c s \frac{\partial m}{\partial x} + (\delta - 1) \frac{\partial f_m^p}{\partial x} \right) \right)_{x=x_{\text{out}}} = 0 \tag{13.57}$$

The entire boundary expression is then set to a prescribed outlet external pressure p_{ext}; this boundary condition selection serves the double duty of both providing a boundary condition for the hyperbolic-parabolic perturbation in the characteristics-bias system and forcing the outlet pressure to approach p_{ext} since ε is set equal to a mesh spacing. The stability, accuracy, and numerical performance of this pressure boundary condition are discussed in Sections 13.6.3, 13.9.

13.6   Finite Element Weak Statement

The finite element solution exists on a partition Ω^h, $\Omega^h \subseteq \Omega$, of Ω, where superscript "h" signifies a spatial discrete approximation. This partition Ω^h has its boundary nodes in $\partial\Omega^h$ on the boundary $\partial\Omega$ of Ω and results from the union of N_e non-overlapping elements Ω_e, $\Omega^h = \bigcup_{e=1}^{N_e} \Omega_e$. With $\eta = 0$ or $\eta = 1$, respectively for shocked or smooth flows, and with $\mathcal{P}_1(\Omega_e)$ and $\mathcal{P}_{1v}(\Omega_e)$ the finite-dimensional function spaces of respectively diagonal square matrix-valued and vector-valued linear polynomials within each Ω_e, for each "t", the corresponding diagonal square matrix-valued and vector-valued finite element discretization spaces employed in this study are defined as

$$\mathcal{S}^1(\Omega^h) \equiv \left\{ w^h \in \mathcal{H}^1(\Omega^h) : w^h\big|_{\Omega_e} \in \mathcal{P}_1(\Omega_e), \forall \Omega_e \in \Omega^h, B(x_{\partial\Omega^h})w^h(x_{\partial\Omega^h}) = 0 \right\}$$

$$\mathcal{S}_v^\eta(\Omega^h) \equiv \left\{ w_v^h \in \mathcal{H}_v^\eta(\Omega^h) : w_v^h\big|_{\Omega_e} \in \mathcal{P}_{1v}(\Omega_e), \forall \Omega_e \in \Omega^h, B(x_{\partial\Omega^h})w_v^h(x_{\partial\Omega^h}, t) = G_v(x_{\partial\Omega^h}, t) \right\}$$

$$(13.58)$$

Based on these spaces, the finite element approximation $q^h \in \mathcal{S}_v^\eta$, is determined for each "t" as the solution of the finite element weak statement associated with (13.53)

$$\int_{\Omega^h} \left(w^h + \varepsilon^h \psi^h s^h \frac{\partial w^h}{\partial x} \right) \left(\frac{\partial q^h}{\partial t} - \phi^h + \frac{\partial f^h}{\partial x} \right) d\Omega$$

$$+ \int_{\Omega^h} \varepsilon^h \psi^h \frac{\partial w^h}{\partial x} \left(\alpha^h c^h \frac{\partial q^h}{\partial x} + s^h (\delta^h - 1) \frac{\partial f^{p^h}}{\partial x} \right) d\Omega = 0, \quad \forall w^h \in \mathcal{S}^1 \qquad (13.59)$$

where $s^h \equiv \operatorname{sgn}(u^h)$ denotes the sign of u^h at the centroid of each element; the linear-momentum discrete weak statement for an outlet node then results from a finite element discretization of (13.57) when an outlet pressure boundary condition is enforced. Since every member w^h of $\mathcal{S}^1$ results from a linear combination of the linearly independent basis functions of this finite dimensional space, statement (13.59) is satisfied for N independent basis functions of the space, where N denotes the dimension of the space. For N mesh nodes within Ω^h, there exist clusters of "master" elements Ω_i^m, each comprising only those adjacent elements that share a mesh node x_i, which implies existence of exactly N master elements. Note that each master element represents a "finite volume" as used in finite volume schemes, which, however, do not employ the following finite element "pyramid" test functions. On each master element Ω_i^m, the discrete test function $w^h \equiv w_i = w_i(x)$, $1 \le i \le N$, will coincide with the pyramid basis function with compact support on Ω_i^m.

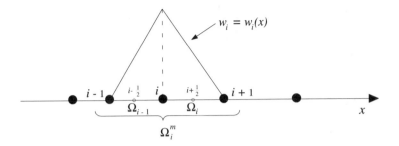

Figure 13.7: Adjacent Finite Elements, Master Element, Pyramid Test Function $w_i(x)$

Such a function equals one at node x_i, zero at all other mesh nodes, and also identically vanishes both on the boundary segments of Ω_i^m not containing x_i, and elsewhere within the computational domain outside Ω_i^m. Statement (13.59) yields a non-linear SUPG formulation, because the controller ψ^h does not remain constant, but varies as a local measure of the discrete solution q^h, as detailed in Chapter 10. As supported by the computational solutions, with subscript "i" denoting numerical value at node i, this measure is established as

$$q^h_{\text{meas}}\Big|_i = \frac{\rho^h_{\text{dim}}}{\rho_{\text{ref}}}\Big|_i + \frac{m^h_{\text{dim}}}{m_{\text{ref}}}\Big|_i + \frac{E^h_{\text{dim}}}{E_{\text{ref}}}\Big|_i = \rho^h\Big|_i + m^h\Big|_i + E^h\Big|_i \qquad (13.60)$$

for a positive m_i^h, this measure becomes the nodal ℓ_1 norm of the solution.

13.6.1 Galerkin Finite Element Expansions and Integrals

The discrete solution $q^h \in \mathcal{S}_v^\eta$ at each time t assumes the form of the following linear combination

$$q^h(x,t) \equiv \sum_{j=1}^N w_j(x) \cdot q^h(x_j,t) \qquad (13.61)$$

of nodal solution values and trial basis functions that coincide with the test functions $w_j(x)$ for a Galerkin formulation. Similarly, the source $\phi = \phi(x, q(x,t))$ and fluxes $f = f(q(x,t))$, $f^q = f^q(q(x,t))$ and $f^p = f^p(q(x,t))$ are discretized through the group expressions

$$\phi^h(x,t) \equiv \sum_{j=1}^N w_j(x) \cdot \phi\left(x_j, q^h(x_j,t)\right), \qquad f^h(x,t) \equiv \sum_{j=1}^N w_j(x) \cdot f\left(q^h(x_j,t)\right)$$

$$f^{q^h}(x,t) \equiv \sum_{j=1}^N w_j(x) \cdot f^q\left(q^h(x_j,t)\right), \qquad f^{p^h}(x,t) \equiv \sum_{j=1}^N w_j(x) \cdot f^p\left(q^h(x_j,t)\right) \qquad (13.62)$$

The notation for the discrete nodal variable and fluxes is then simplified as $q_j(t) \equiv q^h(x_j,t)$, $\phi_j(t) \equiv \phi^h(x_j,t)$, $f_j \equiv f^h(x_j,t)$, $f_j^q \equiv f^{q^h}(x_j,t)$, $f_j^p \equiv f^{p^h}(x_j,t)$. With implied summation on the repeated subscript index j, expansions (13.61)-(13.62) are then inserted into (13.59), which yields the discrete finite element weak statement

$$\int_{\Omega^h} \left(w_i + \varepsilon^h \psi_t^h s^h \frac{\partial w_i}{\partial x} \right) w_j \left. \frac{\partial q^h}{\partial t} \right|_j d\Omega - \int_{\Omega^h} \left(w_i + \varepsilon^h \psi_\phi^h s^h \frac{\partial w_i}{\partial x} \right) w_j \phi_j \, d\Omega$$

$$+ \int_{\Omega^h} \left(w_i + \varepsilon^h \psi^h s^h \frac{\partial w_i}{\partial x} \right) \left(\frac{\partial w_j}{\partial x} f_j \right) d\Omega + \int_{\Omega^h} \varepsilon^h \psi^h \frac{\partial w_i}{\partial x} \frac{\partial w_j}{\partial x} \left(\alpha^h c^h q_j + s^h (\delta^h - 1) f_j^p \right) d\Omega = 0$$

$$(13.63)$$

for $1 \leq i \leq N$, with $\psi_t^h = \psi_\phi^h = \psi^h$ for smooth flows, whereas $\psi_t^h \neq \psi_\phi^h \neq \psi^h$ in the three-element shock region, as amplified in Chapter 10; in respect of ε, this term is set equal to a reference length within each element, typically a measure of the element size, equal to $\Delta x/2$ within each element Ω_e of length Δx, as described in Chapter 10.

Within each element Ω_e, in the upstream-bias integrals

$$\int_{\Omega_e} \psi_{t,\phi}^h \frac{\partial w_i}{\partial x} w_j \, dx, \quad \int_{\Omega_e} \psi^h \frac{\partial w_i}{\partial x} \frac{\partial w_j}{\partial x} \, dx, \quad \int_{\Omega_e} \psi^h \alpha^h c^h \frac{\partial w_i}{\partial x} \frac{\partial w_j}{\partial x} \, dx, \quad \int_{\Omega_e} \psi^h (\delta^h - 1) \frac{\partial w_i}{\partial x} \frac{\partial w_j}{\partial x} \, dx$$

$$(13.64)$$

the functions α^h, δ^h, c^h, ψ^h, and $\psi_{t,\phi}^h$, standing for both ψ_t^h and ψ_ϕ^h, all depend on the variable (x,t), ψ^h and $\psi_{t,\phi}^h$ through an expression like (13.61) and α^h, δ^h, and c^h through

$$\alpha^h = \alpha \left(M \left(q^h(x,t) \right) \right), \quad \delta^h = \delta \left(M \left(q^h(x,t) \right) \right), \quad c^h = c \left(q^h(x,t) \right) \tag{13.65}$$

For the linear basis functions in this paper, each derivative $\partial w_i/\partial x$ becomes a constant, which simplifies (13.64). These integrals directly influence non only the stability, but also the accuracy of the computational solution, especially in the neighbourhood of a discontinuity. Accordingly, within each element where the flow is smooth, these integrals are evaluated by way of a 3-point Gaussian quadrature, with abscissae $\{ -\sqrt{3/5}, 0, \sqrt{3/5} \}$ and weights $\{ 5/9, 8/9, 5/9 \}$, a rule that supplies a 5th-order accurate integration. A 5th-degree continuous polynomial representation of a shock, however, contrasts with its discontinuous nature. As a result, within the three-element shock region, integrals (13.64) are evaluated via a 1-point quadrature, with abscissa $\{ 0 \}$ and weight $\{ 2 \}$, a rule that returns a simple centroidal average for each of these integrals. According to the computational results, in comparison to a uniform use of either the 3-point or 1-point quadrature throughout the computational domain, the use of the separate-quadrature strategy has contributed to non-oscillatory shocks and accurate computation of the jumps in the discontinuous dependent variables.

Since the test and trial functions w_i are prescribed functions of x, the remaining spatial integrations in (13.63) are then exactly carried out. The completion of each indicated spatial integration transforms (13.63) into a system of ordinary differential equations (ODE) in continuum time for determining at each time level t the unknown nodal values $q^h(x_j, t)$.

13.6.2   Discrete Upstream Bias

For a clear comparison between traditional finite difference/volume schemes and the acoustics-convection finite element algorithm (13.63), at any interior node "i" of the representative grid in Figure 13.7, equations (13.63) with $\varepsilon_{i+\frac{1}{2}}^h \equiv (\Delta x_{i+\frac{1}{2}})/2$ can be equivalently recast in difference notation as

$$\left(\frac{\Delta x}{2}\left(\frac{1}{3}+\frac{s\psi_t}{2}\right)\right)_{i-\frac{1}{2}}\left(\frac{dq_{i-1}}{dt}+2\frac{dq_i}{dt}\right)+\left(\frac{\Delta x}{2}\left(\frac{1}{3}-\frac{s\psi_t}{2}\right)\right)_{i+\frac{1}{2}}\left(2\frac{dq_i}{dt}+\frac{dq_{i+1}}{dt}\right)$$

$$-\left(\frac{\Delta x}{2}\left(\frac{1}{3}+\frac{s\psi_\phi}{2}\right)\right)_{i-\frac{1}{2}}(\phi_{i-1}+2\phi_i)-\left(\frac{\Delta x}{2}\left(\frac{1}{3}-\frac{s\psi_\phi}{2}\right)\right)_{i+\frac{1}{2}}(2\phi_i+\phi_{i+1})$$

$$= -(\psi\alpha c)_{i-\frac{1}{2}}(q_i-q_{i-1})+(\psi\alpha c)_{i+\frac{1}{2}}(q_{i+1}-q_i)$$

$$-\frac{1}{2}\left(\left(1+(s\psi)_{i-\frac{1}{2}}\right)(f_i^q-f_{i-1}^q)+\left(1-(s\psi)_{i+\frac{1}{2}}\right)(f_{i+1}^q-f_i^q)\right)$$

$$-\frac{1}{2}\left(\left(1+(s\psi\delta)_{i-\frac{1}{2}}\right)(f_i^p-f_{i-1}^p)+\left(1-(s\psi\delta)_{i+\frac{1}{2}}\right)(f_{i+1}^p-f_i^p)\right) \qquad (13.66)$$

which uniquely couples several time derivatives at each node "i" and features a linear combination of two-point upstream and downstream flux differences. In these finite element equations, the controller ψ^h determines the combination weights of the downstream and upstream expressions, and since ψ^h remains non-negative, these equations induce the appropriate upstream bias, since the upstream weight $1+\psi_{i-\frac{1}{2}}$ always exceeds the downstream weight $1-\psi_{i+\frac{1}{2}}$. As a result, the finite element weak statement (13.63) generates consistent variable-upstream-bias discrete equations that correspond to an upstream-bias discretizations for the original Euler system, within a compact block tri-diagonal matrix statement.

For smooth solutions, these equations will still couple upstream and downstream points even for supersonic flows. Such an algorithm remains mathematically consistent with the physics of supersonic mono-directional wave propagation on the basis of Courant's and Hilbert's classical developments for non-linear hyperbolic systems. These developments show that while waves do propagate along characteristics, smooth solutions can be expanded in Taylor's series within arbitrary regions encircling any given point and along any direction, radiating upstream or downstream from the point.

For a closer comparison with upwind finite-volume schemes, the finite element equations (13.66) can be rearranged to generate the "numerical flux"

$$F_{i+\frac{1}{2}} \equiv \frac{f_i+f_{i+1}}{2}-\psi_{i+\frac{1}{2}}\left[(\alpha c)_{i+\frac{1}{2}}(q_{i+1}-q_i)+\frac{s_{i+\frac{1}{2}}}{2}(f_{i+1}^q-f_i^q)+\frac{(s\delta)_{i+\frac{1}{2}}}{2}(f_{i+1}^p-f_i^p)\right]$$

$$(13.67)$$

which corresponds to the discrete counterpart of the characteristics-bias flux (13.39). By virtue of this numerical flux, equations (13.66) are recast as

$$\left(\frac{\Delta x}{2}\left(\frac{1}{3}+\frac{s\psi_t}{2}\right)\right)_{i-\frac{1}{2}}\left(\frac{dq_{i-1}}{dt}+2\frac{dq_i}{dt}\right)+\left(\frac{\Delta x}{2}\left(\frac{1}{3}-\frac{s\psi_t}{2}\right)\right)_{i+\frac{1}{2}}\left(2\frac{dq_i}{dt}+\frac{dq_{i+1}}{dt}\right)$$

$$-\left(\frac{\Delta x}{2}\left(\frac{1}{3}+\frac{s\psi_\phi}{2}\right)\right)_{i-\frac{1}{2}}(\phi_{i-1}+2\phi_i)-\left(\frac{\Delta x}{2}\left(\frac{1}{3}-\frac{s\psi_\phi}{2}\right)\right)_{i+\frac{1}{2}}(2\phi_i+\phi_{i+1})$$

$$= -\left(F_{i+\frac{1}{2}}-F_{i-\frac{1}{2}}\right) \qquad (13.68)$$

which shows that the finite element weak statement (13.63) naturally leads to a discretely conservative algorithm.

13.6.3   Boundary Equations and Pressure Boundary Condition

The integral statement (13.63) directly yields a set of consistent boundary differential equations, for both unconstrained boundary variables and for pressure, to enforce a pressure boundary condition at a subsonic outlet. These equations do not require any algebraic extrapolation of variables, but rather couple the time derivatives of boundary- and interior-node variables within the boundary cell.

For the linear elements in this study, let N and $N-1$ denote the nodes within the outlet boundary element, with N corresponding to the outlet node. For the discrete finite element equation associated with boundary node x_N, the test function w satisfies the conditions $w(x_{N-1}) = 0$, $w(x_N) = 1$.

The boundary differential equation from (13.63) corresponding to an outlet node becomes

$$\left(\Delta x \left(\frac{1}{3} + \frac{s\psi_t}{2}\right)\right)_{N-\frac{1}{2}} \left(\frac{dq_{N-1}}{dt} + 2\frac{dq_N}{dt}\right)$$

$$-\left(\Delta x \left(\frac{1}{3} + \frac{s\psi_\phi}{2}\right)\right)_{N-\frac{1}{2}} (\phi_{N-1} + 2\phi_N) = -2(\psi ac)_{N-\frac{1}{2}} (q_N - q_{N-1})$$

$$-\left(1 + (s\psi)_{N-\frac{1}{2}}\right)\left(f_N^q - f_{N-1}^q\right) - \left(1 + (s\psi\delta)_{N-\frac{1}{2}}\right)\left(f_N^p - f_{N-1}^p\right) \tag{13.69}$$

This equation directly couples the time derivatives of the solution q at the adjacent boundary and interior nodes x_N and x_{N-1}. A similar equation is then obtained at an inlet, *mutatis mutandis*.

Mechanically, a quasi-one-dimensional flow within a duct is induced by lowering the outlet pressure in comparison to the inlet pressure. Such an outlet pressure boundary condition is naturally enforced within the surface integral that emerges in the momentum-equation weak statement, as shown in the previous section. A linear finite element discretization of (13.57) yields

$$\left(\Delta x \left(\frac{1}{3} + \frac{s\psi_t}{2}\right)\right)_{N-\frac{1}{2}} \left(\frac{dq_{N-1}}{dt} + 2\frac{dq_N}{dt}\right)$$

$$-\left(\Delta x \left(\frac{1}{3} + \frac{s\psi_\phi}{2}\right)\right)_{N-\frac{1}{2}} (\phi_{N-1} + 2\phi_N) = -2(\psi ac)_{N-\frac{1}{2}} (q_N - q_{N-1})$$

$$-\left(1 + (s\psi)_{N-\frac{1}{2}}\right)\left(f_N^q - f_{N-1}^q\right) - (s\psi\delta)_{N-\frac{1}{2}}\left(f_N^p - f_{N-1}^p\right) - \left((2f_{\text{out}}^p - f_N^p) - f_{N-1}^p\right) \tag{13.70}$$

In this equation, f_N^p denotes the outlet-node pressure, as calculated through the equation of state (2.25), whereas, quite significantly, f_{out}^p can correspond to the specified outlet pressure boundary condition. This strategy for imposing an outlet pressure boundary condition remains intrinsically stable. Suppose, for instance, that some numerical perturbation forces f_N^p to decrease below the imposed f_{out}^p. In this case the outlet boundary equation (13.70) induces a negative time rate of change for m_N, which leads to a corresponding reduction in m_N. From the equation of state (2.25), this reduction then leads to an increase in f_N^p, which corresponds to a stable restoration of the imposed pressure condition. A similar conclusion on the stability of (13.70) is achieved by considering a perturbation increase in f_N^p. The results in Section 13.9 confirm the accuracy and stability of this pressure boundary condition procedure.

13.7 Implicit Runge Kutta Time Integration

The finite element equation (13.66) along with appropriate boundary equations and conditions form a system of non-linear ordinary differential equations in the time variable. The discrete characteristics-bias differential system, inclusive of inlet and outlet pressure boundary conditions is cast as

$$
\left(\Delta x \left(\frac{1}{3} - \frac{s\psi_t}{2}\right)\right)_{1+\frac{1}{2}} \left(2\frac{dq_1}{dt} + \frac{dq_2}{dt}\right)
$$

$$
-\left(\Delta x \left(\frac{1}{3} - \frac{s\psi_\phi}{2}\right)\right)_{1+\frac{1}{2}} (2\phi_1 + \phi_2) = 2(\psi\alpha c)_{1+\frac{1}{2}} (q_2 - q_1)
$$

$$
-\left(1 - (s\psi)_{1+\frac{1}{2}}\right)\left(f_2^q - f_1^q\right) - \left(1 - (s\psi\delta)_{1+\frac{1}{2}}\right)\left(f_2^p - f_1^p\right) \tag{13.71}
$$

for the inlet node, $i = 1$, with appropriate components replaced by (13.55)-(13.56)

$$
\left(\frac{\Delta x}{2}\left(\frac{1}{3} + \frac{s\psi_t}{2}\right)\right)_{i-\frac{1}{2}} \left(\frac{dq_{i-1}}{dt} + 2\frac{dq_i}{dt}\right) + \left(\frac{\Delta x}{2}\left(\frac{1}{3} - \frac{s\psi_t}{2}\right)\right)_{i+\frac{1}{2}} \left(2\frac{dq_i}{dt} + \frac{dq_{i+1}}{dt}\right)
$$

$$
-\left(\frac{\Delta x}{2}\left(\frac{1}{3} + \frac{s\psi_\phi}{2}\right)\right)_{i-\frac{1}{2}} (\phi_{i-1} + 2\phi_i) - \left(\frac{\Delta x}{2}\left(\frac{1}{3} - \frac{s\psi_\phi}{2}\right)\right)_{i+\frac{1}{2}} (2\phi_i + \phi_{i+1})
$$

$$
= -(\psi\alpha c)_{i-\frac{1}{2}} (q_i - q_{i-1}) + (\psi\alpha c)_{i+\frac{1}{2}} (q_{i+1} - q_i)
$$

$$
-\frac{1}{2}\left(\left(1 + (s\psi)_{i-\frac{1}{2}}\right)\left(f_i^q - f_{i-1}^q\right) + \left(1 - (s\psi)_{i+\frac{1}{2}}\right)\left(f_{i+1}^q - f_i^q\right)\right)
$$

$$
-\frac{1}{2}\left(\left(1 + (s\psi\delta)_{i-\frac{1}{2}}\right)\left(f_i^p - f_{i-1}^p\right) + \left(1 - (s\psi\delta)_{i+\frac{1}{2}}\right)\left(f_{i+1}^p - f_i^p\right)\right) \tag{13.72}
$$

for the internal nodes, $2 \le i \le N - 1$, and

$$
\left(\Delta x \left(\frac{1}{3} + \frac{s\psi_t}{2}\right)\right)_{N-\frac{1}{2}} \left(\frac{dq_{N-1}}{dt} + 2\frac{dq_N}{dt}\right)
$$

$$
-\left(\Delta x \left(\frac{1}{3} + \frac{s\psi_\phi}{2}\right)\right)_{N-\frac{1}{2}} (\phi_{N-1} + 2\phi_N) = -2(\psi\alpha c)_{N-\frac{1}{2}} (q_N - q_{N-1})
$$

$$
-\left(1 + (s\psi)_{N-\frac{1}{2}}\right)\left(f_N^q - f_{N-1}^q\right) - (s\psi\delta)_{N-\frac{1}{2}}\left(f_N^p - f_{N-1}^p\right)
$$

$$
-\left((2f_{\text{out}}^p - f_N^p) - f_{N-1}^p\right) \tag{13.73}
$$

for the outlet node, $i = N$.

With Q denoting the array of nodal density, momentum, and energy

$$Q \equiv \left\{ \begin{array}{c} \rho_1 \\ m_1 \\ E_1 \\ \rho_2 \\ m_2 \\ E_2 \\ \cdot \\ \cdot \\ \rho_i \\ m_i \\ E_i \\ \cdot \\ \cdot \\ \rho_N \\ m_N \\ E_N \end{array} \right\} \tag{13.74}$$

and $\mathcal{F}$ denoting an array of fluxes, source terms and characteristics-bias expressions, equations (13.71)-(13.73) are abridged as the non-linear differential system

$$\mathcal{M} \frac{dQ(t)}{dt} = \mathcal{F}(t, Q(t)) \tag{13.75}$$

Where the "mass" matrix $\mathcal{M}$ couples the discretization nodal time derivatives. A representative equation at an internal node "i" is thus cast as

$$m_{i,i-1} \frac{dq_{i-1}}{dt} + m_{i,i} \frac{dq_i}{dt} + m_{i,i+1} \frac{dq_{i+1}}{dt} = \mathcal{F}_i(t, Q) \tag{13.76}$$

where $\mathcal{F}_i(t, Q)$ corresponds to the right-hand sides in (13.71)-(13.73).

The differential system (13.75) is integrated through the two-stage implicit Runge-Kutta procedure

$$\begin{aligned} Q_{n+1} - Q_n &= b_1 K_1 + b_2 K_2 \\ \mathcal{M} K_1 &= \Delta t \cdot \mathcal{F}(t_n + c_1 \Delta t, Q_n + a_{11} K_1) \\ \mathcal{M} K_2 &= \Delta t \cdot \mathcal{F}(t_n + c_2 \Delta t, Q_n + a_{21} K_1 + a_{22} K_2) \end{aligned} \tag{13.77}$$

The Runge-Kutta arrays K_1 and K_2 correspond to variations ΔQ_1 and ΔQ_2 of the solution array Q. The first expression in (13.77) thus corresponds to the linear combination

$$Q_{n+1} - Q_n = b_1 \Delta Q_1 + b_2 \Delta Q_2 \tag{13.78}$$

The terminal numerical solution for the second and third expression in (13.77) is then computed using Newton's method, which for the implicit fully-coupled computation of the IRK2 arrays K_i, $1 \leq i \leq 2$, is cast as

$$\left[\mathcal{M} - a_{ii} \Delta t \left(\frac{\partial \mathcal{F}}{\partial Q} \right)^p_{Q^p_i} \right] \left(K^{p+1}_i - K^p_i \right) = \Delta t \, \mathcal{F}(t_n + c_i \Delta t, Q^p_i) - \mathcal{M} K^p_i$$

$$Q_i^p \equiv Q_n + a_{i1}K_1^p + a_{i2}K_2^p \tag{13.79}$$

where $a_{ij} = 0$ for $j > i$, p is the iteration index, and $K_1^p \equiv K_1$ for $i = 2$; for linear finite elements, the Jacobian

$$J_i(Q) \equiv \mathcal{M} - a_{\underline{ii}}\Delta t \left(\frac{\partial \mathcal{F}}{\partial Q}\right)_{Q_i^p}^p \tag{13.80}$$

becomes a block tri-diagonal matrix. Since the Jacobians of the controllers ψ, ψ_t and ψ_ϕ are computationally expensive, these controllers are kept equal to their magnitudes at each time level n, during the determination of the Runge-Kutta arrays K_1 and K_2, a specification that does not require the Jacobians of these controllers; all the other Jacobians have been exactly determined analytically. For all the results documented in the next section, the initial estimate K_i^0 is set equal to the zero array. The practical implementation of Newton's method for (13.79) has allowed a specified maximum number of Newton iterations for the non-linear determination of each IRK array K_i. This number was set to 4, yet the algorithm only employed 4 Newton iterations when the absolute value of the maximum component of the residual $\mathcal{F}$ in (13.75) exceeded the prescribed tolerance of 5.0×10^{-6}, otherwise the algorithm employed fewer iterations, mostly 1. This implementation led to a rapid convergence to steady states, with Courant number reaching 200, as detailed in Section 13.9.

13.8 Reference Exact Solutions

When the source term ϕ in the Euler system

$$\frac{\partial q}{\partial t} + \frac{\partial f(q)}{\partial x} = \phi \tag{13.81}$$

is restricted to the area variation terms, the arrays $q = q(x, t)$, $f = f(q)$ and $\phi = \phi(x, q)$ may be expressed as

$$q \equiv \left\{ \begin{array}{c} \rho \\ m \\ E \end{array} \right\}, \quad f(q) \equiv \left\{ \begin{array}{c} m \\ \dfrac{m^2}{\rho} + p \\ \dfrac{m}{\rho}(E + p) \end{array} \right\}, \quad \phi \equiv -\frac{m}{\rho A}\frac{dA}{dx} \left\{ \begin{array}{c} \rho \\ m \\ E + p \end{array} \right\} \tag{13.82}$$

where ρ, m, E, p, and A respectively denote density, volume-specific linear momentum and total-energy, pressure, and nozzle cross-sectional area. The flow speed u, depends upon m and ρ as $u = m/\rho$. Pressure is then expressed in terms of ρ, m, E through the equation of state

$$p = p(\rho, \epsilon(\rho, m, E)), \quad p = (\gamma - 1)\left(E - \frac{m^2}{2\rho}\right) \tag{13.83}$$

where the first expression applies for an equilibrium gas and the second specifically models a perfect gas.

13.8.1 Smooth-Flow Pressure Integral

The Euler equations may also be expressed as

$$
\begin{cases}
\dfrac{\partial \rho A}{\partial t} + \dfrac{\partial \rho u A}{\partial x} = 0 \\[2mm]
\dfrac{\partial \rho u A}{\partial t} + \dfrac{\partial}{\partial x}\left(\rho u^2 A\right) + A\dfrac{\partial p}{\partial x} = 0 \\[2mm]
\dfrac{\partial E A}{\partial t} + \dfrac{\partial}{\partial x}\left(\rho u A \dfrac{(E+p)}{\rho}\right) = 0
\end{cases}
\tag{13.84}
$$

For a smooth flow, for which the flow-variable partial derivatives exist everywhere within the flow, the combination of the continuity and linear-momentum equations lead to the non-conservation linear-momentum equation

$$
\frac{\partial u}{\partial t} + \frac{\partial}{\partial x}\left(\frac{u^2}{2}\right) + \frac{1}{\rho}\frac{\partial p}{\partial x} = 0
\tag{13.85}
$$

which no longer depends upon the cross sectional area A. This equation along with the continuity and energy equations will lead to an exact pressure integral of the entire system of Euler equations. The conditions for this integral involve a cross sectional area $A = A(x)$ variation that remains sufficiently mild for the average of the speed squared and the square of the average speed to remain accurate representations of each other. This exact integral expresses pressure as a function of density only as $p = p(\rho)$.

In order to determine this pressure integral, begin by expressing the total energy E in terms of the internal and kinetic energies as $E = \rho\epsilon + \rho u^2/2$. Next insert this expression into the energy equation, and simplify the expression through the continuity equation. The resulting energy equation becomes

$$
\rho A\left(\frac{\partial \epsilon}{\partial t} + u\frac{\partial \epsilon}{\partial x}\right) + \rho A\left(\frac{\partial}{\partial t}\left(\frac{u^2}{2}\right) u\frac{\partial}{\partial x}\left(\frac{u^2}{2}\right)\right) + \frac{\partial(Aup)}{\partial x} = 0
\tag{13.86}
$$

which features the substantive derivative of the mass-specific kinetic energy. This derivative is recast in terms of the linear-momentum equation, which leads to the following form of the internal energy equation

$$
\rho A\left(\frac{\partial \epsilon}{\partial t} + u\frac{\partial \epsilon}{\partial x}\right) + p\frac{\partial Au}{\partial x} = 0
\tag{13.87}
$$

The derivative of Au is then recast in terms of the continuity equation. The resulting equation becomes

$$
\rho\left(\frac{\partial \epsilon}{\partial t} + u\frac{\partial \epsilon}{\partial x}\right) = \frac{p}{\rho}\left(\frac{\partial \rho}{\partial t} + u\frac{\partial \rho}{\partial x}\right)
\tag{13.88}
$$

which no longer depends upon the cross-sectional area $A = A(x)$ and can also be cast in substantive-derivative form as

$$
\rho\frac{D\epsilon}{Dt} = \frac{p}{\rho}\frac{D\rho}{Dt}
\tag{13.89}
$$

For an equilibrium gas, the internal energy ϵ depends upon two other thermodynamic variables, hence $\epsilon = \epsilon(p, \rho)$. Equation (13.89), therefore, constrains p to depend upon ρ only as

$$p = p(\rho) \tag{13.90}$$

For a perfect gas, the internal energy ϵ explicitly depends upon p and ρ via the temperature T

$$\varepsilon = c_v T = \frac{c_v}{R} \frac{p}{\rho} = \frac{1}{\gamma - 1} \frac{p}{\rho} \tag{13.91}$$

Equation (13.89), therefore, specializes into

$$\frac{1}{p} \frac{Dp}{Dt} = \frac{\gamma}{\rho} \frac{D\rho}{Dt} \quad \Rightarrow \quad \begin{cases} \dfrac{\partial p}{\partial \rho} = \gamma \dfrac{p}{\rho} \\ \dfrac{D \log p}{Dt} = \dfrac{D \log \rho^{\gamma}}{Dt} \end{cases} \tag{13.92}$$

with solution

$$\frac{p}{\rho^{\gamma}} = \frac{p_o}{\rho_o^{\gamma}} \tag{13.93}$$

which corresponds to the traditional isentropic relationship between p and ρ. More than a simplifying assumption on a gas dynamic process, therefore, this fundamental relationship rather emerges as an exact integral of the quasi-one-dimensional Euler equations for both steady and unsteady smooth, inviscid, and adiabatic flows.

13.8.2 Steady-State Equations

In terms of the primitive variables ρ, u and E, the steady-flow Euler equations become

$$\begin{cases} \dfrac{\partial \rho u A}{\partial x} = 0 \\ \dfrac{\partial}{\partial x}\left(\dfrac{u^2}{2}\right) + \dfrac{1}{\rho} \dfrac{\partial p}{\partial x} = 0 \\ \dfrac{\partial}{\partial x}\left(\rho u A \dfrac{(E + p)}{\rho}\right) = 0 \end{cases} \tag{13.94}$$

These equations lead to the area rule and three exact integrals.

13.8.3 Area Rule

For a smooth flow, results (13.90), (13.92) show that pressure only depends upon density $p = p(\rho)$. The pressure gradient in the linear-momentum equation in (13.94), therefore, explicitly correlates with the density gradient. The expression for this density gradient from the linear-momentum equation thus becomes

$$u \frac{\partial \rho}{\partial x} = -\left(\frac{\rho u^2}{\frac{\partial p}{\partial \rho}\big|_s}\right) \frac{\partial u}{\partial x} \tag{13.95}$$

where subscript "s" indicates an isentropic derivative. From the continuity equation, the same density gradient depends upon the speed and cross-sectional area gradient in the form

$$u\frac{\partial \rho}{\partial x} = -\rho\frac{\partial u}{\partial x} - \frac{\rho u}{A}\frac{\partial A}{\partial x} \tag{13.96}$$

The combination of these two expressions for the density gradient then yields the area-rule expression

$$\frac{1}{A}\frac{\partial A}{\partial x} = -\frac{1}{u}\frac{\partial u}{\partial x}\left(1 - \frac{u^2}{\frac{\partial p}{\partial \rho}\big|_s}\right) = -\frac{1}{u}\frac{\partial u}{\partial x}\left(1 - M^2\right) \tag{13.97}$$

This result shows that besides the speed gradient, the sign of the area gradient depends on the Mach number $M = u/\sqrt{\frac{\partial p}{\partial \rho}\big|_s}$. This number remains less than one as the speed increases from a negligible magnitude. In these conditions the area rule indicates a speed increase takes place along a duct with decreasing cross sectional area. As the speed keeps on increasing, M will eventually equal one. The area rule then shows the cross sectional area no longer decreases for the area gradient vanishes. Accordingly, the cross-sectional area where $M = 1$ corresponds to the nozzle throat. When M exceeds one, the speed can then keep on increasing along a duct with increasing cross-sectional area, since in these conditions the area rule provides a positive area gradient for a positive speed gradient. These considerations reiterate the importance of the Mach number as it critically impacts the correlation between speed and area gradients.

13.8.4   Exact Integrals

The continuity equation in system (13.94) yields

$$u(x)\rho(x)A(x) = u_i\rho_i A_i \tag{13.98}$$

where u_i, ρ_i, and A_i respectively denote the inlet speed, density and nozzle cross sectional area. The second integral is result (13.90)

$$p = p(\rho) \tag{13.99}$$

The third integral emerges from the energy equation as

$$\frac{E(x) + p(x)}{\rho(x)} = \frac{E_i + p_i}{\rho_i} \tag{13.100}$$

Since $E = \rho\epsilon + \rho u^2/2$, this result becomes

$$\epsilon(x) + \frac{u^2(x)}{2} + \frac{p(x)}{\rho(x)} = \epsilon_i + \frac{u_i^2}{2} + \frac{p_i}{\rho_i} \tag{13.101}$$

The specific solution for a perfect gas emerges from inserting expression (13.91) into this integrated energy equation. Along with result (13.93), the complete steady-flow solution of the Euler equation for a perfect gas thus becomes

$$
\begin{cases}
u(x)\rho(x)A(x) = u_i \rho_i A_i \\[2mm]
\dfrac{p(x)}{\rho^\gamma(x)} = \dfrac{p_i}{\rho_i^\gamma} \\[2mm]
\dfrac{u^2(x)}{2} + \dfrac{\gamma}{\gamma-1}\dfrac{p(x)}{\rho(x)} = \dfrac{u_i^2}{2} + \dfrac{\gamma}{\gamma-1}\dfrac{p_i}{\rho_i}
\end{cases}
\tag{13.102}
$$

The direct integration of the linear-momentum equation in system (13.94) yields

$$
\frac{u^2(x)}{2} - \frac{u_i^2}{2} - \int_{x_i}^{x} \frac{1}{\rho}\frac{\partial p}{\partial x}dx = 0
\tag{13.103}
$$

Along with the pressure relationship in system (13.102), this integral yields

$$
\frac{u^2(x)}{2} + \frac{\gamma}{\gamma-1}\frac{p(x)}{\rho(x)} = \frac{u_i^2}{2} + \frac{\gamma}{\gamma-1}\frac{p_i}{\rho_i}
\tag{13.104}
$$

which coincides with the energy equation in (13.102). This outcome is not surprising and does determine the total energy equation because the integration of (13.103) employs the pressure relationship that results from the continuity, linear-momentum and energy equations. The determination of this energy equation through the integration of the linear-momentum equation, however, only remains valid for smooth flows, because both the linear-momentum equation and the pressure relationship used only applies for smooth flows. Instead, the determination of the total energy result from the total-energy equation in system (13.94) applies for both smooth and non-smooth (i.e. shocked) flows because the total energy equation in system (13.94) is in conservation form.

13.8.5   Similarity Solution

The defining flow information consists of the distribution of nozzle cross-sectional area, inlet density and inlet and outlet pressure

$$
A = A(x), \quad \rho_i, \ p_i, \ p_o
$$

These specified quantities uniquely establish a quasi-one dimensional compressible flow within a converging-diverging nozzle. On the basis of this information, a unique steady, similarity solution emerges from system (13.102).

Reference Variables

From the isentropic pressure relationship, the outlet density is established as

$$
\rho_o = \rho_i \left(\frac{p_o}{p_i}\right)^{\frac{1}{\gamma}} \quad \Rightarrow \quad \frac{1}{\rho_o} = \frac{1}{\rho_i}\left(\frac{p_i}{p_o}\right)^{\frac{1}{\gamma}}
\tag{13.105}
$$

From the continuity equation, the outlet speed u_o is expressed in terms of inlet and outlet variables. The insertion of such an expression for u_o in the energy equation then leads to the inlet and outlet speeds in terms of inlet and outlet variables as

$$u_i = \sqrt{\frac{\dfrac{2\gamma}{\gamma-1}\left(\dfrac{p_i}{\rho_i} - \dfrac{p_o}{\rho_o}\right)}{\left(\dfrac{A_i}{A_o}\right)^2 \left(\dfrac{\rho_i}{\rho_o}\right)^2 - 1}}, \quad u_o = \left(\frac{A_i}{A_o}\right)\left(\frac{\rho_i}{\rho_o}\right)\sqrt{\frac{\dfrac{2\gamma}{\gamma-1}\left(\dfrac{p_i}{\rho_i} - \dfrac{p_o}{\rho_o}\right)}{\left(\dfrac{A_i}{A_o}\right)^2 \left(\dfrac{\rho_i}{\rho_o}\right)^2 - 1}} \tag{13.106}$$

The energy equation defines the ratio of stagnation pressure p_{st} and density ρ_{st} as

$$\frac{u^2(x)}{2} + \frac{\gamma}{\gamma-1}\frac{p(x)}{\rho(x)} = \frac{u_i^2}{2} + \frac{\gamma}{\gamma-1}\frac{p_i}{\rho_i} = \frac{\gamma}{\gamma-1}\frac{p_{st}}{\rho_{st}} = c_p T_{st} \tag{13.107}$$

which implicitly defines the stagnation temperature as an adiabatic-flow constant. With reference to the expression for u_i in (13.106), the ratio p_{st}/ρ_{st} then becomes

$$\frac{p_{st}}{\rho_{st}} = \frac{p_i}{\rho_i} + \frac{\dfrac{p_i}{\rho_i} - \dfrac{p_o}{\rho_o}}{\left(\dfrac{A_i}{A_o}\right)^2 \left(\dfrac{\rho_i}{\rho_o}\right)^2 - 1} \tag{13.108}$$

By virtue of the isentropic pressure relationship, the individual stagnation density and pressure then follow as

$$\rho_{st} = \left(\frac{p_i/\rho_i^\gamma}{p_{st}/\rho_{st}}\right)^{\frac{1}{1-\gamma}}, \quad p_{st} = \left(\frac{p_{st}}{\rho_{st}}\right)\rho_{st} \tag{13.109}$$

The maximum speed u_{max} obtains from the energy equation in the limit of a vanishing ratio p/ρ. This ratio and the corresponding maximum speed u_{max} become

$$\lim_{u \to u_{max}} \frac{p}{\rho} = 0, \quad \frac{u_{max}^2}{2} = \frac{\gamma}{\gamma-1}\frac{p_{st}}{\rho_{st}} \quad \Rightarrow u_{max} = \sqrt{2\frac{\gamma}{\gamma-1}\frac{p_{st}}{\rho_{st}}} \tag{13.110}$$

The density and "maximum" nozzle area associated with u_{max} become

$$\lim_{u \to u_{max}} \rho = \lim_{u \to u_{max}} \left(\frac{(p/\rho)}{p_i/\rho_i^\gamma}\right)^{\frac{1}{\gamma-1}} = 0, \quad A_{max} = \lim_{\rho \to 0} u_i \frac{\rho_i A_i}{u_{max}\rho} = \infty \tag{13.111}$$

The nozzle-throat Mach number equals one, hence

$$M_T = 1 \quad \Rightarrow u_T^2 = \gamma\frac{p_T}{\rho_T} \tag{13.112}$$

Through the energy equation, this equality leads to the following expressions for p_T/ρ_T and u_T

$$\frac{p_T}{\rho_T} = \frac{2}{\gamma+1}\frac{p_{st}}{\rho_{st}}, \quad u_T = \sqrt{\frac{2\gamma}{\gamma+1}\frac{p_{st}}{\rho_{st}}} \tag{13.113}$$

Following (13.109), the individual expressions for p_T and ρ_T become

$$\rho_T = \left(\frac{p_{st}/\rho_{st}^\gamma}{p_T/\rho_T}\right)^{\frac{1}{1-\gamma}} = \rho_{st}\left(\frac{2}{\gamma+1}\right)^{\frac{1}{\gamma-1}} , \quad p_T = \frac{2\rho_T}{\gamma+1}\frac{p_{st}}{\rho_{st}} = p_{st}\left(\frac{2}{\gamma+1}\right)^{\frac{\gamma}{\gamma-1}} \quad (13.114)$$

From the continuity equation, the expression for the reference throat area A_T becomes

$$A_T = \frac{\rho_i u_i A_i}{\rho_T u_T} = \frac{\rho_i u_i A_i}{\sqrt{\gamma\rho_{st}p_{st}}}\left(\frac{2}{\gamma+1}\right)^{-\frac{\gamma+1}{2(\gamma-1)}} \quad (13.115)$$

Non-Dimensional Ratios

The ratio of local and stagnation pressure-density ratios will depend on the Mach number M as

$$\frac{p_{st}}{\rho_{st}} = \frac{p}{\rho} + \frac{\gamma-1}{2\gamma}u^2 = \frac{p}{\rho} + \frac{p}{\rho}\frac{\gamma-1}{2}M^2 = \frac{p}{\rho}\left(1+\frac{\gamma-1}{2}M^2\right)$$

$$\Rightarrow \frac{p/\rho}{p_{st}/\rho_{st}} = \frac{T}{T_{st}} = \left(1+\frac{\gamma-1}{2}M^2\right)^{-1} \quad (13.116)$$

With this result and the isentropic pressure relationship, the corresponding expressions for the individual pressure and density ratios become

$$\frac{p}{p_{st}} = \left(1+\frac{\gamma-1}{2}M^2\right)^{-\frac{\gamma}{\gamma-1}} , \quad \frac{\rho}{\rho_{st}} = \left(1+\frac{\gamma-1}{2}M^2\right)^{-\frac{1}{\gamma-1}} \quad (13.117)$$

As shown, the throat speed is expressed as $u_T = \sqrt{\gamma p_T/\rho_T}$. The definition of M then leads to the ratio

$$\frac{u}{u_T} = M\frac{\sqrt{\gamma p/\rho}}{\sqrt{\gamma p_T/\rho_T}} = M\sqrt{\frac{p}{p_T}\frac{\rho_T}{\rho}} \quad \Rightarrow \quad \frac{u}{u_T} = M\sqrt{\frac{1+\gamma}{2+(\gamma-1)M^2}} \quad (13.118)$$

From the continuity equation, the ratio A/A_T follows as

$$\frac{A}{A_T} = \frac{u_T}{u}\frac{\rho_T}{\rho} = \frac{u_T}{u}\frac{\rho_T}{\rho_{st}}\frac{\rho_{st}}{\rho} = \frac{1}{M}\left(\frac{2+(\gamma-1)M^2}{1+\gamma}\right)^{\frac{1+\gamma}{2(\gamma-1)}} \quad (13.119)$$

The distribution of nozzle cross-sectional area $A = A(x)$ is prescribed. The expression for the ratio A/A_T can thus be viewed as the following equation for $M = M(x)$ itself

$$(\gamma-1)M^2 - (\gamma+1)\left(M\frac{A}{A_T}\right)^{\frac{2(\gamma-1)}{1+\gamma}} + 2 = 0 \quad (13.120)$$

which can be solved numerically for M, for each A. With the solutions of this equation, the corresponding pressure, density, and speed follow from (13.117) and (13.118).

13.8.6   Calculation of Shocked Flows

Six quantities are necessary to determine a compressible nozzle flow with embedded normal shock. They consist of the inlet density and pressure ρ_i and p_i, outlet pressure p_o, and inlet, throat, and outlet nozzle areas A_i, A_T, A_o. The complete determination of the flow with embedded normal shock involves the calculation of 13 quantities. They are the inlet and outlet speeds u_i, u_o and outlet density ρ_o, the pre- and post-shock density ρ_1, ρ_2, speed u_1, u_2, and pressure p_1, p_2, the normal-shock nozzle area $A_1 = A_s = A_2$, and the nozzle-throat density ρ_T, speed u_T, and pressure p_T. This determination, therefore, requires 13 equations. These equations involve $M_T = 1$ at the channel throat, the conservation of mass, linear momentum, and total energy across the normal shock, and the pressure relationship as well as the integrated continuity and total energy equations expressed in terms of the inlet and throat conditions, inlet and pre-shock conditions, and post-shock and outlet conditions. Grouped in sets, these 13 equations are

$$
a) \begin{cases} u_i \rho_i A_i = u_T \rho_T A_T \\[4pt] \dfrac{p_i}{\rho_i^\gamma} = \dfrac{p_T}{\rho_T^\gamma} \\[6pt] \dfrac{u_i^2}{2} + \dfrac{\gamma}{\gamma-1}\dfrac{p_i}{\rho_i} = \dfrac{u_T^2}{2} + \dfrac{\gamma}{\gamma-1}\dfrac{p_T}{\rho_T} \\[8pt] u_T^2 = \gamma \dfrac{p_T}{\rho_T} \end{cases}
\quad,\quad
b) \begin{cases} u_i \rho_i A_i = u_1 \rho_1 A_s \\[4pt] \dfrac{p_i}{\rho_i^\gamma} = \dfrac{p_1}{\rho_1^\gamma} \\[6pt] \dfrac{u_i^2}{2} + \dfrac{\gamma}{\gamma-1}\dfrac{p_i}{\rho_i} = \dfrac{u_1^2}{2} + \dfrac{\gamma}{\gamma-1}\dfrac{p_1}{\rho_1} \end{cases}
$$

$$
c) \begin{cases} u_1 \rho_1 = u_2 \rho_2 \\[4pt] \rho_1 u_1^2 + p_1 = \rho u_2^2 + p_2 \\[6pt] \dfrac{u_1^2}{2} + \dfrac{\gamma}{\gamma-1}\dfrac{p_1}{\rho_1} = \dfrac{u_2^2}{2} + \dfrac{\gamma}{\gamma-1}\dfrac{p_2}{\rho_2} \end{cases}
\quad,\quad
d) \begin{cases} u_2 \rho_2 A_s = u_o \rho_o A_o \\[4pt] \dfrac{p_2}{\rho_2^\gamma} = \dfrac{p_o}{\rho_o^\gamma} \\[6pt] \dfrac{u_2^2}{2} + \dfrac{\gamma}{\gamma-1}\dfrac{p_2}{\rho_2} = \dfrac{u_o^2}{2} + \dfrac{\gamma}{\gamma-1}\dfrac{p_o}{\rho_o} \end{cases}
\tag{13.121}
$$

The energy equation in system (13.121)-a) expresses u_i^2 hence M_i in terms of $(p_T/p_i)(\rho_i/\rho_T)$. The continuity equation in the same system expresses $(\rho_i/\rho_T)(A_i/A_T)$ also in terms of $(p_T/p_i)(\rho_i/\rho_T)$. The combination of these two separate results directly yields the following equation for M_i

$$
(\gamma-1)M_i^2 - (\gamma+1)\left(M_i \frac{A_i}{A_T}\right)^{\frac{2(\gamma-1)}{1+\gamma}} + 2 = 0
\tag{13.122}
$$

which can be solved numerically for M_i. The difference between this equation and the formally identical equation (13.119) lies in A_T. In (13.119), A_T represents a reference nozzle throat area, which may or may not coincide with the actual throat area. In (13.122), instead, A_T equals the actual nozzle throat area, because the flow is sonic at the throat for a normal shock to develop.

The energy equations from systems (13.121)-b), c), d) combined with the continuity equations from the same systems yield the following quadratic equation for ρ_i/ρ_o

$$
\left(M_i \frac{A_i}{A_o}\right)^2 \left(\frac{\rho_i}{\rho_o}\right)^2 + \left(\frac{2}{\gamma-1}\frac{p_o}{p_i}\right)\left(\frac{\rho_i}{\rho_o}\right) - \left(M_i^2 + \frac{2}{\gamma-1}\right) = 0
\tag{13.123}
$$

The corresponding physically meaningful solution for ρ_o/ρ_i emerges as

$$\frac{\rho_o}{\rho_i} = \frac{\sqrt{\left[\left(\frac{p_o}{p_i}\right)^2 + M_i^2(\gamma-1)\left(2+(\gamma-1)M_i^2\right)\left(\frac{A_i}{A_o}\right)^2\right] + \frac{p_o}{p_i}}}{2+(\gamma-1)M_i^2} \tag{13.124}$$

The determination of the outlet Mach number M_o results from the continuity equations from systems (13.121)-b), c), d) as

$$M_o = M_i\left(\frac{A_i}{A_o}\right)\sqrt{\left(\frac{p_i}{p_o}\right)\left(\frac{\rho_i}{\rho_o}\right)} \tag{13.125}$$

The corresponding inlet and outlet speeds result from the definition of the Mach number as

$$u_i = M_i\sqrt{\gamma\frac{p_i}{\rho_i}}, \qquad u_o = M_o\sqrt{\gamma\frac{p_o}{\rho_o}} \tag{13.126}$$

Heed that the determination of the outlet density ρ_o as well as the inlet and outlet Mach numbers and speeds do not require the normal-shock Mach numbers.

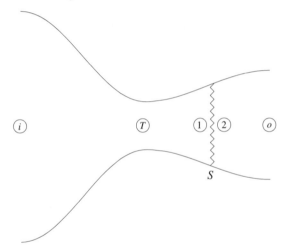

Figure 13.8: Normal Shock Wave

The normal-shock equations in system (13.121)-c) can be recast in terms of pre- and post-shock Mach numbers as

$$\begin{cases} \dfrac{p_2}{p_1}\dfrac{\rho_2}{\rho_1} = \dfrac{M_1^2}{M_2^2} \\[2mm] \dfrac{p_2}{p_1} = \dfrac{1+\gamma M_1^2}{1+\gamma M_2^2} \\[2mm] \dfrac{p_2}{p_1}\dfrac{\rho_1}{\rho_2} = \dfrac{2+(\gamma-1)M_1^2}{2+(\gamma-1)M_2^2} \end{cases} \tag{13.127}$$

The combination of these equations expresses the post-shock Mach number M_2 in terms of only M_1 as

$$M_2^2 = \frac{(\gamma - 1)M_1^2 + 2}{2\gamma M_1^2 - (\gamma - 1)} \tag{13.128}$$

The pressure relationships in systems (13.121)-b), d) lead to the following result in terms of pressure and density ratios

$$\frac{p_2}{p_1}\left(\frac{\rho_1}{\rho_2}\right)^\gamma = \frac{p_o}{p_i}\left(\frac{\rho_i}{\rho_o}\right)^\gamma \tag{13.129}$$

Insertion into this result of the linear-momentum and continuity conditions, from system (13.127), and then of expression (13.128), for M_2, yields the single equation for the pre-shock Mach number M_1

$$M_1^{2\gamma}(\gamma + 1)^{\gamma+1}\left(\frac{p_o}{p_i}\right)\left(\frac{\rho_i}{\rho_o}\right)^\gamma - \left(2 + (\gamma - 1)M_1^2\right)^\gamma\left(2\gamma M_1^2 - (\gamma - 1)\right) = 0 \tag{13.130}$$

which can be solved numerically for M_1. With this solution, the post-shock Mach number M_2 follows from (13.128).

By definition of stagnation states, the stagnation ratios p_{st_1}/ρ_{st_1} and p_{st_2}/ρ_{st_2} are expressed from (13.116) in terms of inlet and outlet variables as

$$\frac{p_{st_1}}{\rho_{st_1}} = \frac{p_i}{\rho_i}\left(1 + \frac{\gamma - 1}{2}M_i^2\right), \quad \frac{p_{st_2}}{\rho_{st_2}} = \frac{p_o}{\rho_o}\left(1 + \frac{\gamma - 1}{2}M_o^2\right) \tag{13.131}$$

The energy equations in systems (13.121)-b), c), d), in particular, show these ratios equal each other

$$\frac{p_{st_1}}{\rho_{st_1}} = \frac{p_{st_2}}{\rho_{st_2}} \Rightarrow T_{st_1} = T_{st_2} \tag{13.132}$$

fundamentally because the flow is adiabatic. With these stagnation ratios, the energy equations in (13.121)-b), d) lead to the calculation of the pre- and post-shock ratios p_1/ρ_1 and p_2/ρ_2

$$\frac{p_1}{\rho_1} = \frac{2p_{st_1}/\rho_{st_1}}{2 + (\gamma - 1)M_1^2}, \quad \frac{p_2}{\rho_2} = \frac{2p_{st_2}/\rho_{st_2}}{2 + (\gamma - 1)M_2^2} \tag{13.133}$$

The individual pre- and post-shock densities and pressures follow from the pre and post-shock pressure relationship as

$$\rho_1 = \left(\frac{p_i/\rho_i^\gamma}{p_1/\rho_1}\right)^{\frac{1}{1-\gamma}}, \quad p_1 = \left(\frac{p_1}{\rho_1}\right)\rho_1, \quad \rho_2 = \left(\frac{p_o/\rho_o^\gamma}{p_2/\rho_2}\right)^{\frac{1}{1-\gamma}}, \quad p_2 = \left(\frac{p_2}{\rho_2}\right)\rho_2 \tag{13.134}$$

The corresponding pre- and post-shock speeds follow from the definition of Mach number as

$$u_1 = M_1\sqrt{\gamma\frac{p_1}{\rho_1}}, \quad u_2 = M_2\sqrt{\gamma\frac{p_2}{\rho_2}} \tag{13.135}$$

The nozzle cross-sectional area A_s, where the normal shock takes place, results from the continuity equation in system (13.121)-b) as

$$A_s = A_i\frac{\rho_i u_i}{\rho_1 u_1} \tag{13.136}$$

The reference post-jump throat speed u_{T_2} and nozzle area A_{T_2} then follow from (13.113) and (13.119) as

$$u_{T_2} = \sqrt{\frac{2}{\gamma+1} \frac{p_{st_2}}{\rho_{st_2}}} = \sqrt{\frac{2}{\gamma+1} \frac{p_{st_1}}{\rho_{st_1}}} = u_{T_1}, \qquad A_{T_2} = A_o M_o \left(\frac{1+\gamma}{2+(\gamma-1)M_o^2}\right)^{\frac{1+\gamma}{2(\gamma-1)}} \qquad (13.137)$$

From (13.114) and (13.132), the pre- and post-shock throat area ratio A_{T_2}/A_{T_1} equals p_{st_1}/p_{st_2} as follows

$$\rho_{T_1} u_{T_1} A_{T_1} = \rho_{T_2} u_{T_2} A_{T_2} \quad \Rightarrow \quad \frac{A_{T_2}}{A_{T_1}} = \frac{\rho_{T_1}}{\rho_{T_2}} \frac{u_{T_1}}{u_{T_2}} = \frac{\rho_{T_1}}{\rho_{T_2}} = \frac{\rho_{st_1}}{\rho_{st_2}} = \frac{p_{st_1}}{p_{st_2}} \qquad (13.138)$$

The results in this section allow a complete calculation of a quasi-one-dimensional compressible flow with embedded normal shock. This calculation is solely based on the prescribed inlet density, nozzle throat area, and inlet and outlet nozzle areas and pressures.

13.9   Computational Results

The characteristics-bias system formulation with continuum acoustics-convection upstream bias has generated physically consistent computational solutions for smooth and shocked mixed subsonic / supersonic flows with cross-sectional area variations, heat as well as mass transfer, shaft work, and wall friction. The 14 different benchmarks in this section encompass one shock-tube flow, one classical heated frictional flow, and several flows with only heat transfer, only mass transfer, mass as well as heat transfer, only shaft work, wall friction, wall friction coupled with heat as well as mass transfer and shaft work, cross-sectional area variations without and with shocks, with shocks and heat transfer, and with wall friction coupled with heat as well as mass transfer and shaft work, with mass added in a direction perpendicular to the x-axis.

The spatial computational domain Ω for these benchmarks is defined as: $\Omega \equiv [x_{\text{in}}, x_{\text{out}}] = [0, 1]$, uniformly discretized through linear finite elements, for five progressively denser grids, as employed in the asymptotic convergence rate studies. These grids respectively contain 50, 100, 200, 400, and 800 linear elements. For each of these grids, every computational solution that admits a steady state rapidly converged, with the norm of the residual driven to the machine zero of 5.0×10^{-14}; the figures documenting the computational solutions in this section correspond to the 200-element grid. For each benchmark, the calculations proceeded with a prescribed constant maximum Courant number $C_{\max}$ defined as

$$C_{\max} \equiv \max\{|u+c|, |u-c|, c\} \frac{\Delta t}{\Delta x} \qquad (13.139)$$

Given Δx and $C_{\max}$ for each benchmark, the algorithm automatically calculates the corresponding Δt as

$$\Delta t = \frac{C_{\max}\Delta x}{\max\{|u+c|, |u-c|, c\}} \qquad (13.140)$$

with variable controller ψ determined as detailed in Chapter 10. Unless otherwise stated, all the computational solutions were obtained with a Courant number set equal to 100 and

$\psi_{\min} = 0.25$. Although $\psi_{\max}$ was set equal to 1.00, the formulation automatically kept ψ essentially equal to $\psi_{\min}$ in regions of solution smoothness and only increased it in the vicinity of a discontinuity.

The formulation has produced essentially non-oscillatory results that also reflect available exact solutions. These computational solutions exhibit rapid convergence rate to a steady state and definite asymptotic convergence rate in both the $\mathcal{H}^0$ and $\mathcal{H}^1$ norms, for continuous solutions, and in the $\mathcal{H}^0$ norm shocked solutions.

13.9.1   Sod's Shock Tube Flow

This benchmark is presented to show that the Galilean invariant characteristics-bias formulation leads to an increased resolution of the computed contact discontinuity. In this flow, the tube is initially divided in two chambers separated by an impermeable diaphragm placed at the midpoint of the tube axis. In these chambers, the non-dimensional initial conditions for the gas are

$$\rho = 1.00, \quad m = 0.00, \quad E = 2.50, \quad 0.0 \le x \le 0.5$$

$$\rho = 0.125, \quad m = 0.00, \quad E = 0.25, \quad 0.5 < x \le 1.0 \tag{13.141}$$

The diaphragm ruptures at $t = 0$ and the computational investigation with $C_{\max} = 1.0$ seeks the solution at the non-dimensional time $t = 0.14152$. At this time station, the exact solution features a normal shock centered at $x = 0.75$, for each of the components of the dependent variables in q, and a contact discontinuity centered at $x = 0.62$, for the distribution of density ρ.

Figures 13.9 a)-d) present the distributions of density at this time level, for several formulations of the mass matrix and magnitudes of the upstream-bias controller ψ. Figure 13.9 a) corresponds to a lumped mass matrix. This solution remains monotone, but diffused, with contact discontinuity and shock spread over several elements; significantly, the lumped-mass formulation for this benchmark has been found to remain devoid of unphysical oscillations only when $\psi \ge 1$, uniformly throughout the computational domain. Figure 13.9 b) is also obtained for a uniform $\psi = 1$, but with a consistent mass matrix, which corresponds to no upstream bias on the time derivative. This distribution remains virtually indistinguishable from the previous one, which shows that a consistent mass-matrix formulation can generate unsteady solutions devoid of unphysical oscillations. The solution in Figure 13.9 c) also corresponds to a consistent mass matrix, but results from a variable ψ, with $\psi_{\min} = 0.25$ and $\psi_{\max} = 0.75$. The upstream bias in this case is no longer uniform, but is non-linearly applied in relation to the variation of the local solution slopes. Accordingly, the resolution of this solution has increased with a crisply captured shock wave and somewhat sharper, but still diffused contact discontinuity. It is the Galilean invariant formulation with a variable ψ and upstream bias on the time derivative that noticeably increases the resolution of the contact discontinuity, which provides another justification the use of this formulation.

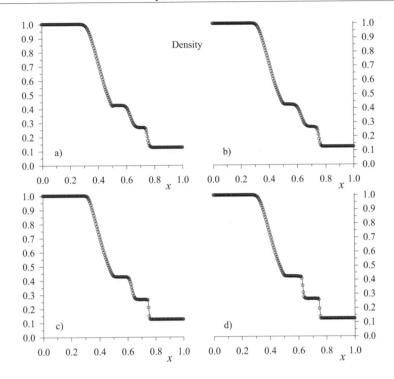

Figure 13.9: Density: a) Linear, Lumped Mass Matrix, b) Linear, Consistent Mass Matrix c) Non-linear, Consistent Mass Matrix, d) Non-linear, Galilean Invariant

The formulation also leads to sharp contact discontinuities in the mass flow and Mach number distributions, as shown in Figures 13.10- 13.11, while it correctly preserves the pressure plateau across this discontinuity, as shown in Figure 13.12.

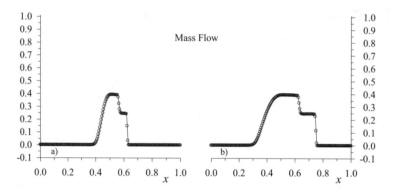

Figure 13.10: Galilean Invariant Formulation, Mass Flow: a) $t=0.07076$, b) $t=0.14152$

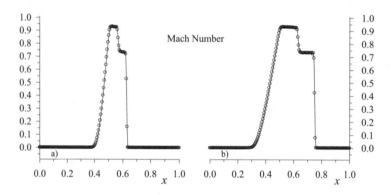

Figure 13.11: Galilean Invariant Formulation, Mach Number: a) $t=0.07076$, b) $t=0.14152$

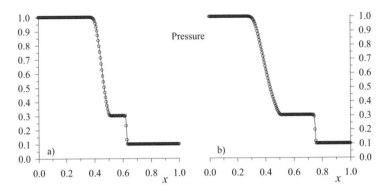

Figure 13.12: Galilean Invariant Formulation, Pressure: a) t=0.07076, b) t=0.14152

These three figures display solutions at two distinct time stations. In particular, they are representative of the essentially non-oscillatory character of this transient solution.

13.9.2 Heat-Transfer Flows

These computations have determined a subsonic and a supersonic duct flow induced by constant heating throughout the length of a smooth-wall cylindrical duct with length $L = 1$m and circular cross section and . The accuracy of the computational predictions is then gauged against the exact Rayleigh-flow results. The dimensional benchmark specifications are the inlet pressure $p_{in} = 100$kPa and temperature $T_{in} = 10^\circ$C, with $(g_T)_{dim} = 105.994$kJ/(kg m) and $M_{in} = 0.5$ for the subsonic flow, and $(g_T)_{dim} = 124.088$kJ/(kg m) and $M_{in} = 3.0$ for the supersonic flow. Employing the sonic state as the reference state, the corresponding non-dimensional specifications as well as boundary conditions are: $\rho_{in} = 2.259$, $E_{in} = 4.756$, $p_{out} = 1.087$, with $g_T = 1.28$, corresponding from Rayleigh flow to $m_{in} = 1.183$ and $M_{out} = 0.929$, for the subsonic flow, and $\rho_{in} = 0.630$, $m_{in} = 1.183$, $E_{in} = 1.553$, with $g_T = 1.40$, corresponding to $p_{out} = 0.899$ and $M_{out} = 1.149$ for the supersonic flow. The initial conditions throughout the duct correspond to a uniform flow from the specified inlet Mach number, a flow that is then advanced in time to a steady state. The corresponding computational results in Figure 13.13

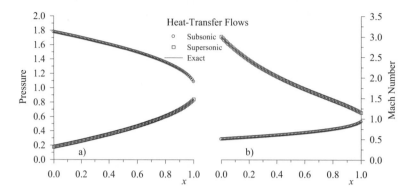

Figure 13.13: Heat-Transfer Flows: a) Pressure, b) Mach Number

reflect the exact Rayleigh-flow solution, with heating inducing an increase or decrease in Mach number, respectively for a subsonic or a supersonic flow. Although the mass flow rate remains unaltered, a constant heating theoretically induces a linear rise in the total enthalpy. This feature is correctly reflected in the computational solution in Figure 13.14, which displays a linear increasing total enthalpy.

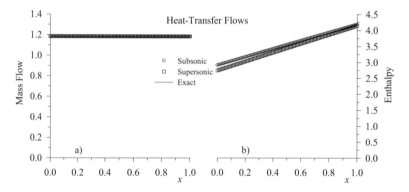

Figure 13.14: Heat-Transfer Flows: a) Mass Flow, b) Total Enthalpy

Figure 13.15 portrays the asymptotic convergence rate for these two benchmarks. As shown, the convergence rate in the $\mathcal{H}^0$ norm achieves an order as high as 2.97, for the subsonic flow. The algorithm features such a desirable rate because as the grid is refined, the characteristics-bias perturbation magnitude also decreases. Accordingly, the computational solution more rapidly approaches the exact solution than it would under sole grid refinement.

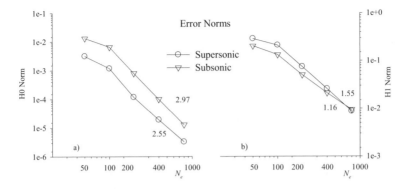

Figure 13.15: Heat-Transfer Flows: a) $\mathcal{H}^0$ Norm, b) $\mathcal{H}^1$ Norm

The convergence rate for the supersonic flow exceeds 2.5, even though it remains less than the subsonic-flow rate. This finding correlates with the absence of the "α" perturbation for supersonic flows, which eliminates the effect of this perturbation from the decreasing error norm. A similar difference in convergence rates between a subsonic and a supersonic flow also affects the $\mathcal{H}^1$ norm, which exceeds 1 and indicates convergence of the solution gradients.

13.9.3 Mass-Transfer Flows

Although infrequently covered in reference Gas Dynamics books, mass-transfer flows remain of technological importance and a related exact steady-state solution can be determined for the case of mass injected in a direction perpendicular to the tube axis. The benchmarks for these flows encompass a subsonic and a supersonic flow in a straight cylindrical duct with circular cross sections, both flows subject to a uniform mass addition. For these benchmarks, the non-dimensional specifications as well as boundary conditions are: $\rho_{in} = 1.808$, $E_{in} = 5.463$, $p_{out} = 1.145$, with $\dot{m}_b = 0.480$, corresponding from the available exact solution to $M_{in} = 0.300$, $m_{in} = 0.697$ and $M_{out} = 0.885$, for the subsonic flow, and $\rho_{in} = 0.412$, $m_{in} = 0.957$, $E_{in} = 1.553$, with $\dot{m}_b = 0.220$, corresponding to $M_{in} = 3.000$, $p_{out} = 0.855$ and $M_{out} = 1.136$ for the supersonic flow. The initial conditions throughout the duct correspond to a uniform flow from the specified inlet Mach number, a flow that is then advanced in time to a steady state.

Presented in Figure 13.16, the computational results are seen to reflet the exact solution, with mass addition inducing an increase or decrease in Mach number, respectively for a subsonic or a supersonic flow.

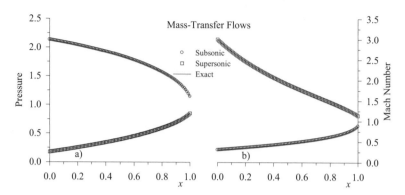

Figure 13.16: Mass-Transfer Flows: a) Pressure, b) Mach Number

Since no heat transfer affects this flow, the total enthalpy remains constant, yet the mass flow rate theoretically increases proportionately to the added mass. This feature is reflected in the computational results in Figure 13.17, which correctly present a constant enthalpy and a linearly increasing mass flow rate.

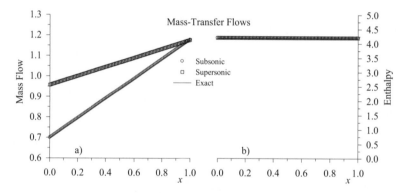

Figure 13.17: Mass-Transfer Flows: a) Mass Flow, b) Total Enthalpy

The available exact solution allowed the calculation of the $\mathcal{H}^0$ and $\mathcal{H}^1$ norms of the error associated with solutions for increasingly denser grids. Indicating solution convergence, these error norms are displayed in Figure 13.18

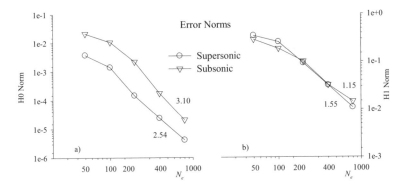

Figure 13.18: Mass-Transfer Flows a) $\mathcal{H}^0$ Norm, b) $\mathcal{H}^1$ Norm

The noted difference between subsonic- and supersonic-flow convergence rates is also reflected in this figure. For the subsonic flow, the asymptotic convergence rate achieves an order as high as 3.10, for the $\mathcal{H}^0$ norm, and as high as 1.55, for the $\mathcal{H}^1$ norm. These norms reflect convergence of not only the solution itself, but also its gradients.

13.9.4 Mass- and Heat-Transfer Flows

As in the case of a combustible mixture injected into a stream, mass addition may also be accompanied by flow heating. For the purpose of this investigation, an exact solution for this challenging case has been obtained and employed to assess the accuracy of the corresponding computational solutions, for both subsonic and supersonic flows. For these flows, the non-dimensional specifications as well as boundary conditions are: $\rho_{in} = 2.254$, $E_{in} = 5.463$, $p_{out} = 1.149$, with $g = 0.820$, $\dot{m}_b = 0.400$, corresponding from the exact solution to $M_{in} = 0.300$, $m_{in} = 0.697$ and $M_{out} = 0.882$, for the subsonic flow, and $\rho_{in} = 0.476$, $m_{in} = 1.029$, $E_{in} = 1.553$, with $g = 0.550$, $\dot{m}_b = 0.150$, corresponding to $M_{in} = 3.000$, $p_{out} = 0.850$ and $M_{out} = 1.141$ for the supersonic flow. As in the previous benchmarks, the initial conditions throughout the duct correspond to a uniform flow from the specified inlet Mach number, a flow that is then advanced in time to a steady state.

The corresponding computational solutions are displayed in Figure 13.19 in terms of static pressure and Mach Number, with mass transfer and heating inducing an increase or decrease in Mach number, respectively for a subsonic or a supersonic flow.

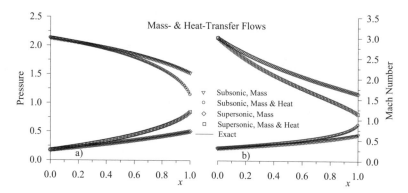

Figure 13.19: Mass-Transfer Flows: a) Pressure, b) Mach Number

As expected, the superposition of heat onto a mass-transfer further increases or decreases the Mach number respectively for subsonic and supersonic flows. As shown in Figure 13.20, the addition of both heat and mass to a flow induces a simultaneous rise in mass flow rate and total enthalpy.

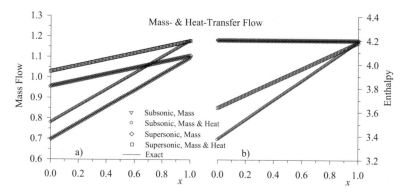

Figure 13.20: Mass-Transfer Flows: a) Mass Flow, b) Total Enthalpy

For constant mass addition and heating, the computational solutions correctly show a linearly increasing mass flow rate and enthalpy.

With reference to the asymptotic convergence rate, the noted difference between subsonic- and supersonic-flow convergence rates is also reflected in these solutions, as shown in Figure 13.21.

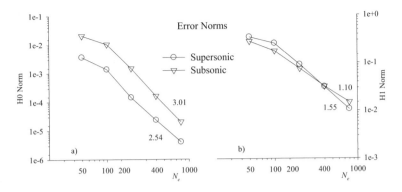

Figure 13.21: Mass-Transfer Flows: a) $\mathcal{H}^0$ Norm, b) $\mathcal{H}^1$ Norm

For the subsonic flow, the asymptotic convergence rate achieves an order as high as 3.01, for the $\mathcal{H}^0$ norm, and as high as 1.55 for the $\mathcal{H}^1$ norm. These norms reflect convergence of not only the solution itself, but also its gradients.

13.9.5 Work Driven Flows

Shaft-work driven flows support turbomachinery, where shaft work is provided to or extracted from a flow respectively through a compressor or a turbine. Both heat transfer and shaft work feature in the same term in the generalized energy equation. The effect of this work on a flow, however, remains markedly different from the effect of heat because, unlike heat, mechanical work also induces a corresponding force in the linear momentum equation.

For the case of a stage efficiency equal to 1, hence $\chi = 1$, also for this kind of flows is it possible to generate a basic steady-state exact solution, which has been determined for this study. On the basis of such a solution, the computational predictions for subsonic and supersonic flows are found in complete agreement with the mechanics of these flows. For the benchmarks for these flows, the non-dimensional specifications as well as boundary conditions are: $\rho_{in} = 2.727$, $E_{in} = 10.442$, $p_{out} = 1.181$, with $w_m = 1.100$, corresponding from the exact solution to $M_{in} = 0.300$, $m_{in} = 1.183$ and $M_{out} = 0.867$, for the subsonic flow, and $\rho_{in} = 0.400$, $m_{in} = 1.183$, $E_{in} = 2.443$, with $w_m = 2.575$ corresponding to $M_{in} = 3.000$, $p_{out} = 0.862$ and $M_{out} = 1.136$ for the supersonic flow. As in the previous benchmarks, the initial conditions throughout the duct correspond to a uniform flow from the specified inlet Mach number, a flow that is then advanced in time to a steady state.

According to basic thermodynamic considerations, extracting work from a subsonic flow decreases static pressure, as expected, but consequently increases the Mach number; conversely when work is extracted from a supersonic flow, it is pressure that increases, while the Mach number decreases. These precise trends are reflected in the computational solutions shown in Figure 13.22, which mirror the exact solutions for both subsonic and supersonic flows.

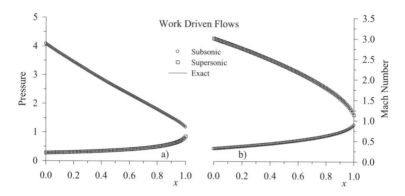

Figure 13.22: Work-Driven Flows: a) Pressure, b) Mach Number

While keeping mass flow unchanged, these flows can only produce work by expending total enthalpy. Such a canonical thermodynamic balance is reflected in the computational solutions graphed in Figure 13.23. Obtained for a constant shaft work, these solutions correctly display a constant mass flow coupled with a linearly decreasing total enthalpy.

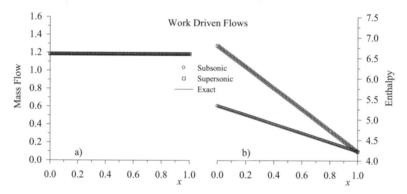

Figure 13.23: Work-Driven Flows a) Mass Flow, b) Total Enthalpy

In respect of the asymptotic convergence rate of these solutions, the noted difference between subsonic- and supersonic-flow convergence rates also manifests itself in these results, graphed in Figure 13.24.

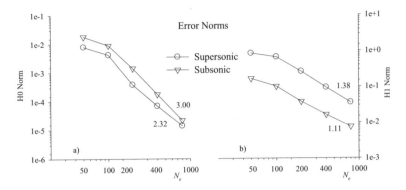

Figure 13.24: Work-Driven Flows: a) $\mathcal{H}^0$ Norm, b) $\mathcal{H}^1$ Norm

For subsonic flows, the asymptotic convergence rate achieves an order as high as 3.00, for the $\mathcal{H}^0$ norm, and an order as high as 1.38 for the $\mathcal{H}^1$ norm. Again, these norms reflect convergence of not only the solution itself, but also its gradients.

13.9.6   Adiabatic Smooth and Shocked Flows with Wall Friction

Wall friction affects any duct flow, even smooth ones, due to viscosity. Directly altering flow momentum, wall friction leads to not only smooth, but also shocked steady flows. This section details computational solutions of frictional subsonic and supersonic flows, without and with shocks, within a straight cylindrical duct with length $L = 1\mathrm{m}$ and circular cross section. Also known as Fanno flows, these flows are described by an exact closed-form solution that depends upon not only the friction factor "f_D", but also the non-dimensional ratio D/L of duct diameter over length; this solution has been employed to asses the accuracy of the computational solutions, as obtained for $f_D = 0.02$, which corresponds to a duct made of cast iron.

The dimensional specifications for the continuous flows are the inlet pressure $p_{\mathrm{in}} = 1.0\mathrm{atm}$ and temperature $T_{\mathrm{in}} = 300\mathrm{K}$, with $D = 4.2\mathrm{cm}$ and $M_{\mathrm{in}} = 0.6$ for the subsonic flow, and $D = 7.5\mathrm{cm}$ as well as $M_{\mathrm{in}} = 2.0$ for the supersonic flow. Employing the sonic state as the reference state, the corresponding non-dimensional specifications as well as boundary conditions are: $\rho_{\mathrm{in}} = 1.575$, $E_{\mathrm{in}} = 4.853$, $p_{\mathrm{out}} = 1.130$, corresponding from Fanno flow to $m_{\mathrm{in}} = 1.183$ and $M_{\mathrm{out}} = 0.900$, for the subsonic flow, and $\rho_{\mathrm{in}} = 0.612$, $m_{\mathrm{in}} = 1.183$, $E_{\mathrm{in}} = 2.164$, corresponding to $p_{\mathrm{out}} = 0.791$ and $M_{\mathrm{out}} = 1.216$ for the supersonic flow. The initial conditions throughout the duct correspond to a uniform flow from the specified inlet Mach number, a flow that is then advanced in time to a steady state.

According to the widely used Fanno-flow solution, friction decreases pressure yet increases the Mach number, in a subsonic flow, but increases pressure yet decreases the Mach number in a supersonic flow. Such trends resonate with those in flows with work extraction because both friction and work extraction develop a force that opposes the flow. These trends are also reflected in the computational solutions in Figure 13.25 that are seen to mirror the exact solutions for both subsonic and supersonic flows.

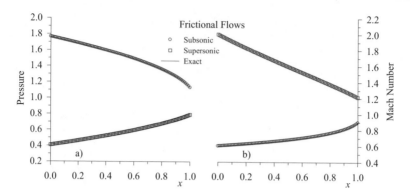

Figure 13.25: Frictional Flows: a) Pressure, b) Mach Number

With reference to the asymptotic convergence rate of these solutions, displayed in Figure 13.26, the noted difference between subsonic- and supersonic-flow convergence rates is also reflected in these results.

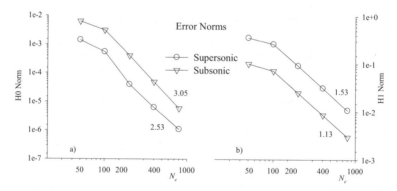

Figure 13.26: Frictional Flows: a) $\mathcal{H}^0$ Norm, b) $\mathcal{H}^1$ Norm

For the subsonic flow, the asymptotic convergence rate achieves an order as high as 3.05, for the $\mathcal{H}^0$ norm, and as high as 1.53 for the $\mathcal{H}^1$ norm subsonic flow. These norms reflect convergence of not only the solution itself, but also its gradients.

The next benchmark involves a steady frictional flow incorporating a normal shock that remains localized in the duct. This benchmark allows testing the capability of the characteristics-bias system formulation to generate an essentially non-oscillatory shocked frictional flow. The accuracy of the computational prediction is then assessed against the exact Fanno-flow solution. For this benchmark with $f_D = 0.02$ and $D = 7.5$cm, the non-dimensional specifications as well as boundary conditions are: $\rho_{\text{in}} = 0.612$, $m_{\text{in}} = 1.183$, $E_{\text{in}} = 2.164$ at the supersonic inlet, and $p_{\text{out}} = 1.368$ at the subsonic outlet. From the exact solution, these specifications correspond to $M_{\text{in}} = 2.000$ and $M_{\text{out}} = 0.759$ with a steady shock that develops at $x = 0.6$ with upstream and downstream shock Mach numbers $M_u = 1.525$ and $M_d = 0.693$. In this case, the initial conditions result from a linear interpolation between the inlet supersonic state and outlet subsonic state. The resulting flow is then advanced in time to a steady state.

Achieved through a Courant number equal to 50, the computational solution is summarized in Figures 13.27-13.28 in terms of pressure, Mach number, density, mass flow, enthalpy and energy. These variations remain essentially non-oscillatory with a crisp normal shock captured in at most two nodes; most importantly, this shock is captured at the theoretically exact location of $x = 0.6$, with accurately calculated pressure ratio and Mach numbers across the shock. In the neighborhood of the outlet, the results remain devoid of any spurious oscillations, which reflects favorably on the enforcement of the pressure boundary condition indicated in Section 13.6.3. With accurately calculated normal-shock and outflow state, this solution reflects the reference exact Fanno-flow solution.

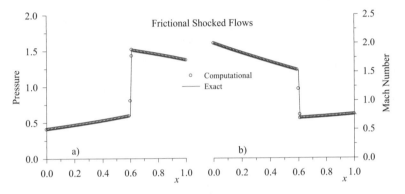

Figure 13.27: Frictional Flows: a) Pressure, b) Mach Number

Despite the normal shock, mass flow remains unchanged and total enthalpy does not vary in an adiabatic flow. Reflecting the physical consistency of the acoustics-convection characteristics-bias formulation, these features are predicted by the computational solution, as shown in Figure 13.28. This solution correctly predicts a constant mass flow and enthalpy that remain undistorted by the shock in the other variables.

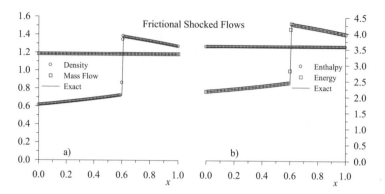

Figure 13.28: Frictional Flows: a) Density and Mass Flow, b) Enthalpy and Energy

The characteristics-bias formulation with acoustics-convection Euler flux-Jacobian decomposition depends on the upstream-bias functions α, δ, ψ. Presented in Figure 13.29 for this shocked frictional flow, the variations of these functions and the nodal ℓ_1 norm of the q^h solution show that α vanishes for $M > 0.6$, while δ, as expected, experiences a sharp decrease as the flow switches from supersonic to subsonic. Following the distribution of the nodal norm, significantly, the formulation leads to $\psi = \psi_{\min}$, for most of the flow, and $\psi = \psi_{\max}$ at the shock. Accordingly, the formulation automatically manages the level of induced upstream-bias, keeping it to a minimum in regions of smooth flow, and focusing an increase only at shocks, in order to generate a solution that remains both accurate and stable.

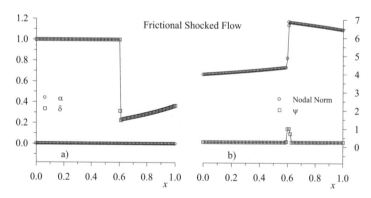

Figure 13.29: Frictional Flows: a) α and δ, b) Nodal Norm and ψ

Also for this shocked flow has the formulation generated an asymptotically converging solution in the $\mathcal{H}^0$ norm, with corresponding convergence-rate curves presented in Figures 13.30 a)-b).

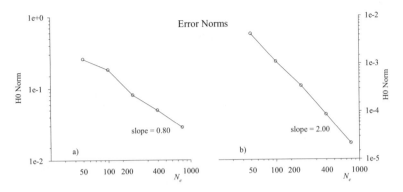

Figure 13.30: Frictional Flows: a) Shocked Solution, b) Continuous Solution Branches

Figure 13.30 a) displays the error norm for the entire solution, including the shock, while Figure 13.30 b) displays the error norm for the union of the continuous subsonic and supersonic solution branches, a union obtained by only considering the computational and exact solutions in the interval $[0.00, 0.56] \cup [0.64, 1.0]$. The shocked solution converges at a rate of 0.8, yet in respect of the continuous solution branches their union is seen to converge at a rate of 2.0. This expected difference in convergence rates between these two solution sets indicates the normal shock dominates the overall convergence rate. The formulation thus generates a computational solution that converges also for a discontinuous flow.

13.9.7 Shocked Flows with Wall Friction, Heat, Mass, and Work Transfer

Difficult to solve analytically in closed form, this problem involves the shocked flow described in the previous section as altered by uniform heat, mass, and work transfer along the entire duct length. Numerically, this problem tests the performance of the algorithm accurately to compute a shocked flow simultaneously subject to wall friction as well as heat, mass, and work transfer; physically, this problem illustrates the quantitative measurable effect of heat, mass, and work transfer on both a duct flow and the position as well as strength of a normal shock within the flow.

The solutions generated by the characteristics-bias formulation for this investigation correspond to the same initial conditions as well as inlet and outlet boundary conditions discussed in the previous section, coupled with different magnitudes of the transfer source terms in the generalized Euler equations. The computed results correspond to the following non-dimensional magnitudes of heat, mass, and work transferred to the flow: $g_T = \pm 1.00$, $\dot{m}_b = \pm 1.00 \times 10^{-3}$, $w_m = \pm 1.00 \times 10^{-2}$, where the positive determination corresponds to heating, mass addition as well as work extraction, while the negative determination corresponds to cooling, mass removal and work provision to the flow.

Computed via a Courant number equal to 50, the solutions achieved for heating and cooling, mass addition and removal, work extraction from and provision to the flow are summarized in Figures 13.31-13.32 in terms of pressure and Mach number variations. As in the previous investigations, these variations remain essentially non-oscillatory for every magnitude of each transfer mechanism, with crisp normal shocks captured in at most two nodes. For all the cases considered of heat, mass and work transfer in this frictional flow, no outflow distortion emerges for any position of the normal shock within the duct, again reflecting favorably on the pressure boundary-condition enforcement procedure in Section 13.6.3.

For the prescribed single constant outlet-pressure boundary condition, the position of the shock is entirely determined by the amount of the three transfer mechanisms. As the figures show, the shock propagates upstream under heating, mass addition and work extraction, but downstream under cooling, mass removal and work provision to the flow. Both mass addition and work extraction have thus the same effect as heating, for together they increase flow entropy, whereas both mass removal and work provision have the same effect as cooling, for the combination of these effects decreases the flow entropy. Conforming to the physics of this flow, the computed solutions can then be discussed in terms of basic heating and cooling. As obtained from Rayleigh-flow theory, heating tends to increase the Mach number for subsonic flows, but to decrease it for supersonic flows.

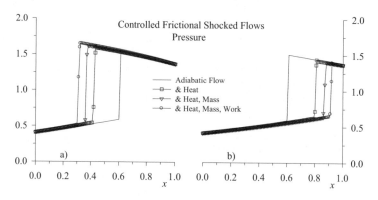

Figure 13.31: Pressure in Shocked Frictional Flows: a) Heat and Mass Added, Work Extracted, b) Heat and Mass Removed, Work Provided

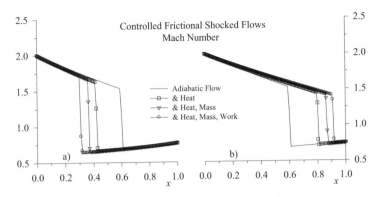

Figure 13.32: Mach Number in Shocked Frictional Flows: a) Heat and Mass Added, Work Extracted, b) Heat and Mass Removed, Work Provided

In comparison to the baseline case of no heat transfer, indicated by the unmarked solid line, heating moves the shock upstream; this perturbation forces a greater drop in Mach number on the supersonic side of the flow, which in turn allows a greater increase in Mach number on the subsonic side of the flow. From a kinetic-theory viewpoint, the addition of heat increases the molecular kinetic energy, which further disturbs a supersonic flow into an upstream-shifted stronger shock, but further energizes a subsonic flow both to negotiate the adverse friction force and meet the imposed pressure boundary condition. This effect is then magnified by mass addition and work extraction.

Conversely, cooling moves the shock downstream in comparison to the baseline case of no heat transfer. As it extracts energy from the flow, cooling forces a smaller decrease in Mach number on the supersonic side of the flow, which in turn correlates with a smaller increase in Mach number on the subsonic side. From a kinetic-theory viewpoint, the removal of heat decreases the molecular kinetic energy, which lessens disturbance of a supersonic flow and thereby sustains it for a longer distance, but also reduces energy of a subsonic flow and thereby decreases its spatial extent. For this reason the shock moves downstream, so as to accommodate a shorter region of subsonic flow that can only negotiate a smaller amount of adverse friction force, but can still meet the prescribed outlet pressure boundary condition. This effect is then strengthened by mass removal and work provision.

These are the first computations known to this author that employ CFD to investigate a straight-duct shocked frictional flow governed by heat, mass and work transfers. As the results indicate, the magnitude of heat, mass, and work transfer directly determine the eventual position and strength of the normal shock. These transfers may thus be employed to control a normal shock.

13.9.8 Shapiro's Benchmark

This benchmark involves the determination of a straight-duct flow that develops under the combined effect of both wall friction and heating. This flow allows comparing the predictions of the characteristics-bias procedure with the results from a trial-and-error method used to solve this problem, as delineated in the classical reference by Shapiro [175]. That method involves several "influence coefficients" to determine the exit Mach number for this flow. Conversely, this formulation solves the generalized Euler equations to determine the entire flow throughout the duct. In this benchmark, the cylindrical duct with circular cross section has a length $L = 1$m and diameter $D = 0.1$m, with a wall-friction factor $f = 0.005$. The benchmark specifies the following conditions: $M_{in} = 0.5$, $p_{in} = 101.5$kPa, $T_{in} = 300$K, outlet pressure ratio $p_{in}/p_{out} = 0.9652$, and outlet-intake stagnation temperature ratio $T_{0out}/T_{0in} = 1.05$, which corresponds to $(g_T)_{dim} = 15.813$kJ/kg as the flow heating per unit mass. Using the inlet stagnation state as reference, these specifications lead to the non-dimensional heating energy $g_T = 0.1749$, along with the non-dimensional magnitudes: $\rho_{in} = 0.8852$, $E_{in} = 2.2550$, and $p_{out} = 0.8137$, corresponding to $m_{in} = 0.5111$. The initial conditions throughout the duct corresponds to a uniform $M_{in} = 0.5$ subsonic flow, which evolved in time towards a final steady state.

The study in [175] calculated the exit Mach number $M_{out} = 0.5293$, a determination that was achieved holding the inlet Mach number equal to 0.5. At a subsonic intake, however, the Mach number is determined by the downstream flow. The computational solution corresponding to an intake Mach number determined by the evolving solution, on the one hand, generated the following results: $M_{in} = 0.4906$, $M_{out} = 0.5206$, which remained insensitive to the magnitude of ψ_{min}; on the other hand, the computational solution corresponding to a fixed intake Mach number of 0.5, exactly as in [175], led to $M_{out} = 0.5292$. In addition to this parameter, the computational solution has determined the entire flow, as displayed in Figure 13.33

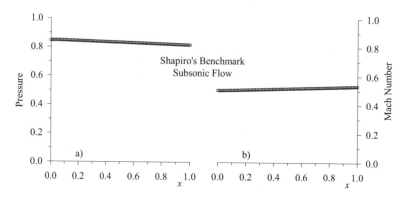

Figure 13.33: Shapiro's Heated Frictional Flow: a) Pressure, b) Mach Number

The Mach number and pressure-ratio variations in these curves remain smooth and follow the mechanics of this flow.

13.9.9 Flows in a Diverging Nozzle

These benchmarks examine the capability of the algorithm to calculate steady isentropic and shocked flows that involve a high-Mach number supersonic region. The nozzle cross-section area distribution features a steep increase in the diverging region, as shown in Figure 13.34, which makes it challenging numerically to compute a non-oscillatory shock located in such a region. The nozzle area ratio distribution for these benchmarks is

$$A(x) = a + b \tanh(8x - 4)$$

$$a = 1.39777, \quad b = 0.34760, \quad A(0) = 1.05041, \quad A(1) = 1.74514 \tag{13.142}$$

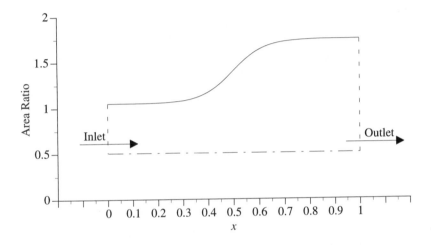

Figure 13.34: Variation of Area Ratio $A(x)/A_*$

The initial conditions for the gas correspond to an $M = 1.26$ uniform supersonic state, which leads to the following initial state throughout the nozzle

$$\rho = 0.50189, \quad m = 0.65187, \quad E = 1.37567 \tag{13.143}$$

The inlet flow is constrained supersonic at $M = 1.26$; Dirichlet boundary conditions are thus enforced on density ρ, linear momentum m and total energy E. An outlet pressure boundary condition is also imposed for the simulation of a shocked flow.

For the given initial conditions, the outlet is already supersonic, hence no boundary conditions are enforced at the outlet. The evolution of the flow from the initial conditions is thus entirely driven by the area source term in the governing Euler equations.

The distributions of density, linear momentum, pressure and enthalpy in Figures 13.35 a)-b) also mirror the corresponding exact solutions, with swift expansions clearly resolved. The algorithm has also correctly held constant the computed enthalpy, which satisfies the steady-adiabatic-flow constant-enthalpy condition. Figure 13.36 a) presents the distributions of total energy E, which remains indistinguishable from the exact solution, and upstream bias controller ψ, with $0.25 \leq \psi < 1.0$. Since the solution is smooth, ψ stays virtually constant with $\psi \simeq 0.25$ for this steady state, which corresponds to a 25% upstream bias.

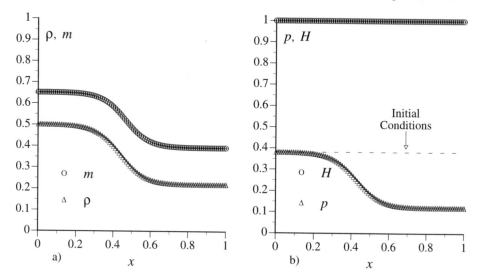

Figure 13.35: Isentropic Flow: a) Density and Momentum, b) Pressure and Enthalpy

Figure 13.36 b) presents the distribution of Mach number M, which also agrees with the exact solution, and the variations of acoustics and pressure-gradient upstream-bias functions α and δ. For a supersonic flow, $\alpha \equiv 0$ and $\delta \equiv 1$. No acoustics-matrix upstream-bias is thus present in this solution and the entire flux divergence receives a uniform upstream-bias approximation.

A normal shock wave features in this steady flow as a result of the subsonic-outlet pressure boundary condition $p/p_{\text{tot}_{in}} = 0.746$, imposed as an impulsive step decrease from the initial conditions, for the entire flow evolution toward steady state. The theoretical solution places the normal shock at the area ratio $A_s = 1.35016$, which corresponds to the interior of the finite element with node coordinates $x = 0.48$ and $x = 0.49$, within the computational domain. The exact shock Mach numbers are $M_{\text{sup}} = 1.71319$ and $M_{\text{sub}} = 0.63717$, which lead to the stagnation pressure and critical-area ratios $p_{\text{tot}_{out}}/p_{\text{tot}_{in}} = A_{*in}/A_{*out} = 0.85022$. The associated outlet area ratio and Mach number are $A_{\text{out}}/A_{*in} = 1.74514$, $M_{\text{out}} = 0.43629$.

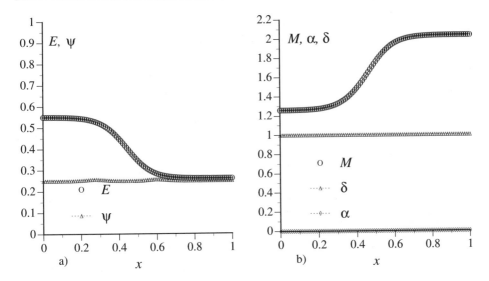

Figure 13.36: Isentropic Flow: a) Energy and Controller, b) Mach Number, α and δ

Figures 13.37 a)-b) present the convergence rate and steady-state speed.

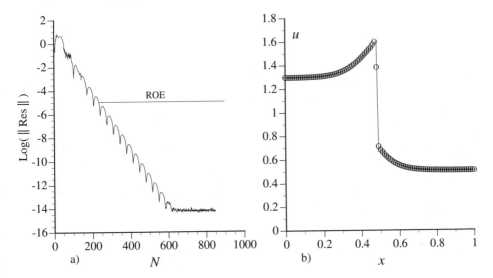

Figure 13.37: Shocked Flow: a) Convergence Rate, b) Speed

The algorithm is certainly capable of driving the maximum residual down to machine zero, with a total residual reduction by 14 orders of magnitude, with $C_{max} = 50$; a residual reduction by 6 orders of magnitude occurs within about 250 time steps. Other initial conditions, closer to the steady state, would presumably lead to faster convergence. These results, however, compare favorably with those computed using Roe's algorithm for this benchmark and with a similar computational solution. This comparison rests on the observation that the acoustics-convection upstream resolution algorithm is not a purely flux-vector splitting algorithm, but it also uses, like Roe's algorithm, a Riemann solver, 'though applied to the acoustics equations. The calculated speed reflects the exact solution, indicated with a solid line, and clearly shows the expected rapid rise preceding the shock as well as an excellent calculated normal shock, captured over only one node.

The distributions of static density and pressure in Figures 13.38 a)-b) also reflect the corresponding exact solutions, with rapid changes in these two variables clearly resolved and normal shock again sharply captured over one internal node. In harmony with the previous benchmark results, also for this problem has the algorithm correctly generated continuous distributions for both linear momentum m and enthalpy H across the normal shock, without any shock distortion. Furthermore, the computed H remains again constant, as has to be the case for a steady adiabatic flow.

Figure 13.39 a) presents the distributions of total energy E, which visually coincides with the

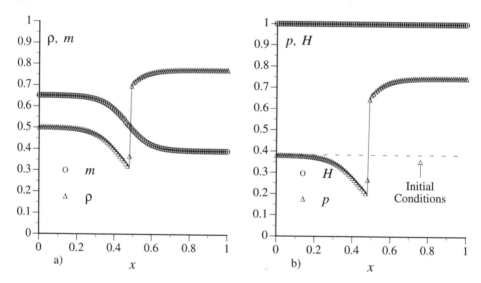

Figure 13.38: Shocked Flow: a) Density and Momentum, b) Pressure and Enthalpy

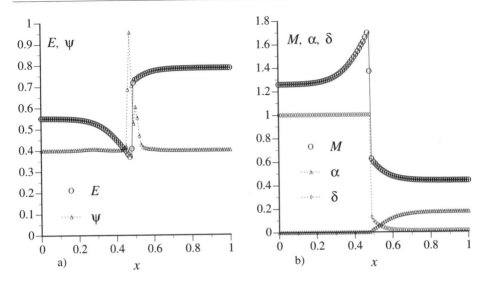

Figure 13.39: Shocked Flow: a) Energy and Controller, b) Mach Number, α and δ

exact solution in solid line, and upstream bias controller ψ, with $0.4 \leq \psi < 1.0$. The controller remains essentially constant over smooth solution regions, with $\psi = 0.4$, which corresponds to a 40% upstream bias. At the shock, ψ rapidly rises and reaches a $\psi \simeq 0.96$ extremum, which induces a 96% upstream-bias. No overshoots are present in this solution, and as Figure 13.39 a) bears out the algorithm succeeds in focusing an increased level of upstream-bias, hence artificial dissipation, at the shock region only, precisely where required for an essentially sharp and non-oscillatory solution. Figure 13.39 b) presents the distribution of Mach number M, which also reflects the exact solution, and the corresponding distributions of the acoustics and pressure-gradient upstream-bias functions α and δ. According to these distributions, the acoustics upstream-bias is present only in the outlet subsonic regions of the flow. This type of upstream-bias vanishes in the supersonic region, including the normal shock, hence it plays no local role in the computation of the shock. Also in this case, therefore, is shock resolution entirely due to the convection and pressure-gradient upstream biases, which occur with the same weight at the supersonic side of the shock, where $\delta = 1$. As M falls across the shock, so does δ, which indicates that the upstream bias in the approximation of the pressure gradient quickly decreases, leading to an essentially centered approximation of this gradient towards the outlet. Also for all the distributions computed for this benchmark are the outlet variations smooth and undistorted; in particular the calculated outlet pressure coincides with the imposed pressure boundary conditions, which again reflects favorably on the surface-integral pressure enforcement strategy.

13.9.10   Adiabatic Smooth and Shocked Flows in a Converging-Diverging Nozzle

This section details the computational solutions for subsonic, critical, supersonic and shocked adiabatic flows within a converging diverging nozzle, employing the stagnation state as a reference. The accuracy of these solutions is then assessed against the corresponding available exact solutions. For these flows, the nozzle non-dimensional cross-sectional area distribution is

$$A(x) = \begin{cases} 1.75 - 0.75\cos\left(\pi(2x-1)\right) & , \ 0 \ \leq \ x \ \leq \ \tfrac{1}{2} \\ 1.25 - 0.25\cos\left(\pi(2x-1)\right) & , \ \tfrac{1}{2} \ \leq \ x \ \leq \ 1 \end{cases} , \quad A(0) = 2.5, \quad A(1) = 1.5$$

(13.144)

Although continuous for every "x", this distribution exhibits a discontinuous curvature at the nozzle throat. At this location, such a geometric characteristic theoretically induces a slope discontinuity on the flow variables, a feature that provides another mechanism for assessing the resolution of the computational solutions.

The first flow involves a completely subsonic flow throughout the nozzle. The reference inlet Mach number is $M = 0.200$, with non-dimensional inlet density and total energy $\rho_{\text{in}} = 0.980$, $E_{\text{in}} = 2.458$ as the two needed boundary conditions at the subsonic inlet; from the exact solution the corresponding mass flow is $m_{\text{in}} = 0.231$. At the subsonic outlet, the pressure boundary condition is $p_{\text{out}} = 0.919$, which corresponds to $M_{\text{out}} = 0.350$. The initial conditions throughout the duct correspond to a uniform flow from the reference inlet Mach number, a flow that is then advanced in time to a steady state.

Achieved with a Courant number equal to 200, the computational solution, shown in Figure 13.40, remains smooth and correctly matches the imposed outlet pressure boundary condition. The flow expansion towards the throat and subsequent compression downstream of the throat are clearly resolved in these computational distributions that reflect the corresponding exact solution.

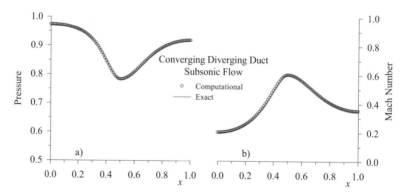

Figure 13.40: Subsonic Nozzle Flow: a) Pressure, b) Mach Number

As shown in Figure 13.41, the computational solutions for denser grids asymptotically converge in both the $\mathcal{H}^0$ and $\mathcal{H}^1$ norms with a converge rate as high as 3.0 in the $\mathcal{H}^0$ norm.

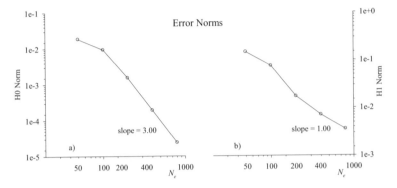

Figure 13.41: Solution Error Norms: a) $\mathcal{H}^0$ Norm, b) $\mathcal{H}^1$ Norm

The next computational solution corresponds to a critical sonic flow at the nozzle throat. The reference inlet Mach number is $M = 0.2395$, with non-dimensional inlet density and total energy $\rho_{in} = 0.972$, $E_{in} = 2.441$ as the two needed boundary conditions at the subsonic inlet; from the exact solution the corresponding mass flow is $m_{in} = 0.274$. At the subsonic outlet, the pressure boundary condition is $p_{out} = 0.881$, which corresponds to $M_{out} = 0.430$. The initial conditions throughout the duct correspond to a uniform flow from the reference inlet Mach number, a flow that is then advanced in time to a steady state.

The eventual steady-state computational solution in Figure 13.42 remains smooth and subsonic. Reflecting the exact solution, this computational solution sharply resolves the slope discontinuity at the throat, which corresponds to a shock of vanishing strength.

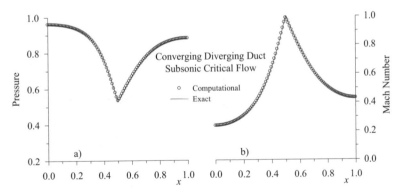

Figure 13.42: Critical Nozzle Flow: a) Pressure, b) Mach Number

This physical slope discontinuity affects the asymptotic convergence rates displayed in Figure 13.43.

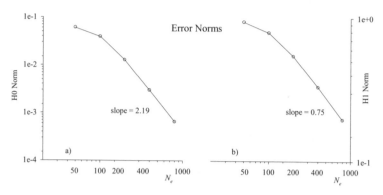

Figure 13.43: Solution Error Norms: a) $\mathcal{H}^0$ Norm, b) $\mathcal{H}^1$ Norm

For this critical flow, the convergence rates equal 0.75 for the $\mathcal{H}^1$ norm and 2.19 for the $\mathcal{H}^0$ norm, rates that both indicate the convergence of this solution as well as its gradients.

For an appropriate back pressure, an isentropic flow throughout the nozzle may emerge that is subsonic upstream of the throat and supersonic downstream, as displayed in Figure 13.44.

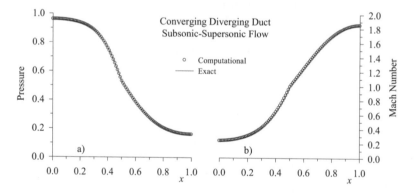

Figure 13.44: Subsonic-Supersonic Isentropic Nozzle Flow: a) Pressure, b) Mach Number

This is the next flow calculated, subject to the previous inlet boundary conditions, but without any outlet flow pressure boundary condition, since the outlet is supersonic; the initial conditions have resulted from a linear variation between the inlet subsonic state and isentropic outlet supersonic state with $M_{\mathrm{out}} = 1.854$ and $p_{\mathrm{out}} = 0.160$. Rapidly achieved with a Courant number equal to 200, the computational solution in remains indistinguishable from the exact solution, with computed outlet pressure and Mach number that reflect the corresponding isentropic magnitudes. The continuous expansion from inlet to outlet remains smooth and the mild slope discontinuity at the throat, induced by the throat discontinuous curvature is sharply resolved.

As the grid is refined, the computational solution for this isentropic flow rapidly converges, as illustrated in Figure 13.45.

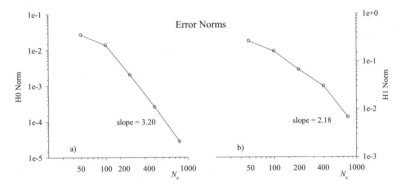

Figure 13.45: Solution Error Norms: a) $\mathcal{H}^0$ Norm, b) $\mathcal{H}^1$ Norm

The convergence rate reaches 3.20 in the $\mathcal{H}^0$ norm and 2.12 in the $\mathcal{H}^1$ norm, indicating convergence of both the solution and solution gradients.

The error norms, hence solution accuracy, depend on the magnitude of $\psi_{\min}$ as documented in Figure 13.46 for this isentropic flow, as calculated on a 200-element grid.

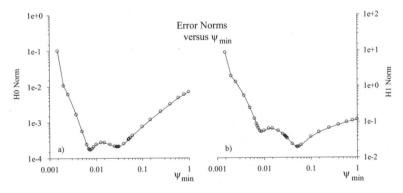

Figure 13.46: Solution Error Norms: a) $\mathcal{H}^0$ Norm, b) $\mathcal{H}^1$ Norm

As $\psi_{\min}$ approaches nought, the error norms rapidly increase and eventually the numerical time integration exponentially diverges, because the corresponding discrete system (13.63)

becomes unstable. As ψ_{min} increases towards unity and then exceeds it, the error norms also rise, in this case because the amount of induced dissipation rises. Accordingly, there exist optimal magnitudes of ψ_{min} at which the error norms reach their individual absolute minimum and correspondingly the computational solution achieves its highest accuracy possible on the prescribed grid.

The final benchmark of this set of adiabatic nozzle flows involves the calculation of a shocked steady flow, with computational-solution accuracy assessed against the available exact solution. For this benchmark, the non-dimensional specifications as well as boundary conditions are: $\rho_{in} = 0.972$, $E_{in} = 2.441$ at the subsonic inlet, and $p_{out} = 0.756$ at the subsonic outlet. From the exact solution, these specifications correspond to $M_{in} = 0.2395$, $m_{in} = 0.274$, and $M_{out} = 0.498$ with a steady shock that develops at $x = 0.750$, with upstream and downstream shock Mach numbers $M_u = 1.600$ and $M_d = 0.669$. The initial conditions throughout the nozzle correspond to a linear interpolation between the inlet and outlet subsonic states.

Achieved with a Courant number equal to 150, the resulting solution is summarized in Figures 13.47-13.48 in terms of pressure, Mach number, density, mass flow, enthalpy and energy. These variations remain essentially non-oscillatory with a crisp normal shock captured in at most one node; most importantly, this shock is captured at the theoretically exact location $x = 0.750$, with accurately calculated pressure ratio and Mach numbers across the shock.

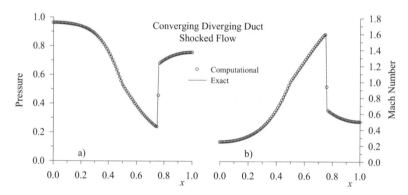

Figure 13.47: Shocked Nozzle Flow: a) Pressure, b) Mach Number

In the neighborhood of the outlet, the results remain devoid of any spurious oscillations, which again reflects favorably on the enforcement of the pressure boundary condition. With accurately calculated normal-shock and outflow states, this solution reflects the available exact solution.

Despite the normal shock, the corresponding mass flow remains constant and the total enthalpy does not change in an adiabatic flow. These important features are correctly predicted by the computational solution, displayed in Figure 13.48, which shows a constant mass flow and enthalpy that remain undistorted by the shock in the other variables.

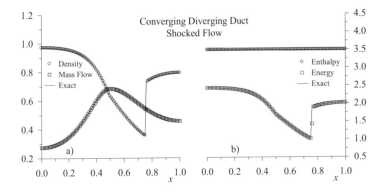

Figure 13.48: Shocked Nozzle Flow: a) Density and Mass Flow, b) Enthalpy and Energy

Presented in Figure 13.49 for this shocked nozzle flow are the variations of the upstream-bias functions "α", "δ" and "ψ" along with the nodal ℓ_1 norm of the q^h solution, variations that are directly determined by the distribution of the Mach number. Following the distribution of the nodal norm, also for this shocked nozzle flow can the formulation lead to $\psi = \psi_{min}$ for most of the flow and focus $\psi = \psi_{max}$ at the shock. Accordingly, the formulation again keeps the upstream bias to a minimum in regions of smooth flow, and focuses an increase of this bias only at shocks, in order to produce a solution that remains accurate and stable.

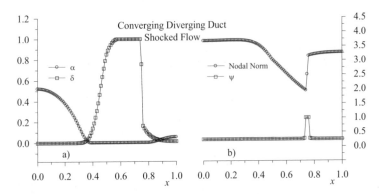

Figure 13.49: Shocked Nozzle Flow: a) α and δ, b) Nodal Norm and ψ

Also for a shocked nozzle flow has the formulation generated an asymptotically convergent solution in the $\mathcal{H}^0$ norm, with corresponding convergence rates presented in Figures 13.50 a)-b).

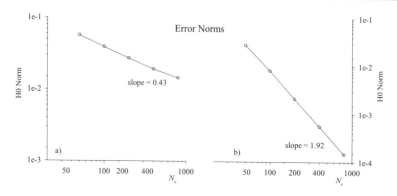

Figure 13.50: Shocked Nozzle Flow: a) Shocked-Solution Norm, b) Continuous-Solution-Branch Norm

Figure 13.50 a) displays the error norm for the entire solution, including the shock, while Figure 13.50 b) displays the error norm for the union of the continuous subsonic and supersonic solution branches, a union obtained by only considering the computational and exact solutions in the interval $[0.00, 0.70] \cup [0.80, 1.0]$. While the shocked solution converges at a rate of 0.43, the continuous solution set is seen to converge at a rate of 1.92. This expected difference in convergence rates between these two solution sets signals the normal shock dominates the overall convergence rate. The formulation has thus produced a computational solution that converges also for a discontinuous flow.

13.9.11   Shocked Nozzle Flows with Wall Friction, Heating, Mass and Work Transfer

Unavailable in analytical closed form, the solution of the gas dynamics problem of a nozzle flow subject to wall friction as well as heat, mass, and work transfer may only be obtained computationally. This section presents the computational solution for two separate nozzle flows: one corresponding to heat transfer only, the other resulting from wall friction as well as heat, mass and work transfer. The reference baseline nozzle flow consists of the shocked frictionless adiabatic flow discussed in the previous section. In the computed results in this section, wall friction exists from nozzle inlet to outlet; heat, mass, and work transfer, however, are only imposed along the diverging part of the nozzle, in order to investigate the effect of these transfers upon the normal shock.

According to the computational results, cooling strengthens a shock and shifts it downstream, whereas heating weakens and can even eliminate a shock from the diverging section of a nozzle. Figures 13.51-13.52 present these results in terms of the Mach number and pressure for both the baseline adiabatic flow, denoted by the unmarked solid line, and the flows corresponding to several magnitudes of heat transfer in the range $-2.0 \leq g_{\mathrm{T}} \leq 1.5$.

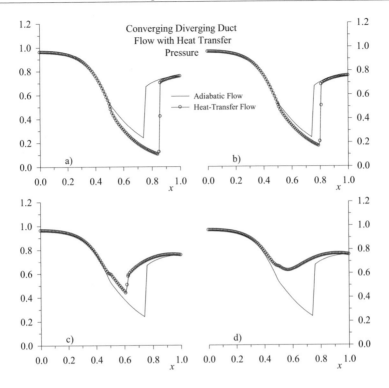

Figure 13.51: Pressure in Nozzle Flow with Heating,
a) $g_T = -2.0$, b) $g_T = -1.0$, c) $g_T = 1.0$, d) $g_T = 1.5$

The results remain essentially non-oscillatory for each magnitude of energy transfer, with crisp normal shocks captured in at most two nodes, without any outflow distortion emerging for any position of the normal shock within the duct. Far from being a numerical artefact, the clearly visible increasing slope discontinuity at the nozzle throat corresponds to both the curvature discontinuity in the nozzle, at this location, and the abrupt rise in local heating, from no heating to the left of the throat, to sizable heating to the right of the throat; the computational results thus succeed in resolving this expected slope discontinuity without spurious oscillations. Consistent with qualitative theoretical deductions, these results also show that in the presence of wall friction or heating, the sonic state develops not at the nozzle throat, but in the diverging duct. Strengthening the available theoretical formulae, the computational results quantitatively locate the sonic state within the diverging duct, for instance at $x = 0.535$ for condition c) in the figure.

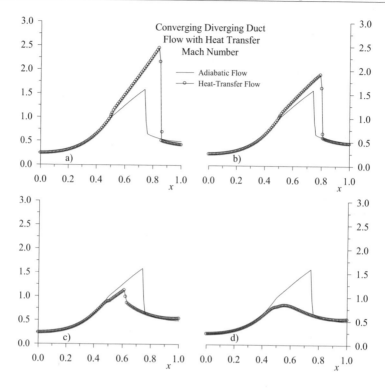

Figure 13.52: Mach Number in Nozzle Flow with Heating,
a) $g_T = -2.0$, b) $g_T = -1.0$, c) $g_T = 1.0$, d) $g_T = 1.5$

For the specified range of heat-transfer magnitudes the normal shock shifts upstream and eventually vanishes from the flow, which becomes totally subsonic. Corresponding to a cooled flow, Figures 13.51, 13.52 a), b) show that supersonic cooling causes the shock to move downstream, with consequent increase of Mach number, decrease of supersonic static pressure, and rise in stagnation-pressure loss across the shock; as the cooling decreases, the shock shifts upstream. With heating, conversely, the normal shock moves upstream towards the throat, as shown in Figures 13.51, 13.52 c), d) with consequent decrease of peak supersonic Mach number, increase of the static pressure minimum, and reduction in stagnation-pressure loss across the shock. A further increase in heating generates a subsonic flow throughout the nozzle. In these conditions, the mass flow rate begins to decrease. According to these results, heat transfer may thus be used to control the location and strength of a normal shock.

The transfer of heat to a flow induces an increase in the flow total enthalpy, but does not alter the mass flow rate of a choked flow. Such fundamental features have been correctly predicted by this solution, as illustrated in Figure 13.53. The results in the figure show a distribution of mass flow that follows that of the reference adiabatic flow and a distribution of total enthalpy that increases linearly from the point of initiation of constant heating.

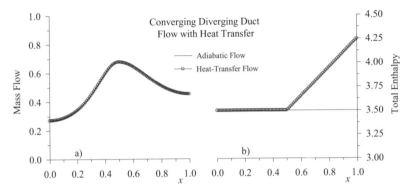

Figure 13.53: Nozzle Flow with Wall Friction and Heating, a) Mach Number, b) Pressure

Similar variations in the position and strength of a normal shock take place when wall friction as well as mass and work transfer are superimposed to heat transfer. The results illustrated in Figures 13.54-13.55 correspond to the baseline adiabatic shocked flow also subject to wall friction with $f_D = 0.02$ and the following transfer magnitudes: $g_T = \pm 1.00$, $\dot{m}_b = \pm 0.10$, $w_m = 0.10$, where the positive determination corresponds to heating, mass addition as well as work extraction, while the negative determination corresponds to cooling, mass removal and work provision to the flow. In agreement with known findings from nozzle-flow analysis, these results show that with or without heating, moderate wall friction minimally affects an accelerating flow within a diverging duct, as indicated by the two unmarked curves in each of the figures.

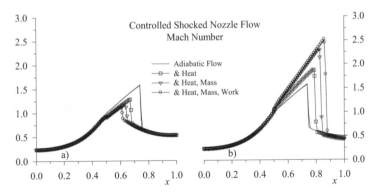

Figure 13.54: Mach Number in Nozzle Flow: a) Heat and Mass Added, Work Extracted,
b) Heat and Mass Removed, Work Provided

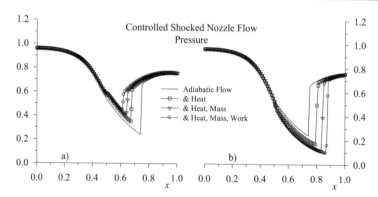

Figure 13.55: Pressure in Nozzle Flow: a) Heat and Mass Added, Work Extracted,
        b) Heat and Mass Removed, Work Provided

As noted in Section 13.9.7, mass addition and work extraction affect the flow analogously to flow heating, whereas mass depletion and work provision affect the flow similarly to flow cooling. With reference to a fixed amount of heating, the addition of mass moves the shock upstream and the subsequent extraction of work further shifts the shock upstream, with consequent reduction of peak supersonic Mach number and stagnation pressure loss across the shock. Conversely, with respect to a fixed amount of cooling, the depletion of mass moves the shock downstream and a subsequent supply of work to the flow further shifts the shock downstream, with consequent increase of peak supersonic Mach number and stagnation pressure loss across the shock. Again, these are the only CFD computations known to this author of shocked nozzle flows subject to wall friction, heat, mass and work transfer. According to these results, also in a nozzle flow will heat, mass, and work transfer exercise significant control on the location and strength of a shock wave.

13.9.12   Convergence to Steady State

The computational solutions in these sections rapidly converged to corresponding steady states for Courant numbers in excess of 50, as allowed by flow features and magnitude of the source terms. In the presence of heat, mass and work transfer in a shocked nozzle flow this number could reach 50; in the absence of all these transfers this number could rise to 150, for shocked adiabatic flows, and 200 for isentropic subsonic and supersonic flows. These magnitudes of Courant number, hence swift convergence, became available because the formulation allowed a specified maximum number of Newton iterations for the non-linear determination of each IRK array K_i, as described in Section 13.7. Although this number was set to 4, the algorithm employed fewer than 4 iterations, mostly 1, at each time station when the Newton-iteration residual fell below the prescribed tolerance of 5.0×10^{-6}.

Representative of the convergence histories of all the computed steady states in this study, Figure 13.56 a) illustrates this process for the shocked nozzle flow with wall friction as well as heat, mass and work transfer, discussed in the previous section, obtained for a Courant number equal to 50, and Figure 13.56 b) presents the corresponding process for the adiabatic shocked nozzle flow presented in Section 13.9.10, obtained for a Courant number equal to 150. In these figures, N_t denotes the number of time cycles and the integers below the convergence curves indicate the number of Newton-iterations employed in the corresponding region of the curves. Essentially 1 iteration was required for the vast majority of each convergence history to determine each K_i at every time station, when the magnitude of the total residual fell below 5.0×10^{-6}.

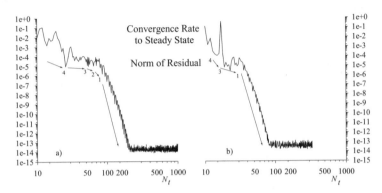

Figure 13.56: Convergence Rate Curves:a) Non-Adiabatic Flow, b) Adiabatic Flow

When the maximum number of Newton iterations per cycle was limited to 1, instead, the calculations could only converge for a reduced Courant number that required more time cycles N_t to converge than in the multiple Newton-iteration mode. Whether in single or multiple Newton-iteration mode, for all the steady states determined in this study, the formulation is capable of driving the total residual to machine zero.

13.10 Computational Performance

This Chapter has formulated a characteristics-bias system for the generalized Euler equations and solved this system through a finite element spatial discretization and an implicit Runge-Kutta time integration, implemented in Newton-iteration mode. The generalized Euler equations feature a source term that models the effects of not only cross-sectional area variations, but also wall friction as well as heat, mass, and work transfer. Since no exact solutions of these equations are available for flows simultaneously influenced by all of these effects, these flows can only be investigated computationally.

The characteristics bias system consists of the generalized equations as augmented with a continuum upstream-bias expression that evolves from a decomposition of the Euler flux Jacobian into physically significant acoustic and convection components. This system induces an upwind representation directly at the differential equation level before the spatial discretization. Owing to this feature, a conventional centered or Galerkin spatial discretization of this system then automatically generates an upwind stable discrete system that does not require any further stabilizing dissipation. Such a continuum upstream bias has been extended to both the source term and the time partial derivatives, which has resulted in a characteristics-bias system that becomes Galilean invariant, a fundamental physical principle satisfied by the original Euler equations. According to the computational results, the extension of the continuum upstream bias to both the source term and time partial derivatives sizably reduces induced upwind dissipation, noticeably increases contact-discontinuity resolution, and beneficially leads to rapid convergence.

The characteristics-bias system has then been cast as an integral weak statement, which has also led to continuum forms of both the DG and SUPG procedures. Unlike the customary SUPG algorithm, however, the characteristics-bias SUPG requires no premultiplication by the transpose of the Euler flux Jacobian matrix. As counterparts of the usual SUPG "τ_{SUPG}" and "δ_{SUPG}" terms, moreover, the parameter products $\varepsilon\psi\alpha$ and $\varepsilon\psi\delta$ in this formulation explicitly depend on the square or cube of the local mesh spacing, which has also contributed to the recorded rapid convergence. The formulation keeps the upstream bias and associated dissipation to a minimum, in regions of smooth flow, and focuses an increase of this bias only at shocks, in order to generate solutions that remain both accurate and stable. Accordingly, the formulation provides a physics based upwind algorithm, but with reduced levels of induced dissipation.

The mechanically significant computational solutions for smooth and shocked flows with cross sectional area variations, wall friction, and heat, mass, as well as work transfer reflect all the available corresponding exact solutions for flows influenced by only one effect, with computed mass flow and total enthalpy correctly remaining constant across normal shocks. The numerical results remain essentially non-oscillatory, with crisp shocks captured in at most one or two nodes. With reference to the available exact solutions, the corresponding computational solutions have converged asymptotically in both the $\mathcal{H}^0$ and $\mathcal{H}^1$ norms, for continuous flows, and in the $\mathcal{H}^0$ norm for shocked flows, with an order as high as 3 in the $\mathcal{H}^0$ norm for continuous solutions. According to the numerical investigations and theoretical considerations in Section 13.9.10, optimal $\psi_{\min}$'s exist that correlate with minimum error norms. Rapid convergence to steady state has also been recorded for booth smooth and shocked flows with Courant numbers ranging between 50 and 200, for the multiple Newton-iteration implementation.

From a physical viewpoint, the computational results have shown that moderate wall friction only marginally affects an accelerating flow in a diverging duct. Conversely, heating, mass addition, and work extraction shift the shock upstream, whereas cooling, mass depletion and work supply shift the shock downstream. From a fluid mechanics standpoint and respectively for frictional and diverging-nozzle flows, the upstream shift can take place in regions of both higher and lower Mach number, and the downstream shift can take place in regions of both lower and higher Mach number. The characteristics-bias formulation has shown that heat, mass and work transfer may be used to control the position as well as strength of a shock. According to the theoretical and computational findings discussed in this paper, this formulation can thus allow comprehensive, efficient and exhaustive investigations of physically realistic and mechanically relevant generalized gas dynamics flows.

Chapter 14

Multi-Dimensional Characteristics and Characteristics-Bias Systems

This chapter develops the multi-dimensional generalizations of the one-dimensional characteristics - bias formulations presented in Chapter 9. Non-linear hyperbolic scalar equations conveniently present the pivotal notion of characteristics for two- and three-dimensional flows. The generalization of this notion to systems classifies systems in hyperbolic, parabolic and elliptic types and leads to multiple characteristic surfaces across which hyperbolic solutions can become discontinuous. The analysis of multi-dimensional characteristic surfaces in the time-space continuum details the propagation of convection and acoustic waves within two- and three-dimensional flows and identifies the streamline and crossflow directions as principal propagation directions. This analysis leads to several physics-based conditions for developing a consistently multi-dimensional and infinite directional characteristics-based approximation. This approximation is induced at the partial differential equation level, in the continuum and before the discretization in space.

The continuum characteristics-bias approximation of the Euler and Navier-Stokes equations is a system of equations that encompasses the corresponding system and induces an authentically multi- dimensional solution-dependent upstream dissipation along not just selected characteristic principal directions, but all directions. This upstream dissipation originates from a multi-dimensional differential hyperbolic-parabolic perturbation within the characteristics-bias system. This system also features an upstream-bias representation of the source term, the divergence of the viscous flux, and the time derivative of the dependent variable, which makes the system Galilean invariant. As a result, the system induces but minimal upwind dissipation.

The development of a characteristics-bias approximation at the differential equation level, prior to the discretization in space, possesses distinctive advantages. The directions of this multi-dimensional upstream bias remain independent of any grid-line direction, and correlate with characteristic directions; furthermore a direct centered approximation of the characteristics-bias system automatically generates a genuinely multi-dimensional consistent upstream-bias approximation of the original Navier-Stokes equations, without any need for additional numerical dissipation terms. This formulation, induces but a minimal amount of upstream dissipation, which leads to essentially non-oscillatory solutions that reflect available exact solutions.

14.1 Hyperbolic Equations and Elliptic, Parabolic, Hyperbolic Systems

This section expresses the solution of a scalar equation in wave-like form. The properties of this form are investigated and then generalized to vector forms, in order to progress with the characteristics analysis of a multi-dimensional hyperbolic system.

14.1.1 Scalar Equations

The multi-dimensional generalization of (9.1) can be cast as

$$\frac{\partial q}{\partial t} + \frac{\partial f_j(q)}{\partial x_j} = 0 \quad \Rightarrow \quad \frac{\partial q}{\partial t} + \frac{\partial f_j(q)}{\partial q}\frac{\partial q}{\partial x_j} = 0 \tag{14.1}$$

with implied summation on repeated subscript indices. Using

$$\lambda_j \equiv \frac{\partial f_j(q)}{\partial q} \tag{14.2}$$

with λ_j a real number, this equation can also be expressed as

$$\frac{\partial q}{\partial t}\frac{1}{\sqrt{1+\lambda_\ell\lambda_\ell}} + \frac{\partial q}{\partial x_j}\frac{\lambda_j}{\sqrt{1+\lambda_\ell\lambda_\ell}} = 0$$

$$\frac{\partial q}{\partial t}\frac{1}{\sqrt{1+\lambda_\ell\lambda_\ell}} + \frac{\partial q}{\partial \overline{x}}\frac{\sqrt{\lambda_j\lambda_j}}{\sqrt{1+\lambda_\ell\lambda_\ell}} = \frac{dq}{ds} = 0, \quad q(\overline{x},0) = q_0(\overline{x}) \tag{14.3}$$

which becomes a one-dimensional equation like (9.1) and indicates that q remains constant with respect to the characteristics-line coordinate s; with reference to $\overline{x}$, this variable represents a coordinate along a multi-dimensional line. The gradients of $(t,\overline{x})$ with respect to s as well as the direction cosines of x_j with respect to $\overline{x}$ and the slope of the corresponding $(t,\overline{x})$ line become

$$\frac{dt}{ds} = \frac{1}{\sqrt{1+\lambda_\ell\lambda_\ell}}, \quad \frac{d\overline{x}}{ds} = \frac{\sqrt{\lambda_j\lambda_j}}{\sqrt{1+\lambda_\ell\lambda_\ell}}, \quad \frac{dx_j}{d\overline{x}} = \frac{\lambda_j}{\sqrt{\lambda_\ell\lambda_\ell}}, \quad \frac{d\overline{x}}{dt} = \sqrt{\lambda_j\lambda_j} \tag{14.4}$$

Since q remains constant along the line defined by these expressions, the corresponding speeds λ_j, $1 \leq j \leq n$, remain constant and thus these expressions can be exactly integrated, which yields the following equations for the characteristic line

$$t = \frac{s}{\sqrt{1+\lambda_\ell\lambda_\ell}}, \quad \overline{x} = \overline{x}_0 + \frac{s\sqrt{\lambda_j\lambda_j}}{\sqrt{1+\lambda_\ell\lambda_\ell}}$$

$$x_j = \frac{\lambda_j}{\sqrt{\lambda_\ell\lambda_\ell}}\overline{x}, \quad \overline{x} = \overline{x}_0 + t\sqrt{\lambda_j\lambda_j} \quad \Rightarrow \quad \overline{x}_0 = x_j\frac{\lambda_j}{\sqrt{\lambda_\ell\lambda_\ell}} - t\sqrt{\lambda_j\lambda_j} \tag{14.5}$$

These results induce the following multi-dimensional unit vector components n_j and characteristic speed λ

$$n_j = \frac{\lambda_j}{\sqrt{\lambda_j \lambda_j}}, \quad \lambda \equiv \lambda_j n_j = \sqrt{\lambda_\ell \lambda_\ell} \tag{14.6}$$

and leads to the following expression for x_0

$$\overline{x}_0 = x_j n_j - \lambda(q)t \tag{14.7}$$

In these results, n_j is constant, because q keeps constant on the characteristic lines. Based on this expression and a constant q, equal to its initial value along the characteristic line, the solution of (14.1) thus becomes

$$q(\overline{x}(s), t(s)) = q(\overline{x}(0), t(0)) = q_0(\overline{x}_0) = q_0\left(x_j n_j - \lambda(q)t\right) = q\left(x_j n_j - \lambda(q)t\right) = q(\boldsymbol{x}, t) \tag{14.8}$$

where $q_0(\overline{x}_0)$ signifies the solution corresponding to the prescribed initial condition and $q(\overline{x}_0)$ indicates the dependence of the solution upon $\overline{x}_0$. This solution thus corresponds to the multi-dimensional generalization of (9.18).

This solution depends on the constant unit vector components, as defined in (14.6). Expression (14.8) still solves (14.1) with arbitrary constant unit-vector components n_j with $\lambda(q)$ still defined as in (14.6). This generalized solution is thus cast as the non-linear wave-like form

$$q = q(x_j n_j - \lambda(q)t) \quad \Rightarrow \quad q = q(\eta_1), \quad \eta_1 \equiv x_j n_j - \lambda(q)t \tag{14.9}$$

where the solution-dependent characteristic speed $\lambda = \lambda(q)$ arises as the solution of an eigenvalue problem. Based on the wave-like form (14.9), the derivatives of q with respect to x_j and t become

$$\frac{\partial q}{\partial x_j} = \frac{\partial q}{\partial \eta_1}\left(n_j - t\frac{\partial \lambda}{\partial x_j}\right) = \frac{\partial q}{\partial \eta_1} n_j - t\frac{\partial q}{\partial \eta_1}\frac{\partial \lambda}{\partial \eta_1}\frac{\partial \eta_1}{\partial x_j} = \frac{\partial q}{\partial \eta_1} n_j - t\frac{\partial \lambda}{\partial \eta_1}\frac{\partial q}{\partial x_j}$$

$$\frac{\partial q}{\partial t} = \frac{\partial q}{\partial \eta_1}\left(-\lambda - t\frac{\partial \lambda}{\partial t}\right) = -\lambda\frac{\partial q}{\partial \eta_1} - t\frac{\partial q}{\partial \eta_1}\frac{\partial \lambda}{\partial \eta_1}\frac{\partial \eta_1}{\partial t} = -\lambda\frac{\partial q}{\partial \eta_1} - t\frac{\partial \lambda}{\partial \eta_1}\frac{\partial q}{\partial t} \tag{14.10}$$

hence

$$\frac{\partial q}{\partial x_j} = \frac{n_j}{1 + t\frac{\partial \lambda}{\partial \eta_1}}\frac{\partial q}{\partial \eta_1}, \quad \frac{\partial q}{\partial t} = -\frac{\lambda}{1 + t\frac{\partial \lambda}{\partial \eta_1}}\frac{\partial q}{\partial \eta_1} \tag{14.11}$$

The insertion of these results into system (14.1) yields the eigenvalue statement

$$\left(-\lambda(q) + \frac{\partial f_j}{\partial q}n_j\right)\frac{\partial q}{\partial \eta_1} = 0 \tag{14.12}$$

This statement implies

$$\lambda(q) = \frac{\partial f_j}{\partial q}n_j \tag{14.13}$$

which corresponds to a multi-dimensional and non-linear characteristic speed.

14.1.2   Non-linear Wave-Like Vector Solutions

Considering the central role of non-linear wave-like solutions, this section explores their salient properties. With implied summation on repeated indices, a non-linear wave-like vector solution is expressed as

$$q = q(\eta_1), \quad \eta_1 \equiv \boldsymbol{x} \cdot \boldsymbol{n} - \lambda(q)t = x_j n_j - \lambda(q)t \tag{14.14}$$

where q now indicates a vector with any number of components, $\boldsymbol{n}$ denotes a space-domain propagation-direction unit vector, independent of $(\boldsymbol{x}, t)$, and $\lambda = \lambda(q)$ indicates a real solution-dependent wave-propagation velocity component along the $\boldsymbol{n}$ direction. For: $\boldsymbol{x} \cdot \boldsymbol{n} - \lambda(q)t$ equal to a constant, the wave-like solution q in (14.14) remains itself constant, which also makes $\lambda = \lambda(q)$ a constant. Accordingly, in the time-space continuum $(\boldsymbol{x}, t)$, the equation

$$\boldsymbol{x} \cdot \boldsymbol{n} - \lambda t = \mathcal{C} \tag{14.15}$$

represents a plane, in 2-D, and a hyperplane, in 3-D, with $\mathcal{C}$ and $(n_1, n_2, -\lambda(q))$, in 2-D, and $(n_1, n_2, n_3, -\lambda(q))$, in 3-D, respectively denoting a constant and a unit vector orthogonal to such a plane as depicted in Figure 14.1 for the representative case of a 2-D flow. The surface that remains tangent to these planes for all $\boldsymbol{n}$'s is then tangent itself to a characteristic surface, for λ corresponding to a characteristic velocity component. Equation (14.15) corresponds in the (x_1, x_2) space to a bundle of parallel lines for a fixed t and different $\mathcal{C}$. For a fixed $\mathcal{C}$ and variable t, the equation represents a $\mathcal{C}$-line propagating in the $\boldsymbol{n}$ direction with velocity component λ.

Next, consider the coordinate transformation

$$\eta_1 = x_j n_j - \lambda(q)t, \quad \eta_2 = \eta_2(x_1, x_2, t), \quad \eta_3 = \eta_3(x_1, x_2, t) \tag{14.16}$$

in 2-D, with η_2 and η_3 chosen so that the reference frame (η_1, η_2, η_3) is also right-handed like (x_1, x_2, t). In 3-D, the corresponding coordinate transformation is

$$\eta_1 = x_j n_j - \lambda(q)t, \quad \eta_2 = \eta_2(x_1, x_2, x_3, t), \quad \eta_3 = \eta_3(x_1, x_2, x_3, t), \quad \eta_4 = \eta_4(x_1, x_2, x_3, t) \tag{14.17}$$

with η_2, η_3, and η_4 chosen so that the reference frame $(\eta_1, \eta_2, \eta_3, \eta_4)$ is also right-handed like (x_1, x_2, x_3, t). With either transformation, $q(x_j n_j - \lambda(q)t) = q(\eta_1)$ and for each t and non-linear q, the η_1 axis then points in the $(n_1, n_2, -\lambda(q))$ direction, in 2-D, and in the $(n_1, n_2, n_3, -\lambda(q))$ direction, in 3-D. In 2-D, this result follows from the partial derivatives of η_1 with respect to (x_1, x_2, t)

$$\left(1 + t\frac{\partial \lambda}{\partial \eta_1}\right)\frac{\partial \eta_1}{\partial t} = -\lambda(q), \quad \left(1 + t\frac{\partial \lambda}{\partial \eta_1}\right)\frac{\partial \eta_1}{\partial x_j} = n_j \tag{14.18}$$

hence

$$\frac{-\lambda}{\frac{\partial \eta_1}{\partial t}} = \frac{n_1}{\frac{\partial \eta_1}{\partial x_1}} = \frac{n_2}{\frac{\partial \eta_1}{\partial x_2}} \tag{14.19}$$

which shows that the η_1-axis direction vector $\left(\frac{\partial \eta_1}{\partial x_1}, \frac{\partial \eta_1}{\partial x_2}, \frac{\partial \eta_1}{\partial t}\right)$ is parallel to $(n_1, n_2, -\lambda(q))$. The corresponding result for 3-D flows follows from the partial derivatives of η_1 with respect to

(x_1, x_2, x_3, t)

$$\left(1 + t\frac{\partial\lambda}{\partial\eta_1}\right)\frac{\partial\eta_1}{\partial t} = -\lambda(q), \quad \left(1 + t\frac{\partial\lambda}{\partial\eta_1}\right)\frac{\partial\eta_1}{\partial x_j} = n_j \tag{14.20}$$

hence

$$\frac{-\lambda}{\frac{\partial\eta_1}{\partial t}} = \frac{n_1}{\frac{\partial\eta_1}{\partial x_1}} = \frac{n_2}{\frac{\partial\eta_1}{\partial x_2}} = \frac{n_3}{\frac{\partial\eta_1}{\partial x_3}} \tag{14.21}$$

which shows that the η_1-axis direction vector $(\frac{\partial\eta_1}{\partial x_1}, \frac{\partial\eta_1}{\partial x_2}, \frac{\partial\eta_1}{\partial x_3}, \frac{\partial\eta_1}{\partial t})$ is parallel to $(n_1, n_2, n_3, -\lambda(q))$.

Since $\boldsymbol{n}$ is a unit vector, hence $n_j n_j = 1$, the second result in (14.18) also leads to the following relations

$$\left(1 + t\frac{\partial\lambda}{\partial\eta_1}\right)\frac{\partial\eta_1}{\partial x_\ell}n_\ell = 1, \quad \frac{\partial\eta_1}{\partial x_j} = \frac{\partial\eta_1}{\partial x_\ell}n_\ell n_j \tag{14.22}$$

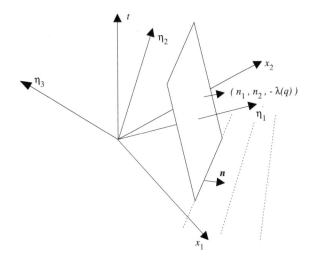

Figure 14.1: Wave Plane and Reference Frames

The partial derivatives of the wave-like solutions (14.14) also satisfy the conditions

$$\frac{\partial q}{\partial t} = -\lambda(q)\frac{\partial q}{\partial x_\ell}n_\ell, \quad \lambda(q)\frac{\partial q}{\partial x_j} = -n_j\frac{\partial q}{\partial t} \tag{14.23}$$

which is proven as follows. The partial derivatives of (14.14) are expressed as

$$\frac{\partial q}{\partial t} = \frac{\partial q}{\partial\eta_1}\left(-t\frac{\partial\lambda}{\partial q}\frac{\partial q}{\partial t} - \lambda(q)\right), \quad \frac{\partial q}{\partial x_j} = \frac{\partial q}{\partial\eta_1}\left(-t\frac{\partial\lambda}{\partial q}\frac{\partial q}{\partial x_j} + n_j\right) \tag{14.24}$$

where $\frac{\partial q}{\partial\eta_1}$ denotes a column array and $\frac{\partial\lambda}{\partial q}$ indicates a row array, which implies that $\frac{\partial q}{\partial\eta_1}\frac{\partial\lambda}{\partial q}$ is a square matrix. These relations then lead to the two systems

$$\left(I + t\frac{\partial q}{\partial\eta_1}\frac{\partial\lambda}{\partial q}\right)\frac{\partial q}{\partial t} = -\lambda(q)\frac{\partial q}{\partial\eta_1}, \quad \left(I + t\frac{\partial q}{\partial\eta_1}\frac{\partial\lambda}{\partial q}\right)\frac{\partial q}{\partial x_j} = n_j\frac{\partial q}{\partial\eta_1} \tag{14.25}$$

where I denotes the identity matrix of appropriate size. Since the square matrix $\frac{\partial q}{\partial \eta_1} \frac{\partial \lambda}{\partial q}$ results from the product of column and row arrays, the rows of this matrix are all linear combinations of the single $\frac{\partial \lambda}{\partial q}$ row; all the eigenvalues of this matrix, but one, accordingly, vanish. With "N" the number of component in q, the only non-vanishing eigenvalue of this matrix has been exactly determined as

$$\lambda_{\eta_1} \equiv \frac{\partial \lambda}{\partial \eta_1} = \sum_{i=1}^{N} \frac{\partial q_i}{\partial \eta_1} \frac{\partial \lambda}{\partial q_i} \tag{14.26}$$

To show that this is the correct eigenvalue, let v_j denote the j-th component of the corresponding eigenvector and express the eigenvalue statement for this matrix in the form

$$\frac{\partial q_i}{\partial \eta_1} \sum_{j=1}^{N} \frac{\partial \lambda}{\partial q_j} v_j = \lambda_{\eta_1} v_i = v_i \sum_{j=1}^{N} \frac{\partial q_j}{\partial \eta_1} \frac{\partial \lambda}{\partial q_j} \tag{14.27}$$

The expression for λ_{η_1} corresponds to an eigenvalue when the determinant of the matrix in the system

$$\frac{\partial q_i}{\partial \eta_1} \sum_{j=1}^{N} \frac{\partial \lambda}{\partial q_j} v_j - \lambda_{\eta_1} v_i = 0 \tag{14.28}$$

vanishes, which takes place when the system equations are linearly dependent. It thus suffices to show that due to the expression for λ_{η_1} the first equation results as a linear combination of the remaining ones. On the one hand, the elaboration of this first equation and multiplicaion by $\frac{\partial \lambda}{\partial q_1}$ yields

$$\frac{\partial q_1}{\partial \eta_1} \frac{\partial \lambda}{\partial q_1} \sum_{j=2}^{N} \frac{\partial \lambda}{\partial q_j} v_j - \frac{\partial \lambda}{\partial q_1} v_1 \sum_{i=2}^{N} \frac{\partial q_i}{\partial \eta_1} \frac{\partial \lambda}{\partial q_i} = 0 \tag{14.29}$$

On the other hand, a linear combination of the remaining $N - 1$ equations, with $\frac{\partial \lambda}{\partial q_i}$ as the linear-combination coefficient for the i-th equation, and subsequent simplification yields

$$\frac{\partial q_1}{\partial \eta_1} \frac{\partial \lambda}{\partial q_1} \sum_{j=2}^{N} \frac{\partial \lambda}{\partial q_j} v_j - \frac{\partial \lambda}{\partial q_1} v_1 \sum_{i=2}^{N} \frac{\partial q_i}{\partial \eta_1} \frac{\partial \lambda}{\partial q_i} = 0 \tag{14.30}$$

which coincides with the previous equation, the first equation in the system. Since this first equation is a linear combination of the others, the determinant of the system matrix vanishses, which shows that λ_{η_1} is the eigenvalue. Consequently, all the eigenvalues of the matrix $\left(I + t \frac{\partial q}{\partial \eta_1} \frac{\partial \lambda}{\partial q}\right)$ equal 1, except one eigenvalue $\tilde{\lambda}$, which is expressed as

$$\tilde{\lambda} = 1 + t \frac{\partial \lambda}{\partial \eta_1} \tag{14.31}$$

As the product of all its matrix eigenvalues, therefore, the determinant of the matrix $\left(I + t \frac{\partial q}{\partial \eta_1} \frac{\partial \lambda}{\partial q}\right)$ equals $\tilde{\lambda}$, and the associated matrix of cofactors is then denoted with B. Systems (14.25) can then be solved to yield

$$\frac{\partial q}{\partial t} = -\frac{\lambda(q)}{1 + t \frac{\partial \lambda}{\partial \eta_1}} B \frac{\partial q}{\partial \eta_1}, \quad \frac{\partial q}{\partial x_j} = \frac{n_j}{1 + t \frac{\partial \lambda}{\partial \eta_1}} B \frac{\partial q}{\partial \eta_1} \tag{14.32}$$

With a finite $\frac{\partial q}{\partial \eta_1}$, if the denominator $1 + t\frac{\partial \lambda}{\partial \eta_1}$ approaches zero, then $\frac{\partial q}{\partial t}$ and $\frac{\partial q}{\partial x_j}$ become unbounded, which indicates that a positive $t = -1/\frac{\partial \lambda}{\partial \eta_1}$ corresponds to a "breaking" time of shock-wave formation. In any region that does not contain shock waves, or in the absence of shock waves, therefore, $\tilde{\lambda} = 1 + t\frac{\partial \lambda}{\partial \eta_1}$ does not vanish, the matrix $\left(I + t\frac{\partial q}{\partial \eta_1}\frac{\partial \lambda}{\partial q} \right)$ is not singular, and the corresponding wave-like solutions are non-singular. Unless stated otherwise, the wave-like solutions considered in the following sections are non-singular. For these solutions, multiplying the first result in (14.32) by n_j and contracting the second with the same n_j lead to the expressions in (14.23). In view of $q = q(\eta_1)$, the first expression in (14.23) can also be expressed as

$$\frac{\partial q}{\partial t} = -\lambda(q)\frac{\partial q}{\partial \eta_1}\frac{\partial \eta_1}{\partial x_\ell}n_\ell \tag{14.33}$$

The results in this section lead to a characteristic solution of non-linear multi-dimensional hyperbolic systems.

14.1.3 Non-Linear Elliptic, Parabolic, Hyperbolic Systems

Let

$$\frac{\partial q}{\partial t} + A_j(q)\frac{\partial q}{\partial x_j} = 0 \tag{14.34}$$

denote a non-linear multi-dimensional system with "n" equations. This system encompasses the particular conservation-law system

$$\frac{\partial q}{\partial t} + \frac{\partial f_j(q)}{\partial x_j} = 0 \quad \Rightarrow \quad \frac{\partial q}{\partial t} + \frac{\partial f_j(q)}{\partial q}\frac{\partial q}{\partial x_j} = 0 \tag{14.35}$$

where the system matrix $A_j(q)$ becomes the Jacobian matrix $\partial f_j/\partial q$.

A solution in wave-like form is expressed as in (14.14) and the associated characteristic velocity component $\lambda = \lambda(q)$ (14.14) is determined by imposing the condition that the non-linear wave-like expression (14.14) solves system (14.34).

For $q = q(\eta_1)$, system (14.34) becomes

$$\frac{\partial q}{\partial t} + A_j(q)\frac{\partial \eta_1}{\partial x_j}\frac{\partial q}{\partial \eta_1} = 0 \tag{14.36}$$

For arbitrary real coefficients $\frac{\partial \eta_1}{\partial x_j}$, the classification of this system as elliptic, parabolic, or hyperbolic depends on the eigenvalues and eigenvectors of the matrix $A_j(q)\frac{\partial \eta_1}{\partial x_j}$. If all the eigenvalues of this matrix are complex, the system is elliptic; if these eigenvalues are mixed, real and complex, the system is of mixed type. If all the eigenvalues are real, but this matrix does not possess "n" linearly independent eigenvectors, the system is parabolic; if all the eigenvalues are real and this matrix possesses "n" linearly independent eigenvectors, the system is hyperbolic. [213] The occurrence of multiple coincident real eigenvalues, accordingly, does not automatically make a system parabolic; on the other hand, when the eigenvalues of this matrix are all real and distinct, the system is hyperbolic, occasionally termed strongly hyperbolic.

The substitution into this system of (14.33) and the second of (14.22), respectively for $\frac{\partial q}{\partial t}$ and $\frac{\partial \eta_1}{\partial x_j}$, leads to

$$\frac{\partial \eta_1}{\partial x_\ell} n_\ell \left(-\lambda(q) \frac{\partial q}{\partial \eta_1} + A_j(q) n_j \frac{\partial q}{\partial \eta_1} \right) = 0 \tag{14.37}$$

According to this result, for not only linear, but also non-linear multi-dimensional systems are the wave eigenvalues determined as a solution of the eigenvalue problem

$$(-\lambda(q)I + A_j(q)n_j) \frac{\partial q}{\partial \eta_1} = 0 \tag{14.38}$$

which leads to

$$\det \left(-\lambda(q)I + A_j(q)n_j \right) = 0 \tag{14.39}$$

The characteristic velocity component λ is thus an eigenvalue of the linear-combination Jacobian matrix $A_j(q)n_j$. For a system there are thus as many characteristic speeds λ as the number of equations in the system.

14.2   Wave-Propagation and Characteristic Surfaces

Among several surfaces associated with a hyperbolic solution, discontinuity surfaces are surfaces across which a hyperbolic solution may become discontinuous. Such discontinuity surfaces are the characteristic surfaces.

14.2.1   Characteristic Surfaces

The characteristic wave-propagation surfaces correspond to information propagation, and therefore are the surfaces that remain tangent to the wave-propagation planes, in 2-D, or hyperplanes, in 3-D, expressed by (14.15), for λ corresponding to a wave-propagation velocity component. The surface

$$q_\ell(x_j n_j - \lambda t) = \text{constant} \tag{14.40}$$

for each scalar component q_ℓ within q, is then itself tangent to a characteristic surface for an appropriate "constant", because of the following results from the second expression in (14.23)

$$\text{grad } q_\ell = -\frac{1}{\lambda}(n_1, n_2, -\lambda) \frac{\partial q_\ell}{\partial t} \quad \text{parallel to} \quad (n_1, n_2, -\lambda) \tag{14.41}$$

in 2-D, and

$$\text{grad } q_\ell = -\frac{1}{\lambda}(n_1, n_2, n_3, -\lambda) \frac{\partial q_\ell}{\partial t} \quad \text{parallel to} \quad (n_1, n_2, n_3, -\lambda) \tag{14.42}$$

in 3-D.
    Let, then

$$\mathcal{F}(x_1, x_2, t) = \text{constant} \tag{14.43}$$

in 2-D, and

$$\mathcal{F}(x_1, x_2, x_3, t) = \text{constant} \tag{14.44}$$

in 3-D, represent the mathematical function of a wave-propagation surface. From vector analysis the vector (grad $\mathcal{F}$) remains perpendicular to the surface at each point. Since the surface must be tangent to the propagation plane (14.15), grad $\mathcal{F}$ must be parallel at each point to the unit vector $(n_1, n_2, -\lambda)$, in 2-D, or $(n_1, n_2, n_3, -\lambda)$, in 3-D, itself perpendicular to plane (14.15). With $n_j n_j = 1$, this condition yields

$$n_1 = \mp \frac{\frac{\partial \mathcal{F}}{\partial x_1}}{\sqrt{\frac{\partial \mathcal{F}}{\partial x_j}\frac{\partial \mathcal{F}}{\partial x_j}}}, \quad n_2 = \mp \frac{\frac{\partial \mathcal{F}}{\partial x_2}}{\sqrt{\frac{\partial \mathcal{F}}{\partial x_j}\frac{\partial \mathcal{F}}{\partial x_j}}}, \quad \lambda = \pm \frac{\frac{\partial \mathcal{F}}{\partial t}}{\sqrt{\frac{\partial \mathcal{F}}{\partial x_j}\frac{\partial \mathcal{F}}{\partial x_j}}} \tag{14.45}$$

in 2-D and

$$n_1 = \mp \frac{\frac{\partial \mathcal{F}}{\partial x_1}}{\sqrt{\frac{\partial \mathcal{F}}{\partial x_j}\frac{\partial \mathcal{F}}{\partial x_j}}}, \quad n_2 = \mp \frac{\frac{\partial \mathcal{F}}{\partial x_2}}{\sqrt{\frac{\partial \mathcal{F}}{\partial x_j}\frac{\partial \mathcal{F}}{\partial x_j}}}, \quad n_3 = \mp \frac{\frac{\partial \mathcal{F}}{\partial x_3}}{\sqrt{\frac{\partial \mathcal{F}}{\partial x_j}\frac{\partial \mathcal{F}}{\partial x_j}}}, \quad \lambda = \pm \frac{\frac{\partial \mathcal{F}}{\partial t}}{\sqrt{\frac{\partial \mathcal{F}}{\partial x_j}\frac{\partial \mathcal{F}}{\partial x_j}}} \tag{14.46}$$

in 3-D. Substitution of either set of expressions into (14.39) thus generates the propagation-surface differential equation for $\mathcal{F}$

$$\det\left(I\frac{\partial \mathcal{F}}{\partial t} + A_j(q)\frac{\partial \mathcal{F}}{\partial x_j}\right) = 0 \tag{14.47}$$

The multiplicity of solutions of this equation implies the existence of several characteristic wave-propagation surfaces.

14.2.2 Characteristic and Discontinuity Surfaces

A discontinuity surface is a surface across which the solution q of (14.34) may become discontinuous. Let η_1 indicate a coordinate in the direction normal to each facet of the discontinuity surface, as shown in Figure 14.1. In terms of the partial derivative with respect to this coordinate direction, therefore, system (14.34) becomes singular, or equivalently, the derivative of q in this direction cannot be determined by the system itself. By expressing the partial derivatives of q in (14.34) in terms of other coordinates η_j as

$$\frac{\partial q}{\partial t} = \frac{\partial q}{\partial \eta_\ell}\frac{\partial \eta_\ell}{\partial t}, \quad \frac{\partial q}{\partial x_j} = \frac{\partial q}{\partial \eta_\ell}\frac{\partial \eta_\ell}{\partial x_j} \tag{14.48}$$

system (14.34) becomes

$$\left(I\frac{\partial \eta_1}{\partial t} + A_j(q)\frac{\partial \eta_1}{\partial x_j}\right)\frac{\partial q}{\partial \eta_1} = -\sum_{\ell=2}^{N_D}\left(I\frac{\partial \eta_\ell}{\partial t} + A_j(q)\frac{\partial \eta_\ell}{\partial x_j}\right)\frac{\partial q}{\partial \eta_\ell} \tag{14.49}$$

where $N_D = 3$, in 2-D, and $N_D = 4$, in 3-D. In this system, $\left(\frac{\partial \eta_1}{\partial x_1}, \frac{\partial \eta_1}{\partial x_2}, \frac{\partial \eta_1}{\partial t}\right)$, in 2-D, and $\left(\frac{\partial \eta_1}{\partial x_1}, \frac{\partial \eta_1}{\partial x_2}, \frac{\partial \eta_1}{\partial x_3}, \frac{\partial \eta_1}{\partial t}\right)$, in 3-D, represent the cartesian components of a vector normal to a facet of each "$\mathcal{F}$ = constant" discontinuity surface. From vector analysis, the components of this normal vector can then be expressed in terms of the gradient of $\mathcal{F}$ as

$$\frac{\partial \eta_1}{\partial t} = k\frac{\partial \mathcal{F}}{\partial t}, \quad \frac{\partial \eta_1}{\partial x_j} = k\frac{\partial \mathcal{F}}{\partial x_j} \tag{14.50}$$

where k is a single scalar constant. With this specification, system (14.49) becomes

$$k \left(I \frac{\partial \mathcal{F}}{\partial t} + A_j(q) \frac{\partial \mathcal{F}}{\partial x_j} \right) \frac{\partial q}{\partial \eta_1} = - \sum_{\ell=2}^{N_D} \left(I \frac{\partial \eta_\ell}{\partial t} + A_j(q) \frac{\partial \eta_\ell}{\partial x_j} \right) \frac{\partial q}{\partial \eta_\ell} \tag{14.51}$$

which becomes singular, hence cannot determine $\frac{\partial q}{\partial \eta_1}$, when the "lhs" Jacobian matrix is itself singular. Accordingly, the governing equation for the discontinuity surface $\mathcal{F}$ is the vanishing determinant

$$\det \left(I \frac{\partial \mathcal{F}}{\partial t} + A_j(q) \frac{\partial \mathcal{F}}{\partial x_j} \right) = 0 \tag{14.52}$$

This result coincides with (14.47), which proves that the discontinuity surfaces are the characteristic surfaces. Since the system cannot determine the derivative of q in the η_1 direction, the boundary conditions for the system must be prescribed on surfaces other than characteristic surfaces.

14.3   Characteristic Cones and Hyperbolic Wave Propagation

The principal propagation directions as well as the flow domain of dependence and range of influence of a flow field point P in the time-space continuum are determined by studying the variation of the characteristic-surface shape versus the characteristic speeds. To investigate these shape changes it suffices to study the shape changes of the corresponding characteristic cones, which are tangent at each flow field point P to the corresponding characteristic surfaces, but are far easier to determine.

14.3.1   Galilean Transformation

The characteristic cones at each flow field point P are readily determined by recasting equations (14.52) through the following Galilean space-time coordinate transformation

$$\begin{cases} X_1 = x_1 - u_1 t \\ X_2 = x_2 - u_2 t \\ \tau = ct \end{cases} \tag{14.53}$$

in 2-D, and

$$\begin{cases} X_1 = x_1 - u_1 t \\ X_2 = x_2 - u_2 t \\ X_3 = x_3 - u_3 t \\ \tau = ct \end{cases} \tag{14.54}$$

in 3-D, where τ now denotes a "space-like" variable corresponding to the distance traveled by a constant-speed acoustic wave "c". With this coordinate transformation, the equation of each eventual cone corresponds to the shape of the cone as recorded by an observer that moves along with a fluid particle with local convection velocity $\boldsymbol{u}$.

By virtue of transformations (14.53), (14.54) the function $\mathcal{F}$ is recast as

$$\mathcal{F}(X_1, X_2, \tau) = \mathcal{F}(X_1(x_1, t), X_2(x_2, t), \tau(t)) \tag{14.55}$$

in 2-D, and

$$\mathcal{F}(X_1, X_2, X_3, \tau) = \mathcal{F}(X_1(x_1, t), X_2(x_2, t), X_3(x_3, t), \tau(t)) \tag{14.56}$$

in 3-D. In 2-D, the partial derivatives of $\mathcal{F}$ with respect to X_1, X_2, τ and x_1, x_2, t are related as

$$\left.\frac{\partial \mathcal{F}}{\partial x_1}\right|_{x_2,t} = \left.\frac{\partial \mathcal{F}}{\partial X_1}\right|_{X_2,\tau}, \quad \left.\frac{\partial \mathcal{F}}{\partial x_2}\right|_{x_1,t} = \left.\frac{\partial \mathcal{F}}{\partial X_2}\right|_{X_1,\tau}$$

$$\left.\frac{\partial \mathcal{F}}{\partial t}\right|_{x_1,x_2} = c\left.\frac{\partial \mathcal{F}}{\partial \tau}\right|_{X_1,X_2,\tau} - u_j\frac{\partial \mathcal{F}}{\partial X_j} \tag{14.57}$$

For a 3-D formulation, the partial derivatives of $\mathcal{F}$ with respect to X_1, X_2, X_3, τ and x_1, x_2, x_3, t are related as

$$\left.\frac{\partial \mathcal{F}}{\partial x_1}\right|_{x_2,x_3,t} = \left.\frac{\partial \mathcal{F}}{\partial X_1}\right|_{X_2,X_3,\tau}, \quad \left.\frac{\partial \mathcal{F}}{\partial x_2}\right|_{x_1,x_3,t} = \left.\frac{\partial \mathcal{F}}{\partial X_2}\right|_{X_1,X_3,\tau}, \quad \left.\frac{\partial \mathcal{F}}{\partial x_3}\right|_{x_1,x_2,t} = \left.\frac{\partial \mathcal{F}}{\partial X_3}\right|_{X_1,X_2,\tau}$$

$$\left.\frac{\partial \mathcal{F}}{\partial t}\right|_{x_1,x_2,x_3} = c\left.\frac{\partial \mathcal{F}}{\partial \tau}\right|_{X_1,X_2,X_3} - u_j\frac{\partial \mathcal{F}}{\partial X_j} \tag{14.58}$$

With either set of partial-derivative relations, therefore, the characteristic-surface equations (14.52) become

$$\mathcal{G} \equiv \det\left(I\left(c\frac{\partial \mathcal{F}}{\partial \tau} - u_j\frac{\partial \mathcal{F}}{\partial X_j}\right) + A_j(q)\frac{\partial \mathcal{F}}{\partial X_j}\right) = 0 \tag{14.59}$$

in 2-D and 3-D, which, despite its convoluted appearance, leads to algebraically simpler operations and results, as shown in the next section.

The equation of the tangent characteristic cone is then directly obtained from these equations upon using the correct multi-dimensional generalization of the Monge-cone [40, 65, 213] ordinary differential system, which yields the characteristic-cone differential equations

$$\frac{d(\frac{\partial \mathcal{F}}{\partial X_1})}{\frac{\partial \mathcal{G}}{\partial X_1} + \frac{\partial \mathcal{G}}{\partial \mathcal{F}}\frac{\partial \mathcal{F}}{\partial X_1}} = \frac{d(\frac{\partial \mathcal{F}}{\partial X_2})}{\frac{\partial \mathcal{G}}{\partial X_2} + \frac{\partial \mathcal{G}}{\partial \mathcal{F}}\frac{\partial \mathcal{F}}{\partial X_2}} = \frac{d(\frac{\partial \mathcal{F}}{\partial \tau})}{\frac{\partial \mathcal{G}}{\partial \tau} + \frac{\partial \mathcal{G}}{\partial \mathcal{F}}\frac{\partial \mathcal{F}}{\partial \tau}} = \frac{dX_1}{-\frac{\partial \mathcal{G}}{\partial(\frac{\partial \mathcal{F}}{\partial X_1})}} = \frac{dX_2}{-\frac{\partial \mathcal{G}}{\partial(\frac{\partial \mathcal{F}}{\partial X_2})}} = \frac{d\tau}{-\frac{\partial \mathcal{G}}{\partial(\frac{\partial \mathcal{F}}{\partial \tau})}} \tag{14.60}$$

in 2-D, and

$$\frac{d(\frac{\partial \mathcal{F}}{\partial X_1})}{\frac{\partial \mathcal{G}}{\partial X_1} + \frac{\partial \mathcal{G}}{\partial \mathcal{F}}\frac{\partial \mathcal{F}}{\partial X_1}} = \frac{d(\frac{\partial \mathcal{F}}{\partial X_2})}{\frac{\partial \mathcal{G}}{\partial X_2} + \frac{\partial \mathcal{G}}{\partial \mathcal{F}}\frac{\partial \mathcal{F}}{\partial X_2}} = \frac{d(\frac{\partial \mathcal{F}}{\partial X_3})}{\frac{\partial \mathcal{G}}{\partial X_3} + \frac{\partial \mathcal{G}}{\partial \mathcal{F}}\frac{\partial \mathcal{F}}{\partial X_3}} = \frac{d(\frac{\partial \mathcal{F}}{\partial \tau})}{\frac{\partial \mathcal{G}}{\partial \tau} + \frac{\partial \mathcal{G}}{\partial \mathcal{F}}\frac{\partial \mathcal{F}}{\partial \tau}} =$$

$$= \frac{dX_1}{-\frac{\partial \mathcal{G}}{\partial(\frac{\partial \mathcal{F}}{\partial X_1})}} = \frac{dX_2}{-\frac{\partial \mathcal{G}}{\partial(\frac{\partial \mathcal{F}}{\partial X_2})}} = \frac{dX_3}{-\frac{\partial \mathcal{G}}{\partial(\frac{\partial \mathcal{F}}{\partial X_3})}} = \frac{d\tau}{-\frac{\partial \mathcal{G}}{\partial(\frac{\partial \mathcal{F}}{\partial \tau})}} \tag{14.61}$$

in 3-D. The solution of these ordinary differential equations for each equation in (14.59) yields the equations of the corresponding characteristic cones.

14.3.2 Characteristic Cones and Wave Propagation Patterns

This section details the general procedure to establish the characteristics cones and then employs this procedure to determine the corresponding cones for the Euler equations. The variation of the shape of these cones with respect to the Mach number allows reaching a deeper understanding of multi-dimensional characteristic wave propagation in the time-space continuum. This analysis leads to specific requirements for a physically consistent multi-dimensional upstream-bias representation.

  The characteristic-surface equation $\mathcal{F}$ in the time-space continuum arises in practice by substitution of (14.45), (14.46) for n_j and λ in the expressions for the eigenvalues of the Jacobian matrix $A_j(q)n_j$. Anticipating the results of Section 17.3 for the Euler system, the three distinct flux-Jacobian eigenvalues are expressed as

$$\lambda_1 = u_j n_j, \quad \lambda_{2,3} = u_j n_j \pm c \tag{14.62}$$

where u_j, $1 \leq j \leq n$, and c respectively denote a Cartesian component of velocity and the local speed of sound. Based on these eigenvalues, the substitution procedure for determining $\mathcal{F}$ yields the Euler characteristic-surface equations

$$\left(\mathcal{F}_t + u_j \mathcal{F}_{x_j}\right)^2 = 0, \quad \left(\mathcal{F}_t + u_j \mathcal{F}_{x_j}\right)^2 = c^2 \mathcal{F}_{x_j} \mathcal{F}_{x_j} \tag{14.63}$$

where each subscript variable denotes differentiation with respect to that variable. In substantive-derivative form, the first of these two equations becomes

$$\left(\frac{D\mathcal{F}}{Dt}\right)^2 = 0 \tag{14.64}$$

which shows that the corresponding characteristic surface remains time-invariant from the perspective of an observer who travels along with a fluid particle.

  The principal propagation directions as well as the flow domain of dependence and range of influence of a flow field point P in the time-space continuum are determined by studying the variation of the characteristic-surface shape versus the Mach number. As indicated in the previous section, in order to investigate these shape changes it suffices to study the shape changes of the corresponding characteristic cones, which are tangent at each flow field point P to the corresponding characteristic surfaces, but are far easier to determine.

  By virtue of the Galilean transformations in Section 14.3, the characteristic-surface equations (14.63) simplify to

$$\mathcal{G} \equiv \left(\frac{\partial \mathcal{F}}{\partial \tau}\right)^2 = 0, \quad \mathcal{G} \equiv \left(\frac{\partial \mathcal{F}}{\partial \tau}\right)^2 - \left(\frac{\partial \mathcal{F}}{\partial X_1}\right)^2 - \left(\frac{\partial \mathcal{F}}{\partial X_2}\right)^2 = 0 \tag{14.65}$$

in 2-D, and

$$\mathcal{G} \equiv \left(\frac{\partial \mathcal{F}}{\partial \tau}\right)^2 = 0, \quad \mathcal{G} \equiv \left(\frac{\partial \mathcal{F}}{\partial \tau}\right)^2 - \left(\frac{\partial \mathcal{F}}{\partial X_1}\right)^2 - \left(\frac{\partial \mathcal{F}}{\partial X_2}\right)^2 - \left(\frac{\partial \mathcal{F}}{\partial X_3}\right)^2 = 0 \tag{14.66}$$

in 3-D. The equation of the tangent characteristic cone is then directly obtained from these equations through the Monge-cone ordinary differential equations. The solution of these ordinary differential equations for each equation in (14.65)-(14.66) is

$$\begin{cases} X_1 - X_{1o} = 0 \\ X_2 - X_{2o} = 0 \end{cases}, \quad (X_1 - X_{1o})^2 + (X_2 - X_{2o})^2 = (\tau - \tau)^2 \tag{14.67}$$

in 2-D, and

$$\begin{cases} X_1 - X_{1o} = 0 \\ X_2 - X_{2o} = 0 \\ X_3 - X_{3o} = 0 \end{cases}, \quad (X_1 - X_{1o})^2 + (X_2 - X_{2o})^2 + (X_3 - X_{3o})^2 = (\tau - \tau_o)^2 \tag{14.68}$$

in 3-D.

By virtue of (14.53)-(14.54), the equation of the first characteristic cone with respect to the fixed reference coordinates (x_1, x_2, t) then becomes

$$\frac{x_1 - x_{1o}}{t - t_o} = u_1, \quad \frac{x_2 - x_{2o}}{t - t_o} = u_2 \tag{14.69}$$

in 2-D, and

$$\frac{x_1 - x_{1o}}{t - t_o} = u_1, \quad \frac{x_2 - x_{2o}}{t - t_o} = u_2, \quad \frac{x_3 - x_{3o}}{t - t_o} = u_3 \tag{14.70}$$

in 3-D. In the time-space continuum, these equations correspond to the single "convection" straight line $S^- S^+$, as displayed in the 2-D case in Figures 14.2-14.4. As time elapses, information waves mono-axially reach and then leave P along this line, which, in the space continuum, projects onto the streamline

$$\frac{x_1 - x_{1o}}{u_1} = \frac{x_2 - x_{2o}}{u_2} \tag{14.71}$$

in 2-D, and

$$\frac{x_1 - x_{1o}}{u_1} = \frac{x_2 - x_{2o}}{u_2} = \frac{x_3 - x_{3o}}{u_3} \tag{14.72}$$

in 3-D, which represents in the time-space continuum a plane that contains the convection line. A spatial discretization that aims to model this propagation mode, therefore, should encompass an upstream bias in the streamline direction.

In the time-space continuum, the equation of the second characteristic cone becomes

$$(x_1 - x_{1o} - u_1(t - t_o))^2 + (x_2 - x_{2o} - u_2(t - t_o))^2 = c^2 (t - t_o)^2 \tag{14.73}$$

in 2-D, and

$$(x_1 - x_{1o} - u_1(t - t_o))^2 + (x_2 - x_{2o} - u_2(t - t_o))^2 \\ + (x_3 - x_{3o} - u_3(t - t_o))^2 = c^2 (t - t_o)^2 \tag{14.74}$$

in 3-D. Hence

$$(x_1 - x_{1o} - u_1(t - t_o))^2 + (x_2 - x_{2o} - u_2(t - t_o))^2 = \|\boldsymbol{u}\|^2 \frac{(t - t_o)^2}{M^2} \tag{14.75}$$

in 2-D, and, in 3-D

$$(x_1 - x_{1o} - u_1(t - t_o))^2 + (x_2 - x_{2o} - u_2(t - t_o))^2$$
$$+ (x_3 - x_{3o} - u_3(t - t_o))^2 = \|\boldsymbol{u}\|^2 \frac{(t - t_o)^2}{M^2} \tag{14.76}$$

Either of these equations corresponds to a characteristic cone with vertex at P that is tangent to the local characteristic surface at P and with a shape that depends upon M. As time elapses, in this case, waves funnel in towards P and out and away from P within the characteristic cone. For a 2-D flow, expression (14.75) also represents the equation in the space continuum a family of circumferences with center coordinates $[x_{1o} + u_1(t - t_o), \ x_{2o} + u_2(t - t_o)]$ and radius $\|\boldsymbol{u}\| (t - t_o)/M$. For a 3-D flow, expression (14.76) also represents the equation of a family of spheres with center coordinates $[x_{1o} + u_1(t - t_o), \ x_{2o} + u_2(t - t_o), \ x_{3o} + u_3(t - t_o)]$ and radius $\|\boldsymbol{u}\| (t - t_o)/M$. For a vanishing c, furthermore, this cone collapses onto the convection line (14.69), which, however, is not the axis of the characteristic cone. Figures 14.2, 14.4 portray this characteristic cone for subsonic and supersonic 2-D flows.

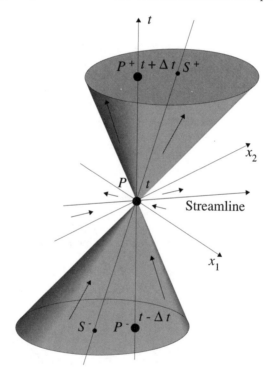

Figure 14.2: Subsonic Characteristic Cone

As M increases, the cone shears in the streamline direction, while its cross section remains an ellipse, with axes that project onto the streamline and crossflow directions. As a fundamental geometric difference between supersonic and subsonic flows, a time axis through P is respectively outside and inside the cone. In this manner, time-independent rays, i.e. lines issuing from the time axis for constant t, that are tangent to this cone only exist when the time axis is not inside the cone, that is for supersonic flows. The lines that issue from the time axis and are tangent to circumferences (14.75) in the (x_1, x_2) plane, and to spheres (14.76) in the (x_1, x_2, x_3) space are then found to be the Mach lines in 2-D, and Mach cones, in 3-D. This finding is not coincidental, because time-invariant rays tangent to the characteristic cone correspond to steady-state (i.e. time- invariant) characteristic lines, which are the Mach lines, in 2-D, and the Mach cone, in 3-D, for the steady Euler equations, as also remarked at the end of Section 17.3. Figure 14.2 presents the subsonic characteristic cone for a 2-D flow. The time axis is inside the cone and no tangent time-invariant lines exist. Accordingly, as time elapses, in this case, the bi-modal funneling of waves within the cone, towards and then away from P, corresponds on any constant-t plane to closed curves encircling P that shrink towards P and then expand away from P. Figure 14.3 illustrates this bi-modal pattern, which corresponds to the familiar subsonic Doppler distribution.

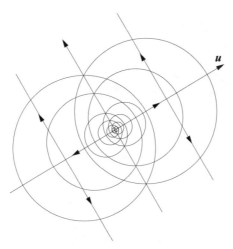

Figure 14.3: Subsonic Wave Propagation

This distribution results from projecting onto the flow plane several sections of the characteristic cone at various time levels and corresponds to pressure pulses emitted by a particle that subsonically travels along a streamline arc. From this perspective, a region of space centered about P is simultaneously the domain of dependence and range of influence for P. A spatial discretization that aims to model this propagation mode, therefore, should encompass an upstream bias along all directions radiating from P.

Figure 14.4 presents the supersonic characteristic cone for a 2-D flow. The time axis is outside the cone and the Mach lines are tangent to the cone on each constant-t plane. As time elapses, in this case, the funneling of waves within the cone, towards and then away from P, corresponds on any constant-t plane to information reaching P within its domain

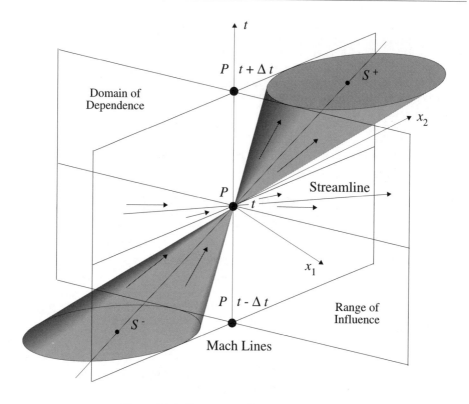

Figure 14.4: Supersonic Characteristic Cone

of dependence, between the two Mach lines upstream of P, and leaving P within its range of influence, between the two Mach lines downstream of P. Figure 14.5 displays this pattern, which corresponds to the familiar supersonic Doppler distribution. Wave propagation is thus mono-axial along the streamline. In the crossflow direction, however, wave propagation remains bi-modal, as illustrated in the figure, for waves propagate both upstream and downstream along this direction due to acoustic-wave propagation. This wave propagation pattern results from projecting onto the flow plane several sections of the characteristic cone at various time levels and corresponds to pressure pulses emitted by a particle that supersonically moves along a streamline arc. A spatial discretization that aims to model this propagation mode, therefore, should encompass a directional bias upstream and downstream of the crossflow direction and an upstream bias along all directions within the domain of dependence of P, with the streamline direction as the upstream-bias principal direction, since the streamline remains the axis of the domain of dependence and region of influence.

The investigation of the characteristic cone shape change versus M, therefore, indicates that for unsteady and steady supersonic flows, the upstream-bias domain of dependence and region of influence for each flow field point P consist of the regions that contain the

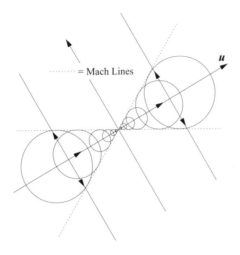

Figure 14.5: Supersonic Wave Propagation

streamline through P and are bound by the Mach lines. For unsteady and steady subsonic flows, both the domain of dependence and region of influence encompass regions encircling P.

14.4 Parabolic Perturbation System

Generalizing the parabolic-perturbation results detailed in Chapter 9, the solution of a multi-dimensional hyperbolic system is represented via the solution of an associated parabolic minimal-perturbation system. By comparison with (9.52) and with n indicating the number of flow dimensions, the generalized parabolic minimal-perturbation system is expressed as

$$\frac{\partial q}{\partial t} + A_j(q)\frac{\partial q}{\partial x_j} - \frac{\partial}{\partial x_i}\left(\varepsilon\psi B_{ij}(q)\frac{\partial q}{\partial x_j}\right) = 0 \tag{14.77}$$

where $\varepsilon\psi$ denotes a positive perturbation parameter and $B_{ij} = B_{ij}(q)$, $1 \leq i, j \leq n$, indicates a non-linear perturbation matrix.

Since both A_j and B_{ij} only depend upon q, even for this system is it possible to express the solution in non-linear wave-like form as

$$q = q(x_j n_j - \lambda(q)t) = q(\eta_1) \tag{14.78}$$

where n_j, $1 \leq j \leq n$, denotes the j-th component of an arbitrary direction unit vector and λ is determined by both A_j and B_{ij}. For a linear system with constant A_j and B_{ij}, λ is also a constant and (14.77) thus becomes

$$-\lambda\frac{dq}{d\eta_1} + A_j n_j\frac{dq}{d\eta_1} - \varepsilon\psi n_i B_{ij} n_j\frac{d^2q}{d\eta_1^2} = 0 \tag{14.79}$$

with λ an eigenvalue of this system. For a Fourier's series expansion of q, an individual series term may be expressed as

$$q_\omega = v \exp(j\omega\eta_1), \quad j \equiv \sqrt{-1} \tag{14.80}$$

and (14.79) becomes

$$\left(-\lambda I + A_j n_j - \varepsilon\psi j\omega n_i B_{ij} n_j \right) q_\omega = 0 \tag{14.81}$$

hence λ corresponds to an eigenvalue of the system matrix $(A_j n_j - \varepsilon\psi j\omega n_i B_{ij} n_j)$.

As for the one-dimensional case, the matrix B_{ij} is determined under the condition that the corresponding differential operator should prevent at each point the unbounded growth of any perturbation δq to the solution; the evolution of infinitesimal perturbations δq at each point is controlled by the linear counterpart of (14.77). The analysis of the time evolution of the energy in the solution of this linear counterpart indicates that (14.77) is stable in the sense indicated when the matrix contraction $n_i B_{ij} n_j$ possesses distinct positive eigenvalues for any n_j, as shown in the following developments. The solution of (14.77) is expressed as

$$q = q_0 + \delta q \tag{14.82}$$

with q_0 that is also a solution of (14.77), which implies that also δq may be expressed in the wave-like form (14.78) and thus depends on η_1 only. With $\overline{q_0}$ denoting a constant local value of q_0 and using a direct multi-dimensional generalization of the expansion procedure presented in Section 9.2.2, the linear counterpart of (14.77) governing the evolution of δq is expressed as

$$\frac{\partial(\delta q)}{\partial t} + A_j(\overline{q_0})\frac{\partial(\delta q)}{\partial x_j} - \frac{\partial}{\partial x_i}\left(\varepsilon\psi B_{ij}(\overline{q_0})\frac{\partial(\delta q)}{\partial x_j} \right) = 0 \tag{14.83}$$

which governs at each point the time evolution of the perturbation (δq). Let the matrix $n_i B_{ij} n_j$ possess distinct eigenvalues for any $\overline{q_0}$, hence any q; under this condition, this matrix may be expressed via the similarity transformation

$$n_i B_{ij}(\overline{q_0}) n_j = X \Lambda X^{-1} \tag{14.84}$$

where the diagonal matrix Λ contains all the eigenvalues of $n_i B_{ij} n_j$ and the constant matrix $X = X(\overline{q_0})$ denotes the associated eigenvector matrix. Next, multiply system (14.83) by X^{-1}, which results in the associated linear system

$$\frac{\partial(X^{-1}\delta q)}{\partial t} + X^{-1}A_j(\overline{q_0})X\frac{\partial(X^{-1}\delta q)}{\partial x_j} - \frac{\partial}{\partial x_i}\left(\varepsilon\psi X^{-1}B_{ij}(\overline{q_0})X\frac{\partial(X^{-1}\delta q)}{\partial x_j} \right) = 0 \tag{14.85}$$

The energy $\mathcal{E}$ in the solution $(X^{-1}\delta q)$ of this system and the time rate of change of this energy may be expressed as

$$\mathcal{E} \equiv \frac{1}{2}\int_\Omega \left(X^{-1}\delta q\right)^T \left(X^{-1}\delta q\right) d\Omega \quad \Rightarrow \quad \frac{d\mathcal{E}}{dt} = \int_\Omega \left(X^{-1}\delta q\right)^T \frac{\partial(X^{-1}\delta q)}{\partial t} d\Omega \tag{14.86}$$

The time derivative of $(X^{-1}\delta q)$ in this integral is then replaced, *à la* Lyapunov, in terms of the system itself. The Cartesian coordinates x_j, $1 \le j \le n$, in the integral containing B_{ij} in the result are then expressed in terms of a non-singular coordinate transformation with curvilinear coordinates η_j, $1 \leq j \leq n$, with η_1 defined as in (14.78). This sequence of operations yields

$$\frac{d\mathcal{E}}{dt} = -\int_\Omega \left(X^{-1}\delta q\right)^T X^{-1} A_j X \frac{\partial \left(X^{-1}\delta q\right)}{\partial x_j} d\Omega + \oint_{\partial\Omega} \left(\varepsilon\psi \left(X^{-1}\delta q\right)^T n_i B_{ij} \frac{\partial \left(X^{-1}\delta q\right)}{\partial x_j}\right) d\Gamma$$

$$-\int_\Omega \varepsilon\psi \frac{\partial \left(X^{-1}\delta q\right)^T}{\partial x_i} X^{-1} B_{ij} X \frac{\partial \left(X^{-1}\delta q\right)}{\partial x_j} d\Omega$$

$$= -\int_\Omega \left(X^{-1}\delta q\right)^T X^{-1} A_j X \frac{\partial \left(X^{-1}\delta q\right)}{\partial x_j} d\Omega + \oint_{\partial\Omega} \left(\varepsilon\psi \left(X^{-1}\delta q\right)^T n_i B_{ij} \frac{\partial \left(X^{-1}\delta q\right)}{\partial x_j}\right) d\Gamma$$

$$-\int_{\Omega_\eta} \frac{\varepsilon\psi(e_{\ell 1}e_{\ell 1})}{\det J} \frac{\partial \left(X^{-1}\delta q\right)^T}{\partial \eta_1} X^{-1} \underbrace{\left(\frac{e_{i1}}{\sqrt{(e_{\ell 1}e_{\ell 1})}}\right)}_{n_i} B_{ij} \underbrace{\left(\frac{e_{j1}}{\sqrt{(e_{\ell 1}e_{\ell 1})}}\right)}_{n_j} X \frac{\partial \left(X^{-1}\delta q\right)}{\partial \eta_1} d\Omega_\eta$$

$$= -\int_\Omega \left(X^{-1}\delta q\right)^T X^{-1} A_j X \frac{\partial \left(X^{-1}\delta q\right)}{\partial x_j} d\Omega + \oint_{\partial\Omega} \left(\varepsilon\psi \left(X^{-1}\delta q\right)^T n_i B_{ij} \frac{\partial \left(X^{-1}\delta q\right)}{\partial x_j}\right) d\Gamma$$

$$-\int_{\Omega_\eta} \frac{\varepsilon\psi(e_{\ell 1}e_{\ell 1})}{\det J} \frac{\partial \left(X^{-1}\delta q\right)^T}{\partial \eta_1} \Lambda \frac{\partial \left(X^{-1}\delta q\right)}{\partial \eta_1} d\Omega_\eta \tag{14.87}$$

where the terms e_{j1}, $1 \leq j \leq n$, correspond to the coordinate-transformation metric data, detailed in Chapter 7. This final integral emerges from (14.84) after identifying the metric-data components $e_{i1}/\sqrt{(e_{\ell 1}e_{\ell 1})}$ with the unit-vector components n_j, $1 \leq j \leq n$, as amplified in Section 14.5.4. In the final expression (14.87), as in the corresponding one-dimensional result, the first domain integral results from the hyperbolic component of the perturbation system, a component that always features in the time rate of change of $\mathcal{E}$, independently of the presence of the perturbation matrix B_{ij}. Expression (14.87) thus allows determining the contribution of this matrix to the rate of change of $\mathcal{E}$; since this expression is valid for any q and q_0, the following considerations apply to any q. The surface integral results from an integration by parts of the domain integral of the second-order term containing B_{ij}; this surface integral remains bounded because of enforced needed parabolic-system boundary conditions on the entire boundary $\partial\Omega$, conditions detailed in Chapters 15-17. The crucial contribution of the Λ matrix, in comparison to the case when this matrix is absent, is the domain integral of the associated quadratic form. Due to the minus sign preceding the integral, it is when this form is positive that the rise of $\mathcal{E}$ is limited, which prevents any unbounded growth of the perturbation δq; additionally, the limiting action of this quadratic form increases with a rise in the magnitude of $\varepsilon\psi$. This quadratic form remains positive when all the diagonal elements of Λ remain positive. As stated at the beginning of this section, accordingly, the perturbation δq will not grow, for any n_j, $1 \leq j \leq n$, when all the eigenvalues of the perturbation matrix $n_i B_{ij} n_j$ are distinct, so that (14.84) exists, and positive, so that the quadratic form associated with Λ in (14.87) remains positive. For any solution $q = q(\boldsymbol{x}, t)$, the parabolic perturbations developed in this book for multi-dimensional parabolic-perturbation equations and systems correspond to a matrix $n_i B_{ij} n_j = n_i B_{ij}(q)n_j$ with distinct and positive eigenvalues.

14.5   Characteristics-Bias Representation

Since for all Mach numbers there simultaneously exist regions of mono-axial and bi-modal propagation, a physically consistent, intrinsically multi-dimensional, and infinite directional upstream formulation for the Euler and Navier-Stokes equations has to provide an upstream approximation suitable for supersonic flows within the mono-axial region, but consistent with subsonic wave propagation within the bi-modal region. This formulation has to involve a streamline upstream approximation that remains bi-modal for subsonic flows and then becomes mono-axial for supersonic flows. For all Mach numbers, a physical upstream approximation must then induce a bi-modal upstream bias along the crossflow direction.

This upstream approximation has to introduce an upstream bias along all directions radiating from each flow field point. The bias in this approximation must change with varying direction and correlate with the directional distribution of the characteristic propagation speeds. Furthermore, the directional distribution of the upstream bias must remain symmetrical like the propagation speeds with respect to both the crossflow and streamline directions.

Traditional discrete approximations of the multi-dimensional system

$$\frac{\partial q}{\partial t} + \frac{\partial f_j(q)}{\partial x_j} - \frac{\partial f_j^v}{\partial x_j} = \phi \tag{14.88}$$

originate from the integral statement

$$\int_{\widehat{\Omega}} \widehat{w} \left(\frac{\partial q}{\partial t} - \phi + \frac{\partial f_j(q)}{\partial x_j} - \frac{\partial f_j^v}{\partial x_j} \right) d\Omega = 0 \tag{14.89}$$

which is equivalent to the governing system (14.88) for arbitrary subdomains $\widehat{\Omega} \subset \Omega$ and arbitrary integrable test functions $\widehat{w}$ with compact support in $\widehat{\Omega}$. For finite volume and element formulations, some upstream-bias approximations can emerge from this statement through numerical flux formulae for f_j and select choices for the test functions.

The characteristics-bias formulation develops the characteristics- bias system

$$G^C - \phi^C + \frac{\partial f_j^C}{\partial x_j} - D^C = 0 \tag{14.90}$$

where G^C, ϕ^C, f_j^C, and D^C respectively correspond to characteristics-bias time derivative, source term, inviscid flux, and viscous-flux divergence that automatically induce a multi-dimensional and infinite directional upstream-bias approximation for the original system. The characteristics-bias representation is formally developed for the time derivative, source term, and viscous-flux divergence so that the resulting characteristics-bias system may be satisfied by an exact solution of the original system, as discussed in Section 14.5.6, and may become Galilean invariant like the original system, which also reduces upwind dissipation, as elaborated in Section 14.6. The characteristics-bias system emerges from the characteristic-bias integral

$$\int_{\widehat{\Omega}} \widehat{w} \left(G^C - \phi^C + \frac{\partial f_j^C}{\partial x_j} - D^C \right) d\Omega = 0 \tag{14.91}$$

associated with (14.89). Since the characteristics-bias terms are developed independently and before any discretization, a genuinely multi-dimensional upstream-bias approximation for the original system on arbitrary grids directly results from a classical centered discretization of this integral statement on any given grid. For finite element formulations, a multi-dimensional upstream-bias approximation of the original system emerges from a classical Galerkin approximation of (14.91).

The following sections show that the divergence of the characteristics-bias inviscid flux f_j^C naturally derives from an upstream-bias integral average as

$$\frac{\partial f_j^C}{\partial x_j} = \frac{\partial f_j}{\partial x_j} - \frac{\partial}{\partial x_i}\left(\varepsilon\psi\sum_{\ell=1}^{L}a_{i\ell}\alpha_\ell A_{\ell j}\frac{\partial q}{\partial x_j}\right) \tag{14.92}$$

In this expression, $A_{\ell j}$ corresponds to a matrix component of the system inviscid flux Jacobian, such that the matrix $A_{\ell j}n_j$ has uniform-sign eigenvalues, where n_j denotes the jth component of a unit vector along an arbitrary wave-propagation direction within conical regions within the flow field. The coefficients α_ℓ denote linear-combination functions, possibly depending upon q. The term $a_{i\ell}$ indicates the ith direction cosine of a unit vector $\boldsymbol{a}_\ell$ along the principal wave-propagation direction of wave "ℓ", a convection or acoustic wave for instance. As detailed in Chapter 10, the positive ε denotes a length scale that measures a local computational-cell size; the positive ψ stands for the "upstream-bias" controller, which controls the amount of induced upstream-bias dissipation. This controller adjusts this dissipation depending on local solution non-smoothness. In regions of discontinuous solution slopes, this controller increases, to preserve numerical stability and capture shocks crisply; in regions of smooth solutions, the controller approaches its minimum, to reduce upstream dissipation. In this manner, the characteristics- bias formulation generates minimal-upstream-dissipation solutions.

It is well known that any discrete approximate solution of a hyperbolic or parabolic system exactly solves not the original equations, but an associate system of equations, also called the modified equations; as a maximum grid spacing within the discrete approximation approaches nought, the modified equations approach the original equations. Likewise, the characteristics-bias system in (14.91) approaches the original system as ε approaches nought, as the computational grid is refined.

14.5.1 Upstream-Bias Time Derivative, Source, Viscous-Flux Divergence

The expression for G^C is defined by the upstream weighted integral average

$$\int_{\widehat{\Omega}}\widehat{w}G^C\,d\Omega \equiv \int_{\widehat{\Omega}}(\widehat{w}+\psi\,\delta\widehat{w})\frac{\partial q}{\partial t}\,d\Omega \tag{14.93}$$

where $\widehat{w} \in \mathcal{H}_0^1(\widehat{\Omega})$ has compact support in $\widehat{\Omega}$ and it thus vanishes on the entire boundary $\partial\widehat{\Omega}$ of $\widehat{\Omega}$. The variation $\delta\widehat{w}$ is then expressed as

$$\delta\widehat{w} = \frac{\partial\widehat{w}}{\partial x_i}\delta x_i, \quad \delta x_i = a_i\varepsilon \tag{14.94}$$

In these two expressions, ε and ψ respectively indicate the length measure and controller, while a_i, $1 \leq i \leq n$, denotes the component of a unit vector $\mathbf{a}$ along a principal characteristic direction; for the Euler and Navier-Stokes equations this direction is the streamline direction. On the basis of this average, via integrating the variation of $\hat{w}$ by parts, the characteristics-bias resolutions of the time-derivative of q emerges as

$$G^C \equiv \frac{\partial q}{\partial t} - \frac{\partial}{\partial x_i}\left[\varepsilon \psi a_i \frac{\partial q}{\partial t}\right] \tag{14.95}$$

Similarly, the formal characteristics-bias resolutions for ϕ and $\frac{\partial f_j^v}{\partial x_j}$ are expressed as

$$\phi^C \equiv \phi - \frac{\partial}{\partial x_i}\left[\varepsilon \psi a_i \, \phi\right], \quad D^C \equiv \frac{\partial f_j^v}{\partial x_j} - \frac{\partial}{\partial x_i}\left[\varepsilon \psi a_i \frac{\partial f_j^v}{\partial x_j}\right] \tag{14.96}$$

14.5.2 Flux Jacobian Decomposition and Upstream-Bias Integral Average

For the development of the characteristics-bias flux f_j^C, consider first the flux Jacobian decomposition (FJD) into L contributions

$$\frac{\partial f_j}{\partial q} = \sum_{\ell=1}^L \alpha_\ell A_{\ell j} \quad \Rightarrow \quad \frac{\partial f_j}{\partial x_j} = \sum_{\ell=1}^L \alpha_\ell A_{\ell j} \frac{\partial q}{\partial x} \tag{14.97}$$

where α_ℓ denotes a linear-combination function, possibly depending upon q, $A_{\ell j}$ corresponds to a flux-Jacobian matrix component such that the matrix $A_{\ell j} n_j$ has uniform-sign eigenvalues within a conical region spanned by n_j, within the flow domain.

An integral average of the Euler flux divergence $\frac{\partial f_j}{\partial x_j}$ as expressed through decomposition (14.97) becomes

$$\int_{\widehat{\Omega}} \hat{w} \frac{\partial f_j}{\partial x_j} \, d\Omega = \int_{\widehat{\Omega}} \sum_{\ell=1}^L \hat{w} \alpha_\ell A_{\ell j} \frac{\partial q}{\partial x_j} \, d\Omega \tag{14.98}$$

The flux f_j^C is therefore defined by way of an upstream-bias integral average as

$$\int_{\widehat{\Omega}} \hat{w} \frac{\partial f_j^C}{\partial x_j} \, d\Omega \equiv \int_{\widehat{\Omega}} \sum_{\ell=1}^L (\hat{w} + \psi \delta_\ell \hat{w}) \, \alpha_\ell A_{\ell j} \frac{\partial q}{\partial x_j} \, d\Omega \tag{14.99}$$

where the rhs integral provides an upstream bias for each matrix component within the FJD in (14.97).

The expression $\delta_\ell \hat{w}$ denotes a directional variation of the test function $\hat{w}$ along the axis of a conical region within the flow domain. This variation induces the appropriate upstream-bias for the test function $\hat{w}$ for each "ℓ" component within (14.99). Depending on the physical significance, magnitude and algebraic sign of the eigenvalues of $A_{\ell j} n_j$, the variation $\delta_\ell \hat{w}$ can vanish or become algebraically positive or negative, which corresponds to an upstream bias respectively in the negative or positive sense of the axis of each conical region.

14.5.3 Characteristics-Bias Flux

The directional variation $\delta_\ell \widehat{w}$ in (14.99) becomes

$$\delta_\ell \widehat{w} = \frac{\partial \widehat{w}}{\partial x_i} \delta_\ell x_i = \frac{\partial \widehat{w}}{\partial x_i} a_{i\ell} \varepsilon, \quad \delta_\ell x_i = a_{i\ell} \varepsilon \tag{14.100}$$

where $a_{i\ell}$ indicates the ith direction cosine of a unit vector $\boldsymbol{a}_\ell$ along the principal wave-propagation direction of wave "ℓ".

With these specifications, the upstream-bias integral average (14.99) becomes

$$\int_{\widehat{\Omega}} \widehat{w} \frac{\partial f_j^C}{\partial x_j} \, d\Omega = \int_{\widehat{\Omega}} \widehat{w} \frac{\partial f_j}{\partial x_j} \, d\Omega + \int_{\widehat{\Omega}} \varepsilon \psi \frac{\partial \widehat{w}}{\partial x_i} \sum_{\ell=1}^{L} a_{i\ell} \alpha_\ell A_{\ell j} \frac{\partial q}{\partial x_j} \, d\Omega \tag{14.101}$$

Considering that $\widehat{w}$ has compact support in $\widehat{\Omega}$, it vanishes on the boundary $\partial \widehat{\Omega}$ of $\widehat{\Omega}$. As a result, integrating (14.101) by parts generates

$$\int_{\widehat{\Omega}} \widehat{w} \left[\frac{\partial f_j^C}{\partial x_j} - \frac{\partial f_j}{\partial x_j} + \frac{\partial}{\partial x_i} \left(\varepsilon \psi \sum_{\ell=1}^{L} a_{i\ell} \alpha_\ell A_{\ell j} \frac{\partial q}{\partial x_j} \right) \right] d\Omega = 0 \tag{14.102}$$

which contains no boundary integrals. Since this integral must vanish for arbitrary test functions $\widehat{w}$ and domains $\widehat{\Omega}$, its integrand must identically equal zero, which generates the following expression for the divergence of the characteristics flux f_j^C

$$\frac{\partial f_j^C}{\partial x_j} = \frac{\partial f_j}{\partial x_j} - \frac{\partial}{\partial x_i} \left(\varepsilon \psi \sum_{\ell=1}^{L} a_{i\ell} \alpha_\ell A_{\ell j} \frac{\partial q}{\partial x_j} \right) \tag{14.103}$$

From (14.103), an expression for the ith component of the characteristics-bias flux may be defined up to a divergence-free flux as

$$f_i^C = f_i - \varepsilon \psi \sum_{\ell=1}^{L} a_{i\ell} \alpha_\ell A_{\ell j} \frac{\partial q}{\partial x_j} \tag{14.104}$$

As a multi-dimensional expression, each cartesian component f_i^C also depends on the derivatives of the solution q along all cartesian coordinate directions. The continuum expression (14.103), or (14.104), thus constitutes a non-discrete multi-dimensional generalization of the various one-dimensional numerical- flux formulae employed in several CFD upwind schemes.

14.5.4 Upstream-Bias Stability Condition

Expression (14.103) exhibits an upstream-bias artificial diffusion, in the form of a second-order differential expression. The associated upstream-bias matrix is

$$\mathcal{A} \equiv n_i \left(\sum_{\ell=1}^{L} a_{i\ell} \alpha_\ell A_{\ell j} \right) n_j \tag{14.105}$$

where n_i indicates the ith direction cosine of a unit vector $\boldsymbol{n}$ along an arbitrary wave - propagation direction. This section shows that for physical consistency of the upstream bias in (14.92), (14.99) and associated mathematical stability of the corresponding second-order differential expression, all the eigenvalues of this upstream-bias matrix must be positive at every flow-field point and for any wave-propagation direction $\boldsymbol{n}$.

In Jacobian form, expression (17.108) becomes

$$\frac{\partial f_j^C}{\partial x_j} = \frac{\partial f_j}{\partial x_j} - \frac{\partial}{\partial x_i}\left[\varepsilon\psi\left(\sum_{\ell=1}^{L} a_{i\ell}\alpha_{\ell}A_{\ell j}\right)\frac{\partial q}{\partial x_j}\right] \tag{14.106}$$

By virtue of a non-singular (x_1, x_2)-space coordinate transformation

$$\eta_1 = x_j n_j - \lambda(q)t, \quad \eta_2 = \eta_2(x_1, x_2, t) \tag{14.107}$$

in 2-D, or a non-singular (x_1, x_2, x_3)-space coordinate transformation

$$\eta_1 = x_j n_j - \lambda(q)t, \quad \eta_2 = \eta_2(x_1, x_2, x_3, t), \quad \eta_3 = \eta_3(x_1, x_2, x_3, t) \tag{14.108}$$

in 3-D, analogous to (14.16)-(14.17), and using the inverse $\{J^{-1}\}_{jk} \equiv \{e_{jk}\}/\det J$ of the associated coordinate-transformation Jacobian $J \equiv \partial \boldsymbol{x}/\partial \boldsymbol{\eta}$, expression (14.106) for each t becomes

$$\frac{\partial f_j^C}{\partial x_j} = \frac{\partial f_j}{\partial x_j} - \frac{1}{\det J}\frac{\partial}{\partial \eta_\ell}\left[\varepsilon\psi e_{i\ell}\left(\sum_{\ell=1}^{L} a_{i\ell}\alpha_{\ell}A_{\ell j}\right)\frac{e_{jk}}{\det J}\frac{\partial q}{\partial \eta_k}\right] \tag{14.109}$$

where the coordinate-transformation metric data e_{jk}, within J, can be brought inside the partial differential operation because of the fundamental equality $\partial e_{i\ell}/\partial \eta_\ell = 0$, for any i and with implied summation on ℓ, as shown in Chapter 7. The same chapter shows that for each ℓ the metric data $e_{i\ell}$ denote the components of a vector that is orthogonal to an $\eta_\ell = $ constant surface. For the hyperbolic and parabolic systems considered in this and the following chapters, systems that encompass the characteristics-bias system, the corresponding solution q in non-linear wave-like form depends upon η_1 only. For these solutions, therefore, $q = q(\eta_1)$ and (14.109) exhibits the upstream-bias expression

$$\mathcal{D} \equiv \frac{1}{\det J}\frac{\partial}{\partial \eta_1}\left[\frac{\varepsilon\psi(e_{\ell 1}e_{\ell 1})}{\det J}\frac{e_{i1}}{\sqrt{(e_{\ell 1}e_{\ell 1})}}\left(\sum_{\ell=1}^{L} a_{i\ell}\alpha_{\ell}A_{\ell j}\right)\frac{e_{j1}}{\sqrt{(e_{\ell 1}e_{\ell 1})}}\frac{\partial q}{\partial \eta_1}\right] \tag{14.110}$$

for which $e_{i1}/\sqrt{(e_{\ell 1}e_{\ell 1})}$ denotes the component n_i of the wave-propagation unit vector $\boldsymbol{n}$. This expression now corresponds to a one-dimensional upstream-bias expression, which, as a result, will be stable when all the eigenvalues of the upstream-bias matrix

$$\mathcal{A} \equiv n_i\left(\sum_{\ell=1}^{L} a_{i\ell}\alpha_{\ell}A_{\ell j}\right)n_j \tag{14.111}$$

remain positive for all propagation directions $\boldsymbol{n}$. This requirement implies a consistent upstream-bias along all directions radiating from any flow field point and thus becomes a fundamental multi-dimensional upstream-bias stability condition.

14.5.5 Incorporation of Flux-Vector and Flux-Difference Splittings

The FJD procedure generalizes and encompasses traditional Flux-Vector and Flux-Difference splitting formulations. Consider the representative Flux Vector Splitting (FVS) of the Euler flux as

$$f_j(q) = f_j^+(q) + f_j^-(q) \tag{14.112}$$

where the Jacobian matrices of $\frac{\partial f_j^+}{\partial q} n_j$ and $\frac{\partial f_j^-}{\partial q} n_j$ respectively possess non-negative and non-positive eigenvalues within a conical region with axis with direction cosines n_j, $1 \leq j \leq 3$. The FJD expression (14.97) encompasses (14.112) with $L = 2$ as

$$\sum_{\ell=1}^{L} \alpha_\ell A_{\ell j} = \frac{\partial f_j^+}{\partial q} + \frac{\partial f_j^-}{\partial q}, \quad \alpha_1 = 1, \quad \alpha_2 = 1 \tag{14.113}$$

The corresponding characteristics-bias flux divergence for this representative flux vector splitting accrues from (14.103) as

$$\frac{\partial f_j^C}{\partial x_j} = \frac{\partial f_j}{\partial x_j} - \frac{\partial}{\partial x_i} \left(\varepsilon \psi \left(a_i^+ \frac{\partial f_j^+}{\partial x_j} - a_i^- \frac{\partial f_j^-}{\partial x_j} \right) \right) \tag{14.114}$$

which generalizes in the continuum the traditional numerical flux formulae for FVS constructions. The associated upstream-bias matrix $\mathcal{A}$ is

$$\mathcal{A} = \sum_{i,j=1}^{3} n_i \left(a_i^+ \frac{\partial f_j^+}{\partial q} + a_i^- \frac{\partial f_j^-}{\partial q} \right) n_j \tag{14.115}$$

where the flux components f_j^+ and f_j^- as well as the direction cosines a_i^+ and a_i^-, frequently $a_i^- = -a_i^+$, should be chosen to satisfy the upstream-bias stability condition on this matrix.

Consider next a representative Flux Difference Splitting (FDS) development, where the inviscid flux Jacobian of f_j is "split" as

$$\frac{\partial f_j}{\partial q} = X_{\underline{j}} \Lambda_j^+ X_{\underline{j}}^{-1} + X_{\underline{j}} \Lambda_j^- X_{\underline{j}}^{-1} \tag{14.116}$$

where $X_{\underline{j}}$ and $\Lambda_j = \Lambda_j^+ + \Lambda_j^-$ denote the right eigenvector matrix and eigenvalue diagonal matrix of the Jacobian, all evaluated at special average values of q, with Λ_j^+ and Λ_j^- respectively containing non-negative and non-positive eigenvalues. The matrices at the rhs of (14.116), therefore, will respectively possess non-negative and non-positive eigenvalues. The FJD expression (14.97) encompasses (14.116) with $L = 2$ as

$$\sum_{\ell=1}^{L} \alpha_\ell A_{\ell j} = X_{\underline{j}} \Lambda_j^+ X_{\underline{j}}^{-1} + X_{\underline{j}} \Lambda_j^- X_{\underline{j}}^{-1}, \quad \alpha_1 = 1, \quad \alpha_2 = 1 \tag{14.117}$$

The corresponding characteristics-bias divergence for this formulation accrues from (14.103) as

$$\frac{\partial f_j^C}{\partial x_j} = \frac{\partial f_j}{\partial x_j} - \frac{\partial}{\partial x_i} \left(\varepsilon \psi \left(a_i^+ \sum_{j=1}^{3} \left(X_{\underline{j}} \Lambda_j^+ X_{\underline{j}}^{-1} \right) + a_i^- \sum_{j=1}^{3} \left(X_{\underline{j}} \Lambda_j^- X_{\underline{j}}^{-1} \right) \right) \frac{\partial q}{\partial x_j} \right) =$$

$$\frac{\partial f_j}{\partial x_j} - \frac{\partial}{\partial x_i} \left(\varepsilon\psi \sum_{j=1}^{3} X_{\underline{j}} \left(a_i^+ \Lambda_j^+ + a_i^- \Lambda_j^- \right) X_{\underline{j}}^{-1} \frac{\partial q}{\partial x_j} \right) \tag{14.118}$$

which generalizes in the continuum the traditional numerical flux formulae for FDS constructions. The associated upstream-bias matrix $\mathcal{A}$ is

$$\mathcal{A} = \sum_{i,j=1}^{3} n_i X_{\underline{j}} \left(a_i^+ \Lambda_j^+ + a_i^- \Lambda_j^- \right) X_{\underline{j}}^{-1} n_j \tag{14.119}$$

where the eigenvalue-matrix components Λ_j^+ and Λ_j^- as well as the direction cosines a_i^+ and a_i^-, frequently $a_i^- = -a_i^+$, should be chosen to satisfy the upstream-bias stability condition on this matrix.

14.5.6   Characteristics-Bias System

The multi-dimensional characteristics-bias system provides a multi-dimensional generalization of the one-dimensional parabolic-perturbation equations described in Chapter 9. Practical computations solve the initial-boundary value problem with a perturbation parameter $\varepsilon\psi$ that remains commensurate with a decreasing, but finite, mesh spacing. As a result, for efficiency's and accuracy's sake, system (14.91) should be solved by an exact solution of the unperturbed system for finite $\varepsilon\psi$, in the case of a Jacobian $\frac{\partial f_j}{\partial q} n_j$ with equal-sign eigenvalues within a streamline conical region, as well as vanishing gradients in the cross-flow direction, and generally admit solutions that should rapidly converge to a weak-solution q of the original system for decreasing $\varepsilon\psi$. It is to promote these properties that the perturbation in the previous sections has been developed from a characteristics-bias representation of the original system. For the reference multi-dimensional system (14.88), the developments in the previous sections yield the characteristics-bias system

$$\frac{\partial q}{\partial t} - \frac{\partial}{\partial x_i} \left(\varepsilon\psi a_i \frac{\partial q}{\partial t} \right) - \phi + \frac{\partial}{\partial x_i} \left(\varepsilon\psi a_i \phi \right)$$

$$-\frac{\partial f_j^v}{\partial x_j} + \frac{\partial}{\partial x_i} \left(\varepsilon\psi \frac{\partial f_j^v}{\partial x_j} \right) + \frac{\partial f_j}{\partial x_j} - \frac{\partial}{\partial x_i} \left(\varepsilon\psi \sum_{\ell=1}^{L} a_{i\ell} \alpha_\ell A_{\ell j} \frac{\partial q}{\partial x_j} \right) = 0 \tag{14.120}$$

This system involves a second-order regularizing perturbation. With reference to (6.21), this perturbation, for viscous flows, leads to the effective Reynolds number

$$\frac{1}{Re^E} \simeq \frac{1}{Re} + \varepsilon\psi \quad \Rightarrow \quad Re^E \simeq \frac{Re}{1 + \varepsilon\psi Re} \tag{14.121}$$

This effective Reynolds number Re^E approaches the reference Reynolds number Re for a sufficiently small magnitude of $\varepsilon\psi$. A current research project consists in further reducing induced upwind diffusion by decreasing $\psi_{\min}$ proportionately to the magnitude of the crossflow deviatoric traction and velocity within a boundary layer. Since

$$\frac{\partial f_j}{\partial x_j} = \sum_{\ell=1}^{L} \alpha_\ell A_{\ell j} \frac{\partial q}{\partial x_j} \tag{14.122}$$

the characteristics-bias system may be more lucidly expressed as

$$\left[I - \frac{\partial}{\partial x_i}\varepsilon\psi a_i\cdot\right]\left(\frac{\partial q}{\partial t} - \phi + \frac{\partial f_j}{\partial x_j} - \frac{\partial f_j^v}{\partial x_j}\right) - \frac{\partial}{\partial x_i}\left(\varepsilon\psi\sum_{\ell=1}^{L}(a_{i\ell} - a_i)\,\alpha_\ell A_{\ell j}\frac{\partial q}{\partial x_j}\right) = 0 \quad (14.123)$$

This "companion" system contains the original system, its derivatives with respect to $\mathbf{x}$ and a compensating second-order term that emerges from the decomposition of the inviscid-flux Jacobian that must satisfy the upstream-bias stability condition for $\mathcal{A}$. This compensating term, however, is only present when the eigenvalues of the flux Jacobian $\frac{\partial f_j}{\partial q}n_j$ do not all have the same algebraic sign; when they do, $a_{i\ell} \equiv a_i$, $1 \le \ell \le L$, and for vanishing solution gradients in the crossflow direction, for the Euler system, the characteristics-bias system becomes

$$\left[I - \frac{\partial}{\partial x_i}\varepsilon\psi a_i\cdot\right]\left(\frac{\partial q}{\partial t} - \phi + \frac{\partial f_j}{\partial x_j} - \frac{\partial f_j^v}{\partial x_j}\right) = 0 \quad (14.124)$$

which only contains the original system and its derivative with respect to $\mathbf{x}$. In these conditions, significantly, an exact solution of the original system will also satisfy this companion system for any finite $\varepsilon\psi$, an important property that is achieved only when the characteristics-bias representation evolves from components of the original system, as developed in the previous sections. When the equations in this system are viewed as quasi-linear equations, independently of one another, and, for the Euler system, the fluid velocity $\mathbf{u}$ is considered constant, the equations are more clearly seen to remain hyperbolic, because of the presence of the mixed derivative with respect to both $\mathbf{x}$ and t. In a suitable neighborhood of a shock, where sufficient dissipation is needed to ensure essential non-oscillatory shock capturing, this mixed derivative will have to be eliminated, so that in the vicinity of a shock the characteristics-bias system becomes a more dissipative parabolic system. In this manner, $\mathcal{H}^1(\Omega)$ solutions of this system may converge to $\mathcal{H}^0(\Omega)$ solution of the unperturbed system.

14.6   Galilean Invariance and Characteristics-Bias Diffusion

Also the multi-dimensional unperturbed Euler and Navier-Stokes systems are Galilean invariant. The characteristics-bias system (14.123) can also satisfy this fundamental physical property, but only when it contains a characteristics-bias term for the time derivative. As for one-dimensional flows, also for multi-dimensional flows is satisfying this property important because in this case the characteristics-bias system reduces the amount of induced characteristics-bias diffusion, as shown in this section.

As stated for one-dimensional flows, the Galilean invariance property reflects the notion that the law of conservation of mass, the second law of mechanics, and the first law of thermodynamics remain independent from the state of motion of flow reference frames that move with constant velocity with respect to one another, called Galilean frames and illustrated in Figure 14.6. The corresponding continuity, linear-momentum, and energy equation thus remain formally invariant with respect to any co-ordinate transformation among Galilean frames.

With $\boldsymbol{V}_B$ the absolute velocity of the moving-reference origin B, and with $\boldsymbol{V}$ and $\boldsymbol{u}$ respectively the absolute and relative velocity of a fluid particle P, the co-ordinate sets (X_1, X_2, X_3) and (x_1, x_2, x_3) as well as velocity of P with respect to any two Galilean frames may be related as

$$X_i = X_{Bi} + x_i = V_{Bi}t + x_i \quad \Rightarrow \quad \left.\frac{\partial X_i}{\partial t}\right|_x = V_{Bi}, \quad u_i = V_i - V_{Bi} \tag{14.125}$$

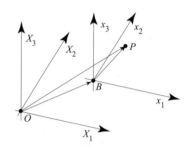

Figure 14.6: Galilean Reference Frames

By virtue of this multi-dimensional coordinate transformation, a function of $(\boldsymbol{X}, t)$ is cast as $\rho(\boldsymbol{X}, t) = \rho(\boldsymbol{X}(\boldsymbol{x}, t), t)$ and the derivatives with respect to $(\boldsymbol{X}, t)$ in the continuity equation, for instance

$$\left.\frac{\partial \rho}{\partial t}\right|_x + \left.\frac{\partial \rho u_i}{\partial x_i}\right|_t = 0 \tag{14.126}$$

correlate with the derivatives with respect to $(\boldsymbol{x}, t)$ as

$$\left.\frac{\partial \rho}{\partial t}\right|_x = \left.\frac{\partial \rho}{\partial X_i}\right|_t \left.\frac{\partial X_i}{\partial t}\right|_x + \left.\frac{\partial \rho}{\partial t}\right|_X, \quad \left.\frac{\partial \rho u_i}{\partial x_i}\right|_t = \left.\frac{\partial \rho u_i}{\partial X_i}\right|_t \tag{14.127}$$

Upon substituting these correlations for the corresponding derivatives, equation (14.126) becomes

$$V_{Bi} \left.\frac{\partial \rho}{\partial X_i}\right|_t + \left.\frac{\partial \rho}{\partial t}\right|_X + \left.\frac{\partial \rho(V_i - V_{Bi})}{\partial X_i}\right|_t = 0 \quad \Rightarrow \quad \left.\frac{\partial \rho}{\partial t}\right|_X + \left.\frac{\partial \rho V_i}{\partial X_i}\right|_t = 0 \tag{14.128}$$

which formally coincides with (14.126) and thus proves Galilean invariance of the continuity equation. The thermodynamic variables and the upstream-bias functions remain unaffected by this co-ordinate transformation because either they do not depend on the velocity of a reference frame or only involve relative velocities with respect to the flow channel, which velocities transform into one another, thereby retaining their form in the transformation. As a result, similar transformations of the other governing equations, also inserting the continuity equation into the linear-momentum equation and these two equations into the energy equation, show that both the linear-momentum and energy equations are Galilean invariant.

Owing to the characteristics-bias expression for the time derivative of q, the multi-dimensional system (14.123) is also Galilean invariant. To show this, it suffices to consider the three-dimensional characteristics-bias continuity equation, with c denoting the local speed of sound, a_i, $a_i^{N_1}$, $a_i^{N_2}$, $1 \le i \le n$, the components of three mutually orthogonal unit vectors respectively in the streamline direction and crossflow plane and α as well as α^N the functions of the upstream bias for the acoustic-wave propagation in the streamline direction and on the crossflow plane. Anticipating the developments of following chapters, this continuity equation is expressed as

$$
\left.\frac{\partial \rho}{\partial t}\right|_x + \left.\frac{\partial \rho u_i}{\partial x_i}\right|_t - \left.\frac{\partial}{\partial x_i}\left(\varepsilon \psi a_i \left.\frac{\partial \rho}{\partial t}\right|_x\right)\right|_t
$$

$$
- \left.\frac{\partial}{\partial x_i}\left[\varepsilon \psi \left(c\left(\alpha a_i a_j + \alpha^N \left(a_i^{N_1} a_j^{N_1} + a_i^{N_2} a_j^{N_2}\right)\right)\right) \left.\frac{\partial \rho}{\partial x_j}\right|_t + a_i \left.\frac{\partial \rho u_j}{\partial x_j}\right|_t\right]\right|_t = 0 \qquad (14.129)
$$

where the second-order terms correspond to the characteristics-bias perturbation expression for the continuity equation. A similar expression is obtained for incompressible and free-surface flows. Transforming the derivatives in this equation following developments (14.126)-(14.128) yields

$$
\left.\frac{\partial \rho}{\partial t}\right|_x + \left.\frac{\partial \rho}{\partial X_i}\right|_t V_{Bi} + \left.\frac{\partial \rho V_i}{\partial X_i}\right|_t - \left.\frac{\partial \rho}{\partial X_i}\right|_t V_{Bi}
$$

$$
- \left.\frac{\partial}{\partial X_i}\left(\varepsilon \psi a_i \left.\frac{\partial \rho}{\partial X_j}\right|_t V_{Bj}\right)\right|_t - \left.\frac{\partial}{\partial X_i}\left(\varepsilon \psi a_i \left.\frac{\partial \rho}{\partial t}\right|_x\right)\right|_t
$$

$$
- \left.\frac{\partial}{\partial X_i}\left[\varepsilon \psi \left(c\left(\alpha a_i a_j + \alpha^N \left(a_i^{N_1} a_j^{N_1} + a_i^{N_2} a_j^{N_2}\right)\right)\right) \left.\frac{\partial \rho}{\partial X_j}\right|_t\right]\right|_t
$$

$$
- \left.\frac{\partial}{\partial X_i}\left[\varepsilon \psi \, a_i \left.\frac{\partial \rho V_j}{\partial X_j}\right|_t\right]\right|_t + \left.\frac{\partial}{\partial X_i}\left(\varepsilon \psi a_i \left.\frac{\partial \rho}{\partial X_j}\right|_t V_{Bj}\right)\right|_t = 0 \qquad (14.130)
$$

with analogous results for the linear-momentum and energy equations. After all the terms containing V_{Bj}, $1 \le j \le n$, cancel one another out, the resulting equation formally coincides with (14.129), which proves Galilean invariance. Not only is this result showing that the characteristics-bias system is Galilean invariant, but also that this system satisfies this fundamental physical property only when a commensurate characteristics-bias expression is also present for the time derivative of q.

Also for multi-dimensional flows, the importance of this fundamental, but seldom, if ever, mentioned, property in CFD relates to its connection with artificial diffusion. In order to highlight this notion it suffices to consider a two-dimensional linear continuity equation with constant positive ε, ψ_t, ψ, c, α, a_i, a_i^N, u_i, $1 \le i \le n$,

$$
\frac{\partial \rho}{\partial t} + u_j \frac{\partial \rho}{\partial x_j} - \frac{\partial}{\partial x_i}\left[\varepsilon \psi_t a_i \frac{\partial \rho}{\partial t} + \varepsilon \psi \left(c \alpha a_i a_j + c \alpha^N a_i^N a_j^N + a_i u_j\right) \frac{\partial \rho}{\partial x_j}\right] = 0 \qquad (14.131)
$$

where $\psi_t = \psi$ for a Galilean invariant formulation. A generalized eigenfunction solution for this equation may be expressed as

$$
\rho(\mathbf{x}, t) = \exp\left(\lambda t\right) \exp\left(\mathbf{j}\, \omega \left(x_j n_j - u_j n_j t\right)\right), \quad \mathbf{j} \equiv \sqrt{-1} \qquad (14.132)
$$

where $\lambda = 0$ corresponds to an exact solution of not only the equation with no bias, hence $\varepsilon\psi = 0$, $\varepsilon\psi_t = 0$, but also the characteristics-bias equation with $\alpha = 0$, $\alpha^N = 0$, and $\psi_t = \psi$, as directly established by substitution. When this substitution is executed for any positive α, α^N, ψ_t, the resulting expression for the dissipation eigenvalue λ is

$$\lambda = -\frac{\varepsilon c \omega^2 \left((M(a_i n_i)^2 + \alpha(a_i n_i)^2 + \alpha^N (a_i^N n_i)^2)\psi - M(a_i n_i)^2 \psi_t \right)(1 + \mathbf{j}\,\omega(a_i n_i)\varepsilon\psi_t)}{1 + \omega^2 (a_i n_i)^2 \varepsilon^2 \psi_t^2}$$

(14.133)

Since the real part of this eigenvalue remains negative, the characteristics bias induces some artificial diffusion, which decreases in proportion to ψ, ψ_t. For $\mathbf{n}$ pointing in the crossflow direction, hence $(a_i^N n_i) = 1$ and $(a_i n_i) = 0$, this eigenvalue becomes independent of ψ_t. For $\mathbf{n}$ pointing in the streamline direction, however, $(a_i^N n_i) = 0$ as well as $(a_i n_i) = 1$ and this result no longer depends upon α^N and thus becomes the already determined one-dimensional expression

$$\lambda = -\frac{\omega c}{2}\lambda_{w\varepsilon} = -\frac{\varepsilon c \omega^2 \left((M + \alpha)\psi - M\psi_t \right)(1 + \mathbf{j}\,\omega\varepsilon\psi_t)}{1 + \omega^2 \varepsilon^2 \psi_t^2}$$

(14.134)

According to this form of λ, the streamline characteristics-bias artificial diffusion reaches a minimum for $\psi_t = \psi$, that is for a Galilean invariant formulation, a minimum that equals zero for $\alpha = 0$. Figure 14.7 displays the variation of the absolute value of the real part of $\lambda_{w\varepsilon}$ for $0 \leq 2\omega\varepsilon \leq \pi$, as obtained for two representative cases: a) $M = 0.3$ hence $\alpha = 0.4$, and b) $M = 1.2$ hence $\alpha \equiv 0$. As this chart shows, for both subsonic and supersonic flows, the induced diffusion rapidly decreases as the characteristics-bias system approaches the Galilean invariant form. For supersonic flows, in particular, this induced diffusion vanishes.

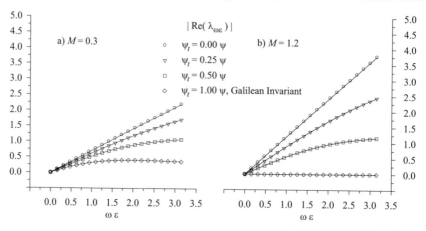

Figure 14.7: Characteristics-Bias Diffusion: a) Subsonic Flow, b) Supersonic Flow

A Galilean invariant characteristics-bias formulation, accordingly, provides a consistent upwinding, but with reduced levels of induced diffusions, which provides another justification for the particular form of upstream bias developed in this formulation for the time derivative.

14.7　Non-Discrete Discontinuous Galerkin (DG) Form

For both a weight function w that has non-vanishing trace and a solution q that is discontinuous on the boundary $\partial\widehat{\Omega}$ of a subdomain $\widehat{\Omega}\subset\Omega$, an integration by parts of both the characteristics-bias expression and the inviscid as well as viscous flux divergence and in (14.123) generates the weak statement

$$\int_{\widehat{\Omega}}w\left(\frac{\partial q}{\partial t}-\phi\right)d\Omega+\int_{\widehat{\Omega}}\varepsilon\psi\frac{\partial w}{\partial x_i}\left[a_i\left(\frac{\partial q}{\partial t}-\phi+\frac{\partial f_j}{\partial x_j}-\frac{\partial f_j^v}{\partial x_j}\right)+\sum_{\ell=1}^{L}\left(a_{i\ell}-a_i\right)\alpha_\ell A_{\ell j}\frac{\partial q}{\partial x_j}\right]d\Omega$$

$$-\int_{\widehat{\Omega}}\frac{\partial w}{\partial x_j}\left(f_j-f_j^v\right)d\Omega$$

$$+\oint_{\partial\widehat{\Omega}}w\left[\left(f_i-f_i^v\right)-\varepsilon\psi\left(a_i\left(\frac{\partial q}{\partial t}-\phi+\frac{\partial f_j}{\partial x_j}-\frac{\partial f_j^v}{\partial x_j}\right)+\sum_{\ell=1}^{L}\left(a_{i\ell}-a_i\right)\alpha_\ell A_{\ell j}\frac{\partial q}{\partial x_j}\right)\right]n_i\,d\Gamma=0$$

$$(14.135)$$

This statement may be viewed as a non-discrete Galilean invariant DG formulation. In respect of the second domain integral, this expression presents a direct counterpart of the "shock-capturing" term employed in current DG algorithms; rather than being a Laplacian expression added to the formulation, this term naturally emerges from the characteristics-bias procedure and system. With reference to the boundary integral, this expression features not only the flux vector $f_j-f_j^v$, but also an expression that corresponds to the characteristics bias resolution of $f_j-f_j^v$; as a result, the entire boundary expression may be viewed as a non-discrete "numerical flux" for $f_j-f_j^v$. Rather than requiring the substitution of $f_j-f_j^v$ with a numerical flux, as takes place in DG developments, the characteristics-bias system procedure directly leads to a surface integral with a specific non-discrete numerical flux, in this case the characteristics-bias resolution flux. A discrete DG algorithm may then result from a finite element discretization of (14.135).

14.8　Non-Discrete Streamline Upwind Petrov-Galerkin (SUPG) Form

For a weight function w with compact support in Ω, a continuous solution q, and I indicating the identity matrix of appropriate rank, an integration by parts of the characteristics-bias expression in (14.123) generates the weak statement

$$\int_{\Omega}\left(w+\varepsilon\psi a_i\frac{\partial w}{\partial x_i}\right)\left(\frac{\partial q}{\partial t}-\phi+\frac{\partial f_j}{\partial x_j}-\frac{\partial f_j^v}{\partial x_j}\right)d\Omega$$

$$+\int_{\Omega}\varepsilon\psi\frac{\partial w}{\partial x_i}\left(\sum_{\ell=1}^{L}\left(a_{i\ell}-a_i\right)\alpha_\ell A_{\ell j}\right)\frac{\partial q}{\partial x_j}\,d\Omega=0\qquad(14.136)$$

The boundary integrals from the integration by parts do not feature in this formulation because either the weight function has vanishing trace on the boundary or the boundary conditions described in Chapters 15-17 are enforced. This statement is recognized as a

Galilean invariant non-discrete SUPG integral statement for the original system (14.88). As an alternative to the reported SUPG formulations, the upstream-bias term in the first domain integral in (14.136) does not require any premultiplication by the transpose of the Euler flux Jacobian matrix, which reduces the number of required calculations; as the counterparts of the τ_{SUPG} and δ_{SUPG} stability parameters, moreover, the characteristics-bias parameters $\varepsilon\psi$ and $\varepsilon\psi(a_{i\ell} - a_i)$ will explicitly depend on the square or cube of a local mesh spacing, respectively for shocked and smooth flows. The second integral in this statement corresponds to the stability and shock-capturing terms of reported SUPG formulations. Unlike the common ad-hoc terms, the stability and shock-capturing terms in this formulation naturally emerge from characteristic wave propagation, for they are directly generated by the characteristics-bias decomposition of the Euler flux divergence $\partial f_j/\partial x_j$. When $a_{i\ell} = a_i$, $1 \leq \ell \leq L$, and for vanishing solution gradients in the crossflow direction for the Euler system, the term $(a_{i\ell} - a_i)$ even identically vanish. In these conditions, the exact solution of the original system can only satisfy this statement when the characteristics-bias formulation is extended to the time derivative as displayed in (14.136). For multi-dimensional flows, the spatially discrete equations in this book derive from a finite element discretization of this integral statement.

Chapter 15

Multi-Dimensional Incompressible Flows

This chapter develops the characteristics-bias procedure for the computational investigation of multi-dimensional incompressible flows. This procedure leads to an eventual continuity equation without any derivatives with respect to time, for consistency with the Navier-Stokes equations, uses the same-order finite elements for pressure, velocity as well as temperature, and solves the coupled continuity, linear-momentum, and energy equations to determine these variables directly.

Within an incompressible-flow field, acoustic and convection waves propagate in infinitely many directions, but along each direction the associated propagation velocity becomes unbounded. It is however possible to investigate characteristic incompressible-flow information propagation for finite propagation speeds. These finite speeds are associated with the slight-compressibility continuity equation that emerges from a direct correlation between pressure and density for a finite polytropic exponent. This slight-compressibility equation, however, is exclusively employed only to derive the second-order characteristics-bias perturbation.

This derivation uses an intrinsically multi-dimensional characteristics analysis, which leads to the spatial distribution of multi-dimensional propagation velocities, investigates the correlation between the time axis and the characteristic surfaces, and shows that among all propagation directions the streamline and crossflow directions are principal propagation directions in the time-space continuum. As an expected result, for any magnitude of pressure and velocity, information propagation remains subsonic, form both upstream and downstream toward any flow-field point. This line of inquiry yields specific conditions for a physically coherent upstream bias formulation that remains consistent with steady and unsteady multi-dimensional acoustic and convection wave propagation. Based on these conditions, the characteristics-bias formulation emerges from a decomposition of the Euler system matrix into crossflow and streamline matrix components that model the physics of multi-dimensional acoustics and convection. Achieved from a limiting process, the eventual continuity equation contains no derivative of pressure.

The discrete equations originate from a Galerkin finite element discretization of the characteristic-bias system. This finite element discretization naturally and automatically leads to consistent boundary differential equations and a new outlet pressure boundary condition that requires no algebraic extrapolation of variables and allows the direct calculation of pressure-driven flows. This formulation directly accommodates an implicit solver, for required Jacobian matrices are determined in a straightforward manner. The finite element equations form a system of algebraic and differential equations. Despite the absence of any pressure time derivative in this system, the equations are directly integrated in time within by way of the implicit non-linearly stable Runge-Kutta algorithm (8.5). This characteristics-bias finite element solver induces but minimal artificial diffusion and generates essentially non-oscillatory solutions.

15.1   Slight-Compressibility System

This section presents the slight-compressibility Euler and Navier-Stokes equations. The Euler equations directly support the development of the characteristics-bias formulation, which is then extended to the Navier-Stokes system with inclusion of the temperature equation.

15.1.1   Slight-Compressibility Continuity Equation

The isentropic correlation between pressure p and density ρ is expressed as

$$p = p(\rho) \tag{15.1}$$

For a perfect gas this result becomes

$$p = p_o \left(\frac{\rho}{\rho_o} \right)^{\gamma} \tag{15.2}$$

where $\gamma = c_p/c_v$ denotes the ratio of specific heats. A general polytropic equation of state emerges as

$$p = p_o \left(\frac{\rho}{\rho_o} \right)^{\kappa \gamma} \tag{15.3}$$

with $\kappa \geq 0$ denoting the polytropic exponent. With this result, the time derivative of density correlates with the time derivative of pressure as

$$\frac{\partial \rho}{\partial t} = \frac{\rho}{\kappa \gamma p} \frac{\partial p}{\partial t} \tag{15.4}$$

which transforms the continuity equation as

$$\frac{\partial \rho}{\partial t} + \frac{\partial \rho u_j}{\partial x_j} = 0 \quad \Rightarrow \quad \frac{\rho}{\kappa \gamma p} \frac{\partial p}{\partial t} + \frac{\partial \rho u_j}{\partial x_j} = 0 \tag{15.5}$$

For an incompressible flow $\rho \simeq$ constant; accordingly this equation transforms into

$$\frac{1}{\kappa \gamma p} \frac{\partial p}{\partial t} + \frac{\partial u_j}{\partial x_j} = 0 \tag{15.6}$$

Unlike Chorin's seminal work, [35, 36], which employs a constant $\kappa \gamma p$, with $\beta \equiv \kappa \gamma p / \rho$, this development allows p in this continuity equation to vary. As a direct result, the eigenvalues of the resulting system depend on a single non-dimensional number, wholly analogous to the Froude and Mach number. For $\kappa = 1$, this equation becomes the slight-compressibility continuity equation, which governs the low-speed flow of a perfect gas. This is the equation employed in the characteristic analysis of incompressible flows and associated development of a characteristics-bias formulation. As noted in Section 4.2.3, $c_p = c_v$ for incompressible flows, hence $\gamma = 1$, which is the magnitude of γ used in this formulation. The corresponding formulation for incompressible flows then results by replacing $\frac{1}{\gamma p}\frac{\partial p}{\partial t}$ with $\frac{1}{\kappa\gamma p}\frac{\partial p}{\partial t}$, showing in Section 15.7 that the corresponding characteristics-bias perturbation in the new system remains stable for any $\kappa > 0$, and then taking the limit for $\kappa \to \infty$. With this limit, the resulting continuity equation achieves a form that does not contain any time derivative of pressure.

15.1.2 Euler and Navier-Stokes Equations

Two-Dimensional Flows

On the basis of the slight-compressibility continuity equation, the corresponding two-dimensional Navier-Stokes equations are expressed as

$$
\begin{cases}
\dfrac{\partial p}{\partial t} + \gamma p \dfrac{\partial u_1}{\partial x_1} + \gamma p \dfrac{\partial u_2}{\partial x_2} = 0 \\[2mm]
\dfrac{\partial u_1}{\partial t} + \dfrac{\partial}{\partial x_1}\left(u_1^2 + \dfrac{p}{\rho}\right) + \dfrac{\partial\left(u_2 u_1\right)}{\partial x_2} = \dfrac{1}{\rho}\dfrac{\partial \tau_{1j}}{\partial x_j} + \phi_1 \\[2mm]
\dfrac{\partial u_2}{\partial t} + \dfrac{\partial\left(u_1 u_2\right)}{\partial x_1} + \dfrac{\partial}{\partial x_2}\left(u_2^2 + \dfrac{p}{\rho}\right) = \dfrac{1}{\rho}\dfrac{\partial \tau_{2j}}{\partial x_j} + \phi_2
\end{cases}
\tag{15.7}
$$

where, with reference to system (4.56) in Chapter 4, ϕ_i and τ_{ij} respectively denote the Cartesian components of the source term and viscous deviatoric stresses. When these terms vanish, these equations reduce to the Euler system, which is cast in matrix form as

$$
\frac{\partial q}{\partial t} + A_j(q)\frac{\partial q}{\partial x_j} = 0
\tag{15.8}
$$

With δ_i^j denoting Kronecker's delta, the matrix for this system is expressed as

$$
A_j(q) = A_i(q)\delta_i^j = \begin{pmatrix} 0 & , & \gamma p \delta_1^j & , & \gamma p \delta_2^j \\ \delta_1^j/\rho & , & u_1\delta_1^j + u_j & , & u_1\delta_2^j \\ \delta_2^j/\rho & , & u_2\delta_1^j & , & u_2\delta_2^j + u_j \end{pmatrix}
\tag{15.9}
$$

Three-Dimensional Flows

For three-dimensional flows, the slight-compressibility Navier-Stokes equations may be expressed as

$$
\begin{cases}
\dfrac{\partial p}{\partial t} + \gamma p \dfrac{\partial u_1}{\partial x_1} + \gamma p \dfrac{\partial u_2}{\partial x_2} + \gamma p \dfrac{\partial u_3}{\partial x_3} = 0 \\[2mm]
\dfrac{\partial u_1}{\partial t} + \dfrac{\partial}{\partial x_1}\left(u_1^2 + \dfrac{p}{\rho} \right) + \dfrac{\partial\,(u_2 u_1)}{\partial x_2} + \dfrac{\partial\,(u_3 u_1)}{\partial x_3} = \dfrac{1}{\rho}\dfrac{\partial \tau_{1j}}{\partial x_j} + \phi_1 \\[2mm]
\dfrac{\partial u_2}{\partial t} + \dfrac{\partial\,(u_1 u_2)}{\partial x_1} + \dfrac{\partial}{\partial x_2}\left(u_2^2 + \dfrac{p}{\rho} \right) + \dfrac{\partial\,(u_3 u_2)}{\partial x_3} = \dfrac{1}{\rho}\dfrac{\partial \tau_{2j}}{\partial x_j} + \phi_2 \\[2mm]
\dfrac{\partial u_3}{\partial t} + \dfrac{\partial\,(u_1 u_3)}{\partial x_1} + \dfrac{\partial\,(u_2 u_3)}{\partial x_2} + \dfrac{\partial}{\partial x_3}\left(u_3^2 + \dfrac{p}{\rho} \right) = \dfrac{1}{\rho}\dfrac{\partial \tau_{3j}}{\partial x_j} + \phi_3
\end{cases}
\tag{15.10}
$$

where, again, ϕ_i and τ_{ij} respectively denote the Cartesian components of the source term and viscous deviatoric stresses. When these terms vanish, these equations reduce to the Euler system, again cast as in (15.8) with system matrix

$$
A_j(q) = A_i(q)\delta_i^j =
\begin{pmatrix}
0 & , & \gamma p \delta_1^j & , & \gamma p \delta_2^j & , & \gamma p \delta_3^j \\
\delta_1^j/\rho & , & u_1 \delta_1^j + u_j & , & u_1 \delta_2^j & , & u_1 \delta_3^j \\
\delta_2^j/\rho & , & u_2 \delta_1^j & , & u_2 \delta_2^j + u_j & , & u_2 \delta_3^j \\
\delta_3^j/\rho & , & u_3 \delta_1^j & , & u_3 \delta_2^j & , & u_3 \delta_3^j + u_j
\end{pmatrix}
\tag{15.11}
$$

15.2 Slight-Compressibility Acoustics Equations

15.2.1 Matrix Form

The matrix form of the acoustics equations for slightly compressible flows is obtained from the slight-compressibility Euler equations

$$
\frac{\partial q}{\partial t} + A_j(q)\frac{\partial q}{\partial x_j} = 0
\tag{15.12}
$$

for a vanishing flow velocity.

Two-Dimensional Flows

Let (v_1, v_2) denote the components of a unit vector $\boldsymbol{v}$ locally parallel to the velocity $\boldsymbol{u}$ and employ the "$\Im$" number

$$
\Im \equiv \frac{\|\boldsymbol{u}\|}{c}, \quad c \equiv \sqrt{\gamma \frac{p}{\rho}}
\tag{15.13}
$$

Upon writing the velocity components (u_1, u_2) in terms of $\Im$ as

$$
(u_1, u_2) = c\Im(v_1, v_2)
\tag{15.14}
$$

the 2-D slight-compressibility system becomes

$$
\frac{\partial}{\partial t}\begin{pmatrix} p \\ u_1 \\ u_2 \end{pmatrix} + \left[\begin{pmatrix} 0 & , & \gamma p \delta_1^j & , & \gamma p \delta_2^j \\ \delta_1^j/\rho & , & 0 & , & 0 \\ \delta_2^j/\rho & , & 0 & , & 0 \end{pmatrix} + c\Im C_j \right] \frac{\partial}{\partial x_j}\begin{pmatrix} p \\ u_1 \\ u_2 \end{pmatrix} = 0 \tag{15.15}
$$

where C_j denotes the completion matrix

$$
C_j \equiv \begin{pmatrix} 0 & , & 0 & , & 0 \\ 0 & , & v_1 \delta_1^j + v_j & , & v_1 \delta_2^j \\ 0 & , & v_2 \delta_1^j & , & v_2 \delta_2^j + v_j \end{pmatrix} \tag{15.16}
$$

For a vanishing $\Im$, these equations then reduce to the acoustics equations

$$
\frac{\partial}{\partial t}\begin{pmatrix} p \\ u_1 \\ u_2 \end{pmatrix} + \begin{pmatrix} 0 & , & \gamma p \delta_1^j & , & \gamma p \delta_2^j \\ \delta_1^j/\rho & , & 0 & , & 0 \\ \delta_2^j/\rho & , & 0 & , & 0 \end{pmatrix} \frac{\partial}{\partial x_j}\begin{pmatrix} p \\ u_1 \\ u_2 \end{pmatrix} = 0 \tag{15.17}
$$

for which A_j^a will denote the acoustics matrix multiplying the gradient of q in (15.17). These equations may also be expressed as

$$
\begin{cases} \dfrac{\partial p}{\partial t} = -\gamma p \dfrac{\partial u_j}{\partial x_j} \\[2mm] \dfrac{\partial u_i}{\partial t} = -\dfrac{1}{\rho}\dfrac{\partial p}{\partial x_i} = -\dfrac{c^2}{\gamma p}\dfrac{\partial p}{\partial x_i} \end{cases} \tag{15.18}
$$

The dependent variables in these equations correspond to those in a flow field that originates from slight perturbations to an otherwise quiescent field. Following the characteristics analysis in Chapter 14, the characteristic speeds λ associated with this system correspond to the eigenvalues of the matrix $A_j^a n_j$, which are

$$
\lambda_1^a = 0, \qquad \lambda_{2,3}^a = \pm\sqrt{\gamma\frac{p}{\rho}} = \pm c \tag{15.19}
$$

Note that these eigenvalues remain independent from the wave-propagation direction unit vector $\boldsymbol{n}$, which corresponds to isotropic wave propagation.

Three-Dimensional Flows

An analogous procedure applies for three-dimensional flows. Let (v_1, v_2, v_3) denote the components of a unit vector $\boldsymbol{v}$ locally parallel to the velocity $\boldsymbol{u}$. Upon expressing the velocity components (u_1, u_2, u_3) in terms of the $\Im$ number as

$$
(u_1, u_2, u_3) = c\Im(v_1, v_2, v_3) \tag{15.20}
$$

the 3-D slight-compressibility system becomes

$$
\frac{\partial}{\partial t}\begin{pmatrix} p \\ u_1 \\ u_2 \\ u_3 \end{pmatrix} + \left[\begin{pmatrix} 0 & , & \gamma p \delta_1^j & , & \gamma p \delta_2^j & , & \gamma p \delta_3^j \\ \delta_1^j/\rho & , & 0 & , & 0 & , & 0 \\ \delta_2^j/\rho & , & 0 & , & 0 & , & 0 \\ \delta_3^j/\rho & , & 0 & , & 0 & , & 0 \end{pmatrix} + c\Im C_j \right] \frac{\partial}{\partial x_j}\begin{pmatrix} p \\ u_1 \\ u_2 \\ u_3 \end{pmatrix} = 0 \tag{15.21}
$$

where C_j denotes the completion matrix

$$
C_j \equiv \begin{pmatrix}
0 & , & 0 & , & 0 & , & 0 \\
0 & , & v_1\delta_1^j + v_j & , & v_1\delta_2^j & , & v_1\delta_3^j \\
0 & , & v_2\delta_1^j & , & v_2\delta_2^j + v_j & , & v_2\delta_3^j \\
0 & , & v_3\delta_1^j & , & v_3\delta_2^j & , & v_3\delta_3^j + v_j
\end{pmatrix}
\tag{15.22}
$$

For a vanishing $\Im$ these equations then reduce to the slight-compressibility acoustics equations

$$
\frac{\partial}{\partial t}\begin{pmatrix} p \\ u_1 \\ u_2 \\ u_3 \end{pmatrix} + \begin{pmatrix}
0 & , & \gamma p\delta_1^j & , & \gamma p\delta_2^j & , & \gamma p\delta_3^j \\
\delta_1^j/\rho & , & 0 & , & 0 & , & 0 \\
\delta_2^j/\rho & , & 0 & , & 0 & , & 0 \\
\delta_3^j/\rho & , & 0 & , & 0 & , & 0
\end{pmatrix} \frac{\partial}{\partial x_j}\begin{pmatrix} p \\ u_1 \\ u_2 \\ u_3 \end{pmatrix} = 0
\tag{15.23}
$$

for which A_j^a will denote the acoustics matrix multiplying the gradient of q in (15.23). These equations may also be expressed as

$$
\begin{cases}
\dfrac{\partial p}{\partial t} = -\gamma p \dfrac{\partial u_j}{\partial x_j} \\[2ex]
\dfrac{\partial u_i}{\partial t} = -\dfrac{1}{\rho}\dfrac{\partial p}{\partial x_i} = -\dfrac{c^2}{\gamma p}\dfrac{\partial p}{\partial x_i}
\end{cases}
\tag{15.24}
$$

The dependent variables in these equations correspond to those in a flow field that originates from slight perturbations to an otherwise quiescent field. The characteristic speeds λ for this system correspond to the eigenvalues of the matrix $A_j^a n_j$, which are

$$
\lambda_{1,2}^a = 0, \qquad \lambda_{3,4}^a = \pm\sqrt{\gamma\frac{p}{\rho}} = \pm c
\tag{15.25}
$$

Note that these eigenvalues remain independent from the wave-propagation direction unit vector $\boldsymbol{n}$.

15.2.2 Streamline and Crossflow Components

Two-Dimensional Flows

For any two mutually perpendicular unit vectors $\boldsymbol{a} = (a_1, a_2)$ and $\boldsymbol{a}^N = (a_1^N, a_2^N)$ within a 2-D flow, along with implied summation on repeated indices, the acoustics matrix product $A_j^a \frac{\partial q}{\partial x_j}$ can be expressed as

$$
A_j^a \frac{\partial q}{\partial x_j} = A_j^a a_j a_k \frac{\partial q}{\partial x_k} + A_j^a a_j^N a_k^N \frac{\partial q}{\partial x_k}
\tag{15.26}
$$

For $\boldsymbol{a}$ parallel to $\boldsymbol{u}$, this expression corresponds to a decomposition of the acoustics matrix product into streamline and crossflow acoustics components. For solutions in wave-like form, shown in Chapter 14, one eigenvalue of each rhs component vanishes; the remaining eigenvalues of these separate components have been determined as

$$\lambda^s_{2,3} = \pm c_s \equiv \pm c a_j n_j \quad , \quad \lambda^N_{2,3} = \pm c_N \equiv \pm c a^N_j n_j \tag{15.27}$$

The two non-vanishing eigenvalues associated with the entire acoustics component at the lhs of (15.26), but as expressed as the rhs combination of streamline and crossflow components have then been determined as

$$\lambda^a_{1,2} = c \left(\left(a_j n_j \right)^2 + \left(a^N_j n_j \right)^2 \right)^{1/2}, \quad \left(a_j n_j \right)^2 + \left(a^N_j n_j \right)^2 = 1 \tag{15.28}$$

which shows that the square of these acoustic eigenvalues equals the sum of the square of the streamline and crossflow acoustic eigenvalues (15.27), as illustrated in Figure 15.1.

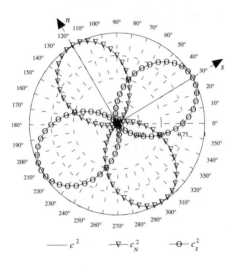

Figure 15.1: Polar Variation of Square of Acoustic Speeds

Three-Dimensional Flows

For any three mutually orthogonal unit vectors $\boldsymbol{a} = (a_1, a_2, a_3)$, $\boldsymbol{a}^{N_1} = (a^{N_1}_1, a^{N_1}_2, a^{N_1}_3)$, and $\boldsymbol{a}^{N_2} = (a^{N_2}_1, a^{N_2}_2, a^{N_2}_3)$ within a 3-D flow, along with implied summation on repeated indices, the acoustics matrix product $A^a_j \frac{\partial q}{\partial x_j}$ can be expressed as

$$A^a_j \frac{\partial q}{\partial x_j} = A^a_j a_j a_k \frac{\partial q}{\partial x_k} + A^a_j a^{N_1}_j a^{N_1}_k \frac{\partial q}{\partial x_k} + A^a_j a^{N_2}_j a^{N_2}_k \frac{\partial q}{\partial x_k} \tag{15.29}$$

For $\boldsymbol{a}$ parallel to $\boldsymbol{u}$, this expression corresponds to a decomposition of the acoustics matrix product into one streamline and two mutually perpendicular crossflow acoustics components.

Two eigenvalues of each rhs component vanish; the remaining eigenvalues of these three separate components have been determined as

$$\lambda_{3,4}^s = \pm c a_j n_j, \quad \lambda_{3,4}^{N_1} = \pm c a_j^{N_1} n_j, \quad \lambda_{3,4}^{N_2} = \pm c a_j^{N_2} n_j \tag{15.30}$$

The two non-vanishing eigenvalues associated with the entire acoustics component at the lhs of (15.29), but as expressed as the rhs combination of streamline and crossflow components have then been determined as

$$\lambda_{3,4}^a = c \left(\left(a_j n_j \right)^2 + \left(a_j^{N_1} n_j \right)^2 + \left(a_j^{N_2} n_j \right)^2 \right)^{1/2}, \quad \left(a_j n_j \right)^2 + \left(a_j^{N_1} n_j \right)^2 + \left(a_j^{N_2} n_j \right)^2 = 1 \tag{15.31}$$

which shows that the square of these acoustic eigenvalues equals the sum of the square of the streamline and crossflow acoustic eigenvalues (15.30). On any plane Π that contains $\boldsymbol{a}$, $\boldsymbol{a}^{N_1}$ and $\boldsymbol{n}$, therefore, acoustic propagation naturally decomposes into two directional components, one along $\boldsymbol{a}$ and the other along $\boldsymbol{a}^{N_1}$, as illustrated in Figure 15.2. The variation

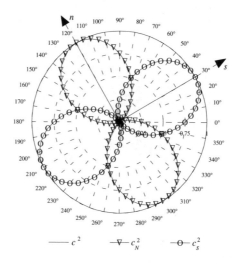

Figure 15.2: Polar Variation of Square of Acoustic Speeds on Plane Π

of the squares of the three principal acoustic speeds is illustrated in Figure 15.3

The planar two-directional acoustic decomposition remains unaltered on plane Π, as this plane spans the entire 3-D space by rotating about the line of $\boldsymbol{a}$. Acoustic propagation, therefore, decomposes into a directional component along $\boldsymbol{a}$ and an isotropic component on any plane orthogonal to $\boldsymbol{a}$. This result also follows from the expression

$$\left(a_j^{N_1} n_j \right)^2 + \left(a_j^{N_2} n_j \right)^2 = 1 - \left(a_j n_j \right)^2 \tag{15.32}$$

from (15.31), which shows that acoustic propagation on the plane of $\boldsymbol{a}^{N_1}$ and $\boldsymbol{a}^{N_2}$ only depends upon $(a_j n_j)$, hence on the angle between $\boldsymbol{a}$ and $\boldsymbol{n}$. For any such angle, the "lhs" of

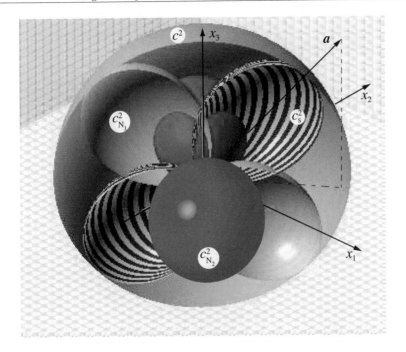

Figure 15.3: Spherical Variation of Square of Three Principal Acoustic Speeds

(15.32) remains constant for any orientation of n with respect to a^{N_1} and a^{N_2}. This result indicates that (15.32) is the equation of a circle, which signifies isotropic acoustic propagation on the plane of a^{N_1} and a^{N_2}, as shown in Figure 15.4

15.2.3   Similarity Transformation

Two-Dimensional Flows

From (15.26), the matrix $A_j^a a_j$ is defined as

$$A_j^a a_j \equiv \begin{pmatrix} 0 & , & \gamma p a_1 & , & \gamma p a_2 \\ a_1/\rho & , & 0 & , & 0 \\ a_2/\rho & , & 0 & , & 0 \end{pmatrix} \tag{15.33}$$

Despite its zero eigenvalue, this matrix features a complete set of eigenvectors X and thus possesses the similarity transformation

$$A_j^a a_j = X \Lambda^a X^{-1} \tag{15.34}$$

with Λ^a defined as

$$\Lambda^a \equiv \begin{pmatrix} c & , & 0 & , & 0 \\ 0 & , & c & , & 0 \\ 0 & , & 0 & , & 0 \end{pmatrix} \tag{15.35}$$

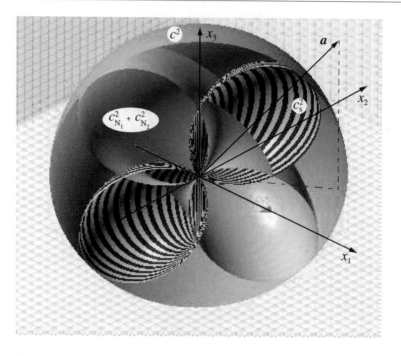

Figure 15.4: Spherical Variation of Square of Crossflow and Streamline Acoustic Speeds

The corresponding eigenvector matrix X and its inverse X^{-1} are

$$X = \begin{pmatrix} \gamma p & , & \gamma p & , & 0 \\ ca_1 & , & -ca_1 & , & -a_2 \\ ca_2 & , & -ca_2 & , & a_1 \end{pmatrix} \tag{15.36}$$

$$X^{-1} = \frac{1}{2\gamma pc} \begin{pmatrix} c & , & \gamma pa_1 & , & \gamma pa_2 \\ c & , & -\gamma pa_1 & , & -\gamma pa_2 \\ 0 & , & -2\gamma pca_2 & , & 2\gamma pca_1 \end{pmatrix} \tag{15.37}$$

Three-Dimensional Flows

From (15.29), the matrix $A_j^a a_j$ is defined as

$$A_j^a a_j \equiv \begin{pmatrix} 0 & , & \gamma pa_1 & , & \gamma pa_2 & , & \gamma pa_3 \\ a_1/\rho & , & 0 & , & 0 & , & 0 \\ a_2/\rho & , & 0 & , & 0 & , & 0 \\ a_3/\rho & , & 0 & , & 0 & , & 0 \end{pmatrix} \tag{15.38}$$

Despite its zero eigenvalues, also this matrix features a complete set of eigenvectors X and thus possesses the similarity transformation

$$A_j^a a_j = X \Lambda^a X^{-1} \tag{15.39}$$

with Λ^a the diagonal matrix of eigenvalues of $A^a_j a_j$. The eigenvector matrix X and its inverse X^{-1} are

$$X = \begin{pmatrix} \gamma p & , & \gamma p & , & 0 & , & 0 \\ ca_1 & , & -ca_1 & , & -a_2 & , & -a_3 \\ ca_2 & , & -ca_2 & , & a_1 & , & 0 \\ ca_3 & , & -ca_3 & , & 0 & , & a_1 \end{pmatrix} \tag{15.40}$$

$$X^{-1} = \frac{1}{2\gamma pc} \begin{pmatrix} c & , & \gamma pa_1 & , & \gamma pa_2 & , & \gamma pa_3 \\ c & , & -\gamma pa_1 & , & -\gamma pa_2 & , & -\gamma pa_3 \\ 0 & , & -2\gamma pca_2 & , & 2\gamma pc(a_1^2 + a_3^2)/a_1 & , & -2\gamma pca_2a_3/a_1 \\ 0 & , & -2\gamma pca_3 & , & -2\gamma pca_2a_3/a_1 & , & 2\gamma pc(a_1^2 + a_2^2)/a_1 \end{pmatrix} \tag{15.41}$$

Since two linearly independent eigenvector are available for the single vanishing eigenvalue, it's always possible to find two linearly independent eigenvectors such that a non-singular X is always available.

15.3 Characteristic Analysis and Velocity Components

For the slight-compressibility Euler system

$$\frac{\partial q}{\partial t} + A_j(q)\frac{\partial q}{\partial x_j} = 0 \tag{15.42}$$

the solution-dependent characteristic velocity components $\lambda = \lambda(q)$ are determined by enforcing the condition that solutions in non-linear wave-like form satisfy this system. With reference to Chapter 14, this condition yields the eigenvalue problem

$$\left(-\lambda(q)I + A_j(q)n_j \right)\frac{\partial q}{\partial \eta_1} = 0 \tag{15.43}$$

For non-trivial solutions $\frac{\partial q}{\partial \eta_1}$, hence $q = q(\eta_1)$, the characteristic velocity components λ are the eigenvalues of the linear combination of system matrices

$$A_j(q)n_j = \begin{pmatrix} 0 & , & \gamma pn_1 & , & \gamma pn_2 \\ n_1/\rho & , & u_1n_1 + u_jn_j & , & u_1n_2 \\ n_2/\rho & , & u_2n_1 & , & u_2n_2 + u_jn_j \end{pmatrix} \tag{15.44}$$

for two-dimensional flows, and

$$A_j(q)n_j = \begin{pmatrix} 0 & , & \gamma pn_1 & , & \gamma pn_2 & , & \gamma pn_3 \\ n_1/\rho & , & u_1n_1 + u_jn_j & , & u_1n_2 & , & u_1n_3 \\ n_2/\rho & , & u_2n_1 & , & u_2n_2 + u_jn_j & , & u_2n_3 \\ n_3/\rho & , & u_3n_1 & , & u_3n_2 & , & u_3n_3 + u_jn_j \end{pmatrix} \tag{15.45}$$

for three-dimensional flows. For both two- and three-dimensional flows, these eigenvalues have been exactly determined in closed form as

$$\lambda^{d_E}_{1,2} = u_jn_j, \qquad \lambda^{d_E}_{3,4} = u_jn_j \pm \sqrt{\left((u_jn_j)^2 + c^2 \right)} \tag{15.46}$$

where superscript d_E signifies dimensional Euler eigenvalues; for two-dimensional flows, the eigenvalue λ_1 is absent. The non-dimensional form of (15.46) follows from division by c, which supplies the $\Im$-number dependent expressions

$$\lambda_{1,2}^{E} = v_j n_j \Im, \qquad \lambda_{3,4}^{E} = v_j n_j \Im \pm \sqrt{\left(1 + \Im^2 \left(v_j n_j\right)^2\right)} \qquad (15.47)$$

where v_1, and v_2, in 2-D, and v_1, v_2, and v_3, in 3-D, denote the components of a unit vector $\boldsymbol{v}$ in the velocity $\boldsymbol{u}$ direction.

As the inner product of the two unit vectors $\boldsymbol{n}$ and $\boldsymbol{v}$, the contraction $v_j n_j$ supplies the cosine of the angle $(\theta - \theta_{\boldsymbol{v}})$ between $\boldsymbol{n}$ and $\boldsymbol{v}$, where θ and $\theta_{\boldsymbol{v}}$ respectively denote the angle between $\boldsymbol{n}$ and the x_1 axis and the angle between $\boldsymbol{v}$ and the x_1 axis in the flow plane, in 2-D, or in any plane Π that contains both $\boldsymbol{n}$ and $\boldsymbol{v}$, in 3-D. The eigenvalues (15.47) thus become

$$\lambda_{1,2}^{E} = \cos\left(\theta - \theta_{\boldsymbol{v}}\right)\Im, \qquad \lambda_{3,4}^{E} = \cos\left(\theta - \theta_{\boldsymbol{v}}\right)\Im \pm \sqrt{\left(1 + \Im^2 \left(\cos\left(\theta - \theta_{\boldsymbol{v}}\right)\right)^2\right)} \qquad (15.48)$$

These expressions, in particular, imply that these Euler eigenvalues achieve their extrema for $\theta = \theta_{\boldsymbol{v}}$, hence when $\boldsymbol{n}$ points in the streamline direction, whereas for $\boldsymbol{n}$ pointing in the crossflow direction, hence $\theta = 90^o + \theta_{\boldsymbol{v}}$, these eigenvalues no longer depend upon $\Im$.

The convection eigenvalues $\lambda_{1,2}^{E}$ vanish when $\cos\left(\theta - \theta_{\boldsymbol{v}}\right) = 0$, hence for $\boldsymbol{n}$ perpendicular to the streamline direction, or, equivalently, pointing in the crossflow direction. On the other hand, the acoustic-convection eigenvalues $\lambda_{3,4}^{E}$ never vanish, for an incompressible flow remains subsonic. A vanishing eigenvalue leads to a steady-flow eigenvalue. Since $\lambda_{3,4}^{E}$ does not vanish, the steady incompressible-flow equations are not hyperbolic. The only real steady characteristic is thus the streamline.

15.3.1 Polar Variation of Characteristic Speeds

Figure 15.5 presents the polar variation of the absolute values of eigenvalues (15.47) for several $\Im$ numbers, in a neighborhood of a flow field point P in the (x_1, x_2) plane, in 2-D, or any plane Π that contains both $\boldsymbol{n}$ and $\boldsymbol{v}$, in 3-D. These variations are obtained for a variable unit vector $\boldsymbol{n} \equiv (\cos\theta, \sin\theta)$ and fixed unit vector $\boldsymbol{v}$, in this representative case inclined by $+30^o$ with respect to the x_1 axis, in 2-D, or a reference x_1 axis on Π, in 3-D.

A collective inspection of all these diagrams reveals three shared features for all $\Im$ numbers. The maximum characteristic speeds occur in the velocity direction, i.e. along a streamline, as noted before. Secondly, all the characteristic speeds are symmetrically distributed about the streamline direction. Thirdly, the eigenvalue pairs $(\|\lambda_1^{E}\|, \|\lambda_2^{E}\|)$ and $(\|\lambda_3^{E}\|, \|\lambda_4^{E}\|)$ remain mirror skew-symmetric with respect to the crossflow direction, in the sense that the curves representative of $\|\lambda_2^{E}\|$ and $\|\lambda_4^{E}\|$ become the respective mirror images of the variations of $\|\lambda_1^{E}\|$ and $\|\lambda_3^{E}\|$ with reference to this direction. The streamline and crossflow directions, therefore, become two fundamental wave-propagation axes.

For vanishing $\Im$ numbers, the acoustic-convection propagation curves in the figure approach two circumferences. The distribution of propagation speeds in this case is therefore isotropic, which corresponds to the direction-invariant propagation of acoustic waves. As the $\Im$ number increases from zero, the curves in the figure progressively become circular asymmetric, which corresponds to anisotropic wave propagation. For $\Im = 0.5$ this anisotropy is

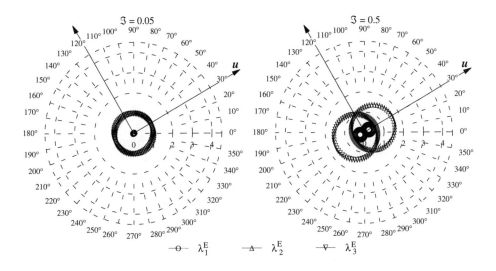

Figure 15.5: Polar Variation of Wave Speeds

already evident and becomes more pronounced for higher $\Im$ numbers. The non-dimensional characteristic speeds then approach 1 in the region about the crossflow direction, which corresponds to essentially acoustic propagation.

Figure 15.7 presents the 3-D spherical variation of the absolute values of eigenvalues (15.47). The surface in this figure was obtained using spherical coordinates (r, θ, ϕ), where r represent the magnitude of these eigenvalues and θ and ϕ correspond to the spherical-coordinate angles depicted in Figure 15.7. With these angles, the cartesian coordinates of n and v are

$$n \equiv (\sin \phi_n \cos \theta_n, \sin \phi_n \sin \theta_n, \cos \phi_n), \quad v \equiv (\sin \phi_v \cos \theta_v, \sin \phi_v \sin \theta_v, \cos \phi_v) \quad (15.49)$$

The corresponding expression for eigenvalues (15.47) become

$$\lambda_{1,2}^E = (\sin \phi_n \cos \theta_n \sin \phi_v \cos \theta_v + \sin \phi_n \sin \theta_n \sin \phi_v \sin \theta_v + \cos \phi_n \cos \phi_v) \Im$$

$$\lambda_{3,4}^E = (\sin \phi_n \cos \theta_n \sin \phi_v \cos \theta_v + \sin \phi_n \sin \theta_n \sin \phi_v \sin \theta_v + \cos \phi_n \cos \phi_v) \Im$$

$$\pm \sqrt{\left(1 + (\sin \phi_n \cos \theta_n \sin \phi_v \cos \theta_v + \sin \phi_n \sin \theta_n \sin \phi_v \sin \theta_v + \cos \phi_n \cos \phi_v)^2 \Im^2\right)}$$

$$(15.50)$$

The surfaces in Figures 15.7 illustrate the space variations of the characteristic speeds. These variations correlate with the 2-D curves, in Figures 15.5-15.6, which result from intersecting these 3-D surfaces with plane Π. These surfaces too show that the extrema of the characteristic speeds occur along the streamline direction, whereas only acoustic propagation exists on the plane orthogonal to velocity at P.

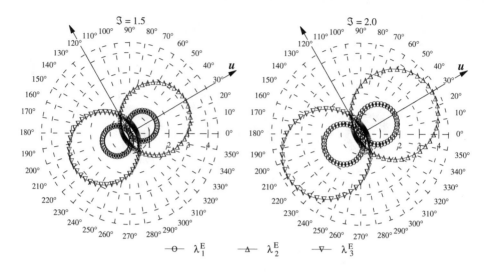

Figure 15.6: Polar Variation of Wave Speeds

For all $\Im$ numbers, the convection eigenvalues $\lambda_{1,2}$ change sign when the $\boldsymbol{n}$-direction shifts from an upstream to a downstream axis with respect to $\boldsymbol{u}$. For this reason, the associated curves cross the polar origin. Pure convective propagation, therefore, remains mono-axial, from upstream to downstream of P, and the axis of this kind of wave propagation is the streamline.

For all $\Im$ numbers the acoustic-convection eigenvalues λ_3^E and λ_4^E respectively remain positive and negative for all directions. For this reason the associated curves contain the polar origin. Acoustic-convection waves, therefore, propagate bi-modally, from both upstream and downstream toward and away from point P, along all directions radiating from P.

15.3.2   Wave Propagation Region and Characteristic Cone

The streamline and crossflow directions are principal directions of acoustic and convection wave propagation. These directions become the axes of distinct wave-propagation regions, which are named the streamline and crossflow wave propagation regions. As Figure 15.8 shows, subsonic acoustic-convection waves propagate bi-modally over the entire plane, hence the Euler eigenvalues do not all have the same algebraic sign. No mono-axial propagation region thus exists.

With reference to Section 14.2, the equations for $\mathcal{F}$ representing the characteristic surfaces in the time-space continuum arise by substitution of (14.45), (14.46) for n_j and λ in the corresponding eigenvalues (15.47). This procedure yields the following partial differential equations for $\mathcal{F}$

$$\left(\mathcal{F}_t + u_j \mathcal{F}_{x_j}\right)^2 = 0, \quad \left(\mathcal{F}_t + u_j \mathcal{F}_{x_j}\right)^2 = \left(u_j \mathcal{F}_{x_j}\right)^2 + c^2 \mathcal{F}_{x_j} \mathcal{F}_{x_j} \tag{15.51}$$

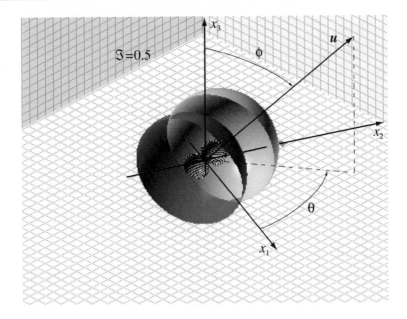

Figure 15.7: 3-D Variation of Wave Speeds

where each subscript variable denotes differentiation with respect to that variable. In substantive-derivative form, the first of these two equations becomes

$$\left(\frac{D\mathcal{F}}{Dt}\right)^2 = 0 \tag{15.52}$$

which shows that the corresponding characteristic surface remains time-invariant from the perspective of an observer who travels along with a fluid particle.

The principal propagation directions as well as the flow domain of dependence and range of influence of a flow field point P in the time-space continuum are determined by studying the variation of the characteristic-surface shape versus the $\Im$ number. As delineated in Section 14.3, in order to investigate these shape changes it suffices to study the shape changes of the corresponding characteristic cones, which are tangent at each flow field point P to the corresponding characteristic surfaces, but are far easier to determine.

By virtue of the Galilean transformations in Section 14.3, the characteristic-surface equations (15.51) simplify to

$$\mathcal{G} \equiv \left(\frac{\partial \mathcal{F}}{\partial \tau}\right)^2 = 0, \quad \mathcal{G} \equiv \left(\frac{\partial \mathcal{F}}{\partial \tau}\right)^2 - \Im^2 \left(v_j \frac{\partial \mathcal{F}}{\partial X_j}\right)^2 - \frac{\partial \mathcal{F}}{\partial X_j}\frac{\partial \mathcal{F}}{\partial X_j} = 0 \tag{15.53}$$

in 2-D and 3-D. The equation of the tangent characteristic cone is then directly obtained from these equations through the Monge-cone ordinary differential equations presented in

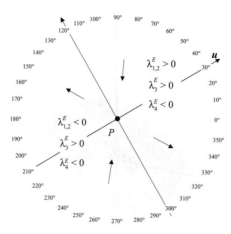

Figure 15.8: Wave Velocity Distribution

Section 14.3. The solution of these ordinary differential equations for each equation in (15.53) is

$$X_j - X_{jo} = 0, \quad (X_j - X_{jo})(X_j - X_{jo}) - \frac{\Im^2}{1+\Im^2}(v_j(X_j - X_{jo}))^2 = (\tau - \tau_o)^2 \qquad (15.54)$$

in 2-D and 3-D.

By virtue of the Galilean transformation, the equation of the first characteristic cone with respect to the fixed reference coordinates (x_1, x_2, t) then becomes

$$\frac{x_1 - x_{1o}}{t - t_o} = u_1, \quad \frac{x_2 - x_{2o}}{t - t_o} = u_2 \qquad (15.55)$$

in 2-D, and

$$\frac{x_1 - x_{1o}}{t - t_o} = u_1, \quad \frac{x_2 - x_{2o}}{t - t_o} = u_2, \quad \frac{x_3 - x_{3o}}{t - t_o} = u_3 \qquad (15.56)$$

in 3-D. In the time-space continuum, these equations correspond to the single "convection" straight line $S^- S^+$, as displayed in the 2-D case in Figure 15.9. As time elapses, therefore, waves reach and then leave P along this line, which, in the space continuum, projects onto the streamline

$$\frac{x_1 - x_{1o}}{u_1} = \frac{x_2 - x_{2o}}{u_2} \qquad (15.57)$$

in 2-D, and

$$\frac{x_1 - x_{1o}}{u_1} = \frac{x_2 - x_{2o}}{u_2} = \frac{x_3 - x_{3o}}{u_3} \qquad (15.58)$$

in 3-D, which represents in the time-space continuum a plane that contains the convection line. A spatial discretization that aims to model this propagation mode, therefore, can receive un upstream bias in the streamline direction.

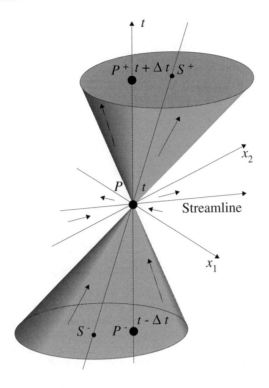

Figure 15.9: Characteristic Cone

In the time-space continuum, the second characteristic equation in (15.54) corresponds to a characteristic cone with vertex at P that is tangent to the local characteristic surface at P and whose shape depends on $\Im$. As time elapses, therefore, waves funnel in towards P and out and away from P within the characteristic cone. For a 2-D flow, this cone also represents in the space continuum a family of ellipses. for a 3-D flow, this cone represents a family of ellipsoids.

For a vanishing c, furthermore, this cone collapses onto the convection line (15.55), which, however, is not the axis of the characteristic cone. Figures 15.9 portrays this characteristic cone for 2-D flows. Significantly, as $\Im$ increases, the cone shears in the streamline direction, while its cross section remains an ellipse, whose axes project onto the streamline and crossflow directions. For all $\Im$, the time axis remains inside the cone. Accordingly, as time elapses, the funneling of waves within the cone, towards and then away from P, corresponds on any constant-t plane to closed curves encircling P that shrink towards P and then expand away from P. Figure 15.10 illustrates this pattern, which corresponds to the familiar Doppler distribution. This distribution results from projecting onto the flow plane several sections of the characteristic cone at various time levels and corresponds to pressure pulses emitted by a particle that moves along a streamline arc. From this perspective, a region of space

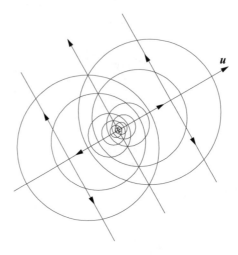

Figure 15.10: Wave Propagation

centered about P is simultaneously the domain of dependence and range of influence for P. A spatial discretization that aims to model this propagation mode, therefore, should receive un upstream bias along all directions radiating from P. The investigation of the characteristic cone shape change versus $\Im$, therefore, indicates that for unsteady and steady subsonic flows, both the domain of dependence and region of influence encompass regions encircling P.

15.3.3   Physical Multi-dimensional Upstream Bias

For all $\Im$ numbers wave propagation remains bi-modal. A physically consistent, intrinsically multi-dimensional, and infinite directional upstream formulation for the incompressible-flow Euler equations, therefore, has to provide an upstream approximation that remains bi-modal. This upstream approximation has to introduce an upstream bias along all directions radiating from each flow field point. The bias in this approximation must change with varying direction and correlate with the directional distribution of the characteristic propagation speeds. The directional distribution of the upstream bias must also remain symmetrical like the propagation speeds with respect to both the crossflow and streamline directions.

The formulation developed in this book first identifies the genuine streamline and crossflow convection and acoustics components within the system matrices. The formulation then establishes a physically consistent upstream approximation for each of these components, along all wave-propagation directions, with an upstream-bias magnitude that correlates with the directional distribution of the characteristic propagation speeds. For all $\Im$ numbers, the approximation remains bi-modal within the domain of dependence and range of influence of each flow-field point.

15.4 Acoustics-Convection Characteristics-Bias Decomposition

This section details a genuinely multi-dimensional characteristics-bias representation for the matrix product $A_j(q)\partial q/\partial x_j$ in the slight-compressibility system

$$\frac{\partial q}{\partial t} + A_j(q)\frac{\partial q}{\partial x_j} = 0 \qquad (15.59)$$

for both two- and three-dimensional flows. This formulation rests on a decomposition of this product into convection, pressure-gradient, and streamline and crossflow acoustic components. For three-dimensional flows, the isotropic acoustic propagation naturally decomposes into three directional propagation modes, the first along the streamline direction and the remaining two along any two mutually perpendicular crossflow directions. These two cross flow propagation combine into a two-dimensional isotropic acoustic propagation on any plane perpendicular to the velocity direction.

15.4.1 2-D Convection and Pressure-Gradient Components

The matrix product $A_j(q)\partial q/\partial x_j$ can be decomposed into convection and pressure-gradient components as

$$A_j(q)\frac{\partial q}{\partial x_j} = A_j^q(q)\frac{\partial q}{\partial x_j} + A_j^p(q)\frac{\partial q}{\partial x_j} \qquad (15.60)$$

where $A_j^q \partial q/\partial x_j$ and $A_j^p \partial q/\partial x_j$ respectively denote the convection and pressure components, defined as

$$A_j^q(q)\frac{\partial q}{\partial x_j} = \begin{pmatrix} 0 & , & \gamma p \delta_1^j & , & \gamma p \delta_2^j \\ 0 & , & u_1\delta_1^j + u_j & , & u_1\delta_2^j \\ 0 & , & u_2\delta_1^j & , & u_2\delta_2^j + u_j \end{pmatrix} \frac{\partial q}{\partial x_j}$$

$$A_j^p(q)\frac{\partial q}{\partial x_j} = \begin{pmatrix} 0 & , & 0 & , & 0 \\ \delta_1^j/\rho & , & 0 & , & 0 \\ \delta_2^j/\rho & , & 0 & , & 0 \end{pmatrix} \frac{\partial q}{\partial x_j} \qquad (15.61)$$

For any $\Im$ number, as Section 15.3 has shown, the eigenvalues of $A_j(q)n_j$

$$\lambda_1^{d_E} = u_j n_j, \quad \lambda_{2,3}^{d_E} = u_j n_j \pm \sqrt{(u_j n_j)^2 + c^2}, \quad c = \sqrt{\gamma\frac{p}{\rho}}$$

$$\lambda_1^E = v_j n_j \Im, \quad \lambda_{2,3}^E = v_j n_j \Im \pm \sqrt{\left(1 + \Im^2 (v_j n_j)^2\right)} \qquad (15.62)$$

have mixed algebraic sign and an upstream approximation for the matrix product along one single direction remains inconsistent with the two-way propagation of acoustic waves. Without the pressure gradient in the momentum equations, however, the speed of sound c would vanish and the corresponding system eigenvalues all have the same algebraic sign. The resulting convection matrix product can then be upstream approximated along one single direction. Since no magnitude of $\Im$ exists that can make λ_2^E and λ_3^E have the same algebraic sign, the system matrix is decomposed as

$$A_j(q)\frac{\partial q}{\partial x_j} = A_j^q \frac{\partial q}{\partial x_j} + A_j^p \frac{\partial q}{\partial x_j} \tag{15.63}$$

This decomposition, however, is insufficient for an accurate multi-dimensional upstream modeling of acoustic waves, for any $\Im$ number within a crossflow wedge region and for low and vanishing $\Im$ numbers within a streamline wedge region.

Concerning the cross-flow wedge region, the eigenvalues (15.63) do not all have the same algebraic sign within such a region. A mono-axial upstream-bias approximation of (15.60) within the cross-flow wedge region, therefore, remains inconsistent with the existing two-way wave propagation in this region.

With regards to the streamline wave propagation region, the eigenvalues associated with the components in (15.63) are the eigenvalues of the matrices

$$A_j^q(q)n_j = \begin{pmatrix} 0 & , & \gamma p n_1 & , & \gamma p n_2 \\ 0 & , & u_1 n_1 + u_j n_j & , & u_1 n_2 \\ 0 & , & u_2 n_1 & , & u_2 n_2 + u_j n_j \end{pmatrix} \tag{15.64}$$

and

$$A_j^p(q)n_j = \begin{pmatrix} 0 & , & 0 & , & 0 \\ n_1/\rho & , & 0 & , & 0 \\ n_2/\rho & , & 0 & , & 0 \end{pmatrix} \tag{15.65}$$

The eigenvalues of these matrices respectively are

$$\lambda_1^q = u_j n_j, \quad \lambda_2^q = 2u_j n_j, \quad \lambda_3^q = 0 \tag{15.66}$$

and

$$\lambda_{1,3}^p = 0 \tag{15.67}$$

which certainly all keep the same algebraic sign for any wave propagation direction, but for vanishing $\Im$ number remain far less than the dominant speed of sound c. For low $\Im$ numbers, therefore, an upstream approximation for the components in (15.63) would inaccurately model the physics of acoustics. This difficulty is resolved by further decomposing the pressure gradient in (15.63) in terms of genuine streamline and crossflow acoustic component, for accurate upstream modeling of acoustic waves throughout the flow field.

15.4.2   2-D Acoustic Components

For arbitrary $\Im$ numbers and corresponding dependent variables p, u_1, and u_2, the matrix product $A_j(q)\frac{\partial q}{\partial x_j}$ may be decomposed as

$$A_j(q)\frac{\partial q}{\partial x_j} = A_j^q(q)\frac{\partial q}{\partial x_j} + A_j^p(q)\frac{\partial q}{\partial x_j} = A_j^q(q)\frac{\partial q}{\partial x_j} + A_j^a(q)\frac{\partial q}{\partial x_j} + A_j^{nc}(q)\frac{\partial q}{\partial x_j} \tag{15.68}$$

where $A_j^a(q)$ denotes the acoustic matrix and A_j^{nc} indicates a non-linear coupling matrix, which completes the non-linear coupling between convection and acoustics within (15.68). All the eigenvalues of the matrix contraction

$$A_j^{nc} n_j \equiv \begin{pmatrix} 0 & , & -\gamma p n_1 & , & -\gamma p n_2 \\ 0 & , & 0 & , & 0 \\ 0 & , & 0 & , & 0 \end{pmatrix} \tag{15.69}$$

identically vanish. No need exists, therefore, to involve A_j^{nc} in the upstream-bias approximation of the system matrix $A_j(q)$. Conversely, the eigenvalues of the acoustic matrix contraction

$$A_j^a n_j \equiv \begin{pmatrix} 0 & , & \gamma p n_1 & , & \gamma p n_2 \\ n_1/\rho & , & 0 & , & 0 \\ n_2/\rho & , & 0 & , & 0 \end{pmatrix} \tag{15.70}$$

are

$$\lambda_1^a = 0, \qquad \lambda_{2,3}^a = \pm c = \pm \sqrt{\gamma \frac{p}{\rho}} \tag{15.71}$$

These eigenvalues coincide with the sound speed and remain independent of the propagation vector $\boldsymbol{n}$ for any $\Im$ number, which signifies isotropic propagation. The acoustics matrix A_j^a is thus used for an upstream-bias approximation of the Euler equations in the low $\Im$-number regime, within the streamline region, and for any $\Im$ number, within the crossflow region.

With reference to Section 15.2.3, for any two mutually perpendicular unit vectors $\boldsymbol{a} = (a_1, a_2)$ and $\boldsymbol{a}^N = (a_1^N, a_2^N)$, along with implied summation on repeated indices, the acoustics component can be expressed as

$$A_j^a \frac{\partial q}{\partial x_j} = A_j^a a_j a_k \frac{\partial q}{\partial x_k} + A_j^a a_j^N a_k^N \frac{\partial q}{\partial x_k}, \quad A_j^a a_j = X \Lambda^{a+} X^{-1} + X \Lambda^{a-} X^{-1} \tag{15.72}$$

For $\boldsymbol{a}$ parallel to $\boldsymbol{u}$, this expression corresponds to a decomposition of the Euler acoustics component into streamline and crossflow acoustics components. The matrices $X \Lambda^{a+} X^{-1}$ and $X \Lambda^{a-} X^{-1}$ respectively correspond to the "forward" and "backward" acoustic-propagation matrix components of $A_j^a a_j$. A bi-modal upstream-bias approximation of $A_j^a a_j$ thus readily follows from instituting a forward and a backward upstream-bias approximation respectively for its forward- and backward- propagation matrices. Similar results then readily follow by replacing $\boldsymbol{a}$ with $\boldsymbol{a}^N$. This bi-modal approximation directly yields both the non-negative-eigenvalue matrices

$$\left| A_j^a a_j \right| \equiv X \left(\Lambda^{a+} - \Lambda^{a-} \right) X^{-1}, \quad \left| A_j^a a_j^N \right| \equiv X_N \left(\Lambda^{a+} - \Lambda^{a-} \right) X_N^{-1} \tag{15.73}$$

and the associated matrix product $\left| A_j^a a_j \right| a_k \partial q / \partial x_k$. The matrices in (15.73) correspond to the streamline and crossflow absolute acoustics matrices. Different choices for the diagonal matrices Λ^{a+}, Λ^{a-} correspond to distinct levels of induced dissipation. With the selection

$$\Lambda^{a+} \equiv \frac{1}{2} \begin{pmatrix} 2c & , & 0 & , & 0 \\ 0 & , & 0 & , & 0 \\ 0 & , & 0 & , & c \end{pmatrix}, \quad \Lambda^{a-} \equiv -\frac{1}{2} \begin{pmatrix} 0 & , & 0 & , & 0 \\ 0 & , & 2c & , & 0 \\ 0 & , & 0 & , & c \end{pmatrix} \tag{15.74}$$

the matrix product of $\left|A_j^a a_j\right| = X\left(\Lambda^{a+} - \Lambda^{a-}\right)X^{-1}$ and the directional derivative $a_k \partial q/\partial x_k$ of q directly leads to the beautifully simple acoustic-field result

$$\left|A_j^a a_j\right| a_k \frac{\partial q}{\partial x_k} = XcIX^{-1}a_k\frac{\partial q}{\partial x_k} = ca_k\frac{\partial q}{\partial x_k} \tag{15.75}$$

The similar result for (a_1^N, a_2^N) replacing (a_1, a_2) is

$$\left|A_j^a a_j^N\right| a_k^N \frac{\partial q}{\partial x_k} = X_N cI X_N^{-1} a_k^N\frac{\partial q}{\partial x_k} = ca_k^N\frac{\partial q}{\partial x_k} \tag{15.76}$$

where I denotes a 3×3 matrix. All the eigenvalues of the streamline and crossflow absolute acoustic matrices, therefore, equal $+c$. The matrix product $A_j(q)\frac{\partial q}{\partial x_j}$ is thus expressed as

$$A_j(q)\frac{\partial q}{\partial x_j} = A_j^q(q)\frac{\partial q}{\partial x_j} + A_j^{nc}\frac{\partial q}{\partial x_j}$$

$$+ \left(X\Lambda^{a+}X^{-1} + X\Lambda^{a-}X^{-1}\right)a_k\frac{\partial q}{\partial x_k} + \left(X_N\Lambda^{a+}X_N^{-1} + X_N\Lambda^{a-}X_N^{-1}\right)a_k^N\frac{\partial q}{\partial x_k} \tag{15.77}$$

15.4.3 Combination of 2-D Acoustics-Convection Decompositions

The previous sections have shown that the system matrix product $A_j(q)\frac{\partial q}{\partial x_j}$ can be equivalently expressed as

$$A_j(q)\frac{\partial q}{\partial x_j} = \begin{cases} A_j^q(q)\dfrac{\partial q}{\partial x_j} + A_j^p(q)\dfrac{\partial q}{\partial x_j}; \\[2mm] A_j^q(q)\dfrac{\partial q}{\partial x_j} + A_j^{nc}\dfrac{\partial q}{\partial x_j} \\[2mm] + \left(X\Lambda^{a+}X^{-1} + X\Lambda^{a-}X^{-1}\right)a_k\dfrac{\partial q}{\partial x_k} \\[2mm] + \left(X_N\Lambda^{a+}X_N^{-1} + X_N\Lambda^{a-}X_N^{-1}\right)a_k^N\dfrac{\partial q}{\partial x_k} \end{cases} \tag{15.78}$$

where the first expression is convenient for a characteristics-bias approximation within any streamline wedge region, for high $\Im$ numbers, and the second expression is needed for a characteristics-bias approximation within any crossflow wedge region and within a streamline wedge region, for low subsonic $\Im$ numbers.

For a two-dimensional upstream-bias approximation throughout the flow field and for all $\Im$ numbers, the system matrix product can thus be cast as a linear combination of these two decompositions, with linear combination parameter α, $0 \leq \alpha \leq 1$

$$A_j(q)\frac{\partial q}{\partial x_j} = (1-\alpha)\left\{A_j^q(q)\frac{\partial q}{\partial x_j} + A_j^p(q)\frac{\partial q}{\partial x_j}\right\} + \alpha\left\{A_j^q(q)\frac{\partial q}{\partial x_j} + A_j^{nc}\frac{\partial q}{\partial x_j}\right\}$$

$$+ \left(X \Lambda^{a+} X^{-1} + X \Lambda^{a-} X^{-1} \right) a_k \frac{\partial q}{\partial x_k} + \left(X_N \Lambda^{a+} X_N^{-1} + X_N \Lambda^{a-} X_N^{-1} \right) a_k^N \frac{\partial q}{\partial x_k} \Bigg\} \qquad (15.79)$$

With $\delta \equiv 1 - \alpha$ and a crossflow function α^N, this combination leads to the following acoustics-convection decomposition

$$A_j(q) \frac{\partial q}{\partial x_j} = \alpha \left(X \Lambda^{a+} X^{-1} + X \Lambda^{a-} X^{-1} \right) a_k \frac{\partial q}{\partial x_k} + \alpha^N \left(X_N \Lambda^{a+} X_N^{-1} + X_N \Lambda^{a-} X_N^{-1} \right) a_k^N \frac{\partial q}{\partial x_k}$$

$$+ A_j(q) \frac{\partial q}{\partial x_j} + \delta A_j(p) \frac{\partial q}{\partial x_j} + \alpha A_j^{nc} \frac{\partial q}{\partial x_j} + (\alpha - \alpha^N) A_j^a a_j^N a_k^N \frac{\partial q}{\partial x_k} \qquad (15.80)$$

A consistent upstream-bias representation results from this combination because of the physical significant of each term. The weights α and α^N respectively for the streamline and crossflow acoustic components in this expression may remain different from each other because the streamline and crossflow characteristic velocity components are different from each other, following the Euler eigenvalues (15.47), which can allow a "differential" upstream bias in the streamline and crossflow directions. Bi-modal wave propagation exists for all $\Im$ numbers in the crossflow direction. Accordingly the crossflow acoustic term $\alpha^N \left(X_N \Lambda^{a+} X_N^{-1} + X_N \Lambda^{a-} X_N^{-1} \right) a_k^N \frac{\partial q}{\partial x_k}$ will receive a bi-modal upstream bias. The associated term $(\alpha - \alpha^N) A_j^a a_j^N a_k^N \frac{\partial q}{\partial x_k}$ need not receive any upstream-bias, for if it did it would obliterate all the crossflow acoustic upstream-bias, which remains fundamental for stability, as indicated by the multi-dimensional characteristics analysis. No need exists to involve the matrix A_j^{nc} in the upstream-bias formulation because this matrix is devoid of physical significance and all its eigenvalues vanish.

The streamline acoustic expression $\alpha \left(X \Lambda^{a+} X^{-1} + X \Lambda^{a-} X^{-1} \right) a_k \frac{\partial q}{\partial x_k}$ accounts for the bi-modal propagation of acoustic waves; this expression is thus employed for an acoustic upstream-bias approximation for low $\Im$ numbers. As $\Im$ increases a greater fraction of the pressure gradient $A_j^p(q) \frac{\partial q}{\partial x_j}$ may be involved in the upstream-bias approximation along with the acoustic matrix; accordingly, as $\Im$ increases, the upstream-bias function α may concurrently decrease. For any magnitude of both pressure and pressure gradient, the convection field uniformly carries information along streamlines; hence, the entire convection term $A_j^q(q) \frac{\partial q}{\partial x}$ can receive an upstream bias along the single streamline direction.

15.4.4   3-D Convection and Pressure-Gradient Components

Also for three-dimensional flows can the matrix product $A_j(q) \frac{\partial q}{\partial x_j}$ be decomposed into convection and pressure-gradient components as

$$A_j(q) \frac{\partial q}{\partial x_j} = A_j^q(q) \frac{\partial q}{\partial x_j} + A_j^p \frac{\partial q}{\partial x_j} \qquad (15.81)$$

where $A_j^q \frac{\partial q}{\partial x_j}$ and $A_j^p \frac{\partial q}{\partial x_j}$ respectively denote the convection and pressure-gradient components

$$A_j^q(q) \frac{\partial q}{\partial x_j} = \begin{pmatrix} 0 & , & \gamma p \delta_1^j & , & \gamma p \delta_2^j & , & \gamma p \delta_3^j \\ 0 & , & u_1 \delta_1^j + u_j & , & u_1 \delta_2^j & , & u_1 \delta_3^j \\ 0 & , & u_2 \delta_1^j & , & u_2 \delta_2^j + u_j & , & u_2 \delta_3^j \\ 0 & , & u_3 \delta_1^j & , & u_3 \delta_2^j & , & u_3 \delta_3^j + u_j \end{pmatrix} \frac{\partial q}{\partial x_j}$$

$$A_j^p \frac{\partial q}{\partial x_j} = \begin{pmatrix} 0 & , & 0 & , & 0 \\ \delta_1^j/\rho & , & 0 & , & 0 \\ \delta_2^j/\rho & , & 0 & , & 0 \\ \delta_3^j/\rho & , & 0 & , & 0 \end{pmatrix} \frac{\partial q}{\partial x_j} \tag{15.82}$$

For any $\Im$ number, the eigenvalues of $A_j(q)n_j$

$$\lambda_{1,2}^{d_E} = u_j n_j, \quad \lambda_{3,4}^{d_E} = u_j n_j \pm \sqrt{(u_j n_j)^2 + c^2}, \quad c = \sqrt{\gamma \frac{p}{\rho}}$$

$$\lambda_{1,2}^{E} = v_j n_j \Im, \quad \lambda_{3,4}^{E} = v_j n_j \Im \pm \sqrt{\left(1 + \Im^2 (v_j n_j)^2\right)} \tag{15.83}$$

have mixed algebraic sign and an upstream approximation for the matrix product along one single direction remains inconsistent with the two-way propagation of acoustic waves. Without the pressure gradient in the momentum equations, however, the speed of sound c would vanish and the corresponding system eigenvalues all have the same algebraic sign. The resulting convection matrix product can then be upstream approximated along one single direction.

The system matrix product can thus be decomposed as

$$A_j^q(q) \frac{\partial q}{\partial x_j} = A_j^q \frac{\partial q}{\partial x_j} A_j^p \frac{\partial q}{\partial x_j} \tag{15.84}$$

Also for three-dimensional flows, however, this decomposition is insufficient for an accurate multi-dimensional upstream modeling of acoustic waves, for any $\Im$ number within the crossflow region and for low and vanishing $\Im$ numbers within any streamline region.

Concerning the crossflow region, the eigenvalues associated with (15.84) do not all have the same algebraic sign within the cross-flow wave propagation region. A mono-axial upstream - bias approximation of (15.84) within the crossflow region, therefore, remains inconsistent with the existing two-way wave propagation in this region.

With regards to the streamline wave propagation region, the eigenvalues associated with the components in (15.84) are the eigenvalues of the matrices

$$A_j^q(q)n_j = \begin{pmatrix} 0 & , & \gamma p n_1 & , & \gamma p n_2 & , & \gamma p n_3 \\ 0 & , & u_1 n_1 + u_j n_j & , & u_1 n_2 & , & u_1 n_3 \\ 0 & , & u_2 n_1 & , & u_2 n_2 + u_j n_j & , & u_2 n_3 \\ 0 & , & u_3 n_1 & , & u_3 n_2 & , & u_3 n_3 + u_j n_j \end{pmatrix} \tag{15.85}$$

$$A_j^p(q)n_j \equiv \begin{pmatrix} 0 & , & 0 & , & 0 \\ n_1/\rho & , & 0 & , & 0 \\ n_2/\rho & , & 0 & , & 0 \\ n_3/\rho & , & 0 & , & 0 \end{pmatrix} \tag{15.86}$$

The eigenvalues of these matrices respectively are

$$\lambda_{1,2}^q = u_j n_j, \quad \lambda_3^q = 2u_j n_j, \quad \lambda_4^q = 0 \tag{15.87}$$

and

$$\lambda_{1,4}^p = 0 \tag{15.88}$$

which certainly all keep the same algebraic sign for any wave propagation direction, but for vanishing $\Im$ number remain far less than the dominant speed of sound c. For low $\Im$ numbers, therefore, an upstream approximation for the components in (15.84) would inaccurately model the physics of acoustics. This difficulty is resolved by further decomposing the pressure gradient in (15.84) in terms of genuine streamline and crossflow acoustic components, for accurate upstream modeling of acoustic waves throughout the flow field.

15.4.5   3-D Acoustic Components

For arbitrary $\Im$ numbers and corresponding dependent variables p, u_1, u_2, and u_3, the matrix product $A_j(q)\frac{\partial q}{\partial x_j}$ may be further decomposed as

$$A_j(q)\frac{\partial q}{\partial x_j} = A_j^q(q)\frac{\partial q}{\partial x_j} + A_j^p(q)\frac{\partial q}{\partial x_j} = A_j^q(q)\frac{\partial q}{\partial x_j} + A_j^a(q)\frac{\partial q}{\partial x_j} + A_j^{nc}(q)\frac{\partial q}{\partial x_j} \qquad (15.89)$$

where $A_j^a(q)$ and $A_j^{nc}(q)$ respectively denote the acoustics and non-linear-coupling matrices. All the eigenvalues of the matrix contraction $A_j^{nc}n_j$

$$A_j^{nc}n_j \equiv \begin{pmatrix} 0 & , & -\gamma p n_1 & , & -\gamma p n_2 & , & -\gamma p n_3 \\ 0 & , & 0 & , & 0 & , & 0 \\ 0 & , & 0 & , & 0 & , & 0 \\ 0 & , & 0 & , & 0 & , & 0 \end{pmatrix} \qquad (15.90)$$

identically vanish. No need exists, therefore, to involve A_j^{nc} in the upstream-bias approximation of the system matrix $A_j(q)$. The eigenvalues of the acoustic matrix contraction

$$A_j^a n_j \equiv \begin{pmatrix} 0 & , & \gamma p n_1 & , & \gamma p n_2 & , & \gamma p \delta_3^j \\ n_1/\rho & , & 0 & , & 0 & , & 0 \\ n_2/\rho & , & 0 & , & 0 & , & 0 \\ n_3/\rho & , & 0 & , & 0 & , & 0 \end{pmatrix} \qquad (15.91)$$

has eigenvalues

$$\lambda_{1,2}^a = 0, \qquad \lambda_{3,4}^a = \pm c \qquad (15.92)$$

These eigenvalues coincide with the sound speed and remain independent of the propagation vector $\boldsymbol{n}$ for any $\Im$ number, which signifies isotropic propagation. The acoustics matrix A_j^a can thus be used for an upstream-bias approximation of the Euler equations in the low $\Im$-number regime, within the streamline region and for any $\Im$ number within the crossflow region.

For any three mutually orthogonal unit vectors $\boldsymbol{a} = (a_1, a_2, a_3)$, $\boldsymbol{a}^{N_1} = (a_1^{N_1}, a_2^{N_1}, a_3^{N_1})$, and $\boldsymbol{a}^{N_2} = (a_1^{N_2}, a_2^{N_2}, a_3^{N_2})$ within a 3-D flow, along with implied summation on repeated indices, the matrix product $A_j^a \frac{\partial q}{\partial x_j}$ can be expressed as

$$A_j^a \frac{\partial q}{\partial x_j} = A_j^a a_j a_k \frac{\partial q}{\partial x_k} + A_j^a a_j^{N_1} a_k^{N_1} \frac{\partial q}{\partial x_k} + A_j^a a_j^{N_2} a_k^{N_2} \frac{\partial q}{\partial x_k} \qquad (15.93)$$

For $\boldsymbol{a}$ parallel to $\boldsymbol{u}$, this expression corresponds to a decomposition of the Euler acoustics component into one streamline and two crossflow acoustics components.

With reference to Section 15.2.3, the matrix contraction $A^a_j a_j$ may be further decomposed as

$$A^a_j a_j = X\Lambda^{a+}X^{-1} + X\Lambda^{a-}X^{-1} \qquad (15.94)$$

The matrices $X\Lambda^{a+}X^{-1}$ and $X\Lambda^{a-}X^{-1}$ respectively correspond to the "forward" and "backward" acoustic-propagation matrix components of $A^a_j a_j$. A bi-modal upstream-bias approximation of $A^a_j a_j$, therefore, readily follows from instituting a forward and a backward upstream-bias approximation respectively for its forward- and backward- propagation matrices. Similar results then readily follow by replacing $\boldsymbol{a}$ with $\boldsymbol{a}^{N_1}$ and $\boldsymbol{a}^{N_2}$ This bi-modal approximation directly yields both the non-negative-eigenvalue matrices

$$\left|A^a_j a_j\right| \equiv X\left(\Lambda^{a+} - \Lambda^{a-}\right)X^{-1}$$

$$\left|A^a_j a^{N_1}_j\right| \equiv X_{N_1}\left(\Lambda^{a+} - \Lambda^{a-}\right)X^{-1}_{N_1}, \quad \left|A^a_j a^{N_2}_j\right| \equiv X_{N_2}\left(\Lambda^{a+} - \Lambda^{a-}\right)X^{-1}_{N_2} \qquad (15.95)$$

and the associated matrix product $\left|A^a_j a_j\right| a_k \partial q/\partial x_k$. The matrices in (15.95) correspond to the streamline and crossflow absolute acoustics matrices. Different forms of Λ^{a+} and Λ^{a-} induce distinct levels of upstream dissipation. With the matrices

$$\Lambda^{a+} \equiv \frac{1}{2}\begin{pmatrix} 2c & , & 0 & , & 0 & , & 0 \\ 0 & , & 0 & , & 0 & , & 0 \\ 0 & , & 0 & , & c & , & 0 \\ 0 & , & 0 & , & 0 & , & c \end{pmatrix}, \quad \Lambda^{a-} \equiv -\frac{1}{2}\begin{pmatrix} 0 & , & 0 & , & 0 & , & 0 \\ 0 & , & 2c & , & 0 & , & 0 \\ 0 & , & 0 & , & c & , & 0 \\ 0 & , & 0 & , & 0 & , & c \end{pmatrix} \qquad (15.96)$$

With these matrices, the matrix product of $\left|A^a_j a_j\right| = X\left(\Lambda^{a+} - \Lambda^{a-}\right)X^{-1}$ with the directional derivative $a_k \partial q/\partial x_k$ of q directly leads to the beautifully simple acoustic-field result

$$\left|A^a_j a_j\right| a_k \frac{\partial q}{\partial x_k} = XcIX^{-1}a_k \frac{\partial q}{\partial x_k} = ca_k \frac{\partial q}{\partial x_k} \qquad (15.97)$$

The similar results for $(a^{N_1}_1, a^{N_1}_2, a^{N_1}_3)$ and $(a^{N_2}_1, a^{N_2}_2, a^{N_2}_3)$ replacing (a_1, a_2, a_3) are

$$\left|A^a_j a^{N_1}_j\right| a^{N_1}_k \frac{\partial q}{\partial x_k} = X_{N_1}cIX^{-1}_{N_1}a^{N_1}_k \frac{\partial q}{\partial x_k} = ca^{N_1}_k \frac{\partial q}{\partial x_k} \qquad (15.98)$$

and

$$\left|A^a_j a^{N_2}_j\right| a^{N_2}_k \frac{\partial q}{\partial x_k} = X_{N_2}cIX^{-1}_{N_2}a^{N_2}_k \frac{\partial q}{\partial x_k} = ca^{N_2}_k \frac{\partial q}{\partial x_k} \qquad (15.99)$$

where I denotes the 4×4 identity matrix. For a vanishing $\Im$, therefore, all the eigenvalues of the streamline and crossflow absolute acoustic matrices coincide with $+c$. The matrix product $A^a_j \frac{\partial q}{\partial x_j}$ may thus be cast as

$$A_j(q)\frac{\partial q}{\partial x_j} = A^q_j \frac{\partial q}{\partial x_j} + A^{nc}_j \frac{\partial q}{\partial x_j} + \left(X\Lambda^{a+}X^{-1} + X\Lambda^{a-}X^{-1}\right)a_k \frac{\partial q}{\partial x_k}$$

$$+ \left(X_{N_1}\Lambda^{a+}X^{-1}_{N_1} + X_{N_1}\Lambda^{a-}X^{-1}_{N_1}\right)a^{N_1}_k \frac{\partial q}{\partial x_k} + \left(X_{N_2}\Lambda^{a+}X^{-1}_{N_2} + X_{N_2}\Lambda^{a-}X^{-1}_{N_2}\right)a^{N_2}_k \frac{\partial q}{\partial x_k} \qquad (15.100)$$

15.4.6 Combination of 3-D Acoustics-Convection Decompositions

The previous sections have shown that the Euler system matrix product $A_j(q)\frac{\partial q}{\partial x_j}$ can be equivalently expressed as

$$
A_j(q)\frac{\partial q}{\partial x_j} = \left\{
\begin{aligned}
& A_j^q(q)\frac{\partial q}{\partial x_j} + A_j^p\frac{\partial q}{\partial x_j}; \\[2mm]
& A_j^q(q)\frac{\partial q}{\partial x_j} + A_j^{nc}\frac{\partial q}{\partial x_j} \\[2mm]
& + \left(X\Lambda^{a+}X^{-1} + X\Lambda^{a-}X^{-1}\right)a_k\frac{\partial q}{\partial x_k} \\[2mm]
& + \left(X_{N_1}\Lambda^{a+}X_{N_1}^{-1} + X_{N_1}\Lambda^{a-}X_{N_1}^{-1}\right)a_k^{N_1}\frac{\partial q}{\partial x_k} \\[2mm]
& + \left(X_{N_2}\Lambda^{a+}X_{N_2}^{-1} + X_{N_2}\Lambda^{a-}X_{N_2}^{-1}\right)a_k^{N_2}\frac{\partial q}{\partial x_k}
\end{aligned}
\right. \tag{15.101}
$$

where the first expression is convenient for a characteristics - bias approximation within the streamline region, for high $\Im$ numbers, and the second expression is needed for a characteristics-bias approximation within the crossflow region and within the streamline region, for low $\Im$ numbers.

For a three-dimensional upstream-bias approximation throughout the flow field and for all $\Im$ numbers, the system matrix product can thus be cast as a linear combination of these two decompositions with linear combination parameter α, $0 \leq \alpha \leq 1$

$$
A_j(q)\frac{\partial q}{\partial x_j} = (1-\alpha)\left\{A_j^q(q)\frac{\partial q}{\partial x_j} + A_j^p\frac{\partial q}{\partial x_j}\right\}
$$

$$
+\alpha\left\{A_j^q(q)\frac{\partial q}{\partial x_j} + A_j^{nc}\frac{\partial q}{\partial x_j} + \left(X\Lambda^{a+}X^{-1} + X\Lambda^{a-}X^{-1}\right)a_k\frac{\partial q}{\partial x_k}\right.
$$

$$
\left. + \left(X_{N_1}\Lambda^{a+}X_{N_1}^{-1} + X_{N_1}\Lambda^{a-}X_{N_1}^{-1}\right)a_k^{N_1}\frac{\partial q}{\partial x_k} + \left(X_{N_2}\Lambda^{a+}X_{N_2}^{-1} + X_{N_2}\Lambda^{a-}X_{N_2}^{-1}\right)a_k^{N_2}\frac{\partial q}{\partial x_k}\right\}
$$

$$\tag{15.102}$$

With $\delta \equiv 1-\alpha$ and a crossflow function α^N, this combination leads to the following acoustics-convection decomposition

$$
A_j(q)\frac{\partial q}{\partial x_j} = \alpha\left(X\Lambda^{a+}X^{-1} + X\Lambda^{a-}X^{-1}\right)a_k\frac{\partial q}{\partial x_k}
$$

$$
+\alpha^N\left(X_{N_1}\Lambda^{a+}X_{N_1}^{-1} + X_{N_1}\Lambda^{a-}X_{N_1}^{-1}\right)a_k^{N_1}\frac{\partial q}{\partial x_k} + \alpha^N\left(X_{N_2}\Lambda^{a+}X_{N_2}^{-1} + X_{N_2}\Lambda^{a-}X_{N_2}^{-1}\right)a_k^{N_2}\frac{\partial q}{\partial x_k}
$$

$$+A_j^q(q)\frac{\partial q}{\partial x_j} + \delta A_j^p \frac{\partial q}{\partial x_j} + \alpha A_j^{nc}\frac{\partial q}{\partial x_j} + (\alpha - \alpha^N)A_j^a\left(a_j^{N_1}a_k^{N_1} + a_j^{N_2}a_k^{N_2}\right)\frac{\partial q}{\partial x_k} \qquad (15.103)$$

As in the two-dimensional formulation, a consistent upstream-bias representation results from this combination because of the physical significant of each term. The weights α and α^N respectively for the streamline and crossflow acoustic components in this expression may remain different from each other because the streamline and crossflow characteristic velocity components are different from each other, following the Euler eigenvalues (15.47), which can allow a "differential" upstream bias in the streamline and crossflow directions. Bi-modal wave propagation exists for all $\Im$ numbers in the crossflow direction. Accordingly the crossflow acoustic term

$$\alpha^N \left(X_{N_1}\Lambda^{a+}X_{N_1}^{-1} + X_{N_1}\Lambda^{a-}X_{N_1}^{-1}\right)a_k^{N_1}\frac{\partial q}{\partial x_k} + \alpha^N \left(X_{N_2}\Lambda^{a+}X_{N_2}^{-1} + X_{N_2}\Lambda^{a-}X_{N_2}^{-1}\right)a_k^{N_2}\frac{\partial q}{\partial x_k}$$

will receive a bi-modal upstream bias. The associated term $(\alpha - \alpha^N)A_j^a a_j^a a_k^N \frac{\partial q}{\partial x_k}$ need not receive any upstream-bias, for if it did it would obliterate all the crossflow acoustic upstream-bias, which remains fundamental for stability, as indicated by the multi-dimensional characteristics analysis. No need exists to involve the matrix A_j^{nc} in the upstream-bias formulation because this matrix is devoid of physical significance and all its eigenvalues vanish.

The streamline acoustic expression $\alpha \left(X\Lambda^{a+}X^{-1} + X\Lambda^{a-}X^{-1}\right)a_k\frac{\partial q}{\partial x_k}$ accounts for the bi-modal propagation of acoustic waves; this expression is thus employed for an acoustic upstream-bias approximation for low $\Im$ numbers. As $\Im$ increases a greater fraction of the pressure gradient $A_j^p(q)\frac{\partial q}{\partial x_j}$ may be involved in the upstream-bias approximation along with the acoustic matrix; accordingly, as $\Im$ increases, the upstream-bias function α may concurrently decrease. For any magnitude of both pressure and pressure gradient, the convection field uniformly carries information along streamlines; hence, the entire convection term $A_j^q(q)\frac{\partial q}{\partial x}$ can receive an upstream bias along the single streamline direction.

15.5 Characteristics-Bias System Matrix

Two-Dimensional Flows

With reference to (14.103), given the physical significance of the terms in decomposition (15.80) and algebraic signs of the corresponding eigenvalues, the associated principal direction unit vectors for these terms are

$$a_1 = -a_2 = a_5 = a_6 = a, \quad a_3 = -a_4 = a^N, \quad a_7 = a_8 = 0 \qquad (15.104)$$

At each flow-field point, a and a^N remain respectively parallel and perpendicular to the local velocity, with a^N obtained by a 90°-degree counterclockwise rotation of a.

With (15.80) and (15.104), the general upstream-bias expression (14.103), expressed in Jacobian-matrix form, leads to the genuinely two-dimensional acoustics-convection characteristics-bias formulation

$$F^C = A_j(q)\frac{\partial q}{\partial x_j} - \frac{\partial}{\partial x_i}\left[\varepsilon\psi\left(c\left(\alpha a_i a_j + \alpha^N a_i^N a_j^N\right)\frac{\partial q}{\partial x_j} + a_i A_j^q(q)\frac{\partial q}{\partial x_j} + a_i\delta A_j^p(q)\frac{\partial q}{\partial x_j}\right)\right]$$
$$(15.105)$$

In this result, the expressions $\left(c\alpha a_i a_j \frac{\partial q}{\partial x_j} + a_i A_j^q(q)\frac{\partial q}{\partial x_j} + a_i \delta A_j^p(q)\frac{\partial q}{\partial x_j} \right)$, $\left(c\alpha^N a_i^N a_j^N \frac{\partial q}{\partial x_j} \right)$ determine the upstream biases within respectively the streamline and crossflow wave propagation regions. Directly corresponding to the physics of acoustics and convection, these two expressions combined then induce a correct upwind bias along all wave propagation regions.

For vanishing $\Im$ numbers, α and α^N will approach 1 whereas δ will approach 0. Under these conditions, (15.105) reduces to

$$F^C = A_j(q)\frac{\partial q}{\partial x_j} - \frac{\partial}{\partial x_i}\left[\varepsilon\psi\left(c\frac{\partial q}{\partial x_i} + a_i A_j^q(q)\frac{\partial q}{\partial x_j} \right) \right] \tag{15.106}$$

which essentially induces only an acoustics upstream bias. Heed that this bias becomes independent of specific propagation directions, for it no longer depends on $\left(\alpha a_i a_j + \alpha^N a_i^N a_j^N \right)$. This bias, therefore, becomes isotropic, in harmony with the isotropic propagation of acoustic waves. In the limit of vacuum flows, hence for unbounded $\Im$, $\alpha \to 0$ and $\delta \to 1$, and (15.105) thus becomes

$$F^C = A_j(q)\frac{\partial q}{\partial x_j} - \frac{\partial}{\partial x_i}\left[\varepsilon\psi\left(c\alpha^N a_i^N a_j^N \frac{\partial q}{\partial x_j} + a_i A_j(q)\frac{\partial q}{\partial x_j} \right) \right] \tag{15.107}$$

which depends on the crossflow component of the absolute acoustics matrix and the entire system matrix product $A_j(q)\partial q/\partial x_j$.

Three-Dimensional Flows

With reference to (14.103), given the physical significance of the terms in decomposition (15.103) and algebraic signs of the corresponding eigenvalues, the associated principal direction unit vectors for these terms are

$$\boldsymbol{a}_1 = -\boldsymbol{a}_2 = \boldsymbol{a}_7 = \boldsymbol{a}_8 = \boldsymbol{a}, \quad \boldsymbol{a}_3 = -\boldsymbol{a}_4 = \boldsymbol{a}^{N_1}, \quad \boldsymbol{a}_5 = -\boldsymbol{a}_6 = \boldsymbol{a}^{N_2}, \quad \boldsymbol{a}_9 = \boldsymbol{a}_{10} = \boldsymbol{0} \tag{15.108}$$

At each flow-field point, $\boldsymbol{a}$ is parallel to the local velocity vector, whereas $\boldsymbol{a}^{N_1}$ and $\boldsymbol{a}^{N_2}$ remain orthogonal to the velocity vector and each other.

With (15.103) and (15.108), the general upstream-bias expression (14.103) leads to the acoustics-convection characteristics-bias formulation

$$F^C = A_j(q)\frac{\partial q}{\partial x_j}$$

$$- \frac{\partial}{\partial x_i}\left[\varepsilon\psi\left(c\left(\alpha a_i a_j + \alpha^N \left(a_i^{N_1} a_j^{N_1} + a_i^{N_2} a_j^{N_2} \right)\right)\frac{\partial q}{\partial x_j} + a_i A_j^q(q)\frac{\partial q}{\partial x_j} + a_i\delta A_j^p(q)\frac{\partial q}{\partial x_j} \right) \right]$$
$$\tag{15.109}$$

In this result, the expressions

$$\left(c\alpha a_i a_j \frac{\partial q}{\partial x_j} + a_i A_j^q(q)\frac{\partial q}{\partial x_j} + a_i\delta A_j^p(q)\frac{\partial q}{\partial x_j} \right), \quad \left(c\alpha^N \left(a_i^{N_1} a_j^{N_1} + a_i^{N_2} a_j^{N_2} \right)\frac{\partial q}{\partial x_j} \right)$$

determine the upstream biases within respectively the streamline and crossflow wave propagation regions. Directly corresponding to the physics of acoustics and convection, these two expressions combined then induce a correct upwind bias along all wave propagation regions.

For vanishing $\Im$ numbers, α and α^N will approach 1 whereas δ will approach 0. Under these conditions (15.109) reduces to

$$F^C = A_j(q)\frac{\partial q}{\partial x_j} - \frac{\partial}{\partial x_i}\left[\varepsilon\psi\left(c\frac{\partial q}{\partial x_i} + a_iA_j^q(q)\frac{\partial q}{\partial x_j}\right)\right] \qquad (15.110)$$

which essentially induces only an acoustics upstream bias that does not depend on any specific propagation direction. This bias, accordingly, becomes isotropic, in harmony with the isotropic propagation of acoustic waves. In the limit of vacuum flows, hence for unbounded $\Im$, it follows that $\alpha \to 0$, $\alpha^N \to 0$ and $\delta \to 1$; expression (15.109) thus becomes

$$F^C = A_j(q)\frac{\partial q}{\partial x_j} - \frac{\partial}{\partial x_i}\left[\varepsilon\psi\left(c\alpha^N\left(a_i^{N_1}a_j^{N_1} + a_i^{N_2}a_j^{N_2}\right)\frac{\partial q}{\partial x_j} + a_iA_j(q)\frac{\partial q}{\partial x_j}\right)\right] \qquad (15.111)$$

which depends on the crossflow component of the absolute acoustics matrix and the entire system matrix product $A_j(q)\partial q/\partial x_j$.

15.5.1 Consistent Multi-Dimensional and Infinite-Directional Upstream Bias

Two-Dimensional Flows

In matrix form, expression (15.105) becomes

$$F^c = A_j(q)\frac{\partial q}{\partial x_j} - \frac{\partial}{\partial x_i}\left[\varepsilon\psi\left(c\left(\alpha a_ia_j + \alpha^N a_i^N a_j^N\right)I + a_iA_j^q(q) + a_i\delta A_j^p(q)\right)\frac{\partial q}{\partial x_j}\right] \qquad (15.112)$$

Such an upstream-bias expression essentially depends upon the five upstream-bias functions a_1, a_2, α, δ, α^N. In order to ensure physical significance for this expression hence for the upstream-bias approximation to decomposition (15.80), these functions are determined by imposing on (15.112) the stringent stability requirement that it should induce an upstream-bias diffusion not just along the principal streamline and crossflow upstream directions, but along all directions $n = (n_1, n_2)$ radiating from any flow-field point. This condition corresponds to stability of the upstream-bias expression.

By virtue of the developments in Sections 14.4, 14.5.4, this expression will be stable when all the eigenvalues of the upstream-bias matrix

$$\mathcal{A} \equiv n_i\left(c\left(\alpha a_ia_j + \alpha^N a_i^N a_j^N\right)I + a_iA_j^q(q) + a_i\delta A_j^p(q)\right)n_j \qquad (15.113)$$

remain positive for all $\Im$ and propagation directions n.

Three-Dimensional Flows

In matrix form, expression (15.109) becomes

$$F^c = A_j(q)\frac{\partial q}{\partial x_j} - \frac{\partial}{\partial x_i}\left[\varepsilon\psi\left(c\left(\alpha a_ia_j + \alpha^N\left(a_i^{N_1}a_j^{N_1} + a_i^{N_2}a_j^{N_2}\right)\right)I + a_iA_j^q(q) + a_i\delta A_j^p(q)\right)\frac{\partial q}{\partial x_j}\right]$$
$$(15.114)$$

This expression only depends upon the six functions a_1, a_2, a_3, α, δ, α^N since $\boldsymbol{a}^{N_1}$ and $\boldsymbol{a}^{N_2}$ are obtained from $\boldsymbol{a}$ as shown in Section 15.5.4. In order to ensure physical significance for the upstream-bias approximation to decomposition (15.103), these functions are determined by imposing on (15.114) the stability requirement that it should induce an upstream-bias diffusion not just along the principal streamline and crossflow upstream directions, but along all directions $\boldsymbol{n} = (n_1, n_2, n_3)$ radiating from any flow-field point. This expression will be stable when all the eigenvalues of the upstream-bias matrix

$$\mathcal{A} \equiv n_i \left(c \left(\alpha a_i a_j + \alpha^N \left(a_i^{N_1} a_j^{N_1} + a_i^{N_2} a_j^{N_2} \right) \right) I + a_i A_j^q(q) + a_i \delta A_j^p(q) \right) n_j \qquad (15.115)$$

remain positive for all $\Im$ and propagation directions $\boldsymbol{n}$.

15.5.2   Upstream-Bias Eigenvalues

Despite the non-linear algebraic complexity of $\mathcal{A}$, all of its eigenvalues have been analytically determined exactly in closed form. Dividing through the speed of sound c, the non-dimensional form of these eigenvalues is

$$\lambda_1 = n_i \left(\alpha a_i a_j + \alpha^N a_i^N a_j^N \right) n_j + n_i a_i v_j n_j \Im$$

$$\lambda_{2,3} = n_i \left(\alpha a_i a_j + \alpha^N a_i^N a_j^N \right) n_j + n_i a_i v_j n_j \Im \pm n_i a_i \sqrt{(v_j n_j \Im)^2 + \delta} \qquad (15.116)$$

for two-dimensional flows, and

$$\lambda_{1,2} = n_i \left(\alpha a_i a_j + \alpha^N \left(a_i^{N_1} a_j^{N_1} + a_i^{N_2} a_j^{N_2} \right) \right) n_j + n_i a_i v_j n_j \Im$$

$$\lambda_{3,4} = n_i \left(\alpha a_i a_j + \alpha^N \left(a_i^{N_1} a_j^{N_1} + a_i^{N_2} a_j^{N_2} \right) \right) n_j$$

$$+ n_i a_i v_j n_j \Im \pm n_i a_i \sqrt{(v_j n_j \Im)^2 + \delta} \qquad (15.117)$$

for three-dimensional flows. In these eigenvalues v_j denotes the jth direction cosine of a unit vector $\boldsymbol{v}$ parallel to the local velocity $\boldsymbol{u}$. Heed that when $\boldsymbol{a} = \boldsymbol{v}$ and $\boldsymbol{n} = \boldsymbol{v}$, the functions α and δ within (15.116) and (15.117) determine the corresponding streamline upstream-bias eigenvalues

$$\lambda_1 = \alpha + \Im, \qquad \lambda_{2,3} = \alpha + \Im \pm \sqrt{\Im^2 + \delta} \qquad (15.118)$$

As shown in the following sections, all of these upstream-bias eigenvalues will remain positive when α and δ are calculated via the streamline-flow expressions, from Chapter 11, and α^N forces $\lambda_{2,3}$, in 2-D, and $\lambda_{3,4}$, in 3-D, to remain positive for all $\boldsymbol{n}$ and Mach numbers.

15.5.3   Conditions on Upstream-Bias Functions and Eigenvalues

Two-Dimensional Flows

The eigenvalues (15.116) are expressed as

$$\lambda_{1,3} = \lambda_{1,3}(\Im, \boldsymbol{n}) \qquad (15.119)$$

to stress their dependence upon both $\Im$ and $\boldsymbol{n}$. The theoretical five conditions for the determination of the five functions $a_1, a_2, \alpha, \delta, \alpha^N$ are

$$\lambda_1\left(\Im, \boldsymbol{n}\right) \geq 0, \;\; \boldsymbol{a} \cdot \boldsymbol{a}^N = 0$$

$$\lambda_2\left(\Im, \boldsymbol{v}\right) + \lambda_3\left(\Im, \boldsymbol{v}\right) = \lambda_2 + \lambda_3, \;\; \delta = 1 - \alpha, \;\; \lambda_{2,3}\left(\Im, \boldsymbol{n}\right) \geq 0 \tag{15.120}$$

where $\lambda_2 + \lambda_3$ now denotes a prescribed streamline upstream-bias eigenvalue expression. The first condition stipulates that the unit vector $\boldsymbol{a}$ is parallel to the velocity vector and the second condition constrains $\boldsymbol{a}$ and $\boldsymbol{a}^N$ to be mutually perpendicular. As a results, these two vectors respectively point along the streamline and crossflow directions. The third and fourth conditions stipulate that α and δ equal the established streamline-flow expressions. For the determined $\boldsymbol{a}$, $\boldsymbol{a}^N$, α and δ, the fifth condition then establishes α^N.

Three-Dimensional Flows

The eigenvalues (15.117) are expressed as

$$\lambda_{1,4} = \lambda_{1,4}\left(\Im, \boldsymbol{n}\right) \tag{15.121}$$

to stress their dependence upon both $\Im$ and $\boldsymbol{n}$. The six conditions for the determination of the six functions a_1, a_2, a_3, α, δ, α^N are

$$\lambda_{1,2}\left(\Im, \boldsymbol{n}\right) \geq 0, \;\; \boldsymbol{a} \cdot \boldsymbol{a}^{N_1} = 0, \;\; \boldsymbol{a} \cdot \boldsymbol{a}^{N_2} = 0$$

$$\lambda_3\left(\Im, \boldsymbol{v}\right) + \lambda_4\left(\Im, \boldsymbol{v}\right) = \lambda_3 + \lambda_4, \;\; \delta = 1 - \alpha, \;\; \lambda_{3,4}\left(\Im, \boldsymbol{n}\right) \geq 0 \tag{15.122}$$

where $\lambda_3 + \lambda_4$ now denotes a prescribed streamline upstream-bias eigenvalue function. The first condition stipulates that the unit vector $\boldsymbol{a}$ is parallel to the velocity vector; the second and third condition then constrain the unit vectors $\boldsymbol{a}$, $\boldsymbol{a}^{N_1}$ and $\boldsymbol{a}^{N_2}$ to be mutually orthogonal to one other. In particular, these three conditions establish that $\boldsymbol{a}$ points in the streamline direction, while $\boldsymbol{a}^{N_1}$ and $\boldsymbol{a}^{N_2}$ point in any two mutually perpendicular crossflow directions. The fourth and fifth conditions again stipulate that α and δ equal the corresponding streamline-flow expressions. For the determined $\boldsymbol{a}$, $\boldsymbol{a}^{N_1}$, $\boldsymbol{a}^{N_2}$, α, and δ, the sixth condition $\lambda_{3,4}\left(\Im, \boldsymbol{n}\right) \geq 0$ then establishes α^N.

15.5.4   Upstream-Bias Vectors $\boldsymbol{a}$, $\boldsymbol{a}^{N_1}$, $\boldsymbol{a}^{N_2}$

Two-Dimensional Flows

These functions are used in actual computations based on the characteristics-bias formulation (15.105). In the eigenvalue λ_1 in (15.116), the components

$$n_i \alpha a_i a_j n_j = \alpha \left(a_j n_j\right)^2, \;\; n_i \alpha^N a_i^N a_j^N n_j = \alpha^N \left(a_j^N n_j\right)^2 \tag{15.123}$$

are already non-negative for non-negative α and α^N. The eigenvalues λ_1, therefore, will remain non-negative for all positive α and α^N, including $\alpha \to 0$ and $\alpha^N \to 0$, when the additional component $n_i a_i v_j n_j \Im$ remains non-negative for all $\Im$. This requirement is met along with the first condition in (15.120) when $\boldsymbol{a} = \boldsymbol{v}$, for

$$n_i a_i v_j n_j \Im = \Im \left(v_j n_j \right)^2 \geq 0 \tag{15.124}$$

This finding is not surprising, for the streamline direction is a principal characteristic direction. Based on $\boldsymbol{a} = \boldsymbol{v}$, the components of $\boldsymbol{a}^N$ are calculated as

$$a_1^N = -a_2, \quad a_2^N = a_1 \tag{15.125}$$

Three-Dimensional Flows

In the eigenvalues $\lambda_{1,2}$ in (15.117), the components

$$n_i \alpha a_i a_j n_j = \alpha \left(a_j n_j \right)^2, \quad n_i \alpha^N a_i^{N_1} a_j^{N_1} n_j = \alpha^{N_1} \left(a_j^{N_1} n_j \right)^2, \quad n_i \alpha^N a_i^{N_2} a_j^{N_2} n_j = \alpha^{N_2} \left(a_j^{N_2} n_j \right)^2 \tag{15.126}$$

are already non-negative for non-negative α and α^N. The eigenvalues $\lambda_{1,2}$, therefore, will remain non-negative for all positive α and α^N, including $\alpha \to 0$ and $\alpha^N \to 0$, when the additional component $n_i a_i v_j n_j \Im$ remains non-negative for all $\Im$. This requirement is met along with the first condition in (15.122) when $\boldsymbol{a} = \boldsymbol{v}$, for

$$n_i a_i v_j n_j \Im = \Im \left(v_j n_j \right)^2 \geq 0 \tag{15.127}$$

This finding is not surprising, for also in a three-dimensional flow is the streamline direction a principal characteristic direction. In any flow region where velocity vanishes, it is then computationally convenient to set $a_1 = 1$, $a_2 = a_3 = 0$. Since $a_1^2 + a_2^2 + a_3^2 = 1$, it follows that either $a_1^2 + a_2^2$ or $a_2^2 + a_3^2$ is always different from zero. For $a_1^2 + a_2^2 > 0$, therefore, the components of $\boldsymbol{a}$, $\boldsymbol{a}^{N_1}$, $\boldsymbol{a}^{N_2}$ can thus be expressed as

$$a_1 = \frac{u_1}{\|\boldsymbol{u}\|}, \quad a_2 = \frac{u_2}{\|\boldsymbol{u}\|}, \quad a_3 = \frac{u_3}{\|\boldsymbol{u}\|}$$

$$a_1^{N_1} = -\frac{a_2}{\sqrt{a_1^2 + a_2^2}}, \quad a_2^{N_1} = \frac{a_1}{\sqrt{a_1^2 + a_2^2}}, \quad a_3^{N_1} = 0$$

$$a_1^{N_2} = -\frac{a_1 a_3}{\sqrt{a_1^2 + a_2^2}}, \quad a_2^{N_2} = -\frac{a_2 a_3}{\sqrt{a_1^2 + a_2^2}}, \quad a_3^{N_2} = \sqrt{a_1^2 + a_2^2} \tag{15.128}$$

With these expressions, $\boldsymbol{a}$, $\boldsymbol{a}^{N_1}$, and $\boldsymbol{a}^{N_2}$ are unit vectors that remain mutually perpendicular with respect to one another. For $a_1^2 + a_2^2 = 0$, instead, the components of $\boldsymbol{a}$, $\boldsymbol{a}^{N_1}$, $\boldsymbol{a}^{N_2}$ can be expressed as

$$a_1 = \frac{u_1}{\|\boldsymbol{u}\|}, \quad a_2 = \frac{u_2}{\|\boldsymbol{u}\|}, \quad a_3 = \frac{u_3}{\|\boldsymbol{u}\|}$$

$$a_1^{N_1} = 0, \quad a_2^{N_1} = -\frac{a_3}{\sqrt{a_2^2 + a_3^2}}, \quad a_3^{N_1} = \frac{a_2}{\sqrt{a_2^2 + a_3^2}}$$

$$a_1^{N_2} = \sqrt{a_2^2 + a_3^2}, \quad a_2^{N_2} = -\frac{a_1 a_2}{\sqrt{a_2^2 + a_3^2}}, \quad a_3^{N_2} = -\frac{a_1 a_3}{\sqrt{a_2^2 + a_3^2}} \tag{15.129}$$

Also with these expressions are $\boldsymbol{a}$, $\boldsymbol{a}^{N_1}$, and $\boldsymbol{a}^{N_2}$ unit vectors that remain mutually perpendicular with respect to one another.

15.5.5   Upstream-Bias Functions α, δ, α^N

With the established upstream-bias unit vectors, the eigenvalues (15.116) and (15.117) remain positive within a conical region about the streamline direction, when α and δ are calculated by way of the streamline-flow expressions

$$\alpha(\Im) = \sqrt{1 + \Im^2} - \Im, \quad \delta = 1 - \alpha \tag{15.130}$$

The eigenvalues remain positive throughout the flow field also depending on α^N.

In 2-D, $\boldsymbol{a} = \boldsymbol{v}$ remains perpendicular to $\boldsymbol{a}^N$, while $n_i a_i$ and $n_i a_i^N$ denote the vector "dot" products between the unit vector $\boldsymbol{n}$ and the unit vectors $\boldsymbol{a}$ and $\boldsymbol{a}^N$ respectively. Accordingly,

$$n_i a_i = \cos\bar{\theta}, \quad n_i a_i^N a_j^N n_j = \sin^2\bar{\theta}, \quad \bar{\theta} \equiv \theta - \theta_v \tag{15.131}$$

where θ and θ_v denote the inclination angles between the x_1 axis and $\boldsymbol{n}$ and $\boldsymbol{v}$ respectively. In 3-D, $\boldsymbol{a} = \boldsymbol{v}$ remains perpendicular to both $\boldsymbol{a}^{N_1}$ and $\boldsymbol{a}^{N_2}$. The expressions $n_i a_i$, $n_i a_i^{N_1}$, and $n_i a_i^{N_2}$ respectively denote the vector "dot" products between the unit vector $\boldsymbol{n}$ and the unit vectors $\boldsymbol{a}$, $\boldsymbol{a}^{N_1}$, and $\boldsymbol{a}^{N_2}$, hence $\boldsymbol{n} = n_i a_i \boldsymbol{a} + n_i a_i^{N_1} \boldsymbol{a}^{N_1} + n_i a_i^{N_2} \boldsymbol{a}^{N_2}$ Accordingly,

$$1 = (n_i a_i)^2 + \left(n_i a_i^{N_1}\right)^2 + \left(n_i a_i^{N_2}\right)^2, \quad n_i a_i = \cos\bar{\theta}$$

$$n_i a_i^{N_1} a_j^{N_1} n_j + n_i a_i^{N_2} a_j^{N_2} n_j = 1 - \cos^2\bar{\theta} = \sin^2\bar{\theta} \tag{15.132}$$

where $\bar{\theta}$ denotes the angle between $\boldsymbol{n}$ and $\boldsymbol{a}$. In terms of this angle $\bar{\theta}$, as results for both two- and three-dimensional flows, the eigenvalues (15.116), (15.117) become

$$\lambda_{1,2} = \alpha \cos^2\bar{\theta} + \alpha^N \sin^2\bar{\theta} + \Im \cos^2\bar{\theta}$$

$$\lambda_{3,4} = \alpha \cos^2\bar{\theta} + \alpha^N \sin^2\bar{\theta} + \Im \cos^2\bar{\theta} \pm \cos\bar{\theta}\sqrt{\left(\Im \cos\bar{\theta}\right)^2 + \delta} \tag{15.133}$$

For $\boldsymbol{n} = \boldsymbol{a}^N$, in 2-D, or $\boldsymbol{n} = b_1 \boldsymbol{a}^{N_1} + b_2 \boldsymbol{a}^{N_2}$, $b_1^2 + b_1^2 = 1$, in 3-D, hence $\bar{\theta} = 90°$, these eigenvalues hence (15.116), (15.117) become

$$\lambda_{1,2} = 0, \qquad \lambda_{3,4} = \alpha^N \tag{15.134}$$

For $\Im \to 0$, propagation is only acoustic and $\alpha^N = 1$. For any $\Im$ number, the specification $\alpha^N = 1$ would certainly lead to non-negative eigenvalues (15.116) and (15.117). In order to decrease the amount of induced dissipation, however, α^N is set equal to a decreasing function of $\Im$. One choice is $\alpha^N = \alpha$. A numerical analysis of eigenvalues (15.116) and (15.117) reveals that with such a choice these eigenvalues remain non-negative for all $\bar{\theta}$ and $\Im$. With such a choice for α^N, the characteristics-bias system becomes independent of $\boldsymbol{a}$, $\boldsymbol{a}^{N_1}$, and $\boldsymbol{a}^{N_2}$. The characteristics-bias formulation in this chapter has employed a formulation in terms of streamline and crossflow components in order to allow use of any suitable expression for α^N.

15.6 Polar Variation of Upstream-Bias

The directional variation of the upstream bias eigenvalues (15.116) is presented in Figures 15.11 - 15.12

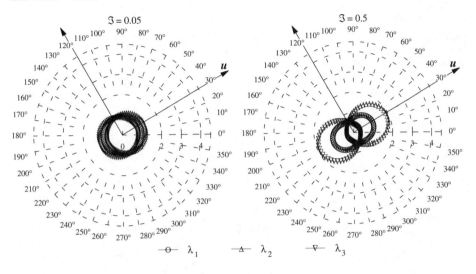

Figure 15.11: Polar Variation of Upstream Bias: $\Im = 0.05, 0.5$

for representative $\Im$ numbers. These variations are obtained for a variable unit vector $n \equiv (\cos\theta, \sin\theta)$ and fixed unit vector $a = v$, in this representative case inclined by $+30^o$ with respect to the x_1 axis on the flow plane, for a 2-D flows, or a reference x_1 axis on any space plane Π that contains both v and n, for a 3-D flow. For any $\Im$ number and wave propagation direction n, as these figures collectively indicate, the characteristics-bias expressions (15.105), (15.109) induce a physically consistent upstream bias because the associated upstream-bias eigenvalues (15.116), (15.117) remain positive and their directional variation mirrors the directional variation of the Euler eigenvalues (15.47). The upstream-bias eigenvalues are symmetrical about the crossflow direction and characteristic streamline, precisely like the Euler eigenvalues (15.47). For $\Im = 0.05$, the directional variation of the upstream-bias eigenvalues in Figure 15.11 correlates with that in Figure 15.5 and thereby corresponds to an isotropic upstream bias, in complete agreement with the isotropic acoustic wave propagation speed in the Euler equations. For increasing $\Im$ numbers, the upstream bias becomes anisotropic, again in agreement with the anisotropic distribution of the Euler eigenvalues (15.47). For $\Im = 0.5$ this anisotropy is already evident and then becomes more marked for increasing $\Im$ numbers as indicated in Figure 15.12, which correlates with Figure 15.6. In particular, the crossflow upstream bias decreases for increasing $\Im$ number.

Next, consider Figures 15.13 - 15.14. These figures compare the directional variations of the representative upstream-bias eigenvalue λ_3 and the corresponding Euler eigenvalue λ_3^E.

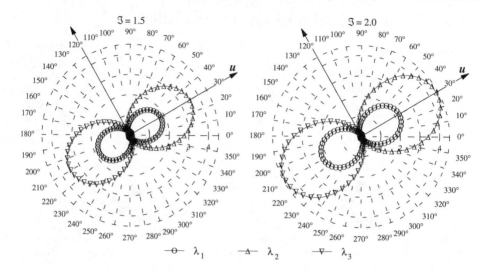

Figure 15.12: Polar Variation of Upstream Bias: $\Im = 1.5, 2.0$

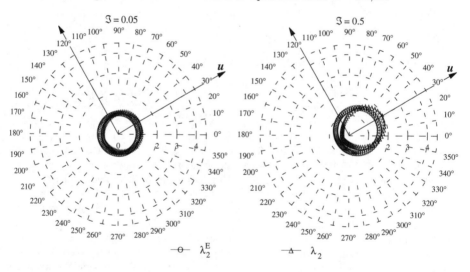

Figure 15.13: Polar Correlation of Characteristic λ_2^E and Upstream λ_2: $\Im = 0.05, 0.5$

This comparison is sufficient to depict the correlation between all the Euler and upstream-bias eigenvalues, for $\lambda_{1,2}$ and $\lambda_{1,2}^E$ are topologically similar to each other, compare Figures 15.6

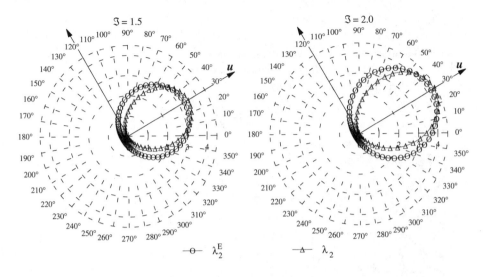

Figure 15.14: Polar Correlation of Characteristic λ_3^E and Upstream λ_3: $\Im = 1.5, 2.0$

and 15.12, while λ_4 and λ_4^E are respectively mirror skew-symmetric to λ_3 and λ_3^E with respect to the crossflow direction.

As Figures 15.13 - 15.14 indicate, λ_3 is symmetrical about the characteristic streamline, precisely like the corresponding characteristic Euler eigenvalue λ_3^E and the corresponding polar curve is topologically similar to the Euler eigenvalue curve. For $\Im = 0.05$, λ_3 and λ_3^E virtually coincide with each other and remain direction invariant, which corresponds to an isotropic upstream bias in correlation with the acoustic speed. For $\Im = 0.5$, Figure 15.13 indicates that λ_3^E is greater than λ_3 in the streamline direction. This results corresponds to minimal streamline upstream diffusion.

As shown in Figure 15.14, the magnitude of the upstream bias reflects the magnitude of the characteristic eigenvalues, within the domain of dependence and range of influence of any flow field point. Outside this region, the upstream-bias eigenvalues are modestly less than the characteristic eigenvalues. In these variations, the upstream-bias eigenvalues are vanishingly small in the cross- flow direction, which, in particular, corresponds to minimal crossflow diffusion.

15.7 Incompressible-Flow Formulation

Following the developments in Chapter 14 and in the previous sections, the characteristics-bias the slight-compressibility continuity equation is expressed as

$$\frac{\partial p}{\partial t} + \gamma p \frac{\partial u_j}{\partial x_j}$$

$$-\frac{\partial}{\partial x_i}\left[\varepsilon\psi\left(a_i\left(\frac{\partial p}{\partial t}+\gamma p\frac{\partial u_j}{\partial x_j}\right)+c\left(\alpha a_i a_j+\alpha^N\left(a_i^{N_1}a_j^{N_1}+a_i^{N_2}a_j^{N_2}\right)\right)\frac{\partial p}{\partial x_j}\right)\right]=0 \quad (15.135)$$

As anticipated in Section 15.1, the corresponding characteristics-bias continuity equation for totally incompressible flows originates from this equation upon replacing $\frac{\partial p}{\partial t}$ with $\frac{1}{\kappa}\frac{\partial p}{\partial t}$. The complete characteristics-bias Euler system becomes

$$\frac{\partial p}{\partial t}+\kappa\gamma p\frac{\partial u_j}{\partial x_j}$$

$$-\kappa\frac{\partial}{\partial x_i}\left[\varepsilon\psi\left(a_i\left(\frac{1}{\kappa}\frac{\partial p}{\partial t}+\gamma p\frac{\partial u_j}{\partial x_j}\right)+c\left(\alpha a_i a_j+\alpha^N\left(a_i^{N_1}a_j^{N_1}+a_i^{N_2}a_j^{N_2}\right)\right)\frac{\partial p}{\partial x_j}\right)\right]=0,$$

$$\frac{\partial u_\ell}{\partial t}+\frac{\partial u_\ell u_j}{\partial x_j}+\frac{\partial(p/\rho)}{\partial x_\ell}$$

$$-\frac{\partial}{\partial x_i}\left[\varepsilon\psi a_i\left(\frac{\partial u_\ell}{\partial t}+\frac{\partial u_\ell u_j}{\partial x_j}+\frac{\partial(p/\rho)}{\partial x_\ell}\right)\right]$$

$$-\frac{\partial}{\partial x_i}\left[\varepsilon\psi\left(c\left(\alpha a_i a_j+\alpha^N\left(a_i^{N_1}a_j^{N_1}+a_i^{N_2}a_j^{N_2}\right)\right)\frac{\partial u_\ell}{\partial x_j}+(\delta-1)a_i\frac{\partial(p/\rho)}{\partial x_\ell}\right)\right]=0 \quad (15.136)$$

Since the polytropic exponent $\kappa>0$ no longer equals one, the eigenvalues of the characteristics-bias expression within this system will now equal the eigenvalues of the matrix product $I_\kappa\mathcal{A}$, where I_κ denotes a positive diagonal matrix with ones on the leading diagonal with the exception of the entry on the first row and column, which equals κ. The eigenvalues of this matrix have been exactly determined for two- and three-dimensional flows. For two-dimensional flows these eigenvalues are

$$\lambda_1^\kappa=n_i\left(\alpha a_i a_j+\alpha^N a_i^N a_j^N\right)n_j+n_i a_i v_j n_j\Im$$

$$\lambda_{2,3}^\kappa=n_i\left(\alpha a_i a_j+\alpha^N a_i^N a_j^N\right)n_j\left(\frac{1+\kappa}{2}\right)+n_i a_i v_j n_j\Im$$

$$\pm\sqrt{\left(n_i\left(\alpha a_i a_j+\alpha^N a_i^N a_j^N\right)n_j\left(\frac{1-\kappa}{2}\right)+n_i a_i v_j n_j\Im\right)^2+\kappa\delta\left(n_i a_i\right)^2} \quad (15.137)$$

which revert to (15.116) for $\kappa=1$. These eigenvalues remain positive for any κ when the following inequality is satisfied

$$n_i\left(\alpha a_i a_j+\alpha^N a_i^N a_j^N\right)n_j\left(\frac{1+\kappa}{2}\right)+n_i a_i v_j n_j\Im$$

$$\geq\sqrt{\left(n_i\left(\alpha a_i a_j+\alpha^N a_i^N a_j^N\right)n_j\left(\frac{1-\kappa}{2}\right)+n_i a_i v_j n_j\Im\right)^2+\kappa\delta\left(n_i a_i\right)^2} \quad (15.138)$$

This inequality leads to

$$\left(n_i\left(\alpha a_i a_j+\alpha^N a_i^N a_j^N\right)n_j\right)^2+2n_i\left(\alpha a_i a_j+\alpha^N a_i^N a_j^N\right)n_j\left(n_i a_i v_j n_j\Im\right)-\delta\left(n_i a_i\right)^2\geq0 \quad (15.139)$$

which no longer depends upon κ and is unconditionally satisfied for the expressions already selected for α, α^N, and δ.

For three-dimensional flows, the eigenvalues of $I_\kappa \mathcal{A}$ are

$$\lambda^\kappa_{1,2} = n_i \left(\alpha a_i a_j + \alpha^N \left(a_i^{N_1} a_j^{N_1} + a_i^{N_2} a_j^{N_2} \right) \right) n_j + n_i a_i v_j n_j \Im$$

$$\lambda^\kappa_{3,4} = n_i \left(\alpha a_i a_j + \alpha^N \left(a_i^{N_1} a_j^{N_1} + a_i^{N_2} a_j^{N_2} \right) \right) n_j \left(\frac{1+\kappa}{2} \right) + n_i a_i v_j n_j \Im$$

$$\pm \sqrt{ \left(n_i \left(\alpha a_i a_j + \alpha^N \left(a_i^{N_1} a_j^{N_1} + a_i^{N_2} a_j^{N_2} \right) \right) n_j \left(\frac{1-\kappa}{2} \right) + n_i a_i v_j n_j \Im \right)^2 + \kappa \delta \left(n_i a_i \right)^2 } \quad (15.140)$$

which revert to (15.117) for $\kappa = 1$. These eigenvalues remain positive for any κ when the following inequality is satisfied

$$n_i \left(\alpha a_i a_j + \alpha^N \left(a_i^{N_1} a_j^{N_1} + a_i^{N_2} a_j^{N_2} \right) \right) n_j \left(\frac{1+\kappa}{2} \right) + n_i a_i v_j n_j \Im$$

$$\geq \sqrt{ \left(n_i \left(\alpha a_i a_j + \alpha^N \left(a_i^{N_1} a_j^{N_1} + a_i^{N_2} a_j^{N_2} \right) \right) n_j \left(\frac{1-\kappa}{2} \right) + n_i a_i v_j n_j \Im \right)^2 + \kappa \delta \left(n_i a_i \right)^2 } \quad (15.141)$$

This inequality leads to

$$\left(n_i \left(\alpha a_i a_j + \alpha^N \left(a_i^{N_1} a_j^{N_1} + a_i^{N_2} a_j^{N_2} \right) \right) n_j \right)^2$$

$$+ 2 n_i \left(\alpha a_i a_j + \alpha^N \left(a_i^{N_1} a_j^{N_1} + a_i^{N_2} a_j^{N_2} \right) \right) n_j \left(n_i a_i v_j n_j \Im \right) - \delta \left(n_i a_i \right)^2 \geq 0 \quad (15.142)$$

which no longer depends upon κ and is unconditionally satisfied for the expressions already selected for α, α^N, and δ. The eigenvalues (15.137), (15.140) thus remain positive for any positive κ, of which the magnitude may be locally optimized for rapid convergence to steady state. The incompressible - flow Euler equations, with $\gamma = 1$, formally result from the limit of system (15.136) for $\kappa \rightarrow \infty$. This positive-eigenvalue condition, very likely either obviates or satisfies the "inf-sup" condition, [18, 19], for the characteristics-bias algorithm allows use of equal-order bi-linear approximation for both pressure and velocity. Following Chapters 4, 14, the characteristics-bias Navier-Stokes system for incompressible flows becomes

$$\frac{1}{\kappa p} \frac{\partial p}{\partial t} - \frac{1}{p} \frac{\partial}{\partial x_i} \left(\varepsilon \psi a_i \frac{1}{\kappa} \frac{\partial p}{\partial t} \right) + \frac{\partial u_j}{\partial x_j} - \frac{1}{p} \frac{\partial}{\partial x_i} \left(\varepsilon \psi a_i p \frac{\partial u_j}{\partial x_j} \right)$$

$$- \frac{1}{p} \frac{\partial}{\partial x_i} \left[\varepsilon \psi \left(c \left(\alpha a_i a_j + \alpha^N \left(a_i^{N_1} a_j^{N_1} + a_i^{N_2} a_j^{N_2} \right) \right) \right) \frac{\partial p}{\partial x_j} \right] = 0,$$

$$\frac{\partial u_\ell}{\partial t} + \frac{\partial u_\ell u_j}{\partial x_j} + \frac{\partial p}{\partial x_\ell} + \frac{Gr}{Re^2} \Delta T \frac{g_\ell}{g} - \frac{1}{Re} \frac{\partial}{\partial x_j} \left(\frac{\partial u_\ell}{\partial x_j} + \frac{\partial u_j}{\partial x_\ell} \right)$$

$$- \frac{\partial}{\partial x_i} \left[\varepsilon \psi a_i \left(\frac{\partial u_\ell}{\partial t} + \frac{\partial u_\ell u_j}{\partial x_j} + \frac{\partial p}{\partial x_\ell} + \frac{Gr}{Re^2} \Delta T \frac{g_\ell}{g} - \frac{1}{Re} \frac{\partial}{\partial x_j} \left(\frac{\partial u_\ell}{\partial x_j} + \frac{\partial u_j}{\partial x_\ell} \right) \right) \right]$$

$$- \frac{\partial}{\partial x_i} \left[\varepsilon \psi \left(c \left(\alpha a_i a_j + \alpha^N \left(a_i^{N_1} a_j^{N_1} + a_i^{N_2} a_j^{N_2} \right) \right) \right) \frac{\partial u_\ell}{\partial x_j} + (\delta - 1) a_i \frac{\partial p}{\partial x_\ell} \right) \right] = 0,$$

$$\frac{\partial \Delta T}{\partial t} + \frac{\partial u_j \Delta T}{\partial x_j} - \frac{1}{Pr Re} \frac{\partial^2 \Delta T}{\partial x_j \partial x_j} - \frac{Ec}{Re} \frac{\partial u_i}{\partial x_j} \left(\frac{\partial u_i}{\partial x_j} + \frac{\partial u_j}{\partial x_i} \right)$$

$$-\frac{\partial}{\partial x_i}\left[\varepsilon\psi a_i\left(\frac{\partial\Delta T}{\partial t}+\frac{\partial u_j\Delta T}{\partial x_j}-\frac{1}{PrRe}\frac{\partial^2\Delta T}{\partial x_j\partial x_j}-\frac{Ec}{Re}\frac{\partial u_i}{\partial x_j}\left(\frac{\partial u_i}{\partial x_j}+\frac{\partial u_j}{\partial x_i}\right)\right)\right]$$

$$-\frac{\partial}{\partial x_i}\left[\varepsilon\psi\left(c\left(\alpha a_i a_j+\alpha^N\left(a_i^{N_1}a_j^{N_1}+a_i^{N_2}a_j^{N_2}\right)\right)\right)\frac{\partial\Delta T}{\partial x_j}\right]=0 \qquad (15.143)$$

from which the two-dimensional equations are obtained by setting $a_1^{N_2}=a_2^{N_2}=0$. The addition of the characteristics-bias temperature equation does not change eigenvalues (15.116), (15.117), but merely contributes another eigenvalue equal to λ_1; in this equation the product of velocity gradients multiplied by Ec/Re corresponds to the viscous dissipation function. System (15.143) supply the characteristics-bias formulation for incompressible flows, a formulation that for $\kappa\to\infty$ does not contain any time pressure derivative in the continuity equation. With and without this time derivative, the formulation has generated essentially non-oscillatory solutions for the velocity, pressure, and temperature fields.

15.8 Variational Statement and Boundary Conditions

With implied summation on repeated subscript indices, $1\le i,j\le n$ and $1\le\ell\le 2$, the incompressible-flow characteristics-bias Navier-Stokes integral systems may be concisely expressed as

$$\int_\Omega w\left(\frac{1}{\kappa p}\frac{\partial p}{\partial t}-\frac{1}{p}\frac{\partial}{\partial x_i}\left(\varepsilon\psi a_i\frac{1}{\kappa}\frac{\partial p}{\partial t}\right)+\frac{\partial u_j}{\partial x_j}-\frac{1}{p}\frac{\partial}{\partial x_i}\left(\varepsilon\psi a_i p\frac{\partial u_j}{\partial x_j}\right)\right)d\Omega$$

$$-\int_\Omega\frac{w}{p}\frac{\partial}{\partial x_i}\left[\varepsilon\psi\left(c\left(\alpha a_i a_j+\alpha^N a_i^{N_\ell}a_j^{N_\ell}\right)\right)\frac{\partial p}{\partial x_j}\right]d\Omega=0,$$

$$\int_\Omega w\left(\frac{\partial q}{\partial t}-\phi+\frac{\partial f_j}{\partial x_j}-\frac{\partial f_j^\nu}{\partial x_j}\right)d\Omega-\int_\Omega w\frac{\partial}{\partial x_i}\left[\varepsilon\psi a_i\left(\frac{\partial q}{\partial t}-\phi+\frac{\partial f_j}{\partial x_j}-\frac{\partial f_j^\nu}{\partial x_j}\right)\right]d\Omega$$

$$-\int_\Omega w\frac{\partial}{\partial x_i}\left[\varepsilon\psi\left(c\left(\alpha a_i a_j+\alpha^N a_i^{N_\ell}a_j^{N_\ell}\right)\frac{\partial q}{\partial x_j}+a_i(\delta-1)\frac{\partial f_j^p}{\partial x_j}\right)\right]d\Omega=0 \qquad (15.144)$$

where the continuity equation has been separated from the linear-momentum and temperature equations and the arrays q, ϕ, f_j, f_j^p, f_j^ν follow from inspection of (15.143). The corresponding weak statement is achieved by integrating by parts the characteristics-bias expression, an operation that yields the integral statements

$$\int_\Omega\left(w+\varepsilon\psi a_i p\frac{\partial}{\partial x_i}\left(\frac{w}{p}\right)\right)\left(\frac{1}{\kappa p}\frac{\partial p}{\partial t}+\frac{\partial u_j}{\partial x_j}\right)d\Omega$$

$$+\int_\Omega\frac{\partial}{\partial x_i}\left(\frac{w}{p}\right)\varepsilon\psi\left(c\left(\alpha a_i a_j+\alpha^N a_i^{N_\ell}a_j^{N_\ell}\right)\right)\frac{\partial p}{\partial x_j}d\Omega$$

$$-\oint_{\partial\Omega}\frac{w}{p}\frac{\partial}{\partial x_i}\left[\varepsilon\psi\left(a_i p\left(\frac{1}{\kappa p}\frac{\partial p}{\partial t}+\frac{\partial u_j}{\partial x_j}\right)+c\left(\alpha a_i a_j+\alpha^N a_i^{N_\ell}a_j^{N_\ell}\right)\right)\frac{\partial p}{\partial x_j}\right]n_i d\Gamma=0,$$

$$\int_\Omega\left(w+\varepsilon\psi a_i\frac{\partial w}{\partial x_i}\right)\left(\frac{\partial q}{\partial t}-\phi+\frac{\partial f_j}{\partial x_j}-\frac{\partial f_j^\nu}{\partial x_j}\right)d\Omega$$

$$
+ \int_\Omega \frac{\partial w}{\partial x_i} \varepsilon \psi \left(c \left(\alpha a_i a_j + \alpha^N a_i^{N_\ell} a_j^{N_\ell} \right) \frac{\partial q}{\partial x_j} + a_i (\delta - 1) \frac{\partial f_j^p}{\partial x_j} \right) d\Omega
$$

$$
- \oint_{\partial \Omega} w \varepsilon \psi \left(a_i \left(\frac{\partial q}{\partial t} - \phi + \frac{\partial (f_j - f_j^\nu)}{\partial x_j} \right) \right) n_i d\Gamma
$$

$$
- \oint_{\partial \Omega} w \varepsilon \psi \left(c \left(\alpha a_i a_j + \alpha^N a_i^{N_\ell} a_j^{N_\ell} \right) \frac{\partial q}{\partial x_j} + a_i (\delta - 1) \frac{\partial f_j^p}{\partial x_j} \right) n_i d\Gamma = 0 \qquad (15.145)
$$

When the divergence of the viscous flux f_j^ν is also integrated by parts, the following equivalent statement is obtained

$$
\int_\Omega \left(w + \varepsilon \psi a_i p \frac{\partial}{\partial x_i} \left(\frac{w}{p} \right) \right) \left(\frac{1}{\kappa p} \frac{\partial p}{\partial t} + \frac{\partial u_j}{\partial x_j} \right) d\Omega
$$

$$
+ \int_\Omega \frac{\partial}{\partial x_i} \left(\frac{w}{p} \right) \varepsilon \psi \left(c \left(\alpha a_i a_j + \alpha^N a_i^{N_\ell} a_j^{N_\ell} \right) \right) \frac{\partial p}{\partial x_j} d\Omega
$$

$$
- \oint_{\partial \Omega} \frac{w}{p} \frac{\partial}{\partial x_i} \left[\varepsilon \psi \left(a_i p \left(\frac{1}{\kappa p} \frac{\partial p}{\partial t} + \frac{\partial u_j}{\partial x_j} \right) + c \left(\alpha a_i a_j + \alpha^N a_i^{N_\ell} a_j^{N_\ell} \right) \right) \frac{\partial p}{\partial x_j} \right] n_i d\Gamma = 0,
$$

$$
\int_\Omega \left(w + \varepsilon \psi a_i \frac{\partial w}{\partial x_i} \right) \left(\frac{\partial q}{\partial t} + \frac{\partial f_j}{\partial x_j} \right) d\Omega + \int_\Omega \frac{\partial w}{\partial x_j} f_j^\nu d\Omega - \oint_{\partial \Omega} w f_j^\nu n_j d\Gamma
$$

$$
+ \int_\Omega \frac{\partial w}{\partial x_i} \varepsilon \psi \left(c \left(\alpha a_i a_j + \alpha^N a_i^{N_\ell} a_j^{N_\ell} \right) \frac{\partial q}{\partial x_j} - a_i \frac{\partial f_j^\nu}{\partial x_j} + a_i (\delta - 1) \frac{\partial f_j^p}{\partial x_j} \right) d\Omega
$$

$$
- \oint_{\partial \Omega} w \varepsilon \psi \left(a_i \left(\frac{\partial q}{\partial t} - \phi + \frac{\partial (f_j - f_j^\nu)}{\partial x_j} \right) \right) n_i d\Gamma
$$

$$
- \oint_{\partial \Omega} w \varepsilon \psi \left(c \left(\alpha a_i a_j + \alpha^N a_i^{N_\ell} a_j^{N_\ell} \right) \frac{\partial q}{\partial x_j} + a_i (\delta - 1) \frac{\partial f_j^p}{\partial x_j} \right) n_i d\Gamma = 0 \qquad (15.146)
$$

where n_i denotes the i-th component of the outward pointing unit vector perpendicular to a boundary facet, as depicted in Figure 15.15; when the boundary corresponds to a wall, then $a_i n_i = 0$ and the corresponding terms in the characteristics-bias surface integral will identically vanish. The $\varepsilon \psi$ surface integral, in any case, is eliminated by enforcing a weak Neumann-type boundary condition for the hyperbolic-parabolic characteristics-bias perturbation. One part of this condition imposes that the original Navier-Stokes system should be satisfied at the boundary; the remaining terms are also set to zero as part of this weak boundary condition. Clearly, the single Neumann-type condition corresponds to the entire boundary expression set to nought. For large $\Im$, $\alpha, \alpha^N \to 0$, $\delta \to 1$ and for either $n_i a_i^N = 0$ or $a_j^N \frac{\partial q}{\partial x_j} = 0$, or both, this boundary condition enforces the Navier-Stokes system on the boundary. With the imposition of this weak boundary condition, system (15.146) becomes

$$
\int_\Omega \left(w + \varepsilon \psi a_i p \frac{\partial}{\partial x_i} \left(\frac{w}{p} \right) \right) \left(\frac{1}{\kappa p} \frac{\partial p}{\partial t} + \frac{\partial u_j}{\partial x_j} \right) d\Omega
$$

$$
+ \int_\Omega \frac{\partial}{\partial x_i} \left(\frac{w}{p} \right) \varepsilon \psi \left(c \left(\alpha a_i a_j + \alpha^N a_i^{N_\ell} a_j^{N_\ell} \right) \right) \frac{\partial p}{\partial x_j} d\Omega = 0,
$$

$$\int_\Omega \left(w + \varepsilon \psi a_i \frac{\partial w}{\partial x_i} \right) \left(\frac{\partial q}{\partial t} - \phi + \frac{\partial f_j}{\partial x_j} \right) d\Omega + \int_\Omega \frac{\partial w}{\partial x_j} f_j^\nu d\Omega - \oint_{\partial \Omega} w f_j^\nu n_j \, d\Gamma$$

$$+ \int_\Omega \frac{\partial w}{\partial x_i} \varepsilon \psi \left(c \left(\alpha a_i a_j + \alpha^N a_i^{N_\ell} a_j^{N_\ell} \right) \frac{\partial q}{\partial x_j} + a_i (\delta - 1) \frac{\partial f_j^p}{\partial x_j} \right) d\Omega = 0 \qquad (15.147)$$

from which, for computational simplicity, the characteristics-bias term for the viscous flux may be omitted without loss of stability. The corresponding weak statement for the continuity equation without time derivative of pressure is

$$\int_\Omega \left(w + \varepsilon \psi a_i p \frac{\partial}{\partial x_i} \left(\frac{w}{p} \right) \right) \frac{\partial u_j}{\partial x_j} \, d\Omega$$

$$+ \int_\Omega \frac{\partial}{\partial x_i} \left(\frac{w}{p} \right) \varepsilon \psi \left(c \left(\alpha a_i a_j + \alpha^N a_i^{N_\ell} a_j^{N_\ell} \right) \right) \frac{\partial p}{\partial x_j} \, d\Omega = 0 \qquad (15.148)$$

For all test functions $w \in \mathcal{H}^1(\Omega)$, the variational formulation seeks a solution $q \in \mathcal{H}^\eta(\Omega)$ with $\eta = 0$ or $\eta = 1$ respectively for discontinuous or smooth flows, that satisfies this weak statement. This statement is subject to prescribed initial conditions $q(x, 0) = q_0(x)$ and boundary conditions on $\partial \Omega \equiv \overline{\Omega} \backslash \Omega$. Synthetically, these boundary conditions are expressed as

$$B(x_{\partial \Omega}) q(x_{\partial \Omega}, t) = G_v(x_{\partial \Omega}, t) \qquad (15.149)$$

where $G_v(x_{\partial \Omega}, t)$ corresponds to the array of prescribed Dirichlet boundary conditions, with a zero entry for each corresponding unconstrained component of q, and $B(x_{\partial \Omega})$ denotes a square diagonal matrix, with a 1 for each diagonal entry, but replaced by zero for each corresponding unconstrained component of q.

At an inlet, Dirichlet boundary conditions may be enforced on the components of velocity u_i, $1 \leq i \leq n$, and on temperature ΔT. These conditions may be cast as

$$\frac{\partial u_i}{\partial t}(x_{\text{in}}, t) = q'_{u_i}(t), \quad 1 \leq i \leq n, \qquad \frac{\partial \Delta T}{\partial t}(x_{\text{in}}, t) = q'_{\Delta T}(t) \qquad (15.150)$$

where $q'_{u_i}(t)$, $1 \leq i \leq n$, and $q'_{\Delta T}(t)$, denote prescribed bounded functions. Since the governing equations determine pressure up to an arbitrary constant, at least one pressure boundary condition is required. If no other pressure boundary conditions are specified, than this single pressure boundary condition is enforced at one point, similarly to (15.150), replacing the continuity equation at that point.

At a solid wall, which corresponds to a streamline, one boundary condition is required for an inviscid flow. This solid-wall boundary condition corresponds to the wall-tangency, i.e no-penetration, condition $\mathbf{u} \cdot \mathbf{n} = 0$. With reference to the decomposition $f_j = f_j^q + f_j^p$ and Figure 15.15-a, the divergence of velocity and the convection flux with components f_j^q within (15.147) are integrated by parts at a wall region, an operation that yields the wall boundary statement

$$\int_\Omega \left[\left(w + \varepsilon \psi a_i \frac{\partial}{\partial x_i} \left(\frac{w}{p} \right) \right) \frac{1}{\kappa} \frac{\partial p}{\partial t} + \varepsilon \psi a_i p \frac{\partial}{\partial x_i} \left(\frac{w}{p} \right) \frac{\partial u_j}{\partial x_j} \right] d\Omega - \int_\Omega \frac{\partial w}{\partial x_j} u_j \, d\Omega$$

$$+ \oint_{\partial \Omega} w \mathbf{u} \cdot \mathbf{n} \, d\Gamma + \int_\Omega \frac{\partial}{\partial x_i} \left(\frac{w}{p} \right) \varepsilon \psi \left(c \left(\alpha a_i a_j + \alpha^N a_i^{N_\ell} a_j^{N_\ell} \right) \right) \frac{\partial p}{\partial x_j} \, d\Omega = 0,$$

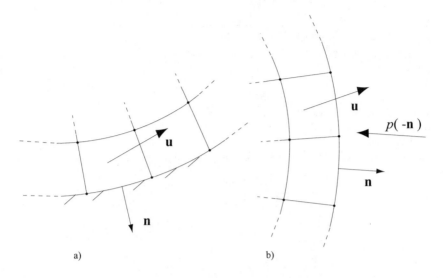

Figure 15.15: Boundary Regions: a) Wall; b) Outlet

$$\int_\Omega \left[\left(w + \varepsilon\psi a_i \frac{\partial w}{\partial x_i} \right) \left(\frac{\partial q}{\partial t} - \phi + \frac{\partial f_j^p}{\partial x_j} \right) + \frac{\partial w}{\partial x_j} f_j^\nu \right] d\Omega - \oint_{\partial\Omega} w f_j^\nu n_j \, d\Gamma$$

$$+ \int_\Omega \frac{\partial w}{\partial x_i} \varepsilon\psi \left(c \left(\alpha a_i a_j + \alpha^N a_i^{N_\ell} a_j^{N_\ell} \right) \frac{\partial q}{\partial x_j} + a_i(\delta - 1) \frac{\partial f_j^p}{\partial x_j} \right) d\Omega$$

$$- \int_\Omega \frac{\partial w}{\partial x_i} \left(f_i^q - \varepsilon\psi a_i \frac{\partial f_j^q}{\partial x_j} \right) d\Omega = - \oint_{\partial\Omega} w q \mathbf{u} \cdot \mathbf{n} \, d\Gamma \qquad (15.151)$$

where $f_j^\nu = 0$ for an inviscid flow. This statement features a generalized "numerical flux" for f_j^q and a wall surface integral that depends on the mass flux $\mathbf{u} \cdot \mathbf{n}$. A wall mass-flux boundary condition is thus directly enforced within this surface integral, with a wall-tangency boundary condition obtained by setting the entire integral to nought. For an inviscid flow, this is the only boundary condition that is required at a solid wall.

For a viscous flow, the two linear-momentum components in statement (15.151) are replaced by the no-slip boundary conditions on u_i, $1 \le i \le n$, and the wall mass-flux surface integrals in the continuity and temperature components in this statement are then eliminated. The wall heat-flux integral in the temperature equation

$$\int_\Omega \left(w + \varepsilon\psi a_i \frac{\partial w}{\partial x_i} \right) \left[\frac{\partial \Delta T}{\partial t} - \phi_E \right] d\Omega - \int_\Omega \frac{\partial w}{\partial x_j} u_j \Delta T d\Omega$$

$$+ \oint_{\partial\Omega} w \Delta T u_j n_j d\Gamma + \int_\Omega \frac{\partial w}{\partial x_i} \varepsilon\psi \left[a_i \frac{\partial(u_j \Delta T)}{\partial x_j} + c \left(\alpha a_i a_j + \alpha^N a_i^{N_\ell} a_j^{N_\ell} \right) \frac{\partial \Delta T}{\partial x_j} \right] d\Omega$$

$$+ \int_\Omega \frac{\partial w}{\partial x_j} \frac{1}{PrRe} \frac{\partial T}{\partial x_j} d\Omega - \oint_{\partial\Omega} w \frac{1}{PrRe} \frac{\partial T}{\partial x_j} n_j d\Gamma = 0 \qquad (15.152)$$

is then used to specify a heat-flux boundary condition, by inserting the prescribed heat-flux $\frac{\partial T}{\partial n} = \frac{\partial T}{\partial x_j} n_j$ in this integral. For a cold- or hot-wall boundary condition, the entire temperature equation (15.152) is then replaced at each hot- or cold-wall point by the Dirichlet boundary condition

$$\frac{\partial \Delta T}{\partial t}(x_{\text{in}}, t) = q'_{\Delta T}(t) \tag{15.153}$$

where $q_{\Delta T}(t)$ denotes a prescribed wall temperature.

At an outlet, a minimal-perturbation temperature boundary condition involves a pre-scribed heat flux This heat-flux boundary condition is enforced by inserting the prescribed magnitude of the heat flux in the corresponding surface integral in the outflow temperature-equation integral

$$\int_\Omega \left(w + \varepsilon \psi a_i \frac{\partial w}{\partial x_i} \right) \left[\frac{\partial \Delta T}{\partial t} - \phi_E + \frac{\partial (u_j \Delta T)}{\partial x_j} \right] d\Omega$$

$$+ \int_\Omega \frac{\partial w}{\partial x_i} \varepsilon \psi \left(c \left(\alpha a_i a_j + \alpha^N a_i^{N_\ell} a_j^{N_\ell} \right) \frac{\partial \Delta T}{\partial x_j} \right) d\Omega$$

$$+ \int_\Omega \frac{\partial w}{\partial x_j} \frac{1}{Pr Re} \frac{\partial T}{\partial x_j} d\Omega - \oint_{\partial \Omega} w \frac{1}{Pr Re} \frac{\partial T}{\partial x_j} n_j \, d\Gamma = 0 \tag{15.154}$$

A vanishing outlet heat flux corresponds to an adiabatic outlet stream.

A similar surface-integral procedure is employed to enforce a specified boundary condition on outlet pressure p and on the deviatoric surface tractions $\tau_{ij} n_j$, $1 \leq i, j \leq n$, for a viscous flow. With reference to the decomposition $f_j = f_j^q + f_j^p$ and Figure 15.15-b, the divergence of the pressure flux with components f_j^p within the momentum equations in (15.147) is integrated by parts at an outlet region, an operation that yields the outflow boundary statement

$$\int_\Omega \left(w + \varepsilon \psi a_k \frac{\partial w}{\partial x_k} \right) \left(\frac{\partial u_i}{\partial t} - \phi_{u_i} + \frac{\partial (u_i u_j)}{\partial x_j} \right) d\Omega$$

$$+ \int_\Omega \frac{\partial w}{\partial x_k} \varepsilon \psi \left(c \left(\alpha a_k a_j + \alpha^N a_k^{N_\ell} a_j^{N_\ell} \right) \frac{\partial u_i}{\partial x_j} + a_k (\delta - 1) \frac{\partial p}{\partial x_i} \right) d\Omega$$

$$- \int_\Omega \frac{\partial w}{\partial x_j} \left(p \delta_i^j - \varepsilon \psi a_j \frac{\partial p}{\partial x_i} - \frac{1}{Re} \tau_{ij} \right) d\Omega + \oint_{\partial \Omega} w \left(p n_i - \frac{1}{Re} \tau_{ij} n_j \right) d\Gamma = 0 \tag{15.155}$$

This statement features an outflow surface integral that depends on the outlet-pressure specific-force components $p n_i$ and surface tractions $\tau_{ij} n_j$, $1 \leq i, j \leq n$. The outlet boundary conditions on pressure and tractions are thus directly enforced by specifying their numerical values within this surface integral.

15.9   Finite Element Galerkin Weak Statement

Since the characteristics-bias system (15.147) is developed independently and before any discretization, a genuinely multi-dimensional upstream-bias approximation for the govern-ing equations (4.56) on arbitrary grids directly results from a classic centered discretization of the characteristics-bias weak statement (15.147) on the prescribed grid. This statement is discretized in space by way of a Galerkin finite element method. This method not only accommodates arbitrary geometries or generates consistent non-extrapolation boundary equations for q, but also retains the ideal surface-integral venues of the integral statement to enforce the boundary conditions described in the previous section.

The finite element approximation exists on a partition Ω^h of Ω, where h signifies spatial discrete approximation. Having its boundary nodes on the boundary $\partial\Omega$ of Ω, this partition Ω^h results from the union of N_e non-overlapping elements Ω_e, $\Omega^h = \bigcup_{e=1}^{N_e} \Omega_e$. With $\eta = 0$ or $\eta = 1$, respectively for shocked or smooth flows, and with $\mathcal{P}_1(\Omega_e)$ and $\mathcal{P}_{1v}(\Omega_e)$ the finite-dimensional function spaces of respectively diagonal square matrix-valued and vector-valued linear polynomials within each Ω_e, for each "t", the corresponding diagonal square matrix-valued and vector-valued finite element discretization spaces employed in this study are defined as

$$\mathcal{S}^1(\Omega^h) \equiv \left\{ w^h \in \mathcal{H}^1(\Omega^h) : w^h\big|_{\Omega_e} \in \mathcal{P}_1(\Omega_e), \forall \Omega_e \in \Omega^h, B(x_{\partial\Omega^h}) w^h(x_{\partial\Omega^h}) = 0 \right\}$$

$$\mathcal{S}_v^\eta(\Omega^h) \equiv \left\{ w_v^h \in \mathcal{H}_v^\eta(\Omega^h) : w_v^h\big|_{\Omega_e} \in \mathcal{P}_{1v}(\Omega_e), \forall \Omega_e \in \Omega^h, B(x_{\partial\Omega^h}) w_v^h(x_{\partial\Omega^h}, t) = G_v(x_{\partial\Omega^h}, t) \right\}$$

$$\tag{15.156}$$

Based on these spaces, the finite element approximation $q^h \in \mathcal{S}_v^\eta$, is determined for each "t" as the solution of the finite element weak statement associated with (15.147), determined as follows.

Using the definition of Ω^h, the finite-element continuity equation may be expressed as

$$\sum_{e=1}^{N_e} \int_{\Omega_e} \left(w^h + \varepsilon^h \psi^h a_i^h p^h \frac{\partial}{\partial x_i} \left(\frac{w^h}{p^h} \right) \right) \left(\frac{1}{\kappa p^h} \frac{\partial p^h}{\partial t} + \frac{\partial u_j^h}{\partial x_j} \right) d\Omega$$

$$+ \sum_{e=1}^{N_e} \int_{\Omega_e} \frac{\partial}{\partial x_i} \left(\frac{w^h}{p^h} \right) \varepsilon^h \psi^h \left(c^h \left(\alpha^h a_i^h a_j^h + \alpha^{N^h} a_i^{N_\ell^h} a_j^{N_\ell^h} \right) \right) \frac{\partial p^h}{\partial x_j} d\Omega = 0 \tag{15.157}$$

Within each finite element Ω_e, for computational simplicity, the discrete pressure p^h in this continuity equation is set equal to a centroidal constant p_e exclusively within both $\varepsilon^h \psi^h a_i^h p^h$ and $\frac{\partial}{\partial x_i}\left(\frac{w^h}{p^h}\right)$; this simplification is not employed within the gradient of pressure in this equation or within any other term in the linear momentum equations. With this simplification, the complete set of characteristics-bias finite element equations can be cast as

$$\sum_{e=1}^{N_e} \int_{\Omega_e} \left(w^h + \varepsilon^h \psi^h a_i^h \frac{\partial w^h}{\partial x_i} \right) \left(\frac{1}{\kappa p^h} \frac{\partial p^h}{\partial t} + \frac{\partial u_j^h}{\partial x_j} \right) d\Omega$$

$$+ \sum_{e=1}^{N_e} \int_{\Omega_e} \frac{\partial w^h}{\partial x_i} \varepsilon^h \psi^h \left(\frac{c^h}{p_e} \left(\alpha^h a_i^h a_j^h + \alpha^{N^h} a_i^{N_\ell^h} a_j^{N_\ell^h} \right) \right) \frac{\partial p^h}{\partial x_j} d\Omega = 0,$$

$$\sum_{e=1}^{N_e} \int_{\Omega_e} \left(w^h + \varepsilon^h \psi^h a_i^h \frac{\partial w^h}{\partial x_i} \right) \left(\frac{\partial q^h}{\partial t} - \phi^h + \frac{\partial f_j^h}{\partial x_j} \right) d\Omega$$

$$+ \sum_{e=1}^{N_e} \int_{\Omega_e} \frac{\partial w^h}{\partial x_j} f_j^{\upsilon^h} d\Omega - \oint_{\partial\Omega^h} w^h f_j^{\upsilon^h} n_j \, d\Gamma +$$

$$+ \sum_{e=1}^{N_e} \int_{\Omega_e} \frac{\partial w^h}{\partial x_i} \varepsilon^h \psi^h \left(c^h \left(\alpha^h a_i^h a_j^h + \alpha^{N^h} a_i^{N_\ell^h} a_j^{N_\ell^h} \right) \frac{\partial q^h}{\partial x_j} + a_i^h (\delta^h - 1) \frac{\partial f_j^{p^h}}{\partial x_j} \right) d\Omega = 0 \quad (15.158)$$

with similar expressions for statements (15.151)-(15.155). The corresponding weak statement for the continuity equation without time derivative of pressure is

$$\sum_{e=1}^{N_e} \int_{\Omega_e} \left(w^h + \varepsilon^h \psi^h a_i^h \frac{\partial w^h}{\partial x_i} \right) \frac{\partial u_j^h}{\partial x_j} d\Omega$$

$$+ \sum_{e=1}^{N_e} \int_{\Omega_e} \frac{\partial w^h}{\partial x_i} \varepsilon^h \psi^h \left(\frac{c^h}{p_e} \left(\alpha^h a_i^h a_j^h + \alpha^{N^h} a_i^{N_\ell^h} a_j^{N_\ell^h} \right) \right) \frac{\partial p^h}{\partial x_j} d\Omega = 0$$

Since every member w^h of $\mathcal{S}^1$ results from a linear combination of the linearly independent basis functions of this finite dimensional space, statement (15.158) is satisfied for N independent basis functions of the space, where N denotes the dimension of the space. For N mesh nodes within Ω^h, there exist clusters of "master" elements Ω_k^M, each cluster comprising only those adjacent elements that share a mesh node $\mathbf{x}_k$, with $1 \leq k \leq N$, where N denotes the total number of not only mesh nodes, but also master elements.

As detailed in Chapter 7, the discrete test function w^h within each master element Ω_k^M will coincide with the "pyramid" basis function $w_k = w_k(\mathbf{x})$, $1 \leq k \leq N$, with compact support on Ω_k^M. Such a function equals one at node $\mathbf{x}_k$, zero at all other mesh nodes and also identically vanishes both on the boundary segments of Ω_k^M not containing $\mathbf{x}_k$ and on the computational domain outside Ω_k^M.

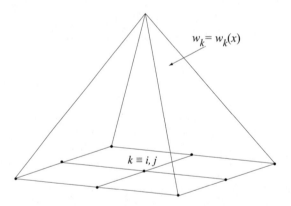

Figure 15.16: Pyramid Test Function for Ω_k^M

The discrete solution q^h and flux f_j^h at each time t assume the form of the following group linear combinations

$$q^h(\mathbf{x}, t) \equiv \sum_{\ell=1}^{N} w_\ell(\mathbf{x}) \cdot q^h(\mathbf{x}_\ell, t), \quad f_j^h(\mathbf{x}, t) \equiv \sum_{\ell=1}^{N} w_\ell(\mathbf{x}) \cdot f_j \left(q^h(\mathbf{x}_\ell, t) \right) \quad (15.159)$$

of time-dependent nodal solution values $q^h(\mathbf{x}_\ell, t)$, to be determined, and trial functions, which coincide with the test functions $w_\ell(\mathbf{x})$ for a Galerkin formulation; an analogous expression applies for ϕ and $f_j^{\nu^h}$. Similarly, the fluxes $f_j^q = f_j^q(q(\mathbf{x}, t))$ and $f_j^p = f_j^p(q(\mathbf{x}, t))$ are discretized through the group expressions

$$f_j^{q^h}(\mathbf{x}, t) \equiv \sum_{\ell=1}^{N} w_\ell(\mathbf{x}) \cdot f_j^q\left(q^h(\mathbf{x}_\ell, t)\right), \quad f_j^{p^h}(\mathbf{x}, t) \equiv \sum_{\ell=1}^{N} w_\ell(\mathbf{x}) \cdot f_j^p\left(q^h(\mathbf{x}_\ell, t)\right) \quad (15.160)$$

The notation for the discrete nodal variable and fluxes is then simplified as $q_\ell(t) \equiv q(\mathbf{x}_\ell, t)$, $f_{j\ell} \equiv f_j\left(q^h(\mathbf{x}_\ell, t)\right)$, $f_{j\ell}^q \equiv f_j^q\left(q^h(\mathbf{x}_\ell, t)\right)$, $f_{j\ell}^p \equiv f_j^p\left(q^h(\mathbf{x}_\ell, t)\right)$ and expansions (15.159)-(15.160) are then inserted into (15.158), which yields the discrete finite element weak statement

$$\sum_{e=1}^{N_e} \int_{\Omega_e} \left(w_k + \varepsilon^h \psi^h a_i^h \frac{\partial w_k}{\partial x_i}\right) \left(\frac{1}{\kappa p^h} w_m \frac{dp_m}{dt} + \frac{\partial w_m}{\partial x_j} u_{jm}\right) d\Omega$$

$$+ \sum_{e=1}^{N_e} \int_{\Omega_e} \frac{\partial w_k}{\partial x_i} \varepsilon^h \psi^h \left(\frac{c^h}{p_e}\left(\alpha^h a_i^h a_j^h + \alpha^{N^h} a_i^{N_\ell^h} a_j^{N_\ell^h}\right)\right) \frac{\partial w_m}{\partial x_j} p_m \, d\Omega = 0,$$

$$\sum_{e=1}^{N_e} \int_{\Omega_e} \left(w_k + \varepsilon^h \psi^h a_i^h \frac{\partial w_k}{\partial x_i}\right) \left(w_m \frac{dq_m}{dt} - w_m \phi_m + \frac{\partial w_m}{\partial x_j} f_{jm}\right) d\Omega$$

$$+ \sum_{e=1}^{N_e} \int_{\Omega_e} \frac{\partial w_k}{\partial x_j} f_j^{\nu^h} \, d\Omega - \oint_{\partial \Omega^h} w_k f_j^{\nu^h} n_j \, d\Gamma +$$

$$+ \sum_{e=1}^{N_e} \int_{\Omega_e} \frac{\partial w_k}{\partial x_i} \frac{\partial w_m}{\partial x_j} \varepsilon^h \psi^h \left[c^h \left(\alpha^h a_i^h a_j^h + \alpha^{N^h} a_i^{N_\ell^h} a_j^{N_\ell^h}\right) q_m + a_i^h (\delta^h - 1) f_{jm}^p\right] d\Omega = 0 \quad (15.161)$$

for $1 \leq k \leq N$. The corresponding finite element weak statement for the continuity equation without time derivative of pressure is

$$\sum_{e=1}^{N_e} \int_{\Omega_e} \left(w_k + \varepsilon^h \psi^h a_i^h \frac{\partial w_k}{\partial x_i}\right) \frac{\partial w_m}{\partial x_j} u_{jm} \, d\Omega$$

$$+ \sum_{e=1}^{N_e} \int_{\Omega_e} \frac{\partial w_k}{\partial x_i} \varepsilon^h \psi^h \left(\frac{c^h}{p_e}\left(\alpha^h a_i^h a_j^h + \alpha^{N^h} a_i^{N_\ell^h} a_j^{N_\ell^h}\right)\right) \frac{\partial w_m}{\partial x_j} p_m \, d\Omega = 0, \quad (15.162)$$

The three-test-function matrices in (7.186) are then needed to discretize the viscous dissipation function in the temperature equation. There are three implied summations with respect to the subscript indices i, j, m. The subscript indices i, j in this expression denote cartesian-axis directions, hence $1 \leq i, j \leq 2$, for two-dimensional flows, or $1 \leq i, j \leq 3$, for three dimensional flows, whereas subscript m indicates a mesh node, hence $1 \leq m \leq N$, although a sum like $\sum_{m=1}^{N} w_m \frac{dq_m}{dt}$ only involves a few neighboring terms because the compact-support test function w_m is only non-zero within a cluster of a few neighboring elements.

While an expansion like the ones in (15.159) for ψ^h, α^h, c^h, $\mathbf{a}^h$, $\mathbf{a}^{N^h}$ and δ^h can be directly accommodated within (15.161), each of these variables in this study has been set equal to a piece-wise constant for computational simplicity, one centroidal constant value per element. Within each element, the terms ψ^h and ε^h are respectively set equal to the controller and reference length developed in Chapter 10. As shown in that chapter, ε^h corresponds to the length of the streamline radius of the generalized ellipse inscribed within each element, since the streamline is a characteristic principal direction.

Since the test and trial functions w_k are prescribed functions of $\mathbf{x}$, the spatial integrations in (15.158) are directly carried out. For arbitrarily shaped elements, these integrations take place as detailed in Chapter 7. Concerning the boundary variables, no extrapolation of variables is needed in this algorithm on a variable that is not constrained via a Dirichlet boundary condition. In this case, instead, the finite element algorithm (15.161) naturally generates for each unconstrained boundary variable a boundary-node ordinary differential equation. For $\kappa \to \infty$, the complete integration with respect to $\mathbf{x}$ transforms (15.161) into a system of continuum-time ordinary algebraic-differential equations (OADE) for determining at each time level t the unknown nodal values $p^h(\mathbf{x}_\ell, t)$, $q^h(\mathbf{x}_\ell, t)$, $1 \le \ell \le N$.

15.10 Implicit Runge-Kutta Time Integration

The finite element equations (15.161) along with appropriate boundary equations and conditions form a system of algebraic and differential equations. The discrete continuity equations provide the algebraic equations and the discrete linear-momentum equations supply the differential equations in the time variable. Although these equations feature no time derivatives of pressure, the implicit Runge-Kutta time integration procedure (8.5) succeeds in simultaneously solving the discrete continuity, linear-momentum, and temperature equations and directly yielding the solution for both pressure, velocity and temperature. The finite element algebraic-differential equations are expressed as

$$\mathcal{M}\frac{dQ(t)}{dt} = \mathcal{F}(t, Q(t)) \tag{15.163}$$

where $\mathcal{M} \equiv \{w_k w_\ell\}$ denotes the mass matrix, $\mathcal{M}\frac{dQ(t)}{dt}$ indicates the corresponding coupling of time derivatives in (15.161), and $\mathcal{F}(t, Q(t))$ represents the remaining terms in (15.161); in this case, only zeros are present in each row of $\mathcal{M}$ that corresponds to a discrete continuity equation. The two-stage diagonally implicit Runge-Kutta algorithms for the numerical time integration of (15.163) is expressed as

$$\begin{aligned}
Q_{n+1} - Q_n &= b_1 K_1 + b_2 K_2 \\
\mathcal{M} K_1 &= \Delta t \cdot \mathcal{F}(t_n + c_1 \Delta t, Q_n + a_{11} K_1) \\
\mathcal{M} K_2 &= \Delta t \cdot \mathcal{F}(t_n + c_2 \Delta t, Q_n + a_{21} K_1 + a_{22} K_2)
\end{aligned} \tag{15.164}$$

where n now denotes a discrete time station. Given the solution Q_n at time t_n, K_1 is computed first, followed by K_2. The solution Q_{n+1} is then determined by way of the first expression in (15.164).

The terminal numerical solution is then determined using Newton's method, which for the implicit fully-coupled computation of the IRK2 arrays K_i, $1 \le i \le 2$, is cast as

$$\left[\mathcal{M} - a_{\underline{ii}}\Delta t \left(\frac{\partial \mathcal{F}}{\partial Q} \right)^\ell_{Q^\ell_i} \right] \left(K^{\ell+1}_i - K^\ell_i \right) = \Delta t\, \mathcal{F}\left(t_n + c_i \Delta t, Q^\ell_i \right) - \mathcal{M} K^\ell_i$$

$$Q_i^\ell \equiv Q_n + a_{i1} K_1^\ell + a_{i2} K_2^\ell \tag{15.165}$$

where $a_{ij} = 0$ for $j > i$, ℓ is the iteration index, and $K_1^\ell \equiv K_1$ for $i = 2$; the Jacobian

$$J_i(Q) \equiv \mathcal{M} - a_{\underline{ii}} \Delta t \left(\frac{\partial \mathcal{F}}{\partial Q}\right)^\ell_{Q_i^\ell} \tag{15.166}$$

has been analytically determined and implemented, leading to a block sparse matrix. For all the results documented in the next section, the initial estimate K_i^0 is set equal to the zero array, while a Gaussian elimination is used with only one iteration executed for (15.165) within each time interval. In this mode, Newton's iteration becomes akin to a classical direct linearized implicit solver.

15.11 Computational Performance

The characteristics-bias finite element algorithm for the direct computational solution of the incompressible-flow Navier-Stokes equations has generated essentially non-oscillatory solutions for not only velocity and temperature, but also pressure for four flows involving: a backward-facing step diffuser, a lid-driven square cavity, natural convection in a square cavity, and the combined effect of natural and forced convection in the same cavity. Each benchmark has employed a non-uniform finite element discretization of Lagrange bilinear elements, consisting of 40 bilinear elements in the transverse and longitudinal directions, for a total of 1600 elements, 1681 nodes and 6724 degrees of freedom.

For each benchmark, the calculations proceeded with a prescribed constant maximum Courant number,

$$C_{\max} \equiv \max\{\|\mathbf{u}\| + c, \|\mathbf{u}\| - c, c\} \frac{\Delta t}{(\Delta \ell)_e} \tag{15.167}$$

For prescribed $(\Delta \ell)_e$ and $C_{\max}$ for each benchmark, the corresponding Δt is determined as

$$\Delta t = \frac{C_{\max}(\Delta \ell)_e}{\max\{\|\mathbf{u}\| + c, \|\mathbf{u}\| - c, c\}} \tag{15.168}$$

Presented in non dimensional form, all the solutions in these validations have asymptotically converged to the preset machine zero of 5×10^{-12}. The streamline fields were developed by placing terminating strips orthogonal to families of streamlines and then connecting the corresponding ends of each closed streamlines. Due to its the characteristics-bias terms, the algorithm provides for local conservation of mass within its order of accuracy. For each of the four benchmarks, the algorithm in 1 Newton-iteration mode, generated a final steady state by using a gradually increasing Courant number from 2, at the beginning of each computational sequence, to 20 in the middle of the sequence, and up to 100 towards steady state. As expected, the maximum allowable Courant number decreases when ψ decreases; on the other hand, the maximum allowable Courant number need not decrease substantially, when more than 1 Newton iteration is employed per time step. The upstream controller ψ was set to $\psi = 0.5$, from beginning to convergence to an initial steady state, and then gradually reduced to $\psi \simeq 0.1$ to lessen the decrease in the effective Reynolds number. A current research project consists in further reducing induced upwind diffusion by decreasing ψ_{min} and ψ_{max} proportionately to the magnitude of velocity and the streamline deviatoric traction within a boundary layer. Each computational final steady state was achieved in about two thousand time steps, with slighter faster convergence achieved with the time derivative of pressure in the continuity equation. For each of the test cases in this section, the characteristics-bias calculations were performed twice: the first time with the pressure time derivative in the continuity equation and the second without this derivative. In both cases, the procedure succeeded for each computational benchmark in generating identical steady states that remain essentially non-oscillatory.

15.11.1 Backward-Facing Step Diffuser Flow

The flow in a diffuser featuring a backward-facing step has been extensively investigated, with reference experimental results available in [9]. In this benchmark, the axial length of the computation domain equals about 1.25 times the reattachment length, in order to test the performance and effect of the outlet pressure and traction boundary condition on the upstream flow. The calculations have determined not only the velocity and pressure fields, but also the temperature distribution within the flow, retaining the viscous-dissipation expression in the temperature equation. Figure 15.17 shows the simplified computational domain for this benchmark.

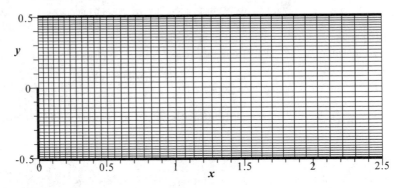

Figure 15.17: Non-uniform 41 × 41 Linear Element Grid

The inlet corresponds to $x = 0$ and $0 \leq y \leq 0.5$ and at this location Dirichlet boundary conditions are enforced on temperature and velocity components, whereas pressure is unconstrained and allowed to vary with the flow. These Dirichlet conditions involve a vanishing transverse velocity component and fully developed temperature and axial-velocity profiles, corresponding to the exact Poiseuille flow solution. For an effective $Re = 178$, the outlet corresponds to $x = 2.5$ and $-0.5 \leq y \leq 0.5$. At this outlet, an adiabatic stream is specified, hence vanishing heat transfer in the temperature equation, along with physically meaningful prescribed longitudinal and transverse surface tractions $\sigma_{1j}n_j$ and $\sigma_{2j}n_j$; since at this outlet $n_1 = 1$ and $n_2 = 0$, these conditions specify pressure $-\sigma_{11} = p = 1.0$ and shear stress $\tau_{21} = \tau_{12}$ from the Poiseuille distribution of u at the flow Reynolds number, conditions that allow the transverse velocity component to vary and lead to a minimally constrained converged solution. As detailed in Section 15.8, these physical conditions require no special boundary equation, but rather specify the surface integrals in the variational statement of the linear-momentum and temperature equations. At the solid walls at the top, bottom, and step site of the computational domain, hence $y = 0.5$, $y = -0.5$, and $-0.5 \leq y \leq 0$ at $x = 0$, the no slip boundary condition is enforced on velocity; the step site and lower wall are treated as adiabatic walls, whereas a cold-wall condition, hence fixed temperature, is specified at the top wall. Quite far from the final steady state, the initial conditions specified fully developed velocity and temperature profiles at the inlet, with temperature variation set to zero at the step, a constant pressure field, with $p = 1$, a vanishing transverse component of velocity $u_2 \equiv v = 0$, and fully developed velocity and temperature profiles for $x > 0$, profiles that satisfy global conservation of mass. These simple initial fields were sufficient to start the calculations and within 1 time step the characteristics-boas procedure for incompressible flows updated these fields to unsteady fields that also satisfy the discretized continuity equation, as follows. The procedure does not require a pressure time derivative in the continuity equation; the system of equations for the first time step was thus solved in Newton iterative mode, but with only five Newton iterations per IRK array K_i, $1 \leq i \leq 2$; at the end of the first time step, accordingly, a field emerges that satisfies the discretized continuity equation; form this field, the calculations efficiently progressed in one-Newton-iteration mode. Corresponding to an eventual $\psi = 0.1$, the final steady state is documented in Figures 15.18-15.23. The velocity vector distribution in Figure 15.18 reflects the expected pattern, with the recirculation zone that extends for 4/5 of the domain length. Toward the outlet, velocity points down, as allowed by the shear-stress boundary condition, while the overall velocity distribution remains uniform, with correct decrease of axial maximum velocity as the effective flow channel widens. These features are reflected in the streamline distribution in Figure 15.19, which clearly portrays the recirculation zone. Normalized by the step height, the reattachment length for this test equals to 4.25, which reflects the result in [9] for $Re = 178$, the effective Reynolds number in this benchmark.

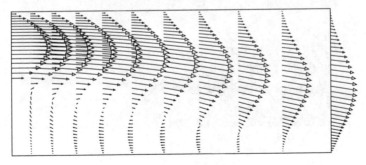

Figure 15.18: Vector Field

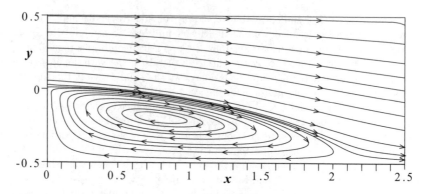

Figure 15.19: Streamline Distribution

These velocity-field features originate from the individual velocity components $u_1 \equiv u$ and $u_2 \equiv v$ displayed in Figures 15.20-15.23. Figure 15.20 portrays the development of an effective flow channel, as the maximum axial velocity progressively decreases, and Figure 15.21 shows that the distribution of u remains monotone, from inlet, within the recirculation zone, and toward the inlet, where the surface-integral outlet boundary conditions promoted an uniform distribution.

The distribution of v displays similar features. Figure 15.20 shows that this velocity component rapidly decreases within the effective flow channel, with minimum above and slightly upstream of the reattachment point. At the outlet, v is not forced to vanish, but is determined by the outlet shear-stress boundary condition, which can be more sharply satisfied by a finer grid in this region; within the flow field, the distribution of v remains monotone.

Figure 15.20: Axial Velocity Contours

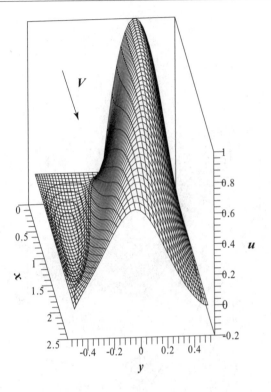

Figure 15.21: Axial Velocity Surface

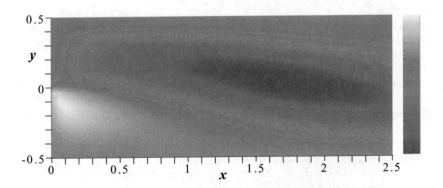

Figure 15.22: Transverse Velocity Contours

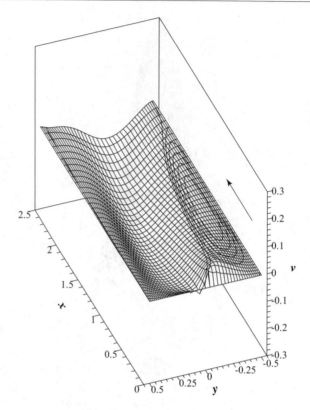

Figure 15.23: Transverse Velocity Surface

The inlet point with coordinates $(0; 0)$ becomes the center of a singular zone where the positive v from the separated flow clashes with the negative v of the downward main flow. The characteristics-bias procedure, nevertheless, negotiates this singularity and leads to an essentially non-oscillatory distribution of v, with a steep variation of this variable in the singular zone, which, as expected, induces a corresponding localized steep variation of pressure.

The distribution of pressure is presented in Figures 15.25-15.24. Directly computed from the coupled characteristics-bias linear-momentum and continuity equations, these distributions show the anticipated abrupt variation of pressure at the singular point, which remains contained within a 4-cell zone. In the remaining field, including the outlet region, pressure remains essentially non-oscillatory, These results support show the ability of the procedure to determine pressure directly and the viability of the surface-integral method to impose a minimally constraining pressure boundary condition.

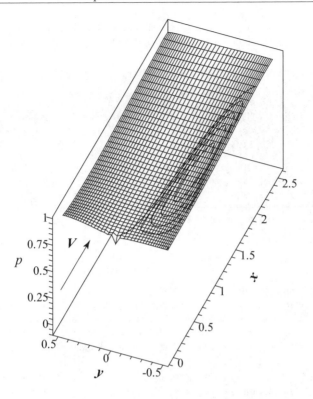

Figure 15.24: Pressure Surface

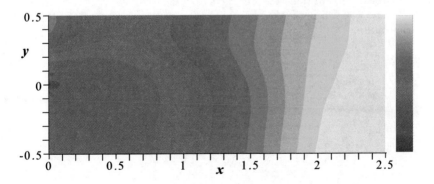

Figure 15.25: Pressure Contours

This study has also determined the temperature field, which is not usually detailed in backward-facing step flow analyses. The calculated distribution of temperature in Figures 15.26-15.27 reflects the flow and the imposed boundary conditions.

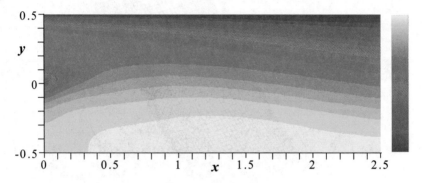

Figure 15.26: Temperature Contours

The inlet temperature profile rapidly merges with the overall temperature distribution, a distribution that becomes less and less dependent on x, downstream of the inlet. This feature echoes the Saint-Venant solid-mechanics principle that away from the ends of a slender bar the distribution of stresses in the bar becomes independent of the particular system of forces at the ends of the bar. The existence of such a feature in this distribution remains consistent with the elliptic nature of the steady-state temperature equation. This temperature distribution remains monotone even at the inlet singularity zone, where the variation from the maximum inlet temperature, down to the vanishing temperature variation at the $(0;0)$ point, and up to the temperatures in the separated flow region takes place without any unphysical oscillation. Downstream of the inlet, the temperature gradually increases in the downward transverse direction, from the cold wall at the top of the channel, to the adiabatic wall, where the thermal energy of convection and internal friction raises the local temperature.

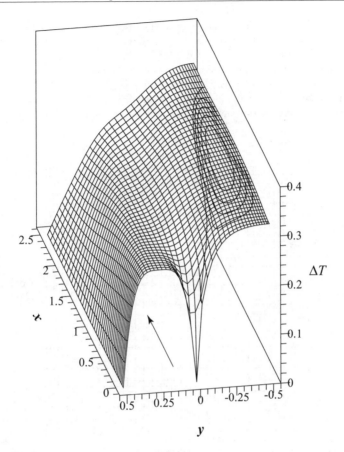

Figure 15.27: Temperature Surface

15.11.2 Natural Convection

The flow induced by natural convection in a square cavity has also been exhaustively investigated experimentally and numerically, with reference numerical results available in [52]. In this benchmark $Ra = 10,000$, $Ec = 1.0$, $Pr = 0.7$, $Re = 1/Pr$. The length of each side of the cavity equals 1 unit and the calculations have determined the velocity, pressure, and temperature fields. The temperature distribution was determined with and without using the viscous dissipation function in the temperature equation and the corresponding temperature fields were found virtually identical, as expected for the selected magnitude of Ra.

Figure 15.28 shows the computational domain for this benchmark. Each side corresponds to a solid wall and the no-slip boundary condition is enforced on the velocity components on every side of the cavity. Dirichlet boundary conditions are imposed on the vertical walls, with $\Delta T = 1$ at $x = 0$ and $\Delta T = 0$ at $x = 1$; at the top and bottom walls a vanishing heat flux is specified. As for pressure, one numerical value for this variable is specified at the $(0; 0)$ corner, with $p = 1$ as a Dirichlet boundary condition replacing the nodal continuity equation at this corner. The initial conditions satisfy the continuity equation as they specified vanishing axial and transverse velocity components, a constant pressure field with $p = 1$ and a linearly varying temperature distribution between the imposed temperature boundary conditions.

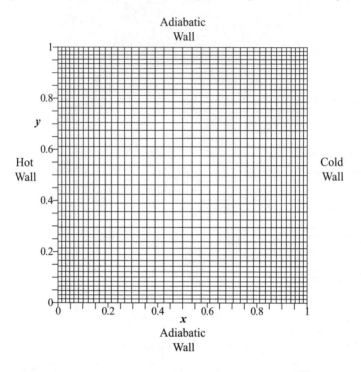

Figure 15.28: Non-uniform 41 × 41 Linear Element Grid

Obtained with only one Newton iteration per time-step cycle and corresponding to an eventual $\psi = 0.15$, the final steady state is documented in Figures 15.29-15.34. The velocity vector distribution in Figure 15.29 correspond to the expected pattern; the streamline distribution in Figure 15.30 reflects the results in [52]. The maximum axial and transverse velocity components reported in this reference are $|u|_{\max} = 16.178$ and $|v|_{\max} = 19.617$, as obtained by way of a $6,561$-node grid. By favorable comparison, the corresponding maxima determined in this study are $|u|_{\max} = 15.922$ and $|v|_{\max} = 18.748$, as generated by way of a $1,681$-node grid.

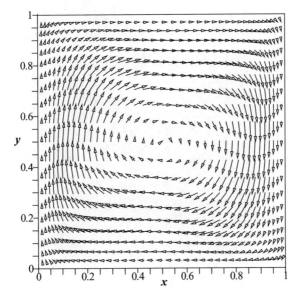

Figure 15.29: Velocity Field

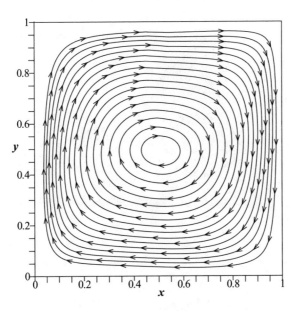

Figure 15.30: Streamline Distribution

The computed distribution of the velocity component $u_1 \equiv u$ is presented in Figures 15.31-15.32. This distribution remains monotone throughout the computational domain.

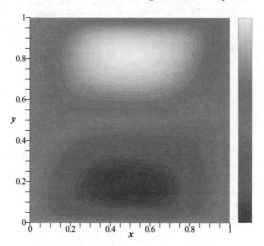

Figure 15.31: Axial Velocity Contours

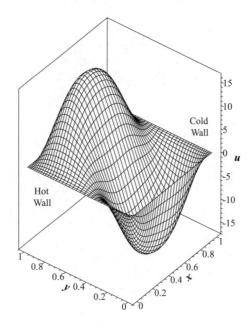

Figure 15.32: Axial Velocity Surface

Figures 15.33-15.34 present the distribution of the transverse velocity v. Also this distribution remains monotone in the entire flow field.

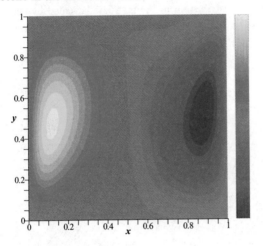

Figure 15.33: Transverse Velocity Contours

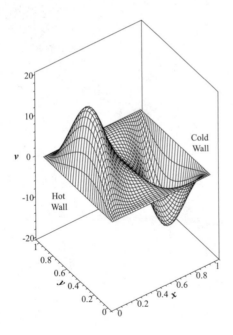

Figure 15.34: Transverse Velocity Surface

The distribution of pressure is presented in Figures 15.36-15.35. Since the buoyancy source term in the transverse momentum equation with $Ra = 10,000$ induces a substantial increase in pressure, the distribution in Figure 15.35 is scaled by the maximum pressure in the field. As shown in the figure, this distribution remains monotone.

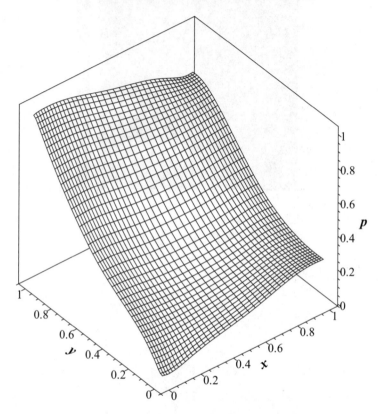

Figure 15.35: Pressure Surface

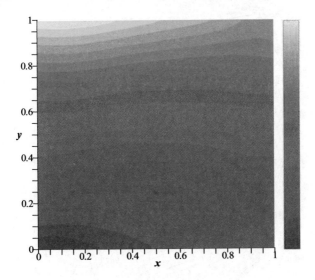

Figure 15.36: Pressure Contours

The calculated distribution of temperature in Figures 15.37-15.38 corresponds to the flow and the imposed boundary conditions. This distribution remains monotone and reflects the results in [52].

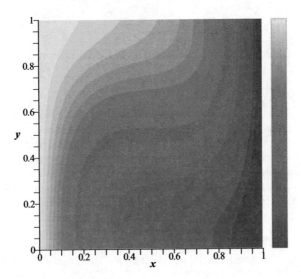

Figure 15.37: Temperature Contours

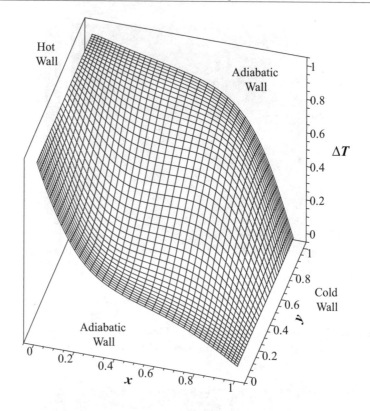

Figure 15.38: Non-uniform 41 × 41 Linear Element Grid

15.11.3 Lid-Driven Cavity Flow

The next test involves lid-driven flow in a square cavity, another well known incompressible flow. Also in this case has the temperature field been determined, using the dissipation function in the temperature equation. Figure 15.39 shows the computational domain for this benchmark and each side corresponds to a solid wall. The transverse velocity component v is set to zero on every side of the cavity; the axial velocity component u is set to zero on the $x = 0$, $x = 1$, and $y = 0$ walls; at the $y = 1$ wall, the boundary condition on this velocity component is $u = 1$. With reference to temperature, Dirichlet boundary conditions are imposed on the vertical walls, with $\Delta T = 0$ at $x = 0$ and $\Delta T = 1$ at $x = 1$; at the top and bottom walls a vanishing heat flux is specified. As for pressure, one numerical value for this variable is specified at the $(0; 0)$ corner, with $p = 1$ as a Dirichlet boundary condition replacing the nodal continuity equation at this corner. The initial conditions involve a constant pressure field with $p = 1$ and a linearly varying temperature distribution between the imposed temperature boundary conditions. The initial conditions for velocity are a vanishing transverse velocity component and an axial velocity component that only vanishes for $y < 1$. With these initial and boundary conditions on velocity, u only depends upon y and as a consequence the initial velocity identically satisfies the continuity equation.

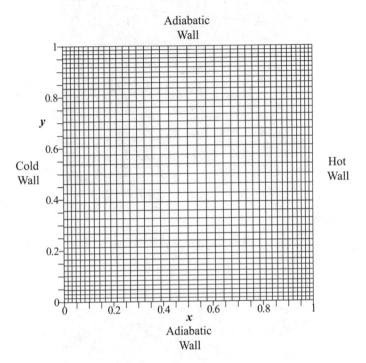

Figure 15.39: Non-uniform 41×41 Linear Element Grid

Corresponding to an effective Reynolds number $Re = 130$ and an eventual $\psi = 0.2$, the final steady state is documented in Figures 15.40-15.49. Figure 15.40 illustrates the known properties of the velocity vector distribution for this flow. The center of the main recirculation zone is asymmetrically located towards the top and right walls of the cavity, while a minute separated-flow region emerges at each bottom corner of the cavity. In the vicinity of the top, left corner, velocity actually points towards this corner, before abruptly turning to follow the lid motion, a well documented phenomenon. These features are reflected in the streamline distribution in Figure 15.41, which shows the main recirculation zone hovering above the secondary corner separated-flow zones. Shown in Figures 15.42-15.45, the corresponding distributions of velocity components u and v, remain monotone, even in the vicinity of the upper corners, which correspond to two singularities.

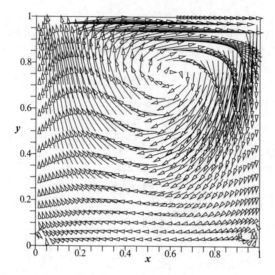

Figure 15.40: Velocity Field

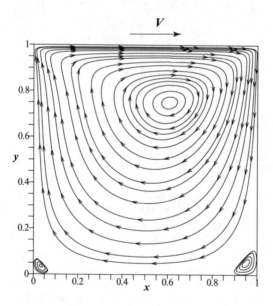

Figure 15.41: Streamline Distribution

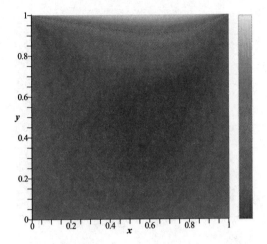

Figure 15.42: Axial Velocity Contours

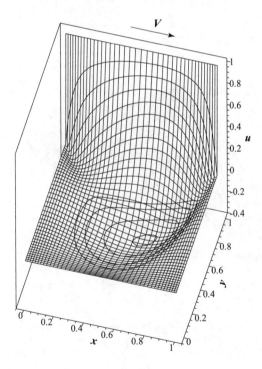

Figure 15.43: Axial Velocity Surface

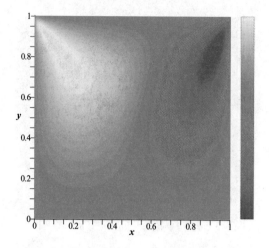

Figure 15.44: Transverse Velocity Contours

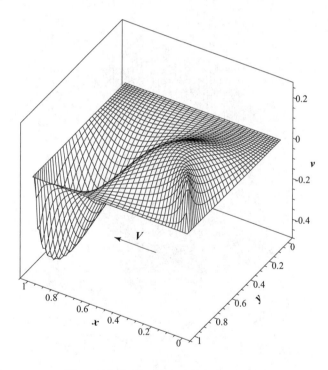

Figure 15.45: Transverse Velocity Surface

Figures 15.46-15.47 present the associated distribution of pressure, which remains smooth, even at the corner singularities. At these two locations, pressure experiences well known abrupt variations, akin to local spikes. These variations are sharply negotiated without any spurious oscillation or distortion. Also these results, accordingly, support the ability of the procedure to determine pressure directly from the coupled continuity and linear-momentum equations.

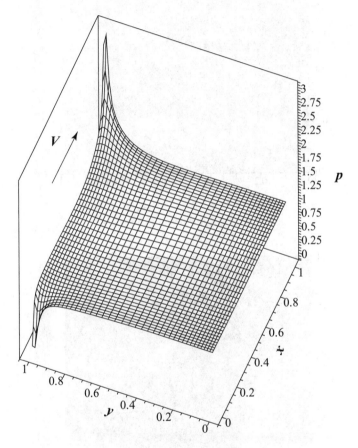

Figure 15.46: Pressure Surface

The calculated distribution of temperature displayed in Figures 15.48-15.49 reflects the flow and the imposed boundary conditions. Also this distribution remains monotone, even at the corner singularity nodes. The maximum temperature occurs at the top right corner where the intense flow shearing increases the local temperature, by means of the dissipation function in the temperature equation, consistently resolved in the finite element discretization.

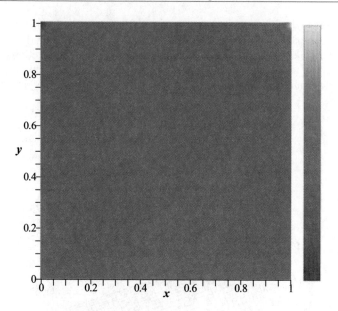

Figure 15.47: Pressure Contours

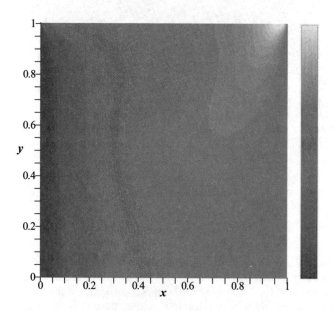

Figure 15.48: Temperature Contours

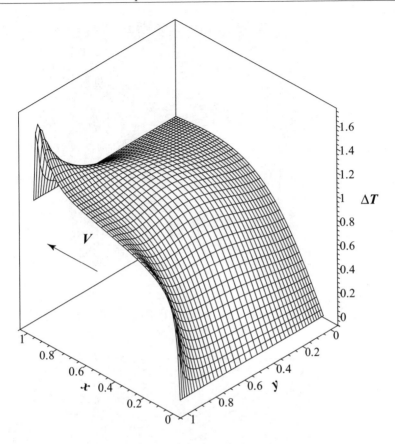

Figure 15.49: Temperature Surface

15.11.4 Lid- and Buoyancy-Driven Cavity Flow

The fourth test case in this chapter investigates a cavity flow induced by both buoyancy and a moving lid. The resulting flow combines several of the features of the flows discussed in the previous two sections. The location of hot and cold wall is selected so as to induce a recirculation zone that opposes the circulatory flow prompted by the sliding lid and for a sufficiently high Rayleigh number Ra two sizable counter-rotating circulatory flows may emerge. This is the kind of flow discussed in this section as obtained with $\psi = 0.2$, $Re = 200$, and $Ra = 10^6$ and corresponding to the same grid, boundary, and initial conditions detailed in the previous section. Figures 15.50-15.59 present the steady-state distributions of velocity, pressure and temperature.

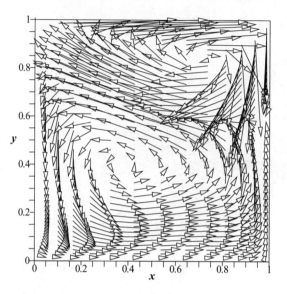

Figure 15.50: Velocity Field

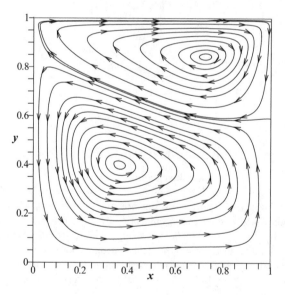

Figure 15.51: Streamline Distribution

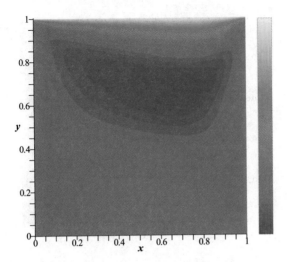

Figure 15.52: Axial Velocity Contours

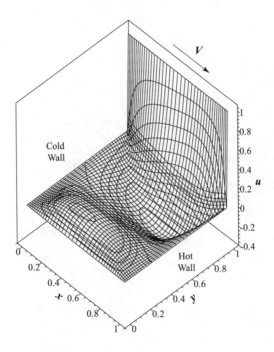

Figure 15.53: Axial Velocity Surface

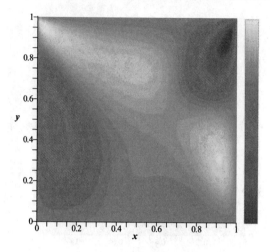

Figure 15.54: Transverse Velocity Contours

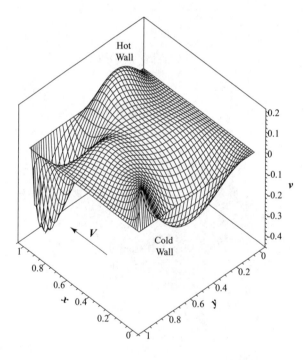

Figure 15.55: Transverse Velocity Surface

The velocity vector distribution in Figure 15.50 shows the anticipated and expected two counter-rotating recirculation zones, which appear of comparable strength, with the buoyancy-induced zone over half of the computational domain. Along the streamline separating these two zones, velocity points upwards, toward the cold wall, before the flow separates in the two zones; in the vicinity of the upper left corner, velocity points toward this corner before the abrupt turn dictated by the sliding lid. These features are reflected in the streamline distribution in Figure 15.51, which shows the extent of the buoyancy circulatory flow, which has pushed the lid-driven recirculation zone upwards and toward the top right corner. Figures 15.52-15.53 graph the corresponding distributions of velocity components u and v. As in the previously discussed flows, these distributions remain monotone, even in the vicinity of the upper corners, which correspond to two singularities.

Figures 15.56-15.57 present the associated distribution of pressure, which again remains smooth, even at the corner singularities. At these two locations, as in the previous lid-driven flow, pressure experiences some rather abrupt variations, akin to local spikes. Once again, these variations are sharply negotiated without any spurious oscillation or distortion.

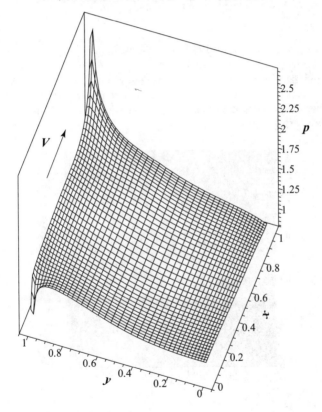

Figure 15.56: Pressure Surface

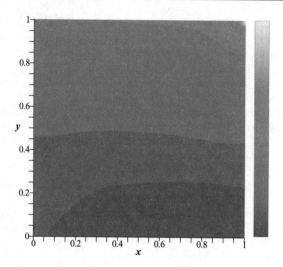

Figure 15.57: Pressure Contours

The calculated distribution of temperature is shown in Figures 15.58-15.59 As in the lid-driven flow, this distribution remains monotone, even at the corner singularity nodes. The maximum temperature occurs at the top right corner where again the intense flow shearing increases the local temperature.

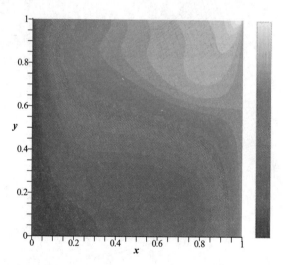

Figure 15.58: Temperature Contours

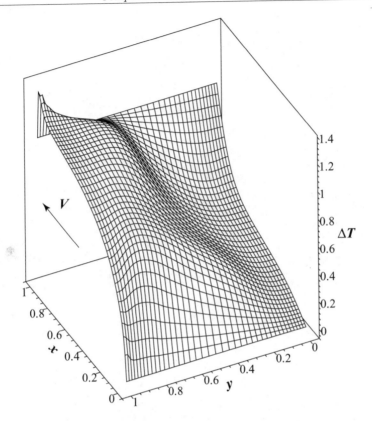

Figure 15.59: Temperature Surface

Chapter 16

Multi-Dimensional Free-Surface Flows

This chapter details the gravity-wave-convection specialization of the characteristics-bias system for investigating depth-averaged free-surface flows. Within a two-dimensional depth-averaged free-surface flow field, gravitational and convection waves propagate in infinitely many directions and along each direction the associated propagation velocity also depends on the Froude number. After stating the governing equations, the chapter describes an intrinsically multi-dimensional characteristics analysis based on a non-linear wave-like form of the solutions of the free-surface equations. This analysis leads to the spatial distribution of multi-dimensional propagation velocities, investigates the correlation between the time axis and the characteristic surfaces, and shows that among all propagation directions the streamline and crossflow directions are principal propagation directions in the time-space continuum.

This line of inquiry yields specific conditions for a physically coherent upstream bias formulation that remains consistent with multi-dimensional gravitational and convection wave propagation. The corresponding gravity-wave-convection flux Jacobian decomposition consists of components that genuinely model the physics of gravity-wave propagation and convection and also feature eigenvalues with uniform algebraic sign. This decomposition leads to a non-discrete upstream bias term that couples the governing equations within the characteristics-bias system. A variational formulation of this system provides several surface integrals as efficient venues for enforcing physically meaningful boundary conditions, such as the wall-tangency and outlet-pressure conditions. A Galerkin finite element spatial discretization and implicit Runge-Kutta time integration of this system generate the computational procedure. The performance of this procedure is assessed for five flows involving: an asymmetric dam-break, a backward-facing step channel, a stream-driven inlet, and two oblique hydraulic jumps. For these flows, the procedure has computed essentially non-oscillatory solutions for not only the free-surface height and velocity, but also temperature.

16.1   Governing Equations

From the developments in Chapter 5 and system (5.185), the non-dimensional depth-averaged free-surface equations can be cast as

$$
\begin{cases}
\dfrac{\partial h}{\partial t} + \dfrac{\partial m_1}{\partial x_1} + \dfrac{\partial m_2}{\partial x_2} = 0 \\[2mm]
\dfrac{\partial m_1}{\partial t} + \dfrac{\partial}{\partial x_1}\left(\dfrac{m_1}{h}m_1 + \dfrac{h^2}{2}\right) + \dfrac{\partial}{\partial x_2}\left(\dfrac{m_2}{h}m_1\right) = \phi_{m_1} + \dfrac{1}{Re}\dfrac{\partial(h\tau_{1j})}{\partial x_j} \\[2mm]
\dfrac{\partial m_2}{\partial t} + \dfrac{\partial}{\partial x_1}\left(\dfrac{m_1}{h}m_2\right) + \dfrac{\partial}{\partial x_2}\left(\dfrac{m_2}{h}m_2 + \dfrac{h^2}{2}\right) = \phi_{m_2} + \dfrac{1}{Re}\dfrac{\partial(h\tau_{2j})}{\partial x_j} \\[2mm]
\dfrac{\partial E}{\partial t} + \dfrac{\partial}{\partial x_1}\left(\dfrac{m_1}{h}\left(E + \dfrac{h^2}{2}\right)\right) + \dfrac{\partial}{\partial x_2}\left(\dfrac{m_2}{h}\left(E + \dfrac{h^2}{2}\right)\right) = \\[2mm]
\qquad\qquad \phi_E + \dfrac{1}{Re}\dfrac{\partial(m_i\tau_{ij})}{\partial x_j} + \dfrac{1}{RePr}\dfrac{\partial}{\partial x_j}\left(h\dfrac{\partial T}{\partial x_j}\right)
\end{cases}
\tag{16.1}
$$

where the array $\phi \equiv \{0, \phi_{m_1}, \phi_{m_2}, \phi_E\}$ is obtained by inspection of the source terms in (5.185). These equations are abridged as the system

$$
\frac{\partial q}{\partial t} + \frac{\partial f_j(q)}{\partial x_j} - \frac{\partial f_j^\nu}{\partial x_j} = \phi
\tag{16.2}
$$

Following the procedure in Chapter 14 to establish the characteristics-bias inviscid flux-divergence, the following developments are based on the inviscid-flow form of this system.

16.2   Gravity-Wave Equations

Let (v_1, v_2) denote the components of a unit vector v locally parallel to the velocity u. Upon writing the velocity components (u_1, u_2) in terms of the Froude number Fr as

$$
(u_1, u_2) = cFr(v_1, v_2)
\tag{16.3}
$$

the free-surface system becomes

$$
\frac{\partial}{\partial t}\begin{pmatrix} h \\ m_1 \\ m_2 \\ E \end{pmatrix} + \left[\begin{pmatrix} 0 & , & \delta_1^j & , & \delta_2^j & , & 0 \\ c^2\delta_1^j & , & 0 & , & 0 & , & 0 \\ c^2\delta_2^j & , & 0 & , & 0 & , & 0 \\ 0 & , & \frac{\delta_1^j}{h}\left(E + \frac{h^2}{2}\right) & , & \frac{\delta_2^j}{h}\left(E + \frac{h^2}{2}\right) & , & 0 \end{pmatrix} + cFrC_j\right]\frac{\partial}{\partial x_j}\begin{pmatrix} h \\ m_1 \\ m_2 \\ E \end{pmatrix}
\tag{16.4}
$$

where δ_i^j and C_j respectively denote Kronecher's delta and the matrix

$$
C_j \equiv \begin{pmatrix} 0 & , & 0 & , & 0 & , & 0 \\ -v_1 u_j & , & v_1\delta_1^j + v_j & , & v_1\delta_2^j & , & 0 \\ -v_2 u_j & , & v_2\delta_1^j & , & v_2\delta_2^j + v_j & , & 0 \\ -\frac{v_j}{h}\left(E + \frac{h^2}{2}\right) + hv_j & , & 0 & , & 0 & , & v_j \end{pmatrix}
\tag{16.5}
$$

For a vanishing Fr these equations then reduce to the gravity-wave equations

$$\frac{\partial}{\partial t}\begin{pmatrix} h \\ m_1 \\ m_2 \\ E \end{pmatrix} + \begin{pmatrix} 0 & , & \delta_1^j & , & \delta_2^j & , & 0 \\ c^2\delta_1^j & , & 0 & , & 0 & , & 0 \\ c^2\delta_2^j & , & 0 & , & 0 & , & 0 \\ 0 & , & \frac{\delta_1^j}{h}\left(E+\frac{h^2}{2}\right) & , & \frac{\delta_2^j}{h}\left(E+\frac{h^2}{2}\right) & , & 0 \end{pmatrix}\frac{\partial}{\partial x_j}\begin{pmatrix} h \\ m_1 \\ m_2 \\ E \end{pmatrix} = 0 \quad (16.6)$$

for which A_j^{gw} will denote the gravity-wave matrix multiplying the gradient of q in (16.6). Heed, in particular, that in this case the energy equation toward steady state is no longer linearly independent from the continuity equation. The dependent variables in these equations correspond to those in a flow field that originates from slight perturbations to an otherwise quiescent field.

Following the characteristics analysis in Chapter 14, the characteristic speeds λ associated with system (16.6) correspond to the eigenvalues of the matrix $A_j^{gw}n_j$, which are

$$\lambda_{1,2}^{gw} = 0, \quad \lambda_{3,4}^{gw} = \pm c \quad (16.7)$$

The eigenvalues $\lambda_{3,4}^{gw}$ in (16.7) correspond to the gravity-wave celerity. These eigenvalues remain independent from the wave-propagation direction unit vector $\boldsymbol{n}$, which corresponds to isotropic propagation.

16.2.1 Streamline and Crossflow Components

For any two mutually perpendicular unit vectors $\boldsymbol{a} = (a_1, a_2)$ and $\boldsymbol{a}^N = (a_1^N, a_2^N)$, along with implied summation on repeated indices, the gravity-wave matrix product $A_j^{gw}\frac{\partial q}{\partial x_j}$ can be expressed as

$$A_j^{gw}\frac{\partial q}{\partial x_j} = A_j^{gw}a_j a_k\frac{\partial q}{\partial x_k} + A_j^{gw}a_j^N a_k^N\frac{\partial q}{\partial x_k} \quad (16.8)$$

For any unit vector $\boldsymbol{a}$, the matrix $A_j^{gw}a_j$ is expressed as

$$A_j^{gw}a_j \equiv \begin{pmatrix} 0 & , & a_1 & , & a_2 & , & 0 \\ c^2 a_1 & , & 0 & , & 0 & , & 0 \\ c^2 a_2 & , & 0 & , & 0 & , & 0 \\ 0 & , & \frac{a_1}{h}\left(E+\frac{h^2}{2}\right) & , & \frac{a_2}{h}\left(E+\frac{h^2}{2}\right) & , & 0 \end{pmatrix} \quad (16.9)$$

For $\boldsymbol{a}$ parallel to $\boldsymbol{u}$, expression (16.8) corresponds to a decomposition into streamline and crossflow gravity-wave components. For wave-like solutions (16.15), two eigenvalues of each component vanish; the remaining eigenvalues of these separate components have been determined as

$$\lambda_{3,4}^s = \pm c_s \equiv \pm ca_j n_j, \quad \lambda_{3,4}^N = \pm c_N \equiv \pm ca_j^N n_j \quad (16.10)$$

The two non-vanishing eigenvalues associated with the entire gravity-wave component at the lhs of (16.8), but as expressed as the rhs combination of streamline and crossflow components have then been determined as

$$\lambda_{3,4}^{gw} = c\left(\left(a_j n_j\right)^2 + \left(a_j^N n_j\right)^2\right)^{1/2}, \quad \left(a_j n_j\right)^2 + \left(a_j^N n_j\right)^2 = 1 \quad (16.11)$$

which shows that the square of the gravity-wave eigenvalues (16.7) equals the sum of the square of the streamline and crossflow gravity-wave eigenvalues (16.10), as illustrated in Figure 16.1.

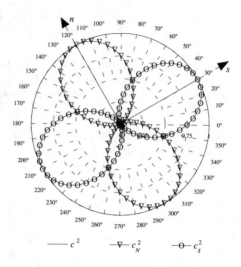

Figure 16.1: Polar Variation of Square of Gravity-Wave Celerities

16.2.2   Similarity Transformation

Despite its zero eigenvalues, $A_j^{gw} a_j$ features a complete set of eigenvectors X and thus possesses the similarity-transformation form

$$A_j^{gw} a_j = X \Lambda^{gw} X^{-1} \tag{16.12}$$

where the diagonal matrix Λ^{gw} is expressed as

$$\Lambda^{gw} \equiv \begin{pmatrix} c & , & 0 & , & 0 & , & 0 \\ 0 & , & -c & , & 0 & , & 0 \\ 0 & , & 0 & , & 0 & , & 0 \\ 0 & , & 0 & , & 0 & , & 0 \end{pmatrix}$$

The associated eigenvector matrix X and its inverse X^{-1} are

$$X = \begin{pmatrix} 1 & , & 1 & , & 0 & , & 0 \\ ca_1 & , & -ca_1 & , & 0 & , & a_2 \\ ca_2 & , & -ca_2 & , & 0 & , & -a_1 \\ \frac{1}{h}\left(E + \frac{h^2}{2}\right) & , & \frac{1}{h}\left(E + \frac{h^2}{2}\right) & , & 1 & , & 0 \end{pmatrix} \tag{16.13}$$

$$X^{-1} = \frac{1}{2c} \begin{pmatrix} c & , & a_1 & , & a_2 & , & 0 \\ c & , & -a_1 & , & -a_2 & , & 0 \\ -\frac{2c}{h}\left(E + \frac{h^2}{2}\right) & , & 0 & , & 0 & , & 2c \\ 0 & , & 2ca_2 & , & -2ca_1 & , & 0 \end{pmatrix} \tag{16.14}$$

16.3   Characteristic Analysis and Velocity Components

The solution of system (16.2) is cast in wave-like form as

$$q = q(\boldsymbol{x} \cdot \boldsymbol{n} - \lambda(q)) \tag{16.15}$$

The solution-dependent velocity component $\lambda = \lambda(q)$ in this solution is determined by requiring this solution to satisfy system (16.2). Chapter 14 shows that this condition yields the eigenvalue problem

$$\left(-\lambda(q)I + \frac{\partial f_j}{\partial q}n_j\right)\frac{\partial q}{\partial \eta_1} = 0 \tag{16.16}$$

For non-trivial solutions $\frac{\partial q}{\partial \eta_1}$, hence $q = q(\eta_1)$, the characteristic velocity components λ are thus the eigenvalues of the linear combination of flux vector Jacobians

$$\frac{\partial f_j(q)}{\partial q}n_j = \begin{pmatrix} 0 & , & n_1 & , & n_2 & , & 0 \\ -u_1 u_j n_j + ghn_1 & , & u_1 n_1 + u_j n_j & , & u_1 n_2 & , & 0 \\ -u_2 u_j n_j + ghn_2 & , & u_2 n_1 & , & u_2 n_2 + u_j n_j & , & 0 \\ -\frac{u_j n_j}{h}\left(E + \frac{h^2}{2}\right) + hu_j n_j & , & \frac{n_1}{h}\left(E + \frac{h^2}{2}\right) & , & \frac{n_2}{h}\left(E + \frac{h^2}{2}\right) & , & u_j n_j \end{pmatrix} \tag{16.17}$$

These eigenvalues have been exactly determined in closed form as

$$\lambda_{1,2}^{fs} = u_j n_j, \qquad \lambda_{3,4}^{fs} = u_j n_j \pm c \tag{16.18}$$

The non-dimensional form of (16.18) follows from division by c, which supplies the Froude-number dependent expressions

$$\lambda_{1,2}^{Fr} = v_j n_j Fr, \qquad \lambda_{3,4}^{Fr} = v_j n_j Fr \pm 1 \tag{16.19}$$

where v_1, and v_2 denote the components of a unit vector $\boldsymbol{v}$ in the velocity $\boldsymbol{u}$ direction.

As the inner product of the two unit vectors $\boldsymbol{n}$ and $\boldsymbol{v}$, the contraction $v_j n_j$ supplies the cosine of the angle $(\theta - \theta_v)$ between $\boldsymbol{n}$ and $\boldsymbol{v}$, where θ and θ_v respectively denote the angle between $\boldsymbol{n}$ and the x_1 axis and the angle between $\boldsymbol{v}$ and the x_1 axis in the flow plane. The eigenvalues (16.19) thus become

$$\lambda_{1,2}^{Fr} = \cos(\theta - \theta_v)Fr, \qquad \lambda_{3,4}^{Fr} = \cos(\theta - \theta_v)Fr \pm 1 \tag{16.20}$$

These expressions, in particular, imply that the free-surface eigenvalues achieve their extrema for $\theta = \theta_v$, hence when $\boldsymbol{n}$ points in the streamline direction, whereas for $\boldsymbol{n}$ pointing in the crossflow direction, hence $\theta = 90° + \theta_v$, these eigenvalues no longer depend upon Fr.

The convection eigenvalues $\lambda_{1,2}$ vanish when $\cos(\theta - \theta_v) = 0$, hence for $\boldsymbol{n}$ perpendicular to the streamline direction, or, equivalently, pointing in the crossflow direction. Since

$\| \cos (\theta - \theta_v) \| \leq 1$, the gravitational-convection eigenvalues $\lambda_{3,4}$ can only vanish for $Fr \geq 1$, hence for supercritical flows. For these flows, $\lambda_{3,4} = 0$ for

$$\mp \cos (\theta - \theta_v) = \pm \sin ((\theta - 90^o) - \theta_v) = \frac{1}{Fr} \tag{16.21}$$

hence for n perpendicular to the Froude lines, for $\pm((\theta - 90^o) - \theta_v)$ corresponds to the angle between a Froude line and the streamline, from the second expression in (16.21). This equation, thus defines the Froude lines. The lines that are perpendicular to the Froude lines are called "conjugate" lines.

The lines that are perpendicular to n for vanishing eigenvalues $\lambda_{1,2}$ and $\lambda_{3,4}$ thus respectively become the streamline and Froude lines. This result is not coincidental, for vanishing eigenvalues $\lambda_{1,4}$ correspond to wave-like supercritical solutions of the steady free-surface equations, for which the streamline and Froude lines are characteristic-wave propagation lines.

16.3.1   Polar Variation of Characteristic Speeds

Figure 16.2 presents the polar variation of the absolute values of eigenvalues (16.19) for subcritical, critical and supercritical Froude numbers, in a neighborhood of a flow field point P in the (x_1, x_2) plane. These variations are obtained for a variable unit vector $n \equiv (\cos \theta, \sin \theta)$ and fixed unit vector v, in this representative case inclined by $+30^o$ with respect to the x_1 axis.

A collective inspection of all these diagrams reveals three shared features for all Froude numbers. The maximum characteristic speeds occur in the velocity direction, i.e. along a streamline, as noted before. Secondly, all the characteristic speeds are symmetrically distributed about the streamline direction. Thirdly, the eigenvalue pairs ($\|\lambda_1\|, \|\lambda_2\|$) and ($\|\lambda_3\|, \|\lambda_4\|$) remain mirror skew-symmetric with respect to the crossflow direction, in the sense that the curves representative of $\|\lambda_2\|$ and $\|\lambda_4\|$ become the respective mirror images of the variations of $\|\lambda_1\|$ and $\|\lambda_3\|$ with reference to this direction. The streamline and crossflow directions become two fundamental wave-propagation axes.

For vanishing Froude numbers, the gravitational-convection propagation curves in the figure approach two circumferences. The distribution of propagation speeds in this case is isotropic, which corresponds to the direction-invariant propagation of gravity waves. As the Froude number increases from zero, the curves in the figure progressively become circular asymmetric, which corresponds to anisotropic wave propagation. For $Fr = 0.5$ this anisotropy is already evident and becomes more pronounced for higher Froude numbers. The non-dimensional characteristic speeds then approach 1 in the region about the crossflow direction, which corresponds to essentially gravitational propagation.

For all Froude numbers, the convection eigenvalues $\lambda_{1,2}$ change sign when the n-direction shifts from an upstream to a downstream axis with respect to u. For this reason, the associated curves cross the polar origin. Pure convective propagation, therefore, remains mono-axial, from upstream to downstream of P and the axis of this type of wave propagation is the streamline.

For subcritical Froude numbers the gravitational-convection eigenvalues λ_3 and λ_4 respectively remain positive and negative for all directions. For this reason the associated

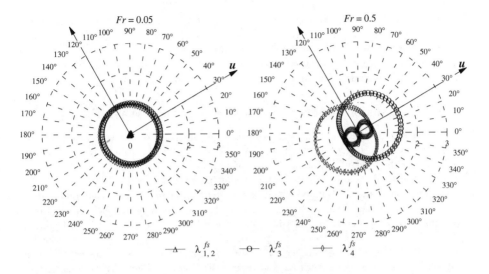

Figure 16.2: Polar Variation of Subcritical Wave Speeds

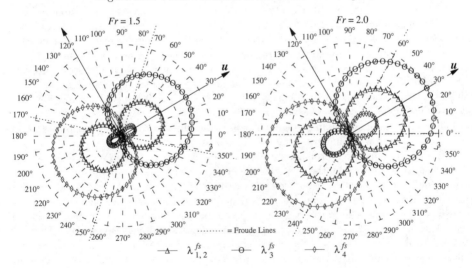

Figure 16.3: Polar Variation of Supercritical Wave Speeds

curves contain the polar origin. For subcritical flows, therefore, gravitational-convection waves propagate bi-modally, from both upstream and downstream toward and away from point P, along all directions radiating from P.

Beginning at the critical state, this pattern drastically changes for supercritical Froude numbers. In this case both λ_3 and λ_4 change algebraic sign when the $\boldsymbol{n}$ direction shifts from upstream to downstream of P along a streamline. For this reason, the associated curves cross the polar origin. For supercritical flows, therefore, gravitational-convection wave propagation becomes mono-axial along a streamline, from upstream to downstream of P.

16.3.2   Regions of Supercritical and Subcritical Wave Propagation

The streamline and crossflow directions are principal directions of free-surface gravity and convection wave propagation. These directions become the axes of distinct wave-propagation regions, named the streamline and crossflow wave propagation regions.

The supercritical mono-axial wave propagation pattern does not extend to the entire (x_1, x_2) plane, but remains confined within two wedge regions about the streamline. About the crossflow direction, there exist two other wedge regions within which gravitational-convection wave propagation remains bi-modal, from both upstream and downstream toward and away from point P, even for supercritical flows. These regions are determined by finding the lines where the free-surface eigenvalues vanish. These lines are the crossflow and conjugate lines, as shown for subcritical and supercritical flows in Figure 16.4, along with the domain of dependence and region of influence of point P and the bi-modal and mono-axial propagation regions on the (x_1, x_2) plane.

As Figure 16.4-a shows, subcritical gravitational-convection waves propagate bi-modally over the entire plane, hence the free-surface eigenvalues do not all have the same algebraic sign. For supercritical flows, Figure 16.4-b shows the streamline wedge regions of mono-axial wave propagation, where the free-surface eigenvalues all have the same algebraic sign. The conjugate lines mark the boundary with the crossflow regions of bi-modal wave propagation, where the free-surface eigenvalues do not all have the same algebraic sign, as in the subcritical-flow situation. In particular, $\lambda_{1,2} < 0$ and $\lambda_{1,2} > 0$ respectively upstream and downstream of the crossflow direction.

The conjugate lines remain perpendicular to the Froude lines while the mono-axial propagation regions grow as the Froude number increases.

For $Fr < 1$, no mono-axial propagation region exists. For $Fr = 1$, the Froude lines coincide with the crossflow direction, the conjugate lines coincide with the streamline direction, hence the mono-axial propagation region consists of the streamline only. As the Froude number increases, the Froude lines approach the streamline and the conjugate lines approach the crossflow direction. The mono-axial propagation regions thus grow as circular sectors of increasing angle, and the bi-modal propagation regions simultaneously shrink. Only for a theoretically infinite Fr will the mono-axial propagation region spread throughout the entire (x_1, x_2) plane. The presence of the cross-flow bi-modal-propagation regions even for supercritical flows is due to gravitation-wave propagation.

16.3.3   Free-Surface Flow Characteristic Cones

With reference to Section 14.2, the equations for $\mathcal{F}$ representing the characteristic surfaces in the time-space continuum arise by substitution of (14.45), (14.46) for n_j and λ in the

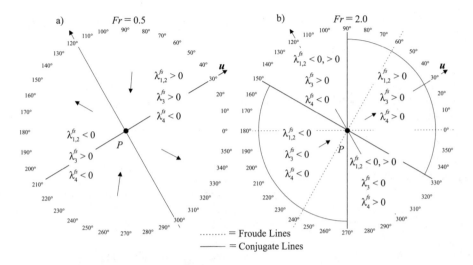

Figure 16.4: Wave Velocity Distribution: a) Subcritical Flows, b) Supercritical Flows

eigenvalues (16.19). This procedure yields the free-surface characteristic-surface equations

$$\left(\mathcal{F}_t + u_j \mathcal{F}_{x_j}\right)^2 = 0, \quad \left(\mathcal{F}_t + u_j \mathcal{F}_{x_j}\right)^2 = c^2 \mathcal{F}_{x_j} \mathcal{F}_{x_j} \tag{16.22}$$

where each subscript variable denotes differentiation with respect to that variable. In substantive-derivative form, the first of these two equations becomes

$$\left(\frac{D\mathcal{F}}{Dt}\right)^2 = 0 \tag{16.23}$$

which shows that the corresponding characteristic surface remains time-invariant from the perspective of an observer who travels along with a fluid particle.

The principal propagation directions as well as the flow domain of dependence and range of influence of a flow field point P in the time-space continuum are determined by studying the variation of the characteristic-surface shape versus the Froude number. As delineated in Section 14.3, in order to investigate these shape changes it suffices to study the shape changes of the corresponding characteristic cones, which are tangent at each flow field point P to the corresponding characteristic surfaces, but are far easier to determine.

By virtue of the Galilean transformations in Section 14.3, the characteristic-surface equations (16.22) simplify to

$$\mathcal{G} \equiv \left(\frac{\partial \mathcal{F}}{\partial \tau}\right)^2 = 0, \quad \mathcal{G} \equiv \left(\frac{\partial \mathcal{F}}{\partial \tau}\right)^2 - \left(\frac{\partial \mathcal{F}}{\partial X_1}\right)^2 - \left(\frac{\partial \mathcal{F}}{\partial X_2}\right)^2 = 0 \tag{16.24}$$

The equation of the tangent characteristic cone is then directly obtained from these equations through the Monge-cone ordinary differential equations presented in Section 14.3. The solution of these ordinary differential equations for each equation in (16.24) is

$$
\begin{cases}
X_1 - X_{1o} = 0 \\
X_2 - X_{2o} = 0
\end{cases}, \quad (X_1 - X_{1o})^2 + (X_2 - X_{2o})^2 = (\tau - \tau)^2
\tag{16.25}
$$

By virtue of the inverse Galilean transformation, the equation of the first characteristic cone with respect to the fixed reference coordinates (x_1, x_2, t) then becomes

$$
\frac{x_1 - x_{1o}}{t - t_o} = u_1, \quad \frac{x_2 - x_{2o}}{t - t_o} = u_2
\tag{16.26}
$$

In the time-space continuum, these equations correspond to the single "convection" straight line $S^- S^+$, as displayed in Figures 16.5, 16.7. As time elapses, therefore, waves reach and then leave P along this line, which, in the space continuum, projects onto the streamline

$$
\frac{x_1 - x_{1o}}{u_1} = \frac{x_2 - x_{2o}}{u_2}
\tag{16.27}
$$

which represents in the time-space continuum a plane that contains the convection line. A spatial discretization that aims to model this propagation mode, therefore, can receive an upstream bias in the streamline direction.

In the time-space continuum, the equation of the second characteristic cone becomes

$$
(x_1 - x_{1o} - u_1(t - t_o))^2 + (x_2 - x_{2o} - u_2(t - t_o))^2 = c^2 (t - t_o)^2
\tag{16.28}
$$

Hence

$$
(x_1 - x_{1o} - u_1(t - t_o))^2 + (x_2 - x_{2o} - u_2(t - t_o))^2 = \|\boldsymbol{u}\|^2 \frac{(t - t_o)^2}{Fr^2}
\tag{16.29}
$$

These equations corresponds to a characteristic cone with vertex at P that is tangent to the local characteristic surface at P and whose shape depends on Fr. As time elapses, therefore, waves funnel in towards P and out and away from P within the characteristic cone. In the space continuum, expression (16.29) represents the equation of a family of circumferences with center coordinates $[x_{1o} + u_1(t - t_o), \ x_{2o} + u_2(t - t_o)]$ and radius $\|\boldsymbol{u}\| (t - t_o)/Fr$.

For a vanishing c, this cone collapses onto the convection line (16.26), which, however, is not the axis of the characteristic cone. Figures 16.5, 16.7 portray this characteristic cone for subcritical and supercritical flows. Significantly, as Fr increases, the cone shears in the streamline direction, while its cross section remains an ellipse, of which the axes project onto the streamline and crossflow directions.

As a fundamental geometric difference between supercritical and subcritical flows, a time axis through P is respectively outside and inside the cone. In this manner, time- independent rays, i.e. lines issuing from the time axis for constant t, that are tangent to this cone only exist when the time axis is not inside the cone, that is for supercritical flows. The lines that issue from the time axis and are tangent to circumferences (16.29) in the (x_1, x_2) plane are then found to be the Froude lines. This finding is not coincidental, because time-invariant rays tangent to the characteristic cone correspond to steady-state (i.e. time-invariant) characteristic lines, which are the Froude lines for the steady free-surface equations.

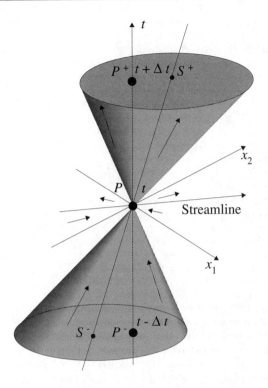

Figure 16.5: Subcritical Characteristic Cone

Figure 16.5 presents the subcritical characteristic cone. The time axis is inside the cone and no tangent time-invariant lines exist. As time elapses, in this case, the funneling of waves within the cone, towards and then away from P, corresponds on any constant-t plane to closed curves encircling P that shrink towards P and then expand away from P. Figure 16.6 illustrates this pattern, which corresponds to the familiar Doppler distribution. This distribution results from projecting onto the flow plane several sections of the characteristic cone at various time levels and corresponds to pressure pulses emitted by a particle that subcritically moves along a streamline arc. From this perspective, a region of space centered about P is simultaneously the domain of dependence and range of influence for P. A spatial discretization that aims to model this propagation mode, therefore, should receive un upstream bias along all directions radiating from P.

Figure 16.7 presents the supercritical characteristic cone. The time axis is outside the cone and the Froude lines are tangent to the cone on each constant-t plane. As time elapses, in this case, the funneling of waves within the cone, towards and then away from P, corresponds on any constant-t plane to information reaching P within its domain of dependence, between the two Froude lines upstream of P, and leaving P within its range of influence, between the two Froude lines downstream of P. Figure 16.8 illustrates this pattern, which corresponds

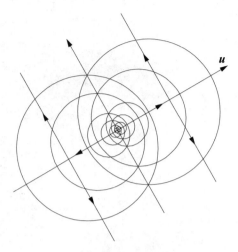

Figure 16.6: Subcritical Wave Propagation

to the familiar Doppler distribution. This distribution results from projecting onto the flow plane several sections of the characteristic cone at various time levels and corresponds to pressure pulses emitted by a particle that supercritically moves along a streamline arc. A spatial discretization that aims to model this propagation mode, therefore, should receive an upstream bias along all directions within the domain of dependence of P, with the streamline direction as the upstream-bias principal direction, since the streamline remains the axis of the domain of dependence and region of influence.

The investigation of the characteristic cone shape change versus Fr, therefore, indicates that for unsteady and steady supercritical flows, the upstream-bias domain of dependence and region of influence for each flow field point P consist of the regions that contain the streamline through P and are bound by the Froude lines. For unsteady and steady subcritical flows, both the domain of dependence and region of influence encompass regions encircling P.

16.3.4   Physical Multi-dimensional Upstream Bias

Since for all Froude numbers there simultaneously exist regions of mono-axial and bi-modal propagation, a physically consistent, intrinsically multi-dimensional, and infinite directional upstream formulation for the free-surface equations has to provide an upstream approximation suitable for supercritical flows within the mono-axial region, but consistent with subcritical wave propagation within the bi-modal region. This formulation has to involve a streamline upstream approximation that remains bi-modal for subcritical flows and then becomes mono-axial for supercritical flows. For all Froude numbers, a physical upstream approximation must then induce a bi-modal upstream bias along the crossflow direction.

This upstream approximation has to introduce an upstream bias along all directions

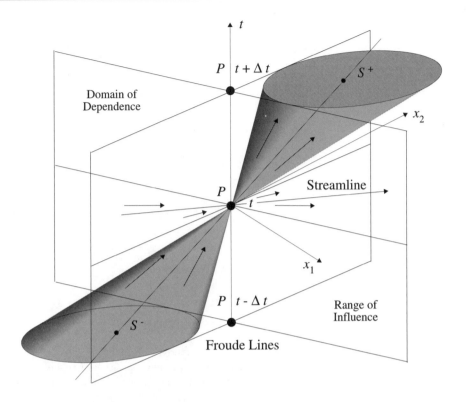

Figure 16.7: Supercritical Characteristic Cone

radiating from each flow field point. The bias in this approximation must change with varying direction and correlate with the directional distribution of the characteristic propagation speeds. Furthermore, the directional distribution of the upstream bias must remain symmetrical like the propagation speeds with respect to both the crossflow and streamline directions.

The formulation developed in the following sections identifies the genuine streamline and crossflow convection and gravitational components within the flux Jacobian matrices. The formulation then establishes a physically consistent upstream approximation for each of these components, along all wave-propagation directions, with an upstream-bias magnitude that correlates with the directional distribution of the characteristic propagation speeds. For increasing Froude numbers, from zero on, the approximation gradually changes from bi-modal in subcritical flows to mono-axial in supercritical flows, within a conical streamline region that encompasses the domain of dependence and range of influence of each flow-field point, while remaining bi-modal for all Froude numbers within an appropriate conical region about the crossflow direction.

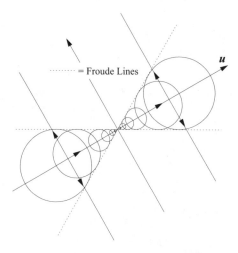

Figure 16.8: Supercritical Wave Propagation

16.4 Gravity-Wave-Convection Flux Divergence Decomposition

This section details a genuinely two-dimensional characteristics-bias formulation for the flux divergence $\partial f_j / \partial x_j$ in (16.2). This formulation rests on a decomposition of this divergence into convection, pressure-gradient, and streamline and crossflow gravity-wave components.

16.4.1 Convection and Pressure-Gradient Components

The flux divergence $\frac{\partial f_j}{\partial x_j}$ can be decomposed into convection and pressure-gradient components as

$$\frac{\partial f_j}{\partial x_j} = \frac{\partial f_j^q}{\partial x_j} + \frac{\partial f_j^p}{\partial x_j} \tag{16.30}$$

where f_j^q and f_j^p respectively denote the convection and pressure flux components, defined as

$$f_j^q(q) \equiv \left\{ \begin{array}{c} m_j \\[4pt] \dfrac{m_j}{h} m_1 \\[4pt] \dfrac{m_j}{h} m_2 \\[4pt] \dfrac{m_j}{h}\left(E + \dfrac{h^2}{2}\right) \end{array} \right\} = \frac{m_j}{h} q^H = \frac{m_j}{h} \cdot \left\{ \begin{array}{c} h \\[4pt] m_1 \\[4pt] m_2 \\[4pt] \left(E + \frac{h^2}{2}\right) \end{array} \right\}, \quad f_j^p(q) \equiv \left\{ \begin{array}{c} 0 \\[4pt] \dfrac{h^2}{2}\delta_1^j \\[4pt] \dfrac{h^2}{2}\delta_2^j \\[4pt] 0 \end{array} \right\} \tag{16.31}$$

For any Froude number, Section 16.3 has shown the non-dimensional eigenvalues of the Jacobian $\frac{\partial f_j}{\partial q} n_j$ are

$$\lambda_{1,2} = v_j n_j Fr, \qquad \lambda_{3,4} = v_j n_j Fr \pm 1 \qquad (16.32)$$

For supercritical flows, the free-surface eigenvalues (16.32), all have the same algebraic sign within the streamline wedge region and the entire flux divergence can be upstream approximated along the streamline principal direction, within this region. For subcritical flows these eigenvalues have mixed algebraic sign and an upstream approximation for the flux divergence along one single direction remains inconsistent with the two-way propagation of gravity waves. Without the pressure gradient in the momentum equations, however, the corresponding flux-Jacobian eigenvalues all have the same algebraic sign within the streamline wedge region and the resulting convection flux divergence can then be upstream approximated along one single direction. The flux divergence can thus be decomposed as the linear combination

$$\frac{\partial f_j}{\partial x_j} = \left[\frac{\partial f_j^q}{\partial x_j} + \beta \frac{\partial f_j^p}{\partial x_j}\right] + \left[(1-\beta)\frac{\partial f_j^p}{\partial x_j}\right], \quad 0 \le \beta \le 1 \qquad (16.33)$$

where the positive pressure-gradient partition function β can be chosen in such a way that all the eigenvalues associated with each of the two components between brackets in (16.33) keep the same algebraic sign within the streamline wedge region, for all Froude numbers. In this manner, these entire components can be upstream approximated along the streamline. The function β will gradually increase toward 1 for increasing Froude number, so that an upstream approximation for the components in (16.33) smoothly approaches and then becomes an upstream approximation for the entire $\frac{\partial f_j}{\partial x_j}$ along the streamline, which also ensures consistency with the corresponding one-dimensional formulation. Decomposition (16.33) is thus used for an upstream approximation of the flux divergence within the streamline wedge region, for subcritical and supercritical flows. This decomposition, however, is insufficient for an accurate multi-dimensional upstream modeling of gravity waves, for any Froude number range within the crossflow wedge region and for low and vanishing Froude numbers within the streamline wedge region.

Concerning the cross-flow wedge region, the eigenvalues associated with (16.33) become the free-surface eigenvalues (16.18) for $\beta = 1$, hence for supercritical flows. As indicated in Section 16.3, these eigenvalues do not all have the same algebraic sign within the cross-flow wave propagation region. A mono-axial upstream-bias approximation of (16.33) within the cross-flow wedge region, therefore, remains inconsistent with the existing two-way wave propagation in this region.

With regards to the streamline wave propagation region, for a Froude number that approaches zero, the free-surface eigenvalues (16.18) can all keep the same algebraic sign only if the gravity-wave celerity contribution vanishes, which corresponds to a vanishing pressure gradient contribution and hence β approaching zero. But for β approaching zero, the eigenvalues associated with the components in (16.33) approach the eigenvalues of the Jacobians

$$
\frac{\partial f_j^q(q)}{\partial q} n_j = \begin{pmatrix} 0 & , & n_1 & , & n_2 & , & 0 \\ -u_1 u_j n_j & , & u_1 n_1 + u_j n_j & , & u_1 n_2 & , & 0 \\ -u_2 u_j n_j & , & u_2 n_1 & , & u_2 n_2 + u_j n_j & , & 0 \\ -\frac{u_j n_j}{h}\left(E + \frac{h^2}{2}\right) + h u_j n_j & , & \frac{n_1}{h}\left(E + \frac{h^2}{2}\right) & , & \frac{n_2}{h}\left(E + \frac{h^2}{2}\right) & , & u_j n_j \end{pmatrix}
$$

(16.34)

and

$$
\frac{\partial f_j^p(q)}{\partial q} n_j = \begin{pmatrix} 0 & , & 0 & , & 0 & , & 0 \\ c^2 n_1 & , & 0 & , & 0 & , & 0 \\ c^2 n_2 & , & 0 & , & 0 & , & 0 \\ 0 & , & 0 & , & 0 & , & 0 \end{pmatrix}
$$

(16.35)

The eigenvalues of these Jacobians respectively are

$$
\lambda_{1,4}^q = u_j n_j
$$

(16.36)

and

$$
\lambda_{1,4}^p = 0
$$

(16.37)

which certainly all keep the same algebraic sign for any wave propagation direction, but for vanishing Froude number remain far less than the dominant gravity-wave celerity c. For low Froude numbers, therefore, an upstream approximation for the components in (16.33) would inaccurately model the physics of gravity-wave propagation. This difficulty is resolved by further decomposing the pressure gradient in (16.33) in terms of genuine streamline and crossflow gravity-wave component, for accurate upstream modeling of gravity waves throughout the flow field.

16.4.2 Gravity-Wave Components

For arbitrary Froude numbers and corresponding dependent variables h, m_1, m_2 and E, the free-surface flux Jacobians in (16.17) can be decomposed as

$$
\frac{\partial f_j}{\partial q} = \frac{\partial f_j^q}{\partial q} + \frac{\partial f_j^p}{\partial q} = \frac{\partial f_j^q}{\partial q} + A_j^{gw} + A_j^{nc}
$$

(16.38)

where, with reference to the gravity-wave equations (16.6) A_j^{gw}, and A_j^{nc} are defined as

$$
A_j^{gw} \equiv \begin{pmatrix} 0 & , & \delta_1^j & , & \delta_2^j & , & 0 \\ c^2 \delta_1^j & , & 0 & , & 0 & , & 0 \\ c^2 \delta_2^j & , & 0 & , & 0 & , & 0 \\ 0 & , & \frac{\delta_1^j}{h}\left(E + \frac{h^2}{2}\right) & , & \frac{\delta_2^j}{h}\left(E + \frac{h^2}{2}\right) & , & 0 \end{pmatrix}
$$

$$
A_j^{nc} \equiv \begin{pmatrix} 0 & , & -\delta_1^j & , & -\delta_2^j & , & 0 \\ 0 & , & 0 & , & 0 & , & 0 \\ 0 & , & 0 & , & 0 & , & 0 \\ 0 & , & -\frac{\delta_1^j}{h}\left(E + \frac{h^2}{2}\right) & , & -\frac{\delta_2^j}{h}\left(E + \frac{h^2}{2}\right) & , & 0 \end{pmatrix}
$$

(16.39)

Heed, in particular that no flux component of $f_j(q)$ exists, of which the Jacobian equals A_j^{gw}. The eigenvalues of the matrix $A_j^{nc} n_j$ have been determined in closed form as

$$\lambda_{1,4}^{nc} = 0 \tag{16.40}$$

which identically vanish for any Fr. The matrix A_j^{nc} can be termed a "non-linear coupling" matrix, for it completes the non-linear coupling between convection and gravity-wave within (16.17) so that the two free-surface eigenvalues $\lambda_{3,4}^{dfs}$ in (16.19) do correspond to the sum of convection velocity components and gravity-wave celerity. Since the matrix A_j^{gw} will be used in the upstream-bias formulation for small Froude numbers only and considering that the eigenvalues in (16.40) identically vanish no need exists to involve A_j^{nc} in the upstream-bias approximation for the flux Jacobian (16.17). The eigenvalues of $A_j^{gw} n_j$ have been exactly determined in closed form as

$$\lambda_{1,2}^{gw} = 0, \qquad \lambda_{3,4}^{gw} = \pm c \tag{16.41}$$

and remain independent of the propagation vector $\boldsymbol{n}$, which signifies isotropic propagation. The matrix A_j^{gw}, therefore, can be termed the "gravity-wave" matrix, for its eigenvalues coincide with the gravity-wave celerity c for any Froude number. This matrix, therefore, can be used for an upstream-bias approximation of the free-surface equations in the low Froude-number regime, within the streamline region, and for any Froude number, within the crossflow region.

For any two mutually perpendicular unit vectors $\boldsymbol{a} = (a_1, a_2)$ and $\boldsymbol{a}^N = (a_1^N, a_2^N)$, along with implied summation on repeated indices, the gravity-wave component within the free-surface flux divergence can be expressed as

$$A_j^{gw} \frac{\partial q}{\partial x_j} = A_j^{gw} a_j a_k \frac{\partial q}{\partial x_k} + A_j^{gw} a_j^N a_k^N \frac{\partial q}{\partial x_k} \tag{16.42}$$

With reference to Section 16.2.2, the matrix $A_j^{gw} a_j$ possesses the similarity-transformation form

$$A_j^{gw} a_j = X \Lambda^{gw} X^{-1} = X \Lambda^{gw+} X^{-1} + X \Lambda^{gw-} X^{-1}, \quad \Lambda^{gw} = \Lambda^{gw+} + \Lambda^{gw-} \tag{16.43}$$

The matrices $X \Lambda^{gw+} X^{-1}$ and $X \Lambda^{gw-} X^{-1}$ respectively correspond to the "forward" and "backward" gravity-wave-propagation matrix components of $A_j^{gw} a_j$. A bi-modal upstream-bias approximation of $A_j^{gw} a_j$, therefore, readily follows from instituting a forward and a backward upstream-bias approximation respectively for the forward- and backward- propagation matrices in (16.43). Results similar to (16.43) then readily follow by replacing $\boldsymbol{a}$ with $\boldsymbol{a}^N$. This bi-modal approximation directly yields both the non-negative-eigenvalue matrix statements

$$\left| A_j^{gw} a_j \right| a_k \frac{\partial q}{\partial x_k} \equiv X \left(\Lambda^{gw+} - \Lambda^{gw-} \right) X^{-1} a_k \frac{\partial q}{\partial x_k}$$

$$\left| A_j^{gw} a_j^N \right| a_k \frac{\partial q}{\partial x_k} \equiv X_N \left(\Lambda^{gw+} - \Lambda^{gw-} \right) X_N^{-1} a_k \frac{\partial q}{\partial x_k} \tag{16.44}$$

The matrices in (16.44) correspond to the streamline and crossflow absolute gravity-wave matrices. Different selection for Λ^{gw+} and Λ^{gw-} induce distinct levels of upstream-bias dissipation. With the forms

$$
\Lambda^{gw+} \equiv \frac{1}{2}
\begin{pmatrix}
2c & , & 0 & , & 0 & , & 0 \\
0 & , & 0 & , & 0 & , & 0 \\
0 & , & 0 & , & c & , & 0 \\
0 & , & 0 & , & 0 & , & c
\end{pmatrix}, \qquad
\Lambda^{gw-} \equiv -\frac{1}{2}
\begin{pmatrix}
0 & , & 0 & , & 0 & , & 0 \\
0 & , & 2c & , & 0 & , & 0 \\
0 & , & 0 & , & c & , & 0 \\
0 & , & 0 & , & 0 & , & c
\end{pmatrix}
\tag{16.45}
$$

the matrix product of $\left| A_j^{gw} a_j \right| = X \left(\Lambda^{gw+} - \Lambda^{gw-} \right) X^{-1}$ and the directional derivative $a_k \partial q / \partial x_k$ of q directly leads to the beautifully simple gravity-wave-field result

$$
\left| A_j^{gw} a_j \right| a_k \frac{\partial q}{\partial x_k} = X c I X^{-1} a_k \frac{\partial q}{\partial x_k} = c a_k \frac{\partial q}{\partial x_k}
\tag{16.46}
$$

The similar expression for (a_1^N, a_2^N) replacing (a_1, a_2) is

$$
\left| A_j^{gw} a_j^N \right| a_k^N \frac{\partial q}{\partial x_k} = X_N c I X_N^{-1} a_k^N \frac{\partial q}{\partial x_k} = c a_k^N \frac{\partial q}{\partial x_k}
\tag{16.47}
$$

where I denotes a 4×4 matrix. For a vanishing Fr, therefore, all the eigenvalues of the streamline and crossflow absolute gravity-wave matrices approach $+c$.

Based on the similarity transformation, the flux divergence can thus be expressed as

$$
\frac{\partial f_j}{\partial x_j} = \frac{\partial f_j^q}{\partial x_j} + A_j^{nc} \frac{\partial q}{\partial x_j}
$$

$$
+ \left(X \Lambda^{gw+} X^{-1} + X \Lambda^{gw-} X^{-1} \right) a_k \frac{\partial q}{\partial x_k} + \left(X_N \Lambda^{gw+} X_N^{-1} + X_N \Lambda^{gw-} X_N^{-1} \right) a_k^N \frac{\partial q}{\partial x_k}
\tag{16.48}
$$

in terms of forward and backward gravity-wave propagation components.

16.4.3   Combination of Flux Divergence Decompositions

The previous sections have shown that the free-surface flux divergence can be equivalently expressed as

$$
\frac{\partial f_j}{\partial x_j} =
\begin{cases}
\left[\dfrac{\partial f_j^q}{\partial x_j} + \beta \dfrac{\partial f_j^p}{\partial x_j} \right] + \left[(1 - \beta) \dfrac{\partial f_j^p}{\partial x_j} \right]; \\[2ex]
\dfrac{\partial f_j^q}{\partial x_j} + A_j^{nc} \dfrac{\partial q}{\partial x_j} \\[2ex]
+ \left(X \Lambda^{a+} X^{-1} + X \Lambda^{a-} X^{-1} \right) a_k \dfrac{\partial q}{\partial x_k} + \left(X_N \Lambda^{a+} X_N^{-1} + X_N \Lambda^{a-} X_N^{-1} \right) a_k^N \dfrac{\partial q}{\partial x_k}
\end{cases}
\tag{16.49}
$$

The first decomposition conveniently leads to a characteristics-bias approximation within the streamline wedge region, for high-subcritical and supercritical Froude numbers, that induces less diffusion than the second, for increasing Froude numbers, but does not represent gravity-wave propagation for low subcritical Froude numbers. The second decomposition leads to a characteristics-bias representation that consistently models gravity-wave propagation within the crossflow and streamline wedge regions, but induces more diffusion than the first for increasing Froude number. A gravity-wave-convection flux divergence decomposition for all Froude numbers for a characteristics-bias representation that not only induces minimal diffusion, but also models gravity-wave propagation is thus established as the following linear combination of these two decompositions, with linear combination parameter α and $0 \leq \alpha, \beta \leq 1$

$$
\frac{\partial f_j}{\partial x_j} = (1 - \alpha) \left\{ \left[\frac{\partial f_j^q}{\partial x_j} + \beta \frac{\partial f_j^p}{\partial x_j} \right] + \left[(1 - \beta) \frac{\partial f_j^p}{\partial x_j} \right] \right\} + \alpha \left\{ \frac{\partial f_j^q(q)}{\partial x_j} + A_j^{nc} \frac{\partial q}{\partial x_j} \right.
$$

$$
\left. + \left(X \Lambda^{a+} X^{-1} + X \Lambda^{a-} X^{-1} \right) a_k \frac{\partial q}{\partial x_k} + \left(X_N \Lambda^{a+} X_N^{-1} + X_N \Lambda^{a-} X_N^{-1} \right) a_k^N \frac{\partial q}{\partial x_k} \right\} \tag{16.50}
$$

Owing to the simplifying parameter $\delta \equiv \beta(1 - \alpha)$ and introducing the crossflow gravity-wave upstream parameter α^N, with $0 \leq \delta$, $\alpha^N \leq 1$, the final form of the gravity-wave-convection flux-divergence decomposition becomes

$$
\frac{\partial f_j(q)}{\partial x_j} = \alpha \left(X \Lambda^{a+} X^{-1} + X \Lambda^{a-} X^{-1} \right) a_k \frac{\partial q}{\partial x_k} + \alpha^N \left(X_N \Lambda^{a+} X_N^{-1} + X_N \Lambda^{a-} X_N^{-1} \right) a_k^N \frac{\partial q}{\partial x_k}
$$

$$
+ \left[\frac{\partial f_j^q}{\partial x_j} + \delta \frac{\partial f_j^p}{\partial x_j} \right] + (1 - \alpha - \delta) \frac{\partial f_j^p}{\partial x_j} + \alpha A_j^{nc} \frac{\partial q}{\partial x_j} + (\alpha - \alpha^N) A_j^a a_j^N a_k^N \frac{\partial q}{\partial x_k} \tag{16.51}
$$

A consistent upstream-bias representation results from this combination because of the physical significant of each term. The weights α and α^N respectively for the streamline and crossflow gravity-wave components in this expression are different from each other because the streamline and crossflow characteristic velocity components remain different from each other, following the system eigenvalues (16.19). For increasing Froude number, furthermore, the pressure gradient term $\delta \frac{\partial f_j^p}{\partial x_j}$ in this decomposition also contributes to the streamline gravity-wave upstream bias. These considerations indicate that the magnitudes of gravity-wave upstream bias for (16.51) along the streamline and crossflow directions will have to differ from each other for varying Fr, a "differential" upstream bias that can be instituted through the distinct weights α and α^N on the streamline and crossflow gravity-wave components.

Bi-modal wave propagation exists for all Froude numbers in the crossflow direction. Accordingly the crossflow gravity-wave term $\alpha^N \left(X_N \Lambda^{a+} X_N^{-1} + X_N \Lambda^{a-} X_N^{-1} \right) a_k^N \frac{\partial q}{\partial x_k}$ will receive a bi-modal upstream bias. The associated term $(\alpha - \alpha^N) A_j^a a_j^N a_k^N \frac{\partial q}{\partial x_k}$ need not receive any upstream-bias, for if it did it would obliterate all the crossflow gravity-wave upstream-bias, which remains fundamental for stability, as indicated by the multi-dimensional characteristics analysis. In respect of the term $\alpha A_j^{nc} \frac{\partial q}{\partial x_j}$, no upstream-bias is needed for this term either, for such an upstream would only add only algebraic complexity without any essential contribution to the upwinding process since all of the eigenvalues of this term vanish, as noted in Section 16.4.2 with reference to (16.40). The magnitude of crossflow upstream bias is controlled by the weight α^N, which does not have to vanish for supercritical flows, because of the presence of the finite bi-modal wave-propagation regions. Nevertheless, α^N can decrease for increasing Fr, because the bi-modal crossflow region shrinks for increasing Froude number.

The streamline gravity-wave term $\alpha\left(X\Lambda^{a+}X^{-1} + X\Lambda^{a-}X^{-1}\right)a_k\frac{\partial q}{\partial x_k}$ accounts for the bimodal streamline propagation of gravity waves; this term is thus employed for a gravity-wave upstream - bias approximation within the streamline wedge region, for low Froude numbers. Since this term already models pressure-induced gravity-wave propagation along each streamline, there is no need to develop an additional upstream bias for the $(1 - \alpha - \delta)\frac{\partial f_j^p}{\partial x_j}$, also considering that the coefficient $(1 - \alpha - \delta)$ of this term vanishes for gravity-wave and supercritical flows and all the associated eigenvalues are identically zero, as noted in Section 16.4.1 with reference to (16.37). As the Froude number rises, besides, an increasing fraction of the pressure gradient receives an upstream bias in the term $\left[\frac{\partial f_j^q}{\partial x_j} + \delta\frac{\partial f_j^p}{\partial x_j}\right]$. This expression is counted as one term because, with reference to (16.33), the eigenvalues associated with this term will all keep the same sign within the streamline wedge region, since $\delta = (1 - \alpha)\beta \leq \beta$.

Based on the two principles of minimal upstream dissipation and consistent infinite-directional upstream bias, the developments in the following sections show that for $Fr = 0$, the functions α and α^N equal 1, whereas the function δ equals 0; for increasing Fr, α rapidly decreases and vanishes, δ rapidly increases and then identically equals 1, for supercritical flows, and α^N monotonically decreases. Employing decomposition (16.51), the following sections establish a computationally efficient characteristics-bias Euler flux divergence.

16.5   Characteristics-Bias Flux-Divergence

With reference to (14.103), given the physical significance of the terms in decomposition (16.51) and algebraic signs of the corresponding eigenvalues, the associated principal direction unit vectors for these terms are

$$a_1 = -a_2 = a_5 = a, \quad a_3 = -a_4 = a^N, \quad a_6 = a_7 = a_8 = 0 \qquad (16.52)$$

At each flow-field point, a and a^N remain respectively parallel and perpendicular to the local velocity, with a^N obtained by a $90°$- degree counterclockwise rotation of a.

With (16.51) and (16.52), the general upstream-bias expression (14.103) directly yields the gravity-wave-convection characteristics flux divergence

$$\frac{\partial f_j^C}{\partial x_j} = \frac{\partial f_j}{\partial x_j} - \frac{\partial}{\partial x_i}\left[\varepsilon\psi\left(c\left(\alpha a_i a_j + \alpha^N a_i^N a_j^N\right)\frac{\partial q}{\partial x_j} + a_i\frac{\partial f_j^q}{\partial x_j} + a_i\delta\frac{\partial f_j^p}{\partial x_j}\right)\right] \qquad (16.53)$$

In this result, the expressions $\left(c\alpha a_i a_j\frac{\partial q}{\partial x_j} + a_i\frac{\partial f_j^q}{\partial x_j} + a_i\delta\frac{\partial f_j^p}{\partial x_j}\right)$, $\left(c\alpha^N a_i^N a_j^N\frac{\partial q}{\partial x_j}\right)$ determine the upstream biases within respectively the streamline and crossflow wave propagation regions. These two combined expressions then induce a correct upwind bias along all wave propagation regions. The operation count for expression (16.53) is then comparable to that of an FVS formulation. The terms in this expression, furthermore, directly correspond to the physics of gravity-wave propagation and convection. Expression (16.53) determines f_i^C itself, up to an arbitrary divergence-free vector, as

$$f_i^C = f_i(q) - \varepsilon\psi\left[c\left(\alpha a_i a_j + \alpha^N a_i^N a_j^N\right)\frac{\partial q}{\partial x_j} + a_i\frac{\partial f_j^q}{\partial x_j} + a_i\delta\frac{\partial f_j^p}{\partial x_j}\right] \qquad (16.54)$$

According to this result, the intrinsic multi-dimensionality of each component f_i^C derives from its dependence upon the entire divergence of f_j^q and f_j^p. The components in this flux remain linearly independent of one another, which avoids the linear-dependence instability in the steady low-Froude-number free-surface equations.

For vanishing Froude numbers, α and α^N will approach 1 whereas δ will approach 0. Under these conditions, (16.53) reduces to

$$\frac{\partial f_j^C}{\partial x_j} = \frac{\partial f_j}{\partial x_j} - \frac{\partial}{\partial x_i}\left[\varepsilon\psi\left(c\frac{\partial q}{\partial x_i} + a_i\frac{\partial f_j^q}{\partial x_j}\right)\right] \tag{16.55}$$

which essentially induces only a gravity-wave upstream bias. Heed that this bias becomes independent of specific propagation directions, for it no longer depends on $\left(\alpha a_i a_j + \alpha^N a_i^N a_j^N\right)$. This bias, therefore, becomes isotropic, in harmony with the isotropic propagation of gravity waves. For supercritical flows, $\alpha = 0$ and $\delta = 1$ and (16.53) thus becomes

$$\frac{\partial f_j^C}{\partial x_j} = \frac{\partial f_j}{\partial x_j} - \frac{\partial}{\partial x_i}\left[\varepsilon\psi\left(c\alpha^N a_i^N a_j^N \frac{\partial q}{\partial x_j} + a_i\frac{\partial f_j}{\partial x_j}\right)\right] \tag{16.56}$$

which depends on the crossflow component of the absolute gravity-wave matrix and the entire divergence of the free-surface inviscid flux vector.

16.5.1 Consistent Multi-Dimensional and Infinite Directional Upstream Bias

In Jacobian form, expression (16.53) becomes

$$\frac{\partial f_j^C}{\partial x_j} = \frac{\partial f_j}{\partial x_j} - \frac{\partial}{\partial x_i}\left[\varepsilon\psi\left(c\left(\alpha a_i a_j + \alpha^N a_i^N a_j^N\right)I + a_i\frac{\partial f_j^q}{\partial q} + a_i\delta\frac{\partial f_j^p}{\partial q}\right)\frac{\partial q}{\partial x_j}\right] \tag{16.57}$$

which essentially depends upon the five upstream-bias functions $a_1, a_2, \alpha, \delta, \alpha^N$. In order to ensure physical significance for the characteristics flux (16.54), hence for the upstream-bias approximation to decomposition (16.51), these functions are determined by imposing on (16.57) the stringent stability requirement that it should induce an upstream-bias diffusion not just along the principal streamline and crossflow upstream directions, but along all directions $\boldsymbol{n} = (n_1, n_2)$ radiating from any flow-field point. This condition corresponds to stability of the upstream-bias expression. By virtue of the developments in Sections 14.4,14.5.4, expression (16.57) will, therefore, be stable when all the eigenvalues of the upstream-bias matrix

$$\mathcal{A} \equiv n_i\left(c\left(\alpha a_i a_j + \alpha^N a_i^N a_j^N\right)I + a_i\frac{\partial f_j^q}{\partial q} + a_i\delta\frac{\partial f_j^p}{\partial q}\right)n_j \tag{16.58}$$

remain positive for all Fr and propagation directions $\boldsymbol{n}$.

16.5.2 Upstream-Bias Eigenvalues and Oblique-Jump Capturing

Despite the formidable non-linear algebraic complexity of $\mathcal{A}$, all of its eigenvalues have been analytically determined exactly in closed form. Dividing by the gravity-wave celerity c, the non-dimensional form of these eigenvalues is

$$\lambda_{1,2} = n_i \left(\alpha a_i a_j + \alpha^N a_i^N a_j^N \right) n_j + n_i a_i v_j n_j Fr$$

$$\lambda_{3,4} = n_i \left(\alpha a_i a_j + \alpha^N a_i^N a_j^N \right) n_j + n_i a_i v_j n_j Fr \pm n_i a_i \sqrt{\delta} \tag{16.59}$$

In these expressions, v_j denotes the jth direction cosine of a unit vector $\boldsymbol{v}$ parallel to the local velocity $\boldsymbol{u}$. Accordingly, the inner product $v_j n_j$ supplies the cosine of the angle between the velocity vector and the unit vector $\boldsymbol{n}$; when this unit vector is orthogonal to an oblique hydraulic-jump facet, these eigenvalues, as a result, depend on the "normal" Froude number $Fr_n = Fr v_j n_j$, which corresponds to the Froude number of the flow in the direction normal to the jump. The function δ from $\lambda_{3,4}$ is thus calculated on the basis of this normal Froude number, so that the resulting multi-dimensional upstream-bias formulation across a jump will sense the supercritical and subcritical flows in the normal direction, as in a one-dimensional flow.

All these eigenvalues must converge to 1 for vanishing Mach number, for an accurate modeling of gravity waves. Heed that for both $\boldsymbol{a} = \boldsymbol{v}$ and $\boldsymbol{n} = \boldsymbol{v}$, the functions α and δ within (16.59) determine the corresponding streamline upstream-bias eigenvalues already established in Chapter 12

$$\lambda_{1,2} = \alpha + Fr, \quad \lambda_{3,4} = \alpha + Fr \pm \sqrt{\delta} \tag{16.60}$$

Rather than prescribing some expressions for α and δ and accepting the resulting variations for these eigenvalues, again physically consistent expressions for the streamline upstream-bias eigenvalues are instead prescribed and the corresponding functions for α and δ determined. As shown in the following sections, all of these upstream-bias eigenvalues will remain positive when α and δ are calculated via the streamline-flow expressions, from Chapter 12, and α^N forces $\lambda_{3,4}$ to remain positive for all $\boldsymbol{n}$ and Froude numbers.

16.5.3   Conditions on Upstream-Bias Functions and Eigenvalues

The eigenvalues (16.59) are expressed as

$$\lambda_{1,4} = \lambda_{1,4} \left(Fr, \boldsymbol{n} \right) \tag{16.61}$$

to stress their dependence upon both Fr and $\boldsymbol{n}$. The five theoretical conditions for the determination of the five functions $a_1, a_2, \alpha, \delta, \alpha^N$ are

$$\lambda_{1,2} \left(Fr, \boldsymbol{n} \right) \geq 0, \quad \boldsymbol{a} \cdot \boldsymbol{a}^N = 0$$

$$\lambda_1 \left(Fr, \boldsymbol{v} \right) = \lambda_1, \quad \lambda_4 \left(Fr, \boldsymbol{v} \right) = \lambda_4, \quad \lambda_{3,4} \left(Fr, \boldsymbol{n} \right) \geq 0 \tag{16.62}$$

where λ_1 and λ_4 now denote prescribed streamline upstream-bias eigenvalues. The first condition stipulates that the unit vector $\boldsymbol{a}$ is parallel to the velocity vector and the second condition determines that $\boldsymbol{a}$ and $\boldsymbol{a}^N$ are mutually perpendicular vectors. In particular, these two conditions specify that the unit vectors $\boldsymbol{a}$ and $\boldsymbol{a}^N$ respectively point along the streamline and crossflow directions. The third and fourth conditions stipulate that the streamline upstream-bias eigenvalues must equal prescribed eigenvalues. The streamline eigenvalues λ_1 and λ_4 will therefore coincide with the eigenvalues λ_1 and λ_2 of the one-dimensional formulation. For the determined $\boldsymbol{a}$, $\boldsymbol{a}^N$, α and δ, the fifth condition then establishes α^N.

16.5.4 Upstream-Bias Functions a, α, and δ

These functions are used in actual computations based on the characteristics flux divergence (16.53). In the eigenvalues $\lambda_{1,2}$ in (16.59), the components

$$n_i \alpha a_i a_j n_j = \alpha \left(a_j n_j\right)^2, \quad n_i \alpha^N a_i^N a_j^N n_j = \alpha^N \left(a_j^N n_j\right)^2 \tag{16.63}$$

are already non-negative for non-negative α and α^N. The eigenvalues $\lambda_{1,2}$ will remain non-negative for all positive α and α^N, including $\alpha \to 0$ and $\alpha^N \to 0$, when the additional component $n_i a_i v_j n_j Fr$ remains non-negative for all Fr. This requirement is met when $\boldsymbol{a} = \boldsymbol{v}$, for

$$n_i a_i v_j n_j Fr = Fr \left(v_j n_j\right)^2 \geq 0 \tag{16.64}$$

This finding is not surprising, for the streamline direction is a principal characteristic direction. Based on $\boldsymbol{a} = \boldsymbol{v}$, the components of $\boldsymbol{a}^N$ are calculated as

$$a_1^N = -a_2, \quad a_2^N = a_1 \tag{16.65}$$

As noted, the expressions for α and δ coincide with the streamline-flow results based on the prescribed streamline eigenvalues λ_1 and λ_4. Restated here, for convenience, the expressions for these two eigenvalues are

$$\lambda_1(Fr) \equiv \begin{cases} 1 - Fr + \dfrac{\varepsilon_{Fr}}{2}(2Fr)^{1/\varepsilon_{Fr}} &, \quad 0 \ \leq\ Fr\ <\ \ \frac{1}{2} \\[2mm] \dfrac{(Fr - \frac{1}{2})^2}{2\varepsilon_{Fr}} + \dfrac{1 + \varepsilon_{Fr}}{2} &, \quad \frac{1}{2} \ \leq\ Fr\ <\ \frac{1}{2} + \varepsilon_{Fr} \\[2mm] Fr &, \quad \frac{1}{2} + \varepsilon_{Fr} \ \leq\ Fr \end{cases} \tag{16.66}$$

$$\lambda_4(Fr) \equiv \begin{cases} 1 - Fr &, \quad 0 \ \leq\ Fr\ \leq\ 1 - \varepsilon_{Fr} \\[2mm] \dfrac{(Fr - 1)^2}{2\varepsilon_{Fr}} + \dfrac{\varepsilon_{Fr}}{2} &, \quad 1 - \varepsilon_{Fr}\ <\ Fr\ <\ 1 + \varepsilon_{Fr} \\[2mm] Fr - 1 &, \quad 1 + \varepsilon_{Fr}\ \leq\ Fr \end{cases} \tag{16.67}$$

where $\varepsilon_{Fr} = \frac{1}{4}$. From the expressions in (16.60), the corresponding functions for both $\alpha = \alpha(Fr)$ and $\delta = \delta(Fr)$ are

$$\alpha(Fr) = \lambda_1(Fr) - Fr, \quad \delta(Fr) = (\lambda_1(Fr) - \lambda_3(Fr))^2 \tag{16.68}$$

16.5.5 Crossflow Upstream Function α^N

For the determination of $\alpha^N = \alpha^N(Fr)$ from $\lambda_{3,4}(Fr, \boldsymbol{n}) \geq 0$ in (16.62), heed that $\boldsymbol{a} = \boldsymbol{v}$ remains perpendicular to $\boldsymbol{a}^N$, while $n_i a_i$ and $n_i a_i^N$ denote the vector "dot" products between the unit vector $\boldsymbol{n}$ and the unit vectors $\boldsymbol{a}$ and $\boldsymbol{a}^N$ respectively. Accordingly,

$$n_i a_i = \cos\bar{\theta}, \quad n_i a_i^N a_j^N n_j = \sin^2\bar{\theta}, \quad \bar{\theta} \equiv \theta - \theta_v \tag{16.69}$$

where θ and θ_v denote the inclination angles between the x_1 axis and $\boldsymbol{n}$ and $\boldsymbol{v}$ respectively. For eigenvalue $\lambda_4 = \lambda_4\,(Fr, \boldsymbol{n})$ in (16.59), the condition $\lambda_4\,(Fr, \boldsymbol{n}) \geq 0$ yields

$$\alpha^N \geq g\left(\bar{\theta}, Fr\right) \equiv \frac{\cos\bar{\theta}\sqrt{\delta} - \cos^2\bar{\theta}\,(\alpha + Fr)}{1 - \cos^2\bar{\theta}} \tag{16.70}$$

For supercritical flows with $Fr \geq 1 + \varepsilon_\lambda$, $\alpha = 0$ and $\delta = 1$, hence (16.70) becomes

$$\alpha^N \geq g\left(\bar{\theta}, Fr\right) = \frac{\cos\bar{\theta} - Fr\cos^2\bar{\theta}}{1 - \cos^2\bar{\theta}} \tag{16.71}$$

and in particular $\alpha^N \geq g_{\max}(Fr)$, where $g_{\max}(Fr)$ denotes the maximum of $g = g(\bar{\theta}, Fr)$ with respect to $\bar{\theta}$, for each Fr. From (16.71), the determination of $g_{\max}(Fr)$ yields

$$\frac{\partial g}{\partial \bar{\theta}} = 0 \quad \Rightarrow \quad \cos^2\bar{\theta} - 2Fr\cos\bar{\theta} + 1 = 0 \tag{16.72}$$

which leads to

$$\cos\bar{\theta}\Big|_{g=g_{\max}} = Fr - \sqrt{Fr^2 - 1}, \qquad g_{\max}(Fr) = \frac{1}{2}\left(Fr - \sqrt{Fr^2 - 1}\right) \tag{16.73}$$

Significantly, the same solution for $g_{\max}(Fr)$ results from the condition $\lambda_3\,(Fr, \boldsymbol{n}) \geq 0$. Consequently,

$$2\alpha^N \geq \left(Fr - \sqrt{Fr^2 - 1}\right), \quad Fr \geq Fr_\lambda \equiv 1 + \varepsilon_\lambda \tag{16.74}$$

and considering that $\lambda_4\,(Fr_\lambda, \boldsymbol{u}) = \varepsilon_\lambda$, an analogous equality is adopted for $2\alpha^N(Fr_\lambda)$, leading to $2\alpha^N(Fr_\lambda) = 2g_{\max}(Fr_\lambda) + \varepsilon_\lambda$. For subcritical flows, a numerical analysis of $g = g(\bar{\theta}, Fr)$ from (16.71) reveals that $g(\bar{\theta}, Fr) < 0.3$ for all $\bar{\theta}$ and $Fr < Fr_\lambda$. Additionally, for an isotropic gravity-wave upstream for vanishing Fr, $\alpha^N(0) = 1$ and $\partial\alpha^N/\partial Fr|_{Fr=0} = \partial\alpha/\partial Fr|_{Fr=0} = \alpha'_0 = -2$, whereas for $Fr \geq Fr_\lambda$ both α^N and its derivative with respect to Fr follow from (16.74). A smooth variation for $\alpha^N = \alpha^N(Fr)$ that satisfies all of these constraints is the composite spline

$$\alpha^N(Fr) \equiv \begin{cases} \left((\alpha^{N'}_{Fr_\lambda} + \alpha'_0)Fr_\lambda - 2\alpha^N_{Fr_\lambda} + 2\right)\left(\dfrac{Fr}{Fr_\lambda}\right)^3 \\[2mm] \quad - \left((\alpha^{N'}_{Fr_\lambda} + 2\alpha'_0)Fr_\lambda - 3\alpha^N_{Fr_\lambda} + 3\right)\left(\dfrac{Fr}{Fr_\lambda}\right)^2 \\[2mm] \quad + \left(\alpha'_0 Fr_\lambda\right)\left(\dfrac{Fr}{Fr_\lambda}\right) + 1, \qquad\qquad\quad 0 \leq Fr < Fr_\lambda \\[4mm] \dfrac{1}{2}\left(1 + \dfrac{2\varepsilon_\lambda}{Fr_\lambda - \sqrt{Fr_\lambda^2 - 1}}\right)\left(Fr - \sqrt{Fr^2 - 1}\right), \quad Fr_\lambda \leq Fr \end{cases} \tag{16.75}$$

where superscript prime "$'$" denotes differentiation with respect to Fr and subscript "Fr_λ" in both $\alpha^{N'}_{Fr_\lambda}$ and $\alpha^N_{Fr_\lambda}$ indicate their respective magnitudes at $Fr = Fr_\lambda$ from the second expression in (16.75).

16.5.6   Variations of α, δ, α^N

The variations of $\alpha = \alpha(Fr)$, $\alpha^N = \alpha^N(Fr)$ and $\delta = \delta(Fr)$ in Figure 16.9 indicate that these three functions as well as their slopes remain continuous for all Froude numbers, with $0 \leq \alpha, \alpha^N, \delta \leq 1$.

These distributions show that the variations of α and α^N remain distinct from each other. The decrease of α^N, hence of the crossflow upstream-bias, is less rapid because this is the only contribution to a crossflow upstream. The function α^N, nevertheless, decreases by 50%, at the critical state, and by 80% for $Fr = 1.8$. For this function, forcing non-negativity of $\lambda_{3,4}$ as opposed to equality to a prescribed positive constant, in particular, ensures minimal cross flow diffusion.

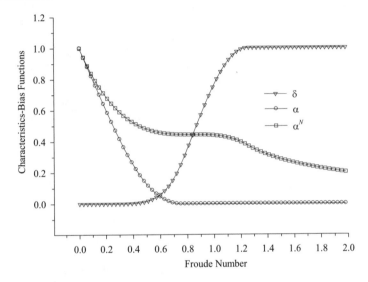

Figure 16.9: Upstream-Bias Functions

Expression (16.75) leads to the conclusions

$$\lim_{Fr \to \infty} \alpha^N(Fr) = 0, \qquad \lim_{Fr \to \infty} \frac{\partial \alpha^N}{\partial Fr} = 0 \qquad (16.76)$$

which indicate that the magnitude of crossflow upstream decreases with increasing Fr. This result agrees with the physics of high-Fr flows, where the bi-modal propagation region narrows about the crossflow direction, as seen in Section 16.3. Convection thereby becomes the prevailing wave propagation mechanism, which therefore reduces the need for gravity-wave crossflow upstream bias.

16.6   Polar Variation of Upstream-Bias

The directional variation of the upstream bias eigenvalues (16.59) is presented in Figures 16.10 - 16.11 for representative subcritical and supercritical Froude numbers. These variations are obtained for a variable unit vector $\boldsymbol{n} \equiv (\cos\theta, \sin\theta)$ and fixed unit vector $\boldsymbol{a} = \boldsymbol{v}$, in this representative case inclined by $+30°$ with respect to the x_1 axis on the flow plane.

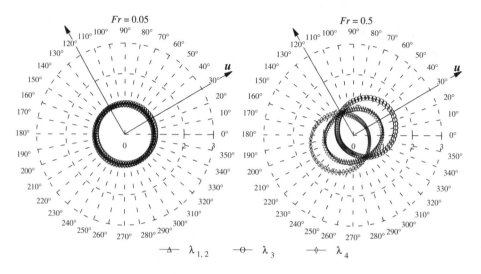

Figure 16.10: Polar Variation of Subcritical Upstream Bias

These figures collectively indicate that the characteristics flux divergence (16.53) induces a physically consistent upstream bias for any Froude number and wave propagation direction $\boldsymbol{n}$, because the associated upstream-bias eigenvalues (16.59) remain positive and their directional variation mirrors the directional variation of the characteristic free-surface eigenvalues (16.19). The upstream-bias eigenvalues are symmetrical about the crossflow direction and characteristic streamline, precisely like the characteristic free-surface eigenvalues (16.19). For $Fr = 0.05$, the directional variation of the upstream-bias eigenvalues in Figure 16.10 correlates with that in Figure 16.2 and thereby corresponds to an isotropic upstream bias, in complete agreement with the isotropic gravity-wave propagation celerity in the free-surface equations. For increasing Froude numbers, the upstream bias becomes anisotropic, again in agreement with the anisotropic distribution of the free-surface eigenvalues (16.18). For $Fr = 0.5$ this anisotropy is already evident and then becomes more marked for supercritical Froude numbers as indicated in Figure 16.13. In particular, the crossflow upstream bias decreases for increasing Froude number.

Figures 16.12 - 16.13 compare the directional variations of the representative upstream-bias eigenvalue λ_3 and the corresponding free-surface eigenvalue λ_3^{Fr}.

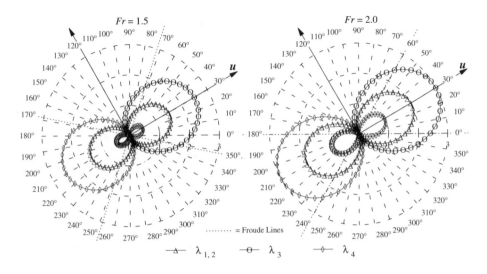

Figure 16.11: Polar Variation of Supercritical Upstream Bias

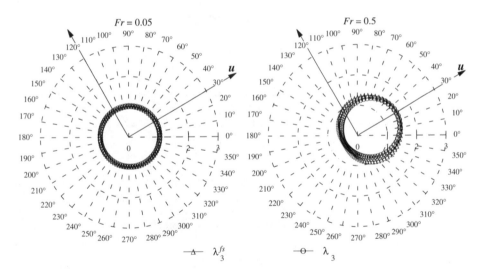

Figure 16.12: Polar Correlation of Subcritical Characteristic λ_3^{fs} and Upstream λ_3

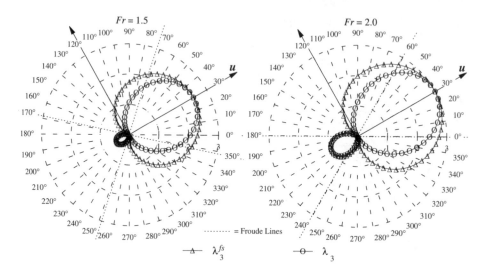

Figure 16.13: Polar Correlation of Supercritical Characteristic λ_3^{fs} and Upstream λ_3

This comparison is sufficient to depict the correlation between all the free-surface and upstream-bias eigenvalues, for $\lambda_{1,2}$ and $\lambda_{1,2}^{Fr}$ are topologically similar to each other, compare Figures 16.3 and 16.13, while λ_4 and λ_4^{Fr} are respectively mirror skew-symmetric to λ_3 and λ_3^{Fr} with respect to the crossflow direction.

As Figures 16.12 - 16.13 indicate, λ_3 is symmetrical about the characteristic streamline, precisely like the corresponding characteristic free-surface eigenvalue λ_3^{Fr} and the corresponding polar curve is topologically similar to the free-surface eigenvalue curve. For $Fr = 0.05$, λ_3 and λ_3^{Fr} virtually coincide with each other and remain direction invariant, which corresponds to an isotropic upstream bias in correlation with the gravity-wave celerity. For $Fr = 0.5$, Figure 16.12 indicates that λ_3^{Fr} is greater than λ_3 in the streamline direction.

For supercritical Froude numbers, λ_3 in the streamline direction coincides with $Fr + 1$. As shown in Figure 16.13, therefore, the magnitude of the upstream bias for supercritical flows is virtually identical to the magnitude of the characteristic eigenvalues, within the domain of dependence and range of influence of any flow field point. Outside this region, the upstream-bias eigenvalues are modestly less than the characteristic eigenvalues. In these variations, the upstream-bias eigenvalues are vanishingly small in the cross- flow direction, which, in particular, corresponds to minimal crossflow diffusion.

16.7   Variational Statement and Boundary Conditions

This section presents the characteristics-bias Euler and Navier-Stokes integral systems and then details a set of boundary conditions.

16.7.1 Characteristics-Bias Euler and Navier-Stokes Systems

With reference to the developments in the previous sections and Chapter 14, the characteristics - bias Euler and Navier-Stokes systems for depth-averaged open-channel flows are formulated as follows.

$$\int_\Omega w \left[\frac{\partial h}{\partial t} + \frac{\partial m_1}{\partial x_1} + \frac{\partial m_2}{\partial x_2} \right] d\Omega - \int_\Omega w \frac{\partial}{\partial x_i} \left[\varepsilon \psi a_i \left[\frac{\partial h}{\partial t} + \frac{\partial m_1}{\partial x_1} + \frac{\partial m_2}{\partial x_2} \right] \right] d\Omega$$

$$- \int_\Omega w \frac{\partial}{\partial x_i} \left[\varepsilon \psi \left[c \left(\alpha a_i a_j + \alpha^N a_i^N a_j^N \right) \frac{\partial h}{\partial x_j} \right] \right] d\Omega = 0 \qquad (16.77)$$

$$\int_\Omega w \left[\frac{\partial m_\ell}{\partial t} - \phi_{m_\ell} + \frac{\partial}{\partial x_1} \left(\frac{m_1}{h} m_\ell \right) + \frac{\partial}{\partial x_2} \left(\frac{m_2}{h} m_\ell \right) + \frac{\partial (h^2/2)}{\partial x_\ell} \right] d\Omega$$

$$- \int_\Omega w \frac{\partial}{\partial x_i} \left[\varepsilon \psi a_i \left[\frac{\partial m_\ell}{\partial t} - \phi_{m_\ell} + \frac{\partial}{\partial x_1} \left(\frac{m_1}{h} m_\ell \right) + \frac{\partial}{\partial x_2} \left(\frac{m_2}{h} m_\ell \right) + \frac{\partial (h^2/2)}{\partial x_\ell} \right] \right] d\Omega$$

$$- \int_\Omega w \frac{\partial}{\partial x_i} \left[\varepsilon \psi \left[c \left(\alpha a_i a_j + \alpha^N a_i^N a_j^N \right) \frac{\partial m_\ell}{\partial x_j} + a_i (\delta - 1) \frac{\partial (h^2/2)}{\partial x_\ell} \right] \right] d\Omega = 0 \qquad (16.78)$$

$$\int_\Omega w \left[\frac{\partial E}{\partial t} - \phi_E + \frac{\partial}{\partial x_1} \left(\frac{m_1}{h} \left(E + \frac{h^2}{2} \right) \right) + \frac{\partial}{\partial x_2} \left(\frac{m_2}{h} \left(E + \frac{h^2}{2} \right) \right) \right] d\Omega$$

$$- \int_\Omega w \frac{\partial}{\partial x_i} \left[\varepsilon \psi a_i \left[\frac{\partial E}{\partial t} - \phi_E + \frac{\partial}{\partial x_1} \left(\frac{m_1}{h} \left(E + \frac{h^2}{2} \right) \right) + \frac{\partial}{\partial x_2} \left(\frac{m_2}{h} \left(E + \frac{h^2}{2} \right) \right) \right] \right] d\Omega$$

$$- \int_\Omega w \frac{\partial}{\partial x_i} \left[\varepsilon \psi \left[c \left(\alpha a_i a_j + \alpha^N a_i^N a_j^N \right) \frac{\partial E}{\partial x_j} \right] \right] d\Omega = 0 \qquad (16.79)$$

The corresponding Navier-Stokes equations are expressed as

$$\int_\Omega w \left[\frac{\partial h}{\partial t} + \frac{\partial m_1}{\partial x_1} + \frac{\partial m_2}{\partial x_2} \right] d\Omega - \int_\Omega w \frac{\partial}{\partial x_i} \left[\varepsilon \psi a_i \left[\frac{\partial h}{\partial t} + \frac{\partial m_1}{\partial x_1} + \frac{\partial m_2}{\partial x_2} \right] \right] d\Omega$$

$$- \int_\Omega w \frac{\partial}{\partial x_i} \left[\varepsilon \psi \left[c \left(\alpha a_i a_j + \alpha^N a_i^N a_j^N \right) \frac{\partial h}{\partial x_j} \right] \right] d\Omega = 0 \qquad (16.80)$$

$$\int_\Omega w \left[\frac{\partial m_\ell}{\partial t} - \phi_{m_\ell} + \frac{\partial}{\partial x_1} \left(\frac{m_1}{h} m_\ell \right) + \frac{\partial}{\partial x_2} \left(\frac{m_2}{h} m_\ell \right) \right] d\Omega$$

$$+ \int_\Omega w \left[\frac{\partial (h^2/2)}{\partial x_\ell} - \frac{1}{Re} \frac{\partial (h \tau_{\ell j})}{\partial x_j} \right] d\Omega$$

$$- \int_\Omega w \frac{\partial}{\partial x_i} \left[\varepsilon \psi a_i \left[\frac{\partial m_\ell}{\partial t} - \phi_{m_\ell} + \frac{\partial}{\partial x_1} \left(\frac{m_1}{h} m_\ell \right) + \frac{\partial}{\partial x_2} \left(\frac{m_2}{h} m_\ell \right) \right] \right] d\Omega$$

$$- \int_\Omega w \frac{\partial}{\partial x_i} \left[\varepsilon \psi a_i \left[\frac{\partial (h^2/2)}{\partial x_\ell} - \frac{1}{Re} \frac{\partial (h \tau_{\ell j})}{\partial x_j} \right] \right] d\Omega$$

$$- \int_\Omega w \frac{\partial}{\partial x_i} \left[\varepsilon \psi \left[c \left(\alpha a_i a_j + \alpha^N a_i^N a_j^N \right) \frac{\partial m_\ell}{\partial x_j} + a_i (\delta - 1) \frac{\partial (h^2/2)}{\partial x_\ell} \right] \right] d\Omega = 0 \qquad (16.81)$$

$$\int_\Omega w \left[\frac{\partial E}{\partial t} - \phi_E + \frac{\partial}{\partial x_1}\left(\frac{m_1}{h}\left(E + \frac{h^2}{2} \right) \right) + \frac{\partial}{\partial x_2}\left(\frac{m_2}{h}\left(E + \frac{h^2}{2} \right) \right) \right] d\Omega$$

$$- \int_\Omega w \left[\frac{1}{Re}\frac{\partial(m_i \tau_{ij})}{\partial x_j} + \frac{1}{Pr\,Re}\frac{\partial}{\partial x_j}\left(h\frac{\partial T}{\partial x_j} \right) \right] d\Omega$$

$$- \int_\Omega w \frac{\partial}{\partial x_i} \left[\varepsilon\psi a_i \left[\frac{\partial E}{\partial t} - \phi_E + \frac{\partial}{\partial x_j}\left(\frac{m_j}{h}\left(E + \frac{h^2}{2} \right) \right) \right] \right] d\Omega$$

$$+ \int_\Omega w \frac{\partial}{\partial x_i} \left[\varepsilon\psi a_i \left[\frac{1}{Re}\frac{\partial(m_\ell \tau_{\ell j})}{\partial x_j} + \frac{1}{Pr\,Re}\frac{\partial}{\partial x_j}\left(h\frac{\partial T}{\partial x_j} \right) \right] \right] d\Omega$$

$$- \int_\Omega w \frac{\partial}{\partial x_i} \left[\varepsilon\psi \left[c\left(\alpha a_i a_j + \alpha^N a_i^N a_j^N \right)\frac{\partial E}{\partial x_j} \right] \right] d\Omega = 0 \qquad (16.82)$$

where $T = E/h - \|\mathbf{V}\|^2/2$. These systems provide the Galilean-invariant characteristics-bias formulation for inviscid and viscous depth-averaged free-surface flows.

16.7.2   Weak Statement and Boundary Conditions

With implied summation on repeated subscript indices, $1 \le i, j \le n$, the characteristics-bias Euler and Navier-Stokes free-surface integral systems may be concisely expressed as

$$\int_\Omega w \left(\frac{\partial q}{\partial t} - \phi + \frac{\partial f_j}{\partial x_j} - \frac{\partial f_j^\nu}{\partial x_j} \right) d\Omega - \int_\Omega w \frac{\partial}{\partial x_i}\left[\varepsilon\psi a_i \left(\frac{\partial q}{\partial t} - \phi + \frac{\partial f_j}{\partial x_j} - \frac{\partial f_j^\nu}{\partial x_j} \right) \right] d\Omega$$

$$- \int_\Omega w \frac{\partial}{\partial x_i}\left[\varepsilon\psi \left(c\left(\alpha a_i a_j + \alpha^N a_i^N a_j^N \right)\frac{\partial q}{\partial x_j} + a_i(\delta - 1)\frac{\partial f_j^p}{\partial x_j} \right) \right] d\Omega = 0 \qquad (16.83)$$

where the arrays ϕ, f_j, f_j^p, f_j^ν follow from inspection of (16.77)-(16.82). The corresponding weak statement is achieved by integrating by parts the characteristics-bias expression, an operation that yields the integral statement

$$\int_\Omega \left(w + \varepsilon\psi a_i \frac{\partial w}{\partial x_i} \right)\left(\frac{\partial q}{\partial t} - \phi + \frac{\partial f_j}{\partial x_j} - \frac{\partial f_j^\nu}{\partial x_j} \right) d\Omega$$

$$+ \int_\Omega \frac{\partial w}{\partial x_i}\varepsilon\psi \left(c\left(\alpha a_i a_j + \alpha^N a_i^N a_j^N \right)\frac{\partial q}{\partial x_j} + a_i(\delta - 1)\frac{\partial f_j^p}{\partial x_j} \right) d\Omega$$

$$- \oint_{\partial\Omega} w\varepsilon\psi \left(a_i\left(\frac{\partial q}{\partial t} - \phi + \frac{\partial(f_j - f_j^\nu)}{\partial x_j} \right) \right) n_i d\Gamma$$

$$- \oint_{\partial\Omega} w\varepsilon\psi \left(c\left(\alpha a_i a_j + \alpha^N a_i^N a_j^N \right)\frac{\partial q}{\partial x_j} + a_i(\delta - 1)\frac{\partial f_j^p}{\partial x_j} \right) n_i d\Gamma = 0 \qquad (16.84)$$

When the divergence of the viscous flux f_j^ν is also integrated by parts, the following equivalent statement is obtained

$$\int_\Omega \left(w + \varepsilon\psi a_i \frac{\partial w}{\partial x_i} \right)\left(\frac{\partial q}{\partial t} + \frac{\partial f_j}{\partial x_j} \right) d\Omega + \int_\Omega \frac{\partial w}{\partial x_j} f_j^\nu d\Omega - \oint_{\partial\Omega} w f_j^\nu n_j \, d\Gamma$$

$$+ \int_\Omega \frac{\partial w}{\partial x_i} \varepsilon \psi \left(c \left(\alpha a_i a_j + \alpha^N a_i^N a_j^N \right) \frac{\partial q}{\partial x_j} - a_i \frac{\partial f_j^\nu}{\partial x_j} + a_i (\delta - 1) \frac{\partial f_j^p}{\partial x_j} \right) d\Omega$$

$$- \oint_{\partial \Omega} w \varepsilon \psi \left(a_i \left(\frac{\partial q}{\partial t} - \phi + \frac{\partial (f_j - f_j^\nu)}{\partial x_j} \right) \right) n_i d\Gamma$$

$$- \oint_{\partial \Omega} w \varepsilon \psi \left(c \left(\alpha a_i a_j + \alpha^N a_i^N a_j^N \right) \frac{\partial q}{\partial x_j} + a_i (\delta - 1) \frac{\partial f_j^p}{\partial x_j} \right) n_i d\Gamma = 0 \qquad (16.85)$$

where n_i denotes the i-th component of the outward pointing unit vector perpendicular to a boundary facet, as depicted in Figure 16.14; when the boundary corresponds to a wall, then $a_i n_i = 0$ and the corresponding terms in the characteristics-bias surface integral will identically vanish. The $\varepsilon \psi$ surface integral, in any case, is eliminated by enforcing a weak Neumann-type boundary condition for the hyperbolic-parabolic characteristics-bias perturbation. One part of this condition imposes that the original Navier-Stokes system should be satisfied at the boundary; the remaining terms are also set to zero as part of this weak boundary condition. Clearly, the single Neumann-type condition corresponds to the entire boundary expression set to nought. For supercritical flows with $Fr \geq 1 + \varepsilon_\lambda$, $\alpha = 0$, $\delta = 1$ and for either $n_i a_i^N = 0$ or $a_j^N \frac{\partial q}{\partial x_j} = 0$, or both, this boundary condition enforces the Navier-Stokes system on the boundary. With the imposition of this weak boundary condition, system (16.85) becomes

$$\int_\Omega \left(w + \varepsilon \psi a_i \frac{\partial w}{\partial x_i} \right) \left(\frac{\partial q}{\partial t} - \phi + \frac{\partial f_j}{\partial x_j} \right) d\Omega + \int_\Omega \frac{\partial w}{\partial x_j} f_j^\nu d\Omega - \oint_{\partial \Omega} w f_j^\nu n_j d\Gamma$$

$$+ \int_\Omega \frac{\partial w}{\partial x_i} \varepsilon \psi \left(c \left(\alpha a_i a_j + \alpha^N a_i^N a_j^N \right) \frac{\partial q}{\partial x_j} + a_i (\delta - 1) \frac{\partial f_j^p}{\partial x_j} \right) d\Omega = 0 \qquad (16.86)$$

from which, for computational simplicity, the characteristics-bias term for the viscous flux may be omitted without loss of stability.

For all test functions $w \in \mathcal{H}^1(\Omega)$, the variational formulation seeks a solution $q \in \mathcal{H}^\eta(\Omega)$ with $\eta = 0$ or $\eta = 1$ respectively for shocked or smooth flows, that satisfies this weak statement. This statement is subject to prescribed initial conditions $q(x,0) = q_0(x)$ and boundary conditions on $\partial \Omega \equiv \overline{\Omega} \backslash \Omega$. Synthetically, these boundary conditions are expressed as

$$B(x_{\partial \Omega}) q(x_{\partial \Omega}, t) = G_v(x_{\partial \Omega}, t) \qquad (16.87)$$

where $G_v(x_{\partial \Omega}, t)$ corresponds to the array of prescribed Dirichlet boundary conditions, with a zero entry for each corresponding unconstrained component of q, and $B(x_{\partial \Omega})$ denotes a square diagonal matrix, with a 1 for each diagonal entry, but replaced by zero for each corresponding unconstrained component of q.

The number of boundary conditions at inlet and outlet depends on the number of negative eigenvalues of the Jacobian $\frac{\partial f_j}{\partial q} n_j$, where n_j denotes the j-th component of the outward pointing normal unit vector $\mathbf{n}$ on $\partial \Omega$.

At a subcritical inlet, only one eigenvalue of $\frac{\partial f_j}{\partial q} n_j$ is positive; as a result only one inlet variable must not be constrained. At such an inlet, Dirichlet boundary conditions are enforced on free-surface height h, total energy E, and transverse linear momentum m_2. For an inviscid flow, no boundary condition is enforced on the axial linear-momentum component

m_1. For a viscous flow, an inlet boundary condition for the axial linear-momentum equation that can be prescribed is a vanishing surface traction $\tau_{1j}n_j$, with τ_{1j} the components of the deviatoric Navier-Stokes stress tensor, a condition effectively enforced by deleting the corresponding surface integral in the statement

$$\int_\Omega \left(w + \varepsilon\psi a_i \frac{\partial w}{\partial x_i} \right) \left(\frac{\partial m_1}{\partial t} - \phi_{m_1} + \frac{\partial}{\partial x_j}\left(\frac{m_1 m_j}{h}\right) + \frac{\partial(h^2/2)}{\partial x_1} \right) d\Omega - \oint_{\partial\Omega} wh\tau_{1j}n_j \, d\Gamma +$$

$$\int_\Omega \frac{\partial w}{\partial x_j} h\tau_{1j} \, d\Omega + \int_\Omega \frac{\partial w}{\partial x_i} \varepsilon\psi \left(c\left(\alpha a_i a_j + \alpha^N a_i^N a_j^N\right)\frac{\partial m_1}{\partial x_j} + a_i(\delta - 1)\frac{\partial(h^2/2)}{\partial x_1}\right) d\Omega = 0 \quad (16.88)$$

Where the inlet flow is supercritical, all the eigenvalues of $\frac{\partial f_j}{\partial q}n_j$ are negative and, correspondingly, Dirichlet boundary conditions are enforced on free-surface height h, energy E, and the components m_1 and m_2 of linear momentum. These conditions may be cast as

$$\frac{\partial h}{\partial t}(x_{\mathrm{in}}, t) = q_h'(t), \quad \frac{\partial E}{\partial t}(x_{\mathrm{in}}, t) = q_E'(t), \quad \frac{\partial m_i}{\partial t}(x_{\mathrm{in}}, t) = q_{m_i}'(t), \quad 1 \le i \le 2 \quad (16.89)$$

where $q_h'(t)$, $q_E'(t)$, $q_{m_i}'(t)$, $1 \le i \le 2$, denote prescribed bounded functions.

At a solid wall, which corresponds to a streamline, $\mathbf{u} \cdot \mathbf{n} = 0$ and, as a result, only one eigenvalue of $\frac{\partial f_j}{\partial q}n_j$ is negative. One boundary condition is thus required at a wall for an inviscid flow. This solid-wall boundary condition corresponds to the wall-tangency, i.e no-penetration, condition $\mathbf{u} \cdot \mathbf{n} = 0$. With reference to the decomposition $f_j = f_j^q + f_j^p$ and Figure 16.14-a,

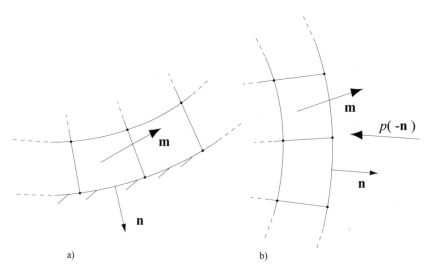

a)   b)

Figure 16.14: Boundary Regions: a) Wall; b) Outlet

the divergence of the convection flux with components f_j^q within (16.86) is integrated by parts at a wall region, an operation that yields the wall boundary statement

$$\int_\Omega \left[\left(w + \varepsilon\psi a_i \frac{\partial w}{\partial x_i} \right) \left(\frac{\partial q}{\partial t} - \phi + \frac{\partial f_j^p}{\partial x_j} \right) + \frac{\partial w}{\partial x_j} f_j^\nu \right] d\Omega - \oint_{\partial\Omega} w f_j^\nu n_j \, d\Gamma$$

$$+ \int_\Omega \frac{\partial w}{\partial x_i} \varepsilon\psi \left(c \left(\alpha a_i a_j + \alpha^N a_i^N a_j^N \right) \frac{\partial q}{\partial x_j} + a_i(\delta - 1) \frac{\partial f_j^p}{\partial x_j} \right) d\Omega$$

$$- \int_\Omega \frac{\partial w}{\partial x_i} \left(f_i^q - \varepsilon\psi a_i \frac{\partial f_j^q}{\partial x_j} \right) d\Omega = - \oint_{\partial\Omega} w q^H \frac{\mathbf{m}}{\rho} \cdot \mathbf{n} \, d\Gamma \qquad (16.90)$$

where $f_j^\nu = 0$ for an inviscid flow. This statement features a generalized "numerical flux" for f_j^q and a wall surface integral that depends on the mass flux $\frac{\mathbf{m}}{\rho} \cdot \mathbf{n}$. A wall mass-flux boundary condition is thus directly enforced within this surface integral, with a wall-tangency boundary condition obtained by setting the entire integral to nought. For an inviscid flow, this is the only boundary condition that is required at a solid wall.

For a viscous flow, the two linear-momentum components in statement (16.90) are replaced by the no-slip boundary conditions on m_1 and m_2; the wall mass-flux surface integrals in the continuity and total-energy components in this statement are then eliminated. The wall heat-flux integral in the energy equation

$$\int_\Omega \left(w + \varepsilon\psi a_i \frac{\partial w}{\partial x_i} \right) \left[\frac{\partial E}{\partial t} - \phi_E \right] d\Omega$$

$$- \int_\Omega \frac{\partial w}{\partial x_j} \frac{m_j}{h} \left(E + \frac{h^2}{2} \right) d\Omega + \oint_{\partial\Omega} w \left(E + \frac{h^2}{2} \right) \frac{m_j}{h} n_j d\Gamma$$

$$+ \int_\Omega \frac{\partial w}{\partial x_i} \varepsilon\psi \left[a_i \frac{\partial}{\partial x_j} \left(\frac{m_j}{h} \left(E + \frac{h^2}{2} \right) \right) + c \left(\alpha a_i a_j + \alpha^N a_i^N a_j^N \right) \frac{\partial E}{\partial x_j} \right] d\Omega$$

$$+ \int_\Omega \frac{\partial w}{\partial x_j} \frac{h}{Re} \left(u_i \tau_{ij} + \frac{1}{Pr} \frac{\partial T}{\partial x_j} \right) d\Omega - \oint_{\partial\Omega} w \frac{h}{Re} \left(u_i \tau_{ij} + \frac{1}{Pr} \frac{\partial T}{\partial x_j} \right) n_j \, d\Gamma = 0 \qquad (16.91)$$

is then used to specify a heat-flux boundary condition, by inserting the prescribed heat-flux $\frac{\partial T}{\partial n} = \frac{\partial T}{\partial x_j} n_j$ in this integral. For a cold- or hot-wall boundary condition on a corresponding wall facet, the heat flux is cast via a temperature variation or the energy equation (16.91) is replaced by a wall surface integral; similarly, for a condition on the axial velocity component u, the axial-momentum equation (16.88) is replaced by a corresponding surface integral. These integrals are

$$\oint_{\partial\Omega} w \left(T - T_w \right) d\Gamma = 0, \quad \oint_{\partial\Omega} w \left(h u_w - m_1 \right) d\Gamma = 0 \qquad (16.92)$$

where T_w and u_w denote prescribed wall quantities.

At a supercritical outlet, all the eigenvalues of $\frac{\partial f_j}{\partial q} n_j$ are positive and thus no boundary conditions are enforced for an inviscid flow; for a viscous flow, the minimal-perturbation boundary conditions that may be prescribed involve an adiabatic outlet stream and vanishing deviatoric work per unit time $u_i \tau_{ij} n_j$ and surface tractions $\tau_{ij} n_j$. The adiabatic-stream condition is enforced by deleting the corresponding surface integral in the outflow energy-equation integral

$$
\int_\Omega \left(w + \varepsilon \psi a_i \frac{\partial w}{\partial x_i} \right) \left[\frac{\partial E}{\partial t} - \phi_E + \frac{\partial}{\partial x_j} \left(\frac{m_j}{h} \left(E + \frac{h^2}{2} \right) \right) \right] d\Omega
$$

$$
+ \int_\Omega \frac{\partial w}{\partial x_i} \varepsilon \psi \left(c \left(\alpha a_i a_j + \alpha^N a_i^N a_j^N \right) \frac{\partial E}{\partial x_j} \right) d\Omega
$$

$$
+ \int_\Omega \frac{\partial w}{\partial x_j} \frac{h}{Re} \left(u_i \tau_{ij} + \frac{1}{Pr} \frac{\partial T}{\partial x_j} \right) d\Omega - \oint_{\partial \Omega} w \frac{h}{Re} \left(u_i \tau_{ij} + \frac{1}{Pr} \frac{\partial T}{\partial x_j} \right) n_j \, d\Gamma = 0 \qquad (16.93)
$$

The vanishing-traction condition is enforced by deleting the corresponding surface integral from the linear-momentum statement

$$
\int_\Omega \left(w + \varepsilon \psi a_k \frac{\partial w}{\partial x_k} \right) \left(\frac{\partial m_i}{\partial t} - \phi_{m_i} + \frac{\partial}{\partial x_j} \left(\frac{m_i m_j}{\rho} \right) + \frac{\partial (h^2/2)}{\partial x_i} \right) d\Omega
$$

$$
+ \int_\Omega \frac{\partial w}{\partial x_k} \varepsilon \psi \left(c \left(\alpha a_k a_j + \alpha^N a_k^N a_j^N \right) \frac{\partial m_i}{\partial x_j} + a_k (\delta - 1) \frac{\partial (h^2/2)}{\partial x_i} \right) d\Omega
$$

$$
+ \int_\Omega \frac{\partial w}{\partial x_j} \frac{h}{Re} \tau_{ij} \, d\Omega - \oint_{\partial \Omega} w \frac{h}{Re} \tau_{ij} n_j \, d\Gamma = 0 \qquad (16.94)
$$

If the traction $\tau_{ij} n_j$ does not vanish at an outlet, its known or prescribed numerical value may be enforced via the corresponding surface integrals in (16.93)-(16.94).

At a subcritical outlet, one eigenvalue of $\frac{\partial f_j}{\partial q} n_j$ is negative and one boundary condition is thus required. A similar surface-integral procedure is employed at such a subcritical outlet region to enforce a specified boundary condition on pressure p for an inviscid flow, and on the deviatoric surface tractions $\tau_{ij} n_j$ for a viscous flow. With reference to the decomposition $f_j = f_j^q + f_j^p$, in Section 16.4.1, and Figure 16.14-b, the divergence of the pressure flux with components f_j^p within the momentum equations (16.94) is integrated by parts at an outlet region, an operation that yields the outflow boundary statement

$$
\int_\Omega \left(w + \varepsilon \psi a_k \frac{\partial w}{\partial x_k} \right) \left(\frac{\partial m_i}{\partial t} - \phi_{m_i} + \frac{\partial}{\partial x_j} \left(\frac{m_j}{h} m_i \right) \right) d\Omega
$$

$$
+ \int_\Omega \frac{\partial w}{\partial x_k} \varepsilon \psi \left(c \left(\alpha a_k a_j + \alpha^N a_k^N a_j^N \right) \frac{\partial m_i}{\partial x_j} + a_k (\delta - 1) \frac{\partial (h^2/2)}{\partial x_i} \right) d\Omega
$$

$$
- \int_\Omega \frac{\partial w}{\partial x_j} \left(\frac{h^2}{2} \delta_i^j - \varepsilon \psi a_j \frac{\partial (h^2/2)}{\partial x_i} - \tau_{ij} \right) d\Omega + \oint_{\partial \Omega} w \left(\frac{h^2}{2} n_i - \frac{h}{Re} \tau_{ij} n_j \right) d\Gamma = 0 \qquad (16.95)
$$

This statement features an outflow surface integral that depends on the outlet-pressure specific-force components $p n_i$, $1 \leq i \leq 3$. With $\tau_{ij} n_j = 0$, the outlet pressure boundary condition is thus directly enforced by specifying the prescribed outlet pressure for p within this surface integral. This strategy for imposing an outlet pressure boundary condition remains intrinsically stable as the following basic considerations on the linear-momentum equation for m_1 indicate; similar conclusions apply to the linear-momentum equation for

m_2. Consider first the case of an outlet with $n_1 < 0$, which implies $\partial w / \partial x_1 < 0$ and $m_1 < 0$. If some computational perturbation induced a decrease in h at the boundary below the imposed magnitude, then m_1 would increase, which would induce a corresponding restoring increase in h from the continuity equation; similar stability conclusions would results by considering a perturbation increase of h at the boundary. Consider the other case of an outlet with $n_1 > 0$, which implies $\partial w / \partial x_1 > 0$ and $m_1 > 0$. If some computational perturbation induced a decrease in h at the boundary, then m_1 would decrease, which again would induce a corresponding restoring increase in h from the continuity equation; similar stability conclusions result by considering a perturbation increase of h at the boundary. The results in Section 16.10 confirm the accuracy and stability of this pressure boundary-condition enforcement procedure.

16.8   Finite Element Galerkin Weak Statement

Since the characteristics flux divergence (16.53), is developed independently and before any discretization, a genuinely multi-dimensional upstream-bias approximation for the governing equations (16.2) on arbitrary grids directly results from a classic centered discretization of the characteristics-bias weak statement on the prescribed grid. To this end a Galerkin finite element method is employed to discretize in space the integral statement (16.86). This method not only accommodates arbitrary geometries or generates consistent non-extrapolation boundary equations for q, but also retains the ideal surface-integral venues of the integral statement to enforce the boundary conditions described in the previous section.

The finite element approximation exists on a partition Ω^h of Ω, where h signifies spatial discrete approximation. Having its boundary nodes on the boundary $\partial\Omega$ of Ω, this partition Ω^h results from the union of N_e non-overlapping elements Ω_e, $\Omega^h = \bigcup_{e=1}^{N_e} \Omega_e$. With $\eta = 0$ or $\eta = 1$, respectively for shocked or smooth flows, and with $\mathcal{P}_1(\Omega_e)$ and $\mathcal{P}_{1v}(\Omega_e)$ the finite-dimensional function spaces of respectively diagonal square matrix-valued and vector-valued linear polynomials within each Ω_e, for each "t", the corresponding diagonal square matrix-valued and vector-valued finite element discretization spaces employed in this study are defined as

$$S^1(\Omega^h) \equiv \left\{ w^h \in \mathcal{H}^1(\Omega^h) : w^h\big|_{\Omega_e} \in \mathcal{P}_1(\Omega_e), \forall\Omega_e \in \Omega^h, B(x_{\partial\Omega^h})w^h(x_{\partial\Omega^h}) = 0 \right\}$$

$$S_v^\eta(\Omega^h) \equiv \left\{ w_v^h \in \mathcal{H}_v^\eta(\Omega^h) : w_v^h\big|_{\Omega_e} \in \mathcal{P}_{1v}(\Omega_e), \forall\Omega_e \in \Omega^h, B(x_{\partial\Omega^h})w_v^h(x_{\partial\Omega^h}, t) = G_v(x_{\partial\Omega^h}, t) \right\}$$

$$(16.96)$$

Based on these spaces, the finite element approximation $q^h \in S_v^\eta$, is determined for each "t" as the solution of the finite element weak statement associated with (16.86)

$$\int_{\Omega^h} \left(w^h + \varepsilon^h \psi^h a_i^h \frac{\partial w^h}{\partial x_i} \right) \left(\frac{\partial q^h}{\partial t} - \phi^h + \frac{\partial f_j^h}{\partial x_j} \right) d\Omega$$

$$+ \int_{\Omega^h} \frac{\partial w^h}{\partial x_j} f_j^{v^h} d\Omega - \oint_{\partial\Omega^h} w^h f_j^{v^h} n_j d\Gamma +$$

$$+ \int_{\Omega^h} \frac{\partial w^h}{\partial x_i} \varepsilon^h \psi^h \left(c^h \left(\alpha^h a_i^h a_j^h + \alpha^{N^h} a_i^{N^h} a_j^{N^h} \right) \frac{\partial q^h}{\partial x_j} + a_i^h (\delta^h - 1) \frac{\partial f_j^{p^h}}{\partial x_j} \right) d\Omega = 0 \qquad (16.97)$$

with similar expressions for statements (16.88)-(16.95). Since every member w^h of $\mathcal{S}^1$ results from a linear combination of the linearly independent basis functions of this finite dimensional space, statement (16.97) is satisfied for N independent basis functions of the space, where N denotes the dimension of the space. For N mesh nodes within Ω^h, there exist clusters of "master" elements Ω_k^M, each cluster comprising only those adjacent elements that share a mesh node $\mathbf{x}_k$, with $1 \leq k \leq N$, where N denotes the total number of not only mesh nodes, but also master elements.

As detailed in Chapter 7, the discrete test function w^h within each master element Ω_k^M will coincide with the "pyramid" basis function $w_k = w_k(\mathbf{x})$, $1 \leq k \leq N$, with compact support on Ω_k^M. Such a function equals one at node $\mathbf{x}_k$, zero at all other mesh nodes and also identically vanishes both on the boundary segments of Ω_k^M not containing $\mathbf{x}_k$ and on the computational domain outside Ω_k^M.

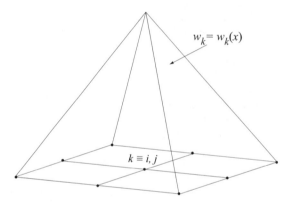

Figure 16.15: Pyramid Test Function for Ω_k^M

The discrete solution q^h and flux f_j^h at each time t assume the form of the following group linear combinations

$$q^h(\mathbf{x}, t) \equiv \sum_{\ell=1}^{N} w_\ell(\mathbf{x}) \cdot q^h(\mathbf{x}_\ell, t), \quad f_j^h(\mathbf{x}, t) \equiv \sum_{\ell=1}^{N} w_\ell(\mathbf{x}) \cdot f_j\left(q^h(\mathbf{x}_\ell, t)\right) \qquad (16.98)$$

of time-dependent nodal solution values $q^h(\mathbf{x}_\ell, t)$, to be determined, and trial functions, which coincide with the test functions $w_\ell(\mathbf{x})$ for a Galerkin formulation; an analogous expression applies for ϕ and $f_j^{\nu^h}$. Similarly, the fluxes $f_j^q = f_j^q(q(\mathbf{x}, t))$ and $f_j^p = f_j^p(q(\mathbf{x}, t))$ are discretized through the group expressions

$$f_j^{q^h}(\mathbf{x}, t) \equiv \sum_{\ell=1}^{N} w_\ell(\mathbf{x}) \cdot f_j^q\left(q^h(\mathbf{x}_\ell, t)\right), \quad f_j^{p^h}(\mathbf{x}, t) \equiv \sum_{\ell=1}^{N} w_\ell(\mathbf{x}) \cdot f_j^p\left(q^h(\mathbf{x}_\ell, t)\right) \qquad (16.99)$$

The notation for the discrete nodal variable and fluxes is then simplified as $q_\ell(t) \equiv q(\mathbf{x}_\ell, t)$, $f_{j\ell} \equiv f_j\left(q^h(\mathbf{x}_\ell, t)\right)$, $f_{j\ell}^q \equiv f_j^q\left(q^h(\mathbf{x}_\ell, t)\right)$, $f_{j\ell}^p \equiv f_j^p\left(q^h(\mathbf{x}_\ell, t)\right)$ and expansions (16.98)-(16.99) are then inserted into (16.97), which yields the discrete finite element weak statement

$$\int_{\Omega^h} \left(w_k + \varepsilon^h \psi^h a_i^h \frac{\partial w_k}{\partial x_i} \right) \left(w_m \frac{dq_m}{dt} - w_m \phi_m + \frac{\partial w_m}{\partial x_j} f_{jm} \right) d\Omega$$

$$+ \int_{\Omega^h} \frac{\partial w_k}{\partial x_j} f_j^{\nu^h} d\Omega - \oint_{\partial\Omega^h} w_k f_j^{\nu^h} n_j \, d\Gamma +$$

$$+ \int_{\Omega^h} \frac{\partial w_k}{\partial x_i} \frac{\partial w_m}{\partial x_j} \varepsilon^h \psi^h \left[c^h \left(\alpha^h a_i^h a_j^h + \alpha^{N^h} a_i^{N^h} a_j^{N^h} \right) q_m + a_i^h (\delta^h - 1) f_{jm}^p \right] d\Omega = 0 \qquad (16.100)$$

for $1 \leq k \leq N$; the three-test-function matrices in (7.186) are then needed to discretize the viscous terms in the linear-momentum and energy equations. There are three implied summations with respect to the subscript indices i, j, m. The subscript indices i, j in this expression denote cartesian-axis directions, hence $1 \leq i, j \leq 2$, for two-dimensional flows, or $1 \leq i, j \leq 3$, for three dimensional flows, whereas subscript m indicates a mesh node, hence $1 \leq m \leq N$, although a sum like $\sum_{m=1}^{N} w_m \frac{dq_m}{dt}$ only involves a few neighboring terms because the compact-support test function w_m is only non-zero within a cluster of a few neighboring elements.

While an expansion like the ones in (16.98) for ψ^h, α^h, c^h, $\mathbf{a}^h$, $\mathbf{a}^{N^h}$ and δ^h can be directly accommodated within (16.100), each of these variables in this study has been set equal to a piece-wise constant for computational simplicity, one centroidal constant value per element. Within each element, the terms ψ^h and ε^h are respectively set equal to the controller and reference length developed in Chapter 10. As shown in that chapter, ε^h corresponds to the length of the streamline radius of the generalized ellipse inscribed within each element, since the streamline is a characteristic principal direction.

Since the test and trial functions w_k are prescribed functions of $\mathbf{x}$, the spatial integrations in (16.97) are directly carried out. For arbitrarily shaped elements, these integrations take place as detailed in Chapter 7. Concerning the boundary variables, no extrapolation of variables is needed in this algorithm on a variable that is not constrained via a Dirichlet boundary condition. In this case, instead, the finite element algorithm (16.100) naturally generates for each unconstrained boundary variable a boundary-node ordinary differential equation. The complete integration with respect to $\mathbf{x}$ transforms (16.100) into a system of continuum-time ordinary differential equations (ODE) for determining at each time level t the unknown nodal values $q^h(\mathbf{x}_\ell, t)$, $1 \leq \ell \leq N$.

16.9 Implicit Runge-Kutta Time Integration

The finite element equations (16.100), along with appropriate boundary equations and conditions, are abridged as the non-linear ODE system

$$\mathcal{M} \frac{dQ(t)}{dt} = \mathcal{F}(t, Q(t)) \qquad (16.101)$$

where $\mathcal{M} \equiv \{w_k w_\ell\}$ denotes the mass matrix, $\mathcal{M}\frac{dQ(t)}{dt}$ indicates the corresponding coupling of time derivatives in (16.100), and $\mathcal{F}(t, Q(t))$ represents the remaining terms in (16.100).

The numerical time integration of (16.101) takes place through a two-stage diagonally implicit Runge-Kutta algorithms (IRK2), as detailed in Chapter 8, expressed as

$$
\begin{aligned}
Q_{n+1} - Q_n &= b_1 K_1 + b_2 K_2 \\
\mathcal{M}K_1 &= \Delta t \cdot \mathcal{F}(t_n + c_1 \Delta t, Q_n + a_{11} K_1) \\
\mathcal{M}K_2 &= \Delta t \cdot \mathcal{F}(t_n + c_2 \Delta t, Q_n + a_{21} K_1 + a_{22} K_2)
\end{aligned} \tag{16.102}
$$

where n now denotes a discrete time station. Given the solution Q_n at time t_n, K_1 is computed first, followed by K_2. The solution Q_{n+1} is then determined by way of the first expression in (16.102).

The terminal numerical solution is then determined using Newton's method, which for the implicit fully-coupled computation of the IRK2 arrays K_i, $1 \leq i \leq 2$, is cast as

$$
\left[\mathcal{M} - a_{\underline{ii}} \Delta t \left(\frac{\partial \mathcal{F}}{\partial Q} \right)^p_{Q_i^p} \right] \left(K_i^{p+1} - K_i^p \right) = \Delta t \, \mathcal{F}(t_n + c_i \Delta t, Q_i^p) - \mathcal{M} K_i^p
$$

$$
Q_i^p \equiv Q_n + a_{i1} K_1^p + a_{i2} K_2^p \tag{16.103}
$$

where $a_{ij} = 0$ for $j > i$, p is the iteration index, and $K_1^p \equiv K_1$ for $i = 2$; the Jacobian

$$
J_i(Q) \equiv \mathcal{M} - a_{\underline{ii}} \Delta t \left(\frac{\partial \mathcal{F}}{\partial Q} \right)^p_{Q_i^p} \tag{16.104}
$$

has been analytically determined and implemented, leading to a block sparse matrix. For all the results documented in the next section, the initial estimate K_i^0 is set equal to the zero array, while a Gaussian elimination is used with only one iteration executed for (16.103) within each time interval. In this mode, Newton's iteration becomes akin to a classical direct linearized implicit solver.

16.10   Computational Performance

The characteristics-bias finite element algorithm detailed in the previous sections has generated essentially non-oscillatory solutions for five flows involving: an asymmetric dam-break, a backward-facing step channel, a stream-driven inlet, and two oblique hydraulic jumps. The asymmetric dam-break problem has employed several grids; the remaining benchmarks have employed a non-uniform finite element discretization of Lagrange bilinear elements, consisting of 40 bilinear elements in the transverse and longitudinal directions, for a total of 1600 elements, 1681 nodes and 6724 degrees of freedom.

For each benchmark, the calculations proceeded with a prescribed constant maximum Courant number,

$$
C_{\max} \equiv \max\{\|\mathbf{u}\| + c, \|\mathbf{u}\| - c, c\} \frac{\Delta t}{(\Delta \ell)_e} \tag{16.105}
$$

For prescribed $(\Delta\ell)_e$ and $C_{\max}$ for each benchmark, the corresponding Δt is determined as

$$\Delta t = \frac{C_{\max}(\Delta\ell)_e}{\max\{\|\mathbf{u}\| + c, \|\mathbf{u}\| - c, c\}} \qquad (16.106)$$

All the solutions in these validations are presented in non dimensional form and have asymptotically converged to the preset machine zero of 5×10^{-12}. For each viscous flows, as mentioned in the previous chapter, the algorithm in 1 Newton-iteration mode generated a final steady state by using a gradually increasing Courant number from 2, at the beginning of each computational sequence, to 20 in the middle of the sequence, and up to 100 towards steady state. As expected, the maximum allowable Courant number decreases when ψ decreases; on the other hand, the maximum allowable Courant number need not decrease substantially, when more than 1 Newton iteration is employed per time step. The upstream controller ψ was set to $\psi = 0.5$, from beginning to convergence to an initial steady state, and then gradually reduced to $\psi \simeq 0.1$ to lessen the decrease in the effective Reynolds number. For these viscous flows, each of the computational final steady state was achieved in about two thousand time steps. The inviscid flow cases rapidly converged each in about two hundred times steps with $C_{\max} = 75$, without employing the increasing Courant-number strategy.

16.10.1   Asymmetric Dam-Break Flow

This section investigates the flow resulting from a the sudden failure of a portion of a dam. The dam is transversally located at the center of a square channel, in which the non-dimensional length of each edge equals 1, and the width of the dam equals 0.05, as shown in Figure 16.16.

The initial conditions involve a vanishing velocity field and a free-surface height that equals 1 and 0.5 respectively upstream and downstream of the dam. In the collapsed-wall region, this height remains transversally constant, but linearly decreases longitudinally from 1 to 0.5. At the "inlet" AB, the free-surface h is set equal to 1; at the outlet CD, this height is set equal to 0.5, but through the average pressure. Along all the remaining walls, the no-penetration boundary condition $\mathbf{m} \cdot \mathbf{n}$ is enforced. The transient solution is obtained for a non-dimensional time in the range $0 \leq t \leq 7.2$ with $C_{\max} = 1$. The solution for this problem is generated using two uniform grids, respectively containing 40×40 and 160×160 bilinear Lagrange elements, as illustrated in Figures 16.17-16.22, with the effect of the controller ψ on solution accuracy investigated on the 40×40-element grid. On this grid, five solutions have been calculated, the first four respectively corresponding to $\psi = 1.00, 0.75, 0.50, 0.25$, and the fifth obtained for a variable ψ, with $0.25 = \psi_{\min} \leq \psi \leq \psi_{\max} = 1.00$.

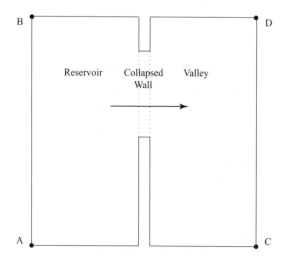

Figure 16.16: Computational Domain

Figure 16.17 present the distribution of free-surface height h resulting from $\psi = 1.00$. Although this solution visibly shows the effect of induced upstream diffusion, it already displays the key features of this solution. The free surface develops a plateau at the entrance of the collapsed-wall region, while it experiences a steep depression at the dam-wall corners. Similar to an acoustic wave, the hydraulic jump propagates as an elliptic front, with jump strength that decreases away from the longitudinal flow propagation direction. These features sharpen as the magnitude of the controller ψ decreases, as shown in Figures 16.18-16.20.

Even though the solution corresponding to $\psi = 0.25$ displays an emerging dispersive mode at the dam-wall corners, a mode due to insufficient upstream diffusion, this solution displays a plateau and hydraulic jump that remain essentially non-oscillatory. The solution generated for a variable ψ, with $0.25 = \psi_{\min} \leq \psi \leq \psi_{\max} = 1.00$, remains essentially non-oscillatory in the flow field as shown in Figure 16.21.

The solution features resulting from distinct magnitudes of the controller ψ reiterate the advantage of a variable ψ that non-linearly depends on the evolving solution. The combination of a variable ψ with a finer grid leads to a solution with refined features. The beneficial effect of this combination is illustrated in Figures 16.22, 16.23 respectively showing the contour and distribution of free-surface height as obtained in the 160×160-element grid. This solution crisply resolves the dam-wall corner depressions, plateau, and curved hydraulic jump.

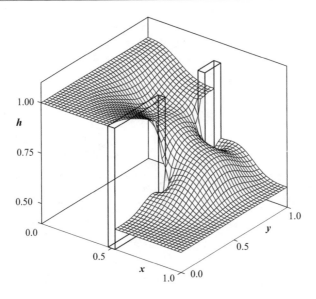

Figure 16.17: Free-Surface Distribution: $\psi = 1.00$

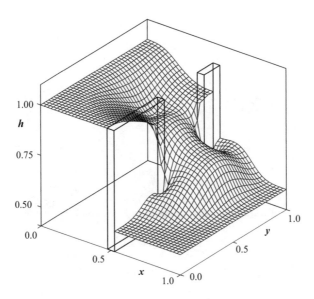

Figure 16.18: Free-Surface Distribution: $\psi = 0.75$

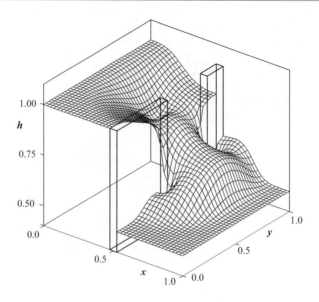

Figure 16.19: Free-Surface Distribution: $\psi = 0.50$

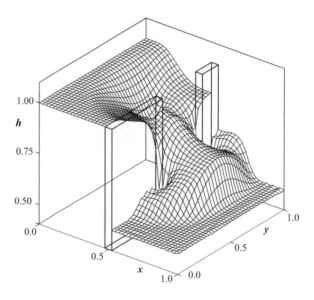

Figure 16.20: Free-Surface Distribution: $\psi = 0.25$

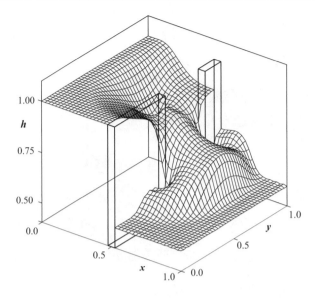

Figure 16.21: Free-Surface Distribution: $0.25 \leq \psi \leq 1.00$

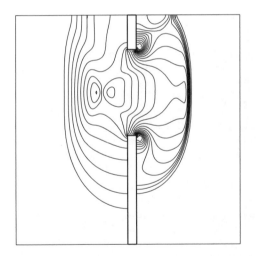

Figure 16.22: Free-Surface Contours: 160×160 Grid, $0.25 \leq \psi \leq 1.00$

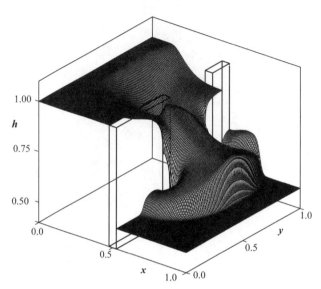

Figure 16.23: Free-Surface Distribution: 160×160 Grid, $0.25 \leq \psi \leq 1.00$

16.10.2   Backward-Facing Step Diffuser Flow

This flow is the free-surface counterpart of the fully enclosed flow described in Section 15.11.1. Also in this benchmark, the axial length of the computation domain equals 5 step lengths. The calculations have determined not only the velocity and pressure fields, but also the temperature distribution within the flow, retaining the viscous-dissipation expression in the temperature equation. Figure 16.24 shows the simplified computational domain for this benchmark. The inlet corresponds to $x = 0$ and $0 \leq y \leq 0.5$ and at this location Dirichlet boundary conditions are enforced on temperature and momentum components, whereas pressure is unconstrained and allowed to vary with the flow. These Dirichlet conditions involve a vanishing transverse momentum component and fully developed temperature and axial-velocity profiles, corresponding to the exact Poiseuille fully-enclosed flow solution. For $Re = 200$, the outlet corresponds to $x = 2.5$ and $-0.5 \leq y \leq 0.5$. At this outlet, an adiabatic stream is specified, hence vanishing heat transfer in the energy equation, along with physically meaningful prescribed longitudinal and transverse surface tractions $\sigma_{1j}n_j$ and $\sigma_{2j}n_j$; since at this outlet $n_1 = 1$ and $n_2 = 0$, these conditions specify pressure $-\sigma_{11} = h^2/2 = 9.0$ and shear stress $\tau_{21} = \tau_{12}$ from the fully-enclosed Poiseuille-flow distribution of u at the flow Reynolds number, conditions that allow the transverse momentum component to vary and lead to a minimally constrained converged solution. As detailed in Section 16.7.2, these physical conditions require no special boundary equation, but rather specify the surface integrals in the variational statement of the linear-momentum and temperature equations.

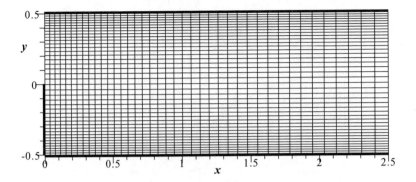

Figure 16.24: Non-uniform 41 × 41 Linear Element Grid

At the solid walls at the top, bottom, and step site of the computational domain, hence $y = 0.5$, $y = -0.5$, and $-0.5 \leq y \leq 0$ at $x = 0$, the no slip boundary condition is enforced on velocity; the step site and lower wall are treated as adiabatic walls, whereas a cold-wall condition, hence fixed temperature, is specified at the top wall. The initial conditions specified fully developed velocity and temperature profiles at the inlet, with temperature variation set to zero at the step, a constant pressure field, with $h^2/2 = 9$, a vanishing transverse component of momentum $m_2 \equiv hv = 0$, and fully developed velocity and temperature profiles for $x > 0$, profiles that satisfy global conservation of mass. Obtained in 1 Newton-iteration mode and corresponding to an eventual $\psi = 0.2$, the final steady state is documented in Figures 16.25-16.33. The velocity vector distribution in Figure 16.25 reflects the expected pattern, with the recirculation zone that extends for almost $1/2$ of the domain length. Toward the outlet, velocity points down, as allowed by the shear-stress boundary condition, while the overall velocity distribution remains uniform, with correct decrease of axial maximum velocity as the effective flow channel widens. These features are reflected in the streamline distribution in Figure 15.19, which clearly portrays the recirculation zone. The reattachment length for this test corresponds to a ratio of this length over the step height equal to 2.5.

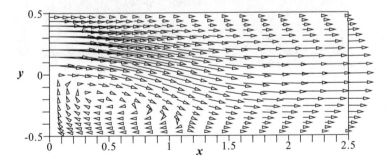

Figure 16.25: Vector Field

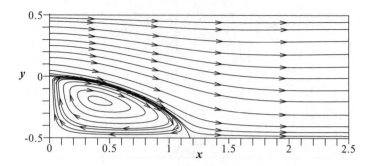

Figure 16.26: Streamline Distribution

These velocity-field features originate from the individual velocity components $u_1 \equiv u$ and $u_2 \equiv v$ displayed in Figures 16.27-16.30. As in the fully-enclosed flow, Figure 16.27 portrays the development of an effective flow channel, as the maximum axial velocity progressively decreases, and Figure 16.28 shows that the distribution of u remains monotone, from inlet, within the recirculation zone, and toward the inlet, where the surface-integral outlet boundary conditions promoted an uniform distribution.

The distribution of v displays similar features. Figure 16.29 shows that this velocity component rapidly decreases within the effective flow channel, with minimum above and slightly upstream of the reattachment point. At the outlet, v is not forced to vanish, but is determined by the outlet shear-stress boundary condition, which can be more sharply satisfied by a finer grid in this region; within the flow field, the distribution of v remains monotone.

Figure 16.27: Axial Velocity Contours

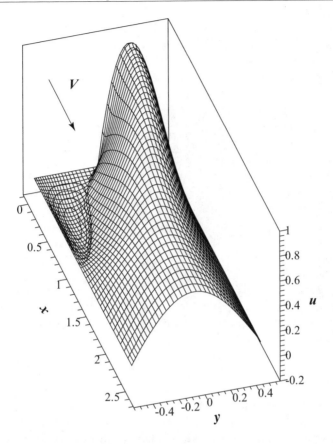

Figure 16.28: Axial Velocity Surface

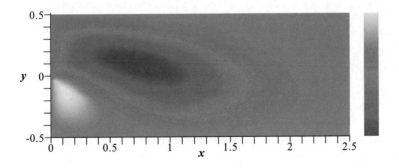

Figure 16.29: Transverse Velocity Contours

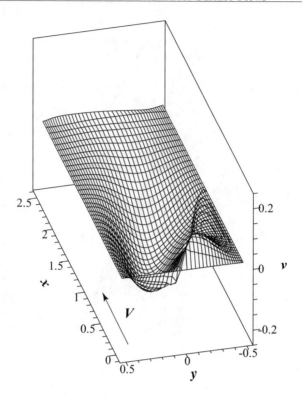

Figure 16.30: Transverse Velocity Surface

Also in this case is the inlet point with coordinates $(0;0)$ the center of a singular zone, where the positive v from the separated flow tends to oppose the negative v of the downward main flow. The characteristics-bias procedure, nevertheless, negotiates this singularity and leads to an essentially non-oscillatory distribution of v, with a steep variation of this variable in the singular zone, which, as expected, induces a corresponding localized steep variation of free-surface height.

The distribution of this height is presented in Figure 16.31. This distribution shows the anticipated abrupt variation of height at the singular point, which remains contained within a 4-cell zone, and indicates that a finer grid in the inlet region would increase the computational accuracy. In the remaining field, including the outlet region, the free-surface height remains essentially non-oscillatory

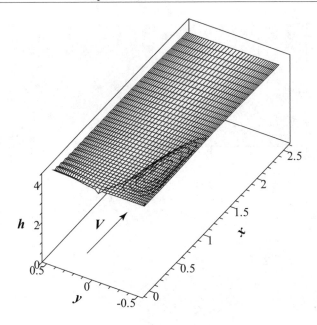

Figure 16.31: Free-Surface Height

· This investigation has also determined the temperature field, which is not usually considered in free-surface flow analyses. The calculated distribution of temperature in Figures 16.32-16.33 reflects the flow and the imposed boundary conditions. The inlet temperature profile rapidly merges with the overall temperature distribution, a distribution that becomes less and less dependent on x, downstream of the inlet. This temperature distribution remains monotone even at the inlet singularity zone, where the variation from the maximum inlet temperature, down to the vanishing temperature variation at the $(0;0)$ point, and up to the temperatures in the separated flow region takes place without any unphysical oscillation. Downstream of the inlet, the temperature gradually increases in the downward transverse direction, from the cold wall at the top of the channel, to the adiabatic wall, where the thermal energy of convection and internal friction raises the local temperature.

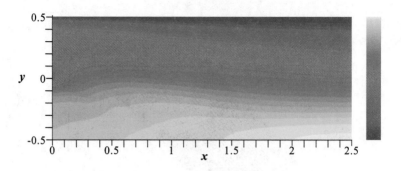

Figure 16.32: Temperature Contours

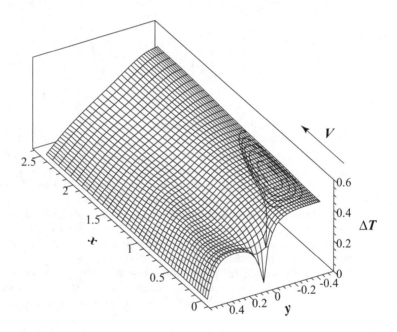

Figure 16.33: Temperature Surface

16.10.3   Stream-Driven Inlet Flow

The next test involves a flow induced by a stream flowing past the entrance of a square inlet, a model that represents some harbors. On the basis of this representation, the flow is expected to remain periodic with cyclical variations of the free surface height. Figure 16.34 shows the computational domain for this test; the left, bottom, and right sides correspond to solid walls, whereas the top side corresponds to a flowing stream. The transverse momentum component m_2 is set to zero on every side of the inlet; the axial momentum component m_1 is set to zero on the $x = 0$, $x = 1$, and $y = 0$ walls. Along the entire $y = 1$ wall, a boundary condition is enforced on the axial component of velocity, as described in Section 16.7.2; the condition on this velocity component corresponds to a $Fr = 0.4$. Concerning the temperature variation ΔT, this variable is constrained to zero, at the cold side, and to one at the hot side. The computational procedure monitors the boundary magnitude of Fr; whenever this number locally exceeds one, the procedure automatically enforces a Dirichlet boundary condition on all the components of the depended variable q, a constraint equal to the numerical value that q attains when Fr first exceeds one. The initial conditions involve a constant free-surface height field with $h = 1$ and a linearly varying temperature field with $\Delta T = x$. The initial conditions for velocity are a vanishing transverse velocity component and an axial velocity component that only vanishes for $y < 1$. The initial field for the total energy E is then calculated form these distributions of h, $\mathbf{V}$, and ΔT, so that the resulting field also satisfies the boundary conditions on ΔT.

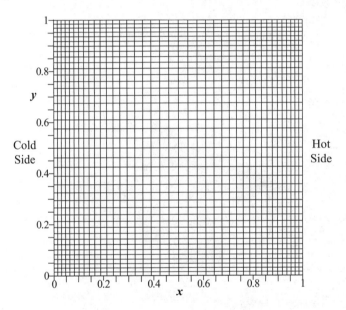

Figure 16.34: Non-uniform 41 × 41 Linear Element Grid

Corresponding to an effective Reynolds number $Re = 100$ and an eventual $\psi = 0.375$, the solution, as expected, remains periodic, with a magnitude of the maximum residual of the order of 3.8×10^{-5}, as recorded during 50 periods of the flow with $C_{\max} = 1$, subsequent to the initial transient. Throughout these periods, the solution remains essentially non-oscillatory. The state corresponding to the middle of the 50-th period is documented in Figures 16.35-16.44. Figures 16.35-16.36 present the distribution of free-surface height.

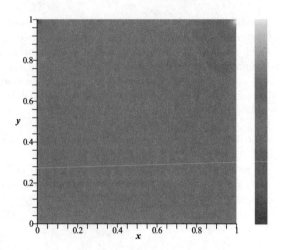

Figure 16.35: Free-Surface Height Contours

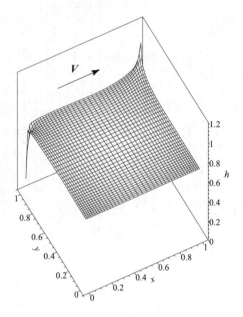

Figure 16.36: Free-Surface Height

This distribution remains essentially non-oscillatory, with clearly defined spikes at the upper corners of the domain, where the solution for this variable is expected to be singular.

The steep drop in h at the upper left corner leads to a local supercritical flow with $Fr = 1.08$; remaining subcritical, the right-corner spike represents the increase in water height in this location of the inlet. Away form the corner singularities, the distribution of free-surface height remains monotone.

Figure 16.37 presents the velocity vector distribution for this flow. The center of the main recirculation zone is asymmetrically located towards the top and right walls of the inlet. In the vicinity of the stream, left corner, velocity actually points towards this corner, before sharply turning to follow the outer stream. These features are reflected in the streamline distribution in Figure 16.38, which shows the associated recirculation zone. Normalized by the outer stream speed, the corresponding distributions of velocity components u and v, Figures 16.39-16.42, remain monotone, even in the vicinity of the stream singularity corners. These results, in particular, document the effectiveness of the surface-integral procedure to enforce a prescribed velocity magnitude.

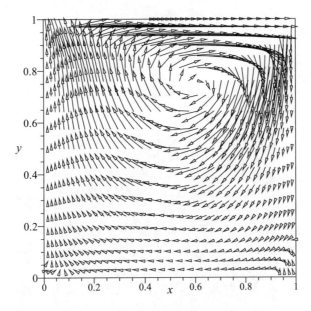

Figure 16.37: Velocity Field

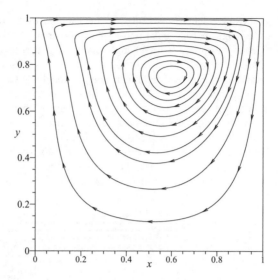

Figure 16.38: Streamline Distribution

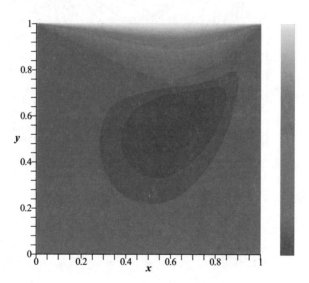

Figure 16.39: Axial Velocity Contours

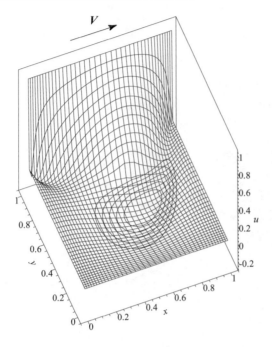

Figure 16.40: Axial Velocity Surface

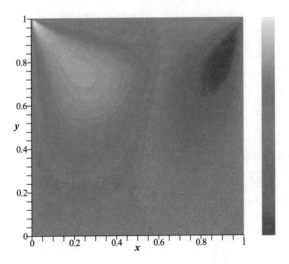

Figure 16.41: Transverse Velocity Contours

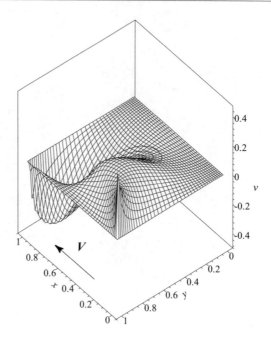

Figure 16.42: Transverse Velocity Surface

Resulting from convection and conduction, the calculated distribution of temperature in Figures 16.43-16.44 reflects the flow and the imposed boundary conditions. The temperature profile at the stream side slowly increases towards the left corner, before rapidly rising to the hot-side magnitude, within a localized thermal boundary layer. The convection of fluid towards the hot side thus limits the increase in temperature along the stream side. Away from this side, the temperature distribution remains monotone, from the cold to the hot side, with reduced dependence upon y towards the opposite side of the domain.

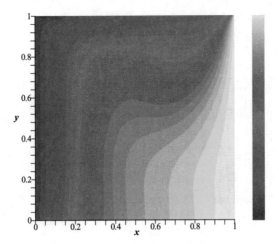

Figure 16.43: Temperature-Variation Contours

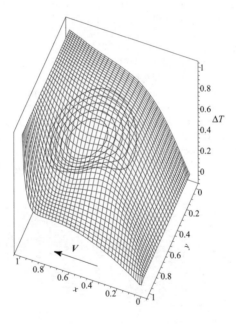

Figure 16.44: Temperature-Variation Surface

16.10.4   Fr=2.5 Hydraulic-Jump Reflection

This test involves an oblique hydraulic-jump configuration, with grid illustrated in Figure 16.45.

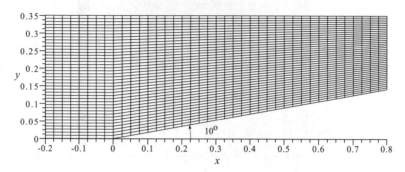

Figure 16.45: Non-uniform 41×41 Linear Element Grid

The supercritical upstream flow corresponds to a free-stream Froude number $Fr_\infty = 2.5$, hence the inlet boundary conditions constrain free-surface height h, longitudinal as well as transversal linear momentum components m_1 and m_2 and total energy E. The outlet remains supercritical, hence no boundary conditions are enforced at this boundary. At the solid upper and lower walls, the inviscid wall-tangency boundary condition is enforced using the method in Section 16.7.2. An initially uniform supercritical $Fr = 2.5$ flow is deflected by 10° by a side wall and the final steady state is computationally achieved in about two hundred time steps with $\psi = 0.35$, by advancing the solution in time.

The wall-tangency boundary condition on the whole upper and lower boundary walls leads to an oblique hydraulic jump shock that propagates toward the upper wall and is then reflected downward towards the outlet. As shown in the velocity vector and streamline distributions in Figures 16.46-16.47, the velocity vector abruptly turns across the oblique hydraulic jumps.

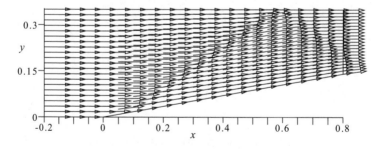

Figure 16.46: Vector Field

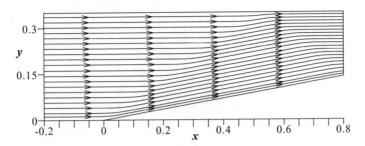

Figure 16.47: Streamline Distribution

With maximum upstream-bias activated as described in Section 10.2, after Figure 10.5, at the geometric singularity of the jump initiation point, the free-surface height distribution in Figures 17.35 and 17.36 corresponds to an essentially non-oscillatory solution with crisply calculated incident and reflected jumps. In particular, the algorithm allows the reflected jump to cross the outflow boundary unperturbed, without any spurious distortion. Significantly, this computational solution mirrors the available exact solution, with three juxtaposed plateaus connected by two oblique jumps.

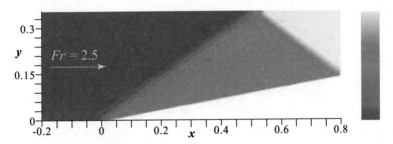

Figure 16.48: Free-Surface Height Contours

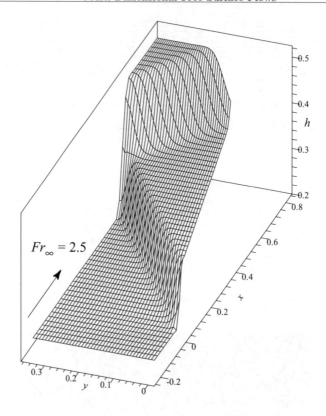

Figure 16.49: Free-Surface Height

A similar essentially non-oscillatory field is displayed in the Froude-number distribution in Figures 16.50-16.51. The incident and reflected jumps are crisply calculated, with a reflected jump that can cross the outflow boundary without any spurious distortion.

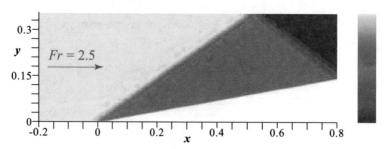

Figure 16.50: Froude Number Contours

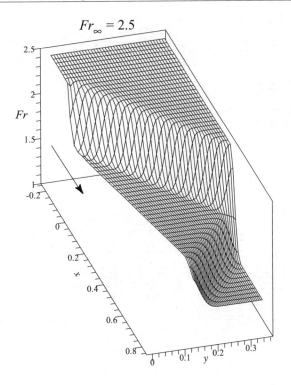

Figure 16.51: Froude Number Surface

16.10.5   Fr=5.0 Supercritical Flow

This final test determines a supercritical flow with free stream Froude number $Fr = 5.0$, to demonstrate the capability of the characteristics-bias procedure to calculate a strong oblique hydraulic jump. The sharpness of the captured hydraulic jumps directly depends on the magnitude of ψ. A decrease in the magnitude of the controller may lead to a solution that may tend lose its essentially non-oscillatory character, as shown in this section. The solution corresponds to the same grid as well as initial and boundary conditions described in the previous section, and is obtained for the magnitude $\psi = 0.2$ for the controller.

As shown in the velocity vector and streamline distributions in Figures 16.52-16.53, the free stream Froude number leads to an oblique hydraulic jump that remains close to the deflecting wall and does not reach the upper wall in the computational domain. As these charts indicate, the velocity vector abruptly turns across the oblique jump.

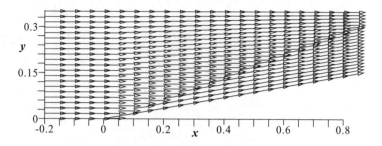

Figure 16.52: Velocity Field

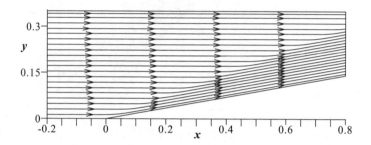

Figure 16.53: Streamline Distribution

The free-surface height distribution in Figures 16.54-16.55, with maximum upstream-bias at the jump initiation corner, remains essentially non - oscillatory. Event though this character may be lost with any reduction of ψ, as implied by the mild undulations at the edges of the jump, this jump is crisply captured and allowed to cross the outlet boundary undistorted. Also this solution mirrors the available exact solution, with two juxtaposed plateaus connected by the oblique jump.

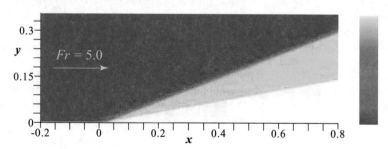

Figure 16.54: Free-Surface Height Contours

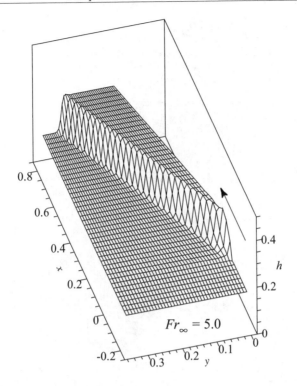

Figure 16.55: Free Surface Height

A similar essentially non-oscillatory field is displayed in the Froude-number distribution in Figures 16.56-16.57. The jump is crisply calculated and can cross outflow boundary without any spurious distortion.

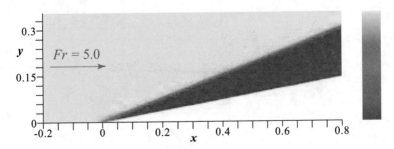

Figure 16.56: Froude Number Contours

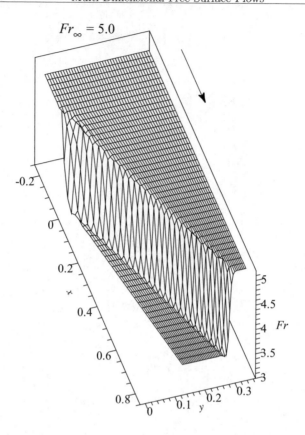

Figure 16.57: Froude Number Surface

Chapter 17

Multi-Dimensional Compressible Flows

This chapter expounds the multi-dimensional acoustics-convection upstream resolution algorithm and documents its computational performance for subsonic, transonic, supersonic and hypersonic smooth and shocked flows. Developed for the two- and three-dimensional Euler and Navier-Stokes equations, this formulation introduces an intrinsically infinite directional and multi-dimensional upstream bias along characteristic directions. This upstream bias is directly induced at the differential equation level, before any discretization, within a characteristics-bias system associated with the Euler and Navier-Stoles equations. For any Mach number, this system induces consistent upwinding along all wave-propagation directions radiating from any flow-field point. An integral formulation of this companion system then directly yields a generalized non-discrete SUPG formulation.

The characteristics-bias system relies upon an intrinsically multi-dimensional characteristics analysis based on a non-linear wave-like form of the solutions of the Euler equations. This analysis recognizes that within a compressible flow field, acoustic and convection waves propagate in infinitely many directions and along each direction the associated propagation velocity also depends on the Mach number. The results of this characteristics analysis lead to the spatial distribution of multi-dimensional propagation velocities and shows that among all propagation directions the streamline and crossflow directions are principal propagation directions in the time-space continuum. This line of inquiry yields specific conditions for a physically coherent upstream bias formulation that remains consistent with multidimensional acoustic and convection wave propagation.

The formulation developed in this chapter decomposes the multi-dimensional flux Jacobian into components that genuinely model the physics of acoustics and convection and feature eigenvalues with uniform algebraic sign. The formulation then establishes a physically consistent upstream approximation for each of these components, with an upstream-bias magnitude that correlates with the directional distribution of the characteristic propagation speeds. This formulation eliminates the unstable linear-dependence problem in steady low-Mach-number flows and satisfies by design the upstream-bias stability condition. For increasing Mach numbers from zero, the approximation gradually changes from bi-modal in subsonic flows to mono-axial in supersonic flows, within a streamline conical region that encompasses the domain of dependence and range of influence of each flow-field point; simultaneously, the approximation remains bi-modal for all Mach numbers within an appropriate conical region about the crossflow direction.

A traditional Galerkin finite element discretization on arbitrary grids of the characteristics-bias system automatically and directly generates a consistent and genuinely multi - dimensional upstream-bias approximation of the Euler and Navier-Stokes equations. The associated discrete infinite - directional upstream-bias remains independent of the direction of the coordinate axes as well as orientation of each computational-cell side, which obviates the need for rotated stencils. This approximation requires data only from the computational cells shared by each grid node and also reduces to a consistent upstream approximation of the acoustics equations, for vanishing Mach number. The operation count for this algorithm is comparable to that of a simple flux vector splitting algorithm. The upstream principal directions are continuously updated and high-rate convergence to machine zero achieved without additional shock-capturing terms, data filtering or loss of essential monotonicity.

17.1   Euler and Navier-Stokes Systems

With respect to an inertial Cartesian reference frame and implied summation on repeated subscript indices, the Navier-Stokes conservation law system

$$\frac{\partial q}{\partial t} + \frac{\partial f_j(q)}{\partial x_j} - \frac{\partial f_j^\nu}{\partial x_j} = 0 \tag{17.1}$$

consists of the continuity, momentum and total-energy equations. In the absence of viscous stresses and heat conduction, the viscous vector components f_j^ν vanishes and this system reduces to the Euler system. This system becomes a hyperbolic system because the flux vector Jacobian $(\partial f_j/\partial q)n_j$ possesses real eigenvalues and a complete set of eigenvectors for any unit vector $\boldsymbol{n}$, with direction cosines n_j providing the Cartesian components of $\boldsymbol{n}$ along the coordinate axes. With reference to Figure 17.1, for a representative 2-D flow field, the unit vector $\boldsymbol{n}$ indicates the propagation direction of plane waves with speeds that equal the eigenvalues of $\frac{\partial f_j(q)}{\partial q}n_j$, as elaborated in Section 17.3.

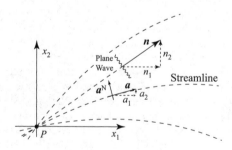

Figure 17.1: Reference Unit Vectors

The figure also displays the unit vectors $\mathbf{a}$ and $\mathbf{a}^N$ respectively pointing in the streamline and crossflow directions, vectors that provide the two principal directions for the decomposition developed in Section 17.4.

17.1.1 2-D Formulation

For two-dimensional flows, subscript j varies in the range $1 \leq j \leq 2$, while the independent variable $(\boldsymbol{x}, t) \equiv (x_1, x_2, t)$ in (17.1) varies in the Cartesian-product domain $D \equiv \Omega \times [t_o, t_f]$, $[t_o, t_f] \in \Re^+$, $\Omega \subset \Re^2$, with $\Re$ denoting the real-number field. The dependent-variable array $q = q(\boldsymbol{x}, t)$ and viscous and inviscid flux "vector" components f_j^ν, and $f_j = f_j(q)$ are then defined as

$$
q \equiv \left\{ \begin{array}{c} \rho \\ m_1 \\ m_2 \\ E \end{array} \right\}, \quad
f_1 \equiv \left\{ \begin{array}{c} m_1 \\ \dfrac{m_1}{\rho} m_1 + p \\ \dfrac{m_1}{\rho} m_2 \\ \dfrac{m_1}{\rho}(E + p) \end{array} \right\}, \quad
f_2 \equiv \left\{ \begin{array}{c} m_2 \\ \dfrac{m_2}{\rho} m_1 \\ \dfrac{m_2}{\rho} m_2 + p \\ \dfrac{m_2}{\rho}(E + p) \end{array} \right\}, \quad
f_j^\nu \equiv \left\{ \begin{array}{c} 0 \\ \sigma_{1j}^\mu \\ \sigma_{2j}^\mu \\ \dfrac{m_i}{\rho}\sigma_{ij}^\mu - q_j^F \end{array} \right\}
\tag{17.2}
$$

17.1.2 3-D Formulation

For three-dimensional flows, subscript j varies in the range $1 \leq j \leq 3$ and the independent variable $(\boldsymbol{x}, t) \equiv (x_1, x_2, x_3, t)$ varies in the domain $D \equiv \Omega \times [t_o, t_f]$, $[t_o, t_f] \in \Re^+$, $\Omega \subset \Re^3$. The dependent-variable array $q = q(\boldsymbol{x}, t)$ and viscous and inviscid flux "vector" components f_j^ν and $f_j = f_j(q)$ are then defined as

$$
q \equiv \left\{ \begin{array}{c} \rho \\ m_1 \\ m_2 \\ m_3 \\ E \end{array} \right\}, \quad
f_j^\nu \equiv \left\{ \begin{array}{c} 0 \\ \sigma_{1j}^\mu \\ \sigma_{2j}^\mu \\ \sigma_{3j}^\mu \\ \dfrac{m_i}{\rho}\sigma_{ij}^\mu - q_j^F \end{array} \right\}
\tag{17.3}
$$

$$
f_1 \equiv \left\{ \begin{array}{c} m_1 \\ \dfrac{m_1}{\rho} m_1 + p \\ \dfrac{m_1}{\rho} m_2 \\ \dfrac{m_1}{\rho} m_3 \\ \dfrac{m_1}{\rho}(E + p) \end{array} \right\}, \quad
f_2 \equiv \left\{ \begin{array}{c} m_2 \\ \dfrac{m_2}{\rho} m_1 \\ \dfrac{m_2}{\rho} m_2 + p \\ \dfrac{m_2}{\rho} m_3 \\ \dfrac{m_2}{\rho}(E + p) \end{array} \right\}, \quad
f_3 \equiv \left\{ \begin{array}{c} m_3 \\ \dfrac{m_3}{\rho} m_1 \\ \dfrac{m_3}{\rho} m_2 \\ \dfrac{m_3}{\rho} m_3 + p \\ \dfrac{m_3}{\rho}(E + p) \end{array} \right\}
\tag{17.4}
$$

17.1.3 Dependent Variables

In the array q, the variables ρ, m_1, m_2, m_3, and E, respectively denote static density and volume-specific linear momentum components and total energy. Concerning the viscous and inviscid fluxes, the variables σ_{ij}^{μ}, q_j^F and p respectively indicate the deviatoric stress-tensor components, the Fourier heat conduction flux component and static pressure. The Eulerian flow velocity $\boldsymbol{u}$, with cartesian components u_j, $1 \le j \le 3$, is then defined as $\boldsymbol{u} \equiv \boldsymbol{m}/\rho$.

17.1.4 Constitutive Relations

With reference to Chapter 2, the components $\tau_{ij} \equiv \sigma_{ij}^{\mu}$ of the deviatoric stress tensor are expressed as

$$\tau_{ij} = \mu \left(\frac{\partial u_i}{\partial x_j} + \frac{\partial u_j}{\partial x_i} \right) + \overline{\lambda} \frac{\partial u_\ell}{\partial x_\ell} \delta_i^j, \quad \overline{\lambda} = -\frac{2}{3}\mu + \eta_B \tag{17.5}$$

where μ and $\overline{\lambda}$ respectively denote the first and second coefficient of viscosity, with η_B the bulk viscosity. For mono-atomic gases η_B vanishes, while for other fluids, like air, the two coefficients of viscosity are classically related by specifying equality between mechanical and thermodynamic pressure, which leads to $\overline{\lambda} = -2\mu/3$, traditionally known as Stokes' hypothesis. The components q_j^F of the Fourier heat flux vector are expressed as

$$q_i^F = -k \frac{\partial T}{\partial x_i} \tag{17.6}$$

where T indicates static temperature and k denotes the coefficient of thermal conductivity.

17.1.5 Equations of State and Speed of Sound

For any homogeneous equilibrium gas, pressure can be expressed as a function of two other thermodynamic variables. They are density ρ and mass-specific internal energy ϵ, in this case, since they are readily available from the Euler system (17.1): ρ directly from the continuity equation in the system, and ϵ from q as

$$\epsilon \equiv \frac{E}{\rho} - \frac{1}{2\rho^2} \left(m_1^2 + m_2^2 + m_3^2 \right) \tag{17.7}$$

The pressure equation of state thus becomes

$$p = p(\rho, \epsilon) = p \left(\rho, \frac{E}{\rho} - \frac{1}{2\rho^2} \left(m_1^2 + m_2^2 + m_3^2 \right) \right) \tag{17.8}$$

According to this expression, the Jacobian derivatives of p with respect to q, for the Jacobian $\partial f_j / \partial q$ of $f_j(q)$, are not all independent of one another. The Jacobian derivatives of (17.8) with respect to m_1, m_2, m_3 and E in fact satisfy the constraints

$$\frac{\partial p}{\partial m_1} = -\frac{m_1}{\rho} \frac{\partial p}{\partial E}, \quad \frac{\partial p}{\partial m_2} = -\frac{m_2}{\rho} \frac{\partial p}{\partial E}, \quad \frac{\partial p}{\partial m_3} = -\frac{m_3}{\rho} \frac{\partial p}{\partial E} \tag{17.9}$$

as obtained by expressing the derivatives of p with respect to m_1, m_2, m_3 and E in terms of the thermodynamic derivative of p with respect to ϵ, from the first expression in (17.8). In the following sections, for simplicity, the abridged notation

$$p_\rho \equiv \frac{\partial p}{\partial \rho}, \quad p_{m_1} \equiv \frac{\partial p}{\partial m_1}, \quad p_{m_2} \equiv \frac{\partial p}{\partial m_2}, \quad p_{m_3} \equiv \frac{\partial p}{\partial m_3}, \quad p_E \equiv \frac{\partial p}{\partial E} \tag{17.10}$$

will denote the Jacobian derivatives of pressure. In terms of these derivatives, the square of the speed of sound c for general equations of state can be expressed as

$$c^2 \equiv \left.\frac{\partial p}{\partial \rho}\right|_S = p_\rho + p_E \left(\frac{E+p}{\rho} - \frac{1}{\rho^2}\left(m_1^2 + m_2^2 + m_3^2\right) \right) \tag{17.11}$$

This result, in particular, allows expressing the mass-specific total enthalpy H as

$$H = \frac{E+p}{\rho} = \frac{c^2\left(1 + p_E M^2\right) - p_\rho}{p_E} \tag{17.12}$$

where $M \equiv \|\boldsymbol{u}\|/c$ denotes the Mach number.

Chapter 3 details the calculation of pressure and its derivative for high-temperature reacting gases. For a perfect gas, the specific expressions for (17.7) and (17.8) follow from the internal energy and equation of state

$$\epsilon = c_v T = \frac{R}{\gamma - 1}T, \quad p = \rho R T \tag{17.13}$$

where c_v, T, R and γ respectively denote the constant-volume specific heat, static temperature, gas constant and specific-heat ratio. The elimination of T from these two expressions, along with (17.7), leads to the following familiar expressions for the equation of state for p in terms of q

$$p = (\gamma - 1)\rho\epsilon = (\gamma - 1)\left(E - \frac{1}{2\rho}\left(m_1^2 + m_2^2 + m_3^2\right) \right) \tag{17.14}$$

With this equation of state, the corresponding Jacobian derivatives of pressure become

$$p_\rho = (\gamma - 1)\frac{m_1^2 + m_2^2 + m_3^2}{2\rho^2}, \quad p_{m_i} = -(\gamma - 1)\frac{m_i}{\rho}, \quad p_E = (\gamma - 1) \tag{17.15}$$

which satisfy (2.22). From (17.11), the perfect-gas speed of sound can be expressed as

$$c^2 = (\gamma - 1)\left[\frac{E+p}{\rho} - \frac{m_1^2 + m_2^2 + m_3^2}{2\rho^2} \right] \tag{17.16}$$

directly in terms of the dependent variables.

17.2 Acoustics Equations and Streamline and Crossflow Components

This section identifies the acoustics equations embedded within the multi-dimensional Euler equations. Such an identification determines the acoustic-wave propagation matrix that then leads to an accurate upstream-bias modelling of acoustics-wave propagation.

17.2.1   Acoustics Systems

Two-Dimensional Flows

The 2-D Euler equations

$$
\begin{cases}
\dfrac{\partial \rho}{\partial t} + \dfrac{\partial m_1}{\partial x_1} + \dfrac{\partial m_2}{\partial x_2} = 0 \\[2mm]
\dfrac{\partial m_1}{\partial t} + \dfrac{\partial}{\partial x_1}\left(\dfrac{m_1}{\rho} m_1 + p\right) + \dfrac{\partial}{\partial x_2}\left(\dfrac{m_2}{\rho} m_1\right) = 0 \\[2mm]
\dfrac{\partial m_2}{\partial t} + \dfrac{\partial}{\partial x_1}\left(\dfrac{m_1}{\rho} m_2\right) + \dfrac{\partial}{\partial x_2}\left(\dfrac{m_2}{\rho} m_2 + p\right) = 0 \\[2mm]
\dfrac{\partial E}{\partial t} + \dfrac{\partial}{\partial x_1}\left(\dfrac{m_1}{\rho}(E + p)\right) + \dfrac{\partial}{\partial x_2}\left(\dfrac{m_2}{\rho}(E + p)\right) = 0
\end{cases}
\tag{17.17}
$$

are abridged as the system

$$
\frac{\partial q}{\partial t} + \frac{\partial f_j(q)}{\partial x_j} = 0
\tag{17.18}
$$

Let (v_1, v_2) denote the components of a unit vector $\boldsymbol{v}$ locally parallel to the velocity $\boldsymbol{u}$. Upon writing the velocity components (u_1, u_2) in terms of the Mach number M as

$$
(u_1, u_2) = cM(v_1, v_2)
\tag{17.19}
$$

and using the pressure derivative identities (17.9), the Euler system becomes

$$
\frac{\partial}{\partial t}\begin{pmatrix}\rho \\ m_1 \\ m_2 \\ E\end{pmatrix} + \left[\begin{pmatrix}0 & , & \delta_1^j & , & \delta_2^j & , & 0 \\ p_\rho \delta_1^j & , & 0 & , & 0 & , & p_E \delta_1^j \\ p_\rho \delta_2^j & , & 0 & , & 0 & , & p_E \delta_2^j \\ 0 & , & H\delta_1^j & , & H\delta_2^j & , & 0\end{pmatrix} + cMC_j\right]\frac{\partial}{\partial x_j}\begin{pmatrix}\rho \\ m_1 \\ m_2 \\ E\end{pmatrix}
\tag{17.20}
$$

where δ_i^j and C_j respectively denote Kronecher's delta and the matrix

$$
C_j \equiv \begin{pmatrix}
0 & , & 0 & , & 0 & , & 0 \\
-v_1 u_j & , & v_1\delta_1^j + v_j - v_1 p_E \delta_1^j & , & v_1\delta_2^j - v_2 p_E \delta_1^j & , & 0 \\
-v_2 u_j & , & v_2\delta_1^j - v_1 p_E \delta_2^j & , & v_2\delta_2^j + v_j - v_2 p_E \delta_2^j & , & 0 \\
v_j\left(p_\rho - H\right) & , & -v_j u_1 p_E & , & -v_j u_2 p_E & , & v_j\left(1 + p_E\right)
\end{pmatrix}
\tag{17.21}
$$

For a vanishing Mach Number, these equations reduce to the acoustics equations

$$
\frac{\partial}{\partial t}\begin{pmatrix}\rho \\ m_1 \\ m_2 \\ E\end{pmatrix} + \begin{pmatrix}0 & , & \delta_1^j & , & \delta_2^j & , & 0 \\ p_\rho \delta_1^j & , & 0 & , & 0 & , & p_E \delta_1^j \\ p_\rho \delta_2^j & , & 0 & , & 0 & , & p_E \delta_2^j \\ 0 & , & H\delta_1^j & , & H\delta_2^j & , & 0\end{pmatrix}\frac{\partial}{\partial x_j}\begin{pmatrix}\rho \\ m_1 \\ m_2 \\ E\end{pmatrix} = 0
\tag{17.22}
$$

for which A_j^a will denote the acoustics matrix multiplying the gradient of q in (17.22). The dependent variables in these equations correspond to those in a flow field that originates from slight perturbations to an otherwise quiescent field. Chapter 1 has developed the following relationship between the derivatives of density and energy

$$\left(\frac{c^2 - p_\rho}{p_E}\right)\frac{\partial\rho}{\partial x_i} = \frac{\partial E}{\partial x_i} - \frac{m_j}{\rho}\frac{\partial m_j}{\partial x_i} \tag{17.23}$$

In terms of the Mach number this result becomes

$$\left(\frac{c^2 - p_\rho}{p_E}\right)\frac{\partial\rho}{\partial x_i} = \frac{\partial E}{\partial x_i} - cv_j M\frac{\partial m_j}{\partial x_i} \tag{17.24}$$

For acoustic flows with $M \to 0$, this result simplifies as

$$\frac{\partial E}{\partial x_i} = \left(\frac{c^2 - p_\rho}{p_E}\right)\frac{\partial\rho}{\partial x_i}, \quad \frac{\partial\rho}{\partial x_i} = \left(\frac{p_E}{c^2 - p_\rho}\right)\frac{\partial E}{\partial x_i} \tag{17.25}$$

On the basis of (17.25) the acoustics equations become

$$\begin{cases} \dfrac{\partial\rho}{\partial t} = -\dfrac{\partial m_j}{\partial x_j} \\[2mm] \dfrac{\partial m_i}{\partial t} = -c^2\dfrac{\partial\rho}{\partial x_i} \end{cases} \tag{17.26}$$

Heed, in particular, that the energy equation toward steady state is no longer linearly independent from the continuity equation. This linear-dependence phenomenon explains the widely reported convergence difficulties towards steady state experienced in the CFD simulation of incompressible, i.e. low-Mach-number, flows with a compressible flow formulation. The infinite-directional algorithm detailed in this chapter incorporates a convenient upstream-bias expression to eliminate this steady-state linear-dependence phenomenon.

Following the characteristics analysis in Chapter 14, the characteristic speeds λ associated with this system correspond to the eigenvalues of the matrix $A_j^a n_j$, which are

$$\lambda_{1,2}^a = 0, \quad \lambda_{3,4}^a = \pm\left(p_\rho + p_E H\right)^{1/2} = \pm c = \pm\left(\left.\frac{\partial p}{\partial\rho}\right|_S\right)^{1/2} \quad \Rightarrow \quad H = \frac{c^2 - p_\rho}{p_E} \tag{17.27}$$

Note that these eigenvalues remain independent from the wave-propagation direction unit vector n, which corresponds to isotropic propagation. Both c and the eigenvalues $\lambda_{3,4}^a$ in (17.27) correspond to isentropic speed of sound (17.11).

Three-Dimensional Flows

The 3-D Euler equations

$$
\begin{cases}
\dfrac{\partial \rho}{\partial t} + \dfrac{\partial m_1}{\partial x_1} + \dfrac{\partial m_2}{\partial x_2} + \dfrac{\partial m_3}{\partial x_3} = 0 \\[2mm]
\dfrac{\partial m_1}{\partial t} + \dfrac{\partial}{\partial x_1}\left(\dfrac{m_1}{\rho}m_1 + p\right) + \dfrac{\partial}{\partial x_2}\left(\dfrac{m_2}{\rho}m_1\right) + \dfrac{\partial}{\partial x_3}\left(\dfrac{m_3}{\rho}m_1\right) = 0 \\[2mm]
\dfrac{\partial m_2}{\partial t} + \dfrac{\partial}{\partial x_1}\left(\dfrac{m_1}{\rho}m_2\right) + \dfrac{\partial}{\partial x_2}\left(\dfrac{m_2}{\rho}m_2 + p\right) + \dfrac{\partial}{\partial x_3}\left(\dfrac{m_3}{\rho}m_2\right) = 0 \\[2mm]
\dfrac{\partial m_3}{\partial t} + \dfrac{\partial}{\partial x_1}\left(\dfrac{m_1}{\rho}m_3\right) + \dfrac{\partial}{\partial x_2}\left(\dfrac{m_2}{\rho}m_3\right) + \dfrac{\partial}{\partial x_3}\left(\dfrac{m_3}{\rho}m_3 + p\right) = 0 \\[2mm]
\dfrac{\partial E}{\partial t} + \dfrac{\partial}{\partial x_1}\left(\dfrac{m_1}{\rho}(E+p)\right) + \dfrac{\partial}{\partial x_2}\left(\dfrac{m_2}{\rho}(E+p)\right) + \dfrac{\partial}{\partial x_3}\left(\dfrac{m_3}{\rho}(E+p)\right) = 0
\end{cases}
\tag{17.28}
$$

are abridged as the system

$$
\frac{\partial q}{\partial t} + \frac{\partial f_j(q)}{\partial x_j} = 0
\tag{17.29}
$$

Similarly to the two-dimensional case, let (v_1, v_2, v_3) denote the components of a unit vector $\boldsymbol{v}$ locally parallel to the velocity $\boldsymbol{u}$. Upon expressing the velocity components (u_1, u_2, u_3) in terms of the Mach number M as

$$
(u_1, u_2, u_3) = cM(v_1, v_2, v_3)
\tag{17.30}
$$

and using the pressure derivative identities (17.9), the Euler system becomes

$$
\frac{\partial}{\partial t}
\begin{pmatrix} \rho \\ m_1 \\ m_2 \\ m_3 \\ E \end{pmatrix}
+
\left[
\begin{pmatrix}
0 & , & \delta_1^j & , & \delta_2^j & , & \delta_3^j & , & 0 \\
p_\rho \delta_1^j & , & 0 & , & 0 & , & 0 & , & p_E \delta_1^j \\
p_\rho \delta_2^j & , & 0 & , & 0 & , & 0 & , & p_E \delta_2^j \\
p_\rho \delta_3^j & , & 0 & , & 0 & , & 0 & , & p_E \delta_3^j \\
0 & , & H\delta_1^j & , & H\delta_2^j & , & H\delta_3^j & , & 0
\end{pmatrix}
+ cMC_j
\right]
\frac{\partial}{\partial x_j}
\begin{pmatrix} \rho \\ m_1 \\ m_2 \\ m_3 \\ E \end{pmatrix}
= 0
\tag{17.31}
$$

where C_j denotes the completion matrix

$$
C_j \equiv
$$

$$
\begin{pmatrix}
0 & , & 0 & , & 0 & , & 0 & , & 0 \\
-v_1 u_j & , & v_1\delta_1^j + v_j - v_1 p_E \delta_1^j & , & v_1\delta_2^j - v_2 p_E \delta_1^j & , & v_1\delta_3^j - v_3 p_E \delta_1^j & , & 0 \\
-v_2 u_j & , & v_2\delta_1^j - v_1 p_E \delta_2^j & , & v_2\delta_2^j + v_j - v_2 p_E \delta_2^j & , & v_2\delta_3^j - v_3 p_E \delta_2^j & , & 0 \\
-v_3 u_j & , & v_3\delta_1^j - v_1 p_E \delta_3^j & , & v_3\delta_2^j - v_2 p_E \delta_3^j & , & v_3\delta_3^j + v_j - v_3 p_E \delta_3^j & , & 0 \\
v_j\left(p_\rho - H\right) & , & -v_j u_1 p_E & , & -v_j u_2 p_E & , & -v_j u_3 p_E & , & v_j\left(1 + p_E\right)
\end{pmatrix}
\tag{17.32}
$$

For a vanishing M these equations reduce to the acoustics equations

$$\frac{\partial}{\partial t}\begin{pmatrix} \rho \\ m_1 \\ m_2 \\ m_3 \\ E \end{pmatrix} + \begin{pmatrix} 0 & , & \delta_1^j & , & \delta_2^j & , & \delta_3^j & , & 0 \\ p_\rho\delta_1^j & , & 0 & , & 0 & , & 0 & , & p_E\delta_1^j \\ p_\rho\delta_2^j & , & 0 & , & 0 & , & 0 & , & p_E\delta_2^j \\ p_\rho\delta_3^j & , & 0 & , & 0 & , & 0 & , & p_E\delta_3^j \\ 0 & , & H\delta_1^j & , & H\delta_2^j & , & H\delta_3^j & , & 0 \end{pmatrix} \frac{\partial}{\partial x_j}\begin{pmatrix} \rho \\ m_1 \\ m_2 \\ m_3 \\ E \end{pmatrix} = 0 \qquad (17.33)$$

for which A_j^a will again denote the acoustics matrix multiplying the gradient of q in (17.33). The dependent variables in these equations correspond to those in a flow field that originates from slight perturbations to an otherwise quiescent field. Heed, in particular, that also for three-dimensional flows is the energy equation toward steady state no longer linearly independent from the continuity equation.

Following the characteristics analysis in Chapter 14, the characteristic speeds λ for this system correspond to the eigenvalues of the matrix $A_j^a n_j$, which are

$$\lambda_{0,2}^a = 0, \quad \lambda_{3,4}^a = \pm\left(p_\rho + p_E H\right)^{1/2} = \pm c = \pm\left(\left.\frac{\partial p}{\partial \rho}\right|_S\right)^{1/2} \quad \Rightarrow \quad H = \frac{c^2 - p_\rho}{p_E} \qquad (17.34)$$

Note that these eigenvalues remain again independent from the wave-propagation direction unit vector $\boldsymbol{n}$. Both c and the eigenvalues $\lambda_{3,4}^a$ in (17.34) correspond to isentropic speed of sound (2.30).

17.2.2 Streamline and Crossflow Components

Two-Dimensional Flows

For any two mutually perpendicular unit vectors $\boldsymbol{a} = (a_1, a_2)$ and $\boldsymbol{a}^N = (a_1^N, a_2^N)$ within a 2-D flow, along with implied summation on repeated indices, the acoustic matrix product $A_j^a\frac{\partial q}{\partial x_j}$ may be expressed as

$$A_j^a\frac{\partial q}{\partial x_j} = A_j^a a_j a_k \frac{\partial q}{\partial x_k} + A_j^a a_j^N a_k^N \frac{\partial q}{\partial x_k} \qquad (17.35)$$

For any unit vector $\boldsymbol{a}$, the matrix $A_j^a a_j$ is expressed as

$$A_j^a a_j \equiv \begin{pmatrix} 0 & , & a_1 & , & a_2 & , & 0 \\ p_\rho a_1 & , & 0 & , & 0 & , & p_E a_1 \\ p_\rho a_2 & , & 0 & , & 0 & , & p_E a_2 \\ 0 & , & \dfrac{c^2 - p_\rho}{p_E}a_1 & , & \dfrac{c^2 - p_\rho}{p_E}a_2 & , & 0 \end{pmatrix} \qquad (17.36)$$

For $\boldsymbol{a}$ parallel to $\boldsymbol{u}$, expression (17.35) corresponds to a decomposition of the Euler acoustics component into streamline and crossflow acoustics components. For wave-like solutions (17.53), two eigenvalues of each component vanish; the remaining eigenvalues of these separate components have been determined as

$$\lambda_{3,4}^s = \pm c_s = \pm ca_j n_j, \quad \lambda_{3,4}^N = \pm c_N = \pm ca_j^N n_j \qquad (17.37)$$

The two non-vanishing eigenvalues associated with the entire acoustics component at the lhs of (17.35), but as expressed as the rhs combination of streamline and crossflow components have then been determined as

$$\lambda_{3,4}^{a} = c\left(\left(a_j n_j \right)^2 + \left(a_j^N n_j \right)^2 \right)^{1/2}, \quad \left(a_j n_j \right)^2 + \left(a_j^N n_j \right)^2 = 1 \tag{17.38}$$

which shows that the square of the acoustic eigenvalues (17.27) equals the sum of the square of the streamline and crossflow acoustic eigenvalues (17.37), as illustrated in Figure 17.2.

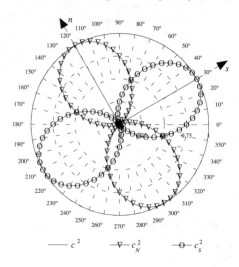

Figure 17.2: Polar Variation of Square of Acoustic Speeds

Three-Dimensional Flows

For any three mutually orthogonal unit vectors $\boldsymbol{a} = (a_1, a_2, a_3)$, $\boldsymbol{a}^{N_1} = (a_1^{N_1}, a_2^{N_1}, a_3^{N_1})$, and $\boldsymbol{a}^{N_2} = (a_1^{N_2}, a_2^{N_2}, a_3^{N_2})$ within a 3-D flow, along with implied summation on repeated indices, the acoustic matrix product $A_j^a \frac{\partial q}{\partial x_j}$ may be expressed as

$$A_j^a \frac{\partial q}{\partial x_j} = A_j^a a_j a_k \frac{\partial q}{\partial x_k} + A_j^a a_j^{N_1} a_k^{N_1} \frac{\partial q}{\partial x_k} + A_j^a a_j^{N_2} a_k^{N_2} \frac{\partial q}{\partial x_k} \tag{17.39}$$

For any vector $\boldsymbol{a}$, the matrix $A_j^a a_j$ is expressed as

$$A_j^a a_j \equiv \begin{pmatrix} 0 & , & a_1 & , & a_2 & , & a_3 & , & 0 \\ p_\rho a_1 & , & 0 & , & 0 & , & 0 & , & p_E a_1 \\ p_\rho a_2 & , & 0 & , & 0 & , & 0 & , & p_E a_2 \\ p_\rho a_3 & , & 0 & , & 0 & , & 0 & , & p_E a_3 \\ 0 & , & \dfrac{c^2 - p_\rho}{p_E} a_1 & , & \dfrac{c^2 - p_\rho}{p_E} a_2 & , & \dfrac{c^2 - p_\rho}{p_E} a_3 & , & 0 \end{pmatrix} \tag{17.40}$$

For $\boldsymbol{a}$ parallel to $\boldsymbol{u}$, expression (17.39) corresponds to a decomposition of the Euler acoustics component into one streamline and two crossflow acoustics components. For wave-like solutions (17.53), three eigenvalues of each component vanish; the remaining eigenvalues of these three separate components have been determined as

$$\lambda_{3,4}^s = \pm c a_j n_j, \quad \lambda_{3,4}^{N_1} = \pm c a_j^{N_1} n_j, \quad \lambda_{3,4}^{N_2} = \pm c a_j^{N_2} n_j \tag{17.41}$$

The two non-vanishing eigenvalues associated with the entire acoustics component at the lhs of (17.39), but as expressed as the rhs combination of streamline and crossflow components have then been determined as

$$\lambda_{3,4}^a = c \left(\left(a_j n_j \right)^2 + \left(a_j^{N_1} n_j \right)^2 + \left(a_j^{N_2} n_j \right)^2 \right)^{1/2}, \quad \left(a_j n_j \right)^2 + \left(a_j^{N_1} n_j \right)^2 + \left(a_j^{N_2} n_j \right)^2 = 1 \tag{17.42}$$

which shows that the square of the acoustic eigenvalues (17.34) equals the sum of the square of the streamline and crossflow acoustic eigenvalues (17.41). On any plane Π that contains $\boldsymbol{a}$, $\boldsymbol{a}^{N_1}$ and $\boldsymbol{n}$, therefore, acoustic propagation naturally decomposes into two directional components, one along $\boldsymbol{a}$ and the other along $\boldsymbol{a}^{N_1}$, as illustrated in Figure 17.3.

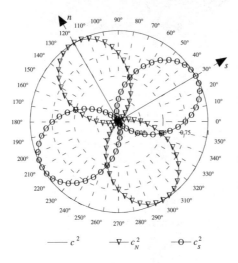

Figure 17.3: Polar Variation of Square of Acoustic Speeds on Plane Π

The variation of the squares of the three principal acoustic speeds is illustrated in Figure 17.4. The planar two-directional acoustic decomposition remains unaltered on plane Π, as this plane spans the entire 3-D space by rotating about the line of $\boldsymbol{a}$. Acoustic propagation thus decomposes into a directional component along $\boldsymbol{a}$ and an isotropic component on any plane orthogonal to $\boldsymbol{a}$. This result also follows from the expression

$$\left(a_j^{N_1} n_j \right)^2 + \left(a_j^{N_2} n_j \right)^2 = 1 - \left(a_j n_j \right)^2 \tag{17.43}$$

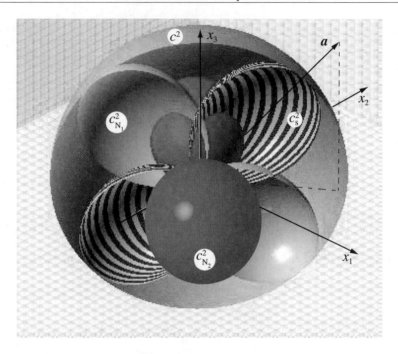

Figure 17.4: Spherical Variation of Square of Three Principal Acoustic Speeds

from (17.42), which shows that acoustic propagation on the plane of $\boldsymbol{a}^{N_1}$ and $\boldsymbol{a}^{N_2}$ only depends upon $(a_j n_j)$, hence on the angle between $\boldsymbol{a}$ and $\boldsymbol{n}$. For any such angle, the "lhs" of (17.43) remains constant for any orientation of $\boldsymbol{n}$ with respect to $\boldsymbol{a}^{N_1}$ and $\boldsymbol{a}^{N_2}$, as $\boldsymbol{n}$ rotates about $\boldsymbol{a}$. This result indicates that (17.43) is the equation of a circle, which signifies isotropic acoustic propagation on the plane of $\boldsymbol{a}^{N_1}$ and $\boldsymbol{a}^{N_2}$, as shown in Figure 17.5

17.2.3   Similarity Transformation

Two-Dimensional Flows

Despite its zero eigenvalues, $A_j^a a_j$ features a complete set of eigenvectors X and thus possess the similarity-transformation form

$$A_j^a a_j = X \Lambda^a X^{-1} \tag{17.44}$$

The corresponding diagonal matrix Λ^a is

$$\Lambda^a \equiv \begin{pmatrix} c & , & 0 & , & 0 & , & 0 \\ 0 & , & -c & , & 0 & , & 0 \\ 0 & , & 0 & , & 0 & , & 0 \\ 0 & , & 0 & , & 0 & , & 0 \end{pmatrix} \tag{17.45}$$

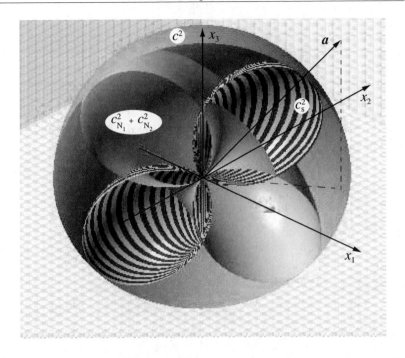

Figure 17.5: Spherical Variation of Square of Crossflow and Streamline Acoustic Speeds

The associated eigenvector matrix X and its inverse X^{-1} for general equations of state (17.8) are

$$X = \begin{pmatrix} p_E & , & p_E & , & p_E & , & 0 \\ cp_E a_1 & , & -cp_E a_1 & , & 0 & , & -a_2 \\ cp_E a_2 & , & -cp_E a_2 & , & 0 & , & a_1 \\ c^2 - p_\rho & , & c^2 - p_\rho & , & -p_\rho & , & 0 \end{pmatrix} \qquad (17.46)$$

$$X^{-1} = \frac{1}{2c^2 p_E} \begin{pmatrix} p_\rho & , & ca_1 & , & ca_2 & , & p_E \\ p_\rho & , & -ca_1 & , & -ca_2 & , & p_E \\ 2\left(c^2 - p_\rho\right) & , & 0 & , & 0 & , & -2p_E \\ 0 & , & -2c^2 p_E a_2 & , & 2c^2 p_E a_1 & , & 0 \end{pmatrix} \qquad (17.47)$$

Three-Dimensional Flows

Also for three-dimensional flows, despite its zero eigenvalues, will $A_j^a a_j$ feature a complete set of eigenvectors X and it thus possesses the similarity-transformation form

$$A_j^a a_j = X \Lambda^a X^{-1} \qquad (17.48)$$

where the diagonal matrix Λ^a is

$$\Lambda^a \equiv \begin{pmatrix} c & , & 0 & , & 0 & , & 0 & , & 0 \\ 0 & , & -c & , & 0 & , & 0 & , & 0 \\ 0 & , & 0 & , & 0 & , & 0 & , & 0 \\ 0 & , & 0 & , & 0 & , & 0 & , & 0 \\ 0 & , & 0 & , & 0 & , & 0 & , & 0 \end{pmatrix} \qquad (17.49)$$

where the corresponding eigenvector matrix X and its inverse X^{-1} for general equations of state (17.8) are

$$X = \begin{pmatrix} p_E & , & p_E & , & p_E & , & 0 & , & 0 \\ cp_E a_1 & , & -cp_E a_1 & , & 0 & , & -a_2 & , & -a_1 a_3 \\ cp_E a_2 & , & -cp_E a_2 & , & 0 & , & a_1 & , & -a_2 a_3 \\ cp_E a_3 & , & -cp_E a_3 & , & 0 & , & 0 & , & a_1^2 + a_2^2 \\ c^2 - p_\rho & , & c^2 - p_\rho & , & -p_\rho & , & 0 & , & 0 \end{pmatrix} \qquad (17.50)$$

$$X^{-1} = \frac{1}{2c^2 p_E} \begin{pmatrix} p_\rho & , & ca_1 & , & ca_2 & , & ca_3 & , & p_E \\ p_\rho & , & -ca_1 & , & -ca_2 & , & -ca_3 & , & p_E \\ 2\left(c^2 - p_\rho\right) & , & 0 & , & 0 & , & 0 & , & -2p_E \\ 0 & , & -\dfrac{2c^2 p_E a_2}{a_1^2 + a_2^2} & , & \dfrac{2c^2 p_E a_1}{a_1^2 + a_2^2} & , & 0 & , & 0 \\ 0 & , & -\dfrac{2c^2 p_E a_1 a_3}{a_1^2 + a_2^2} & , & -\dfrac{2c^2 p_E a_2 a_3}{a_1^2 + a_2^2} & , & 1 & , & 0 \end{pmatrix}$$

$$(17.51)$$

Since two linearly independent eigenvector are available for the single vanishing eigenvalue, it's always possible to find two linearly independent eigenvectors such that a non-singular X is always available.

17.3 Characteristic Analysis and Velocity Components

Within a multi-dimensional flow field, acoustic and convection waves propagate in infinitely many directions; along each direction the associated propagation velocity also depends on the Mach number. As an essential prerequisite for the developments in Section 17.4, this section presents an intrinsically multi-dimensional characteristics analysis based on a non-linear wave-like form of the Euler solution q. This analysis leads to the spatial distribution of multi-dimensional propagation velocities and shows that among all propagation directions the streamline and crossflow directions are principal propagation directions. This line of enquiry is highlighted because it yields specific conditions for a physically coherent upstream bias formulation.

17.3.1 Characteristic Velocity Components

For the multi-dimensional Euler system

$$\frac{\partial q}{\partial t} + \frac{\partial f_j(q)}{\partial x_j} = 0 \qquad (17.52)$$

the non-linear wave-like form of q is expressed as

$$q = q(\eta_1), \quad \eta_1 = \boldsymbol{x} \cdot \boldsymbol{n} - \lambda(q)t = x_j n_j - \lambda(q)t \qquad (17.53)$$

where $\boldsymbol{n}$ denotes a space-domain propagation-direction unit vector, independent of $(\boldsymbol{x}, t)$, and $\lambda = \lambda(q)$ indicates a wave-propagation velocity component along the $\boldsymbol{n}$ direction. This solution-dependent velocity component for this system is determined by enforcing the condition that the non-linear wave-like solutions (17.53) satisfy this system. For linear and non-linear $\lambda = \lambda(q)$, Chapter 14 shows that this condition yields the eigenvalue problem

$$\left(-\lambda(q)I + \frac{\partial f_j}{\partial q} n_j \right) \frac{\partial q}{\partial \eta_1} = 0 \qquad (17.54)$$

For non-trivial solutions $\frac{\partial q}{\partial \eta_1}$, hence $q = q(\eta_1)$, the characteristic velocity components λ are thus the eigenvalues of the linear combination of flux vector Jacobians

$$\frac{\partial f_j(q)}{\partial q} n_j =$$

$$\begin{pmatrix} 0 & , & n_1 & , & n_2 & , & 0 \\ -u_1 u_j n_j + p_\rho n_1 & , & u_1 n_1 + u_j n_j + p_{m_1} n_1 & , & u_1 n_2 + p_{m_2} n_1 & , & p_E n_1 \\ -u_2 u_j n_j + p_\rho n_2 & , & u_2 n_1 + p_{m_1} n_2 & , & u_2 n_2 + u_j n_j + p_{m_2} n_2 & , & p_E n_2 \\ u_j n_j \left(p_\rho - H \right) & , & H n_1 + u_j n_j p_{m_1} & , & H n_2 + u_j n_j p_{m_2} & , & u_j n_j \left(1 + p_E \right) \end{pmatrix}$$
$$(17.55)$$

in 2-D, and

$$\frac{\partial f_j(q)}{\partial q} n_j =$$

$$\begin{pmatrix} 0 & , & n_1 & , & n_2 & , & n_3 & , & 0 \\ -u_1 u_j n_j + p_\rho n_1 & , & u_1 n_1 + u_j n_j + p_{m_1} n_1 & , & u_1 n_2 + p_{m_2} n_1 & , & u_1 n_3 + p_{m_3} n_1 & , & p_E n_1 \\ -u_2 u_j n_j + p_\rho n_2 & , & u_2 n_1 + p_{m_1} n_2 & , & u_2 n_2 + u_j n_j + p_{m_2} n_2 & , & u_2 n_3 + p_{m_3} n_2 & , & p_E n_2 \\ -u_3 u_j n_j + p_\rho n_3 & , & u_3 n_1 + p_{m_1} n_3 & , & u_3 n_2 + p_{m_2} n_3 & , & u_3 n_3 + u_j n_j + p_{m_3} n_3 & , & p_E n_3 \\ u_j n_j \left(p_\rho - H \right) & , & H n_1 + u_j n_j p_{m_1} & , & H n_2 + u_j n_j p_{m_2} & , & H n_3 + u_j n_j p_{m_3} & , & u_j n_j \left(1 + p_E \right) \end{pmatrix}$$

$$(17.56)$$

in 3-D. For general equilibrium equations of state, these eigenvalues have been exactly determined in closed form as

$$\lambda_{1,2}^{d_E} = u_j n_j, \quad \lambda_{3,4}^{d_E} = u_j n_j \pm \left(p_\rho + p_E \left(H - u_j u_j \right) \right)^{1/2} \qquad (17.57)$$

where superscript d_E signifies dimensional Euler eigenvalues. These eigenvalues correspond to a 2-D flow. A 3-D flow simply induces another eigenvalue $\lambda_0^{d_E}$ that coincides with $\lambda_{1,2}^{d_E}$. Of interest, eigenvalues $\lambda_{3,4}^{d_E}$ directly incorporate a sound speed expression that coincides with the isentropic partial derivative of pressure (2.30), a result that is achieved without having to make the assumption of an isentropic flow. Through (2.30) these equilibrium-gas eigenvalues become

$$\lambda_{1,2}^{d_E} = u_j n_j, \quad \lambda_{3,4}^{d_E} = u_j n_j \pm c \qquad (17.58)$$

which have the same familiar form as the perfect-gas eigenvalues. The non-dimensional form of (17.58) follows from division by c, which supplies the Mach-number dependent expressions

$$\lambda_{1,2}^{E} = v_j n_j M, \quad \lambda_{3,4}^{E} = v_j n_j M \pm 1 \tag{17.59}$$

where v_1, and v_2, in 2-D, and v_1, v_2, and v_3, in 3-D, denote the components of a unit vector $\boldsymbol{v}$ in the velocity $\boldsymbol{u}$ direction.

As the inner product of the two unit vectors $\boldsymbol{n}$ and $\boldsymbol{v}$, the contraction $v_j n_j$ supplies the cosine of the angle $(\theta - \theta_v)$ between $\boldsymbol{n}$ and $\boldsymbol{v}$, where θ and θ_v respectively denote the angle between $\boldsymbol{n}$ and the x_1 axis and the angle between $\boldsymbol{v}$ and the x_1 axis in the flow plane, in 2-D, or in any plane Π that contains both $\boldsymbol{n}$ and $\boldsymbol{v}$, in 3-D. The eigenvalues (17.59) thus become

$$\lambda_{1,2}^{E} = \cos\left(\theta - \theta_v\right) M, \quad \lambda_{3,4}^{E} = \cos\left(\theta - \theta_v\right) M \pm 1 \tag{17.60}$$

These expressions, in particular, imply that the Euler eigenvalues achieve their extrema for $\theta = \theta_v$, hence when $\boldsymbol{n}$ points in the streamline direction, whereas for $\boldsymbol{n}$ pointing in the crossflow direction, hence $\theta = 90^\circ + \theta_v$, these eigenvalues no longer depend upon M.

The convection eigenvalues $\lambda_{1,2}^{E}$ vanish when $\cos\left(\theta - \theta_v\right) = 0$, hence for $\boldsymbol{n}$ perpendicular to the streamline direction, or, equivalently, pointing in the crossflow direction. Since $\| \cos\left(\theta - \theta_v\right) \| \leq 1$, the acoustic-convection eigenvalues $\lambda_{3,4}^{E}$ can only vanish for $M \geq 1$, hence for supersonic flows. For these flows, $\lambda_{3,4}^{E} = 0$ for

$$\mp \cos\left(\theta - \theta_v\right) = \pm \sin\left(\left(\theta - 90^\circ\right) - \theta_v\right) = \frac{1}{M} \tag{17.61}$$

hence for $\boldsymbol{n}$ perpendicular to the Mach lines, for $\pm\left(\left(\theta - 90^\circ\right) - \theta_v\right)$ corresponds to the angle between a Mach line and the streamline, from the well known second expression in (17.61). This equation, thus defines the Mach lines, in 2-D, and the Mach cone, in 3-D. The lines that are perpendicular to the Mach lines will be called "conjugate" lines.

The lines that are perpendicular to $\boldsymbol{n}$ for vanishing eigenvalues $\lambda_{1,2}^{E}$ and $\lambda_{3,4}^{E}$ thus respectively become the streamline and Mach lines. This result is not coincidental, for vanishing eigenvalues $\lambda_{1,4}^{E}$ correspond to wave-like solutions of the steady Euler equations, for which the streamline and Mach lines are characteristic-wave propagation lines.

17.3.2   Polar Variation of Characteristic Speeds

As a novel way of visualizing the Euler eigenvalues, Figures 17.6-17.7 present the polar variation of the absolute values of eigenvalues (17.59) for subsonic, sonic and supersonic Mach numbers, in a neighborhood of a flow field point P in the (x_1, x_2) plane, in 2-D, or any plane Π that contains both $\boldsymbol{n}$ and $\boldsymbol{v}$, in 3-D. These variations are obtained for a variable unit vector $\boldsymbol{n} \equiv (\cos\theta, \sin\theta)$ and fixed unit vector $\boldsymbol{v}$, in this representative case inclined by $+30^\circ$ with respect to the x_1 axis, in 2-D, or a reference x_1 axis on Π, in 3-D.

A collective inspection of all these diagrams reveals three shared features for all Mach numbers. The maximum characteristic speeds occur in the velocity direction, i.e. along a streamline, as noted before. Secondly, all the characteristic speeds are symmetrically distributed about the streamline direction. Thirdly, the eigenvalue pairs $(\|\lambda_1^{E}\|, \|\lambda_2^{E}\|)$ and

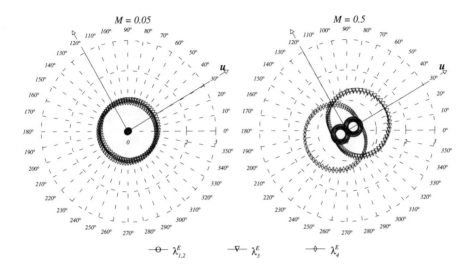

Figure 17.6: Polar Variation of Subsonic Wave Speeds

$(\|\lambda_3^E\|, \|\lambda_4^E\|)$ remain mirror skew-symmetric with respect to the crossflow direction, in the sense that the curves representative of $\|\lambda_2^E\|$ and $\|\lambda_4^E\|$ become the respective mirror images of the variations of $\|\lambda_1^E\|$ and $\|\lambda_3^E\|$ with reference to this direction. The streamline and crossflow directions thus become two fundamental wave-propagation axes.

For vanishing Mach numbers, the acoustic-convection propagation curves in the figure approach two circumferences. The distribution of propagation speeds in this case is isotropic, which corresponds to the direction-invariant propagation of acoustic waves. As the Mach number increases from zero, the curves in the figure progressively become circular asymmetric, which corresponds to anisotropic wave propagation. For $M = 0.5$ this anisotropy is already evident and becomes more pronounced for higher Mach numbers. The non-dimensional characteristic speeds then approach 1 in the region about the crossflow direction, which corresponds to essentially acoustic propagation.

Figures 17.8-17.9 present the 3-D spherical variation of the absolute values of eigenvalues (17.59) for subsonic and supersonic flows. The surfaces in these figures were drawn using spherical coordinates (r, θ, ϕ), where r represent the magnitude of these eigenvalues and θ and ϕ correspond to the spherical-coordinate angles depicted in Figure 17.8. With these angles, the unit vectors n and v may be expressed in Cartesian-coordinate form as

$$n \equiv (\sin\phi_n \cos\theta_n, \sin\phi_n \sin\theta_n, \cos\phi_n), \quad v \equiv (\sin\phi_v \cos\theta_v, \sin\phi_v \sin\theta_v, \cos\phi_v) \quad (17.62)$$

The corresponding expressions for eigenvalues (17.59) become

$$\lambda_{1,2}^E = (\sin\phi_n \cos\theta_n \sin\phi_v \cos\theta_v + \sin\phi_n \sin\theta_n \sin\phi_v \sin\theta_v + \cos\phi_n \cos\phi_v) M$$

$$\lambda_{3,4}^E = (\sin\phi_n \cos\theta_n \sin\phi_v \cos\theta_v + \sin\phi_n \sin\theta_n \sin\phi_v \sin\theta_v + \cos\phi_n \cos\phi_v) M \pm 1$$

$$(17.63)$$

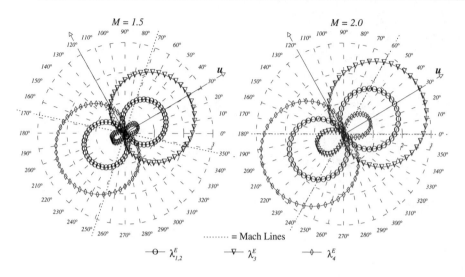

Figure 17.7: Polar Variation of Supersonic Wave Speeds

The surfaces in Figures 17.8-17.9 illustrate the space variations of the characteristic speeds. These variations correlate with the 2-D curves, in Figures 17.6-17.7, which result from intersecting these 3-D surfaces with plane Π. These surfaces too show that the extrema of the characteristic speeds occur along the streamline direction, whereas only acoustic propagation exists on the plane orthogonal to velocity at P.

For all Mach numbers, the convection eigenvalues $\lambda_{1,2}$ change sign when the $\boldsymbol{n}$-direction shifts from an upstream to a downstream axis with respect to $\boldsymbol{u}$. For this reason, the associated curves cross the polar origin. Pure convective propagation remains mono-axial, from upstream to downstream of P, and the axis of this type of wave propagation is the streamline.

For subsonic Mach numbers the acoustic-convection eigenvalues λ_3^E and λ_4^E respectively remain positive and negative for all directions. For this reason the associated curves contain the polar origin. For subsonic flows acoustic-convection waves propagate bi-modally, from both upstream and downstream toward and away from point P, along all directions radiating from P.

Beginning at the sonic state, this pattern drastically changes for supersonic Mach numbers. In this case both λ_3^E and λ_4^E change algebraic sign when the $\boldsymbol{n}$ shifts from upstream to downstream of P along a streamline. For this reason, the associated curves cross the polar origin. For supersonic flows acoustic-convection wave propagation becomes mono-axial along a streamline, from upstream to downstream of P.

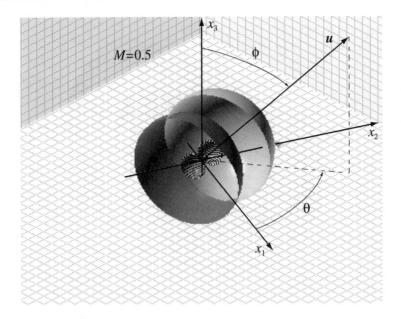

Figure 17.8: 3-D Variation of Subsonic Wave Speeds

17.3.3   Regions of Supersonic and Subsonic Wave Propagation

The streamline and crossflow directions are principal directions of gas dynamic acoustic and convection wave propagation. These directions become the axes of distinct wave-propagation regions, which are named the streamline and crossflow wave propagation regions.

The supersonic mono-axial wave propagation pattern does not extend to the entire (x_1, x_2) plane, in 2-D, or the entire (x_1, x_2, x_3) space in 3-D, but remains confined within two conical regions about the streamline. About the crossflow direction, in fact, there exist two other conical regions within which wave propagation remains bi-modal, from both upstream and downstream toward and away from point P, even for supersonic flows, essentially due to acoustic propagation. These regions are determined by finding the lines where the Euler eigenvalues vanish. These lines are the crossflow and conjugate lines, as shown for subsonic and supersonic flows in Figure 17.10, along with the domain of dependence and region of influence of point P and the bi-modal and mono-axial propagation regions on the (x_1, x_2) plane, in 2-D, or on plane Π, in 3-D.

As Figure 17.10-a shows, subsonic acoustic-convection waves propagate bi-modally over the entire plane, hence the Euler eigenvalues do not all have the same algebraic sign. For supersonic flows, Figure 17.10-b shows the streamline conical regions of mono-axial wave propagation, where the Euler eigenvalues all have the same algebraic sign. The conjugate lines mark the boundary with the crossflow regions of bi-modal wave propagation, where the Euler eigenvalues do not all have the same algebraic sign, as in the subsonic-flow situation.

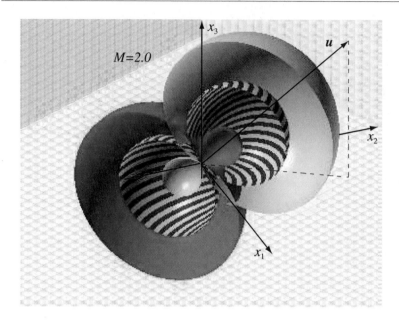

Figure 17.9: 3-D Variation of Supersonic Wave Speeds

In particular, $\lambda_{1,2} < 0$ and $\lambda_{1,2} > 0$ respectively upstream and downstream of the crossflow direction. The sign of an eigenvalue in any region of the diagram indicates the propagation direction along an axis with unit vector $\boldsymbol{n}$ pointing from P into that region.

Since the conjugate lines remain perpendicular to the Mach lines, they directly determine the growth of the mono-axial propagation regions as the Mach number increases. For $M < 1$, no mono-axial propagation region exists. For $M = 1$, the Mach lines coincide with the crossflow direction, the conjugate lines coincide with the streamline direction, hence the mono-axial propagation region consists of the streamline only. As the Mach number increases, the Mach lines approach the streamline and the conjugate lines approach the crossflow direction. The mono-axial propagation regions thus grow as circular sectors of increasing angle, and the bi-modal propagation regions simultaneously shrink. Only for a theoretically infinite M will the mono-axial propagation region spread throughout the entire (x_1, x_2) plane, in 2-D, or the entire (x_1, x_2, x_3) space, in 3-D. The presence, perhaps unexpected, of these bi-modal-propagation regions even for supersonic flows is due to crossflow acoustic propagation.

As seen, pure acoustic propagation remains bi-modal about the crossflow direction, while pure convection is bi-modal or mono-axial about the streamline direction, depending on the presence of either a subsonic or supersonic flow. The following sections develop an upstream-bias formulation with magnitude of upstream bias that reflects this distribution of characteristic speeds.

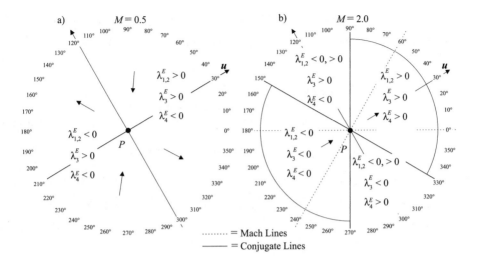

Figure 17.10: Wave Velocity Distribution: a) Subsonic Flows, b) Supersonic Flows

17.4 Acoustics-Convection Euler Flux Divergence Decomposition

This section details an intrinsically multi-dimensional characteristics-bias representation for the flux divergence $\partial f_j / \partial x_j$. This formulation rests on a decomposition of this divergence into convection, pressure-gradient, and streamline as well as crossflow acoustic components. As a guiding principle in formulating this particular Euler flux Jacobian decomposition, the form of the eventual characteristics-bias flux divergence should minimally depart from the form of the Euler flux divergence, both for efficiency of implementation and accuracy of numerical computations. The acoustics-convection flux Jacobian decomposition consists of components that genuinely model the physics of multi-dimensional acoustics and convection. The characteristic analysis of the acoustics components shows that the three-dimensional isotropic acoustic propagation naturally decomposes into three directional propagations, the first along the streamline direction and the remaining two along any two mutually perpendicular crossflow directions; these two cross flow propagation modes, in particular, combine into a two-dimensional isotropic acoustic propagation on any plane perpendicular to the velocity direction. This formulation eliminates an unstable linear-dependence problem in steady low-Mach-number flows and satisfies by design the upstream-bias stability condition. As the Mach number increases, the formulation smoothly approaches and then becomes an upstream-bias approximation of the entire flux divergence, along the principal streamline direction.

17.4.1   2-D Convection and Pressure-Gradient Components

For supersonic flows, the Euler eigenvalues (17.59) associated with the Jacobian $\frac{\partial f_j}{\partial q} n_j$

$$\lambda_{1,2}^E = v_j n_j M, \quad \lambda_{3,4}^E = v_j n_j M \pm 1 \tag{17.64}$$

all have the same algebraic sign within a streamline wedge region so that within this region the entire flux divergence $\frac{\partial f_j}{\partial x_j}$ can be upstream approximated along the streamline principal direction. For subsonic flows these eigenvalues have mixed algebraic sign and an upstream approximation for the flux divergence along one single direction remains inconsistent with the two-way propagation of acoustic waves. For subsonic flows it is the pressure gradient in the momentum equations that induces mixed-sign eigenvalues. Accordingly, by suitably decreasing the contribution from this gradient, the resulting flux-Jacobian eigenvalues all have the same algebraic sign within a streamline wedge region and the resulting convection flux divergence can then be upstream approximated along one single direction. In order to isolate the pressure contribution, the Euler flux is expressed in terms of convection and pressure fluxes as $f_j \equiv f_j^q + f_j^p$, where f_j^q and f_j^p indicate the arrays

$$f_j^q(q) \equiv \left\{ \begin{array}{c} m_j \\ \dfrac{m_j}{\rho} m_1 \\ \dfrac{m_j}{\rho} m_2 \\ \dfrac{m_j}{\rho}(E+p) \end{array} \right\} = \dfrac{m_j}{\rho} q^H = \dfrac{m_j}{\rho} \cdot \left\{ \begin{array}{c} \rho \\ m_1 \\ m_2 \\ E+p \end{array} \right\}, \quad f_j^p(q) \equiv \left\{ \begin{array}{c} 0 \\ p\delta_1^j \\ p\delta_2^j \\ 0 \end{array} \right\} \tag{17.65}$$

Based on these arrays, a flux-divergence decomposition with variable pressure-gradient contribution is then obtained through the linear combination

$$\frac{\partial f_j}{\partial x_j} = \left[\frac{\partial f_j^q}{\partial x_j} + \beta \frac{\partial f_j^p}{\partial x_j} \right] + \left[(1-\beta) \frac{\partial f_j^p}{\partial x_j} \right], \quad 0 \le \beta \le 1 \tag{17.66}$$

where the positive pressure-gradient partition function $\beta = \beta(M)$ can be chosen in such a way that all the eigenvalues associated with each of the two components between brackets in (17.66) keep the same algebraic sign within the streamline wedge region, for all Mach numbers. In this manner, these entire components can be upstream approximated along the streamline. This choice for β is possible because the eigenvalues of a matrix are continuous functions of the matrix entries and hence all the eigenvalues for the components in (17.66) will continuously depend upon β. The function β will gradually increase toward 1 for increasing Mach number, so that an upstream approximation for the components in (17.66) smoothly approaches and then becomes an upstream approximation for the entire $\frac{\partial f_j}{\partial x_j}$ along the streamline, which also ensures consistency with the one-dimensional formulation. Decomposition (17.66) is thus used for an upstream approximation of the flux divergence within the streamline wedge region, for subsonic and supersonic flows. This decomposition, however, is insufficient for an accurate multi-dimensional upstream modeling of acoustic waves, for any Mach number range within the crossflow wedge region and for low and vanishing Mach numbers within the streamline wedge region.

Concerning the cross-flow wedge region, the eigenvalues associated with (17.66) become the Euler eigenvalues (17.58) for $\beta = 1$. As indicated in Section 17.3, these eigenvalues do not all have the same algebraic sign within the cross-flow wave propagation region. A mono-axial upstream-bias approximation of (17.66) within the cross-flow wedge region remains inconsistent with the existing two-way wave propagation in this region.

With regards to the streamline wave propagation region, for a Mach number that approaches zero, the Euler eigenvalues (17.58) can all keep the same algebraic sign only if the sound speed contribution vanishes, which corresponds to a vanishing pressure gradient contribution and hence β approaching zero. But for β approaching zero, the acoustics components are obliterated and the eigenvalues associated with the components in (17.66) approach the eigenvalues of the Jacobians

$$
\frac{\partial f_j^q(q)}{\partial q} n_j =
\begin{pmatrix}
0 & , & n_1 & , & n_2 & , & 0 \\
-u_1 u_j n_j & , & u_1 n_1 + u_j n_j & , & u_1 n_2 & , & 0 \\
-u_2 u_j n_j & , & u_2 n_1 & , & u_2 n_2 + u_j n_j & , & 0 \\
u_j n_j \left(p_\rho - H \right) & , & H n_1 + u_j n_j p_{m_1} & , & H n_2 + u_j n_j p_{m_2} & , & u_j n_j \left(1 + p_E \right)
\end{pmatrix}
\tag{17.67}
$$

and

$$
\frac{\partial f_j^p(q)}{\partial q} n_j =
\begin{pmatrix}
0 & , & 0 & , & 0 & , & 0 \\
p_\rho n_1 & , & p_{m_1} n_1 & , & p_{m_2} n_1 & , & p_E n_1 \\
p_\rho n_2 & , & p_{m_1} n_2 & , & p_{m_2} n_2 & , & p_E n_2 \\
0 & , & 0 & , & 0 & , & 0
\end{pmatrix}
\tag{17.68}
$$

Using the pressure-derivative identity (2.22) the eigenvalues of these Jacobians respectively are

$$
\lambda_{1,3}^q = u_j n_j, \quad \lambda_4^q = u_j n_j (1 + p_E)
\tag{17.69}
$$

and

$$
\lambda_{1,3}^p = 0, \quad \lambda_4^p = -u_j n_j p_E
\tag{17.70}
$$

which certainly all keep the same algebraic sign for any wave propagation direction, but for vanishing Mach number they also vanish rather than approaching the dominant speed of sound c; considering that all of the pressure eigenvalues vanish in these conditions, an upstream-bias approximation of the isolated pressure gradient is unnecessary, for it would not represent acoustic or any other propagation. For low Mach numbers an upstream approximation for the components in (17.66) would inaccurately model the physics of acoustics. This difficulty is resolved by further decomposing the two-dimensional pressure gradient in (17.66) in terms of genuine streamline and crossflow acoustic components, for accurate upstream modeling of acoustic waves throughout the flow field.

17.4.2 2-D Acoustic Components

For arbitrary Mach numbers and corresponding dependent variables ρ, m_1, m_2 and E, the Euler flux Jacobians in (17.55) can be decomposed as

$$
\frac{\partial f_j}{\partial q} = \frac{\partial f_j^q}{\partial q} + \frac{\partial f_j^p}{\partial q} = \frac{\partial f_j^q}{\partial q} + A_j^a + A_j^{nc}
\tag{17.71}
$$

where, with reference to the acoustics equations (17.22), the matrices $\frac{\partial f_j^q}{\partial q}$, $\frac{\partial f_j^p}{\partial q}$, A_j^a, and A_j^{nc} are defined as

$$
\frac{\partial f_j^q}{\partial q} = \begin{pmatrix}
0 & , & \delta_1^j & , & \delta_2^j & , & 0 \\
-u_1 u_j & , & u_1\delta_1^j + u_j & , & u_1\delta_2^j & , & 0 \\
-u_2 u_j & , & u_2\delta_1^j & , & u_2\delta_2^j + u_j & , & 0 \\
u_j\left(p_\rho - H\right) & , & H\delta_1^j + u_j p_{m_1} & , & H\delta_2^j + u_j p_{m_2} & , & u_j\left(1 + p_E\right)
\end{pmatrix}
$$

$$
\frac{\partial f_j^p}{\partial q} = \begin{pmatrix}
0 & , & 0 & , & 0 & , & 0 \\
p_\rho\delta_1^j & , & p_{m_1}\delta_1^j & , & p_{m_2}\delta_1^j & , & p_E\delta_1^j \\
p_\rho\delta_2^j & , & p_{m_1}\delta_2^j & , & p_{m_2}\delta_2^j & , & p_E\delta_2^j \\
0 & , & 0 & , & 0 & , & 0
\end{pmatrix}
$$

$$
A_j^a \equiv \begin{pmatrix}
0 & , & \delta_1^j & , & \delta_2^j & , & 0 \\
p_\rho\delta_1^j & , & 0 & , & 0 & , & p_E\delta_1^j \\
p_\rho\delta_2^j & , & 0 & , & 0 & , & p_E\delta_2^j \\
0 & , & \dfrac{c^2 - p_\rho}{p_E}\delta_1^j & , & \dfrac{c^2 - p_\rho}{p_E}\delta_2^j & , & 0
\end{pmatrix}
$$

$$
A_j^{nc} \equiv \begin{pmatrix}
0 & , & -\delta_1^j & , & -\delta_2^j & , & 0 \\
0 & , & p_{m_1}\delta_1^j & , & p_{m_2}\delta_1^j & , & 0 \\
0 & , & p_{m_1}\delta_2^j & , & p_{m_2}\delta_2^j & , & 0 \\
0 & , & -\dfrac{c^2 - p_\rho}{p_E}\delta_1^j & , & -\dfrac{c^2 - p_\rho}{p_E}\delta_1^j & , & 0
\end{pmatrix} \tag{17.72}
$$

Heed, in particular that no flux component of $f_j(q)$ exists, of which the Jacobian equals A_j^a. The eigenvalues of the matrix $A_j^{nc}n_j$ have been determined in closed form as

$$
\lambda_{1,3}^{nc} = 0, \quad \lambda_4^{nc} = -cMp_E v_j n_j \tag{17.73}
$$

which become infinitesimal for vanishing M. The matrix A_j^{nc} can be termed a "non-linear coupling" matrix, for it completes the non-linear coupling between convection and acoustics within (17.71) so that the two Euler eigenvalues $\lambda_{3,4}^{d_E}$ in (17.58) do correspond to the sum of convection and acoustic speeds. Since the matrix A_j^a will be used in the upstream-bias formulation for small Mach numbers only and considering that the eigenvalues in (17.73) vanish for these Mach numbers and also for n pointing in the crossflow direction, for which $v_j n_j = 0$, no need exists to involve A_j^{nc} in the upstream-bias approximation of the flux Jacobian (17.55).

The eigenvalues of $A_j^a n_j$ have been exactly determined in closed form as

$$
\lambda_{1,2}^a = 0, \quad \lambda_{3,4}^a = \pm c \tag{17.74}
$$

and remain independent of the propagation vector n, which signifies isotropic propagation. The matrix A_j^a can thus be termed the "acoustics" matrix, for its eigenvalues equal the speed of sound c. This matrix is thus used for an upstream-bias approximation of the Euler equations in the low Mach-number regime, within the streamline region, and for any Mach number, within the crossflow region.

For any two mutually perpendicular unit vectors $\boldsymbol{a} = (a_1, a_2)$ and $\boldsymbol{a}^N = (a_1^N, a_2^N)$, along with implied summation on repeated indices, the acoustics component within the Euler flux divergence can be expressed as

$$A_j^a \frac{\partial q}{\partial x_j} = A_j^a a_j a_k \frac{\partial q}{\partial x_k} + A_j^a a_j^N a_k^N \frac{\partial q}{\partial x_k} \tag{17.75}$$

For $\boldsymbol{a}$ parallel to $\boldsymbol{u}$, this expression corresponds to a decomposition of the Euler acoustics component into streamline and crossflow acoustics components.

With reference to Section 17.2.3, the two-dimensional-flow similarity transformation for these components may be expressed as

$$A_j^a a_j = X \Lambda^a X^{-1} = X \Lambda^{a+} X^{-1} + X \Lambda^{a-} X^{-1}, \quad \Lambda^a = \Lambda^{a+} + \Lambda^{a-}$$

$$A_j^a a_j^N = X_N \Lambda^a X_N^{-1} = X_N \Lambda^{a+} X_N^{-1} + X_N \Lambda^{a-} X_N^{-1} \tag{17.76}$$

where the matrices $X \Lambda^{a+} X^{-1}$ and $X \Lambda^{a-} X^{-1}$ respectively correspond to the "forward" and "backward" acoustic-propagation matrix components of $A_j^a a_j$. Accordingly, a bi-modal upstream-bias approximation of $A_j^a a_j$ readily follows from instituting a forward and a backward upstream-bias approximation respectively for the forward- and backward- propagation matrices in (17.76). This bi-modal approximation directly yields both the non-negative-eigenvalue matrices

$$\left| A_j^a a_j \right| \equiv X \left(\Lambda^{a+} - \Lambda^{a-} \right) X^{-1}, \quad \left| A_j^a a_j^N \right| \equiv X_N \left(\Lambda^{a+} - \Lambda^{a-} \right) X_N^{-1} \tag{17.77}$$

and the associated matrix product $\left| A_j^a a_j \right| a_k \partial q / \partial x_k$. The matrices in (17.77) correspond to the streamline and crossflow absolute acoustics matrices, which depend on the diagonal matrices Λ^{a+} and Λ^{a-}, with different choices inducing distinct levels of upstream-bias dissipation. With the following matrices

$$\Lambda^{a+} \equiv \frac{1}{2} \begin{pmatrix} 2c & , & 0 & , & 0 & , & 0 \\ 0 & , & 0 & , & 0 & , & 0 \\ 0 & , & 0 & , & c & , & 0 \\ 0 & , & 0 & , & 0 & , & c \end{pmatrix}, \quad \Lambda^{a-} \equiv -\frac{1}{2} \begin{pmatrix} 0 & , & 0 & , & 0 & , & 0 \\ 0 & , & 2c & , & 0 & , & 0 \\ 0 & , & 0 & , & c & , & 0 \\ 0 & , & 0 & , & 0 & , & c \end{pmatrix} \tag{17.78}$$

the matrix product of $\left| A_j^a a_j \right| = X \left(\Lambda^{a+} - \Lambda^{a-} \right) X^{-1}$ and the directional derivative $a_k \partial q / \partial x_k$ of q directly leads to the beautifully simple acoustic-field result

$$\left| A_j^a a_j \right| a_k \frac{\partial q}{\partial x_k} = X c I X^{-1} a_k \frac{\partial q}{\partial x_k} = c a_k \frac{\partial q}{\partial x_k} \tag{17.79}$$

The similar result for (a_1^N, a_2^N) replacing (a_1, a_2) is

$$\left| A_j^a a_j^N \right| a_k^N \frac{\partial q}{\partial x_k} = X_N c I X_N^{-1} a_k^N \frac{\partial q}{\partial x_k} = c a_k^N \frac{\partial q}{\partial x_k} \tag{17.80}$$

where I denotes a 4×4 matrix. For a vanishing M, all the eigenvalues of the streamline and crossflow absolute acoustic matrices thus approach $+c$.

On the basis of the acoustics similarity transformation, the Euler flux divergence may thus be expressed as

$$\frac{\partial f_j}{\partial x_j} = \frac{\partial f_j^q}{\partial x_j} + A_j^{nc}\frac{\partial q}{\partial x_j}$$

$$+ \left(X\Lambda^{a+}X^{-1} + X\Lambda^{a-}X^{-1}\right)a_k\frac{\partial q}{\partial x_k} + \left(X_N\Lambda^{a+}X_N^{-1} + X_N\Lambda^{a-}X_N^{-1}\right)a_k^N\frac{\partial q}{\partial x_k} \qquad (17.81)$$

For a and a^N respectively pointing in the streamline and crossflow directions, this expression corresponds to a decomposition in terms of streamline and crossflow acoustic components.

17.4.3 Combination of 2-D Acoustics-Convection Characteristics-Bias Decompositions

The previous sections have shown that the Euler flux divergence can be equivalently expressed as

$$\frac{\partial f_j}{\partial x_j} = \begin{cases} \left[\dfrac{\partial f_j^q}{\partial x_j} + \beta\dfrac{\partial f_j^p}{\partial x_j}\right] + \left[(1-\beta)\dfrac{\partial f_j^p}{\partial x_j}\right]; \\[3mm] \dfrac{\partial f_j^q}{\partial x_j} + A_j^{nc}\dfrac{\partial q}{\partial x_j} \\[3mm] + \left(X\Lambda^{a+}X^{-1} + X\Lambda^{a-}X^{-1}\right)a_k\dfrac{\partial q}{\partial x_k} + \left(X_N\Lambda^{a+}X_N^{-1} + X_N\Lambda^{a-}X_N^{-1}\right)a_k^N\dfrac{\partial q}{\partial x_k} \end{cases}$$

$$(17.82)$$

The first decomposition will conveniently lead to a characteristics-bias approximation within the streamline wedge region, for high-subsonic and supersonic Mach numbers that induces less diffusion than the second, for increasing Mach numbers, but does not represent acoustic propagation for low subsonic Mach numbers. The second decomposition will lead to a characteristics-bias approximation that consistently models acoustic propagation within the crossflow and streamline wedge regions, but induces more diffusion than the first for increasing Mach number. An acoustics-convection flux divergence decomposition for all Mach numbers for a characteristics-bias resolution that not only induces minimal diffusion, but also models acoustic propagation is thus established as the following linear combination of these two decompositions, with linear combination parameter α and $0 \le \alpha, \beta \le 1$

$$\frac{\partial f_j}{\partial x_j} = (1-\alpha)\left\{\left[\frac{\partial f_j^q}{\partial x_j} + \beta\frac{\partial f_j^p}{\partial x_j}\right] + \left[(1-\beta)\frac{\partial f_j^p}{\partial x_j}\right]\right\} + \alpha\left\{\frac{\partial f_j^q(q)}{\partial x_j} + A_j^{nc}\frac{\partial q}{\partial x_j}\right.$$

$$\left. + \left(X\Lambda^{a+}X^{-1} + X\Lambda^{a-}X^{-1}\right)a_k\frac{\partial q}{\partial x_k} + \left(X_N\Lambda^{a+}X_N^{-1} + X_N\Lambda^{a-}X_N^{-1}\right)a_k^N\frac{\partial q}{\partial x_k}\right\} \qquad (17.83)$$

Owing to the simplifying parameter $\delta \equiv \beta(1-\alpha)$ and introducing the crossflow acoustic upstream parameter α^N, with $0 \le \delta$, $\alpha^N \le 1$, the final form of the acoustics-convection flux-divergence decomposition becomes

$$\frac{\partial f_j(q)}{\partial x_j} = \alpha\left(X\Lambda^{a+}X^{-1} + X\Lambda^{a-}X^{-1}\right)a_k\frac{\partial q}{\partial x_k} + \alpha^N\left(X_N\Lambda^{a+}X_N^{-1} + X_N\Lambda^{a-}X_N^{-1}\right)a_k^N\frac{\partial q}{\partial x_k}$$

$$+ \left[\frac{\partial f_j^q}{\partial x_j} + \delta \frac{\partial f_j^p}{\partial x_j} \right] + (1 - \alpha - \delta) \frac{\partial f_j^p}{\partial x_j} + \alpha A_j^{nc} \frac{\partial q}{\partial x_j} + (\alpha - \alpha^N) A_j^a a_j^N a_k^N \frac{\partial q}{\partial x_k} \qquad (17.84)$$

A consistent upstream-bias representation results from this combination because of the physical significant of each term. The weights α and α^N respectively for the streamline and crossflow acoustic components in this expression are different from each other because the streamline and crossflow characteristic velocity components remain different from each other, following the Euler eigenvalues (17.59). For increasing Mach number, furthermore, the pressure gradient term $\delta \frac{\partial f_j^p}{\partial x_j}$ in this decomposition also contributes to the streamline acoustic upstream bias. These considerations indicate that the magnitudes of acoustic upstream bias for (17.84) along the streamline and crossflow directions will have to differ from each other for varying M, a "differential" upstream bias that can be instituted through the distinct weights α and α^N on the streamline and crossflow acoustic components.

Bi-modal wave propagation exists for all Mach numbers in the crossflow direction. Accordingly the crossflow acoustic term $\alpha^N \left(X_N \Lambda^{a+} X_N^{-1} + X_N \Lambda^{a-} X_N^{-1} \right) a_k^N \frac{\partial q}{\partial x_k}$ will receive a bi-modal upstream bias. The associated term $(\alpha - \alpha^N) A_j^a a_j^N a_k^N \frac{\partial q}{\partial x_k}$ need not receive any upstream-bias, for if it did it would obliterate all the crossflow acoustic upstream-bias, which remains fundamental for stability, as indicated by the multi-dimensional characteristics analysis. In respect of the term $\alpha A_j^{nc} \frac{\partial q}{\partial x_j}$, no upstream-bias is needed for this term either, for such an upstream of this term would essentially add only algebraic complexity to the formulation, since its coefficient α rapidly decreases and then vanishes for supersonic and high subsonic Mach numbers; three out of four eigenvalues of the Jacobian of this term vanish, as discussed in Section 17.4.2 with reference to (17.73), and the product of α and the remaining fourth eigenvalue remains negligible for $\alpha > 0$; additionally all the eigenvalues of this term vanish in the crossflow direction. The magnitude of crossflow upstream bias is controlled by the weight α^N, which does not have to vanish for supersonic flows, because of the presence of the finite bi-modal wave-propagation regions. Nevertheless, α^N can decrease for increasing M, because the bi-modal crossflow region shrinks for increasing Mach number.

The streamline acoustic term $\alpha \left(X \Lambda^{a+} X^{-1} + X \Lambda^{a-} X^{-1} \right) a_k \frac{\partial q}{\partial x_k}$ accounts for the bi-modal streamline propagation of acoustic waves; this term is thus employed for an acoustics upstream - bias approximation within the streamline wedge region, for low Mach numbers. Since this term already models pressure-induced acoustic-wave propagation along each streamline, there is no need to develop an additional upstream bias for the $(1 - \alpha - \delta) \frac{\partial f_j^p}{\partial x_j}$, also considering that the coefficient $(1 - \alpha - \delta)$ of this term vanishes for acoustic and supersonic flows and three out of four eigenvalues of the Jacobian of this term identically vanish, as discussed in Section 17.4.1 with reference to (17.70). As the Mach number rises, besides, an increasing fraction of the pressure gradient receives an upstream bias in the term $\left[\frac{\partial f_j^q}{\partial x_j} + \delta \frac{\partial f_j^p}{\partial x_j} \right]$. This expression is counted as one term because, with reference to (17.66), the eigenvalues associated with this term will all keep the same sign within the streamline wedge region, since $\delta = (1 - \alpha) \beta \leq \beta$.

Based on the two principles of minimal upstream dissipation and consistent infinite-directional upstream bias, the developments in the following sections show that for $M = 0$, the functions α and α^N equal 1, whereas the function δ equals 0; for increasing M, α rapidly decreases and vanishes, δ rapidly increases and then identically equals 1, for supersonic flows, and α^N monotonically decreases. Employing decomposition (17.84), the following sections establish a computationally efficient characteristics-bias Euler flux divergence.

17.4.4   3-D Convection and Pressure-Gradient Components

Also for three-dimensional flows is the Euler flux f_j decomposed as $f_j = f_j^q + f_j^p$, where the convection and pressure fluxes are defined as

$$
f_j^q(q) \equiv \left\{ \begin{array}{c} m_j \\ \dfrac{m_j}{\rho}m_1 \\ \dfrac{m_j}{\rho}m_2 \\ \dfrac{m_j}{\rho}m_3 \\ \dfrac{m_j}{\rho}(E+p) \end{array} \right\} = \frac{m_j}{\rho}q^H = \frac{m_j}{\rho}\cdot\left\{ \begin{array}{c} \rho \\ m_1 \\ m_2 \\ m_3 \\ E+p \end{array} \right\}, \quad f_j^p(q) \equiv \left\{ \begin{array}{c} 0 \\ p\delta_1^j \\ p\delta_2^j \\ p\delta_3^j \\ 0 \end{array} \right\} \qquad (17.85)
$$

For any Mach number, Section 17.3 has shown that the non-dimensional Euler eigenvalues of the Jacobian $\frac{\partial f_j}{\partial q}n_j$ are

$$
\lambda_{0,2}^E = v_j n_j M, \quad \lambda_{3,4}^E = v_j n_j M \pm 1 \qquad (17.86)
$$

For supersonic flows, these eigenvalues all have the same algebraic sign within the streamline region and the entire flux divergence can be upstream approximated along the streamline principal direction, within this region. For subsonic flows these eigenvalues have mixed algebraic sign and an upstream approximation for the flux divergence along one single direction remains inconsistent with the two-way propagation of acoustic waves. Without the pressure gradient in the momentum equations, however, the corresponding flux-Jacobian eigenvalues all have the same algebraic sign within the streamline region and the resulting convection flux divergence can then be upstream approximated along one single direction. Also for three-dimensional flows, accordingly, can the flux divergence be decomposed as the linear combination

$$
\frac{\partial f_j}{\partial x_j} = \left[\frac{\partial f_j^q}{\partial x_j} + \beta\frac{\partial f_j^p}{\partial x_j}\right] + \left[(1-\beta)\frac{\partial f_j^p}{\partial x_j}\right], \quad 0 \le \beta \le 1 \qquad (17.87)
$$

where the positive pressure-gradient partition function $\beta = \beta(M)$ is chosen in such a way that all the eigenvalues associated with each of the two components between brackets in (17.87) keep the same algebraic sign within the streamline region, for all Mach numbers. In this manner, these entire components can be upstream approximated along the streamline. The function β will gradually increase toward 1 for increasing Mach number, so that an upstream approximation for the components in (17.87) smoothly approaches and then becomes an upstream approximation for the entire $\frac{\partial f_j}{\partial x_j}$ along the streamline, which also ensures consistency with the one-dimensional formulation. Decomposition (17.87) is thus used for an upstream approximation of the flux divergence within the streamline region, for subsonic and supersonic flows. This decomposition, however, is insufficient for an accurate multi-dimensional upstream modeling of acoustic waves, for any Mach number range within the crossflow region and for low and vanishing Mach numbers within the streamline region, for the same reasons stated for the case of two-dimensional flows. For β approaching zero, the eigenvalues associated with the components in (17.87) approach the eigenvalues of the Jacobians

$$\frac{\partial f_j^q(q)}{\partial q} n_j =$$

$$\begin{pmatrix} 0 & , & n_1 & , & n_2 & , & n_3 & , & 0 \\ -u_1 u_j n_j & , & u_1 n_1 + u_j n_j & , & u_1 n_2 & , & u_1 n_3 & , & 0 \\ -u_2 u_j n_j & , & u_2 n_1 & , & u_2 n_2 + u_j n_j & , & u_2 n_3 & , & 0 \\ -u_3 u_j n_j & , & u_3 n_1 & , & u_3 n_2 & , & u_3 n_3 + u_j n_j & , & 0 \\ u_j n_j \left(p_\rho - H \right) & , & H n_1 + u_j n_j p_{m_1} & , & H n_2 + u_j n_j p_{m_2} & , & H n_3 + u_j n_j p_{m_3} & , & u_j n_j \left(1 + p_E \right) \end{pmatrix}$$

$$(17.88)$$

and

$$\frac{\partial f_j^p(q)}{\partial q} n_j = \begin{pmatrix} 0 & , & 0 & , & 0 & , & 0 & , & 0 \\ p_\rho n_1 & , & p_{m_1} n_1 & , & p_{m_2} n_1 & , & p_{m_3} n_1 & , & p_E n_1 \\ p_\rho n_2 & , & p_{m_1} n_2 & , & p_{m_2} n_2 & , & p_{m_3} n_2 & , & p_E n_2 \\ p_\rho n_3 & , & p_{m_1} n_3 & , & p_{m_2} n_3 & , & p_{m_3} n_3 & , & p_E n_3 \\ 0 & , & 0 & , & 0 & , & 0 & , & 0 \end{pmatrix} \qquad (17.89)$$

Using the pressure-derivative identity (2.22) the eigenvalues of these Jacobians respectively are

$$\lambda_{0,3}^q = u_j n_j, \quad \lambda_4^q = u_j n_j (1 + p_E) \qquad (17.90)$$

and

$$\lambda_{0,3}^p = 0, \quad \lambda_4^p = -u_j n_j p_E \qquad (17.91)$$

which certainly all keep the same algebraic sign for any wave propagation direction, but for vanishing Mach number they also vanish instead of converging to the dominant speed of sound c. For low Mach numbers, an upstream approximation for the components in (17.87) would thus inaccurately model the physics of acoustics. Again, this predicament is resolved by further decomposing the three-dimensional pressure gradient in (17.87) in terms of genuine streamline and crossflow acoustic components, for accurate upstream modeling of acoustic waves throughout a three-dimensional flow field.

17.4.5   3-D Acoustic Components

For arbitrary Mach numbers and corresponding dependent variables ρ, m_1, m_2, m_3, and E, the Euler flux Jacobians in (17.56) can be decomposed as

$$\frac{\partial f_j}{\partial q} = \frac{\partial f_j^q}{\partial q} + \frac{\partial f_j^p}{\partial q} = \frac{\partial f_j^q}{\partial q} + A_j^a + A_j^{nc} \qquad (17.92)$$

where, with reference to the acoustics equations (17.33), the matrices $\frac{\partial f_j^q}{\partial q}$, $\frac{\partial f_j^p}{\partial q}$, A_j^a, and A_j^{nc} are defined as

$$
\frac{\partial f_j^q}{\partial q} = \begin{pmatrix}
0 & , & \delta_1^j & , & \delta_2^j & , & \delta_3^j & , & 0 \\
-u_1 u_j & , & u_1 \delta_1^j + u_j & , & u_1 \delta_2^j & , & u_1 \delta_3^j & , & 0 \\
-u_2 u_j & , & u_2 \delta_1^j & , & u_2 \delta_2^j + u_j & , & u_2 \delta_3^j & , & 0 \\
-u_3 u_j & , & u_3 \delta_1^j & , & u_3 \delta_2^j & , & u_3 \delta_3^j + u_j & , & 0 \\
u_j \left(p_\rho - H \right) & , & H\delta_1^j + u_j p_{m_1} & , & H\delta_2^j + u_j p_{m_2} & , & H\delta_3^j + u_j p_{m_3} & , & u_j \left(1 + p_E \right)
\end{pmatrix}
$$

$$
\frac{\partial f_j^p}{\partial q} = \begin{pmatrix}
0 & , & 0 & , & 0 & , & 0 & , & 0 \\
p_\rho \delta_1^j & , & p_{m_1} \delta_1^j & , & p_{m_2} \delta_1^j & , & p_{m_3} \delta_1^j & , & p_E \delta_1^j \\
p_\rho \delta_2^j & , & p_{m_1} \delta_2^j & , & p_{m_2} \delta_2^j & , & p_{m_3} \delta_2^j & , & p_E \delta_2^j \\
p_\rho \delta_3^j & , & p_{m_1} \delta_3^j & , & p_{m_2} \delta_3^j & , & p_{m_3} \delta_3^j & , & p_E \delta_3^j \\
0 & , & 0 & , & 0 & , & 0 & , & 0
\end{pmatrix}
$$

$$
A_j^a \equiv \begin{pmatrix}
0 & , & \delta_1^j & , & \delta_2^j & , & \delta_3^j & , & 0 \\
p_\rho \delta_1^j & , & 0 & , & 0 & , & 0 & , & p_E \delta_1^j \\
p_\rho \delta_2^j & , & 0 & , & 0 & , & 0 & , & p_E \delta_2^j \\
p_\rho \delta_3^j & , & 0 & , & 0 & , & 0 & , & p_E \delta_3^j \\
0 & , & \dfrac{c^2 - p_\rho}{p_E}\delta_1^j & , & \dfrac{c^2 - p_\rho}{p_E}\delta_2^j & , & \dfrac{c^2 - p_\rho}{p_E}\delta_3^j & , & 0
\end{pmatrix}
$$

$$
A_j^{nc} \equiv \begin{pmatrix}
0 & , & -\delta_1^j & , & -\delta_2^j & , & -\delta_3^j & , & 0 \\
0 & , & p_{m_1} \delta_1^j & , & p_{m_2} \delta_1^j & , & p_{m_3} \delta_1^j & , & 0 \\
0 & , & p_{m_1} \delta_2^j & , & p_{m_2} \delta_2^j & , & p_{m_3} \delta_2^j & , & 0 \\
0 & , & p_{m_1} \delta_3^j & , & p_{m_2} \delta_3^j & , & p_{m_3} \delta_3^j & , & 0 \\
0 & , & -\dfrac{c^2 - p_\rho}{p_E}\delta_1^j & , & -\dfrac{c^2 - p_\rho}{p_E}\delta_2^j & , & -\dfrac{c^2 - p_\rho}{p_E}\delta_3^j & , & 0
\end{pmatrix} \tag{17.93}
$$

Heed, in particular that no flux component of $f_j(q)$ exists, of which the Jacobian equals A_j^a. The eigenvalues of the matrix $A_j^{nc} n_j$ have been determined in closed form as

$$
\lambda_{0,3}^{nc} = 0, \quad \lambda_4^{nc} = -cM p_E v_j n_j \tag{17.94}
$$

which become infinitesimal for vanishing M. Again, the matrix A_j^{nc} can be termed a "non-linear coupling" matrix, for it completes the non-linear coupling between convection and acoustics within (17.56) so that the two Euler eigenvalues $\lambda_{3,4}^{d_E}$ in (17.58) do correspond to the sum of convection and acoustic speeds. Since the matrix A_j^a will be used in the upstream-bias formulation for small Mach numbers only and considering that the eigenvalues in (17.94) vanish for these Mach numbers and also for n pointing in the crossflow direction, for which $v_j n_j = 0$, no need exists to involve A_j^{nc} in the upstream-bias approximation of the flux Jacobian (17.56).

The eigenvalues of $A_j^a n_j$ have been exactly determined in closed form as

$$
\lambda_{0,2}^a = 0, \quad \lambda_{3,4}^a = \pm c \tag{17.95}
$$

and remain independent of the propagation vector n, which signifies isotropic propagation. The matrix A_j^a can be thus termed the "acoustics" matrix, for its eigenvalues equal the speed of sound c. This matrix can be thus used for an upstream-bias approximation of the Euler equations in the low Mach-number regime, within the streamline region, and for any Mach number, within the crossflow region.

For any three mutually orthogonal unit vectors $\boldsymbol{a} = (a_1, a_2, a_3)$, $\boldsymbol{a}^{N_1} = (a_1^{N_1}, a_2^{N_1}, a_3^{N_1})$, and $\boldsymbol{a}^{N_2} = (a_1^{N_2}, a_2^{N_2}, a_3^{N_2})$, along with implied summation on repeated indices, the acoustics component within the Euler flux divergence can be expressed as

$$A_j^a \frac{\partial q}{\partial x_j} = A_j^a a_j a_k \frac{\partial q}{\partial x_k} + A_j^a a_j^{N_1} a_k^{N_1} \frac{\partial q}{\partial x_k} + A_j^a a_j^{N_2} a_k^{N_2} \frac{\partial q}{\partial x_k} \qquad (17.96)$$

For $\boldsymbol{a}$ parallel to $\boldsymbol{u}$, this expression corresponds to a decomposition of the Euler acoustics component into one streamline and two crossflow acoustics components.

With reference to Section 17.2.3, the three-dimensional-flow similarity transformation for these components may be expressed as

$$A_j^a a_j = X\Lambda^a X^{-1} = X\Lambda^{a+} X^{-1} + X\Lambda^{a-} X^{-1}, \quad \Lambda^a = \Lambda^{a+} + \Lambda^{a-}$$

$$A_j^a a_j^{N_1} = X_{N_1} \Lambda^a X_{N_1}^{-1} = X_{N_1} \Lambda^{a+} X_{N_1}^{-1} + X_{N_1} \Lambda^{a-} X_{N_1}^{-1}$$

$$A_j^a a_j^{N_2} = X_{N_2} \Lambda^a X_{N_2}^{-1} = X_{N_2} \Lambda^{a+} X_{N_2}^{-1} + X_{N_2} \Lambda^{a-} X_{N_2}^{-1} \qquad (17.97)$$

where the matrices $X\Lambda^{a+} X^{-1}$ and $X\Lambda^{a-} X^{-1}$ respectively correspond to the "forward" and "backward" acoustic-propagation matrix components of $A_j^a a_j$. Accordingly, a bi-modal upstream-bias approximation of $A_j^a a_j$ readily follows from instituting a forward and a backward upstream-bias approximation respectively for the forward- and backward- propagation matrices in (17.97). This bi-modal approximation directly yields both the non-negative-eigenvalue matrices

$$\left| A_j^a a_j \right| \equiv X \left(\Lambda^{a+} - \Lambda^{a-} \right) X^{-1}$$

$$\left| A_j^a a_j^{N_1} \right| \equiv X_{N_1} \left(\Lambda^{a+} - \Lambda^{a-} \right) X_{N_1}^{-1}, \quad \left| A_j^a a_j^{N_2} \right| \equiv X_{N_2} \left(\Lambda^{a+} - \Lambda^{a-} \right) X_{N_2}^{-1} \qquad (17.98)$$

and the associated matrix product $\left| A_j^a a_j \right| a_k \partial q / \partial x_k$. The matrices in (17.98) correspond to the streamline and crossflow absolute acoustics matrices. As in the two-dimensional-flow case, these absolute matrices depend on the diagonal matrices Λ^{a+} and Λ^{a-}, with different forms inducing distinct magnitudes of upstream-bias dissipation. With the following matrices

$$\Lambda^{a+} \equiv \frac{1}{2} \begin{pmatrix} 2c & , & 0 & , & 0 & , & 0 & , & 0 \\ 0 & , & 0 & , & 0 & , & 0 & , & 0 \\ 0 & , & 0 & , & c & , & 0 & , & 0 \\ 0 & , & 0 & , & 0 & , & c & , & 0 \\ 0 & , & 0 & , & 0 & , & 0 & , & c \end{pmatrix}, \quad \Lambda^{a-} \equiv -\frac{1}{2} \begin{pmatrix} 0 & , & 0 & , & 0 & , & 0 & , & 0 \\ 0 & , & 2c & , & 0 & , & 0 & , & 0 \\ 0 & , & 0 & , & c & , & 0 & , & 0 \\ 0 & , & 0 & , & 0 & , & c & , & 0 \\ 0 & , & 0 & , & 0 & , & 0 & , & c \end{pmatrix} \qquad (17.99)$$

the matrix product of $\left| A_j^a a_j \right| = X \left(\Lambda^{a+} - \Lambda^{a-} \right) X^{-1}$ and the directional derivative $a_k \partial q / \partial x_k$ of q directly leads to the beautifully simple acoustic-field result

$$\left| A_j^a a_j \right| a_k \frac{\partial q}{\partial x_k} = XcIX^{-1} a_k \frac{\partial q}{\partial x_k} = ca_k \frac{\partial q}{\partial x_k} \qquad (17.100)$$

The similar results for $(a_1^{N_1}, a_2^{N_1}, a_3^{N_1})$ and $(a_1^{N_2}, a_2^{N_2}, a_3^{N_2})$ replacing (a_1, a_2, a_3) are

$$\left| A_j^a a_j^{N_1} \right| a_k^{N_1} \frac{\partial q}{\partial x_k} = X_{N_1} cI X_{N_1}^{-1} a_k^{N_1} \frac{\partial q}{\partial x_k} = c a_k^{N_1} \frac{\partial q}{\partial x_k} \tag{17.101}$$

and

$$\left| A_j^a a_j^{N_2} \right| a_k^{N_2} \frac{\partial q}{\partial x_k} = X_{N_2} cI X_{N_2}^{-1} a_k^{N_2} \frac{\partial q}{\partial x_k} = c a_k^{N_2} \frac{\partial q}{\partial x_k} \tag{17.102}$$

where I denotes the 5×5 identity matrix. For a vanishing M, all the eigenvalues of the streamline and crossflow absolute acoustic matrices thus approach $+c$.

On the basis of the acoustics similarity transformation, the Euler flux divergence may thus be expressed as

$$\frac{\partial f_j}{\partial x_j} = \frac{\partial f_j^q}{\partial x_j} + A_j^{nc} \frac{\partial q}{\partial x_j} + \left(X \Lambda^{a+} X^{-1} + X \Lambda^{a-} X^{-1} \right) a_k \frac{\partial q}{\partial x_k}$$

$$+ \left(X_{N_1} \Lambda^{a+} X_{N_1}^{-1} + X_{N_1} \Lambda^{a-} X_{N_1}^{-1} \right) a_k^{N_1} \frac{\partial q}{\partial x_k} + \left(X_{N_2} \Lambda^{a+} X_{N_2}^{-1} + X_{N_2} \Lambda^{a-} X_{N_2}^{-1} \right) a_k^{N_2} \frac{\partial q}{\partial x_k} \tag{17.103}$$

For $\boldsymbol{a}$, $\boldsymbol{a}^{N_1}$, and $\boldsymbol{a}^{N_2}$ respectively pointing in the streamline and two mutually perpendicular crossflow directions, this expression provides a decomposition in terms of three-dimensional streamline and crossflow acoustic components.

17.4.6   Combination of 3-D Acoustics-Convection Characteristics-Bias Decompositions

The previous sections have shown that the Euler flux divergence can be equivalently expressed as

$$\frac{\partial f_j}{\partial x_j} = \begin{cases} \left[\dfrac{\partial f_j^q}{\partial x_j} + \beta \dfrac{\partial f_j^p}{\partial x_j} \right] + \left[(1 - \beta) \dfrac{\partial f_j^p}{\partial x_j} \right]; \\[2mm] \dfrac{\partial f_j^q}{\partial x_j} + A_j^{nc} \dfrac{\partial q}{\partial x_j} \\[2mm] + \left(X \Lambda^{a+} X^{-1} + X \Lambda^{a-} X^{-1} \right) a_k \dfrac{\partial q}{\partial x_k} \\[2mm] + \left(X_{N_1} \Lambda^{a+} X_{N_1}^{-1} + X_{N_1} \Lambda^{a-} X_{N_1}^{-1} \right) a_k^{N_1} \dfrac{\partial q}{\partial x_k} \\[2mm] + \left(X_{N_2} \Lambda^{a+} X_{N_2}^{-1} + X_{N_2} \Lambda^{a-} X_{N_2}^{-1} \right) a_k^{N_2} \dfrac{\partial q}{\partial x_k} \end{cases} \tag{17.104}$$

As for two-dimensional flows, the first decomposition will conveniently lead to a characteristics - bias approximation within the streamline wedge region, for high-subsonic and supersonic Mach numbers that induces less diffusion than the second, for increasing Mach numbers, but does not represent acoustic propagation for low subsonic Mach numbers. The second decomposition will lead to a characteristics-bias approximation that consistently models acoustic propagation within the crossflow and streamline wedge regions, but induces more diffusion than the first for increasing Mach number. An acoustics-convection flux divergence decomposition for all Mach numbers for a characteristics-bias resolution that not only induces minimal diffusion, but also models acoustic propagation is thus established as the following linear combination of these two decompositions, with linear combination parameter α and $0 \leq \alpha, \beta \leq 1$

$$\frac{\partial f_j}{\partial x_j} = (1 - \alpha) \left\{ \left[\frac{\partial f_j^q}{\partial x_j} + \beta \frac{\partial f_j^p}{\partial x_j} \right] + \left[(1 - \beta) \frac{\partial f_j^p}{\partial x_j} \right] \right\}$$

$$+\alpha \left\{ \frac{\partial f_j^q(q)}{\partial x_j} + A_j^{nc} \frac{\partial q}{\partial x_j} + \left(X\Lambda^{a+} X^{-1} + X\Lambda^{a-} X^{-1} \right) a_k \frac{\partial q}{\partial x_k} \right.$$

$$\left. + \left(X_{N_1} \Lambda^{a+} X_{N_1}^{-1} + X_{N_1} \Lambda^{a-} X_{N_1}^{-1} \right) a_k^{N_1} \frac{\partial q}{\partial x_k} + \left(X_{N_2} \Lambda^{a+} X_{N_2}^{-1} + X_{N_2} \Lambda^{a-} X_{N_2}^{-1} \right) a_k^{N_2} \frac{\partial q}{\partial x_k} \right\}$$

$$(17.105)$$

Owing to the simplifying parameter $\delta \equiv \beta(1 - \alpha)$ and introducing the crossflow acoustic upstream parameter α^N, with $0 \leq \delta, \alpha^N \leq 1$, the final form of the acoustics-convection flux-divergence decomposition becomes

$$\frac{\partial f_j(q)}{\partial x_j} = \alpha \left(X\Lambda^{a+} X^{-1} + X\Lambda^{a-} X^{-1} \right) a_k \frac{\partial q}{\partial x_k}$$

$$+\alpha^N \left(X_{N_1} \Lambda^{a+} X_{N_1}^{-1} + X_{N_1} \Lambda^{a-} X_{N_1}^{-1} \right) a_k^{N_1} \frac{\partial q}{\partial x_k} + \alpha^N \left(X_{N_2} \Lambda^{a+} X_{N_2}^{-1} + X_{N_2} \Lambda^{a-} X_{N_2}^{-1} \right) a_k^{N_2} \frac{\partial q}{\partial x_k}$$

$$+ \left[\frac{\partial f_j^q}{\partial x_j} + \delta \frac{\partial f_j^p}{\partial x_j} \right] + (1 - \alpha - \delta) \frac{\partial f_j^p}{\partial x_j} + \alpha A_j^{nc} \frac{\partial q}{\partial x_j} + (\alpha - \alpha^N) A_j^a \left(a_j^{N_1} a_k^{N_1} + a_j^{N_2} a_k^{N_2} \right) \frac{\partial q}{\partial x_k} \quad (17.106)$$

As for two-dimensional flows, a consistent upstream-bias representation results from this combination because of the physical significant of each term. The weights α and α^N respectively for the streamline and crossflow acoustic components in this expression are different from each other because the streamline and crossflow characteristic velocity components remain different from each other, following the Euler eigenvalues (17.59). For increasing Mach number, furthermore, the pressure gradient term $\delta \frac{\partial f_j^p}{\partial x_j}$ in this decomposition also contributes to the streamline acoustic upstream bias. These considerations indicate that the magnitudes of acoustic upstream bias for (17.106) along the streamline and crossflow directions will have to differ from each other for varying M, a "differential" upstream bias that can be instituted through the distinct weights α and α^N on the streamline and crossflow acoustic components.

Bi-modal wave propagation region exists for all Mach numbers in the crossflow plane perpendicular to each streamline. Accordingly the crossflow acoustics expression

$$\alpha^N \left(X_{N_1} \Lambda^{a+} X_{N_1}^{-1} + X_{N_1} \Lambda^{a-} X_{N_1}^{-1} \right) a_k^{N_1} \frac{\partial q}{\partial x_k} + \alpha^N \left(X_{N_2} \Lambda^{a+} X_{N_2}^{-1} + X_{N_2} \Lambda^{a-} X_{N_2}^{-1} \right) a_k^{N_2} \frac{\partial q}{\partial x_k}$$

will receive a bi-modal upstream bias. The associated term $(\alpha - \alpha^N) A_j^a (a_j^{N_1} a_k^{N_1} + a_j^{N_2} a_k^{N_2}) \frac{\partial q}{\partial x_k}$ need not receive any upstream-bias, for if it did it would obliterate all the crossflow acoustic upstream-bias, which remains fundamental for stability, as indicated by the multi-dimensional characteristics analysis. In respect of the term $\alpha A_j^{nc} \frac{\partial q}{\partial x_j}$, no upstream-bias is needed for this term either, for such an upstream of this term would essentially add only algebraic complexity to the formulation, since its coefficient α rapidly decreases and then vanishes for supersonic and high subsonic Mach numbers; three out of four eigenvalues of the Jacobian of this term vanish, as discussed in Section 17.4.5 with reference to (17.94), and the product of α and the remaining fourth eigenvalue remains negligible for $\alpha > 0$; additionally all the eigenvalues of this term vanish in the crossflow direction. The magnitude of crossflow upstream bias is controlled by the weight α^N, which does not have to vanish for supersonic flows, because of the presence of the finite bi-modal wave-propagation regions. Nevertheless, α^N can decrease for increasing M, because the bi-modal crossflow region shrinks for increasing Mach number.

The streamline acoustic term $\alpha \left(X\Lambda^{a+} X^{-1} + X\Lambda^{a-} X^{-1} \right) a_k \frac{\partial q}{\partial x_k}$ accounts for the bi-modal streamline propagation of acoustic waves; this term is thus employed for an acoustics upstream - bias approximation within the streamline wedge region, for low Mach numbers. Since this term already models pressure-induced acoustic-wave propagation along each streamline, there is no need to develop an additional upstream bias for the $(1 - \alpha - \delta) \frac{\partial f_j^p}{\partial x_j}$, also considering that the coefficient $(1 - \alpha - \delta)$ of this term vanishes for acoustic and supersonic flows and three out of four eigenvalues of the Jacobian of this term identically vanish, as discussed in Section 17.4.4 with reference to (17.91). Besides, as the Mach number rises, an increasing fraction of the pressure gradient receives an upstream bias in the term $\left[\frac{\partial f_j^q}{\partial x_j} + \delta \frac{\partial f_j^p}{\partial x_j} \right]$. This expression is counted as one term because, with reference to (17.87), the eigenvalues associated with this term will all keep the same sign within the streamline wedge region, since $\delta = (1 - \alpha)\beta \leq \beta$.

Based on the two principles of minimal upstream dissipation and consistent infinite-directional upstream bias, the developments in the following sections show that for $M = 0$, the functions α and α^N equal 1, whereas the function δ equals 0; for increasing M, α rapidly decreases and vanishes, δ rapidly increases and then identically equals 1, for supersonic flows, and α^N monotonically decreases. Employing decomposition (17.106), the following sections establish a physically meaningful three-dimensional characteristics-bias Euler flux divergence.

17.5   Characteristics-Bias Euler Flux Divergence

This section presents the characteristics-bias Euler flux divergence for both two- and three-dimensional flows.

Two-Dimensional Flows

With reference to (14.103), given the physical significance of the terms in decomposition (17.84) and algebraic signs of the corresponding principal-direction eigenvalues, the associated principal direction unit vectors for these terms are

$$a_1 = -a_2 = a_5 = a, \quad a_3 = -a_4 = a^N, \quad a_6 = a_7 = a_8 = 0 \qquad (17.107)$$

At each flow-field point, a and a^N remain respectively parallel and perpendicular to the local velocity, with a^N obtained by a 90°- degree counterclockwise rotation of a. With (17.84) and (17.107) the general upstream-bias expression (14.103) in Chapter 14 directly yields the acoustics-convection characteristics flux divergence

$$\frac{\partial f_j^C}{\partial x_j} = \frac{\partial f_j}{\partial x_j} - \frac{\partial}{\partial x_i}\left[\varepsilon\psi\left(c\left(\alpha a_i a_j + \alpha^N a_i^N a_j^N\right)\frac{\partial q}{\partial x_j} + a_i\frac{\partial f_j^q}{\partial x_j} + a_i\delta\frac{\partial f_j^p}{\partial x_j}\right)\right] \qquad (17.108)$$

In this result, the expressions $\left(c\alpha a_i a_j\frac{\partial q}{\partial x_j} + a_i\frac{\partial f_j^q}{\partial x_j} + a_i\delta\frac{\partial f_j^p}{\partial x_j}\right)$ and $\left(c\alpha^N a_i^N a_j^N\frac{\partial q}{\partial x_j}\right)$ determine the upstream biases within respectively the streamline and crossflow wave propagation regions. These two combined expressions then induce a correct upwind bias along all wave propagation regions. The operation count for expression (17.108) is comparable to that of an FVS formulation. The terms in this expression, furthermore, directly correspond to the physics of acoustics and convection. Expression (17.108) determines f_i^C itself, up to an arbitrary divergence-free vector, as

$$f_i^C = f_i(q) - \varepsilon\psi\left[c\left(\alpha a_i a_j + \alpha^N a_i^N a_j^N\right)\frac{\partial q}{\partial x_j} + a_i\frac{\partial f_j^q}{\partial x_j} + a_i\delta\frac{\partial f_j^p}{\partial x_j}\right] \qquad (17.109)$$

According to this result, the intrinsic two-dimensionality of each component f_i^C derives from its dependence upon the entire divergence of f_j^q and f_j^p. The components within this flux, in particular, remain linearly independent of one another, which avoids the linear-dependence instability in the steady low-Mach-number Euler equations.

For vanishing Mach numbers, α and α^N will approach 1 whereas δ will approach 0. Under these conditions, (17.108) reduces to

$$\frac{\partial f_j^C}{\partial x_j} = \frac{\partial f_j}{\partial x_j} - \frac{\partial}{\partial x_i}\left[\varepsilon\psi\left(c\frac{\partial q}{\partial x_i} + a_i\frac{\partial f_j^q}{\partial x_j}\right)\right] \qquad (17.110)$$

which essentially induces only an acoustics upstream bias. Heed that this bias becomes independent of any specific propagation direction, for it no longer depends on $\left(\alpha a_i a_j + \alpha^N a_i^N a_j^N\right)$. This bias, as a result, becomes isotropic, in harmony with the isotropic propagation of acoustic waves. For supersonic flows, $\alpha = 0$ and $\delta = 1$ and (17.108) thus becomes

$$\frac{\partial f_j^C}{\partial x_j} = \frac{\partial f_j}{\partial x_j} - \frac{\partial}{\partial x_i}\left[\varepsilon\psi\left(c\alpha^N a_i^N a_j^N\frac{\partial q}{\partial x_j} + a_i\frac{\partial f_j}{\partial x_j}\right)\right] \qquad (17.111)$$

which depends not only on the entire divergence of the Euler inviscid flux vector, but also on the crossflow component of the absolute acoustics matrix. This component is needed because the Euler flux Jacobian eigenvalues remain of mixed algebraic sign within a conical flow region about the crossflow direction, due to acoustic propagation.

3-D Characteristics-Bias Euler Flux

With reference to (14.103), given the physical significance of the terms in decomposition (17.106) and algebraic signs of the corresponding principal-direction eigenvalues, the associated principal - direction unit vectors for these terms are

$$a_1 = -a_2 = a_7 = a, \quad a_3 = -a_4 = a^{N_1}, \quad a_5 = -a_6 = a^{N_2}, \quad a_8 = a_9 = a_{10} = 0 \quad (17.112)$$

At each flow-field point, a is parallel to the local velocity vector, whereas the mutually orthogonal unit vectors a^{N_1} and a^{N_2} remain perpendicular to velocity. With (17.106) and (17.112), the general upstream-bias expression (14.103) directly yields the three-dimensional acoustics-convection characteristics flux divergence

$$\frac{\partial f_j^C}{\partial x_j} = \frac{\partial f_j}{\partial x_j} - \frac{\partial}{\partial x_i} \left[\varepsilon \psi \left(c \left(\alpha a_i a_j + \alpha^N \left(a_i^{N_1} a_j^{N_1} + a_i^{N_2} a_j^{N_2} \right) \right) \frac{\partial q}{\partial x_j} + a_i \frac{\partial f_j^q}{\partial x_j} + a_i \delta \frac{\partial f_j^p}{\partial x_j} \right) \right]$$
$$(17.113)$$

In this result, the expressions $\left(c\alpha a_i a_j \frac{\partial q}{\partial x_j} + a_i \frac{\partial f_j^q}{\partial x_j} + a_i \delta \frac{\partial f_j^p}{\partial x_j} \right)$ and $\left(c\alpha^N \left(a_i^{N_1} a_j^{N_1} + a_i^{N_2} a_j^{N_2} \right) \frac{\partial q}{\partial x_j} \right)$ determine the upstream biases within respectively the streamline and crossflow wave propagation regions. These two expressions combined then induce a correct upwind bias along all wave propagation regions. The terms in this expression, furthermore, directly correspond to the physics of acoustics and convection.

Expression (17.113) determines f_i^C itself, up to an arbitrary divergence-free vector, as

$$f_i^C = f_i(q) - \varepsilon \psi \left[c \left(\alpha a_i a_j + \alpha^N \left(a_i^N a_j^{N_1} + a_i^{N_2} a_j^{N_2} \right) \right) \frac{\partial q}{\partial x_j} + a_i \frac{\partial f_j^q}{\partial x_j} + a_i \delta \frac{\partial f_j^p}{\partial x_j} \right] \quad (17.114)$$

According to this result, the intrinsic three-dimensionality of each component f_i^C derives from its dependence upon the entire divergence of f_j^q and f_j^p. The components in this flux remain linearly independent of one another, which avoids the linear-dependence instability in the steady low-Mach-number Euler equations.

For vanishing Mach numbers, α and α^N will approach 1 whereas δ will approach 0. Under these conditions, (17.113) reduces to

$$\frac{\partial f_j^C}{\partial x_j} = \frac{\partial f_j}{\partial x_j} - \frac{\partial}{\partial x_i} \left[\varepsilon \psi \left(c \frac{\partial q}{\partial x_i} + a_i \frac{\partial f_j^q}{\partial x_j} \right) \right] \quad (17.115)$$

which essentially induces only an acoustics upstream bias. Heed that this bias becomes independent of any specific propagation direction, for it no longer depends on the directional expression $\left(\alpha a_i a_j + \alpha^N \left(a_i^{N_1} a_j^{N_1} + a_i^{N_2} a_j^{N_2} \right) \right)$. This bias, as a result, becomes isotropic, in harmony with the isotropic propagation of acoustic waves. For supersonic flows, $\alpha = 0$ and $\delta = 1$ and (17.113) thus becomes

$$\frac{\partial f_j^C}{\partial x_j} = \frac{\partial f_j}{\partial x_j} - \frac{\partial}{\partial x_i} \left[\varepsilon \psi \left(c\alpha^N \left(a_i^{N_1} a_j^{N_1} + a_i^{N_2} a_j^{N_2} \right) \frac{\partial q}{\partial x_j} + a_i \frac{\partial f_j}{\partial x_j} \right) \right] \quad (17.116)$$

which depends on the crossflow component of the absolute acoustics matrix and the entire divergence of the Euler inviscid flux vector.

17.5.1   Consistent Multi-Dimensional and Infinite-Directional Upstream Bias

This section describes the conditions that allow the characteristics-bias Euler flux divergence to provide a consistent infinite-directional and intrinsically multi-dimensional upstream formulation.

Two-Dimensional Flows

In Jacobian form, the characteristics-bias Euler flux divergence becomes

$$\frac{\partial f_j^C}{\partial x_j} = \frac{\partial f_j}{\partial x_j} - \frac{\partial}{\partial x_i}\left[\varepsilon\psi\left(c\left(\alpha a_i a_j + \alpha^N a_i^N a_j^N\right)I + a_i\frac{\partial f_j^q}{\partial q} + a_i\delta\frac{\partial f_j^p}{\partial q}\right)\frac{\partial q}{\partial x_j}\right] \quad (17.117)$$

which essentially depends upon the five upstream-bias functions $a_1, a_2, \alpha, \delta, \alpha^N$. In order to ensure physical significance for this characteristic-bias expression, these functions are determined by imposing on (17.117) the stringent stability requirement of an upstream-bias dissipation not just along the principal streamline and crossflow directions, but along all directions $\boldsymbol{n} = (n_1, n_2)$ radiating from any flow-field point, a condition that corresponds to stability of the upstream-bias expression. By virtue of the developments in Sections 14.4,14.5.4, expression (17.117) will be stable when all the eigenvalues of the upstream-bias matrix

$$\mathcal{A} \equiv n_i\left(c\left(\alpha a_i a_j + \alpha^N a_i^N a_j^N\right)I + a_i\frac{\partial f_j^q}{\partial q} + a_i\delta\frac{\partial f_j^p}{\partial q}\right)n_j \quad (17.118)$$

remain positive for all Mach numbers M and propagation directions $\boldsymbol{n}$.

Three-Dimensional Flows

An analogous conclusion applies for 3-D flows. In Jacobian form, the characteristics-bias Euler flux divergence becomes

$$\frac{\partial f_j^C}{\partial x_j} = \frac{\partial f_j}{\partial x_j} - \frac{\partial}{\partial x_i}\left[\varepsilon\psi\left(c\left(\alpha a_i a_j + \alpha^N\left(a_i^{N_1}a_j^{N_1} + a_i^{N_2}a_j^{N_2}\right)\right)I + a_i\frac{\partial f_j^q}{\partial q} + a_i\delta\frac{\partial f_j^p}{\partial q}\right)\frac{\partial q}{\partial x_j}\right]$$
$$(17.119)$$

which essentially depend upon the six upstream-bias functions $a_1,\ a_2,\ a_3,\ \alpha,\ \delta,\ \alpha^N$, since $\boldsymbol{a}^{N_1}$ and $\boldsymbol{a}^{N_2}$ are directly obtained from $\boldsymbol{a}$ as shown in Section 17.5.4. Again, in order to ensure physical significance for this characteristics-bias expression, these functions are determined by imposing on expression (17.119) the stringent stability requirement of an upstream-bias dissipation not just along the principal streamline and crossflow directions, but along all three-dimensional directions $\boldsymbol{n} = (n_1, n_2, n_3)$ radiating from any flow-field point. This expression will be stable when all the eigenvalues of the upstream-bias matrix

$$\mathcal{A} \equiv n_i\left(c\left(\alpha a_i a_j + \alpha^N\left(a_i^{N_1}a_j^{N_1} + a_i^{N_2}a_j^{N_2}\right)\right)I + a_i\frac{\partial f_j^q}{\partial q} + a_i\delta\frac{\partial f_j^p}{\partial q}\right)n_j \quad (17.120)$$

remain positive for all Mach numbers M and propagation directions $\boldsymbol{n}$.

17.5.2 Upstream-Bias Eigenvalues and Oblique-Shock Capturing

Despite the formidable non-linear algebraic complexity of $\mathcal{A}$, all of its eigenvalues have been analytically determined exactly in closed form for general equilibrium equations of state and for two- and three-dimensional flows. Dividing through the speed of sound c, the non-dimensional form of these eigenvalues is

$$\lambda_{1,2} = n_i\left(\alpha a_i a_j + \alpha^N a_i^N a_j^N\right)n_j + n_i a_i v_j n_j M$$

$$\lambda_{3,4} = n_i \left(\alpha a_i a_j + \alpha^N a_i^N a_j^N \right) n_j$$

$$+ n_i a_i \left(1 + \frac{1-\delta}{2} p_E \right) v_j n_j M \pm n_i a_i \sqrt{\left(\frac{1-\delta}{2} p_E v_j n_j M \right)^2 + \delta} \qquad (17.121)$$

for two dimensional flows, and

$$\lambda_{0,2} = n_i \left(\alpha a_i a_j + \alpha^N \left(a_i^{N_1} a_j^{N_1} + a_i^{N_2} a_j^{N_2} \right) \right) n_j + n_i a_i v_j n_j M$$

$$\lambda_{3,4} = n_i \left(\alpha a_i a_j + \alpha^N \left(a_i^{N_1} a_j^{N_1} + a_i^{N_2} a_j^{N_2} \right) \right) n_j +$$

$$+ n_i a_i \left(1 + \frac{1-\delta}{2} p_E \right) v_j n_j M \pm n_i a_i \sqrt{\left(\frac{1-\delta}{2} p_E v_j n_j M \right)^2 + \delta} \qquad (17.122)$$

for three-dimensional flows.

In these expressions, v_j denotes the jth direction cosine of a unit vector $\boldsymbol{v}$ parallel to the local velocity $\boldsymbol{u}$. Accordingly, the inner product $v_j n_j$ supplies the cosine of the angle between the velocity vector and the unit vector $\boldsymbol{n}$; when this unit vector is orthogonal to an oblique-shock facet, these eigenvalues, as a result, depend on the "normal" Mach number $M_n = M v_j n_j$, which corresponds to the Mach number of the flow in the direction normal to the shock. The function δ from $\lambda_{3,4}$ is thus calculated on the basis of this normal Mach number, so that the resulting multi-dimensional upstream-bias formulation across a shock will sense the supersonic and subsonic flows in the normal direction, as in a one-dimensional flow.

All these eigenvalues must converge to 1 for vanishing Mach number, for an accurate modeling of acoustic waves. Heed that for both $\boldsymbol{a} = \boldsymbol{v}$ and $\boldsymbol{n} = \boldsymbol{v}$, the functions α and δ within (17.121) and (17.122) determine the corresponding streamline upstream-bias eigenvalues already established for one-dimensional flows in Chapter 13

$$\lambda_{1,2} = \alpha + M, \quad \lambda_{3,4} = \alpha + \left(1 + \frac{1-\delta}{2} p_E \right) M \pm \sqrt{\left(\frac{1-\delta}{2} p_E M \right)^2 + \delta} \qquad (17.123)$$

Rather than prescribing some expressions for α and δ and accepting the resulting variations for these eigenvalues, again physically consistent expressions for the streamline upstream-bias eigenvalues are instead prescribed and the corresponding functions for α and δ determined. As shown in the following sections, all of these upstream-bias eigenvalues will remain positive when α and δ are calculated via the streamline-flow expressions, from Chapter 13, and α^N forces $\lambda_{3,4}$ to remain positive for all $\boldsymbol{n}$ and Mach numbers.

17.5.3   Conditions on Upstream-Bias Functions and Eigenvalues

Two-Dimensional Flows

The eigenvalues (17.121) are expressed as

$$\lambda_{1,4} = \lambda_{1,4} (M, \boldsymbol{n}) \qquad (17.124)$$

to stress their dependence upon both M and $\boldsymbol{n}$. The five theoretical conditions for the determination of the five functions $a_1, a_2, \alpha, \delta, \alpha^N$ are

$$\lambda_{1,2} (M, \boldsymbol{n}) \geq 0, \quad \boldsymbol{a} \cdot \boldsymbol{a}^N = 0$$

$$\lambda_1 (M, \boldsymbol{v}) = \lambda_1, \quad \lambda_4 (M, \boldsymbol{v}) = \lambda_4, \quad \lambda_{3,4} (M, \boldsymbol{n}) \geq 0 \tag{17.125}$$

where λ_1 and λ_4 now denote prescribed streamline upstream-bias eigenvalues. The first condition stipulates that the unit vector $\boldsymbol{a}$ is parallel to velocity and the second condition determines that $\boldsymbol{a}$ and $\boldsymbol{a}^N$ are mutually perpendicular. In particular, these two conditions constrain the unit vectors $\boldsymbol{a}$ and $\boldsymbol{a}^N$ respectively to point along the streamline and cross-flow directions. The third and fourth conditions stipulate that the streamline upstream-bias eigenvalues must equal prescribed eigenvalues. The streamline eigenvalues λ_1 and λ_4 will coincide with the eigenvalues λ_1 and λ_2 of the one-dimensional formulation. These eigenvalues will thus lead to the same expressions for α and δ as in the 1-D formulation. For the determined $\boldsymbol{a}$, $\boldsymbol{a}^N$, α and δ, the fifth condition then establishes α^N.

Three-Dimensional Flows

In this case, the eigenvalues (17.122) are expressed as

$$\lambda_{0,4} = \lambda_{0,4} (M, \boldsymbol{n}) \tag{17.126}$$

The six conditions for the determination of the six functions a_1, a_2, a_3, α, δ, α^N are

$$\lambda_{0,2} (M, \boldsymbol{n}) \geq 0, \quad \boldsymbol{a} \cdot \boldsymbol{a}^{N_1} = 0, \quad \boldsymbol{a} \cdot \boldsymbol{a}^{N_2} = 0$$

$$\lambda_{0,1} (M, \boldsymbol{v}) = \lambda_1, \quad \lambda_4 (M, \boldsymbol{v}) = \lambda_4, \quad \lambda_{3,4} (M, \boldsymbol{n}) \geq 0 \tag{17.127}$$

where λ_1 and λ_4 again denote prescribed streamline upstream-bias eigenvalues. The first condition stipulates that the unit vector $\boldsymbol{a}$ is parallel to velocity and the second and third condition determine that $\boldsymbol{a}$, $\boldsymbol{a}^{N_1}$, and $\boldsymbol{a}^{N_2}$ are mutually perpendicular to one another. In particular, these three conditions constrain the unit vector $\boldsymbol{a}$ to point in the streamline direction and the unit vectors $\boldsymbol{a}^{N_1}$ and $\boldsymbol{a}^{N_2}$ to point in any two mutually perpendicular crossflow directions. The fourth and fifth conditions stipulate that the streamline upstream-bias eigenvalues must equal prescribed eigenvalues, and thereby determine α and δ as in the 1-D formulation. For the determined $\boldsymbol{a}$, $\boldsymbol{a}^{N_1}$, $\boldsymbol{a}^{N_2}$, α, and δ, the sixth condition then establishes α^N.

17.5.4 Upstream-Bias Functions a, α, and δ

These functions are used in actual computations based on the characteristics flux divergence (17.108) and (17.113)

Two-Dimensional Flows

In the eigenvalues $\lambda_{1,2}$ in (17.121), the components

$$n_i \alpha a_i a_j n_j = \alpha \left(a_j n_j \right)^2, \quad n_i \alpha^N a_i^N a_j^N n_j = \alpha^N \left(a_j^N n_j \right)^2 \tag{17.128}$$

are already non-negative for non-negative α and α^N. The eigenvalues $\lambda_{1,2}$ will thus remain non-negative for all positive α and α^N, including $\alpha \to 0$ and $\alpha^N \to 0$, when the additional component $n_i a_i v_j n_j M$ remains non-negative for all M. This requirement is met when $\boldsymbol{a} = \boldsymbol{v}$, for

$$n_i a_i v_j n_j M = M \left(v_j n_j \right)^2 \geq 0 \tag{17.129}$$

This finding is not surprising, for the streamline direction is a principal characteristic direction. The unit vector $\boldsymbol{a}^N$ then equals the vector obtained by rotating $\boldsymbol{a}$ by 90^0 counterclockwise, with components

$$a_1^N = -a_2, \quad a_2^N = a_1 \tag{17.130}$$

Three-Dimensional Flows

In the eigenvalues $\lambda_{0,2}$ in (17.122), the components

$$n_i \alpha a_i a_j n_j = \alpha \left(a_j n_j \right)^2, \quad n_i \alpha^N a_i^{N_1} a_j^{N_1} n_j = \alpha^{N_1} \left(a_j^{N_1} n_j \right)^2, \quad n_i \alpha^N a_i^{N_2} a_j^{N_2} n_j = \alpha^{N_2} \left(a_j^{N_2} n_j \right)^2 \tag{17.131}$$

are already non-negative for non-negative α and α^N. The eigenvalues $\lambda_{0,2}$ will thus remain non-negative for all positive α and α^N, including $\alpha \to 0$ and $\alpha^N \to 0$, when the additional component $n_i a_i v_j n_j M$ remains non-negative for all M. This requirement is met along with the first condition in (17.127) when $\boldsymbol{a} = \boldsymbol{v}$, for

$$n_i a_i v_j n_j M = M \left(v_j n_j \right)^2 \geq 0 \tag{17.132}$$

Again, this finding is not surprising, for also within a three-dimensional flow is the streamline direction a principal characteristic direction. In any flow region where velocity vanishes, it is then computationally convenient to set $a_1 = 1$, $a_2 = a_3 = 0$. Since $a_1^2 + a_2^2 + a_3^2 = 1$, it follows that either $a_1^2 + a_2^2$ or $a_2^2 + a_3^2$ is always different from zero. For $a_1^2 + a_2^2 > 0$, accordingly, the components of $\boldsymbol{a}$, $\boldsymbol{a}^{N_1}$, $\boldsymbol{a}^{N_2}$ may be expressed as

$$a_1 = \frac{u_1}{\|\boldsymbol{u}\|}, \quad a_2 = \frac{u_2}{\|\boldsymbol{u}\|}, \quad a_3 = \frac{u_3}{\|\boldsymbol{u}\|}$$

$$a_1^{N_1} = -\frac{a_2}{\sqrt{a_1^2 + a_2^2}}, \quad a_2^{N_1} = \frac{a_1}{\sqrt{a_1^2 + a_2^2}}, \quad a_3^{N_1} = 0$$

$$a_1^{N_2} = -\frac{a_1 a_3}{\sqrt{a_1^2 + a_2^2}}, \quad a_2^{N_2} = -\frac{a_2 a_3}{\sqrt{a_1^2 + a_2^2}}, \quad a_3^{N_2} = \sqrt{a_1^2 + a_2^2} \tag{17.133}$$

With these expressions, $\boldsymbol{a}$, $\boldsymbol{a}^{N_1}$, and $\boldsymbol{a}^{N_2}$ are unit vectors that remain mutually perpendicular with respect to one another. For $a_2^2 + a_3^2 > 0$, instead, the components of $\boldsymbol{a}$, $\boldsymbol{a}^{N_1}$, $\boldsymbol{a}^{N_2}$ may be expressed as

$$a_1 = \frac{u_1}{\|\boldsymbol{u}\|}, \quad a_2 = \frac{u_2}{\|\boldsymbol{u}\|}, \quad a_3 = \frac{u_3}{\|\boldsymbol{u}\|}$$

$$a_1^{N_1} = 0, \quad a_2^{N_1} = -\frac{a_3}{\sqrt{a_2^2 + a_3^2}}, \quad a_3^{N_1} = \frac{a_2}{\sqrt{a_2^2 + a_3^2}}$$

$$a_1^{N_2} = \sqrt{a_2^2 + a_3^2}, \quad a_2^{N_2} = -\frac{a_1 a_2}{\sqrt{a_2^2 + a_3^2}}, \quad a_3^{N_2} = -\frac{a_1 a_3}{\sqrt{a_2^2 + a_3^2}} \tag{17.134}$$

With these expressions too $\boldsymbol{a}$, $\boldsymbol{a}^{N_1}$, and $\boldsymbol{a}^{N_2}$ are unit vectors that remain mutually perpendicular with respect to one another.

Determination of α and δ

As noted, the expressions for α and δ coincide with the one-dimensional-flow results based on the prescribed streamline eigenvalues λ_1 and λ_4. Restated here, for convenience, the expressions for these two eigenvalues are

$$\lambda_1(M) \equiv \begin{cases} 1 - M + \frac{\varepsilon_M}{2}(2M)^{1/\varepsilon_M} & , \quad 0 \leq M < \frac{1}{2} \\[2mm] \frac{(M - \frac{1}{2})^2}{2\varepsilon_M} + \frac{1 + \varepsilon_M}{2} & , \quad \frac{1}{2} \leq M < \frac{1}{2} + \varepsilon_M \\[2mm] M & , \quad \frac{1}{2} + \varepsilon_M \leq M \end{cases} \tag{17.135}$$

$$\lambda_4(M) \equiv \begin{cases} 1 - M & , \quad 0 \leq M \leq 1 - \varepsilon_M \\[2mm] \frac{(M - 1)^2}{2\varepsilon_M} + \frac{\varepsilon_M}{2} & , \quad 1 - \varepsilon_M < M < 1 + \varepsilon_M \\[2mm] M - 1 & , \quad 1 + \varepsilon_M \leq M \end{cases} \tag{17.136}$$

where $\varepsilon_M = \frac{1}{4}$. From the expressions in (17.123), the corresponding functions for both $\alpha = \alpha(M)$ and $\delta = \delta(M)$ in terms of these streamline eigenvalues are

$$\alpha(M) = \lambda_1(M) - M, \quad \delta(M) = \frac{(\lambda_1(M) - \lambda_4(M))(\lambda_1(M) - \lambda_4(M) + p_E M)}{1 + p_E M(\lambda_1(M) - \lambda_4(M))} \tag{17.137}$$

17.5.5   Crossflow Upstream-Bias Function α^N

The condition $\lambda_{3,4}(M, \boldsymbol{n}) \geq 0$ in (17.125), (17.127) determines $\alpha^N = \alpha^N(M)$ for both two- and three-dimensional flows.

Two-Dimensional Flows

Heed that $\boldsymbol{a} = \boldsymbol{v}$ remains perpendicular to $\boldsymbol{a}^N$, while $n_i a_i$ and $n_i a_i^N$ denote the vector "dot" products between the unit vector $\boldsymbol{n}$ and the unit vectors $\boldsymbol{a}$ and $\boldsymbol{a}^N$ respectively. Accordingly,

$$n_i a_i = \cos\overline{\theta}, \quad n_i a_i^N a_j^N n_j = \sin^2\overline{\theta}, \quad \overline{\theta} \equiv \theta - \theta_v \tag{17.138}$$

where θ and θ_v denote the inclination angles between the x_1 axis and $\boldsymbol{n}$ and $\boldsymbol{v}$ respectively.

Three-Dimensional Flows

Heed that $\boldsymbol{a} = \boldsymbol{v}$ remains perpendicular to both $\boldsymbol{a}^{N_1}$ and $\boldsymbol{a}^{N_2}$. Moreover, the expressions $n_i a_i$, $n_i a_i^{N_1}$, and $n_i a_i^{N_2}$ respectively denote the vector "dot" products between the unit vector $\boldsymbol{n}$ and the unit vectors $\boldsymbol{a}$, $\boldsymbol{a}^{N_1}$, and $\boldsymbol{a}^{N_2}$, hence

$$\boldsymbol{n} = n_i a_i \boldsymbol{a} + n_i a_i^{N_1} \boldsymbol{a}^{N_1} + n_i a_i^{N_2} \boldsymbol{a}^{N_2} \tag{17.139}$$

Accordingly,

$$1 = \left(n_i a_i\right)^2 + \left(n_i a_i^{N_1}\right)^2 + \left(n_i a_i^{N_2}\right)^2, \quad n_i a_i = \cos\bar{\theta}$$

$$n_i a_i^{N_1} a_j^{N_1} n_j + n_i a_i^{N_2} a_j^{N_2} n_j = 1 - \cos^2\bar{\theta} = \sin^2\bar{\theta} \tag{17.140}$$

where $\bar{\theta}$ denotes the angle between $\boldsymbol{n}$ and $\boldsymbol{a}$.

Determination of α^N

For both two- and three-dimensional flows, the condition $\lambda_4\left(M, \boldsymbol{n}\right) \geq 0$ leads to the following single constraint on α^N

$$\alpha^N \geq g\left(\bar{\theta}, M\right) \equiv \frac{\cos\bar{\theta}\sqrt{\left(\frac{1-\delta}{2}p_E \cos\bar{\theta}M\right)^2 + \delta - \cos^2\bar{\theta}\left(\alpha + \left(1 + \frac{1-\delta}{2}p_E\right)M\right)}}{1 - \cos^2\bar{\theta}} \tag{17.141}$$

For supersonic flows with $M \geq 1 + \varepsilon_M$, $\alpha = 0$ and $\delta = 1$, hence (17.141) becomes

$$\alpha^N \geq g\left(\bar{\theta}, M\right) = \frac{\cos\bar{\theta} - M\cos^2\bar{\theta}}{1 - \cos^2\bar{\theta}} \tag{17.142}$$

and in particular $\alpha^N \geq g_{\max}(M)$, where $g_{\max}(M)$ denotes the maximum of $g = g(\bar{\theta}, M)$ with respect to $\bar{\theta}$, for each M. From (17.142), the determination of $g_{\max}(M)$ yields

$$\frac{\partial g}{\partial \bar{\theta}} = 0 \quad \Rightarrow \quad \cos^2\bar{\theta} - 2M\cos\bar{\theta} + 1 = 0 \tag{17.143}$$

which leads to

$$\cos\bar{\theta}\Big|_{g=g_{\max}} = M - \sqrt{M^2 - 1}, \quad g_{\max}(M) = \frac{1}{2}\left(M - \sqrt{M^2 - 1}\right) \tag{17.144}$$

Significantly, the same solution for $g_{\max}(M)$ results from the condition $\lambda_3\left(M, \boldsymbol{n}\right) \geq 0$. Consequently,

$$2\alpha^N \geq \left(M - \sqrt{M^2 - 1}\right), \quad M \geq M_M \equiv 1 + \varepsilon_M \tag{17.145}$$

and considering that $\lambda_4\left(M_M, \boldsymbol{u}\right) = \varepsilon_M$, an analogous equality is adopted for $2\alpha^N(M_M)$, leading to $2\alpha^N(M_M) = 2g_{\max}(M_M) + \varepsilon_M$.

For subsonic flows, a numerical analysis of $g = g(\bar{\theta}, M)$ from (17.141) reveals that $g(\bar{\theta}, M) < 0.3$ for all $\bar{\theta}$ and $M < M_M$. Additionally, for an isotropic acoustic upstream for vanishing M, $\alpha^N(0) = 1$ and $\partial\alpha^N/\partial M|_{M=0} = \partial\alpha/\partial M|_{M=0} = \alpha'_0 = -2$, whereas for

$M \geq M_M$ both α^N and its derivative with respect to M can be calculated from (17.145). A smooth variation for $\alpha^N = \alpha^N(M)$ that satisfies all of these constraints is the composite spline

$$\alpha^N(M) \equiv \begin{cases} \left((\alpha_M^{N'} + \alpha_0')M_M - 2\alpha_M^N + 2\right)\left(\dfrac{M}{M_M}\right)^3 \\ \quad - \left((\alpha_M^{N'} + 2\alpha_0')M_M - 3\alpha_M^N + 3\right)\left(\dfrac{M}{M_M}\right)^2 \\ \quad + \left(\alpha_0' M_M\right)\left(\dfrac{M}{M_M}\right) + 1, \qquad 0 \leq M < M_M \\ \dfrac{1}{2}\left(1 + \dfrac{2\varepsilon_M}{M_M - \sqrt{M_M^2 - 1}}\right)\left(M - \sqrt{M^2 - 1}\right), \quad M_M \leq M \end{cases} \tag{17.146}$$

where superscript prime "$'$" denotes differentiation with respect to M and subscript "M" in both $\alpha_M^{N'}$ and α_M^N indicate their respective magnitudes at $M = M_M$ from the second expression in (17.146).

17.5.6   Variations of α, δ, α^N

The variations of $\alpha = \alpha(M)$, $\alpha^N = \alpha^N(M)$ and $\delta = \delta(M)$ in Figure 17.11 indicate that

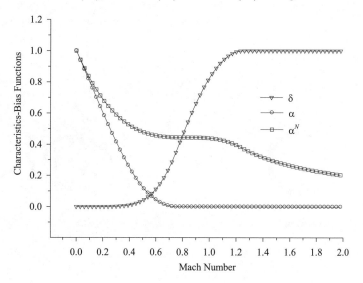

Figure 17.11: Upstream-Bias Functions

these three functions as well as their slopes remain continuous for all Mach numbers, with $0 \leq \alpha, \alpha^N, \delta \leq 1$.

These distributions show that the variations of α and α^N remain distinct from each other. The decrease of α^N, hence of the crossflow upstream-bias is less rapid because this is the only contribution to a crossflow upstream. The function α^N, nevertheless, decreases by 50%, at the sonic state, and by 80% for $M = 1.8$. For this function, forcing non-negativity of $\lambda_{3,4}$ as opposed to equality to a prescribed positive constant, in particular, ensures minimal cross flow diffusion. Expression (17.146) leads to the results

$$\lim_{M \to \infty} \alpha^N(M) = 0, \quad \lim_{M \to \infty} \frac{\partial \alpha^N}{\partial M} = 0 \qquad (17.147)$$

which indicate that the magnitude of crossflow upstream decreases with increasing M. This result agrees with the physics of high-M flows, where the bi-modal propagation region narrows about the crossflow direction, as seen in Section 17.3. Convection thereby becomes the prevailing wave propagation mechanism, which thus reduces the need for acoustic crossflow upstream bias.

17.6   Polar Variation of Upstream-Bias

The directional variation of the upstream bias eigenvalues (17.121), (17.122) is presented in Figures 17.12 - 17.13 for representative subsonic and supersonic Mach numbers. These variations are obtained for a variable unit vector $\boldsymbol{n} \equiv (\cos\theta, \sin\theta)$ and fixed unit vector $\boldsymbol{a} = \boldsymbol{v}$, in this representative case inclined by $+30°$ with respect to the x_1 axis on the flow plane, in 2-D, or a reference x_1 axis on any plane Π that contains both $\boldsymbol{v}$ and $\boldsymbol{n}$.

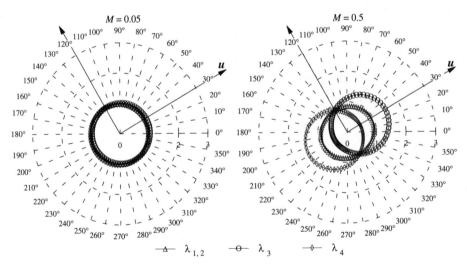

Figure 17.12: Polar Variation of Subsonic Upstream Bias

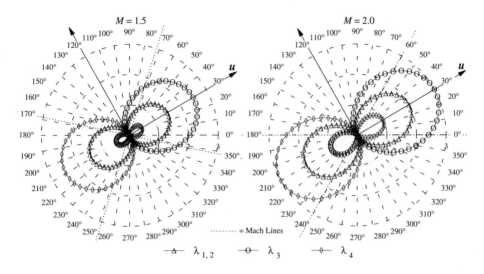

Figure 17.13: Polar Variation of Supersonic Upstream Bias

These figures collectively indicate that the characteristics flux divergence expressions (17.108), (17.113) induce a physically consistent upstream bias because, for any Mach number and wave propagation direction $\boldsymbol{n}$, the associated upstream-bias eigenvalues (17.121), (17.122) remain positive and their directional variation mirrors the directional variation of the characteristic Euler eigenvalues (17.58). The upstream-bias eigenvalues, moreover, are symmetrical about the crossflow direction and characteristic streamline, precisely like the characteristic Euler eigenvalues (17.58). For $M = 0.05$, the directional variation of the upstream-bias eigenvalues in Figure 17.12 correlates with that in Figure 17.6 and thereby corresponds to an isotropic upstream bias, in complete agreement with the isotropic acoustic wave propagation speed in the Euler equations. For increasing Mach numbers, the upstream bias becomes anisotropic, again in agreement with the anisotropic distribution of the Euler eigenvalues (17.58). For $M = 0.5$ this anisotropy is already evident and then becomes more marked for supersonic Mach numbers as indicated in Figure 17.13. In particular, the crossflow upstream bias decreases for increasing Mach number.

Figures 17.14 - 17.15 compare the directional variations of the representative upstream-bias eigenvalue λ_3 and the corresponding Euler eigenvalue λ_3^E. This comparison is sufficient to depict the correlation between all the Euler and upstream-bias eigenvalues, for $\lambda_{1,2}$ and $\lambda_{1,2}^E$ are topologically similar to each other, compare Figures 17.7 and 17.13, while λ_4 and λ_4^E are respectively mirror skew-symmetric to λ_3 and λ_3^E with respect to the crossflow direction.

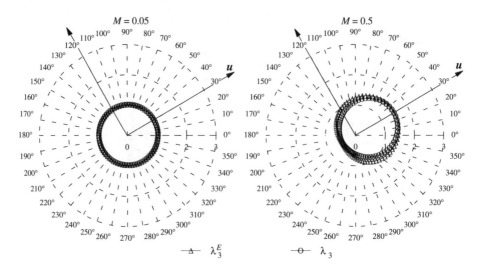

Figure 17.14: Polar Correlation of Subsonic Characteristic λ_3^E and Upstream λ_3

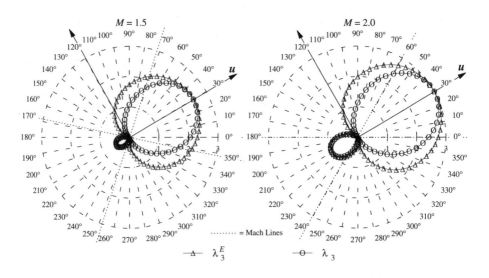

Figure 17.15: Polar Correlation of Supersonic Characteristic λ_3^E and Upstream λ_3

As Figures 17.14 - 17.15 indicate, λ_3 is symmetrical about the characteristic streamline, precisely like the corresponding characteristic Euler eigenvalue λ_3^E and the corresponding polar curve is topologically similar to the Euler eigenvalue curve. For $M = 0.05$, λ_3 and λ_3^E virtually coincide with each other and remain direction invariant, which corresponds to an isotropic upstream bias in correlation with the acoustic speed. For $M = 0.5$, Figure 17.14 indicates that λ_3^E is greater than λ_3 in the streamline direction. This results corresponds to minimal streamline upstream diffusion.

For supersonic Mach numbers, λ_3 in the streamline direction coincides with $M + 1$. As shown in Figure 17.15, the magnitude of the upstream bias for supersonic flows is thus virtually identical to the magnitude of the characteristic eigenvalues, within the domain of dependence and range of influence of any flow field point. Outside this region, the upstream-bias eigenvalues are modestly less than the characteristic eigenvalues. In these variations, the upstream-bias eigenvalues are vanishingly small in the cross-flow direction, which, in particular, corresponds to minimal crossflow diffusion.

17.7   Variational Statement and Boundary Conditions

This section presents the characteristics-bias Euler and Navier-Stokes integral systems and then details a set of boundary conditions.

17.7.1   Characteristics-Bias Euler and Navier-Stokes Systems

With reference to the developments in the previous sections and Chapter 14, the characteristics - bias Euler and Navier-Stokes systems for two- and three-dimensional flows are formulated as follows.

Two-Dimensional Flows

The characteristics-bias Euler equations are cast as

$$
\int_\Omega w \left(\frac{\partial \rho}{\partial t} + \frac{\partial m_1}{\partial x_1} + \frac{\partial m_2}{\partial x_2} \right) d\Omega - \int_\Omega w \frac{\partial}{\partial x_i} \left(\varepsilon \psi a_i \left(\frac{\partial \rho}{\partial t} + \frac{\partial m_1}{\partial x_1} + \frac{\partial m_2}{\partial x_2} \right) \right) d\Omega
$$

$$
- \int_\Omega w \frac{\partial}{\partial x_i} \left(\varepsilon \psi \left(c \left(\alpha a_i a_j + \alpha^N a_i^N a_j^N \right) \frac{\partial \rho}{\partial x_j} \right) \right) d\Omega = 0 \qquad (17.148)
$$

$$
\int_\Omega w \left(\frac{\partial m_\ell}{\partial t} + \frac{\partial}{\partial x_1} \left(\frac{m_1}{\rho} m_\ell \right) + \frac{\partial}{\partial x_2} \left(\frac{m_2}{\rho} m_\ell \right) + \frac{\partial p}{\partial x_\ell} \right) d\Omega
$$

$$
- \int_\Omega w \frac{\partial}{\partial x_i} \left(\varepsilon \psi a_i \left(\frac{\partial m_\ell}{\partial t} + \frac{\partial}{\partial x_1} \left(\frac{m_1}{\rho} m_\ell \right) + \frac{\partial}{\partial x_2} \left(\frac{m_2}{\rho} m_\ell \right) + \frac{\partial p}{\partial x_\ell} \right) \right) d\Omega
$$

$$
- \int_\Omega w \frac{\partial}{\partial x_i} \left(\varepsilon \psi \left(c \left(\alpha a_i a_j + \alpha^N a_i^N a_j^N \right) \frac{\partial m_\ell}{\partial x_j} + a_i (\delta - 1) \frac{\partial p}{\partial x_\ell} \right) \right) d\Omega = 0 \qquad (17.149)
$$

$$
\int_\Omega w \left(\frac{\partial E}{\partial t} + \frac{\partial}{\partial x_1} \left(\frac{m_1}{\rho} (E + p) \right) + \frac{\partial}{\partial x_2} \left(\frac{m_2}{\rho} (E + p) \right) \right) d\Omega
$$

$$
- \int_\Omega w \frac{\partial}{\partial x_i} \left(\varepsilon \psi a_i \left(\frac{\partial E}{\partial t} + \frac{\partial}{\partial x_1} \left(\frac{m_1}{\rho} (E + p) \right) + \frac{\partial}{\partial x_2} \left(\frac{m_2}{\rho} (E + p) \right) \right) \right) d\Omega
$$

$$-\int_\Omega w \frac{\partial}{\partial x_i}\left(\varepsilon\psi\left(c\left(\alpha a_i a_j + \alpha^N a_i^N a_j^N\right)\frac{\partial E}{\partial x_j}\right)\right) d\Omega = 0 \tag{17.150}$$

The corresponding Navier-Stokes equations are expressed as

$$\int_\Omega w \left(\frac{\partial \rho}{\partial t} + \frac{\partial m_1}{\partial x_1} + \frac{\partial m_2}{\partial x_2}\right) d\Omega - \int_\Omega w \frac{\partial}{\partial x_i}\left(\varepsilon\psi a_i \left(\frac{\partial \rho}{\partial t} + \frac{\partial m_1}{\partial x_1} + \frac{\partial m_2}{\partial x_2}\right)\right) d\Omega$$

$$-\int_\Omega w \frac{\partial}{\partial x_i}\left(\varepsilon\psi\left(c\left(\alpha a_i a_j + \alpha^N a_i^N a_j^N\right)\frac{\partial \rho}{\partial x_j}\right)\right) d\Omega = 0 \tag{17.151}$$

$$\int_\Omega w \left(\frac{\partial m_\ell}{\partial t} + \frac{\partial}{\partial x_1}\left(\frac{m_1}{\rho}m_\ell\right) + \frac{\partial}{\partial x_2}\left(\frac{m_2}{\rho}m_\ell\right) + \frac{\partial p}{\partial x_\ell} - \frac{1}{Re}\frac{\partial \sigma_{\ell j}^\mu}{\partial x_j}\right) d\Omega$$

$$-\int_\Omega w \frac{\partial}{\partial x_i}\left(\varepsilon\psi a_i \left(\frac{\partial m_\ell}{\partial t} + \frac{\partial}{\partial x_1}\left(\frac{m_1}{\rho}m_\ell\right) + \frac{\partial}{\partial x_2}\left(\frac{m_2}{\rho}m_\ell\right) + \frac{\partial p}{\partial x_\ell} - \frac{1}{Re}\frac{\partial \sigma_{\ell j}^\mu}{\partial x_j}\right)\right) d\Omega$$

$$-\int_\Omega w \frac{\partial}{\partial x_i}\left(\varepsilon\psi\left(c\left(\alpha a_i a_j + \alpha^N a_i^N a_j^N\right)\frac{\partial m_\ell}{\partial x_j} + a_i(\delta - 1)\frac{\partial p}{\partial x_\ell}\right)\right) d\Omega = 0 \tag{17.152}$$

$$\int_\Omega w \left(\frac{\partial E}{\partial t} + \frac{\partial}{\partial x_1}\left(\frac{m_1}{\rho}(E+p)\right) + \frac{\partial}{\partial x_2}\left(\frac{m_2}{\rho}(E+p)\right)\right) d\Omega$$

$$-\int_\Omega w \left(\frac{1}{Re}\frac{\partial}{\partial x_j}\left(\frac{m_i}{\rho}\sigma_{ij}^\mu\right) + \frac{1}{EcPrRe}\frac{\partial}{\partial x_j}\left(\frac{\partial T}{\partial x_j}\right)\right) d\Omega$$

$$-\int_\Omega w \frac{\partial}{\partial x_i}\left(\varepsilon\psi a_i \left(\frac{\partial E}{\partial t} + \frac{\partial}{\partial x_j}\left(\frac{m_j}{\rho}(E+p)\right)\right)\right) d\Omega$$

$$+\int_\Omega w \frac{\partial}{\partial x_i}\left(\varepsilon\psi a_i \left(\frac{1}{Re}\frac{\partial}{\partial x_j}\left(\frac{m_\ell}{\rho}\sigma_{\ell j}^\mu\right) + \frac{1}{EcPrRe}\frac{\partial}{\partial x_j}\left(\frac{\partial T}{\partial x_j}\right)\right)\right) d\Omega$$

$$-\int_\Omega w \frac{\partial}{\partial x_i}\left(\varepsilon\psi\left(c\left(\alpha a_i a_j + \alpha^N a_i^N a_j^N\right)\frac{\partial E}{\partial x_j}\right)\right) d\Omega = 0 \tag{17.153}$$

These systems provide the Galilean-invariant characteristics-bias formulation for two-dimensional inviscid and viscous compressible flows.

Three-Dimensional Flows

The characteristics-bias Euler equations are cast as

$$\int_\Omega w \left(\frac{\partial \rho}{\partial t} + \frac{\partial m_1}{\partial x_1} + \frac{\partial m_2}{\partial x_2} + \frac{\partial m_3}{\partial x_3}\right) d\Omega$$

$$-\int_\Omega w \frac{\partial}{\partial x_i}\left(\varepsilon\psi a_i \left(\frac{\partial \rho}{\partial t} + \frac{\partial m_1}{\partial x_1} + \frac{\partial m_2}{\partial x_2} + \frac{\partial m_3}{\partial x_3}\right)\right) d\Omega$$

$$-\int_\Omega w \frac{\partial}{\partial x_i}\left(\varepsilon\psi\left(c\left(\alpha a_i a_j + \alpha^N \left(a_i^{N_1} a_j^{N_1} + a_i^{N_2} a_j^{N_2}\right)\right)\frac{\partial \rho}{\partial x_j}\right)\right) d\Omega = 0 \tag{17.154}$$

$$\int_\Omega w \left(\frac{\partial m_\ell}{\partial t} + \frac{\partial}{\partial x_1}\left(\frac{m_1}{\rho}m_\ell\right) + \frac{\partial}{\partial x_2}\left(\frac{m_2}{\rho}m_\ell\right) + \frac{\partial}{\partial x_3}\left(\frac{m_3}{\rho}m_\ell\right) + \frac{\partial p}{\partial x_\ell} \right) d\Omega$$

$$- \int_\Omega w \frac{\partial}{\partial x_i} \left(\varepsilon\psi a_i \left(\frac{\partial m_\ell}{\partial t} + \frac{\partial}{\partial x_1}\left(\frac{m_1}{\rho}m_\ell\right) + \frac{\partial}{\partial x_2}\left(\frac{m_2}{\rho}m_\ell\right) + \frac{\partial}{\partial x_3}\left(\frac{m_3}{\rho}m_\ell\right) + \frac{\partial p}{\partial x_\ell} \right) \right) d\Omega$$

$$- \int_\Omega w \frac{\partial}{\partial x_i} \left(\varepsilon\psi \left(c\left(\alpha a_i a_j + \alpha^N \left(a_i^{N_1} a_j^{N_1} + a_i^{N_2} a_j^{N_2}\right)\right) \frac{\partial m_\ell}{\partial x_j} + a_i(\delta - 1)\frac{\partial p}{\partial x_\ell} \right) \right) d\Omega = 0$$

$$(17.155)$$

$$\int_\Omega w \left(\frac{\partial E}{\partial t} + \frac{\partial}{\partial x_1}\left(\frac{m_1}{\rho}(E+p)\right) + \frac{\partial}{\partial x_2}\left(\frac{m_2}{\rho}(E+p)\right) + \frac{\partial}{\partial x_3}\left(\frac{m_3}{\rho}(E+p)\right) \right) d\Omega$$

$$- \int_\Omega w \frac{\partial}{\partial x_i} \left(\varepsilon\psi a_i \left(\frac{\partial E}{\partial t} + \frac{\partial}{\partial x_1}\left(\frac{m_1}{\rho}(E+p)\right) + \frac{\partial}{\partial x_2}\left(\frac{m_2}{\rho}(E+p)\right) + \frac{\partial}{\partial x_3}\left(\frac{m_3}{\rho}(E+p)\right) \right) \right) d\Omega$$

$$- \int_\Omega w \frac{\partial}{\partial x_i} \left(\varepsilon\psi \left(c\left(\alpha a_i a_j + \alpha^N \left(a_i^{N_1} a_j^{N_1} + a_i^{N_2} a_j^{N_2}\right)\right) \frac{\partial E}{\partial x_j} \right) \right) d\Omega = 0 \qquad (17.156)$$

The corresponding Navier-Stokes equations are expressed as

$$\int_\Omega w \left(\frac{\partial \rho}{\partial t} + \frac{\partial m_1}{\partial x_1} + \frac{\partial m_2}{\partial x_2} + \frac{\partial m_3}{\partial x_3} \right) d\Omega$$

$$- \int_\Omega w \frac{\partial}{\partial x_i} \left(\varepsilon\psi a_i \left(\frac{\partial \rho}{\partial t} + \frac{\partial m_1}{\partial x_1} + \frac{\partial m_2}{\partial x_2} + \frac{\partial m_3}{\partial x_3} \right) \right) d\Omega$$

$$- \int_\Omega w \frac{\partial}{\partial x_i} \left(\varepsilon\psi \left(c\left(\alpha a_i a_j + \alpha^N \left(a_i^{N_1} a_j^{N_1} + a_i^{N_2} a_j^{N_2}\right)\right) \frac{\partial \rho}{\partial x_j} \right) \right) d\Omega = 0 \qquad (17.157)$$

$$\int_\Omega w \left(\frac{\partial m_\ell}{\partial t} + \frac{\partial}{\partial x_1}\left(\frac{m_1}{\rho}m_\ell\right) + \frac{\partial}{\partial x_2}\left(\frac{m_2}{\rho}m_\ell\right) + \frac{\partial}{\partial x_3}\left(\frac{m_3}{\rho}m_\ell\right) + \frac{\partial p}{\partial x_\ell} - \frac{1}{Re}\frac{\partial \sigma_{\ell j}^\mu}{\partial x_j} \right) d\Omega$$

$$- \int_\Omega w \frac{\partial}{\partial x_i} \left(\varepsilon\psi a_i \left(\frac{\partial m_\ell}{\partial t} + \frac{\partial}{\partial x_j}\left(\frac{m_j}{\rho}m_\ell\right) + \frac{\partial p}{\partial x_\ell} - \frac{1}{Re}\frac{\partial \sigma_{\ell j}^\mu}{\partial x_j} \right) \right) d\Omega$$

$$- \int_\Omega w \frac{\partial}{\partial x_i} \left(\varepsilon\psi \left(c\left(\alpha a_i a_j + \alpha^N \left(a_i^{N_1} a_j^{N_1} + a_i^{N_2} a_j^{N_2}\right)\right) \frac{\partial m_\ell}{\partial x_j} + a_i(\delta - 1)\frac{\partial p}{\partial x_\ell} \right) \right) d\Omega = 0$$

$$(17.158)$$

$$\int_\Omega w \left(\frac{\partial E}{\partial t} + \frac{\partial}{\partial x_1}\left(\frac{m_1}{\rho}(E+p)\right) + \frac{\partial}{\partial x_2}\left(\frac{m_2}{\rho}(E+p)\right) + \frac{\partial}{\partial x_3}\left(\frac{m_3}{\rho}(E+p)\right) \right) d\Omega$$

$$- \int_\Omega w \left(\frac{1}{Re}\frac{\partial}{\partial x_j}\left(\frac{m_i}{\rho}\sigma_{ij}^\mu\right) + \frac{1}{EcPrRe}\frac{\partial}{\partial x_j}\left(\frac{\partial T}{\partial x_j}\right) \right) d\Omega$$

$$- \int_\Omega w \frac{\partial}{\partial x_i} \left(\varepsilon\psi a_i \left(\frac{\partial E}{\partial t} + \frac{\partial}{\partial x_j}\left(\frac{m_j}{\rho}(E+p)\right) \right) \right) d\Omega$$

$$+ \int_\Omega w \frac{\partial}{\partial x_i} \left(\varepsilon\psi a_i \left(\frac{1}{Re}\frac{\partial}{\partial x_j}\left(\frac{m_\ell}{\rho}\sigma_{\ell j}^\mu\right) + \frac{1}{EcPrRe}\frac{\partial}{\partial x_j}\left(\frac{\partial T}{\partial x_j}\right) \right) \right) d\Omega$$

$$- \int_\Omega w \frac{\partial}{\partial x_i} \left(\varepsilon\psi \left(c\left(\alpha a_i a_j + \alpha^N \left(a_i^{N_1} a_j^{N_1} + a_i^{N_2} a_j^{N_2}\right)\right) \frac{\partial E}{\partial x_j} \right) \right) d\Omega = 0 \qquad (17.159)$$

These systems provide the Galilean-invariant characteristics-bias formulation for three-dimensional inviscid and viscous compressible flows.

17.7.2   Weak Statement and Boundary Conditions

For two- and three-dimensional flows, with implied summation on repeated subscript indices, $1 \leq i,j \leq n$, the characteristics-bias Euler and Navier-Stokes integral systems may be concisely expressed as

$$\int_\Omega w \left(\frac{\partial q}{\partial t} + \frac{\partial f_j}{\partial x_j} - \frac{\partial f_j^\nu}{\partial x_j} \right) d\Omega - \int_\Omega w \frac{\partial}{\partial x_i} \left[\varepsilon \psi a_i \left(\frac{\partial q}{\partial t} + \frac{\partial f_j}{\partial x_j} - \frac{\partial f_j^\nu}{\partial x_j} \right) \right] d\Omega$$

$$- \int_\Omega w \frac{\partial}{\partial x_i} \left[\varepsilon \psi \left(c \left(\alpha a_i a_j + \alpha^N a_i^{N\ell} a_j^{N\ell} \right) \frac{\partial q}{\partial x_j} + a_i (\delta - 1) \frac{\partial f_j^p}{\partial x_j} \right) \right] d\Omega = 0 \qquad (17.160)$$

with $\ell = 1$, for two-dimensional flows, and $1 \leq \ell \leq 2$ for three-dimensional flows. The corresponding weak statement is achieved by integrating by parts the characteristics-bias expression, an operation that yields the integral statement

$$\int_\Omega \left(w + \varepsilon \psi a_i \frac{\partial w}{\partial x_i} \right) \left(\frac{\partial q}{\partial t} + \frac{\partial f_j}{\partial x_j} - \frac{\partial f_j^\nu}{\partial x_j} \right) d\Omega$$

$$+ \int_\Omega \frac{\partial w}{\partial x_i} \varepsilon \psi \left(c \left(\alpha a_i a_j + \alpha^N a_i^{N\ell} a_j^{N\ell} \right) \frac{\partial q}{\partial x_j} + a_i (\delta - 1) \frac{\partial f_j^p}{\partial x_j} \right) d\Omega$$

$$- \oint_{\partial \Omega} w \varepsilon \psi \left(a_i \left(\frac{\partial q}{\partial t} + \frac{\partial (f_j - f_j^\nu)}{\partial x_j} \right) + c \left(\alpha a_i a_j + \alpha^N a_i^{N\ell} a_j^{N\ell} \right) \frac{\partial q}{\partial x_j} + a_i (\delta - 1) \frac{\partial f_j^p}{\partial x_j} \right) n_i d\Gamma = 0$$

$$(17.161)$$

When the divergence of the viscous flux f_j^ν is also integrated by parts, the following equivalent statement is obtained

$$\int_\Omega \left(w + \varepsilon \psi a_i \frac{\partial w}{\partial x_i} \right) \left(\frac{\partial q}{\partial t} + \frac{\partial f_j}{\partial x_j} \right) d\Omega + \int_\Omega \frac{\partial w}{\partial x_j} f_j^\nu d\Omega - \oint_{\partial \Omega} w f_j^\nu n_j \, d\Gamma$$

$$+ \int_\Omega \frac{\partial w}{\partial x_i} \varepsilon \psi \left(c \left(\alpha a_i a_j + \alpha^N a_i^{N\ell} a_j^{N\ell} \right) \frac{\partial q}{\partial x_j} - a_i \frac{\partial f_j^\nu}{\partial x_j} + a_i (\delta - 1) \frac{\partial f_j^p}{\partial x_j} \right) d\Omega$$

$$- \oint_{\partial \Omega} w \varepsilon \psi \left(a_i \left(\frac{\partial q}{\partial t} + \frac{\partial (f_j - f_j^\nu)}{\partial x_j} \right) + c \left(\alpha a_i a_j + \alpha^N a_i^{N\ell} a_j^{N\ell} \right) \frac{\partial q}{\partial x_j} + a_i (\delta - 1) \frac{\partial f_j^p}{\partial x_j} \right) n_i d\Gamma = 0$$

$$(17.162)$$

where n_i denotes the i-th component of the outward pointing unit vector perpendicular to a boundary facet, as depicted in Figure 17.16; when the boundary corresponds to a wall, then $a_i n_i = 0$ and the corresponding terms in the characteristics-bias surface integral will identically vanish. The $\varepsilon \psi$ surface integral, in any case, is eliminated by enforcing a weak Neumann-type boundary condition for the hyperbolic-parabolic characteristics-bias perturbation. One part of this condition imposes that the original Navier-Stokes system should be satisfied at the boundary; the remaining terms are also set to zero as part of this weak boundary condition. Clearly, the single Neumann-type condition corresponds to the entire boundary expression set to nought. For supersonic flows with $M \geq 1 + \varepsilon_M$, $\alpha = 0$, $\delta = 1$ and for either $n_i a_i^{N\ell} = 0$ or $a_j^{N\ell} \frac{\partial q}{\partial x_j} = 0$, or both, this boundary condition enforces the Navier-Stokes system on the boundary. With the imposition of this weak boundary condition, system (17.162) becomes

$$\int_{\Omega} \left(w + \varepsilon \psi a_i \frac{\partial w}{\partial x_i} \right) \left(\frac{\partial q}{\partial t} + \frac{\partial f_j}{\partial x_j} \right) d\Omega + \int_{\Omega} \frac{\partial w}{\partial x_j} f_j^{\nu} d\Omega - \oint_{\partial \Omega} w f_j^{\nu} n_j \, d\Gamma$$

$$+ \int_{\Omega} \frac{\partial w}{\partial x_i} \varepsilon \psi \left(c \left(\alpha a_i a_j + \alpha^N a_i^{N\ell} a_j^{N\ell} \right) \frac{\partial q}{\partial x_j} + a_i (\delta - 1) \frac{\partial f_j^p}{\partial x_j} \right) d\Omega = 0 \qquad (17.163)$$

from which, for computational simplicity, the characteristics-bias term for the viscous flux may be omitted without loss of stability. Significantly, this statement corresponds to a generalized SUPG statement, which, however, does not require pre-multiplication of the Euler system by the transpose of the Euler Jacobian, while it automatically provides the characteristics-based counterparts of the SUPG "shock capturing" term and stability parameters, in this case $\varepsilon \psi$, $\boldsymbol{a}$, $\boldsymbol{a}^N$, α, α^N, and δ.

For all test functions $w \in \mathcal{H}^1(\Omega)$, the variational formulation seeks a solution $q \in \mathcal{H}^{\eta}(\Omega)$ with $\eta = 0$ or $\eta = 1$ respectively for shocked or smooth flows, that satisfies this weak statement. This statement is subject to prescribed initial conditions $q(x, 0) = q_0(x)$ and boundary conditions on $\partial \Omega \equiv \overline{\Omega} \backslash \Omega$. Synthetically, these boundary conditions are expressed as

$$B(x_{\partial \Omega}) q(x_{\partial \Omega}, t) = G_v(x_{\partial \Omega}, t) \qquad (17.164)$$

where $G_v(x_{\partial \Omega}, t)$ corresponds to the array of prescribed Dirichlet boundary conditions, with a zero entry for each corresponding unconstrained component of q, and $B(x_{\partial \Omega})$ denotes a square diagonal matrix, with a 1 for each diagonal entry, but replaced by zero for each corresponding unconstrained component of q.

The number of boundary conditions at inlet and outlet depends on the number of negative eigenvalues of the Jacobian $\frac{\partial f_j}{\partial q} n_j$, where n_j denotes the j-th component of the outward pointing normal unit vector $\mathbf{n}$ on $\partial \Omega$. A negative eigenvalue of this Jacobian signifies a wave propagation in the direction opposite $\mathbf{n}$, hence toward the flow domain, a propagation that carries information from the flow outside the domain Ω, which information in embodied in a corresponding boundary condition.

At a subsonic inlet, only one eigenvalue of $\frac{\partial f_j}{\partial q} n_j$ is positive; as a result only one inlet variable must not be constrained. At such an inlet, Dirichlet boundary conditions are enforced on density ρ, total energy E, and transverse linear momentum components m_2, m_3. For an inviscid flow, no boundary condition is enforced on the axial linear-momentum component m_1. For a viscous flow, an inlet boundary condition for the axial linear-momentum equation that can be prescribed is a vanishing surface traction $\tau_{1j} n_j$, with τ_{1j} the components of the deviatoric Navier-Stokes stress tensor, a condition effectively enforced by deleting the corresponding surface integral in the statement

$$\int_{\Omega} \left(w + \varepsilon \psi a_i \frac{\partial w}{\partial x_i} \right) \left(\frac{\partial m_1}{\partial t} + \frac{\partial}{\partial x_j} \left(\frac{m_1 m_j}{\rho} \right) + \frac{\partial p}{\partial x_1} \right) d\Omega + \int_{\Omega} \frac{\partial w}{\partial x_j} \tau_{1j} d\Omega - \oint_{\partial \Omega} w \tau_{1j} n_j \, d\Gamma +$$

$$\int_{\Omega} \frac{\partial w}{\partial x_i} \varepsilon \psi \left(c \left(\alpha a_i a_j + \alpha^N a_i^{N\ell} a_j^{N\ell} \right) \frac{\partial m_1}{\partial x_j} + a_i (\delta - 1) \frac{\partial p}{\partial x_1} \right) d\Omega = 0 \qquad (17.165)$$

Where the inlet flow is supersonic, all the eigenvalues of $\frac{\partial f_j}{\partial q} n_j$ are negative and, correspondingly, Dirichlet boundary conditions are enforced on density ρ, energy E, and all the components m_1, m_2 and m_3 of linear momentum. These conditions may be cast as

$$\frac{\partial \rho}{\partial t}(x_{\text{in}}, t) = q'_{\rho}(t), \quad \frac{\partial E}{\partial t}(x_{\text{in}}, t) = q'_{E}(t), \quad \frac{\partial m_i}{\partial t}(x_{\text{in}}, t) = q'_{m_i}(t), \quad 1 \leq i \leq n \qquad (17.166)$$

where $q'_{\rho}(t)$, $q'_{E}(t)$, $q'_{m_i}(t)$, $1 \leq i \leq n$, denote prescribed bounded functions.

At a solid wall, which corresponds to a streamline, $\mathbf{u} \cdot \mathbf{n} = 0$ and, as a result, only one eigenvalue of $\frac{\partial f_j}{\partial q} n_j$ is negative. One boundary condition is thus required at a wall for an inviscid flow. This solid-wall boundary condition corresponds to the wall-tangency, i.e no-penetration, condition $\mathbf{u} \cdot \mathbf{n} = 0$. With reference to the decomposition $f_j = f_j^q + f_j^p$ and Figure 17.16-a,

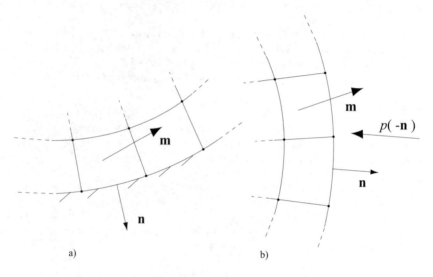

Figure 17.16: Boundary Regions: a) Wall; b) Outlet

the divergence of the convection flux with components f_j^q within (17.163) is integrated by parts at a wall region, an operation that yields the wall boundary statement

$$\int_{\Omega} \left[\left(w + \varepsilon \psi a_i \frac{\partial w}{\partial x_i} \right) \left(\frac{\partial q}{\partial t} + \frac{\partial f_j^p}{\partial x_j} \right) + \frac{\partial w}{\partial x_j} f_j^\nu \right] d\Omega - \oint_{\partial \Omega} w f_j^\nu n_j \, d\Gamma$$

$$+ \int_{\Omega} \frac{\partial w}{\partial x_i} \varepsilon \psi \left(c \left(\alpha a_i a_j + \alpha^N a_i^{N_\ell} a_j^{N_\ell} \right) \frac{\partial q}{\partial x_j} + a_i (\delta - 1) \frac{\partial f_j^p}{\partial x_j} \right) d\Omega$$

$$- \int_{\Omega} \frac{\partial w}{\partial x_i} \left(f_i^q - \varepsilon \psi a_i \frac{\partial f_j^q}{\partial x_j} \right) d\Omega = - \oint_{\partial \Omega} w q^H \frac{\mathbf{m}}{\rho} \cdot \mathbf{n} \, d\Gamma \qquad (17.167)$$

where $f_j^v = 0$ for an inviscid flow. This statement features a generalized "numerical flux" for f_j^q and a wall surface integral that depends on the mass flux $\frac{\mathbf{m}}{\rho} \cdot \mathbf{n}$. A wall mass-flux boundary condition is thus directly enforced within this surface integral, with a wall-tangency boundary condition obtained by setting the entire integral to nought. For an inviscid flow, this is the only boundary condition that is required at a solid wall.

For a viscous flow, the three linear-momentum components in statement (17.167) are replaced by the no-slip boundary conditions on m_1, m_2 and m_3; the wall mass-flux surface integrals in the continuity and total-energy components in this statement are then eliminated. The wall heat-flux integral in the energy equation

$$\int_\Omega \left[\left(w + \varepsilon \psi a_i \frac{\partial w}{\partial x_i} \right) \frac{\partial E}{\partial t} - \frac{\partial w}{\partial x_j} \left(m_j \frac{E+p}{\rho} - \varepsilon \psi a_j \frac{\partial}{\partial x_i} \left(m_i \frac{E+p}{\rho} \right) - u_i \tau_{ij} - k \frac{\partial T}{\partial x_j} \right) \right] d\Omega$$

$$+ \int_\Omega \frac{\partial w}{\partial x_i} \varepsilon \psi \left(c \left(\alpha a_i a_j + \alpha^N a_i^{N_\ell} a_j^{N_\ell} \right) \frac{\partial E}{\partial x_j} \right) d\Omega + \oint_{\partial\Omega} w \left(m_j \frac{E+p}{\rho} - u_i \tau_{ij} - k \frac{\partial T}{\partial x_j} \right) n_j \, d\Gamma = 0$$

$$(17.168)$$

is then used to specify a heat-flux boundary condition, by inserting the prescribed heat-flux $\frac{\partial T}{\partial n} = \frac{\partial T}{\partial x_j} n_j$ in this integral. For a cold- or hot-wall boundary condition, the heat flux is cast via a temperature variation or the energy equation (17.168) is locally replaced by the wall surface integral

$$\oint_{\partial\Omega} w \left(\frac{p(q)}{\rho R} - T_w \right) d\Gamma = 0 \qquad (17.169)$$

where T_w denotes a prescribed wall temperature.

At a supersonic outlet, all the eigenvalues of $\frac{\partial f_j}{\partial q} n_j$ are positive and thus no boundary conditions are enforced for an inviscid flow; for a viscous flow, the minimal-perturbation boundary conditions that may be prescribed involve an adiabatic outlet stream and vanishing deviatoric work per unit time $u_i \tau_{ij} n_j$ and surface tractions $\tau_{ij} n_j$. The adiabatic-stream condition is enforced by deleting the corresponding surface integral in the outflow energy-equation integral

$$\int_\Omega \left(w + \varepsilon \psi a_i \frac{\partial w}{\partial x_i} \right) \left(\frac{\partial E}{\partial t} + \frac{\partial}{\partial x_j} \left(m_j \frac{E+p}{\rho} \right) \right) d\Omega + \int_\Omega \frac{\partial w}{\partial x_j} \left(u_i \tau_{ij} + k \frac{\partial T}{\partial x_j} \right) d\Omega$$

$$+ \int_\Omega \frac{\partial w}{\partial x_i} \varepsilon \psi \left(c \left(\alpha a_i a_j + \alpha^N a_i^{N_\ell} a_j^{N_\ell} \right) \frac{\partial E}{\partial x_j} \right) d\Omega - \oint_{\partial\Omega} w \left(u_i \tau_{ij} n_j + k \frac{\partial T}{\partial n} \right) d\Gamma = 0 \quad (17.170)$$

The vanishing-traction condition is enforced by deleting the corresponding surface integral from the linear-momentum statement

$$\int_\Omega \left(w + \varepsilon \psi a_k \frac{\partial w}{\partial x_k} \right) \left(\frac{\partial m_i}{\partial t} + \frac{\partial}{\partial x_j} \left(\frac{m_i m_j}{\rho} \right) + \frac{\partial p}{\partial x_i} \right) d\Omega + \int_\Omega \frac{\partial w}{\partial x_j} \tau_{ij} \, d\Omega$$

$$+ \int_\Omega \frac{\partial w}{\partial x_k} \varepsilon \psi \left(c \left(\alpha a_k a_j + \alpha^N a_k^{N_\ell} a_j^{N_\ell} \right) \frac{\partial m_i}{\partial x_j} + a_k (\delta - 1) \frac{\partial p}{\partial x_i} \right) d\Omega - \oint_{\partial\Omega} w \tau_{ij} n_j \, d\Gamma = 0$$

$$(17.171)$$

If the traction $\tau_{ij} n_j$ does not vanish at an outlet, its known or prescribed numerical value may be enforced via the corresponding surface integrals in (17.170)-(17.171).

At a subsonic outlet, one eigenvalue of $\frac{\partial f_j}{\partial q} n_j$ is negative and one boundary condition is thus required. A similar surface-integral procedure is employed at such a subsonic outlet region to enforce a specified boundary condition on pressure p for an inviscid flow, and on the deviatoric surface tractions $\tau_{ij} n_j$ for a viscous flow. With reference to the decomposition $f_j = f_j^q + f_j^p$ in Sections 17.4.1, 17.4.4, and Figure 17.16-b, the divergence of the pressure flux with components f_j^p within the momentum equations (17.171) is integrated by parts at an outlet region, an operation that yields the outflow boundary statement

$$\int_\Omega \left(w + \varepsilon\psi a_k \frac{\partial w}{\partial x_k} \right) \left(\frac{\partial m_i}{\partial t} + \frac{\partial}{\partial x_j} \left(\frac{m_j}{\rho} m_i \right) \right) d\Omega$$

$$+ \int_\Omega \frac{\partial w}{\partial x_k} \varepsilon\psi \left(c \left(\alpha a_k a_j + \alpha^N a_k^{N\ell} a_j^{N\ell} \right) \frac{\partial m_i}{\partial x_j} + a_k(\delta - 1)\frac{\partial p}{\partial x_i} \right) d\Omega$$

$$- \int_\Omega \frac{\partial w}{\partial x_j} \left(p\delta_i^j - \varepsilon\psi a_j \frac{\partial p}{\partial x_i} - \tau_{ij} \right) d\Omega + \oint_{\partial\Omega} w \left(pn_i - \tau_{ij} n_j \right) d\Gamma = 0 \qquad (17.172)$$

This statement features an outflow surface integral that depends on the outlet-pressure specific-force components pn_i, $1 \le i \le 3$; on a facet where p is not constrained, this integral is formed using the variable p as determined by the local evolving solution q. With $\tau_{ij} n_j = 0$, the outlet pressure boundary condition is thus directly enforced by specifying the prescribed outlet pressure for p within this surface integral on a corresponding outlet facet. This strategy for imposing an outlet pressure boundary condition remains intrinsically stable as the following basic considerations on the linear-momentum equation for m_1 indicate; similar conclusions apply to the linear-momentum equations for m_2 and m_3. Consider first the case of an outlet with $n_1 < 0$, which implies $\partial w/\partial x_1 < 0$ and $m_1 < 0$. If some computational perturbation induced a decrease in m_1 at the boundary, then $\|m_1\|$ would increase hence p from (17.14) would decrease, which through the boundary domain integral of pressure in (17.172) would induce a restoring increase in m_1; similar stability conclusions would results by considering a perturbation increase of m_1 at the boundary. Consider the other case of an outlet with $n_1 > 0$, which implies $\partial w/\partial x_1 > 0$ and $m_1 > 0$. If some computational perturbation induced a decrease in m_1 at the boundary, then p from (17.14) would increase, which through the boundary domain integral of pressure in (17.172) would induce a restoring increase in m_1; similar stability conclusions result by considering a perturbation increase of m_1 at the boundary. The results in Section 17.10 confirm the accuracy and stability of this pressure boundary-condition enforcement procedure.

17.8   Finite Element Galerkin Weak Statement

Since the characteristics flux divergence (17.108), (17.113) is developed independently and before any discretization, a genuinely multi-dimensional upstream-bias approximation for the governing equations (17.1) on arbitrary grids directly results from a classic centered discretization of the characteristics-bias weak statement on the prescribed grid. To this end a Galerkin finite element method is employed to discretize in space the integral statement

(17.163). This method not only accommodates arbitrary geometries or generates consistent non-extrapolation boundary equations for q, but also retains the ideal surface-integral venues of the integral statement to enforce the boundary conditions described in the previous section.

The finite element approximation exists on a partition Ω^h of Ω, where h signifies spatial discrete approximation. Having its boundary nodes on the boundary $\partial\Omega$ of Ω, this partition Ω^h results from the union of N_e non-overlapping elements Ω_e, $\Omega^h = \bigcup_{e=1}^{N_e} \Omega_e$. With $\eta = 0$ or $\eta = 1$, respectively for shocked or smooth flows, and with $\mathcal{P}_1(\Omega_e)$ and $\mathcal{P}_{1v}(\Omega_e)$ the finite-dimensional function spaces of respectively diagonal square matrix-valued and vector-valued linear polynomials within each Ω_e, for each "t", the corresponding diagonal square matrix-valued and vector-valued finite element discretization spaces employed in this study are defined as

$$ \mathcal{S}^1(\Omega^h) \equiv \left\{ w^h \in \mathcal{H}^1(\Omega^h) : w^h \big|_{\Omega_e} \in \mathcal{P}_1(\Omega_e), \forall \Omega_e \in \Omega^h, B(x_{\partial\Omega^h}) w^h(x_{\partial\Omega^h}) = 0 \right\} $$

$$ \mathcal{S}_v^{\eta}(\Omega^h) \equiv \left\{ w_v^h \in \mathcal{H}_v^{\eta}(\Omega^h) : w_v^h \big|_{\Omega_e} \in \mathcal{P}_{1v}(\Omega_e), \forall \Omega_e \in \Omega^h, B(x_{\partial\Omega^h}) w_v^h(x_{\partial\Omega^h}, t) = G_v(x_{\partial\Omega^h}, t) \right\} $$

$$ (17.173) $$

Based on these spaces, the finite element approximation $q^h \in \mathcal{S}_v^{\eta}$, is determined for each "t" as the solution of the finite element weak statement associated with (17.163)

$$ \int_{\Omega^h} \left(w^h + \varepsilon^h \psi^h a_i^h \frac{\partial w^h}{\partial x_i} \right) \left(\frac{\partial q^h}{\partial t} + \frac{\partial f_j^h}{\partial x_j} \right) d\Omega + \int_{\Omega^h} \frac{\partial w^h}{\partial x_j} f_j^{\nu^h} d\Omega - \oint_{\partial\Omega^h} w^h f_j^{\nu^h} n_j \, d\Gamma + $$

$$ + \int_{\Omega^h} \frac{\partial w^h}{\partial x_i} \varepsilon^h \psi^h \left(c^h \left(\alpha^h a_i^h a_j^h + \alpha^{N^h} a_i^{N_\ell} a_j^{N_\ell} \right) \frac{\partial q^h}{\partial x_j} + a_i^h (\delta^h - 1) \frac{\partial f_j^{p^h}}{\partial x_j} \right) d\Omega = 0 \quad (17.174) $$

with similar expressions for statements (17.165)-(17.172). Since every member w^h of $\mathcal{S}^1$ results from a linear combination of the linearly independent basis functions of this finite dimensional space, statement (17.174) is satisfied for N independent basis functions of the space, where N denotes the dimension of the space. For N mesh nodes within Ω^h, there exist clusters of "master" elements Ω_k^M, each cluster comprising only those adjacent elements that share a mesh node $\mathbf{x}_k$, with $1 \le k \le N$, where N denotes the total number of not only mesh nodes, but also master elements.

As detailed in Chapter 7, the discrete test function w^h within each master element Ω_k^M will coincide with the "pyramid" basis function $w_k = w_k(\mathbf{x})$, $1 \le k \le N$, with compact support on Ω_k^M. Such a function equals one at node $\mathbf{x}_k$, zero at all other mesh nodes and also identically vanishes both on the boundary segments of Ω_k^M not containing $\mathbf{x}_k$ and on the computational domain outside Ω_k^M.

The discrete solution q^h and flux f_j^h at each time t assume the form of the following group linear combinations

$$ q^h(\mathbf{x}, t) \equiv \sum_{\ell=1}^{N} w_\ell(\mathbf{x}) \cdot q^h(\mathbf{x}_\ell, t), \quad f_j^h(\mathbf{x}, t) \equiv \sum_{\ell=1}^{N} w_\ell(\mathbf{x}) \cdot f_j\left(q^h(\mathbf{x}_\ell, t) \right) \quad (17.175) $$

of time-dependent nodal solution values $q^h(\mathbf{x}_\ell, t)$, to be determined, and trial functions, which coincide with the test functions $w_\ell(\mathbf{x})$ for a Galerkin formulation; an analogous expression applies for $f_j^{\nu^h}$. Similarly, the fluxes $f_j^q = f_j^q(q(\mathbf{x}, t))$ and $f_j^p = f_j^p(q(\mathbf{x}, t))$ are

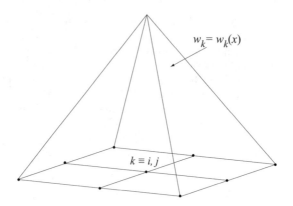

Figure 17.17: Pyramid Test Function for Ω_k^M

discretized through the group expressions

$$f_j^{q^h}(\mathbf{x}, t) \equiv \sum_{\ell=1}^{N} w_\ell(\mathbf{x}) \cdot f_j^q \left(q^h(\mathbf{x}_\ell, t) \right), \qquad f_j^{p^h}(\mathbf{x}, t) \equiv \sum_{\ell=1}^{N} w_\ell(\mathbf{x}) \cdot f_j^p \left(q^h(\mathbf{x}_\ell, t) \right) \qquad (17.176)$$

The notation for the discrete nodal variable and fluxes is then simplified as $q_\ell(t) \equiv q(\mathbf{x}_\ell, t)$, $f_{j\ell} \equiv f_j \left(q^h(\mathbf{x}_\ell, t) \right)$, $f_{j\ell}^q \equiv f_j^q \left(q^h(\mathbf{x}_\ell, t) \right)$, $f_{j\ell}^p \equiv f_j^p \left(q^h(\mathbf{x}_\ell, t) \right)$ and expansions (17.175)-(17.176) are then inserted into (17.174), which yields the discrete finite element weak statement

$$\int_{\Omega^h} \left(w_k + \varepsilon^h \psi^h a_i^h \frac{\partial w_k}{\partial x_i} \right) \left(w_m \frac{dq_m}{dt} + \frac{\partial w_m}{\partial x_j} f_{jm} \right) d\Omega + \int_{\Omega^h} \frac{\partial w_k}{\partial x_j} f_j^{\nu^h} d\Omega - \oint_{\partial \Omega^h} w_k f_j^{\nu^h} n_j d\Gamma +$$

$$+ \int_{\Omega^h} \frac{\partial w_k}{\partial x_i} \frac{\partial w_m}{\partial x_j} \varepsilon^h \psi^h \left[c^h \left(\alpha^h a_i^h a_j^h + \alpha^{N^h} a_i^{N_\ell^h} a_j^{N_\ell^h} \right) q_m + a_i^h (\delta^h - 1) f_{jm}^p \right] d\Omega = 0 \qquad (17.177)$$

for $1 \le k \le N$; the three-test-function matrices in (7.186) are then needed to discretize the viscous dissipation expression in the energy equation. There are three implied summations with respect to the subscript indices i, j, m. The subscript indices i, j in this expression denote cartesian-axis directions, hence $1 \le i, j \le 2$, for two-dimensional flows, or $1 \le i, j \le 3$, for three dimensional flows, whereas subscript m indicates a mesh node, hence $1 \le m \le N$, although a sum like $\sum_{m=1}^{N} w_m \frac{dq_m}{dt}$ only involves a few neighboring terms because the compact-support test function w_m is only non-zero within a cluster of a few neighboring elements.

While an expansion like the ones in (17.175) for ψ^h, α^h, c^h, $\mathbf{a}^h$, $\mathbf{a}^{N^h_\ell}$ and δ^h can be directly accommodated within (17.177), each of these variables in this study has been set equal to a piece-wise constant for computational simplicity, one centroidal constant value per element. Within each element, the terms ψ^h and ε^h are respectively set equal to the controller and reference length developed in Chapter 10. As shown in that chapter, ε^h corresponds to the length of the streamline radius of the generalized ellipse inscribed within each element, since the streamline is a characteristic principal direction.

Since the test and trial functions w_k are prescribed functions of $\mathbf{x}$, the spatial integrations in (17.174) are directly carried out. For arbitrarily shaped elements, these integrations take place as detailed in Chapter 7. Concerning the boundary variables, no extrapolation of variables is needed in this algorithm on a variable that is not constrained via a Dirichlet boundary condition. In this case, instead, the finite element algorithm (17.177) naturally generates for each unconstrained boundary variable a boundary-node ordinary differential equation. The complete integration with respect to $\mathbf{x}$ transforms (17.177) into a system of continuum-time ordinary differential equations (ODE) for determining at each time level t the unknown nodal values $q^h(\mathbf{x}_\ell, t)$, $1 \leq \ell \leq N$.

17.9   Implicit Runge-Kutta Time Integration

The finite element equations (17.177), along with appropriate boundary equations and conditions, are abridged as the non-linear ODE system

$$\mathcal{M}\frac{dQ(t)}{dt} = \mathcal{F}(t, Q(t)) \tag{17.178}$$

where $\mathcal{M} \equiv \{w_k w_\ell\}$ denotes the mass matrix, $\mathcal{M}\frac{dQ(t)}{dt}$ indicates the corresponding coupling of time derivatives in (17.177), and $\mathcal{F}(t, Q(t))$ represents the remaining terms in (17.177).

The numerical time integration of (17.178) takes place through a two-stage diagonally implicit Runge-Kutta algorithms (IRK2), as detailed in Chapter 8, expressed as

$$
\begin{aligned}
Q_{n+1} - Q_n &= b_1 K_1 + b_2 K_2 \\
\mathcal{M}K_1 &= \Delta t \cdot \mathcal{F}(t_n + c_1 \Delta t, Q_n + a_{11} K_1) \\
\mathcal{M}K_2 &= \Delta t \cdot \mathcal{F}(t_n + c_2 \Delta t, Q_n + a_{21} K_1 + a_{22} K_2)
\end{aligned}
\tag{17.179}
$$

where n now denotes a discrete time station. Given the solution Q_n at time t_n, K_1 is computed first, followed by K_2. The solution Q_{n+1} is then determined by way of the first expression in (17.179).

The terminal numerical solution is then determined using Newton's method, which for the implicit fully-coupled computation of the IRK2 arrays K_i, $1 \leq i \leq 2$, is cast as

$$\left[\mathcal{M} - a_{\underline{ii}}\Delta t \left(\frac{\partial \mathcal{F}}{\partial Q} \right)^p_{Q^p_i} \right] \left(K_i^{p+1} - K_i^p \right) = \Delta t\, \mathcal{F}(t_n + c_i \Delta t, Q_i^p) - \mathcal{M}K_i^p$$

$$Q_i^p \equiv Q_n + a_{i1} K_1^p + a_{i2} K_2^p \tag{17.180}$$

where $a_{ij} = 0$ for $j > i$, p is the iteration index, and $K_1^p \equiv K_1$ for $i = 2$; the Jacobian

$$J_i(Q) \equiv \mathcal{M} - a_{\underline{ii}}\Delta t \left(\frac{\partial \mathcal{F}}{\partial Q} \right)^p_{Q^p_i} \tag{17.181}$$

has been analytically determined and implemented, leading to a block sparse matrix. For all the results documented in the next section, the initial estimate K_i^0 is set equal to the zero array, while a Gaussian elimination is used with only one iteration executed for (17.180) within each time interval. In this mode, Newton's iteration becomes akin to a classical direct linearized implicit solver.

17.10 Computational Results

The Acoustics-Convection Upstream Resolution Algorithm has generated essentially non-oscillatory results for subsonic, transonic, supersonic, hypersonic flows, encompassing oblique and interacting shocks that reflect exact solutions. The benchmarks discussed in this section include 5 flows: a chemically-reacting hypersonic flow about a blunt body, a subsonic flow and a transonic flow about a symmetrical airfoil, and two supersonic inlet flows. In order to determine the coarse-grid performance of the algorithm and ultimate accuracy of quadrilateral elements, each benchmark has employed a finite element discretization of Lagrange bilinear elements without any MUSCL-type local extrapolation of variables; for these computational tests, with the exception of the hypersonic-flow study, the body-fitted grid consists of 40 bilinear elements in the transverse and longitudinal directions, for a total of 1600 elements, 1681 nodes and 6724 degrees of freedom. The computational efficiency of the procedure has remained comparable to that of a conventional centered algorithm for the characteristics-bias system with upstream directions continuously updated without any filtering or freezing, with high-rate convergence of the residual norm to machine zero.

For each benchmark, the calculations proceeded with a prescribed constant maximum Courant number $C_{\max} = 100$. Considering the definition of the Courant number,

$$C_{\max} \equiv \max\{\|\mathbf{u}\| + c, \|\mathbf{u}\| - c, c\}\frac{\Delta t}{(\Delta\ell)_e} \tag{17.182}$$

for prescribed $(\Delta\ell)_e$ and $C_{\max}$ for each benchmark, the corresponding Δt is determined as

$$\Delta t = \frac{C_{\max}(\Delta\ell)_e}{\max\{\|\mathbf{u}\| + c, \|\mathbf{u}\| - c, c\}} \tag{17.183}$$

All the solutions in these validations are presented in non dimensional form, with pressure p made dimensionless through the corresponding inlet stagnation (i.e. total) pressure.

17.10.1 Hypersonic Blunt-Body Flow

This study computationally investigates the formation of a bow shock and shock layer about a blunt axisymmetric sphere-biconic body in a hypersonic flow with $M_\infty = 8$, for two sets of thermodynamic freestream conditions: one keeping a perfect-gas flow, the other leading to high-temperature gas effects. The algorithm for these investigations featured altered acoustics-convection characteristics-bias parameters, to assess the impact on solution quality resulting from any variations of the optimal characteristics-bias parameters. These modifications involved $\alpha^N = 0$, $\delta = 0.0$, $\alpha = 1.0$, and c in the acoustic term replaced by the local magnitude $\|\boldsymbol{u}\|$. When these modifications are not used, the corresponding computational solutions display sharper viscous layers and crisper bow shocks. The configuration consists of a frontal hemispherical blunt body with radius $R_N = 0.5$ in, a $10.5°$ forecone, and a $7.0°$ aft cone section with axial lengths 12.32 in and 20.00 in respectively. The body-fitted mesh for this study contains 64×74 bilinear elements with non-dimensional $\Delta x_{\min} = \mathcal{O}(10^{-5}) = \mathcal{O}(Re^{-1})$ in the nosetip region. Figure 17.18 displays a detail of this mesh about the frontal blunt body.

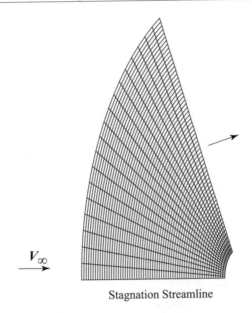

Stagnation Streamline

Figure 17.18: Hypersonic Blunt-Body Flows, Mesh Detail

At the inlet, this computational study has imposed Dirichlet constraints on density, momentum components and specific total energy. Vanishing radial momentum is enforced along the centerline, and both momentum components are set to zero at the blunt-body stagnation point. Along the aerodynamic surface, vanishing momentum flux is imposed for the inviscid flow simulation, whereas the no-slip condition is enforced for the viscous flow prediction. The Rankine-Hugoniot conditions, along with Newtonian surface pressure distribution and available bow-shock location estimates are used to determine the initial field [7].

For the freestream thermodynamic conditions $p_\infty = 600.12$ Pa and $T_\infty = 54.31$ K, the perfect-gas viscous flow about this configuration is determined at the reference $Re_\infty = 500,000$, based on the blunt-body nose radius, with the boundary-layer initial condition obtained via linear interpolation to zero of the momentum components in a narrow strip region about the aerodynamic surface. The machine-zero steady state was achieved for $\psi = 0.50$ in 800 time steps at a maximum Courant number of 200. For an adiabatic wall, the boundary layer strongly interacts with the exterior essentially inviscid flow. The computed shock stand-off distance $\delta_{\text{SD}} = 0.075$ in, agrees to within 1.5% with available correlations [6]. Figures 17.19-17.21 summarize the steady solution in terms of pressure contours and stagnation-line distributions of pressure and temperature. The displayed shock thickness in the contour fields results from the use of a significantly expanded axial scale, a use needed to make these fields visually discernible. The stagnation-line distributions use an x-axis with origin at the blunt-body stagnation point and distances in inches. These distributions display a limited post-shock two-node loss of monotonicity, as induced by the alterations in the optimal characteristics-bias parameters. Away from the post-shock region, the distributions remain essentially non-oscillatory in the shock compression and subsequent expansion regions. The computed stagnation-point peak values for pressure and temperature are $p_\infty = 55,383.87$ Pa and $T_\infty = 832.11$ K. The associated inviscid flow solution is obtained at a maximum Courant number of 500, with maximum convergence rate occurring at a maximum Courant number of 200. For the previous magnitude of ψ, the essentially non-oscillatory solution leads to a stand-off distance $\delta_{\mathrm{SD}} = 0.068$ in, which agrees to within 4% with available correlations [6], with a peak pressure $p_\infty = 55,383.87$ Pa. The subsequent flow expansion about the spherical nose is also smooth and clearly defined. For a free-stream temperature of $T_\infty = 54.31$ K, the post-shock and stagnation-point temperature values of $T = 773.2$ K and $T = 832.11$ K remain consistent with a non-reacting gas flow.

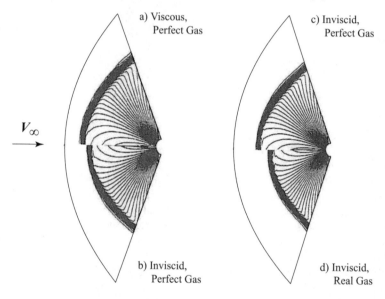

Figure 17.19: Hypersonic Blunt-Body Flows, Pressure Contours

For representative atmospheric free stream conditions, however, the temperature rise across the bow shock triggers non-negligible high-temperature gas effects. This study then performed two additional inviscid-flow calculations corresponding to the free stream conditions $T_\infty = 221.1$ K and $p_\infty = 2.72 \times 10^{-2}$ atm, with perfect-gas equation of state and then with high-temperature gas effects modeled as described in Chapter 3. Starting from the perfect-gas steady-state solution as an initial field, the final steady state with $\psi = 0.5$ was achieved in 400 time steps at a maximum Courant number of 200. For this chemically reacting gas flow, following the methods of Chapter 3, the continuous and accurate computation of the pressure derivatives for the implicit implementation contributes to the observed monotone convergence. Figure 17.19-d) graphs the pressure contours for this more realistic case, as contrasted with the associated perfect-gas distribution. As indicated in this figure, real-gas effects lead to a decrease of the shock stand-off distance to $\delta_{\mathrm{SD}} = 0.056$ in, almost 18% less than the associated perfect-gas value. The distribution of pressure along the stagnation streamline and aerodynamic surface is presented in Figure 17.20. For the reference freestream conditions, the equilibrium air pressure distribution does not radically differ from that corresponding to the perfect-gas case, with the exception of the shock location.

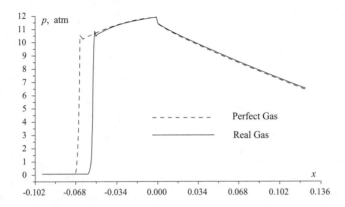

Figure 17.20: Hypersonic Blunt-Body Flows, Stagnation-Streamline Pressure

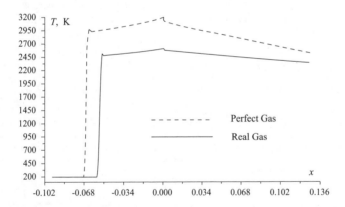

Figure 17.21: Hypersonic Blunt-Body Flows, Stagnation-Streamline Temperature

Conversely, as indicated in Figure 17.21, the temperature fields are significantly different. For the same free-stream conditions, the computed post-shock temperature for the perfect-gas simulation is $T = 3149.9$ K, whereas the reacting-flow magnitude is $T = 2595.2$ K, a decrease of almost 18% accounted for by the excitation of vibrational modes and eventual formation of nitric oxide and atomic oxygen.

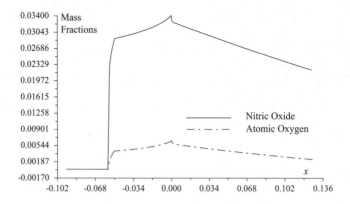

Figure 17.22: Hypersonic Blunt-Body Flows, Stagnation-Streamline Mass Fractions

As indicated in Figure 17.22, these species are formed immediately after the shock, with peak production at the stagnation point. The mass fraction of nitric oxide is almost six times larger than the oxygen mass fraction, while only negligible traces of atomic nitrogen are present. Reflecting the existing knowledge that high-temperature gas effects for realistic free stream conditions significantly influence the air mixture chemical composition and temperature field, these computational results show that the even a simplified form of the finite element acoustics-convection formulation coupled with the implicit Runge-Kutta time integration leads to essentially non-oscillatory solutions for challenging flow fields with chemical reactions.

17.10.2   Airfoil Flows

The next three computational tests employ the full characteristics-bias formulation without any alteration of the associated parameters. The computational test in this section corresponds to a subsonic critical flow about a 3% thick symmetrical airfoil, with stretched grid illustrated in Figure 17.23.

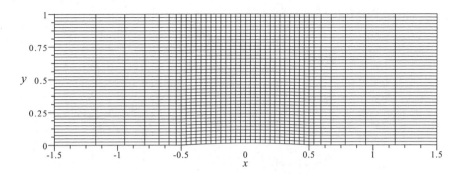

Figure 17.23: Airfoil Flows: Computational Grid

The subsonic inlet corresponds to a subsonic free-stream Mach number $M_\infty = 0.87$, hence the inlet boundary conditions only constrain density ρ, transverse linear momentum component m_2 and total energy E. The outlet, consisting of both the downstream exit and upper side, remains subsonic, hence static pressure is constrained at this boundary. At the lower surface, the inviscid wall-tangency boundary condition is enforced according to the method delineated in Section 17.7.2. A pressure drop is imposed upon an initially quiescent field and the final steady state is computationally achieved by advancing the solution in time.

The Mach-number distribution and flooded contours in Figures 17.24 and 17.25 for $\psi = 0.5$ portray a non-oscillatory solution with sharply resolved drops in Mach number at the airfoil leading and trailing edges and undistorted capturing of a vanishingly small supersonic region over the airfoil surface, towards the trailing edge.

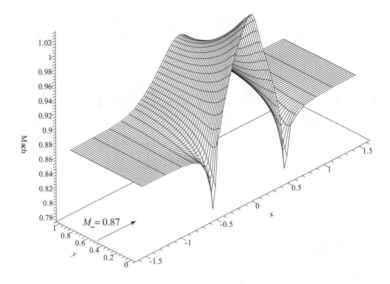

Figure 17.24: $M_\infty = 0.87$ Airfoil Critical Subsonic Flow, Mach Number Contours

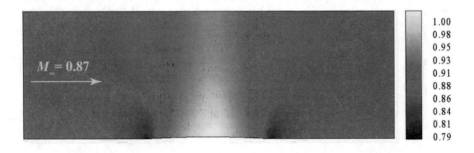

Figure 17.25: $M_\infty = 0.87$ Airfoil Critical Subsonic Flow, Mach Number Distribution

Despite the stretched grid, a similar essentially non-oscillatory field is portrayed in the pressure distribution and flooded contours in Figures 17.26 and 17.27. Although this distribution indicates increased accuracy would result from a locally refined grid, the pressure peaks at the airfoil leading and trailing edges remain undistorted. In particular the calculated pressure in the outlet region remains smooth and the calculated outlet pressure at the $x = 1.5$ outlet coincides with the imposed pressure boundary conditions, which reflects favorably on the surface-integral pressure enforcement strategy delineated in Section 17.7.2.

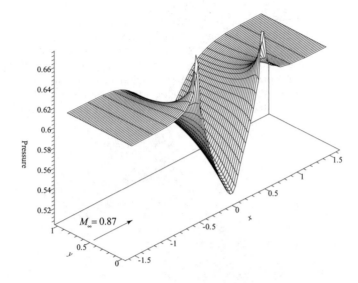

Figure 17.26: $M_\infty = 0.87$ Airfoil Critical Subsonic Flow, Pressure Distribution

Figure 17.27: $M_\infty = 0.87$ Airfoil Critical Subsonic Flow, Pressure Contours

The second computational test corresponds to a transonic flow about the same airfoil. The subsonic inlet corresponds to a subsonic free-stream Mach number $M_\infty = 0.87$, hence the inlet boundary conditions only constrain density ρ, transverse linear momentum component m_2 and total energy E. As in the first test, the outlet, consisting of both the downstream exit and portions of the upper side of the grid, remain subsonic, hence static pressure is constrained at these boundary segments; when the local Mach number exceeds one, the algorithm no longer enforces an outlet pressure boundary condition. At the lower surface, the inviscid wall-tangency boundary condition is enforced according to the method delineated in Section 17.7.2. A pressure drop is imposed upon an initial field corresponding to the steady-state flow of the previous test and the final steady state is computationally achieved by advancing the solution in time. The imposed pressure drop was selected to induce a supersonic region that extends past the upper boundary, in order to test the capability of the algorithm to allow a captured shock to cross a boundary with minimal reflection.

The Mach-number distribution and flooded contours in Figures 17.28 and 17.29 present an essentially non-oscillatory solution with crisply calculated compressions and ex-

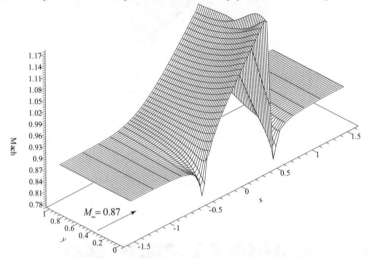

Figure 17.28: $M_\infty = 0.87$ Airfoil Transonic Flow, Mach Number Distribution

Figure 17.29: $M_\infty = 0.87$ Airfoil Transonic Flow, Mach Number Contours

pansions, at the airfoil leading and trailing edges, and with a supersonic pocket terminated by a slightly curved sharp shock adjacent to the airfoil trailing edge. In particular this two-dimensional shock is captured within two to three nodes.

Despite the stretched grid, a similar essentially non-oscillatory field is displayed in the pressure distribution and flooded contours in Figures 17.26 and 17.27.

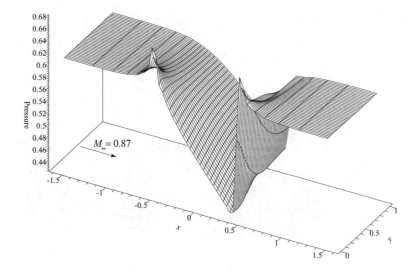

Figure 17.30: $M_\infty = 0.87$ Airfoil Transonic Flow, Pressure Distribution

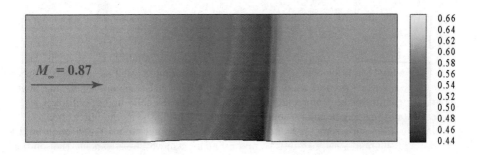

Figure 17.31: $M_\infty = 0.87$ Airfoil Transonic Flow, Pressure Contours

This distribution indicates increased accuracy would result from a locally refined grid. Nevertheless the pressure peaks at the airfoil leading and trailing edges remain undistorted and the slightly curved shock in this variable is also sharply captured within two to three nodes. The calculated subsonic distribution downstream of the shock remains smooth. In particular the calculated outlet pressure at the $x = 1.5$ outlet coincides with the imposed pressure boundary conditions, which again reflects favorably on the surface-integral pressure enforcement strategy delineated in Section 17.7.2.

17.10.3   Inlet Flows

The next computational tests encompass two supersonic inlet flows. One of these flows involves an oblique shock reflection case, with grid illustrated in Figure 17.32.

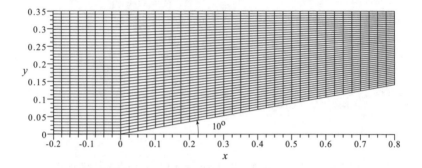

Figure 17.32: $M_\infty = 2.40$ Shock Reflection, Computational Grid

The supersonic inlet corresponds to a free-stream Mach number $M_\infty = 2.40$, hence the inlet boundary conditions constrain density ρ, longitudinal as well as transversal linear momentum components m_1 and m_2 and total energy E. The outlet remains supersonic, hence no boundary conditions are enforced at this boundary. At the solid upper and lower walls, the inviscid wall-tangency boundary condition is enforced using the method in Section 17.7.2. An initially uniform supersonic $M = 2.40$ shockless flow is subject to a 10° deflection by the lower wall and the final steady state is computationally achieved by advancing the solution in time.

It is the wall-tangency boundary condition on the whole upper and lower boundary walls that induces emergence of an oblique shock at the lower ramp initiation point, a shock that propagates toward the upper wall and is then reflected downward towards the outlet. As shown in the velocity vector and streamline distributions in Figures 17.33-17.34, the velocity vector abruptly turns across the oblique hydraulic jumps.

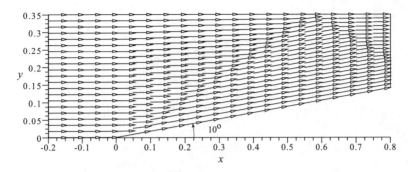

Figure 17.33: $M_\infty = 2.40$ Shock Reflection, Vector Field

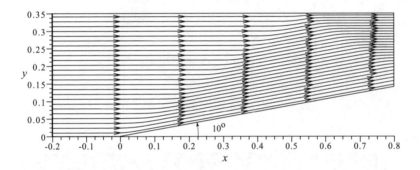

Figure 17.34: $M_\infty = 2.40$ Shock Reflection, Streamline Distribution

A localized maximum upstream-bias is activated at the geometric singularity of the ramp initiation point, as described in Section 10.2 after Figure 10.5, while $\psi_{min} = 0.2$ and $\psi_{max} = 0.35$. The Mach-number distribution and flooded contours in Figures 17.35 and 17.36 present an essentially non-oscillatory solution with crisply calculated incident and reflected shocks. In particular, the algorithm allows the reflected shock to cross the outflow boundary unperturbed, without any spurious distortion. Significantly, this computational solution mirrors the available exact solution, with three juxtaposed plateaus connected by two oblique shocks. The calculated Mach numbers in the plateaus downstream of the two shocks are $M_2 = 2.00$ and $M_3 = 1.64$; the shock inclination angles are $\theta_2 = 33^\circ$ and $\theta_3 = -29^\circ$. Not only for Mach number and shock angles, but also for pressure do the computed results reflect the corresponding exact values.

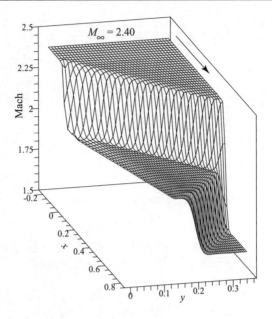

Figure 17.35: $M_\infty = 2.40$ Shock Reflection, Mach Number Distribution

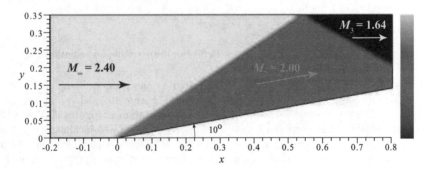

Figure 17.36: $M_\infty = 2.40$ Shock Reflection, Mach Number Contours

A similar essentially non-oscillatory field is displayed in the pressure distribution and flooded contours in Figures 17.37 and 17.38. The incident and reflected pressure shocks are crisply calculated, with a reflected shock that can cross the outflow boundary without any spurious distortion.

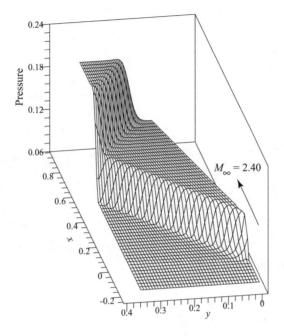

Figure 17.37: $M_\infty = 2.40$ Shock Reflection, Pressure Distribution

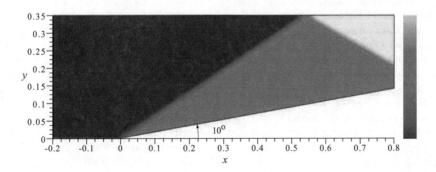

Figure 17.38: $M_\infty = 2.40$ Shock Reflection, Pressure Contours

The final test case in this study involves the asymmetric interaction of two oblique shocks, with non-uniform grid illustrated in Figure 17.39. The supersonic inlet corresponds to a free-stream Mach number $M_\infty = 2.40$, hence the inlet boundary conditions constrain density ρ, longitudinal and transversal linear momentum components m_1 and m_2 and total energy E. The solid wall specification only extends to $x = 0.4$; beyond this station, the upper and lower boundaries correspond to outlets, to allow the reflected shocks to cross the computational boundaries with minimal further reflection. The entire outlet remains supersonic, hence no boundary conditions are enforced at this boundary. At the solid upper and lower walls, only extending to $x = 0.4$, the inviscid wall-tangency boundary condition is enforced using the method in Section 17.7.2. An initially uniform supersonic $M = 2.40$ shockless flow is subject to a 5° deflection by the lower wall and a 3° deflection by the upper wall; the final steady state is computationally achieved by advancing the solution in time.

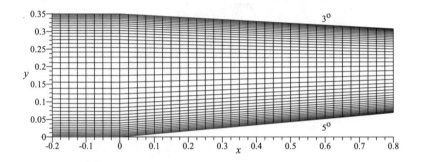

Figure 17.39: $M_\infty = 2.40$ Shock-on-Shock Interaction, Computational Grid

The wall-tangency boundary conditions on the upper and lower boundary walls induce emergence of two separate oblique shocks of different strengths, shocks that propagate toward each other, interact, and reflect away from each other, towards the outlet. With maximum upstream bias at the ramp corners, determined from the method in Section 10.2, and $0.2\psi = 0.35$, the Mach-number distribution and flooded contours in Figures 17.40 and 17.41 present an essentially non-oscillatory solution with crisply calculated incident and reflected shocks. Significantly, this computational solution mirrors the available exact solution, with four juxtaposed plateaus connected by four oblique shocks. Since the ramp deflection angles remain somewhat moderate and close to each other, the velocity magnitudes and Mach numbers remain essentially constant across the slip line that originates at the reflected-shock intersection point. Nevertheless, the reflected shock interaction rotates the emerging velocity vector to the equilibrium angle of 2° with respect to the horizontal direction. This angle corresponds to a deflection of 3°, across the lower reflected shock, and a deflection angle of -5°, across the upper reflected shock. These calculated velocity and deflection angles mirror the exact values. The calculated Mach numbers in the plateaus downstream of the four shocks are $M_2 = 2.20$, $M_3 = 2.28$, $M_4 \simeq M_5 = 2.08$; the shock inclination angles are $\theta_2 = 28.53^o$, $\theta_3 = -26.90^o$, $\theta_4 = 29.52$ and $\theta_5 = -30.15$. Not only for Mach number and shock angles but also for pressure do these computed results coincide with the corresponding exact values.

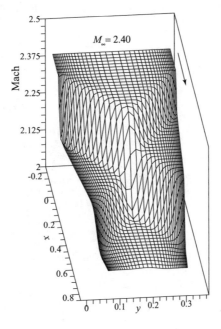

Figure 17.40: $M_\infty = 2.40$ Shock-on-Shock Interaction, Mach Number Distribution

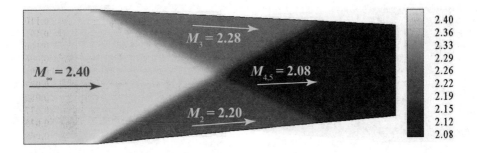

Figure 17.41: $M_\infty = 2.40$ Shock-on-Shock Interaction, Mach Number Contours

A similar essentially non-oscillatory field is displayed in the pressure distribution and flooded contours in Figures 17.42 and 17.43. The incident and reflected pressure shocks are crisply calculated, with reflected shocks that can cross the outflow boundary essentially unperturbed without spurious distortion.

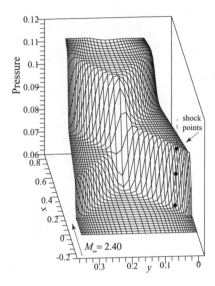

Figure 17.42: $M_\infty = 2.40$ Shock-on-Shock Interaction, Pressure Distribution

Figure 17.43: $M_\infty = 2.40$ Shock-on-Shock Interaction, Pressure Contours

17.10.4 Convergence to Steady State

For all the flow cases discussed, the algorithm has rapidly calculated a steady state, as exemplified by Figure 17.44 for the shock reflection and interaction problems.

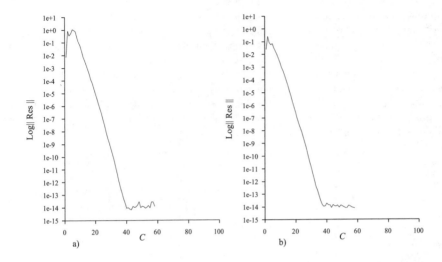

Figure 17.44: Convergence Rate: a) Shock Reflection, b) Shock Interaction

These curves document the high-rate convergence rate of the algorithm, with a reduction of the residual norm to 1×10^{-14}, hence machine zero, achieved in about 45 cycles "C" at a constant maximum Courant number equal to 100. The recorded essentially monotone decrease of the residual is seen to correspond to an exponential convergence rate.

17.11 Computational Performance

The characteristics-bias procedure relies upon the physics and mathematics of multi-dimensional characteristic acoustics and convection for general equations of state. It generates the upstream bias at the differential equation level, before any discrete approximation, by way of a characteristics-bias system and associated decomposition of the Euler flux divergence into convection and streamline and crossflow acoustic components. A natural finite element discretization of the characteristics-bias system directly provides an accurate and genuinely multi-dimensional upstream-bias approximation of the Euler equations, in the form of a non-linear combination of upstream diffusive and downstream anti-diffusive flux differences, with greater bias on the upstream diffusive flux difference.

Along the infinite directions of wave propagation, the formulation induces anisotropic and variable-strength consistent upwinding that correlates with the spatial distribution of characteristic velocities. It furthermore induces minimal artificial dissipation and yields correct essentially non-oscillatory solutions, while the upstream directions are continuously updated, and rapid convergence is achieved. The study in this chapter has implemented the algorithm using a linear approximation of fluxes within quadrilateral cells without any MUSCL-type local extrapolation of variables.

According to the solution-driven numerical values of the upstream-bias magnitude, the computed smooth and shocked solutions resulted from a mostly centered discretization. The algorithm succeeds in both reducing artificial dissipation in regions of smooth flow, for higher accuracy, and focusing an increased level of upstream-bias, hence artificial dissipation, at the shock regions only, precisely where required for stability and sharp shock capturing. This characteristics-bias algorithm, however, admits a straightforward implicit implementation, features a computational simplicity that parallels a traditional centered discretization, and rationally eliminates superfluous artificial diffusion. The computed solutions for subsonic, transonic, supersonic, and hypersonic flows remain essentially non-oscillatory and reflect available reference solutions.

Bibliography

[1] Akin J. E., Tezduyar T. E., "Calculation of the Advective Limit of the SUPG Stabilization Parameter for Linear and Higher Order Elements", *Comput. Methods Appl. Mech. Engr.*, 193:1909-1922, 2004.

[2] Alcrudo F., Garcia-Navarro P., "A High-Resolution Godunov-type Scheme in Finite Volumes for the 2-D Shallow-Water Equations", *Int. Journ. Numer. Meth. in Fluids*, 16:489-505, 1993.

[3] Aldrighetti E., Zanolli P., "A High Resolution Scheme for Flows in Open Channels with Arbitrary Cross Section", *Int. Journ. Numer. Meth. in Fluids*, 47:817-824, 2005.

[4] Aliabadi S. K., Tezduyar T. E., "Parallel Fluid Dynamics Computations in Aerospace Applications", *Int. Journ. Numer. Meth. in Fluids*, 21:783-805, 1995.

[5] Ames, W. F. *Numerical Methods for Partial Differential Equations*, 3rd Ed., Academic Press, Boston, MA, 1992.

[6] Anderson J. D. Jr., *Modern Compressible Flow with Historical Perspective*, McGraw-Hill, New York, NY, 1982.

[7] Anderson J. D. Jr., *Hypersonic and High Temperature Gas Dynamics*, Mc Graw-Hill, New York, NY, 1989.

[8] Anderson W. K., Thomas J. L., van Leer B., "A Comparison of Finite Volume Flux Vector Splittings for the Euler Equations", Tech. Paper AIAA 85-0122, 23rd Aerospace Sciences Meeting, Reno, NV, 1985.

[9] Armaly B. F., Durst F., Pereira J. C. F., Schonung B., "Experimental and Theoretical Investigation of Backward Facing Step Flow", *Journal of Fluid Mechanics*, 172, 473-496, 1983.

[10] Arminjon P., Dervieux A. "Construction of TVD-Like Artificial Viscosities on Two-Dimensional Arbitrary FEM Grids", *Journal of Computational Physics*, 106, 176-198, 1993.

[11] Baker A. J., *Finite Element Computational Fluid Mechanics*, Hemisphere Publishers, Washington, DC, 1983.

[12] Baker A. J., Kim J. W., "A Taylor Weak Statement Algorithm for Hyperbolic Conservation Laws", *International Journal for Numerical Methods in Fluids*, 7, 489-520, 1987.

[13] Barth T., "Numerical Aspects of Computing Viscous High Reynolds Number Flows on Unstructured Meshes", AIAA-91-0721, 1991.

[14] Batchelor G. K. *An Introduction to Fluid Dynamics* Cambridge University Press, Cambridge, 1970.

[15] Bathe K.-J., Wilson E.L. *Numerical Methods in Finite Element Analysis*, Prentice-Hall, 1976.

[16] Bathe K.-J. *Finite Element Procedures in Engineering Analysis*, Prentice-Hall, 1982.

[17] Bathe K.-J. *Finite Element Procedures*, Prentice-Hall, 1996.

[18] Bathe K.-J., Hendriana D., Brezzi F., Sangalli G., "Inf-sup Testing of Upwind Methods", *International Journal for Numerical Methods in Engineering*, 48, 745-760, 2000.

[19] Bathe K.-J., "The inf-sup Condition and its Evaluation for Mixed Finite Element Methods", *Computers & Structures*, 79, 243-252, 2001.

[20] Bathe K.-J., Pontaza J. P., "A Flow-Condition-Based Interpolation Mixed Finite Element Procedure for Higher Reynolds Number Fluid Flows", *Mathematical Models and Methods in Applied Sceinces*, 12, 4, 525-539, 2002.

[21] Bathe K.-J., Zhang H., "Finite Element Developments for General Fluid Flows with Structural Interactions", *International Journal for Numerical Methods in Engineering*, 60, 1, 213-232, 2004.

[22] Beam R. M., Warming R. F., "An Implicit Finite- Difference Algorithm for Hyperbolic Systems in Conservation-Law Form", *Journal of Computational Physics*, 22, 87-110, 1976.

[23] Berzins M., "Modified Mass Matrices and Positivity Preservation for Hyperbolic and Parabolic PDEs", *Commun. Mumer Meth. Engnr.*, 00:1-6, 2000.

[24] Boris J. P., Book D. L., "Flux Corrected Transport: I. SHASTA, a Fluid Transport Algorithm that Works", *Journal of Computational Physics*, 11, 38-69, 1973.

[25] Brenan K. E., Campbell S. L., Petzold L. R., *Numerical Solution of Initial-Value Problems in Differential-Algebraic Equations*, North-Holland, Amsterdam, 1989.

[26] Butcher J. C., *The Numerical Analysis of Ordinary Differential Equations: Runge-Kutta and General Linear Methods*, John Wiley, New York, NY, 1987.

[27] K. Burrage, "A Special Family of Runge-Kutta Methods for Solving Stiff Differential Equations", *BIT*, 18, 22-41, 1978.

[28] K. Burrage, J. C. Butcher, "Non-linear Stability for Implicit Runge-Kutta Methods", SIAM *Journal of Numerical Analysis*, 16:46-57, 1979.

[29] Carette J.-C., Deconinck H., Paillere H., Roe P. L., "Multidimensional Upwinding: Its Relation to Finite Elements", *Inter. Journ. Numer. Meth. in Fluids*, 20, 935-955, 1995.

[30] Carey G. F., Oden J. T., *Finite Elements*, Vols. I-VI, Prentice Hall, 1981 - 1986.

[31] Carey G. F., *Computational Grids: Generation, Adaptation and Solution Strategies*, Taylor & Francis, 1997.

[32] Catabriga L., Coutinho A. L.G.A., "Implicit SUPG Solution of the Euler Equations Using Edge-Based Data Structures", *Computer Methods in Applied Mechanics and Engineering*, 191, 3477-3490, 2002.

[33] Chanson H., *The Hydraulics of Open Channel Flow: An Introduction*, 2nd Ed. Elsevier, 2004.

[34] Chaudhry M. H., *Open Channel Flow* , Prentice-Hall, N.J., 1993.

[35] Chorin A. J., " A Numerical Method for Solving Incompressible Viscous Flow Problems", *J. of Comput. Phys.*, 2, 12-16, 1967.

[36] Chorin A. J., "Numerical Solution of the Navier-Stokes Equations", *Math. Comput.*, 22, 745-762, 1968.

[37] Chorin A. J., Marsden J. E. *A Mathematical Introduction to Fluid Mechanics*, Springer-Verlag, Heidelberg, 1992.

[38] Cockburn B., Shu C.-W., "The Runge-Kutta Discontinuous Galerkin Method for Conservation Laws V", *Journal of Computational Physics*, 141, 199-224, 1998.

[39] Coirier W., van Leer B., "Numerical Flux Formulas for the Euler and Navier-Stokes Equations: II. Progress in Flux-Vector Splitting", AIAA-91-1566, 1991.

[40] Courant R., Hilbert D., *Methods of Mathematical Physics*, John Wiley & Sons, New York, NY, 1989.

[41] Christie I., Mitchell A. R., "Upwinding of High Order Galerkin Methods in Conduction-Convection Problems", *International Journal for Numerical Methods in Engineering*, 12, 1764-1771, 1978.

[42] Crossley A. J., Wright N.G., "Time Accurate Local Time Stepping for the Unsteady Shallow Water Equations", *International Journal for Numerical Methods in Fluids*, 48:775-799, 2005.

[43] Crouzeix M., "Sur la B- Stabilite des Methodes de Runge Kutta", *Numerische Mathematik*, 32:75-82, 1979.

[44] Dadone A., and Grossman B., "A Rotated Upwind Scheme for the Euler Equations", AIAA 91-0635, 1991.

[45] Davis S.F., "A Rotationally-Biased Upwind Difference Scheme for the Euler Equations", *Journal of Computational Physics*, 56, 65-92, 1984.

[46] Debnath L., Mikusinski P., *Introduction to Hilbert Spaces with Applications*, Academic Press, Inc., 1990.

[47] Deconinck H., "Beyond the Riemann Problem: Part II", in *Algorithmic Trends in CFD*, Springer Verlag, 1993.

[48] Deconinck H., Roe P.L., Struijs R. "A Multidimensional Generalization of Roe's Flux difference Splitter for the Euler Equations", *Computers and Fluids*, 22, 215-222, 1993.

[49] Dekker K., Verwer J.G., *Stability of Runge-Kutta Methods for Stiff Non-Linear Differential Equations*, Elsevier Publishers, Amsterdam, 1984.

[50] Demkowicz L., Oden J. T., Rachowicz W., Hardy O., "An h-p Taylor-Galerkin Finite Element Method for Compressible Euler Equations", *Computer Methods in Applied Mechanics and Engineering*, 88, 363-396, 1991.

[51] Desideri J.-A., Glinsky N., Hettena E., "Hypersonic Reacting Flow Computations", *Computers and Fluids*, Vol. 18, No. 2, 151-182, 1990.

[52] De Vahl Davis G., "Natural Convection of Air in a Square Cavity: A Benchmark Numerical Solution", *International Journal for Numerical Methods in Fluids*, 3, 249-264, 1983.

[53] Dieudonne' J., *Foundations of Modern Analysis*, Academic Press, 1969.

[54] Dolejší V., Feistauer M., "A Semi-Implicit Discontinuous Galerkinn Finite Element Method for the Numerical Solution of Inviscid Compressible Flow", *Journal of Computational Physics*, 198, 727-746, 2004.

[55] Donea J., "A Taylor-Galerkin Method for Convective Transport Problems", *International Journal for Numerical Methods in Engineering*, Vol. 20, 101-119, 1984.

[56] Donea J., Quartapelle L., Selmin V., "An Analysis of Time Discretization in the Finite Element Solution of Hyperbolic Problems", *Journal of Computational Physics*, 70, 463-499, 1987.

[57] J. Emsley *The Elements*, Oxford University Press, New York, NY, 1998.

[58] Eskilsson C., Sherwin S. J., "A Triangular Spectral/hp Discontinuous Galerkin Method for Modelling 2D Shallow Water Equations", *International Journal for Numerical Methods in Fluids*, 45:605-623, 2004.

[59] Faber T. E. *Fluid Dynamics for Physicists* , Cambridge University Press, Cambridge, 1995.

[60] Ferziger J. H., Perić M. *Computational Methods for Fluid Dynamics*, Springer Verlag, Heidelberg, 1997.

[61] Fidkowski K. J., Oliver T. A., Lu. J., Darmofal D. L., "*p*-Multigrid Solution of High-Order Discontinuous Galerkin Discretizations of the Compressible Navier-Stokes Equations", *Journal of Computaional Physics*, 207, 92-113, 2005.

[62] Fletcher C. A. J. *Computational Techniques for Fluid Dynamics*, Vol. 1 & 2, Springer-Verlag, Heidelberg, 1991.

[63] Frankel T. *The Geometry of Physics*, Cambridge University Press, Cambridge, 1997.

[64] Gaitonde D., Shang J., "The Performance of Flux-Split Algorithms in High- Speed Viscous Flows", AIAA-92-0186, 1992.

[65] Garabedian P. R., *Partial Differential Equations*, Chelsea Publishing Company, New York, NY, 1986.

[66] Garcia-Navarro P., Vazquez-Cendon M., "On Numerical Treatment of the Source Terms in the Shallow Water Equations", *Computers and Fluids*, 29:951 979, 2000.

[67] R. Glowinski and J. Periaux, "Numerical Methods for Nonlinear Problems in Fluid Dynamics", in *Supercomputing*, A. Lichnewsky, C. Saguez Eds., North-Holland, 1987.

[68] Godlewski E., Raviart P.-A. *Numerical Approximation of Hyperbolic Systems of Conservation Laws*, Springer-Verlag, Heidelberg, 1996.

[69] Gnoffo P. A., Gupta R.N., Shinn J. L., "Conservation Equations and Physical Models for Hypersonic Air Flows in Thermal and Chemical Nonequilibrium", NASA Technical Paper 2868, 1989.

[70] Greiner W., Neise L., Stöcker H. *Thermodynamics and Statistical Mechanics*, Springer, New York, NY, 1997.

[71] Gresho P. M., Lee R. L., Sani R. L., "Finite Elements in Fluids III", 325, John Wiley, 1978.

[72] Gresho P. M., Sani R. L. *Incompressible Flow and the Finite Element Method* , John Wiley, New York, NY, 1999.

[73] Grossman B., Cinnella P., "Flux-Split Algorithms for Flows with Non-equilibrium Chemistry and Vibrational Relaxation", *Journ. of Comp. Physics*, 88, 131-168, 1990.

[74] Gunzburger M. D. *Finite Element Methods for Viscous Incompresisble Flows*, Academic Press, Boston, MA, 1989.

[75] Gupta R. N., Yos J. M., Thompson R. A., Lee K.-P., "A Review of Reaction Rates and Thermodynamic and Transport Properties for an 11-Species Air Model for Chemical and Thermal Nonequilibrium Calculations to 30,000K", NASA Reference Publication 1232, 1990.

[76] Hartman R., Houston P., "Adaptive Discontinuous Galerkin Finite Element Methods for the Compressible Euler Equations", *Journal of Computational Physics*, 183(2):508-532, 2002.

[77] Hendriana D., Bathe K.-J. "On Upwind Methods for Parabolic Finite Elements in Incompressible Flows", *International Journal for Numerical Methods in Engineering*, 47, 317-340, 2000.

[78] Hendriana D., Bathe K.-J. "On a Parabolic Quadrilateral Finite Element for Compressible Flows", *Computer Methods in Applied Mechanics and Enngineering*, 186, 1-22, 2000.

[79] Hill P., Peterson C.*Mechanics and Thermodynamics of Propulsion*, Addison Wesley, Reading, MA, 1992.

[80] C. Hirsch, *Numerical Computation of Internal and External Flows*, Vol. 1, 2 John Wiley & Sons, New York, NY, 1991.

[81] Hoffman J., "Development of an Algorithm for the Three-Dimensional Fully-Coupled Navier-Stokes Equations with Finite Rate Chemistry", AIAA- 89-0670, 1989.

[82] Horn R. A., Johnson C. R., *Matrix Analysis*, Cambridge, 1991.

[83] Horn R. A., Johnson C. R., *Topics in Matrix Analysis*, Cambridge University Press, New York, NY, 1991.

[84] Hubbard J.H., West B. H., *Differential Equations: A Dynamical Systems Approach*, Vols. 1-4, Springer-Verlag, Heidelberg, 1995.

[85] Hughes T. J. R., Brooks A., "A Theoretical Framework for Petrov-Galerkin Methods with Discontinuous Weighting Functions: Application to the Streamline-Upwind Procedure", Finite Elements in Fluids IV, 47-65, John Wiley, 1982.

[86] Hughes T. J. R., "Recent Progress in the Development and Understanding of SUPG Methods with Special Reference to the Compressible Euler and Navier-Stokes Equations", *International Journal for Numerical Methods in Fluids*, 7, 11, 1987.

[87] Hughes T. J. R., "The Finite Element Method: Linear Static and Dynamic Finite Element Analysis", Prentice-Hall, 1987.

[88] Hussaini M. Y., Kumar A., Salas M.D., ed., *Algorithmic Trends in Computational Fluid Dynamics*, Springer-Verlag, 1993.

[89] Hutson V., Pym J. S., *Applications of Functional Analysis and Operator Theory*, Academic Press, New York, NY, 1980.

[90] Iannelli J., Baker A.J., "A Stiffly-Stable Implicit Runge-Kutta Algorithm for CFD Applications", Tech. Paper AIAA 88-0115, 26th Aerospace Sciences Meeting, Reno NV, 1988.

[91] Iannelli J., Baker A.J., "An Implicit and Stiffly Stable Finite Element CFD Algorithm for Aerodynamic Applications", Tech. Paper AIAA-89-0656, 27th Aerospace Sciences Meeting, Reno, NV, 1989.

[92] Iannelli J., Baker A.J., "An Intrinsically N-Dimensional Generalized Flux Vector Splitting Implicit Finite Element Euler Algorithm", AIAA 91-0123, 29th Aerospace Sciences Meeting, Reno, NV, 1991.

[93] Iannelli J., Baker A. J., "A Globally Well-Posed Finite Element Algorithm for Aerodynamics Applications", International Journal for Numerical Methods in Fluids, Vol. 12, 407-441, 1991.

[94] Iannelli J., Baker A. J., "A Non-linearly Stable Implicit Finite Element CFD Algorithm for Hypersonic Aerodynamics", *International Journal for Numerical Methods in Engineering*, 34, 419-441, 1992.

[95] Iannelli J., "A CFD Euler Solver from a Physical Acoustics- Convection Flux Jacobian Decomposition", *Int. J. Numer. Meth. Fluids* 31, 821-860, 1999.

[96] Iannelli J., "A Multi-Dimensional Acoustics-Convection Upstream Resolution Euler Solver", Keynote Presentation, 3rd M.I.T. Conference on Computational Fluid and Solid Mechanics, Massachusetts Institute of Technology, Cambridge, MS, USA, June 14-17, 2005.

[97] John F. *Partial Differential Equations*, Springer-Verlag, Heidelberg, 1971.

[98] Johnson C., *Numerical Solution of Partial Differential Equations by the Finite Element Method*, Cambridge, 1987.

[99] Johnson C., Szepessy A., "On the Convergence of a Finite Element Method for a Nonlinear Hyperbolic Conservation Law", *Mathematics of Computation*, 49, 180, 427-444, 1987.

[100] Johnson C., Szepessy A., Hansbo P., "On the Convergence of Shock-Capturing Streamline Diffusion Finite Element Methods for Hyperbolic Conservation Laws", *Mathematics of Computation*, 54, 189, 107-129, 1990.

[101] Johnson G. M., "Relaxation Solution of the Full Euler Equations", Springer-Verlag, Lecture Notes in Physics, 170, 273-279, 1982.

[102] Jorgenson P., Turkel E., "Central Difference TVD Schemes for Time Dependent and Steady State Problems", *Journ. of Comp. Phys.*, 107, 297-308, 1993.

[103] Keener J. P., *Principles of Applied Mathematics*, Addison-Wesley, New York, NY, 1988.

[104] Kevorkian J., *Partial Differential Equations: Analytical Solutions Techniques*, Wadsworth & Brooks/Cole, Pacific Grove, CA, 1990.

[105] Kohno H., Bathe K.-J. "Insight into the Flow-Condition-Based Interpolation Finite Element Approach: Solution of Steady-State Advection-Diffusion Problems", *International Journal for Numerical Methods in Engineering*, 63, 2, 197-217, 2005.

[106] Kopal Z., *Numerical Analysis* , John Wiley, New York, NY, 1955.

[107] Kreiss H.O., Lorenz J., *Initial-Boundary Value Problems and the Navier-Stokes Equations*, Academic Press, New York, NY, 1989.

[108] Kronzon Y., *et al.*, "Gas Flows in Rocket Motors", Tech. Report AL-TR-89-011, AFSC, AL/LSCF, Edwards AFB, CA, 1989.

[109] Ladyzhenskaya A., "On the Construction of Discontinuous Solutions of Quasi-Linear Hyperbolic Equations as Limits to the Solutions of the Respective Parabolic Equations when the Viscosity Coefficient is Approaching Zero", *Doklady Akad. Nauk USSR*, 3 291-295, 1956.

[110] Landau L. D., Lifshitz E. M. *Fluid Mechanics*, Pergamn Press, Oxford, 1987.

[111] Lafon F., Osher S., "Essentially Non-Oscillatory Postprocessing Filtering Methods", Algorithmic Trends in Computational Fluid Dynamics, Springer-Verlag, 1993.

[112] Lambert J. D., *Computational Methods in Ordinary Differential Equations*, John Wiley & Sons, New York, NY, 1983.

[113] Laney C. B. *Computational Gas Dynamics*, Cambridge University Press, Cambridge, 1998.

[114] Laurien E., Böhle M., Holthoff H., Wiesbaum J., Lieseberg A., "Finite-Element Algorithm for Chemically Reacting Hypersonic Flows", AIAA-92- 0754, 1992.

[115] Lax P. D., "Weak Solutions of Nonlinear Hyperbolic Equations and Their Numerical Computation", *Comm. Pure Appl. Math.*, 7, 159-193, 1954.

[116] Lax P. D., " Hyperbolic Systems of Conservation Laws II", *Comm. Pure Appl. Math.*, 10, 537-566, 1957.

[117] Lax P. D., *Hyperbolic Systems of Conservation Laws and the Mathematical Theory of Shock Waves*, 3rd Ed. SIAM, Philadelphia, PA, 1990.

[118] LeBeau G. J., Tezduyar T. E., "Finite Element Computation of Compressible Flows with the SUPG Formulation", FED- 123, Advances in Finite Element Analysis in Fluid Dynamics, ASME, 1991.

[119] LeBeau G. J., Ray S. E., Aliabadi S. K., Tezduyar T. E., "SUPG Finite Element Computation of Compressible Flows With the Entropy and Conservation Variables Formulations ", *Computer Methods in Applied Mechanics and Engineering*, 104, 3, 397-498, 1993.

[120] LeVeque R. J., *Numerical Methods for Conservation Laws*, Birkhauser Verlag, Boston, MS, 1990.

[121] LeVeque R. J., *Finite Volume Methods for Hyperbolic Problems*, Cambridge University Press, Cambridge, 2002.

[122] Levy D.W., Powell K., van Leer B., "An Implementation of a Grid-Independent Upwind Scheme for the Euler Equations", AIAA 89-1931-CP, 1989.

[123] Levy D.W., Powell K.G., van Leer B., "Use of a Rotated Riemann Solver for the Two-Dimensional Euler Equations", *Journal of Computational Physics*, 106, 201-214, 1993.

[124] Liou M.-S., van Leer B.,"Choice of Implicit and Explicit Operators for The Upwind Differencing Method", AIAA 88-0624, 26th Aerospace Sciences Meeting, Reno, NV, 1988.

[125] Liou M. S., van Leer B., "Splitting of Inviscid Fluxes for Real Gases", NASA Tech. Memorandum 100856 ICOMP-88-7, 1988.

[126] Liou M., "A Newton/Upwind Method and Numerical Study of Shock Wave/Boundary Layer Interactions", *Int. Journ. Num. Meth. in Fluids*, 9, 747-761, 1989.

[127] Liou M. S., van Leer B., and Shuen J. S., "Splitting of Inviscid Fluxes for Real Gases", *Journal of Computational Physics*, 87, 1-24, 1990.

[128] Liou M.-S., Steffen C.J., "A New Flux Splitting Scheme", *Journal of Computational Physics*, 107, 23-29, 1993.

[129] Liou M.-S., "A Sequel to AUSM: AUSM+, *Journal of Computational Physics*, 129, 364-382, 1996.

[130] Liska R., Wendroff B., "Two-Dimensional Shallow Water Equations by Composite Schemes", *International Journal for Numerical Methods in Fluids*, 30:461-479, 1999.

[131] Lombard C. K., Oliger J., Yang J. Y., "A Natural Conservative Flux Difference Splitting for the Hyperbolic Systems of Gasdynamics", Tech. Paper AIAA 82-0976, 3rd Joint Thermophysics, Fluids, Plasma and Heat Transfer Conference, St. Louis, Missouri, 1982.

[132] Luo H., Baum J .D., Löhner R., Cabello J., "Adaptive Edge-Based Finite Element Schemes for the Euler and Navier-Stokes Equations on Unstructured Grids", AIAA 93-0336, 1993.

[133] MacCormack R.W., "Algorithmic Trends in CFD in the 1990's for Aerospace Flow Field Calculations", Algorithmic Trends in Computational Fluid Dynamics, Springer-Verlag, 1993.

[134] Malvern L. E., *Introduction to the Mechanics of a Continuous Medium*, Prentice-Hall, 1969.

[135] Majda A., Osher S., "Numerical Viscosity and The Entropy Condition", *Comm. Pure Appl. Math.*, 32, 797-838, 1979.

[136] Meintjes K., Morgan A. P., "Element Variables and the Solution of Complex Chemical Equilibrium Problems", *Combust. Sci. and Tech.*, 68, 35-48, 1989.

[137] Milne W. E., "A Note on the Numerical Integration of Differential Equations", *J. Res. Nat. Bur. Standards*, 43, 537-542, 1949.

[138] Mitchell C. R., Walters R. W., "K-Exact Reconstruction for the Navier Stokes Equations on Arbitrary Grids", AIAA 93-0536, 1993.

[139] Molvik G., Merkle C., "A Set of Strongly Coupled, Upwind Algorithms for Computing Flows in Chemical Nonequilibrium", AIAA-89- 0199, 1989.

[140] Morton E. W., Parrott A. K., "Generalized Galerkin Methods for First-Order Hyperbolic Equations", *Journ. Comp. Phys.*, 36, 249-270, 1980.

[141] Morton K. W., Mayers D. F., *Numerical Solution of Partial Differential Equations*, Cambridge University Press, Cambridge, 1994.

[142] Nørsett S. P., "Runge-Kutta Methods with a Multiple Real Eigenvalue Only", *BIT*, 16, 388-393, 1976.

[143] Oden J. T., "Formulation and Application of Certain Primal and Mixed Finite Element Models of Finite Deformations of Elastic Bodies," in: R. Glowinski and J. L. Lions, eds., Computing Methods in Applied Sciences and Engineering, Springer, Berlin, 1974.

[144] Oden J. T., Reddy J. N.*An Introduction to the Mathematical Theory of Finite Elements*, John Wiley, New York, NY, 1976.

[145] Oden J. T., Demkowicz L., Rachowicz L., Westermann L., "A Posteriori Error Analysis in Finite Elements: The Element Residual Method for Symmetrizable Problems with Application to Compressible Euler and Navier-Stokes Equations", *Comput. Methods Appl. Mech. Engrg.*, 82, 183-203, 1990.

[146] Oden J. T., *Finite Elements of Nonlinear Continua*, Dover, 2006.

[147] Olejnik O., "On the Uniqueness of a Generalised Solution to the Cauchy Problem for a Nonlinear System of Equations", Arising in Mechanics, *Usp. Mat. Nauk.*, 12(6): 169-176, 1957.

[148] Olejnik O. A., "Uniqueness and Stability of the Generalized Solutions of the Cauchy Problem for a Quasi-Linear Equation", *Amer. Math. Soc. Transl.*, (2)33, 285-290, 1964.

[149] Olejnik O. A., "Construction of a Generalized Solution of the Cauchy Problem for a Quasi-Linear Equation of First Order by the Introduction of Vanishing Viscosity", *Amer. Math. Soc. Transl.*, 2, 26, 1963.

[150] Olejnik O. A. "Discontinuous Solutions of Non-Linear Differential Equations", *Amer. Math. Soc. Transl.*, 2, 26, 1963.

[151] Ortega J. M. *Numerical Analysis, A Second Course*, SIAM, Philadelhia, PA, 1992.

[152] Paillere H., Deconinck H., Struijs R., Roe P. L., Mesaros L.M., Muller J.D., "Computations of Inviscid Compressible Flows Using Fluctuation-Splitting on Triangular Meshes", AIAA 93-3301, 1993.

[153] Pandolfi M., D'Ambrosio D., "Numerical Instabilities in Upwind Methods: Analysis and Cures for the 'Carbuncle' Phenomenon", *Journal of Computational Physics*, 166, 271-301, 2001.

[154] Panton R. L. *Incompressible Flow*, John Wiley, New York, NY, 1996.

[155] Park C., "On Convergence of Computation of Chemically Reacting Flows", Tech. Paper, AIAA 85-0247, 23rd Aerospace Sciences Meeting, Reno, NV, 1985.

[156] Park C., Yoon S., "A Fully-Coupled Implicit Method for Thermo-Chemical Non-Equilibrium Air at Sub-Orbital Flight Speeds", Tech. Paper, AIAA 89-1974, 9th Computational Fluid Dynamics Conference, Buffalo, New York, 1989.

[157] Parpia I. H., "A Planar Oblique Wave Model for the Euler Equations" AIAA-91-1545-CP, 1991.

[158] Pepper, D. W., Heinrich J. C. *The Finite Element Method: Basic Concepts and Applications*, Taylor and Francis, New York, NY, 1992.

[159] Petrovsky I. G. *Lectures on Partial Differential Equations* 4th Ed., Interscience Publishers, New York, NY, 1964.

[160] Pironneau O. *Finite Element Methods for Fluids*, John Wiley, New York, NY, 1989.

[161] Powell K., van Leer B., "A Genuinely Multi-Dimensional Upwind Cell-Vertex Scheme for the Euler Equations", NASA Technical Memorandum 102029, ICOMP-89-13, 1989.

[162] Prabhu R. K., Stewart J. R., and Thareja R. R., "A Navier-Stokes Solver for High Speed Equilibrium Flows and Application to Blunt Bodies", AIAA 89-0668, 27th Aerospace Sciences Meeting, Reno, NV, 1989.

[163] Rasmussen M., *Hypersonic Flow*, John Wiley, New York, NY, 1994.

[164] Richtmyer R. D., Morton K. W., *Difference Methods for Initial Value Problems*, Interscience Publishers, John Wiley & Sons, New York, NY, 1967.

[165] Rider W. J., "Extension of High-Resolution Schemes to Multiple Dimensions," AIAA 93-0877, 1993.

[166] Rizzetta D. P., Mach K. D., "Comparative Numerical Study of Hypersonic Compression Ramp Flows" Tech. Paper AIAA-89-1877, 1989.

[167] D. P. Rizzetta, "Numerical Simulation of a Supersonic Inlet", AIAA-91-0128, 1991.

[168] Roe P. L., "Approximate Riemann Solvers, Parameter Vectors, and Difference Schemes", *Journal of Computational Physics*, 43, 357-372, 1981.

[169] Roe P. L., "Sonic Flux Formulae", *SIAM J. Sci. Statistic. Comp.*, 13, 611-630, 1992.

[170] Roe P. L., "Beyond the Riemann Problem: Part I", in *Algorithmic Trends in CFD*, Springer Verlag, 1993.

[171] Roe P. L., "Local Reducation of Certain Wave Operators to One-Dimensional Form", ICASE Report No. 94-96, ICASE NASA Langley Research Center, Hampton, VA, 1994.

[172] Roe P. L., "Sock Capturing", in Handbook of Shock Waves, 787-876, Vol. 1 Academic Press, Boston, MA, 2001.

[173] Rumsey C. L., van Leer B., Roe P. L., "A Multidimensional Flux Function with Applications to the Euler and Navier-Stokes Equations", *Journal of Computational Physics*, 105, 306-323, 1993.

[174] Salençon J., *Mécanique des Milieux Continus*, Tome 1-3, Éditions del' Ècole Polytechnique, Paris, 2000.

[175] Shapiro A., *The Dynamics and Thermodynamics of Compressible Fluid Flow*, Vols. 1, 2, The Ronald Press Company, NY, 1954.

[176] Simpson B., "Unsteady Three-Dimensional Thin Layer Navier- Stokes Solutions on Dynamic Blocked Grids", Tech. Report AFATL-TR-89-19, ABAB, AF Armament LAB, Eglin AFB, FL, 1989.

[177] Smoller J., *Shock Waves and Reaction-Diffusion Equations*, Springer-Verlag, 1983.

[178] Sod G. A., *Numerical Methods in Fluid Dynamics*, Cambridge University Press, New York, NY, 1985.

[179] Steger J. L., Warming R. F., "Flux Vector Splitting of the Inviscid Gasdynamics Equations with Applications to Finite Difference Methods", *Journal of Computational Physics*, 40, 263-293, 1981.

[180] Strang G., Fix G. J., *An Analysis of the Finite Element Method*, Prentice-Hall, Englewood Cliffs, NJ, 1973.

[181] Tadmor E., "Local Error Estimates for Discontinuous Solutions of Nonlinear Hyperbolic Equations", *Siam J. Numer. Anal.*, Vol. 28, No. 4, pp. 891-906, 1991.

[182] Tadmor E., "Approximate Solutions of Nonlinear Conservattion Laws", in Advanced Numerical Approximation of Numerical Hyperbolic Equations, 1-149, Lecture Notes in Mathematics, 1697, Springer-Verlag, Heidelberg, 1998.

[183] Tannehill J. C., Mugge P. H., "Improved Curve Fits for the Thermodynamic Properties of Equilibrium Air Suitable for Numerical Computation Using Time-Dependent or Shock-Capturing Methods", NASA CR-2470, 1974.

[184] Tannehill J. C., Anderson D. A., Pletcher R. H. *Computational Fluid Mechanics and Heat Transfer*, Taylor & Francis, Washington DC, 1997.

[185] Tezduyar T., Osawa J., "Finite Element Stabilization Parameters Computed from Element Matrices and Vectors", *Computer Methods in Applied Mechanics and Engineering*, 190, 411-430, 2000.

[186] Tezduyar T., "Stabilization Parameters and Local Length Scales in SUPG and PSPG Formulations", Paper-81508, Proceedings of the Fifth World Congress on Computational Mechanics,, Vienna, Austria, 2002.

[187] Tezduyar T., Sathte S., "Stabilization Parameters in SUPG and PSPG Formulations", *Journal of Computational and Applied mechanics*, 4, 4, 71-88, 2003.

[188] Thomas J. L., van Leer B., Walters R. W., "Implicit Flux-Split Schemes for the Euler Equations", Tech. Paper AIAA 85-1680, 18th Fluid Dynamics, Plasmadynamics and Lasers Conference, Cincinnati, OH, 1985.

[189] Taylor A. III, Ng W., Walters R., "Upwind Relaxation Methods for the Navier-Stokes Equations Using Inner Iterations", *Journ. of Comp. Physics*, 99, 68- 78, 1992.

[190] Taylor A. III, Ng W., Walters R., "An Improved Upwind Finite Volume Relaxation Method for High Speed Viscous Flows", *Journ of Comp. Physics*, 99, 159-168, 1992.

[191] Taylor M. E., "First-Order Hyperbolic Systems with a Small Viscosity Term", *Comm. Pure Appl. Math.*, 31, 707-786, 1978.

[192] Toro E. F., *Rieman Solvers and Numerical Methods for Fluid Dynamics*, 2nd Ed., Springer-Verlag, Berlin, 1999.

[193] Toro E. F., *Shock Capturing Methods for Free Surface Shallow Flows*, John Wiley, New York, 2001.

[194] van der Vegt J. J. W., van der Ven H., "Discontinuous Galerkin Finite Element Method with Anisotropic Local Grid Refinement for Inviscid Compressible Flows", *Journal of Computational Physics*, 141, 46-77, 1998.

[195] vanLeer B., "Towards the Ultimate Conservative Difference Scheme. II. Monotonicity and Conservation Combined in a Second Order Scheme", *Journal of Computational Physics*, 14, 361-70, 1974.

[196] vanLeer B., "Towards the Ultimate Conservative Difference Scheme. III. Upstream-Centered Finite-Difference Schemes for Ideal Compressible Flows", *Journal of Computational Physics*, 23, 263-275, 1977.

[197] vanLeer B., "Towards the Ultimate Conservative Difference Scheme. IV. A new approach to numerical convection", *Journal of Computational Physics*, 23, 276-279, 1977.

[198] vanLeer B., "Towards the Ultimate Conservative Difference Scheme. V. A Second-Order Sequel to Godunov's Method", *Journal of Computational Physics*, 32, 101-136, 1979.

[199] vanLeer B., "Flux-Vector Splitting for the Euler Equations", Lecture Notes in Physics, 170, 507-512, 1982.

[200] vanLeer B., "Progress in Multi-Dimensional Upwind Differencing", NASA ICASE Report No. 92-43, 1992.

[201] Venkatakrishnan V., "On the Accuracy of Limiters and Convergence to Steady State Solutions", AIAA 93-0880, 1993.

[202] Vichnevetsky, Bowles J. B. *Fourier Analysis of Numerical Approximations of Hyperbolic Equations*, SIAM, Philadelphia, PA, 1982.

[203] Vincenti W. G., C. H. Kruger, Jr., *Introduction to Physical Gas Dynamics*, John Wiley and Sons, Inc., New York, London, Sydney, 1965.

[204] Vinokur M., Montagne' J. L., "Generalized Flux-Vector Splitting and Roe Average for an Equilibrium Real Gas", *Journal of Computational Physics*, 89, 276-300, 1990.

[205] Walters R., Cinnella P., Slack D., Halt D., "Characteristic-Based Algorithms for Flows in Thermo-Chemical Nonequilibrium", AIAA-90-0393, 1990.

[206] Warming R. F., Hyett B. J., "The Modified Equation Approach to the Stability and Accuracy Analysis of Finite-Difference Methods", *Journal of Computational Physics*, 14, 159-179, 1974.

[207] White F. M., *Viscous Fluid Flow*, 3rd ed., McGraw Hill, 2005.

[208] Whitham G. B., *Linear and Non-Linear Waves*, John Wiley & Sons, 1974.

[209] Yang H.Q., Przekwas A. J., "A Comparative Study of Advanced Shock-Capturing Schemes Applied to Burgers' Equation", *Journal of Computational Physics*, 102, 139-159, 1992.

[210] Yee H. C., Klopfer G. H., Montagne J. L., "High Resolution Shock-Capturing Schemes for Inviscid and Viscous Hypersonic Flows", *Journal of Computational Physics*, 88, 31-61, 1990.

[211] Young D. M., Gregory R. T. *A Survey of Numerical Mathematics* Vol. 1 & 2, Addison Wesley, Reading, MA, 1973.

[212] Young N., *An Introduction to Hilbert Spaces*, Cambridge Mathematical Textbooks, 1988.

[213] Zauderer E., *Partial Differential Equations of Applied Mathematics*, John Wiley, New York, NY, 1989.

[214] Zienkiewicz O. C., . Taylor R. L, *The Finite Element Method*, Vols. 1, 3, Butterworth & Heinemann, 2000.

[215] Zucrow M. J., Hoffman J. D. *Gas Dynamics*, John Wiley & Sons, NY, 1976.

Index

Printing: Krips bv, Meppel
Binding: Stürtz, Würzburg